Rational (Reciprocal) Function

$$f(x) = \frac{1}{x}$$

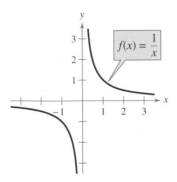

Domain: $(-\infty, 0) \cup (0, \infty)$
Range: $(-\infty, 0) \cup (0, \infty)$
No intercepts
Decreasing on $(-\infty, 0)$ and $(0, \infty)$
Odd function
Origin symmetry
Vertical asymptote: y-axis
Horizontal asymptote: x-axis

Exponential Function

$$f(x) = a^x, \ a > 1$$

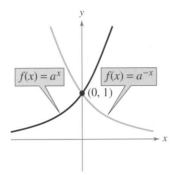

Domain: $(-\infty, \infty)$
Range: $(0, \infty)$
Intercept: $(0, 1)$
Increasing on $(-\infty, \infty)$
  for $f(x) = a^x$
Decreasing on $(-\infty, \infty)$
  for $f(x) = a^{-x}$
Horizontal asymptote: x-axis
Continuous

Logarithmic Function

$$f(x) = \log_a x, \ a > 1$$

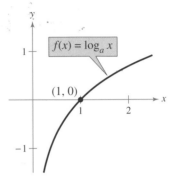

Domain: $(0, \infty)$
Range: $(-\infty, \infty)$
Intercept: $(1, 0)$
Increasing on $(0, \infty)$
Vertical asymptote: y-axis
Continuous
Reflection of graph of $f(x) = a^x$
  in the line $y = x$

Sine Function

$$f(x) = \sin x$$

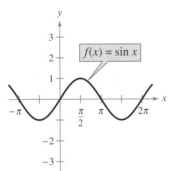

Domain: $(-\infty, \infty)$
Range: $[-1, 1]$
Period: 2π
x-intercepts: $(n\pi, 0)$
y-intercept: $(0, 0)$
Odd function
Origin symmetry

Cosine Function

$$f(x) = \cos x$$

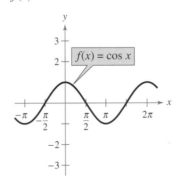

Domain: $(-\infty, \infty)$
Range: $[-1, 1]$
Period: 2π
x-intercepts: $\left(\frac{\pi}{2} + n\pi, 0\right)$
y-intercept: $(0, 1)$
Even function
y-axis symmetry

Tangent Function

$$f(x) = \tan x$$

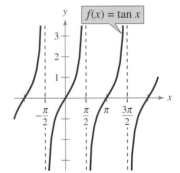

Domain: all $x \neq \frac{\pi}{2} + n\pi$
Range: $(-\infty, \infty)$
Period: π
x-intercepts: $(n\pi, 0)$
y-intercept: $(0, 0)$
Vertical asymptotes:
$$x = \frac{\pi}{2} + n\pi$$
Odd function
Origin symmetry

Cosecant Function

$f(x) = \csc x$

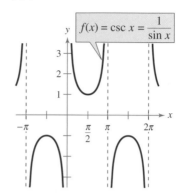

Domain: all $x \neq n\pi$
Range: $(-\infty, -1] \cup [1, \infty)$
Period: 2π
No intercepts
Vertical asymptotes: $x = n\pi$
Odd function
Origin symmetry

Secant Function

$f(x) = \sec x$

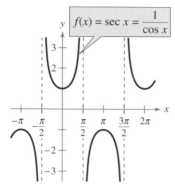

Domain: all $x \neq \dfrac{\pi}{2} + n\pi$

Range: $(-\infty, -1] \cup [1, \infty)$
Period: 2π
y-intercept: $(0, 1)$
Vertical asymptotes:

$$x = \frac{\pi}{2} + n\pi$$

Even function
y-axis symmetry

Cotangent Function

$f(x) = \cot x$

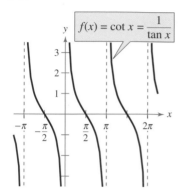

Domain: all $x \neq n\pi$
Range: $(-\infty, \infty)$
Period: π

x-intercepts: $\left(\dfrac{\pi}{2} + n\pi, 0\right)$

Vertical asymptotes: $x = n\pi$
Odd function
Origin symmetry

Inverse Sine Function

$f(x) = \arcsin x$

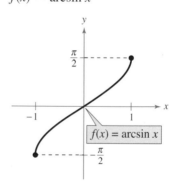

Domain: $[-1, 1]$

Range: $\left[-\dfrac{\pi}{2}, \dfrac{\pi}{2}\right]$

Intercept: $(0, 0)$
Odd function
Origin symmetry

Inverse Cosine Function

$f(x) = \arccos x$

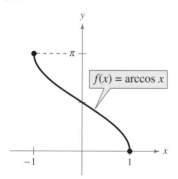

Domain: $[-1, 1]$
Range: $[0, \pi]$

y-intercept: $\left(0, \dfrac{\pi}{2}\right)$

Inverse Tangent Function

$f(x) = \arctan x$

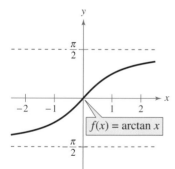

Domain: $(-\infty, \infty)$

Range: $\left(-\dfrac{\pi}{2}, \dfrac{\pi}{2}\right)$

Intercept: $(0, 0)$
Horizontal asymptotes:

$$y = \pm\frac{\pi}{2}$$

Odd function
Origin symmetry

Trigonometry

Eighth Edition

Ron Larson
The Pennsylvania State University
The Behrend College

With the assistance of
David C. Falvo
The Pennsylvania State University
The Behrend College

BROOKS/COLE
CENGAGE Learning™

Australia • Brazil • Japan • Korea • Mexico • Singapore • Spain • United Kingdom • United States

BROOKS/COLE
CENGAGE Learning™

Trigonometry, Eighth Edition
Ron Larson

Publisher: Charlie VanWagner

Acquiring Sponsoring Editor: Gary Whalen

Development Editor: Stacy Green

Assistant Editor: Cynthia Ashton

Editorial Assistant: Guanglei Zhang

Associate Media Editor: Lynh Pham

Marketing Manager: Myriah FitzGibbon

Marketing Coordinator: Angela Kim

Marketing Communications Manager: Katy Malatesta

Content Project Manager: Susan Miscio

Senior Art Director: Jill Ort

Senior Print Buyer: Diane Gibbons

Production Editor: Carol Merrigan

Text Designer: Walter Kopek

Rights Acquiring Account Manager, Photos: Don Schlotman

Photo Researcher: Prepress PMG

Cover Designer: Harold Burch

Cover Image: Richard Edelman/Woodstock Graphics Studio

Compositor: Larson Texts, Inc.

For product information and technology assistance, contact us at
Cengage Learning Customer & Sales Support, 1-800-354-9706

For permission to use material from this text or product, submit all requests online at **www.cengage.com/permissions.**
Further permissions questions can be emailed to
permissionrequest@cengage.com

Library of Congress Control Number: 2009930254

Student Edition:

ISBN-13: 978-1-4390-4907-5

ISBN-10: 1-4390-4907-6

Brooks/Cole
10 Davis Drive
Belmont, CA 94002-3098
USA

Cengage Learning is a leading provider of customized learning solutions with office locations around the globe, including Singapore, the United Kingdom, Australia, Mexico, Brazil, and Japan. Locate your local office at: **international.cengage.com/region**

Cengage Learning products are represented in Canada by Nelson Education, Ltd.

For your course and learning solutions, visit **www.cengage.com**

Purchase any of our products at your local college store or at our preferred online store **www.ichapters.com**

Printed in China
3 4 5 6 7 14 13 12 11

Contents

chapter 3

Additional Topics in Trigonometry   281

chapter 4

Complex Numbers   337

chapter 5

Exponential and Logarithmic Functions   375

A Word from
the Author

Welcome to the Eighth Edition of *Trigonometry*! We are proud to offer you a new and revised version of our textbook. With this edition, we have listened to you, our users, and have incorporated many of your suggestions for improvement.

8th 7th 6th 5th

4th 3rd 2nd 1st

In the Eighth Edition, we continue to offer instructors and students a text that is pedagogically sound, mathematically precise, and still comprehensible. There are many changes in the mathematics, art, and design; the more significant changes are noted here.

- **New Chapter Openers** Each *Chapter Opener* has three parts, *In Mathematics, In Real Life*, and *In Careers*. *In Mathematics* describes an important mathematical topic taught in the chapter. *In Real Life* tells students where they will encounter this topic in real-life situations. *In Careers* relates application exercises to a variety of careers.

- **New Study Tips and Warning/Cautions** Insightful information is given to students in two new features. The *Study Tip* provides students with useful information or suggestions for learning the topic. The *Warning/Caution* points out common mathematical errors made by students.

- **New Algebra Helps** *Algebra Help* directs students to sections of the textbook where they can review algebra skills needed to master the current topic.

- **New Side-by-Side Examples** Throughout the text, we present solutions to many examples from multiple perspectives—algebraically, graphically, and numerically. The side-by-side format of this pedagogical feature helps students to see that a problem can be solved in more than one way and to see that different methods yield the same result. The side-by-side format also addresses many different learning styles.

- *New Capstone Exercises* *Capstones* are conceptual problems that synthesize key topics and provide students with a better understanding of each section's concepts. Capstone exercises are excellent for classroom discussion or test prep, and teachers may find value in integrating these problems into their reviews of the section.

- *New Chapter Summaries* The *Chapter Summary* now includes an explanation and/or example of each objective taught in the chapter.

- *Revised Exercise Sets* The exercise sets have been carefully and extensively examined to ensure they are rigorous and cover all topics suggested by our users. Many new skill-building and challenging exercises have been added.

For the past several years, we've maintained an independent website—**CalcChat.com**—that provides free solutions to all odd-numbered exercises in the text. Thousands of students using our textbooks have visited the site for practice and help with their homework. For the Eighth Edition, we were able to use information from CalcChat.com, including which solutions students accessed most often, to help guide the revision of the exercises.

I hope you enjoy the Eighth Edition of *Trigonometry*. As always, I welcome comments and suggestions for continued improvements.

Ron Larson

Acknowledgments

I would like to thank the many people who have helped me prepare the text and the supplements package. Their encouragement, criticisms, and suggestions have been invaluable.

Thank you to all of the instructors who took the time to review the changes in this edition and to provide suggestions for improving it. Without your help, this book would not be possible.

Reviewers

Chad Pierson, *University of Minnesota-Duluth*; Sally Shao, *Cleveland State University*; Ed Stumpf, *Central Carolina Community College*; Fuzhen Zhang, *Nova Southeastern University*; Dennis Shepherd, *University of Colorado, Denver*; Rhonda Kilgo, *Jacksonville State University*; C. Altay Özgener, *Manatee Community College Bradenton*; William Forrest, *Baton Rouge Community College*; Tracy Cook, *University of Tennessee Knoxville*; Charles Hale, *California State Poly University Pomona*; Samuel Evers, *University of Alabama*; Seongchun Kwon, *University of Toledo*; Dr. Arun K. Agarwal, *Grambling State University*; Hyounkyun Oh, *Savannah State University*; Michael J. McConnell, *Clarion University*; Martha Chalhoub, *Collin County Community College*; Angela Lee Everett, *Chattanooga State Tech Community College*; Heather Van Dyke, *Walla Walla Community College*; Gregory Buthusiem, *Burlington County Community College*; Ward Shaffer, *College of Coastal Georgia*; Carmen Thomas, *Chatham University*

My thanks to David Falvo, The Behrend College, The Pennsylvania State University, for his contributions to this project. My thanks also to Robert Hostetler, The Behrend College, The Pennsylvania State University, and Bruce Edwards, University of Florida, for their significant contributions to previous editions of this text.

I would also like to thank the staff at Larson Texts, Inc. who assisted with proofreading the manuscript, preparing and proofreading the art package, and checking and typesetting the supplements.

On a personal level, I am grateful to my spouse, Deanna Gilbert Larson, for her love, patience, and support. Also, a special thanks goes to R. Scott O'Neil. If you have suggestions for improving this text, please feel free to write to me. Over the past two decades I have received many useful comments from both instructors and students, and I value these comments very highly.

Ron Larson

Ron Larson

Supplements

Supplements for the Instructor

Annotated Instructor's Edition   This AIE is the complete student text plus point-of-use annotations for the instructor, including extra projects, classroom activities, teaching strategies, and additional examples. Answers to even-numbered text exercises, Vocabulary Checks, and Explorations are also provided.

Complete Solutions Manual   This manual contains solutions to all exercises from the text, including Chapter Review Exercises and Chapter Tests.

Instructor's Companion Website   This free companion website contains an abundance of instructor resources.

PowerLecture™ with ExamView®   The CD-ROM provides the instructor with dynamic media tools for teaching Trigonometry. PowerPoint® lecture slides and art slides of the figures from the text, together with electronic files for the test bank and a link to the Solution Builder, are available. The algorithmic ExamView allows you to create, deliver, and customize tests (both print and online) in minutes with this easy-to-use assessment system. Enhance how your students interact with you, your lecture, and each other.

Solutions Builder   This is an electronic version of the complete solutions manual available via the PowerLecture and Instructor's Companion Website. It provides instructors with an efficient method for creating solution sets to homework or exams that can then be printed or posted.

Supplements for the Student

Student Companion Website   This free companion website contains an abundance of student resources.

Instructional DVDs   Keyed to the text by section, these DVDs provide comprehensive coverage of the course—along with additional explanations of concepts, sample problems, and applications—to help students review essential topics.

Student Study and Solutions Manual   This guide offers step-by-step solutions for all odd-numbered text exercises, Chapter and Cumulative Tests, and Practice Tests with solutions.

Premium eBook   The Premium eBook offers an interactive version of the textbook with search features, highlighting and note-making tools, and direct links to videos or tutorials that elaborate on the text discussions.

Enhanced WebAssign   Enhanced WebAssign is designed for you to do your homework online. This proven and reliable system uses pedagogy and content found in Larson's text, and then enhances it to help you learn Trigonometry more effectively. Automatically graded homework allows you to focus on your learning and get interactive study assistance outside of class.

Prerequisites

In Mathematics

Functions show how one variable is related to another variable.

In Real Life

Functions are used to estimate values, simulate processes, and discover relationships. You can model the enrollment rate of children in preschool and estimate the year in which the rate will reach a certain number. This estimate can be used to plan for future needs, such as adding teachers and buying books. (See Exercise 113, page 83.)

Jose Luis Pelaez/Getty Images

IN CAREERS

There are many careers that use functions. Several are listed below.

- Roofing Contractor
  Exercise 131, page 55

- Sociologist
  Exercise 80, page 101

- Automotive Engineer
  Exercise 61, page 108

- Demographer
  Exercises 67 and 68, page 109

What you should learn

- Represent and classify real numbers.
- Order real numbers and use inequalities.
- Find the absolute values of real numbers and find the distance between two real numbers.
- Evaluate algebraic expressions.
- Use the basic rules and properties of algebra.

Why you should learn it

Real numbers are used to represent many real-life quantities. For example, in Exercises 83–88 on page 13, you will use real numbers to represent the federal deficit.

Real Numbers

Real numbers are used in everyday life to describe quantities such as age, miles per gallon, and population. Real numbers are represented by symbols such as

$$-5, 9, 0, \frac{4}{3}, 0.666\ldots, 28.21, \sqrt{2}, \pi, \text{ and } \sqrt[3]{-32}.$$

Here are some important **subsets** (each member of subset B is also a member of set A) of the real numbers. The three dots, called *ellipsis points*, indicate that the pattern continues indefinitely.

$$\{1, 2, 3, 4, \ldots\} \quad \text{Set of natural numbers}$$
$$\{0, 1, 2, 3, 4, \ldots\} \quad \text{Set of whole numbers}$$
$$\{\ldots, -3, -2, -1, 0, 1, 2, 3, \ldots\} \quad \text{Set of integers}$$

A real number is **rational** if it can be written as the ratio p/q of two integers, where $q \neq 0$. For instance, the numbers

$$\frac{1}{3} = 0.3333\ldots = 0.\overline{3}, \frac{1}{8} = 0.125, \text{ and } \frac{125}{111} = 1.126126\ldots = 1.\overline{126}$$

are rational. The decimal representation of a rational number either repeats $\left(\text{as in } \frac{173}{55} = 3.1\overline{45}\right)$ or terminates $\left(\text{as in } \frac{1}{2} = 0.5\right)$. A real number that cannot be written as the ratio of two integers is called **irrational.** Irrational numbers have infinite nonrepeating decimal representations. For instance, the numbers

$$\sqrt{2} = 1.4142135\ldots \approx 1.41 \quad \text{and} \quad \pi = 3.1415926\ldots \approx 3.14$$

are irrational. (The symbol $\approx$ means "is approximately equal to.") Figure P.1 shows subsets of real numbers and their relationships to each other.

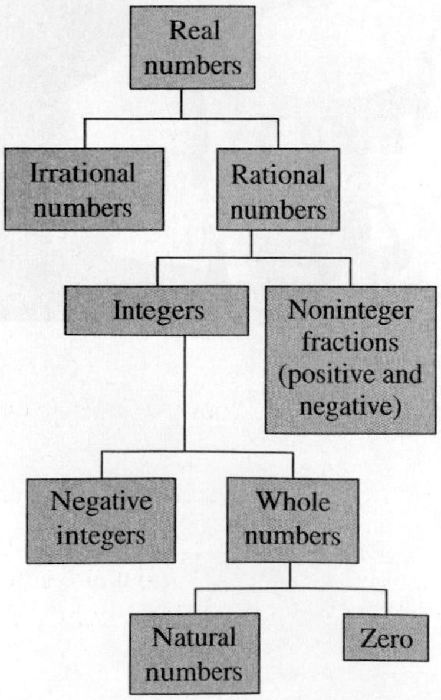

Example 1   Classifying Real Numbers

Determine which numbers in the set

$$\left\{-13, -\sqrt{5}, -1, -\frac{1}{3}, 0, \frac{5}{8}, \sqrt{2}, \pi, 7\right\}$$

are (a) natural numbers, (b) whole numbers, (c) integers, (d) rational numbers, and (e) irrational numbers.

Solution

a. Natural numbers: $\{7\}$

b. Whole numbers: $\{0, 7\}$

c. Integers: $\{-13, -1, 0, 7\}$

d. Rational numbers: $\left\{-13, -1, -\frac{1}{3}, 0, \frac{5}{8}, 7\right\}$

e. Irrational numbers: $\left\{-\sqrt{5}, \sqrt{2}, \pi\right\}$

FIGURE **P.1**   Subsets of real numbers

CHECK Point   Now try Exercise 11.

Real numbers are represented graphically on the **real number line.** When you draw a point on the real number line that corresponds to a real number, you are **plotting** the real number. The point 0 on the real number line is the **origin.** Numbers to the right of 0 are positive, and numbers to the left of 0 are negative, as shown in Figure P.2. The term **nonnegative** describes a number that is either positive or zero.

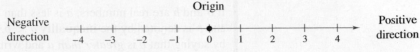

FIGURE P.2   The real number line

As illustrated in Figure P.3, there is a *one-to-one correspondence* between real numbers and points on the real number line.

Every real number corresponds to exactly one point on the real number line.

Every point on the real number line corresponds to exactly one real number.

FIGURE P.3   One-to-one correspondence

Example 2   Plotting Points on the Real Number Line

Plot the real numbers on the real number line.

a. $-\dfrac{7}{4}$

b. 2.3

c. $\dfrac{2}{3}$

d. -1.8

Solution

All four points are shown in Figure P.4.

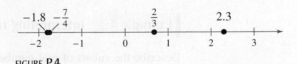

FIGURE P.4

a. The point representing the real number $-\frac{7}{4} = -1.75$ lies between -2 and -1, but closer to -2, on the real number line.

b. The point representing the real number 2.3 lies between 2 and 3, but closer to 2, on the real number line.

c. The point representing the real number $\frac{2}{3} = 0.666\ldots$ lies between 0 and 1, but closer to 1, on the real number line.

d. The point representing the real number -1.8 lies between -2 and -1, but closer to -2, on the real number line. Note that the point representing -1.8 lies slightly to the left of the point representing $-\frac{7}{4}$.

CHECK Point   Now try Exercise 17.

Ordering Real Numbers

One important property of real numbers is that they are *ordered*.

> ### Definition of Order on the Real Number Line
>
> If a and b are real numbers, a is less than b if $b - a$ is positive. The **order** of a and b is denoted by the **inequality** $a < b$. This relationship can also be described by saying that b is *greater than* a and writing $b > a$. The inequality $a \le b$ means that a is *less than or equal to* b, and the inequality $b \ge a$ means that b is *greater than or equal to* a. The symbols $<$, $>$, $\le$, and $\ge$ are *inequality symbols*.

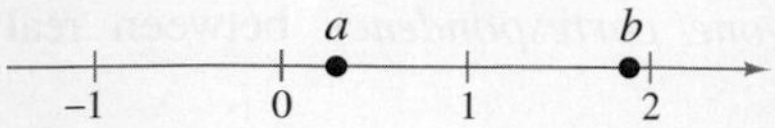

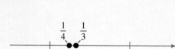

FIGURE **P.5** $a < b$ if and only if a lies to the left of b.

Geometrically, this definition implies that $a < b$ if and only if a lies to the *left* of b on the real number line, as shown in Figure P.5.

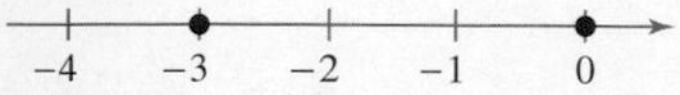

FIGURE **P.6**

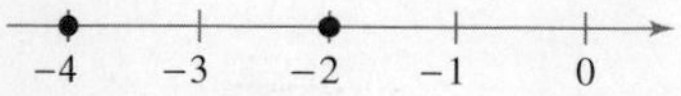

FIGURE **P.7**

FIGURE **P.8**

FIGURE **P.9**

Example 3   Ordering Real Numbers

Place the appropriate inequality symbol ($<$ or $>$) between the pair of real numbers.

a. $-3, 0$ **b.** $-2, -4$ **c.** $\dfrac{1}{4}, \dfrac{1}{3}$ **d.** $-\dfrac{1}{5}, -\dfrac{1}{2}$

Solution

a. Because -3 lies to the left of 0 on the real number line, as shown in Figure P.6, you can say that -3 is *less than* 0, and write $-3 < 0$.

b. Because -2 lies to the right of -4 on the real number line, as shown in Figure P.7, you can say that -2 is *greater than* -4, and write $-2 > -4$.

c. Because $\frac{1}{4}$ lies to the left of $\frac{1}{3}$ on the real number line, as shown in Figure P.8, you can say that $\frac{1}{4}$ is *less than* $\frac{1}{3}$, and write $\frac{1}{4} < \frac{1}{3}$.

d. Because $-\frac{1}{5}$ lies to the right of $-\frac{1}{2}$ on the real number line, as shown in Figure P.9, you can say that $-\frac{1}{5}$ is *greater than* $-\frac{1}{2}$, and write $-\frac{1}{5} > -\frac{1}{2}$.

CHECK*Point*   Now try Exercise 25.

Example 4   Interpreting Inequalities

Describe the subset of real numbers represented by each inequality.

a. $x \le 2$ **b.** $-2 \le x < 3$

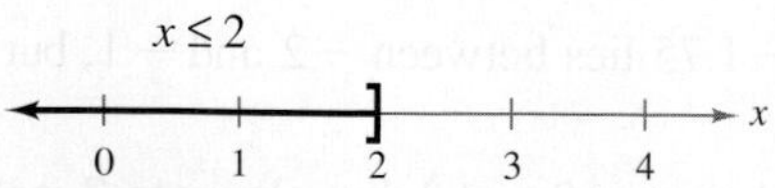

Solution

a. The inequality $x \le 2$ denotes all real numbers less than or equal to 2, as shown in Figure P.10.

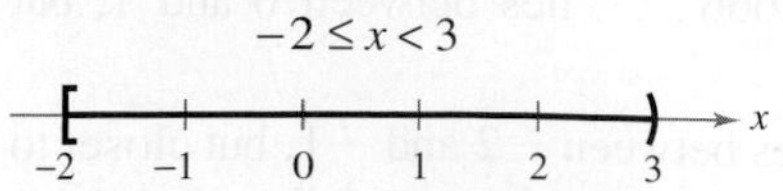

b. The inequality $-2 \le x < 3$ means that $x \ge -2$ *and* $x < 3$. This "double inequality" denotes all real numbers between -2 and 3, including -2 but not including 3, as shown in Figure P.11.

CHECK*Point*   Now try Exercise 31.

$x \le 2$

FIGURE **P.10**

$-2 \le x < 3$

FIGURE **P.11**

Inequalities can be used to describe subsets of real numbers called **intervals.** In the bounded intervals below, the real numbers a and b are the **endpoints** of each interval. The endpoints of a closed interval are included in the interval, whereas the endpoints of an open interval are not included in the interval.

Study Tip

The reason that the four types of intervals at the right are called *bounded* is that each has a finite length. An interval that does not have a finite length is *unbounded* (see below).

Bounded Intervals on the Real Number Line

Notation	Interval Type	Inequality	Graph
$[a, b]$	Closed	$a \le x \le b$	
(a, b)	Open	$a < x < b$	
$[a, b)$		$a \le x < b$	
$(a, b]$		$a < x \le b$	

 WARNING/CAUTION

Whenever you write an interval containing ∞ or $-\infty$, always use a parenthesis and never a bracket. This is because ∞ and $-\infty$ are never an endpoint of an interval and therefore are not included in the interval.

The symbols ∞, **positive infinity,** and $-\infty$, **negative infinity,** do not represent real numbers. They are simply convenient symbols used to describe the unboundedness of an interval such as $(1, \infty)$ or $(-\infty, 3]$.

Unbounded Intervals on the Real Number Line

Notation	Interval Type	Inequality	Graph
$[a, \infty)$		$x \ge a$	
(a, ∞)	Open	$x > a$	
$(-\infty, b]$		$x \le b$	
$(-\infty, b)$	Open	$x < b$	
$(-\infty, \infty)$	Entire real line	$-\infty < x < \infty$	

Example 5   Using Inequalities to Represent Intervals

Use inequality notation to describe each of the following.

a. c is at most 2.   **b.** m is at least -3.   **c.** All x in the interval $(-3, 5]$

Solution

a. The statement "c is at most 2" can be represented by $c \le 2$.

b. The statement "m is at least -3" can be represented by $m \ge -3$.

c. "All x in the interval $(-3, 5]$" can be represented by $-3 < x \le 5$.

CHECK*Point*   Now try Exercise 45.

Example 6   **Interpreting Intervals**

Give a verbal description of each interval.

a. $(-1, 0)$ **b.** $[2, \infty)$ **c.** $(-\infty, 0)$

Solution

a. This interval consists of all real numbers that are greater than -1 and less than 0.

b. This interval consists of all real numbers that are greater than or equal to 2.

c. This interval consists of all negative real numbers.

CHECK Point   Now try Exercise 41.

Absolute Value and Distance

The **absolute value** of a real number is its *magnitude*, or the distance between the origin and the point representing the real number on the real number line.

Definition of Absolute Value

If a is a real number, then the absolute value of a is

$$|a| = \begin{cases} a, & \text{if } a \geq 0 \\ -a, & \text{if } a < 0 \end{cases}.$$

Notice in this definition that the absolute value of a real number is never negative. For instance, if $a = -5$, then $|-5| = -(-5) = 5$. The absolute value of a real number is either positive or zero. Moreover, 0 is the only real number whose absolute value is 0. So, $|0| = 0$.

Example 7   **Finding Absolute Values**

a. $|-15| = 15$ **b.** $\left|\dfrac{2}{3}\right| = \dfrac{2}{3}$

c. $|-4.3| = 4.3$ **d.** $-|-6| = -(6) = -6$

CHECK Point   Now try Exercise 51.

Example 8   **Evaluating the Absolute Value of a Number**

Evaluate $\dfrac{|x|}{x}$ for (a) $x > 0$ and (b) $x < 0$.

Solution

a. If $x > 0$, then $|x| = x$ and $\dfrac{|x|}{x} = \dfrac{x}{x} = 1$.

b. If $x < 0$, then $|x| = -x$ and $\dfrac{|x|}{x} = \dfrac{-x}{x} = -1$.

CHECK Point   Now try Exercise 59.

The **Law of Trichotomy** states that for any two real numbers a and b, *precisely* one of three relationships is possible:

$$a = b, \quad a < b, \quad \text{or} \quad a > b.$$ Law of Trichotomy

Example 9   Comparing Real Numbers

Place the appropriate symbol ($<$, $>$, or $=$) between the pair of real numbers.

a. $|-4|$ ▓ $|3|$ **b.** $|-10|$ ▓ $|10|$ **c.** $-|-7|$ ▓ $|-7|$

Solution

a. $|-4| > |3|$ because $|-4| = 4$ and $|3| = 3$, and 4 is greater than 3.

b. $|-10| = |10|$ because $|-10| = 10$ and $|10| = 10$.

c. $-|-7| < |-7|$ because $-|-7| = -7$ and $|-7| = 7$, and -7 is less than 7.

CHECK*Point*   Now try Exercise 61.   ▄

Properties of Absolute Values

1. $|a| \geq 0$

2. $|-a| = |a|$

3. $|ab| = |a||b|$

4. $\left|\dfrac{a}{b}\right| = \dfrac{|a|}{|b|}, \quad b \neq 0$

Absolute value can be used to define the distance between two points on the real number line. For instance, the distance between -3 and 4 is

$$|-3 - 4| = |-7|$$
$$= 7$$

as shown in Figure P.12.

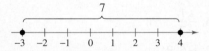

FIGURE P.12   The distance between -3 and 4 is 7.

Distance Between Two Points on the Real Number Line

Let a and b be real numbers. The **distance between a and b** is

$$d(a, b) = |b - a| = |a - b|.$$

Example 10   Finding a Distance

Find the distance between -25 and 13.

Solution

The distance between -25 and 13 is given by

$$|-25 - 13| = |-38| = 38.$$ Distance between -25 and 13

The distance can also be found as follows.

$$|13 - (-25)| = |38| = 38$$ Distance between -25 and 13

CHECK*Point*   Now try Exercise 67.   ▄

Algebraic Expressions

One characteristic of algebra is the use of letters to represent numbers. The letters are **variables,** and combinations of letters and numbers are **algebraic expressions.** Here are a few examples of algebraic expressions.

$$5x, \qquad 2x - 3, \qquad \frac{4}{x^2 + 2}, \qquad 7x + y$$

Definition of an Algebraic Expression

An **algebraic expression** is a collection of letters (**variables**) and real numbers (**constants**) combined using the operations of addition, subtraction, multiplication, division, and exponentiation.

The **terms** of an algebraic expression are those parts that are separated by *addition.* For example,

$$x^2 - 5x + 8 = x^2 + (-5x) + 8$$

has three terms: x^2 and $-5x$ are the **variable terms** and 8 is the **constant term.** The numerical factor of a term is called the **coefficient.** For instance, the coefficient of $-5x$ is -5, and the coefficient of x^2 is 1.

Example 11   Identifying Terms and Coefficients

Algebraic Expression	Terms	Coefficients
a. $5x - \dfrac{1}{7}$	$5x, -\dfrac{1}{7}$	$5, -\dfrac{1}{7}$
b. $2x^2 - 6x + 9$	$2x^2, -6x, 9$	$2, -6, 9$
c. $\dfrac{3}{x} + \dfrac{1}{2}x^4 - y$	$\dfrac{3}{x}, \dfrac{1}{2}x^4, -y$	$3, \dfrac{1}{2}, -1$

CHECK*Point*▶   Now try Exercise 89.

To **evaluate** an algebraic expression, substitute numerical values for each of the variables in the expression, as shown in the next example.

Example 12   Evaluating Algebraic Expressions

Expression	Value of Variable	Substitute	Value of Expression
a. $-3x + 5$	$x = 3$	$-3(3) + 5$	$-9 + 5 = -4$
b. $3x^2 + 2x - 1$	$x = -1$	$3(-1)^2 + 2(-1) - 1$	$3 - 2 - 1 = 0$
c. $\dfrac{2x}{x + 1}$	$x = -3$	$\dfrac{2(-3)}{-3 + 1}$	$\dfrac{-6}{-2} = 3$

Note that you must substitute the value for *each* occurrence of the variable.

CHECK*Point*▶   Now try Exercise 95.

When an algebraic expression is evaluated, the **Substitution Principle** is used. It states that "If $a = b$, then a can be replaced by b in any expression involving a." In Example 12(a), for instance, 3 is *substituted* for x in the expression $-3x + 5$.

Basic Rules of Algebra

There are four arithmetic operations with real numbers: *addition*, *multiplication*, *subtraction*, and *division*, denoted by the symbols $+$, $\times$ or $\cdot$, $-$, and $\div$ or $/$. Of these, addition and multiplication are the two primary operations. Subtraction and division are the inverse operations of addition and multiplication, respectively.

Definitions of Subtraction and Division

Subtraction: Add the opposite.   **Division:** Multiply by the reciprocal.

$$a - b = a + (-b) \qquad\qquad \text{If } b \neq 0, \text{ then } a/b = a\left(\frac{1}{b}\right) = \frac{a}{b}.$$

In these definitions, $-b$ is the **additive inverse** (or opposite) of b, and $1/b$ is the **multiplicative inverse** (or reciprocal) of b. In the fractional form a/b, a is the **numerator** of the fraction and b is the **denominator.**

Because the properties of real numbers below are true for variables and algebraic expressions as well as for real numbers, they are often called the **Basic Rules of Algebra.** Try to formulate a verbal description of each property. For instance, the first property states that *the order in which two real numbers are added does not affect their sum.*

Basic Rules of Algebra

Let a, b, and c be real numbers, variables, or algebraic expressions.

Property		*Example*
Commutative Property of Addition:	$a + b = b + a$	$4x + x^2 = x^2 + 4x$
Commutative Property of Multiplication:	$ab = ba$	$(4 - x)x^2 = x^2(4 - x)$
Associative Property of Addition:	$(a + b) + c = a + (b + c)$	$(x + 5) + x^2 = x + (5 + x^2)$
Associative Property of Multiplication:	$(ab)c = a(bc)$	$(2x \cdot 3y)(8) = (2x)(3y \cdot 8)$
Distributive Properties:	$a(b + c) = ab + ac$	$3x(5 + 2x) = 3x \cdot 5 + 3x \cdot 2x$
	$(a + b)c = ac + bc$	$(y + 8)y = y \cdot y + 8 \cdot y$
Additive Identity Property:	$a + 0 = a$	$5y^2 + 0 = 5y^2$
Multiplicative Identity Property:	$a \cdot 1 = a$	$(4x^2)(1) = 4x^2$
Additive Inverse Property:	$a + (-a) = 0$	$5x^3 + (-5x^3) = 0$
Multiplicative Inverse Property:	$a \cdot \dfrac{1}{a} = 1, \quad a \neq 0$	$(x^2 + 4)\left(\dfrac{1}{x^2 + 4}\right) = 1$

Because subtraction is defined as "adding the opposite," the Distributive Properties are also true for subtraction. For instance, the "subtraction form" of $a(b + c) = ab + ac$ is $a(b - c) = ab - ac$. Note that the operations of subtraction and division are neither commutative nor associative. The examples

$$7 - 3 \neq 3 - 7 \quad \text{and} \quad 20 \div 4 \neq 4 \div 20$$

show that subtraction and division are not commutative. Similarly

$$5 - (3 - 2) \neq (5 - 3) - 2 \quad \text{and} \quad 16 \div (4 \div 2) \neq (16 \div 4) \div 2$$

demonstrate that subtraction and division are not associative.

Example 13 Identifying Rules of Algebra

Identify the rule of algebra illustrated by the statement.

a. $(5x^3)2 = 2(5x^3)$

b. $\left(4x + \dfrac{1}{3}\right) - \left(4x + \dfrac{1}{3}\right) = 0$

c. $7x \cdot \dfrac{1}{7x} = 1, \quad x \neq 0$

d. $(2 + 5x^2) + x^2 = 2 + (5x^2 + x^2)$

Solution

a. This statement illustrates the Commutative Property of Multiplication. In other words, you obtain the same result whether you multiply $5x^3$ by 2, or 2 by $5x^3$.

b. This statement illustrates the Additive Inverse Property. In terms of subtraction, this property simply states that when any expression is subtracted from itself the result is 0.

c. This statement illustrates the Multiplicative Inverse Property. Note that it is important that x be a nonzero number. If x were 0, the reciprocal of x would be undefined.

d. This statement illustrates the Associative Property of Addition. In other words, to form the sum

$$2 + 5x^2 + x^2$$

it does not matter whether 2 and $5x^2$, or $5x^2$ and x^2 are added first.

CHECK*Point* ▶ Now try Exercise 101.

Study Tip

Notice the difference between the *opposite of a number* and a *negative number*. If a is already negative, then its opposite, $-a$, is positive. For instance, if $a = -5$, then

$$-a = -(-5) = 5.$$

Properties of Negation and Equality

Let a, b, and c be real numbers, variables, or algebraic expressions.

Property	Example
1. $(-1)a = -a$	$(-1)7 = -7$
2. $-(-a) = a$	$-(-6) = 6$
3. $(-a)b = -(ab) = a(-b)$	$(-5)3 = -(5 \cdot 3) = 5(-3)$
4. $(-a)(-b) = ab$	$(-2)(-x) = 2x$
5. $-(a + b) = (-a) + (-b)$	$-(x + 8) = (-x) + (-8)$
	$\quad\quad\quad\quad = -x - 8$
6. If $a = b$, then $a \pm c = b \pm c$.	$\frac{1}{2} + 3 = 0.5 + 3$
7. If $a = b$, then $ac = bc$.	$4^2 \cdot 2 = 16 \cdot 2$
8. If $a \pm c = b \pm c$, then $a = b$.	$1.4 - 1 = \frac{7}{5} - 1 \implies 1.4 = \frac{7}{5}$
9. If $ac = bc$ and $c \neq 0$, then $a = b$.	$3x = 3 \cdot 4 \implies x = 4$

The "or" in the Zero-Factor Property includes the possibility that either or both factors may be zero. This is an **inclusive or,** and it is the way the word "or" is generally used in mathematics.

Properties of Zero

Let a and b be real numbers, variables, or algebraic expressions.

1. $a + 0 = a$ and $a - 0 = a$ 2. $a \cdot 0 = 0$

3. $\dfrac{0}{a} = 0, \quad a \neq 0$ 4. $\dfrac{a}{0}$ is undefined.

5. **Zero-Factor Property:** If $ab = 0$, then $a = 0$ or $b = 0$.

Properties and Operations of Fractions

Let a, b, c, and d be real numbers, variables, or algebraic expressions such that $b \neq 0$ and $d \neq 0$.

1. **Equivalent Fractions:** $\dfrac{a}{b} = \dfrac{c}{d}$ if and only if $ad = bc$.

2. **Rules of Signs:** $-\dfrac{a}{b} = \dfrac{-a}{b} = \dfrac{a}{-b}$ and $\dfrac{-a}{-b} = \dfrac{a}{b}$

3. **Generate Equivalent Fractions:** $\dfrac{a}{b} = \dfrac{ac}{bc}, \quad c \neq 0$

4. **Add or Subtract with Like Denominators:** $\dfrac{a}{b} \pm \dfrac{c}{b} = \dfrac{a \pm c}{b}$

5. **Add or Subtract with Unlike Denominators:** $\dfrac{a}{b} \pm \dfrac{c}{d} = \dfrac{ad \pm bc}{bd}$

6. **Multiply Fractions:** $\dfrac{a}{b} \cdot \dfrac{c}{d} = \dfrac{ac}{bd}$

7. **Divide Fractions:** $\dfrac{a}{b} \div \dfrac{c}{d} = \dfrac{a}{b} \cdot \dfrac{d}{c} = \dfrac{ad}{bc}, \quad c \neq 0$

In Property 1 of fractions, the phrase "if and only if" implies two statements. One statement is: If $a/b = c/d$, then $ad = bc$. The other statement is: If $ad = bc$, where $b \neq 0$ and $d \neq 0$, then $a/b = c/d$.

Example 14   Properties and Operations of Fractions

a. Equivalent fractions: $\dfrac{x}{5} = \dfrac{3 \cdot x}{3 \cdot 5} = \dfrac{3x}{15}$ **b.** Divide fractions: $\dfrac{7}{x} \div \dfrac{3}{2} = \dfrac{7}{x} \cdot \dfrac{2}{3} = \dfrac{14}{3x}$

c. Add fractions with unlike denominators: $\dfrac{x}{3} + \dfrac{2x}{5} = \dfrac{5 \cdot x + 3 \cdot 2x}{3 \cdot 5} = \dfrac{11x}{15}$

CHECK*Point* Now try Exercise 119.

If a, b, and c are integers such that $ab = c$, then a and b are **factors** or **divisors** of c. A **prime number** is an integer that has exactly two positive factors—itself and 1—such as 2, 3, 5, 7, and 11. The numbers 4, 6, 8, 9, and 10 are **composite** because each can be written as the product of two or more prime numbers. The number 1 is neither prime nor composite. The **Fundamental Theorem of Arithmetic** states that every positive integer greater than 1 can be written as the product of prime numbers in precisely one way (disregarding order). For instance, the *prime factorization* of 24 is $24 = 2 \cdot 2 \cdot 2 \cdot 3$.

P.1   EXERCISES

See www.CalcChat.com for worked-out solutions to odd-numbered exercises.

VOCABULARY: Fill in the blanks.

1. A real number is _____ if it can be written as the ratio $\frac{p}{q}$ of two integers, where $q \neq 0$.

2. _____ numbers have infinite nonrepeating decimal representations.

3. The point 0 on the real number line is called the _____.

4. The distance between the origin and a point representing a real number on the real number line is the _____ _____ of the real number.

5. A number that can be written as the product of two or more prime numbers is called a _____ number.

6. An integer that has exactly two positive factors, the integer itself and 1, is called a _____ number.

7. An algebraic expression is a collection of letters called _____ and real numbers called _____.

8. The _____ of an algebraic expression are those parts separated by addition.

9. The numerical factor of a variable term is the _____ of the variable term.

10. The _____ _____ states that if $ab = 0$, then $a = 0$ or $b = 0$.

SKILLS AND APPLICATIONS

In Exercises 11–16, determine which numbers in the set are (a) natural numbers, (b) whole numbers, (c) integers, (d) rational numbers, and (e) irrational numbers.

11. $\left\{ -9, -\frac{7}{2}, 5, \frac{2}{3}, \sqrt{2}, 0, 1, -4, 2, -11 \right\}$

12. $\left\{ \sqrt{5}, -7, -\frac{7}{3}, 0, 3.12, \frac{5}{4}, -3, 12, 5 \right\}$

13. $\{2.01, 0.666 \ldots, -13, 0.010110111 \ldots, 1, -6\}$

14. $\{2.3030030003 \ldots, 0.7575, -4.63, \sqrt{10}, -75, 4\}$

15. $\left\{ -\pi, -\frac{1}{3}, \frac{6}{3}, \frac{1}{2}\sqrt{2}, -7.5, -1, 8, -22 \right\}$

16. $\left\{ 25, -17, -\frac{12}{5}, \sqrt{9}, 3.12, \frac{1}{2}\pi, 7, -11.1, 13 \right\}$

In Exercises 17 and 18, plot the real numbers on the real number line.

17. (a) 3   (b) $\frac{7}{2}$ (c) $-\frac{5}{2}$ (d) -5.2

18. (a) 8.5   (b) $\frac{4}{3}$ (c) -4.75 (d) $-\frac{8}{3}$

In Exercises 19–22, use a calculator to find the decimal form of the rational number. If it is a nonterminating decimal, write the repeating pattern.

19. $\frac{5}{8}$ 20. $\frac{1}{3}$

21. $\frac{41}{333}$ 22. $\frac{6}{11}$

In Exercises 23 and 24, approximate the numbers and place the correct symbol (< or >) between them.

23.

24.

In Exercises 25–30, plot the two real numbers on the real number line. Then place the appropriate inequality symbol (< or >) between them.

25. $-4, -8$ 26. $-3.5, 1$

27. $\frac{3}{2}, 7$ 28. $1, \frac{16}{3}$

29. $\frac{5}{6}, \frac{2}{3}$ 30. $-\frac{8}{7}, -\frac{3}{7}$

In Exercises 31–42, (a) give a verbal description of the subset of real numbers represented by the inequality or the interval, (b) sketch the subset on the real number line, and (c) state whether the interval is bounded or unbounded.

31. $x \leq 5$ 32. $x \geq -2$

33. $x < 0$ 34. $x > 3$

35. $[4, \infty)$ 36. $(-\infty, 2)$

37. $-2 < x < 2$ 38. $0 \leq x \leq 5$

39. $-1 \leq x < 0$ 40. $0 < x \leq 6$

41. $[-2, 5)$ 42. $(-1, 2]$

In Exercises 43–50, use inequality notation and interval notation to describe the set.

43. y is nonnegative.

44. y is no more than 25.

45. x is greater than -2 and at most 4.

46. y is at least -6 and less than 0.

47. t is at least 10 and at most 22.

48. k is less than 5 but no less than -3.

49. The dog's weight W is more than 65 pounds.

50. The annual rate of inflation r is expected to be at least 2.5% but no more than 5%.

In Exercises 51–60, evaluate the expression.

51. $|-10|$

52. $|0|$

53. $|3 - 8|$

54. $|4 - 1|$

55. $|-1| - |-2|$

56. $-3 - |-3|$

57. $\dfrac{-5}{|-5|}$

58. $-3|-3|$

59. $\dfrac{|x + 2|}{x + 2}, \quad x < -2$

60. $\dfrac{|x - 1|}{x - 1}, \quad x > 1$

In Exercises 61–66, place the correct symbol (<, >, or =) between the two real numbers.

61. $|-3|$ ▨ $-|-3|$

62. $|-4|$ ▨ $|4|$

63. -5 ▨ $-|5|$

64. $-|-6|$ ▨ $|-6|$

65. $-|-2|$ ▨ $-|2|$

66. $-(-2)$ ▨ -2

In Exercises 67–72, find the distance between a and b.

67. $a = 126, b = 75$

68. $a = -126, b = -75$

69. $a = -\frac{5}{2}, b = 0$

70. $a = \frac{1}{4}, b = \frac{11}{4}$

71. $a = \frac{16}{5}, b = \frac{112}{75}$

72. $a = 9.34, b = -5.65$

In Exercises 73–78, use absolute value notation to describe the situation.

73. The distance between x and 5 is no more than 3.

74. The distance between x and −10 is at least 6.

75. y is at least six units from 0.

76. y is at most two units from a.

77. While traveling on the Pennsylvania Turnpike, you pass milepost 57 near Pittsburgh, then milepost 236 near Gettysburg. How many miles do you travel during that time period?

78. The temperature in Bismarck, North Dakota was 60°F at noon, then 23°F at midnight. What was the change in temperature over the 12-hour period?

BUDGET VARIANCE   In Exercises 79–82, the accounting department of a sports drink bottling company is checking to see whether the actual expenses of a department differ from the budgeted expenses by more than $500 or by more than 5%. Fill in the missing parts of the table, and determine whether each actual expense passes the "budget variance test."

		Budgeted Expense, b	Actual Expense, a	$|a - b|$	$0.05b$
79.	Wages	$112,700	$113,356	▨	▨
80.	Utilities	$9,400	$9,772	▨	▨
81.	Taxes	$37,640	$37,335	▨	▨
82.	Insurance	$2,575	$2,613	▨	▨

FEDERAL DEFICIT   In Exercises 83–88, use the bar graph, which shows the receipts of the federal government (in billions of dollars) for selected years from 1996 through 2006. In each exercise you are given the expenditures of the federal government. Find the magnitude of the surplus or deficit for the year.   (Source: U.S. Office of Management and Budget)

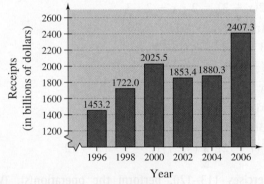

Year	Receipts	Expenditures	$|Receipts - Expenditures|$
83. 1996	▨	$1560.6 billion	▨
84. 1998	▨	$1652.7 billion	▨
85. 2000	▨	$1789.2 billion	▨
86. 2002	▨	$2011.2 billion	▨
87. 2004	▨	$2293.0 billion	▨
88. 2006	▨	$2655.4 billion	▨

In Exercises 89–94, identify the terms. Then identify the coefficients of the variable terms of the expression.

89. $7x + 4$

90. $6x^3 - 5x$

91. $\sqrt{3}x^2 - 8x - 11$

92. $3\sqrt{3}x^2 + 1$

93. $4x^3 + \dfrac{x}{2} - 5$

94. $3x^4 - \dfrac{x^2}{4}$

In Exercises 95–100, evaluate the expression for each value of x. (If not possible, state the reason.)

Expression	Values	
95. $4x - 6$	(a) $x = -1$	(b) $x = 0$
96. $9 - 7x$	(a) $x = -3$	(b) $x = 3$
97. $x^2 - 3x + 4$	(a) $x = -2$	(b) $x = 2$
98. $-x^2 + 5x - 4$	(a) $x = -1$	(b) $x = 1$
99. $\dfrac{x + 1}{x - 1}$	(a) $x = 1$	(b) $x = -1$
100. $\dfrac{x}{x + 2}$	(a) $x = 2$	(b) $x = -2$

In Exercises 101–112, identify the rule(s) of algebra illustrated by the statement.

101. $x + 9 = 9 + x$

102. $2\left(\frac{1}{2}\right) = 1$

103. $\dfrac{1}{h + 6}(h + 6) = 1, \quad h \neq -6$

104. $(x + 3) - (x + 3) = 0$

105. $2(x + 3) = 2 \cdot x + 2 \cdot 3$

106. $(z - 2) + 0 = z - 2$

107. $1 \cdot (1 + x) = 1 + x$

108. $(z + 5)x = z \cdot x + 5 \cdot x$

109. $x + (y + 10) = (x + y) + 10$

110. $x(3y) = (x \cdot 3)y = (3x)y$

111. $3(t - 4) = 3 \cdot t - 3 \cdot 4$

112. $\frac{1}{7}(7 \cdot 12) = \left(\frac{1}{7} \cdot 7\right)12 = 1 \cdot 12 = 12$

In Exercises 113–120, perform the operation(s). (Write fractional answers in simplest form.)

113. $\frac{3}{16} + \frac{5}{16}$

114. $\frac{6}{7} - \frac{4}{7}$

115. $\frac{5}{8} - \frac{5}{12} + \frac{1}{6}$

116. $\frac{10}{11} + \frac{6}{33} - \frac{13}{66}$

117. $12 \div \frac{1}{4}$

118. $-\left(6 \cdot \frac{4}{8}\right)$

119. $\dfrac{2x}{3} - \dfrac{x}{4}$

120. $\dfrac{5x}{6} \cdot \dfrac{2}{9}$

EXPLORATION

In Exercises 121 and 122, use the real numbers A, B, and C shown on the number line. Determine the sign of each expression.

121. (a) $-A$

(b) $B - A$

122. (a) $-C$

(b) $A - C$

123. CONJECTURE

(a) Use a calculator to complete the table.

n	1	0.5	0.01	0.0001	0.000001
$5/n$					

(b) Use the result from part (a) to make a conjecture about the value of $5/n$ as n approaches 0.

124. CONJECTURE

(a) Use a calculator to complete the table.

n	1	10	100	10,000	100,000
$5/n$					

(b) Use the result from part (a) to make a conjecture about the value of $5/n$ as n increases without bound.

TRUE OR FALSE? In Exercises 125–128, determine whether the statement is true or false. Justify your answer.

125. If $a > 0$ and $b < 0$, then $a - b > 0$.

126. If $a > 0$ and $b < 0$, then $ab > 0$.

127. If $a < b$, then $\dfrac{1}{a} < \dfrac{1}{b}$, where $a \neq 0$ and $b \neq 0$.

128. Because $\dfrac{a + b}{c} = \dfrac{a}{c} + \dfrac{b}{c}$, then $\dfrac{c}{a + b} = \dfrac{c}{a} + \dfrac{c}{b}$.

129. THINK ABOUT IT Consider $|u + v|$ and $|u| + |v|$, where $u \neq 0$ and $v \neq 0$.

(a) Are the values of the expressions always equal? If not, under what conditions are they unequal?

(b) If the two expressions are not equal for certain values of u and v, is one of the expressions always greater than the other? Explain.

130. THINK ABOUT IT Is there a difference between saying that a real number is positive and saying that a real number is nonnegative? Explain.

131. THINK ABOUT IT Because every even number is divisible by 2, is it possible that there exist any even prime numbers? Explain.

132. THINK ABOUT IT Is it possible for a real number to be both rational and irrational? Explain.

133. WRITING Can it ever be true that $|a| = -a$ for a real number a? Explain.

134. CAPSTONE Describe the differences among the sets of natural numbers, whole numbers, integers, rational numbers, and irrational numbers.

What you should learn

- Identify different types of equations.
- Solve linear equations in one variable and equations that lead to linear equations.
- Solve quadratic equations by factoring, extracting square roots, completing the square, and using the Quadratic Formula.
- Solve polynomial equations of degree three or greater.
- Solve equations involving radicals.
- Solve equations with absolute values.

Why you should learn it

Linear equations are used in many real-life applications. For example, in Exercises 155 and 156 on page 27, linear equations can be used to model the relationship between the length of a thigh bone and the height of a person, helping researchers learn about ancient cultures.

P.2 SOLVING EQUATIONS

Equations and Solutions of Equations

An **equation** in x is a statement that two algebraic expressions are equal. For example

$$3x - 5 = 7, \quad x^2 - x - 6 = 0, \quad \text{and} \quad \sqrt{2x} = 4$$

are equations. To **solve** an equation in x means to find all values of x for which the equation is true. Such values are **solutions.** For instance, $x = 4$ is a solution of the equation

$$3x - 5 = 7$$

because $3(4) - 5 = 7$ is a true statement.

The solutions of an equation depend on the kinds of numbers being considered. For instance, in the set of rational numbers, $x^2 = 10$ has no solution because there is no rational number whose square is 10. However, in the set of real numbers, the equation has the two solutions $x = \sqrt{10}$ and $x = -\sqrt{10}$.

An equation that is true for *every* real number in the *domain* of the variable is called an **identity.** The domain is the set of all real numbers for which the equation is defined. For example

$$x^2 - 9 = (x + 3)(x - 3) \qquad \text{Identity}$$

is an identity because it is a true statement for any real value of x. The equation

$$\frac{x}{3x^2} = \frac{1}{3x} \qquad \text{Identity}$$

where $x \neq 0$, is an identity because it is true for any nonzero real value of x.

An equation that is true for just *some* (or even none) of the real numbers in the domain of the variable is called a **conditional equation.** For example, the equation

$$x^2 - 9 = 0 \qquad \text{Conditional equation}$$

is conditional because $x = 3$ and $x = -3$ are the only values in the domain that satisfy the equation. The equation $2x - 4 = 2x + 1$ is conditional because there are no real values of x for which the equation is true.

Linear Equations in One Variable

Definition of a Linear Equation

A **linear equation in one variable** x is an equation that can be written in the standard form

$$ax + b = 0$$

where a and b are real numbers with $a \neq 0$.

HISTORICAL NOTE

British Museum

This ancient Egyptian papyrus, discovered in 1858, contains one of the earliest examples of mathematical writing in existence. The papyrus itself dates back to around 1650 B.C., but it is actually a copy of writings from two centuries earlier. The algebraic equations on the papyrus were written in words. Diophantus, a Greek who lived around A.D. 250, is often called the Father of Algebra. He was the first to use abbreviated word forms in equations.

A linear equation in one variable, written in standard form, always has *exactly one* solution. To see this, consider the following steps.

$$ax + b = 0 \qquad \text{Original equation, with } a \neq 0$$

$$ax = -b \qquad \text{Subtract } b \text{ from each side.}$$

$$x = -\frac{b}{a} \qquad \text{Divide each side by } a.$$

To solve a conditional equation in x, isolate x on one side of the equation by a sequence of **equivalent** (and usually simpler) **equations,** each having the same solution(s) as the original equation. The operations that yield equivalent equations come from the Substitution Principle and the Properties of Equality studied in Section P.1.

Generating Equivalent Equations

An equation can be transformed into an *equivalent equation* by one or more of the following steps.

	Given Equation	Equivalent Equation
1. Remove symbols of grouping, combine like terms, or simplify fractions on one or both sides of the equation.	$2x - x = 4$	$x = 4$
2. Add (or subtract) the same quantity to (from) *each* side of the equation.	$x + 1 = 6$	$x = 5$
3. Multiply (or divide) *each* side of the equation by the same *nonzero* quantity.	$2x = 6$	$x = 3$
4. Interchange the two sides of the equation.	$2 = x$	$x = 2$

Study Tip

After solving an equation, you should check each solution in the original equation. For instance, you can check the solution of Example 1(a) as follows.

$$3x - 6 = 0 \qquad \text{Write original equation.}$$

$$3(2) - 6 \overset{?}{=} 0 \qquad \text{Substitute 2 for } x.$$

$$0 = 0 \qquad \text{Solution checks. } \checkmark$$

Try checking the solution of Example 1(b).

Example 1   Solving a Linear Equation

a. $3x - 6 = 0$ Original equation

$\qquad 3x = 6$ Add 6 to each side.

$\qquad\quad x = 2$ Divide each side by 3.

b. $5x + 4 = 3x - 8$ Original equation

$\quad 2x + 4 = -8$ Subtract $3x$ from each side.

$\quad 2x = -12$ Subtract 4 from each side.

$\qquad x = -6$ Divide each side by 2.

CHECK*Point*   Now try Exercise 15.

To solve an equation involving fractional expressions, find the least common denominator (LCD) of all terms and multiply every term by the LCD. This process will clear the original equation of fractions and produce a simpler equation.

Study Tip

An equation with a *single fraction* on each side can be cleared of denominators by **cross multiplying.** To do this, multiply the left numerator by the right denominator and the right numerator by the left denominator as follows.

$$\frac{a}{b} = \frac{c}{d} \qquad \text{Original equation}$$

$$ad = cb \qquad \text{Cross multiply.}$$

Example 2   An Equation Involving Fractional Expressions

Solve $\dfrac{x}{3} + \dfrac{3x}{4} = 2$.

Solution

$\dfrac{x}{3} + \dfrac{3x}{4} = 2$	Write original equation.
$(12)\dfrac{x}{3} + (12)\dfrac{3x}{4} = (12)2$	Multiply each term by the LCD of 12.
$4x + 9x = 24$	Divide out and multiply.
$13x = 24$	Combine like terms.
$x = \dfrac{24}{13}$	Divide each side by 13.

The solution is $x = \frac{24}{13}$. Check this in the original equation.

CHECK*Point* Now try Exercise 23.

When multiplying or dividing an equation by a *variable* quantity, it is possible to introduce an extraneous solution. An **extraneous solution** is one that does not satisfy the original equation. Therefore, it is essential that you check your solutions.

Example 3   An Equation with an Extraneous Solution

Solve $\dfrac{1}{x-2} = \dfrac{3}{x+2} - \dfrac{6x}{x^2-4}$.

Solution

The LCD is $x^2 - 4$, or $(x+2)(x-2)$. Multiply each term by this LCD.

$$\frac{1}{x-2}(x+2)(x-2) = \frac{3}{x+2}(x+2)(x-2) - \frac{6x}{x^2-4}(x+2)(x-2)$$

$$x + 2 = 3(x-2) - 6x, \qquad x \neq \pm 2$$

$$x + 2 = 3x - 6 - 6x$$

$$x + 2 = -3x - 6$$

$$4x = -8 \quad \Longrightarrow \quad x = -2 \qquad \text{Extraneous solution}$$

In the original equation, $x = -2$ yields a denominator of zero. So, $x = -2$ is an extraneous solution, and the original equation has *no solution*.

CHECK*Point* Now try Exercise 35.

Study Tip

Recall that the least common denominator of two or more fractions consists of the product of all prime factors in the denominators, with each factor given the highest power of its occurrence in any denominator. For instance, in Example 3, by factoring each denominator you can determine that the LCD is $(x+2)(x-2)$.

Quadratic Equations

A **quadratic equation** in x is an equation that can be written in the general form

$$ax^2 + bx + c = 0$$

where a, b, and c are real numbers, with $a \neq 0$. A quadratic equation in x is also known as a **second-degree polynomial equation** in x.

You should be familiar with the following four methods of solving quadratic equations.

Solving a Quadratic Equation

Factoring: If $ab = 0$, then $a = 0$ or $b = 0$.

Example:
$$x^2 - x - 6 = 0$$
$$(x - 3)(x + 2) = 0$$
$$x - 3 = 0 \implies x = 3$$
$$x + 2 = 0 \implies x = -2$$

Square Root Principle: If $u^2 = c$, where $c > 0$, then $u = \pm\sqrt{c}$.

Example:
$$(x + 3)^2 = 16$$
$$x + 3 = \pm 4$$
$$x = -3 \pm 4$$
$$x = 1 \quad \text{or} \quad x = -7$$

Completing the Square: If $x^2 + bx = c$, then

$$x^2 + bx + \left(\frac{b}{2}\right)^2 = c + \left(\frac{b}{2}\right)^2 \qquad \text{Add } \left(\frac{b}{2}\right)^2 \text{ to each side.}$$

$$\left(x + \frac{b}{2}\right)^2 = c + \frac{b^2}{4}.$$

Example:
$$x^2 + 6x = 5$$
$$x^2 + 6x + 3^2 = 5 + 3^2 \qquad \text{Add } \left(\frac{6}{2}\right)^2 \text{ to each side.}$$
$$(x + 3)^2 = 14$$
$$x + 3 = \pm\sqrt{14}$$
$$x = -3 \pm \sqrt{14}$$

Quadratic Formula: If $ax^2 + bx + c = 0$, then $x = \dfrac{-b \pm \sqrt{b^2 - 4ac}}{2a}$.

Example:
$$2x^2 + 3x - 1 = 0$$
$$x = \frac{-3 \pm \sqrt{3^2 - 4(2)(-1)}}{2(2)}$$
$$= \frac{-3 \pm \sqrt{17}}{4}$$

Study Tip

The Square Root Principle is also referred to as *extracting square roots*.

Study Tip

You can solve every quadratic equation by completing the square or using the Quadratic Formula.

| **Example 4** | **Solving a Quadratic Equation by Factoring** |

a. $2x^2 + 9x + 7 = 3$ Original equation

$2x^2 + 9x + 4 = 0$ Write in general form.

$(2x + 1)(x + 4) = 0$ Factor.

$2x + 1 = 0$ ⟹ $x = -\dfrac{1}{2}$ Set 1st factor equal to 0.

$x + 4 = 0$ ⟹ $x = -4$ Set 2nd factor equal to 0.

The solutions are $x = -\frac{1}{2}$ and $x = -4$. Check these in the original equation.

b. $6x^2 - 3x = 0$ Original equation

$3x(2x - 1) = 0$ Factor.

$3x = 0$ ⟹ $x = 0$ Set 1st factor equal to 0.

$2x - 1 = 0$ ⟹ $x = \dfrac{1}{2}$ Set 2nd factor equal to 0.

The solutions are $x = 0$ and $x = \frac{1}{2}$. Check these in the original equation.

CHECK *Point*   Now try Exercise 49.

Note that the method of solution in Example 4 is based on the Zero-Factor Property from Section P.1. Be sure you see that this property works *only* for equations written in general form (in which the right side of the equation is zero). So, all terms must be collected on one side *before* factoring. For instance, in the equation $(x - 5)(x + 2) = 8$, it is *incorrect* to set each factor equal to 8. Try to solve this equation correctly.

| **Example 5** | **Extracting Square Roots** |

Solve each equation by extracting square roots.

a. $4x^2 = 12$ **b.** $(x - 3)^2 = 7$

Solution

a. $4x^2 = 12$ Write original equation.

$x^2 = 3$ Divide each side by 4.

$x = \pm\sqrt{3}$ Extract square roots.

When you take the square root of a variable expression, you must account for both positive and negative solutions. So, the solutions are $x = \sqrt{3}$ and $x = -\sqrt{3}$. Check these in the original equation.

b. $(x - 3)^2 = 7$ Write original equation.

$x - 3 = \pm\sqrt{7}$ Extract square roots.

$x = 3 \pm \sqrt{7}$ Add 3 to each side.

The solutions are $x = 3 \pm \sqrt{7}$. Check these in the original equation.

CHECK *Point*   Now try Exercise 65.

When solving quadratic equations by completing the square, you must add $(b/2)^2$ to *each side* in order to maintain equality. If the leading coefficient is *not* 1, you must divide each side of the equation by the leading coefficient *before* completing the square, as shown in Example 7.

Example 6 Completing the Square: Leading Coefficient Is 1

Solve $x^2 + 2x - 6 = 0$ by completing the square.

Solution

$$x^2 + 2x - 6 = 0 \qquad \text{Write original equation.}$$

$$x^2 + 2x = 6 \qquad \text{Add 6 to each side.}$$

$$x^2 + 2x + 1^2 = 6 + 1^2 \qquad \text{Add } 1^2 \text{ to each side.}$$

$$\underbrace{}_{(\text{half of } 2)^2}$$

$$(x + 1)^2 = 7 \qquad \text{Simplify.}$$

$$x + 1 = \pm\sqrt{7} \qquad \text{Take square root of each side.}$$

$$x = -1 \pm \sqrt{7} \qquad \text{Subtract 1 from each side.}$$

The solutions are $x = -1 \pm \sqrt{7}$. Check these in the original equation.

CHECK*Point* Now try Exercise 73.

Example 7 Completing the Square: Leading Coefficient Is Not 1

$$3x^2 - 4x - 5 = 0 \qquad \text{Original equation}$$

$$3x^2 - 4x = 5 \qquad \text{Add 5 to each side.}$$

$$x^2 - \frac{4}{3}x = \frac{5}{3} \qquad \text{Divide each side by 3.}$$

$$x^2 - \frac{4}{3}x + \left(-\frac{2}{3}\right)^2 = \frac{5}{3} + \left(-\frac{2}{3}\right)^2 \qquad \text{Add } \left(-\frac{2}{3}\right)^2 \text{ to each side.}$$

$$\underbrace{\phantom{x^2 - \frac{4}{3}x + \left(-\frac{2}{3}\right)^2}}_{\left(\text{half of } -\frac{4}{3}\right)^2}$$

$$x^2 - \frac{4}{3}x + \frac{4}{9} = \frac{19}{9} \qquad \text{Simplify.}$$

$$\left(x - \frac{2}{3}\right)^2 = \frac{19}{9} \qquad \text{Perfect square trinomial}$$

$$x - \frac{2}{3} = \pm\frac{\sqrt{19}}{3} \qquad \text{Extract square roots.}$$

$$x = \frac{2}{3} \pm \frac{\sqrt{19}}{3} \qquad \text{Solutions}$$

CHECK*Point* Now try Exercise 77.

When using the Quadratic Formula, remember that *before* the formula can be applied, you must first write the quadratic equation in general form.

Example 8   The Quadratic Formula: Two Distinct Solutions

Use the Quadratic Formula to solve $x^2 + 3x = 9$.

Solution

$$x^2 + 3x = 9$$
Write original equation.

$$x^2 + 3x - 9 = 0$$
Write in general form.

$$x = \frac{-b \pm \sqrt{b^2 - 4ac}}{2a}$$
Quadratic Formula

$$x = \frac{-3 \pm \sqrt{(3)^2 - 4(1)(-9)}}{2(1)}$$
Substitute $a = 1$, $b = 3$, and $c = -9$.

$$x = \frac{-3 \pm \sqrt{45}}{2}$$
Simplify.

$$x = \frac{-3 \pm 3\sqrt{5}}{2}$$
Simplify.

The equation has two solutions:

$$x = \frac{-3 + 3\sqrt{5}}{2} \quad \text{and} \quad x = \frac{-3 - 3\sqrt{5}}{2}.$$

Check these in the original equation.

CHECK *Point* Now try Exercise 87.

Example 9   The Quadratic Formula: One Solution

Use the Quadratic Formula to solve $8x^2 - 24x + 18 = 0$.

Solution

$$8x^2 - 24x + 18 = 0$$
Write original equation.

$$4x^2 - 12x + 9 = 0$$
Divide out common factor of 2.

$$x = \frac{-b \pm \sqrt{b^2 - 4ac}}{2a}$$
Quadratic Formula

$$x = \frac{-(-12) \pm \sqrt{(-12)^2 - 4(4)(9)}}{2(4)}$$
Substitute $a = 4$, $b = -12$, and $c = 9$.

$$x = \frac{12 \pm \sqrt{0}}{8} = \frac{3}{2}$$
Simplify.

This quadratic equation has only one solution: $x = \frac{3}{2}$. Check this in the original equation.

CHECK *Point* Now try Exercise 91.

Note that Example 9 could have been solved without first dividing out a common factor of 2. Substituting $a = 8$, $b = -24$, and $c = 18$ into the Quadratic Formula produces the same result.

Polynomial Equations of Higher Degree

The methods used to solve quadratic equations can sometimes be extended to solve polynomial equations of higher degree.

> ⚠ **WARNING/CAUTION**
>
> A common mistake that is made in solving equations such as the equation in Example 10 is to divide each side of the equation by the variable factor x^2. This loses the solution $x = 0$. When solving an equation, always write the equation in general form, then factor the equation and set each factor equal to zero. Do not divide each side of an equation by a variable factor in an attempt to simplify the equation.

Example 10 Solving a Polynomial Equation by Factoring

Solve $3x^4 = 48x^2$.

Solution

First write the polynomial equation in general form with zero on one side, factor the other side, and then set each factor equal to zero and solve.

$$3x^4 = 48x^2 \qquad \text{Write original equation.}$$

$$3x^4 - 48x^2 = 0 \qquad \text{Write in general form.}$$

$$3x^2(x^2 - 16) = 0 \qquad \text{Factor out common factor.}$$

$$3x^2(x + 4)(x - 4) = 0 \qquad \text{Write in factored form.}$$

$$3x^2 = 0 \quad \Longrightarrow \quad x = 0 \qquad \text{Set 1st factor equal to 0.}$$

$$x + 4 = 0 \quad \Longrightarrow \quad x = -4 \qquad \text{Set 2nd factor equal to 0.}$$

$$x - 4 = 0 \quad \Longrightarrow \quad x = 4 \qquad \text{Set 3rd factor equal to 0.}$$

You can check these solutions by substituting in the original equation, as follows.

Check

$$3(0)^4 = 48(0)^2 \qquad \qquad \text{0 checks.} ✓$$

$$3(-4)^4 = 48(-4)^2 \qquad \qquad \text{-4 checks.} ✓$$

$$3(4)^4 = 48(4)^2 \qquad \qquad \text{4 checks.} ✓$$

So, you can conclude that the solutions are $x = 0$, $x = -4$, and $x = 4$.

CHECK*Point* Now try Exercise 113.

Example 11 Solving a Polynomial Equation by Factoring

Solve $x^3 - 3x^2 - 3x + 9 = 0$.

Solution

$$x^3 - 3x^2 - 3x + 9 = 0 \qquad \text{Write original equation.}$$

$$x^2(x - 3) - 3(x - 3) = 0 \qquad \text{Factor by grouping.}$$

$$(x - 3)(x^2 - 3) = 0 \qquad \text{Distributive Property}$$

$$x - 3 = 0 \quad \Longrightarrow \quad x = 3 \qquad \text{Set 1st factor equal to 0.}$$

$$x^2 - 3 = 0 \quad \Longrightarrow \quad x = \pm\sqrt{3} \qquad \text{Set 2nd factor equal to 0.}$$

The solutions are $x = 3$, $x = \sqrt{3}$, and $x = -\sqrt{3}$. Check these in the original equation.

CHECK*Point* Now try Exercise 119.

Equations Involving Radicals

Operations such as squaring each side of an equation, raising each side of an equation to a rational power, and multiplying each side of an equation by a variable quantity all can introduce extraneous solutions. So, when you use any of these operations, checking your solutions is crucial.

Example 12   Solving Equations Involving Radicals

a. $\sqrt{2x + 7} - x = 2$ Original equation

$\qquad \sqrt{2x + 7} = x + 2$ Isolate radical.

$\qquad 2x + 7 = x^2 + 4x + 4$ Square each side.

$\qquad 0 = x^2 + 2x - 3$ Write in general form.

$\qquad 0 = (x + 3)(x - 1)$ Factor.

$\qquad x + 3 = 0 \implies x = -3$ Set 1st factor equal to 0.

$\qquad x - 1 = 0 \implies x = 1$ Set 2nd factor equal to 0.

By checking these values, you can determine that the only solution is $x = 1$.

b. $\sqrt{2x - 5} - \sqrt{x - 3} = 1$ Original equation

$\qquad \sqrt{2x - 5} = \sqrt{x - 3} + 1$ Isolate $\sqrt{2x - 5}$.

$\qquad 2x - 5 = x - 3 + 2\sqrt{x - 3} + 1$ Square each side.

$\qquad 2x - 5 = x - 2 + 2\sqrt{x - 3}$ Combine like terms.

$\qquad x - 3 = 2\sqrt{x - 3}$ Isolate $2\sqrt{x - 3}$.

$\qquad x^2 - 6x + 9 = 4(x - 3)$ Square each side.

$\qquad x^2 - 10x + 21 = 0$ Write in general form.

$\qquad (x - 3)(x - 7) = 0$ Factor.

$\qquad x - 3 = 0 \implies x = 3$ Set 1st factor equal to 0.

$\qquad x - 7 = 0 \implies x = 7$ Set 2nd factor equal to 0.

The solutions are $x = 3$ and $x = 7$. Check these in the original equation.

CHECK *Point*   Now try Exercise 129.

> ### Study Tip
>
> When an equation contains two radicals, it may not be possible to isolate both. In such cases, you may have to raise each side of the equation to a power at *two* different stages in the solution, as shown in Example 12(b).

Example 13   Solving an Equation Involving a Rational Exponent

$(x - 4)^{2/3} = 25$ Original equation

$\sqrt[3]{(x - 4)^2} = 25$ Rewrite in radical form.

$(x - 4)^2 = 15{,}625$ Cube each side.

$x - 4 = \pm 125$ Take square root of each side.

$x = 129, \quad x = -121$ Add 4 to each side.

CHECK *Point*   Now try Exercise 137.

Equations with Absolute Values

To solve an equation involving an absolute value, remember that the expression inside the absolute value signs can be positive or negative. This results in *two* separate equations, each of which must be solved. For instance, the equation

$$|x - 2| = 3$$

results in the two equations $x - 2 = 3$ and $-(x - 2) = 3$, which implies that the equation has two solutions: $x = 5$ and $x = -1$.

Example 14   Solving an Equation Involving Absolute Value

Solve $|x^2 - 3x| = -4x + 6$.

Solution

Because the variable expression inside the absolute value signs can be positive or negative, you must solve the following two equations.

First Equation

$x^2 - 3x = -4x + 6$	Use positive expression.
$x^2 + x - 6 = 0$	Write in general form.
$(x + 3)(x - 2) = 0$	Factor.
$x + 3 = 0 \implies x = -3$	Set 1st factor equal to 0.
$x - 2 = 0 \implies x = 2$	Set 2nd factor equal to 0.

Second Equation

$-(x^2 - 3x) = -4x + 6$	Use negative expression.
$x^2 - 7x + 6 = 0$	Write in general form.
$(x - 1)(x - 6) = 0$	Factor.
$x - 1 = 0 \implies x = 1$	Set 1st factor equal to 0.
$x - 6 = 0 \implies x = 6$	Set 2nd factor equal to 0.

Check

$\left|(-3)^2 - 3(-3)\right| \overset{?}{=} -4(-3) + 6$	Substitute -3 for x.
$18 = 18$	-3 checks. ✔
$\left|(2)^2 - 3(2)\right| \overset{?}{=} -4(2) + 6$	Substitute 2 for x.
$2 \neq -2$	2 does not check.
$\left|(1)^2 - 3(1)\right| \overset{?}{=} -4(1) + 6$	Substitute 1 for x.
$2 = 2$	1 checks. ✔
$\left|(6)^2 - 3(6)\right| \overset{?}{=} -4(6) + 6$	Substitute 6 for x.
$18 \neq -18$	6 does not check.

The solutions are $x = -3$ and $x = 1$.

CHECK*Point* Now try Exercise 151.

P.2   EXERCISES

See www.CalcChat.com for worked-out solutions to odd-numbered exercises.

VOCABULARY: Fill in the blanks.

1. An _____ is a statement that equates two algebraic expressions.

2. A linear equation in one variable is an equation that can be written in the standard form _____.

3. When solving an equation, it is possible to introduce an _____ solution, which is a value that does not satisfy the original equation.

4. The four methods that can be used to solve a quadratic equation are _____, _____, _____, and the _____.

SKILLS AND APPLICATIONS

In Exercises 5–12, determine whether the equation is an identity or a conditional equation.

5. $4(x + 1) = 4x + 4$

6. $2(x - 3) = 7x - 1$

7. $-6(x - 3) + 5 = -2x + 10$

8. $3(x + 2) - 5 = 3x + 1$

9. $4(x + 1) - 2x = 2(x + 2)$

10. $x^2 + 2(3x - 2) = x^2 + 6x - 4$

11. $3 + \dfrac{1}{x + 1} = \dfrac{4x}{x + 1}$

12. $\dfrac{5}{x} + \dfrac{3}{x} = 24$

In Exercises 13–26, solve the equation and check your solution.

13. $x + 11 = 15$

14. $7 - x = 19$

15. $7 - 2x = 25$

16. $7x + 2 = 23$

17. $8x - 5 = 3x + 20$

18. $7x + 3 = 3x - 17$

19. $4y + 2 - 5y = 7 - 6y$

20. $3(x + 3) = 5(1 - x) - 1$

21. $x - 3(2x + 3) = 8 - 5x$

22. $9x - 10 = 5x + 2(2x - 5)$

23. $\dfrac{3x}{8} - \dfrac{4x}{3} = 4$

24. $\dfrac{x}{5} - \dfrac{x}{2} = 3 + \dfrac{3x}{10}$

25. $\frac{3}{2}(z + 5) - \frac{1}{4}(z + 24) = 0$

26. $0.60x + 0.40(100 - x) = 50$

In Exercises 27–42, solve the equation and check your solution. (If not possible, explain why.)

27. $x + 8 = 2(x - 2) - x$

28. $8(x + 2) - 3(2x + 1) = 2(x + 5)$

29. $\dfrac{100 - 4x}{3} = \dfrac{5x + 6}{4} + 6$

30. $\dfrac{17 + y}{y} + \dfrac{32 + y}{y} = 100$

31. $\dfrac{5x - 4}{5x + 4} = \dfrac{2}{3}$

32. $\dfrac{15}{x} - 4 = \dfrac{6}{x} + 3$

33. $3 = 2 + \dfrac{2}{z + 2}$

34. $\dfrac{1}{x} + \dfrac{2}{x - 5} = 0$

35. $\dfrac{x}{x + 4} + \dfrac{4}{x + 4} + 2 = 0$

36. $\dfrac{7}{2x + 1} - \dfrac{8x}{2x - 1} = -4$

37. $\dfrac{2}{(x - 4)(x - 2)} = \dfrac{1}{x - 4} + \dfrac{2}{x - 2}$

38. $\dfrac{1}{x - 2} + \dfrac{3}{x + 3} = \dfrac{4}{x^2 + x - 6}$

39. $\dfrac{3}{x^2 - 3x} + \dfrac{4}{x} = \dfrac{1}{x - 3}$

40. $\dfrac{6}{x} - \dfrac{2}{x + 3} = \dfrac{3(x + 5)}{x^2 + 3x}$

41. $(x + 2)^2 + 5 = (x + 3)^2$

42. $(2x + 1)^2 = 4(x^2 + x + 1)$

In Exercises 43–46, write the quadratic equation in general form.

43. $2x^2 = 3 - 8x$

44. $13 - 3(x + 7)^2 = 0$

45. $\frac{1}{5}(3x^2 - 10) = 18x$

46. $x(x + 2) = 5x^2 + 1$

In Exercises 47–58, solve the quadratic equation by factoring.

47. $6x^2 + 3x = 0$

48. $9x^2 - 4 = 0$

49. $x^2 - 2x - 8 = 0$

50. $x^2 - 10x + 9 = 0$

51. $x^2 - 12x + 35 = 0$

52. $4x^2 + 12x + 9 = 0$

53. $3 + 5x - 2x^2 = 0$

54. $2x^2 = 19x + 33$

55. $x^2 + 4x = 12$

56. $\frac{1}{8}x^2 - x - 16 = 0$

57. $x^2 + 2ax + a^2 = 0$, a is a real number

58. $(x + a)^2 - b^2 = 0$, a and b are real numbers

In Exercises 59–70, solve the equation by extracting square roots.

59. $x^2 = 49$

60. $x^2 = 32$

61. $3x^2 = 81$

62. $9x^2 = 36$

63. $(x - 12)^2 = 16$

64. $(x + 13)^2 = 25$

65. $(x + 2)^2 = 14$

66. $(x - 5)^2 = 30$

67. $(2x - 1)^2 = 18$

68. $(2x + 3)^2 - 27 = 0$

69. $(x - 7)^2 = (x + 3)^2$

70. $(x + 5)^2 = (x + 4)^2$

In Exercises 71–80, solve the quadratic equation by completing the square.

71. $x^2 + 4x - 32 = 0$

72. $x^2 + 6x + 2 = 0$

73. $x^2 + 12x + 25 = 0$

74. $x^2 + 8x + 14 = 0$

75. $8 + 4x - x^2 = 0$

76. $9x^2 - 12x = 14$

77. $2x^2 + 5x - 8 = 0$

78. $4x^2 - 4x - 99 = 0$

79. $5x^2 - 15x + 7 = 0$

80. $3x^2 + 9x + 5 = 0$

In Exercises 81–98, use the Quadratic Formula to solve the equation.

81. $2x^2 + x - 1 = 0$

82. $25x^2 - 20x + 3 = 0$

83. $2 + 2x - x^2 = 0$

84. $x^2 - 10x + 22 = 0$

85. $x^2 + 14x + 44 = 0$

86. $6x = 4 - x^2$

87. $x^2 + 8x - 4 = 0$

88. $4x^2 - 4x - 4 = 0$

89. $12x - 9x^2 = -3$

90. $16x^2 + 22 = 40x$

91. $9x^2 + 24x + 16 = 0$

92. $16x^2 - 40x + 5 = 0$

93. $28x - 49x^2 = 4$

94. $3x + x^2 - 1 = 0$

95. $8t = 5 + 2t^2$

96. $25h^2 + 80h + 61 = 0$

97. $(y - 5)^2 = 2y$

98. $\left(\frac{5}{7}x - 14\right)^2 = 8x$

In Exercises 99–104, use the Quadratic Formula to solve the equation. (Round your answer to three decimal places.)

99. $0.1x^2 + 0.2x - 0.5 = 0$

100. $2x^2 - 2.50x - 0.42 = 0$

101. $-0.067x^2 - 0.852x + 1.277 = 0$

102. $-0.005x^2 + 0.101x - 0.193 = 0$

103. $422x^2 - 506x - 347 = 0$

104. $-3.22x^2 - 0.08x + 28.651 = 0$

In Exercises 105–112, solve the equation using any convenient method.

105. $x^2 - 2x - 1 = 0$

106. $11x^2 + 33x = 0$

107. $(x + 3)^2 = 81$

108. $x^2 - 14x + 49 = 0$

109. $x^2 - x - \frac{11}{4} = 0$

110. $3x + 4 = 2x^2 - 7$

111. $4x^2 + 2x + 4 = 2x + 8$

112. $a^2x^2 - b^2 = 0$, a and b are real numbers, $a \neq 0$

In Exercises 113–126, find all real solutions of the equation. Check your solutions in the original equation.

113. $2x^4 - 50x^2 = 0$

114. $20x^3 - 125x = 0$

115. $x^4 - 81 = 0$

116. $x^6 - 64 = 0$

117. $x^3 + 216 = 0$

118. $9x^4 - 24x^3 + 16x^2 = 0$

119. $x^3 - 3x^2 - x + 3 = 0$

120. $x^3 + 2x^2 + 3x + 6 = 0$

121. $x^4 + x = x^3 + 1$

122. $x^4 - 2x^3 = 16 + 8x - 4x^3$

123. $x^4 - 4x^2 + 3 = 0$

124. $36t^4 + 29t^2 - 7 = 0$

125. $x^6 + 7x^3 - 8 = 0$

126. $x^6 + 3x^3 + 2 = 0$

In Exercises 127–154, find all solutions of the equation. Check your solutions in the original equation.

127. $\sqrt{2x} - 10 = 0$

128. $7\sqrt{x} - 6 = 0$

129. $\sqrt{x - 10} - 4 = 0$

130. $\sqrt{5 - x} - 3 = 0$

131. $\sqrt{2x + 5} + 3 = 0$

132. $\sqrt{3 - 2x} - 2 = 0$

133. $\sqrt[3]{2x + 1} + 8 = 0$

134. $\sqrt[3]{4x - 3} + 2 = 0$

135. $\sqrt{5x - 26} + 4 = x$

136. $\sqrt{x + 5} = \sqrt{2x - 5}$

137. $(x - 6)^{3/2} = 8$

138. $(x + 3)^{3/2} = 8$

139. $(x + 3)^{2/3} = 5$

140. $(x^2 - x - 22)^{4/3} = 16$

141. $3x(x - 1)^{1/2} + 2(x - 1)^{3/2} = 0$

142. $4x^2(x - 1)^{1/3} + 6x(x - 1)^{4/3} = 0$

143. $x = \dfrac{3}{x} + \dfrac{1}{2}$

144. $\dfrac{4}{x + 1} - \dfrac{3}{x + 2} = 1$

145. $\dfrac{20 - x}{x} = x$

146. $4x + 1 = \dfrac{3}{x}$

147. $\dfrac{x}{x^2 - 4} + \dfrac{1}{x + 2} = 3$

148. $\dfrac{x + 1}{3} - \dfrac{x + 1}{x + 2} = 0$

149. $|2x - 1| = 5$

150. $|13x + 1| = 12$

151. $|x| = x^2 + x - 3$

152. $|x^2 + 6x| = 3x + 18$

153. $|x + 1| = x^2 - 5$

154. $|x - 10| = x^2 - 10x$

ANTHROPOLOGY In Exercises 155 and 156, use the following information. The relationship between the length of an adult's femur (thigh bone) and the height of the adult can be approximated by the linear equations

$y = 0.432x - 10.44$ Female

$y = 0.449x - 12.15$ Male

where y is the length of the femur in inches and x is the height of the adult in inches (see figure).

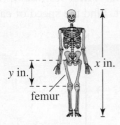

155. An anthropologist discovers a femur belonging to an adult human female. The bone is 16 inches long. Estimate the height of the female.

156. From the foot bones of an adult human male, an anthropologist estimates that the person's height was 69 inches. A few feet away from the site where the foot bones were discovered, the anthropologist discovers a male adult femur that is 19 inches long. Is it likely that both the foot bones and the thigh bone came from the same person?

157. OPERATING COST   A delivery company has a fleet of vans. The annual operating cost C per van is $C = 0.32m + 2500$, where m is the number of miles traveled by a van in a year. What number of miles will yield an annual operating cost of $10,000?

158. FLOOD CONTROL   A river has risen 8 feet above its flood stage. The water begins to recede at a rate of 3 inches per hour. Write a mathematical model that shows the number of feet above flood stage after t hours. If the water continually recedes at this rate, when will the river be 1 foot above its flood stage?

159. GEOMETRY   The hypotenuse of an isosceles right triangle is 5 centimeters long. How long are its sides?

160. GEOMETRY   An equilateral triangle has a height of 10 inches. How long is one of its sides? (*Hint:* Use the height of the triangle to partition the triangle into two congruent right triangles.)

161. PACKAGING An open box with a square base (see figure) is to be constructed from 84 square inches of material. The height of the box is 2 inches. What are the dimensions of the box? (*Hint:* The surface area is $S = x^2 + 4xh$.)

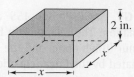

162. FLYING SPEED Two planes leave simultaneously from Chicago's O'Hare Airport, one flying due north and the other due east (see figure). The northbound plane is flying 50 miles per hour faster than the eastbound plane. After 3 hours, the planes are 2440 miles apart. Find the speed of each plane.

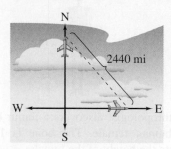

163. VOTING POPULATION The total voting-age population P (in millions) in the United States from 1990 through 2006 can be modeled by

$$P = \frac{182.17 - 1.542t}{1 - 0.018t}, \quad 0 \le t \le 16$$

where t represents the year, with $t = 0$ corresponding to 1990. (Source: U.S. Census Bureau)

(a) In which year did the total voting-age population reach 200 million?

(b) Use the model to predict the year in which the total voting-age population will reach 241 million. Is this prediction reasonable? Explain.

164. AIRLINE PASSENGERS An airline offers daily flights between Chicago and Denver. The total monthly cost C (in millions of dollars) of these flights is $C = \sqrt{0.2x + 1}$, where x is the number of passengers (in thousands). The total cost of the flights for June is 2.5 million dollars. How many passengers flew in June?

EXPLORATION

TRUE OR FALSE? In Exercises 165 and 166, determine whether the statement is true or false. Justify your answer.

165. An equation can never have more than one extraneous solution.

166. When solving an absolute value equation, you will always have to check more than one solution.

167. THINK ABOUT IT What is meant by *equivalent equations*? Give an example of two equivalent equations.

168. Solve $3(x + 4)^2 + (x + 4) - 2 = 0$ in two ways.

(a) Let $u = x + 4$, and solve the resulting equation for u. Then solve the u-solution for x.

(b) Expand and collect like terms in the equation, and solve the resulting equation for x.

(c) Which method is easier? Explain.

THINK ABOUT IT In Exercises 169–172, write a quadratic equation that has the given solutions. (There are many correct answers.)

169. -3 and 6

170. -4 and -11

171. $1 + \sqrt{2}$ and $1 - \sqrt{2}$

172. $-3 + \sqrt{5}$ and $-3 - \sqrt{5}$

In Exercises 173 and 174, consider an equation of the form $x + |x - a| = b$, where a and b are constants.

173. Find a and b when the solution of the equation is $x = 9$. (There are many correct answers.)

174. WRITING Write a short paragraph listing the steps required to solve this equation involving absolute values, and explain why it is important to check your solutions.

In Exercises 175 and 176, consider an equation of the form $x + \sqrt{x - a} = b$, where a and b are constants.

175. Find a and b when the solution of the equation is $x = 20$. (There are many correct answers.)

176. WRITING Write a short paragraph listing the steps required to solve this equation involving radicals, and explain why it is important to check your solutions.

177. Solve each equation, given that a and b are not zero.

(a) $ax^2 + bx = 0$ (b) $ax^2 - ax = 0$

178. CAPSTONE

(a) Explain the difference between a conditional equation and an identity.

(b) Give an example of an absolute value equation that has only one solution.

(c) State the Quadratic Formula in words.

(d) Does raising each side of an equation to the nth power always yield an equivalent equation? Explain.

P.3 # THE CARTESIAN PLANE AND GRAPHS OF EQUATIONS

What you should learn

- Plot points in the Cartesian plane.
- Use the Distance Formula to find the distance between two points.
- Use the Midpoint Formula to find the midpoint of a line segment.
- Use a coordinate plane to model and solve real-life problems.
- Sketch graphs of equations.
- Find x- and y-intercepts of graphs of equations.
- Use symmetry to sketch graphs of equations.
- Find equations of and sketch graphs of circles.

Why you should learn it

The graph of an equation can help you see relationships between real-life quantities. For example, in Exercise 120 on page 42, a graph can be used to estimate the life expectancies of children who are born in the years 2005 and 2010.

The Cartesian Plane

Just as you can represent real numbers by points on a real number line, you can represent ordered pairs of real numbers by points in a plane called the **rectangular coordinate system,** or the **Cartesian plane,** named after the French mathematician René Descartes (1596–1650).

The Cartesian plane is formed by using two real number lines intersecting at right angles, as shown in Figure P.13. The horizontal real number line is usually called the **x-axis,** and the vertical real number line is usually called the **y-axis.** The point of intersection of these two axes is the **origin,** and the two axes divide the plane into four parts called **quadrants.**

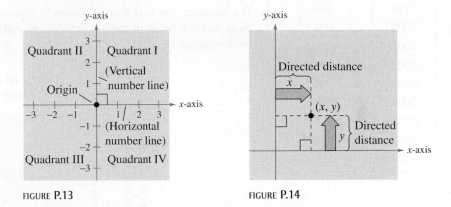

FIGURE P.13   FIGURE P.14

Each point in the plane corresponds to an **ordered pair** (x, y) of real numbers x and y, called **coordinates** of the point. The **x-coordinate** represents the directed distance from the y-axis to the point, and the **y-coordinate** represents the directed distance from the x-axis to the point, as shown in Figure P.14.

The notation (x, y) denotes both a point in the plane and an open interval on the real number line. The context will tell you which meaning is intended.

Example 1   Plotting Points in the Cartesian Plane

Plot the points $(-1, 2)$, $(3, 4)$, $(0, 0)$, $(3, 0)$, and $(-2, -3)$.

Solution

To plot the point $(-1, 2)$, imagine a vertical line through -1 on the x-axis and a horizontal line through 2 on the y-axis. The intersection of these two lines is the point $(-1, 2)$. The other four points can be plotted in a similar way, as shown in Figure P.15.

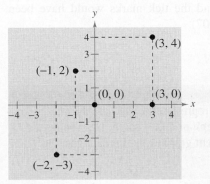

FIGURE P.15

CHECK*Point*   Now try Exercise 11.

The beauty of a rectangular coordinate system is that it allows you to *see* relationships between two variables. It would be difficult to overestimate the importance of Descartes's introduction of coordinates in the plane. Today, his ideas are in common use in virtually every scientific and business-related field.

Example 2 Sketching a Scatter Plot

From 1994 through 2007, the numbers N (in millions) of subscribers to a cellular telecommunication service in the United States are shown in the table, where t represents the year. Sketch a scatter plot of the data. (Source: CTIA-The Wireless Association)

Solution

To sketch a *scatter plot* of the data shown in the table, you simply represent each pair of values by an ordered pair (t, N) and plot the resulting points, as shown in Figure P.16. For instance, the first pair of values is represented by the ordered pair (1994, 24.1). Note that the break in the t-axis indicates that the numbers between 0 and 1994 have been omitted.

Year, t	Subscribers, N
1994	24.1
1995	33.8
1996	44.0
1997	55.3
1998	69.2
1999	86.0
2000	109.5
2001	128.4
2002	140.8
2003	158.7
2004	182.1
2005	207.9
2006	233.0
2007	255.4

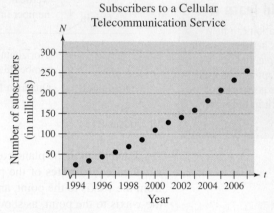

FIGURE **P.16**

CHECK*Point* Now try Exercise 29.

In Example 2, you could have let $t = 1$ represent the year 1994. In that case, the horizontal axis would not have been broken, and the tick marks would have been labeled 1 through 14 (instead of 1994 through 2007).

TECHNOLOGY

The scatter plot in Example 2 is only one way to represent the data graphically. You could also represent the data using a bar graph or a line graph. If you have access to a graphing utility, try using it to represent graphically the data given in Example 2.

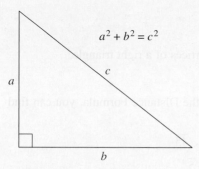

$$a^2 + b^2 = c^2$$

FIGURE **P.17**

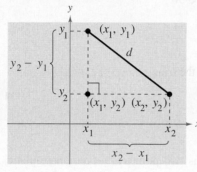

FIGURE **P.18**

The Distance Formula

Recall from the Pythagorean Theorem that, for a right triangle with hypotenuse of length c and sides of lengths a and b, you have

$$a^2 + b^2 = c^2 \qquad \text{Pythagorean Theorem}$$

as shown in Figure P.17. (The converse is also true. That is, if $a^2 + b^2 = c^2$, then the triangle is a right triangle.)

Suppose you want to determine the distance d between two points (x_1, y_1) and (x_2, y_2) in the plane. With these two points, a right triangle can be formed, as shown in Figure P.18. The length of the vertical side of the triangle is $|y_2 - y_1|$, and the length of the horizontal side is $|x_2 - x_1|$. By the Pythagorean Theorem, you can write

$$d^2 = |x_2 - x_1|^2 + |y_2 - y_1|^2$$

$$d = \sqrt{|x_2 - x_1|^2 + |y_2 - y_1|^2} = \sqrt{(x_2 - x_1)^2 + (y_2 - y_1)^2}.$$

This result is the **Distance Formula.**

The Distance Formula

The distance d between the points (x_1, y_1) and (x_2, y_2) in the plane is

$$d = \sqrt{(x_2 - x_1)^2 + (y_2 - y_1)^2}.$$

Example 3   Finding a Distance

Find the distance between the points $(-2, 1)$ and $(3, 4)$.

Algebraic Solution

Let $(x_1, y_1) = (-2, 1)$ and $(x_2, y_2) = (3, 4)$. Then apply the Distance Formula.

$$d = \sqrt{(x_2 - x_1)^2 + (y_2 - y_1)^2} \qquad \text{Distance Formula}$$

$$= \sqrt{[3 - (-2)]^2 + (4 - 1)^2} \qquad \text{Substitute for } x_1, y_1, x_2, \text{ and } y_2.$$

$$= \sqrt{(5)^2 + (3)^2} \qquad \text{Simplify.}$$

$$= \sqrt{34} \qquad \text{Simplify.}$$

$$\approx 5.83 \qquad \text{Use a calculator.}$$

So, the distance between the points is about 5.83 units. You can use the Pythagorean Theorem to check that the distance is correct.

$$d^2 \overset{?}{=} 3^2 + 5^2 \qquad \text{Pythagorean Theorem}$$

$$\left(\sqrt{34}\right)^2 \overset{?}{=} 3^2 + 5^2 \qquad \text{Substitute for } d.$$

$$34 = 34 \qquad \text{Distance checks. } \checkmark$$

Graphical Solution

Use centimeter graph paper to plot the points $A(-2, 1)$ and $B(3, 4)$. Carefully sketch the line segment from A to B. Then use a centimeter ruler to measure the length of the segment.

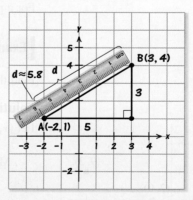

FIGURE **P.19**

The line segment measures about 5.8 centimeters, as shown in Figure P.19. So, the distance between the points is about 5.8 units.

CHECK *Point*   Now try Exercise 41(a) and (b).

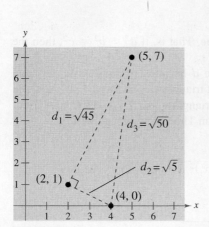

FIGURE P.20

Example 4 Verifying a Right Triangle

Show that the points $(2, 1)$, $(4, 0)$, and $(5, 7)$ are vertices of a right triangle.

Solution

The three points are plotted in Figure P.20. Using the Distance Formula, you can find the lengths of the three sides as follows.

$$d_1 = \sqrt{(5-2)^2 + (7-1)^2} = \sqrt{9+36} = \sqrt{45}$$

$$d_2 = \sqrt{(4-2)^2 + (0-1)^2} = \sqrt{4+1} = \sqrt{5}$$

$$d_3 = \sqrt{(5-4)^2 + (7-0)^2} = \sqrt{1+49} = \sqrt{50}$$

Because

$$(d_1)^2 + (d_2)^2 = 45 + 5 = 50 = (d_3)^2$$

you can conclude by the Pythagorean Theorem that the triangle must be a right triangle.

CHECK Point Now try Exercise 51.

The Midpoint Formula

To find the **midpoint** of the line segment that joins two points in a coordinate plane, you can simply find the average values of the respective coordinates of the two endpoints using the **Midpoint Formula.**

The Midpoint Formula

The midpoint of the line segment joining the points (x_1, y_1) and (x_2, y_2) is given by the Midpoint Formula

$$\text{Midpoint} = \left(\frac{x_1 + x_2}{2}, \frac{y_1 + y_2}{2} \right).$$

For a proof of the Midpoint Formula, see Proofs in Mathematics on page 130.

Example 5 Finding a Line Segment's Midpoint

Find the midpoint of the line segment joining the points $(-5, -3)$ and $(9, 3)$.

Solution

Let $(x_1, y_1) = (-5, -3)$ and $(x_2, y_2) = (9, 3)$.

$$\text{Midpoint} = \left(\frac{x_1 + x_2}{2}, \frac{y_1 + y_2}{2} \right) \qquad \text{Midpoint Formula}$$

$$= \left(\frac{-5 + 9}{2}, \frac{-3 + 3}{2} \right) \qquad \text{Substitute for } x_1, y_1, x_2, \text{ and } y_2.$$

$$= (2, 0) \qquad \text{Simplify.}$$

The midpoint of the line segment is $(2, 0)$, as shown in Figure P.21.

CHECK Point Now try Exercise 41(c).

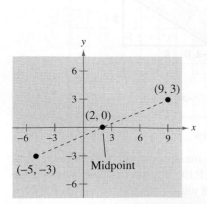

FIGURE P.21

Applications

Example 6   Finding the Length of a Pass

A football quarterback throws a pass from the 28-yard line, 40 yards from the sideline. The pass is caught by the wide receiver on the 5-yard line, 20 yards from the same sideline, as shown in Figure P.22. How long is the pass?

Solution

You can find the length of the pass by finding the distance between the points (40, 28) and (20, 5).

$$d = \sqrt{(x_2 - x_1)^2 + (y_2 - y_1)^2}$$ Distance Formula

$$= \sqrt{(40 - 20)^2 + (28 - 5)^2}$$ Substitute for x_1, y_1, x_2, and y_2.

$$= \sqrt{400 + 529}$$ Simplify.

$$= \sqrt{929}$$ Simplify.

$$\approx 30$$ Use a calculator.

So, the pass is about 30 yards long.

CHECK*Point*   Now try Exercise 57.

In Example 6, the scale along the goal line does not normally appear on a football field. However, when you use coordinate geometry to solve real-life problems, you are free to place the coordinate system in any way that is convenient for the solution of the problem.

Example 7   Estimating Annual Revenue

Barnes & Noble had annual sales of approximately $5.1 billion in 2005, and $5.4 billion in 2007. Without knowing any additional information, what would you estimate the 2006 sales to have been?   (Source: Barnes & Noble, Inc.)

Solution

One solution to the problem is to assume that sales followed a linear pattern. With this assumption, you can estimate the 2006 sales by finding the midpoint of the line segment connecting the points (2005, 5.1) and (2007, 5.4).

$$\text{Midpoint} = \left(\frac{x_1 + x_2}{2}, \frac{y_1 + y_2}{2} \right)$$ Midpoint Formula

$$= \left(\frac{2005 + 2007}{2}, \frac{5.1 + 5.4}{2} \right)$$ Substitute for x_1, x_2, y_1 and y_2.

$$= (2006, 5.25)$$ Simplify.

So, you would estimate the 2006 sales to have been about $5.25 billion, as shown in Figure P.23. (The actual 2006 sales were about $5.26 billion.)

CHECK*Point*   Now try Exercise 59.

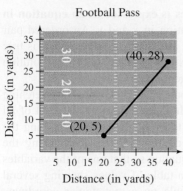

Football Pass

FIGURE **P.22**

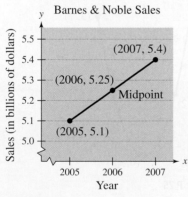

Barnes & Noble Sales

FIGURE **P.23**

The Graph of an Equation

Earlier in this section, you used a coordinate system to represent graphically the relationship between two quantities. There, the graphical picture consisted of a collection of points in a coordinate plane (see Example 2).

Frequently, a relationship between two quantities is expressed as an **equation in two variables.** For instance, $y = 7 - 3x$ is an equation in x and y. An ordered pair (a, b) is a **solution** or **solution point** of an equation in x and y if the equation is true when a is substituted for x, and b is substituted for y. For instance, $(1, 4)$ is a solution of $y = 7 - 3x$ because $4 = 7 - 3(1)$ is a true statement.

In the remainder of this section you will review some basic procedures for sketching the graph of an equation in two variables. The **graph of an equation** is the set of all points that are solutions of the equation. The basic technique used for sketching the graph of an equation is the **point-plotting method.** To sketch a graph using the point-plotting method, first, if possible, rewrite the equation so that one of the variables is isolated on one side of the equation. Next, make a table of values showing several solution points. Then plot the points from your table on a rectangular coordinate system. Finally, connect the points with a smooth curve or line.

Example 8   Sketching the Graph of an Equation

Sketch the graph of $y = x^2 - 2$.

Solution

Because the equation is already solved for y, begin by constructing a table of values.

x	-2	-1	0	1	2	3
$y = x^2 - 2$	2	-1	-2	-1	2	7
(x, y)	$(-2, 2)$	$(-1, -1)$	$(0, -2)$	$(1, -1)$	$(2, 2)$	$(3, 7)$

Next, plot the points given in the table, as shown in Figure P.24. Finally, connect the points with a smooth curve, as shown in Figure P.25.

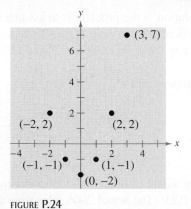

FIGURE **P.24**

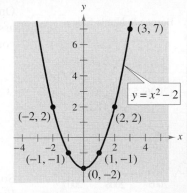

FIGURE **P.25**

CHECK *Point* ▸ Now try Exercise 65.

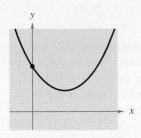

No *x*-intercepts; one *y*-intercept

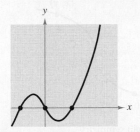

Three *x*-intercepts; one *y*-intercept

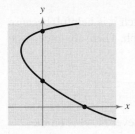

One *x*-intercept; two *y*-intercepts

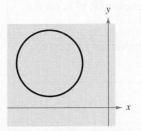

No intercepts

FIGURE **P.26**

TECHNOLOGY

To graph an equation involving *x* and *y* on a graphing utility, use the following procedure.

1. Rewrite the equation so that *y* is isolated on the left side.

2. Enter the equation into the graphing utility.

3. Determine a *viewing window* that shows all important features of the graph.

4. Graph the equation.

Intercepts of a Graph

It is often easy to determine the solution points that have zero as either the *x*-coordinate or the *y*-coordinate. These points are called **intercepts** because they are the points at which the graph intersects or touches the *x*- or *y*-axis. It is possible for a graph to have no intercepts, one intercept, or several intercepts, as shown in Figure P.26.

Note that an *x*-intercept can be written as the ordered pair $(x, 0)$ and a *y*-intercept can be written as the ordered pair $(0, y)$. Some texts denote the *x*-intercept as the *x*-coordinate of the point $(a, 0)$ [and the *y*-intercept as the *y*-coordinate of the point $(0, b)$] rather than the point itself. Unless it is necessary to make a distinction, we will use the term *intercept* to mean either the point or the coordinate.

Finding Intercepts

1. To find *x*-intercepts, let *y* be zero and solve the equation for *x*.

2. To find *y*-intercepts, let *x* be zero and solve the equation for *y*.

| **Example 9** | **Finding *x*- and *y*-Intercepts** |

Find the *x*- and *y*-intercepts of the graph of $y = x^3 - 4x$.

Solution

Let $y = 0$. Then

$$0 = x^3 - 4x = x(x^2 - 4)$$

has solutions $x = 0$ and $x = \pm 2$.

x-intercepts: $(0, 0)$, $(2, 0)$, $(-2, 0)$

Let $x = 0$. Then

$$y = (0)^3 - 4(0)$$

has one solution, $y = 0$.

y-intercept: $(0, 0)$ See Figure P.27.

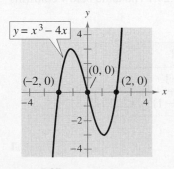

$y = x^3 - 4x$

FIGURE **P.27**

CHECK*Point* ▶ Now try Exercise 69.

Symmetry

Graphs of equations can have **symmetry** with respect to one of the coordinate axes or with respect to the origin. Symmetry with respect to the *x*-axis means that if the Cartesian plane were folded along the *x*-axis, the portion of the graph above the *x*-axis would coincide with the portion below the *x*-axis. Symmetry with respect to the *y*-axis or the origin can be described in a similar manner, as shown in Figure P.28.

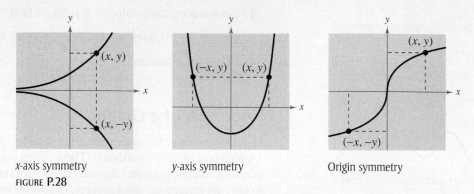

x-axis symmetry
FIGURE **P.28**

y-axis symmetry

Origin symmetry

Knowing the symmetry of a graph *before* attempting to sketch it is helpful, because then you need only half as many solution points to sketch the graph. There are three basic types of symmetry, described as follows.

Graphical Tests for Symmetry

1. A graph is **symmetric with respect to the *x*-axis** if, whenever (x, y) is on the graph, $(x, -y)$ is also on the graph.

2. A graph is **symmetric with respect to the *y*-axis** if, whenever (x, y) is on the graph, $(-x, y)$ is also on the graph.

3. A graph is **symmetric with respect to the origin** if, whenever (x, y) is on the graph, $(-x, -y)$ is also on the graph.

Example 10   Testing for Symmetry

The graph of $y = x^2 - 2$ is symmetric with respect to the *y*-axis because the point $(-x, y)$ is also on the graph of $y = x^2 - 2$. (See Figure P.29.) The table below confirms that the graph is symmetric with respect to the *y*-axis.

x	-3	-2	-1	1	2	3
y	7	2	-1	-1	2	7
(x, y)	$(-3, 7)$	$(-2, 2)$	$(-1, -1)$	$(1, -1)$	$(2, 2)$	$(3, 7)$

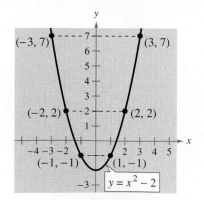

FIGURE **P.29** *y*-axis symmetry

CHECK*Point* Now try Exercise 79.

Algebraic Tests for Symmetry

1. The graph of an equation is symmetric with respect to the *x*-axis if replacing *y* with $-y$ yields an equivalent equation.

2. The graph of an equation is symmetric with respect to the *y*-axis if replacing *x* with $-x$ yields an equivalent equation.

3. The graph of an equation is symmetric with respect to the origin if replacing *x* with $-x$ and *y* with $-y$ yields an equivalent equation.

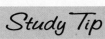

FIGURE **P.30**

Example 11   Using Symmetry as a Sketching Aid

Use symmetry to sketch the graph of

$$x - y^2 = 1.$$

Solution

Of the three tests for symmetry, the only one that is satisfied is the test for *x*-axis symmetry because $x - (-y)^2 = 1$ is equivalent to $x - y^2 = 1$. So, the graph is symmetric with respect to the *x*-axis. Using symmetry, you only need to find the solution points above the *x*-axis and then reflect them to obtain the graph, as shown in Figure P.30.

y	$x = y^2 + 1$	(x, y)
0	1	$(1, 0)$
1	2	$(2, 1)$
2	5	$(5, 2)$

Study Tip

Notice that when creating the table in Example 11, it is easier to choose *y*-values and then find the corresponding *x*-values of the ordered pairs.

CHECK*Point*   Now try Exercise 95.

Example 12   Sketching the Graph of an Equation

Sketch the graph of

$$y = |x - 1|.$$

Solution

This equation fails all three tests for symmetry and consequently its graph is not symmetric with respect to either axis or to the origin. The absolute value sign indicates that *y* is always nonnegative. Create a table of values and plot the points as shown in Figure P.31. From the table, you can see that $x = 0$ when $y = 1$. So, the *y*-intercept is $(0, 1)$. Similarly, $y = 0$ when $x = 1$. So, the *x*-intercept is $(1, 0)$.

x	-2	-1	0	1	2	3	4
$y = |x - 1|$	3	2	1	0	1	2	3
(x, y)	$(-2, 3)$	$(-1, 2)$	$(0, 1)$	$(1, 0)$	$(2, 1)$	$(3, 2)$	$(4, 3)$

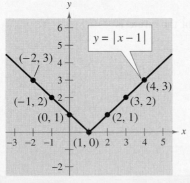

FIGURE **P.31**

CHECK*Point*   Now try Exercise 99.

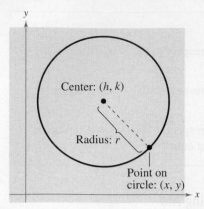

FIGURE P.32

Throughout this course, you will learn to recognize several types of graphs from their equations. For instance, you will learn to recognize that the graph of a second-degree equation of the form

$$y = ax^2 + bx + c$$

is a parabola (see Example 8). The graph of a **circle** is also easy to recognize.

Circles

Consider the circle shown in Figure P.32. A point (x, y) is on the circle if and only if its distance from the center (h, k) is r. By the Distance Formula,

$$\sqrt{(x - h)^2 + (y - k)^2} = r.$$

By squaring each side of this equation, you obtain the **standard form of the equation of a circle.**

Standard Form of the Equation of a Circle

The point (x, y) lies on the circle of **radius** r and **center** (h, k) if and only if

$$(x - h)^2 + (y - k)^2 = r^2.$$

 **WARNING/CAUTION**

Be careful when you are finding h and k from the standard equation of a circle. For instance, to find the correct h and k from the equation of the circle in Example 13, rewrite the quantities $(x + 1)^2$ and $(y - 2)^2$ using subtraction.

$$(x + 1)^2 = [x - (-1)]^2,$$

$$(y - 2)^2 = [y - (2)]^2$$

So, $h = -1$ and $k = 2$.

From this result, you can see that the standard form of the equation of a circle *with its center at the origin,* $(h, k) = (0, 0)$, is simply

$$x^2 + y^2 = r^2. \qquad \text{Circle with center at origin}$$

Example 13 Finding the Equation of a Circle

The point $(3, 4)$ lies on a circle whose center is at $(-1, 2)$, as shown in Figure P.33. Write the standard form of the equation of this circle.

Solution

The radius of the circle is the distance between $(-1, 2)$ and $(3, 4)$.

$$r = \sqrt{(x - h)^2 + (y - k)^2} \qquad \text{Distance Formula}$$

$$= \sqrt{[3 - (-1)]^2 + (4 - 2)^2} \qquad \text{Substitute for } x, y, h, \text{ and } k.$$

$$= \sqrt{4^2 + 2^2} \qquad \text{Simplify.}$$

$$= \sqrt{16 + 4} \qquad \text{Simplify.}$$

$$= \sqrt{20} \qquad \text{Radius}$$

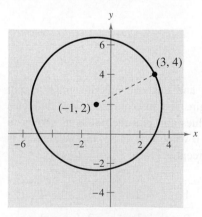

FIGURE P.33

Using $(h, k) = (-1, 2)$ and $r = \sqrt{20}$, the equation of the circle is

$$(x - h)^2 + (y - k)^2 = r^2 \qquad \text{Equation of circle}$$

$$[x - (-1)]^2 + (y - 2)^2 = \left(\sqrt{20}\right)^2 \qquad \text{Substitute for } h, k, \text{ and } r.$$

$$(x + 1)^2 + (y - 2)^2 = 20. \qquad \text{Standard form}$$

CHECK*Point* Now try Exercise 107.

P.3   EXERCISES

See www.CalcChat.com for worked-out solutions to odd-numbered exercises.

VOCABULARY: Fill in the blanks.

1. An ordered pair of real numbers can be represented in a plane called the rectangular coordinate system or the _____ plane.

2. The _____ _____ is a result derived from the Pythagorean Theorem.

3. Finding the average values of the respective coordinates of the two endpoints of a line segment in a coordinate plane is also known as using the _____ _____.

4. An ordered pair (a, b) is a _____ of an equation in x and y if the equation is true when a is substituted for x, and b is substituted for y.

5. The set of all solution points of an equation is the _____ of the equation.

6. The points at which a graph intersects or touches an axis are called the _____ of the graph.

7. A graph is symmetric with respect to the _____ if, whenever (x, y) is on the graph, $(-x, y)$ is also on the graph.

8. The equation $(x - h)^2 + (y - k)^2 = r^2$ is the standard form of the equation of a _____ with center _____ and radius _____.

SKILLS AND APPLICATIONS

In Exercises 9 and 10, approximate the coordinates of the points.

9. 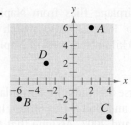 **10.**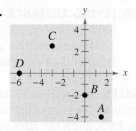

In Exercises 11–14, plot the points in the Cartesian plane.

11. $(-4, 2)$, $(-3, -6)$, $(0, 5)$, $(1, -4)$

12. $(0, 0)$, $(3, 1)$, $(-2, 4)$, $(1, -1)$

13. $(3, 8)$, $(0.5, -1)$, $(5, -6)$, $(-2, 2.5)$

14. $\left(1, -\frac{1}{3}\right)$, $\left(\frac{3}{4}, 3\right)$, $(-3, 4)$, $\left(-\frac{4}{3}, -\frac{3}{2}\right)$

In Exercises 15–18, find the coordinates of the point.

15. The point is located three units to the left of the y-axis and four units above the x-axis.

16. The point is located eight units below the x-axis and four units to the right of the y-axis.

17. The point is located five units below the x-axis and the coordinates of the point are equal.

18. The point is on the x-axis and 12 units to the left of the y-axis.

In Exercises 19–28, determine the quadrant(s) in which (x, y) is located so that the condition(s) is (are) satisfied.

19. $x > 0$ and $y < 0$ **20.** $x < 0$ and $y < 0$

21. $x = -4$ and $y > 0$ **22.** $x > 2$ and $y = 3$

23. $y < -5$ **24.** $x > 4$

25. $x < 0$ and $-y > 0$ **26.** $-x > 0$ and $y < 0$

27. $xy > 0$ **28.** $xy < 0$

In Exercises 29 and 30, sketch a scatter plot of the data.

29. NUMBER OF STORES   The table shows the number y of Wal-Mart stores for each year x from 2000 through 2007.   (Source: Wal-Mart Stores, Inc.)

Year, x	Number of stores, y
2000	4189
2001	4414
2002	4688
2003	4906
2004	5289
2005	6141
2006	6779
2007	7262

30. METEOROLOGY   The following data points (x, y) represent the lowest temperatures on record y (in degrees Fahrenheit) in Duluth, Minnesota, for each month x, where $x = 1$ represents January.   (Source: NOAA)

$(1, -39)$, $(2, -39)$, $(3, -29)$, $(4, -5)$, $(5, 17)$, $(6, 27)$, $(7, 35)$, $(8, 32)$, $(9, 22)$, $(10, 8)$, $(11, -23)$, $(12, -34)$

In Exercises 31–38, find the distance between the points. (*Note:* In each case, the two points lie on the same horizontal or vertical line.)

31. $(6, -3), (6, 5)$ **32.** $(1, 4), (8, 4)$

33. $(-3, -1), (2, -1)$ **34.** $(-3, -4), (-3, 6)$

35. $(-2, 6), (3, 6)$ **36.** $(8, 5), (8, 20)$

37. $(-5, 4), (-5, -1)$ **38.** $(1, 3), (1, -2)$

In Exercises 39 and 40, (a) find the length of each side of the right triangle, and (b) show that these lengths satisfy the Pythagorean Theorem.

39.

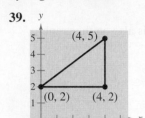

40.

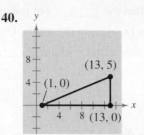

In Exercises 41–50, (a) plot the points, (b) find the distance between the points, and (c) find the midpoint of the line segment joining the points.

41. $(1, 1), (9, 7)$ **42.** $(1, 12), (6, 0)$

43. $(-4, 10), (4, -5)$ **44.** $(-7, -4), (2, 8)$

45. $(-1, 2), (5, 4)$ **46.** $(2, 10), (10, 2)$

47. $\left(\frac{1}{2}, 1\right), \left(-\frac{5}{2}, \frac{4}{3}\right)$ **48.** $\left(-\frac{1}{3}, -\frac{1}{3}\right), \left(-\frac{1}{6}, -\frac{1}{2}\right)$

49. $(6.2, 5.4), (-3.7, 1.8)$ **50.** $(-16.8, 12.3), (5.6, 4.9)$

In Exercises 51–54, show that the points form the vertices of the indicated polygon.

51. Right triangle: $(4, 0), (2, 1), (-1, -5)$

52. Right triangle: $(-1, 3), (3, 5), (5, 1)$

53. Isosceles triangle: $(1, -3), (3, 2), (-2, 4)$

54. Isosceles triangle: $(2, 3), (4, 9), (-2, 7)$

ADVERTISING In Exercises 55 and 56, use the graph below, which shows the average costs (in thousands of dollars) of a 30-second television spot during the Super Bowl from 2000 to 2008. (Source: Nielson Media and TNS Media Intelligence)

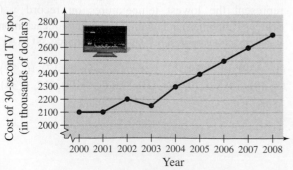

55. Estimate the percent increase in the average cost of a 30-second spot from Super Bowl XXXIV in 2000 to Super Bowl XXXVIII in 2004.

56. Estimate the percent increase in the average cost of a 30-second spot from Super Bowl XXXIV in 2000 to Super Bowl XLII in 2008.

57. SPORTS A soccer player passes the ball from a point that is 18 yards from the endline and 12 yards from the sideline. The pass is received by a teammate who is 42 yards from the same endline and 50 yards from the same sideline, as shown in the figure. How long is the pass?

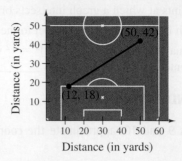

58. FLYING DISTANCE An airplane flies from Naples, Italy in a straight line to Rome, Italy, which is 120 kilometers north and 150 kilometers west of Naples. How far does the plane fly?

SALES In Exercises 59 and 60, use the Midpoint Formula to estimate the sales of Big Lots, Inc. and Dollar Tree Stores, Inc. in 2005, given the sales in 2003 and 2007. Assume that the sales followed a linear pattern. (Source: Big Lots, Inc.; Dollar Tree Stores, Inc)

59. Big Lots

Year	Sales (in millions)
2003	$4174
2007	$4656

60. Dollar Tree

Year	Sales (in millions)
2003	$2800
2007	$4243

In Exercises 61–64, determine whether each point lies on the graph of the equation.

Equation	Points	
61. $y = \sqrt{x + 4}$	(a) $(0, 2)$	(b) $(5, 3)$
62. $y = \sqrt{5 - x}$	(a) $(1, 2)$	(b) $(5, 0)$

Equation   *Points*

63. $y = x^2 - 3x + 2$ (a) $(2, 0)$ (b) $(-2, 8)$

64. $2x - y - 3 = 0$ (a) $(1, 2)$ (b) $(1, -1)$

In Exercises 65 and 66, complete the table. Use the resulting solution points to sketch the graph of the equation.

65. $y = \frac{3}{4}x - 1$

x	-2	0	1	$\frac{4}{3}$	2
y					
(x, y)					

66. $y = 5 - x^2$

x	-2	-1	0	1	2
y					
(x, y)					

In Exercises 67–78, find the x- and y-intercepts of the graph of the equation.

67. $y = 16 - 4x^2$ **68.** $y = (x + 3)^2$

69. $y = 5x - 6$ **70.** $y = 8 - 3x$

71. $y = \sqrt{x + 4}$ **72.** $y = \sqrt{2x - 1}$

73. $y = |3x - 7|$ **74.** $y = -|x + 10|$

75. $y = 2x^3 - 4x^2$ **76.** $y = x^4 - 25$

77. $y^2 = 6 - x$ **78.** $y^2 = x + 1$

In Exercises 79–82, assume that the graph has the indicated type of symmetry. Sketch the complete graph of the equation. To print an enlarged copy of the graph, go to the website *www.mathgraphs.com.*

79.
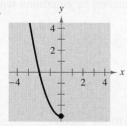
y-axis symmetry

80.
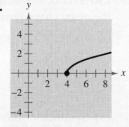
x-axis symmetry

81.
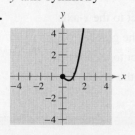
Origin symmetry

82.
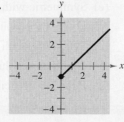
y-axis symmetry

In Exercises 83–90, use the algebraic tests to check for symmetry with respect to both axes and the origin.

83. $x^2 - y = 0$ **84.** $x - y^2 = 0$

85. $y = x^3$ **86.** $y = x^4 - x^2 + 3$

87. $y = \dfrac{x}{x^2 + 1}$ **88.** $y = \dfrac{1}{x^2 + 1}$

89. $xy^2 + 10 = 0$ **90.** $xy = 4$

In Exercises 91–102, use symmetry to sketch the graph of the equation.

91. $y = -3x + 1$ **92.** $y = 2x - 3$

93. $y = x^2 - 2x$ **94.** $y = -x^2 - 2x$

95. $y = x^3 + 3$ **96.** $y = x^3 - 1$

97. $y = \sqrt{x - 3}$ **98.** $y = \sqrt{1 - x}$

99. $y = |x - 6|$ **100.** $y = 1 - |x|$

101. $x = y^2 - 1$ **102.** $x = y^2 - 5$

In Exercises 103–110, write the standard form of the equation of the circle with the given characteristics.

103. Center: $(0, 0)$; radius: 6

104. Center: $(0, 0)$; radius: 8

105. Center: $(2, -1)$; radius: 4

106. Center: $(-7, -4)$; radius: 7

107. Center: $(-1, 2)$; solution point: $(0, 0)$

108. Center: $(3, -2)$; solution point: $(-1, 1)$

109. Endpoints of a diameter: $(0, 0), (6, 8)$

110. Endpoints of a diameter: $(-4, -1), (4, 1)$

In Exercises 111–116, find the center and radius of the circle, and sketch its graph.

111. $x^2 + y^2 = 25$

112. $x^2 + y^2 = 16$

113. $(x - 1)^2 + (y + 3)^2 = 9$

114. $x^2 + (y - 1)^2 = 1$

115. $\left(x - \frac{1}{2}\right)^2 + \left(y - \frac{1}{2}\right)^2 = \frac{9}{4}$

116. $(x - 2)^2 + (y + 3)^2 = \frac{16}{9}$

117. DEPRECIATION   A hospital purchases a new magnetic resonance imaging machine for \$500,000. The depreciated value y (reduced value) after t years is given by $y = 500{,}000 - 40{,}000t$, $0 \le t \le 8$. Sketch the graph of the equation.

118. CONSUMERISM   You purchase an all-terrain vehicle (ATV) for \$8000. The depreciated value y after t years is given by $y = 8000 - 900t$, $0 \le t \le 6$. Sketch the graph of the equation.

119. ELECTRONICS The resistance y (in ohms) of 1000 feet of solid copper wire at 68 degrees Fahrenheit can be approximated by the model $y = \dfrac{10{,}770}{x^2} - 0.37$, $5 \le x \le 100$, where x is the diameter of the wire in mils (0.001 inch). (Source: American Wire Gage)

(a) Complete the table.

x	5	10	20	30	40	50
y						

x	60	70	80	90	100
y					

(b) Use the table of values in part (a) to sketch a graph of the model. Then use your graph to estimate the resistance when $x = 85.5$.

(c) Use the model to confirm algebraically the estimate you found in part (b).

(d) What can you conclude in general about the relationship between the diameter of the copper wire and the resistance?

120. POPULATION STATISTICS The table shows the life expectancies of a child (at birth) in the United States for selected years from 1920 to 2000. (Source: U.S. National Center for Health Statistics)

Year	Life expectancy, y
1920	54.1
1930	59.7
1940	62.9
1950	68.2
1960	69.7
1970	70.8
1980	73.7
1990	75.4
2000	77.0

A model for the life expectancy during this period is $y = -0.0025t^2 + 0.574t + 44.25$, $20 \le t \le 100$, where y represents the life expectancy and t is the time in years, with $t = 20$ corresponding to 1920.

(a) Sketch a scatter plot of the data.

(b) Graph the model for the data and compare the scatter plot and the graph.

(c) Determine the life expectancy in 1948 both graphically and algebraically.

(d) Use the graph of the model to estimate the life expectancies of a child for the years 2005 and 2010.

(e) Do you think this model can be used to predict the life expectancy of a child 50 years from now? Explain.

EXPLORATION

TRUE OR FALSE? In Exercises 121–124, determine whether the statement is true or false. Justify your answer.

121. In order to divide a line segment into 16 equal parts, you would have to use the Midpoint Formula 16 times.

122. The points $(-8, 4)$, $(2, 11)$, and $(-5, 1)$ represent the vertices of an isosceles triangle.

123. A graph is symmetric with respect to the x-axis if, whenever (x, y) is on the graph, $(-x, y)$ is also on the graph.

124. A graph of an equation can have more than one y-intercept.

125. THINK ABOUT IT What is the y-coordinate of any point on the x-axis? What is the x-coordinate of any point on the y-axis?

126. THINK ABOUT IT When plotting points on the rectangular coordinate system, is it true that the scales on the x- and y-axes must be the same? Explain.

127. PROOF Prove that the diagonals of the parallelogram in the figure intersect at their midpoints.

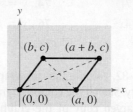

128. CAPSTONE Match the equation or equations with the given characteristic.

(i) $y = 3x^3 - 3x$ (ii) $y = (x + 3)^2$

(iii) $y = 3x - 3$ (iv) $y = \sqrt[3]{x}$

(v) $y = 3x^2 + 3$ (vi) $y = \sqrt{x + 3}$

(vii) $x^2 + y^2 = 9$

(a) Symmetric with respect to the y-axis

(b) Three x-intercepts

(c) Symmetric with respect to the x-axis

(d) $(-2, 1)$ is a point on the graph

(e) Symmetric with respect to the origin

(f) Graph passes through the origin

(g) Equation of a circle

P.4   LINEAR EQUATIONS IN TWO VARIABLES

What you should learn

- Use slope to graph linear equations in two variables.
- Find the slope of a line given two points on the line.
- Write linear equations in two variables.
- Use slope to identify parallel and perpendicular lines.
- Use slope and linear equations in two variables to model and solve real-life problems.

Why you should learn it

Linear equations in two variables can be used to model and solve real-life problems. For instance, in Exercise 129 on page 55, you will use a linear equation to model student enrollment at the Pennsylvania State University.

Courtesy of Pennsylvania State University

Using Slope

The simplest mathematical model for relating two variables is the **linear equation in two variables** $y = mx + b$. The equation is called *linear* because its graph is a line. (In mathematics, the term *line* means *straight line*.) By letting $x = 0$, you obtain

$$y = m(0) + b \qquad \text{Substitute 0 for } x.$$
$$= b.$$

So, the line crosses the y-axis at $y = b$, as shown in Figure P.34. In other words, the y-intercept is $(0, b)$. The steepness or slope of the line is m.

$$y = mx + b$$

Slope ⌐——⌐ y-Intercept

The **slope** of a nonvertical line is the number of units the line rises (or falls) vertically for each unit of horizontal change from left to right, as shown in Figure P.34 and Figure P.35.

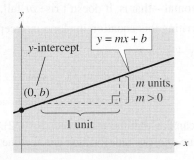

Positive slope, line rises.
FIGURE P.34

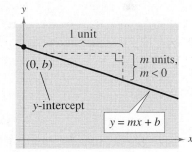

Negative slope, line falls.
FIGURE P.35

A linear equation that is written in the form $y = mx + b$ is said to be written in **slope-intercept form.**

The Slope-Intercept Form of the Equation of a Line

The graph of the equation

$$y = mx + b$$

is a line whose slope is m and whose y-intercept is $(0, b)$.

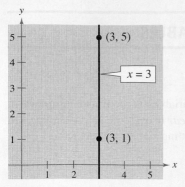

FIGURE P.36 Slope is undefined.

Once you have determined the slope and the y-intercept of a line, it is a relatively simple matter to sketch its graph. In the next example, note that none of the lines is vertical. A vertical line has an equation of the form

$$x = a. \qquad \text{Vertical line}$$

The equation of a vertical line cannot be written in the form $y = mx + b$ because the slope of a vertical line is undefined, as indicated in Figure P.36.

Example 1 Graphing a Linear Equation

Sketch the graph of each linear equation.

a. $y = 2x + 1$

b. $y = 2$

c. $x + y = 2$

Solution

a. Because $b = 1$, the y-intercept is $(0, 1)$. Moreover, because the slope is $m = 2$, the line *rises* two units for each unit the line moves to the right, as shown in Figure P.37.

b. By writing this equation in the form $y = (0)x + 2$, you can see that the y-intercept is $(0, 2)$ and the slope is zero. A zero slope implies that the line is horizontal—that is, it doesn't rise *or* fall, as shown in Figure P.38.

c. By writing this equation in slope-intercept form

$$x + y = 2 \qquad \text{Write original equation.}$$

$$y = -x + 2 \qquad \text{Subtract } x \text{ from each side.}$$

$$y = (-1)x + 2 \qquad \text{Write in slope-intercept form.}$$

you can see that the y-intercept is $(0, 2)$. Moreover, because the slope is $m = -1$, the line *falls* one unit for each unit the line moves to the right, as shown in Figure P.39.

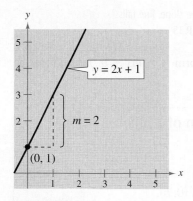

When *m* is positive, the line rises.
FIGURE P.37

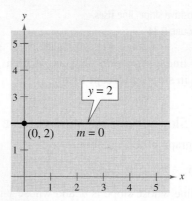

When *m* is 0, the line is horizontal.
FIGURE P.38

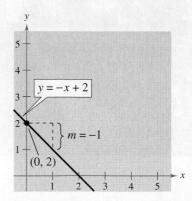

When *m* is negative, the line falls.
FIGURE P.39

CHECKPoint Now try Exercise 17.

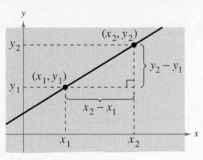

FIGURE **P.40**

Finding the Slope of a Line

Given an equation of a line, you can find its slope by writing the equation in slope-intercept form. If you are not given an equation, you can still find the slope of a line. For instance, suppose you want to find the slope of the line passing through the points (x_1, y_1) and (x_2, y_2), as shown in Figure P.40. As you move from left to right along this line, a change of $(y_2 - y_1)$ units in the vertical direction corresponds to a change of $(x_2 - x_1)$ units in the horizontal direction.

$$y_2 - y_1 = \text{the change in } y = \text{rise}$$

and

$$x_2 - x_1 = \text{the change in } x = \text{run}$$

The ratio of $(y_2 - y_1)$ to $(x_2 - x_1)$ represents the slope of the line that passes through the points (x_1, y_1) and (x_2, y_2).

$$\text{Slope} = \frac{\text{change in } y}{\text{change in } x}$$

$$= \frac{\text{rise}}{\text{run}}$$

$$= \frac{y_2 - y_1}{x_2 - x_1}$$

The Slope of a Line Passing Through Two Points

The **slope** m of the nonvertical line through (x_1, y_1) and (x_2, y_2) is

$$m = \frac{y_2 - y_1}{x_2 - x_1}$$

where $x_1 \neq x_2$.

When this formula is used for slope, the *order of subtraction* is important. Given two points on a line, you are free to label either one of them as (x_1, y_1) and the other as (x_2, y_2). However, once you have done this, you must form the numerator and denominator using the same order of subtraction.

$$m = \frac{y_2 - y_1}{x_2 - x_1} \qquad m = \frac{y_1 - y_2}{x_1 - x_2} \qquad m = \frac{y_2 - y_1}{x_1 - x_2}$$

Correct   Correct   Incorrect

For instance, the slope of the line passing through the points $(3, 4)$ and $(5, 7)$ can be calculated as

$$m = \frac{7 - 4}{5 - 3} = \frac{3}{2}$$

or, reversing the subtraction order in both the numerator and denominator, as

$$m = \frac{4 - 7}{3 - 5} = \frac{-3}{-2} = \frac{3}{2}.$$

Example 2 Finding the Slope of a Line Through Two Points

Find the slope of the line passing through each pair of points.

a. $(-2, 0)$ and $(3, 1)$ **b.** $(-1, 2)$ and $(2, 2)$

c. $(0, 4)$ and $(1, -1)$ **d.** $(3, 4)$ and $(3, 1)$

Solution

a. Letting $(x_1, y_1) = (-2, 0)$ and $(x_2, y_2) = (3, 1)$, you obtain a slope of

$$m = \frac{y_2 - y_1}{x_2 - x_1} = \frac{1 - 0}{3 - (-2)} = \frac{1}{5}.$$ See Figure P.41.

b. The slope of the line passing through $(-1, 2)$ and $(2, 2)$ is

$$m = \frac{2 - 2}{2 - (-1)} = \frac{0}{3} = 0.$$ See Figure P.42.

c. The slope of the line passing through $(0, 4)$ and $(1, -1)$ is

$$m = \frac{-1 - 4}{1 - 0} = \frac{-5}{1} = -5.$$ See Figure P.43.

d. The slope of the line passing through $(3, 4)$ and $(3, 1)$ is

$$m = \frac{1 - 4}{3 - 3} = \frac{-3}{0}.$$ See Figure P.44.

Because division by 0 is undefined, the slope is undefined and the line is vertical.

Study Tip

In Figures P.41 to P.44, note the relationships between slope and the orientation of the line.

a. Positive slope: line rises from left to right

b. Zero slope: line is horizontal

c. Negative slope: line falls from left to right

d. Undefined slope: line is vertical

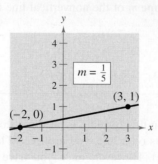

FIGURE **P.41**

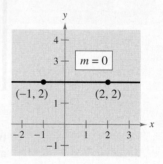

FIGURE **P.42**

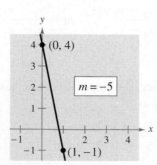

FIGURE **P.43**

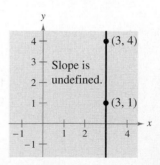

FIGURE **P.44**

CHECK*Point* Now try Exercise 31.

Writing Linear Equations in Two Variables

If (x_1, y_1) is a point on a line of slope m and (x, y) is *any other* point on the line, then

$$\frac{y - y_1}{x - x_1} = m.$$

This equation, involving the variables x and y, can be rewritten in the form

$$y - y_1 = m(x - x_1)$$

which is the **point-slope form** of the equation of a line.

Point-Slope Form of the Equation of a Line

The equation of the line with slope m passing through the point (x_1, y_1) is

$$y - y_1 = m(x - x_1).$$

The point-slope form is most useful for *finding* the equation of a line. You should remember this form.

Example 3   Using the Point-Slope Form

Find the slope-intercept form of the equation of the line that has a slope of 3 and passes through the point $(1, -2)$.

Solution

Use the point-slope form with $m = 3$ and $(x_1, y_1) = (1, -2)$.

$y - y_1 = m(x - x_1)$	Point-slope form
$y - (-2) = 3(x - 1)$	Substitute for m, x_1, and y_1.
$y + 2 = 3x - 3$	Simplify.
$y = 3x - 5$	Write in slope-intercept form.

The slope-intercept form of the equation of the line is $y = 3x - 5$. The graph of this line is shown in Figure P.45.

CHECK*Point*   Now try Exercise 51.

The point-slope form can be used to find an equation of the line passing through two points (x_1, y_1) and (x_2, y_2). To do this, first find the slope of the line

$$m = \frac{y_2 - y_1}{x_2 - x_1}, \quad x_1 \neq x_2$$

and then use the point-slope form to obtain the equation

$$y - y_1 = \frac{y_2 - y_1}{x_2 - x_1}(x - x_1). \qquad \text{Two-point form}$$

This is sometimes called the **two-point form** of the equation of a line.

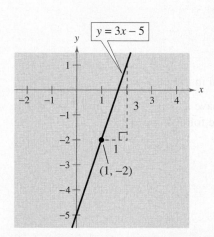

$y = 3x - 5$

FIGURE P.45

Study Tip

When you find an equation of the line that passes through two given points, you only need to substitute the coordinates of one of the points in the point-slope form. It does not matter which point you choose because both points will yield the same result.

Parallel and Perpendicular Lines

Slope can be used to decide whether two nonvertical lines in a plane are parallel, perpendicular, or neither.

Parallel and Perpendicular Lines

1. Two distinct nonvertical lines are **parallel** if and only if their slopes are equal. That is, $m_1 = m_2$.

2. Two nonvertical lines are **perpendicular** if and only if their slopes are negative reciprocals of each other. That is, $m_1 = -1/m_2$.

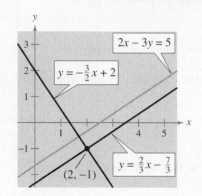

FIGURE **P.46**

Example 4   Finding Parallel and Perpendicular Lines

Find the slope-intercept forms of the equations of the lines that pass through the point $(2, -1)$ and are (a) parallel to and (b) perpendicular to the line $2x - 3y = 5$.

Solution

By writing the equation of the given line in slope-intercept form

$$2x - 3y = 5 \qquad \text{Write original equation.}$$
$$-3y = -2x + 5 \qquad \text{Subtract } 2x \text{ from each side.}$$
$$y = \tfrac{2}{3}x - \tfrac{5}{3} \qquad \text{Write in slope-intercept form.}$$

you can see that it has a slope of $m = \tfrac{2}{3}$, as shown in Figure P.46.

a. Any line parallel to the given line must also have a slope of $\tfrac{2}{3}$. So, the line through $(2, -1)$ that is parallel to the given line has the following equation.

$$y - (-1) = \tfrac{2}{3}(x - 2) \qquad \text{Write in point-slope form.}$$
$$3(y + 1) = 2(x - 2) \qquad \text{Multiply each side by 3.}$$
$$3y + 3 = 2x - 4 \qquad \text{Distributive Property}$$
$$y = \tfrac{2}{3}x - \tfrac{7}{3} \qquad \text{Write in slope-intercept form.}$$

b. Any line perpendicular to the given line must have a slope of $-\tfrac{3}{2}$ (because $-\tfrac{3}{2}$ is the negative reciprocal of $\tfrac{2}{3}$). So, the line through $(2, -1)$ that is perpendicular to the given line has the following equation.

$$y - (-1) = -\tfrac{3}{2}(x - 2) \qquad \text{Write in point-slope form.}$$
$$2(y + 1) = -3(x - 2) \qquad \text{Multiply each side by 2.}$$
$$2y + 2 = -3x + 6 \qquad \text{Distributive Property}$$
$$y = -\tfrac{3}{2}x + 2 \qquad \text{Write in slope-intercept form.}$$

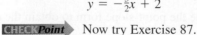 Now try Exercise 87.

Notice in Example 4 how the slope-intercept form is used to obtain information about the graph of a line, whereas the point-slope form is used to write the equation of a line.

TECHNOLOGY

On a graphing utility, lines will not appear to have the correct slope unless you use a viewing window that has a square setting. For instance, try graphing the lines in Example 4 using the standard setting $-10 \le x \le 10$ and $-10 \le y \le 10$. Then reset the viewing window with the square setting $-9 \le x \le 9$ and $-6 \le y \le 6$. On which setting do the lines $y = \tfrac{2}{3}x - \tfrac{5}{3}$ and $y = -\tfrac{3}{2}x + 2$ appear to be perpendicular?

Applications

In real-life problems, the slope of a line can be interpreted as either a *ratio* or a *rate*. If the *x*-axis and *y*-axis have the same unit of measure, then the slope has no units and is a **ratio**. If the *x*-axis and *y*-axis have different units of measure, then the slope is a **rate** or **rate of change.**

Example 5   Using Slope as a Ratio

The maximum recommended slope of a wheelchair ramp is $\frac{1}{12}$. A business is installing a wheelchair ramp that rises 22 inches over a horizontal length of 24 feet. Is the ramp steeper than recommended?   (Source: *Americans with Disabilities Act Handbook*)

Solution

The horizontal length of the ramp is 24 feet or $12(24) = 288$ inches, as shown in Figure P.47. So, the slope of the ramp is

$$\text{Slope} = \frac{\text{vertical change}}{\text{horizontal change}} = \frac{22 \text{ in.}}{288 \text{ in.}} \approx 0.076.$$

Because $\frac{1}{12} \approx 0.083$, the slope of the ramp is not steeper than recommended.

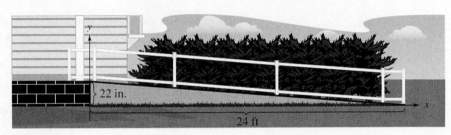

22 in.

24 ft

FIGURE **P.47**

CHECK*Point*   Now try Exercise 115.

Example 6   Using Slope as a Rate of Change

A kitchen appliance manufacturing company determines that the total cost in dollars of producing *x* units of a blender is

$$C = 25x + 3500. \qquad \text{Cost equation}$$

Describe the practical significance of the *y*-intercept and slope of this line.

Solution

The *y*-intercept $(0, 3500)$ tells you that the cost of producing zero units is $3500. This is the *fixed cost* of production—it includes costs that must be paid regardless of the number of units produced. The slope of $m = 25$ tells you that the cost of producing each unit is $25, as shown in Figure P.48. Economists call the cost per unit the *marginal cost*. If the production increases by one unit, then the "margin," or extra amount of cost, is $25. So, the cost increases at a rate of $25 per unit.

CHECK*Point*   Now try Exercise 119.

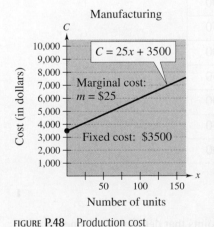

Manufacturing

$C = 25x + 3500$

Marginal cost:
$m = \$25$

Fixed cost: $3500

FIGURE **P.48**   Production cost

Most business expenses can be deducted in the same year they occur. One exception is the cost of property that has a useful life of more than 1 year. Such costs must be *depreciated* (decreased in value) over the useful life of the property. If the *same amount* is depreciated each year, the procedure is called *linear* or *straight-line depreciation*. The *book value* is the difference between the original value and the total amount of depreciation accumulated to date.

Example 7   Straight-Line Depreciation

A college purchased exercise equipment worth $12,000 for the new campus fitness center. The equipment has a useful life of 8 years. The salvage value at the end of 8 years is $2000. Write a linear equation that describes the book value of the equipment each year.

Solution

Let V represent the value of the equipment at the end of year t. You can represent the initial value of the equipment by the data point (0, 12,000) and the salvage value of the equipment by the data point (8, 2000). The slope of the line is

$$m = \frac{2000 - 12,000}{8 - 0} = -\$1250$$

which represents the annual depreciation in *dollars per year*. Using the point-slope form, you can write the equation of the line as follows.

$V - 12,000 = -1250(t - 0)$	Write in point-slope form.
$V = -1250t + 12,000$	Write in slope-intercept form.

The table shows the book value at the end of each year, and the graph of the equation is shown in Figure P.49.

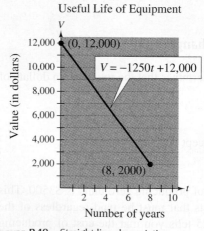

Useful Life of Equipment

$V = -1250t + 12,000$

FIGURE P.49   Straight-line depreciation

Year, t	Value, V
0	12,000
1	10,750
2	9500
3	8250
4	7000
5	5750
6	4500
7	3250
8	2000

CHECK*Point*   Now try Exercise 121.

In many real-life applications, the two data points that determine the line are often given in a disguised form. Note how the data points are described in Example 7.

Example 8   Predicting Sales

The sales for Best Buy were approximately $35.9 billion in 2006 and $40.0 billion in 2007. Using only this information, write a linear equation that gives the sales (in billions of dollars) in terms of the year. Then predict the sales for 2010.   (Source: Best Buy Company, Inc.)

Solution

Let $t = 6$ represent 2006. Then the two given values are represented by the data points $(6, 35.9)$ and $(7, 40.0)$. The slope of the line through these points is

$$m = \frac{40.0 - 35.9}{7 - 6}$$

$$= 4.1.$$

Using the point-slope form, you can find the equation that relates the sales y and the year t to be

$$y - 35.9 = 4.1(t - 6) \qquad \text{Write in point-slope form.}$$

$$y = 4.1t + 11.3. \qquad \text{Write in slope-intercept form.}$$

According to this equation, the sales for 2010 will be

$$y = 4.1(10) + 11.3 = 41 + 11.3 = \$52.3 \text{ billion. (See Figure P.50.)}$$

CHECK**Point**   Now try Exercise 129.

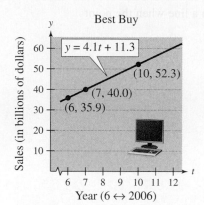

y = 4.1t + 11.3

(10, 52.3)

(7, 40.0)

(6, 35.9)

Best Buy

Sales (in billions of dollars)

Year (6 ↔ 2006)

FIGURE **P.50**

The prediction method illustrated in Example 8 is called **linear extrapolation.** Note in Figure P.51 that an extrapolated point does not lie between the given points. When the estimated point lies between two given points, as shown in Figure P.52, the procedure is called **linear interpolation.**

Because the slope of a vertical line is not defined, its equation cannot be written in slope-intercept form. However, every line has an equation that can be written in the **general form**

$$Ax + By + C = 0 \qquad \text{General form}$$

where A and B are not both zero. For instance, the vertical line given by $x = a$ can be represented by the general form $x - a = 0$.

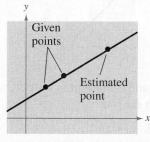

Given points

Estimated point

Linear extrapolation

FIGURE **P.51**

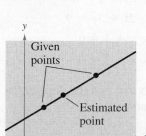

Given points

Estimated point

Linear interpolation

FIGURE **P.52**

Summary of Equations of Lines

1. General form: $Ax + By + C = 0$

2. Vertical line: $x = a$

3. Horizontal line: $y = b$

4. Slope-intercept form: $y = mx + b$

5. Point-slope form: $y - y_1 = m(x - x_1)$

6. Two-point form: $y - y_1 = \dfrac{y_2 - y_1}{x_2 - x_1}(x - x_1)$

P.4 EXERCISES

See www.CalcChat.com for worked-out solutions to odd-numbered exercises.

VOCABULARY

In Exercises 1–7, fill in the blanks.

1. The simplest mathematical model for relating two variables is the _____ equation in two variables $y = mx + b$.

2. For a line, the ratio of the change in y to the change in x is called the _____ of the line.

3. Two lines are _____ if and only if their slopes are equal.

4. Two lines are _____ if and only if their slopes are negative reciprocals of each other.

5. When the x-axis and y-axis have different units of measure, the slope can be interpreted as a _____.

6. The prediction method _____ _____ is the method used to estimate a point on a line when the point does not lie between the given points.

7. Every line has an equation that can be written in _____ form.

8. Match each equation of a line with its form.
   (a) $Ax + By + C = 0$ (i) Vertical line
   (b) $x = a$ (ii) Slope-intercept form
   (c) $y = b$ (iii) General form
   (d) $y = mx + b$ (iv) Point-slope form
   (e) $y - y_1 = m(x - x_1)$ (v) Horizontal line

SKILLS AND APPLICATIONS

In Exercises 9 and 10, identify the line that has each slope.

9. (a) $m = \frac{2}{3}$
   (b) m is undefined.
   (c) $m = -2$

10. (a) $m = 0$
    (b) $m = -\frac{3}{4}$
    (c) $m = 1$

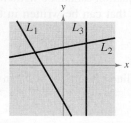

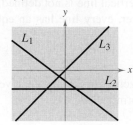

In Exercises 11 and 12, sketch the lines through the point with the indicated slopes on the same set of coordinate axes.

Point	Slopes
11. $(2, 3)$	(a) 0 (b) 1 (c) 2 (d) -3
12. $(-4, 1)$	(a) 3 (b) -3 (c) $\frac{1}{2}$ (d) Undefined

In Exercises 13–16, estimate the slope of the line.

13.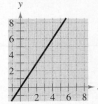

14.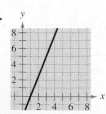

15.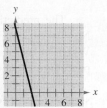

16.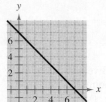

In Exercises 17–28, find the slope and y-intercept (if possible) of the equation of the line. Sketch the line.

17. $y = 5x + 3$
18. $y = x - 10$
19. $y = -\frac{1}{2}x + 4$
20. $y = -\frac{3}{2}x + 6$
21. $5x - 2 = 0$
22. $3y + 5 = 0$
23. $7x + 6y = 30$
24. $2x + 3y = 9$
25. $y - 3 = 0$
26. $y + 4 = 0$
27. $x + 5 = 0$
28. $x - 2 = 0$

In Exercises 29–40, plot the points and find the slope of the line passing through the pair of points.

29. $(0, 9), (6, 0)$
30. $(12, 0), (0, -8)$
31. $(-3, -2), (1, 6)$
32. $(2, 4), (4, -4)$
33. $(5, -7), (8, -7)$
34. $(-2, 1), (-4, -5)$
35. $(-6, -1), (-6, 4)$
36. $(0, -10), (-4, 0)$
37. $\left(\frac{11}{2}, -\frac{4}{3}\right), \left(-\frac{3}{2}, -\frac{1}{3}\right)$
38. $\left(\frac{7}{8}, \frac{3}{4}\right), \left(\frac{5}{4}, -\frac{1}{4}\right)$
39. $(4.8, 3.1), (-5.2, 1.6)$
40. $(-1.75, -8.3), (2.25, -2.6)$

In Exercises 41–50, use the point on the line and the slope m of the line to find three additional points through which the line passes. (There are many correct answers.)

41. $(2, 1)$, $m = 0$ **42.** $(3, -2)$, $m = 0$

43. $(5, -6)$, $m = 1$ **44.** $(10, -6)$, $m = -1$

45. $(-8, 1)$, m is undefined.

46. $(1, 5)$, m is undefined.

47. $(-5, 4)$, $m = 2$ **48.** $(0, -9)$, $m = -2$

49. $(7, -2)$, $m = \frac{1}{2}$ **50.** $(-1, -6)$, $m = -\frac{1}{2}$

In Exercises 51–64, find the slope-intercept form of the equation of the line that passes through the given point and has the indicated slope m. Sketch the line.

51. $(0, -2)$, $m = 3$ **52.** $(0, 10)$, $m = -1$

53. $(-3, 6)$, $m = -2$ **54.** $(0, 0)$, $m = 4$

55. $(4, 0)$, $m = -\frac{1}{3}$ **56.** $(8, 2)$, $m = \frac{1}{4}$

57. $(2, -3)$, $m = -\frac{1}{2}$ **58.** $(-2, -5)$, $m = \frac{3}{4}$

59. $(6, -1)$, m is undefined.

60. $(-10, 4)$, m is undefined.

61. $\left(4, \frac{5}{2}\right)$, $m = 0$ **62.** $\left(-\frac{1}{2}, \frac{3}{2}\right)$, $m = 0$

63. $(-5.1, 1.8)$, $m = 5$ **64.** $(2.3, -8.5)$, $m = -2.5$

In Exercises 65–78, find the slope-intercept form of the equation of the line passing through the points. Sketch the line.

65. $(5, -1)$, $(-5, 5)$ **66.** $(4, 3)$, $(-4, -4)$

67. $(-8, 1)$, $(-8, 7)$ **68.** $(-1, 4)$, $(6, 4)$

69. $\left(2, \frac{1}{2}\right)$, $\left(\frac{1}{2}, \frac{5}{4}\right)$ **70.** $(1, 1)$, $\left(6, -\frac{2}{3}\right)$

71. $\left(-\frac{1}{10}, -\frac{3}{5}\right)$, $\left(\frac{9}{10}, -\frac{9}{5}\right)$ **72.** $\left(\frac{3}{4}, \frac{3}{2}\right)$, $\left(-\frac{4}{3}, \frac{7}{4}\right)$

73. $(1, 0.6)$, $(-2, -0.6)$ **74.** $(-8, 0.6)$, $(2, -2.4)$

75. $(2, -1)$, $\left(\frac{1}{3}, -1\right)$ **76.** $\left(\frac{1}{5}, -2\right)$, $(-6, -2)$

77. $\left(\frac{7}{3}, -8\right)$, $\left(\frac{7}{3}, 1\right)$ **78.** $(1.5, -2)$, $(1.5, 0.2)$

In Exercises 79–82, determine whether the lines are parallel, perpendicular, or neither.

79. $L_1: y = \frac{1}{3}x - 2$ **80.** $L_1: y = 4x - 1$

   $L_2: y = \frac{1}{3}x + 3$ $L_2: y = 4x + 7$

81. $L_1: y = \frac{1}{2}x - 3$ **82.** $L_1: y = -\frac{4}{5}x - 5$

   $L_2: y = -\frac{1}{2}x + 1$ $L_2: y = \frac{5}{4}x + 1$

In Exercises 83–86, determine whether the lines L_1 and L_2 passing through the pairs of points are parallel, perpendicular, or neither.

83. $L_1: (0, -1), (5, 9)$ **84.** $L_1: (-2, -1), (1, 5)$

   $L_2: (0, 3), (4, 1)$ $L_2: (1, 3), (5, -5)$

85. $L_1: (3, 6), (-6, 0)$ **86.** $L_1: (4, 8), (-4, 2)$

   $L_2: (0, -1), \left(5, \frac{7}{3}\right)$ $L_2: (3, -5), \left(-1, \frac{1}{3}\right)$

In Exercises 87–96, write the slope-intercept forms of the equations of the lines through the given point (a) parallel to the given line and (b) perpendicular to the given line.

87. $4x - 2y = 3$, $(2, 1)$ **88.** $x + y = 7$, $(-3, 2)$

89. $3x + 4y = 7$, $\left(-\frac{2}{3}, \frac{7}{8}\right)$ **90.** $5x + 3y = 0$, $\left(\frac{7}{8}, \frac{3}{4}\right)$

91. $y + 3 = 0$, $(-1, 0)$ **92.** $y - 2 = 0$, $(-4, 1)$

93. $x - 4 = 0$, $(3, -2)$ **94.** $x + 2 = 0$, $(-5, 1)$

95. $x - y = 4$, $(2.5, 6.8)$

96. $6x + 2y = 9$, $(-3.9, -1.4)$

In Exercises 97–102, use the *intercept form* to find the equation of the line with the given intercepts. The intercept form of the equation of a line with intercepts $(a, 0)$ and $(0, b)$ is

$$\frac{x}{a} + \frac{y}{b} = 1, \quad a \neq 0, \ b \neq 0.$$

97. x-intercept: $(2, 0)$ **98.** x-intercept: $(-3, 0)$

   y-intercept: $(0, 3)$ y-intercept: $(0, 4)$

99. x-intercept: $\left(-\frac{1}{6}, 0\right)$ **100.** x-intercept: $\left(\frac{2}{3}, 0\right)$

   y-intercept: $\left(0, -\frac{2}{3}\right)$ y-intercept: $(0, -2)$

101. Point on line: $(1, 2)$

   x-intercept: $(c, 0)$

   y-intercept: $(0, c)$, $c \neq 0$

102. Point on line: $(-3, 4)$

   x-intercept: $(d, 0)$

   y-intercept: $(0, d)$, $d \neq 0$

GRAPHICAL ANALYSIS In Exercises 103–106, identify any relationships that exist among the lines, and then use a graphing utility to graph the three equations in the same viewing window. Adjust the viewing window so that the slope appears visually correct—that is, so that parallel lines appear parallel and perpendicular lines appear to intersect at right angles.

103. (a) $y = 2x$ (b) $y = -2x$ (c) $y = \frac{1}{2}x$

104. (a) $y = \frac{2}{3}x$ (b) $y = -\frac{3}{2}x$ (c) $y = \frac{2}{3}x + 2$

105. (a) $y = -\frac{1}{2}x$ (b) $y = -\frac{1}{2}x + 3$ (c) $y = 2x - 4$

106. (a) $y = x - 8$ (b) $y = x + 1$ (c) $y = -x + 3$

In Exercises 107–110, find a relationship between x and y such that (x, y) is equidistant (the same distance) from the two points.

107. $(4, -1)$, $(-2, 3)$ **108.** $(6, 5)$, $(1, -8)$

109. $\left(3, \frac{5}{2}\right)$, $(-7, 1)$ **110.** $\left(-\frac{1}{2}, -4\right)$, $\left(\frac{7}{2}, \frac{5}{4}\right)$

111. SALES The following are the slopes of lines representing annual sales y in terms of time x in years. Use the slopes to interpret any change in annual sales for a one-year increase in time.

(a) The line has a slope of $m = 135$.

(b) The line has a slope of $m = 0$.

(c) The line has a slope of $m = -40$.

112. REVENUE The following are the slopes of lines representing daily revenues y in terms of time x in days. Use the slopes to interpret any change in daily revenues for a one-day increase in time.

(a) The line has a slope of $m = 400$.

(b) The line has a slope of $m = 100$.

(c) The line has a slope of $m = 0$.

113. AVERAGE SALARY The graph shows the average salaries for senior high school principals from 1996 through 2008. (Source: Educational Research Service)

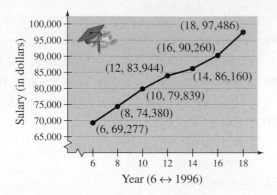

Year (6 ↔ 1996)

(a) Use the slopes of the line segments to determine the time periods in which the average salary increased the greatest and the least.

(b) Find the slope of the line segment connecting the points for the years 1996 and 2008.

(c) Interpret the meaning of the slope in part (b) in the context of the problem.

114. SALES The graph shows the sales (in billions of dollars) for Apple Inc. for the years 2001 through 2007. (Source: Apple Inc.)

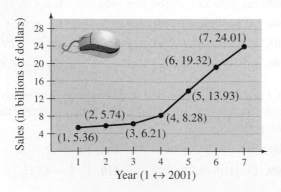

Year (1 ↔ 2001)

(a) Use the slopes of the line segments to determine the years in which the sales showed the greatest increase and the least increase.

(b) Find the slope of the line segment connecting the points for the years 2001 and 2007.

(c) Interpret the meaning of the slope in part (b) in the context of the problem.

115. ROAD GRADE You are driving on a road that has a 6% uphill grade (see figure). This means that the slope of the road is $\frac{6}{100}$. Approximate the amount of vertical change in your position if you drive 200 feet.

116. ROAD GRADE From the top of a mountain road, a surveyor takes several horizontal measurements x and several vertical measurements y, as shown in the table (x and y are measured in feet).

x	300	600	900	1200	1500	1800	2100
y	-25	-50	-75	-100	-125	-150	-175

(a) Sketch a scatter plot of the data.

(b) Use a straightedge to sketch the line that you think best fits the data.

(c) Find an equation for the line you sketched in part (b).

(d) Interpret the meaning of the slope of the line in part (c) in the context of the problem.

(e) The surveyor needs to put up a road sign that indicates the steepness of the road. For instance, a surveyor would put up a sign that states "8% grade" on a road with a downhill grade that has a slope of $-\frac{8}{100}$. What should the sign state for the road in this problem?

RATE OF CHANGE In Exercises 117 and 118, you are given the dollar value of a product in 2010 and the rate at which the value of the product is expected to change during the next 5 years. Use this information to write a linear equation that gives the dollar value V of the product in terms of the year t. (Let $t = 10$ represent 2010.)

	2010 Value	*Rate*
117.	$2540	$125 decrease per year
118.	$156	$4.50 increase per year

119. DEPRECIATION The value V of a molding machine t years after it is purchased is

$$V = -4000t + 58,500, \quad 0 \le t \le 5.$$

Explain what the V-intercept and the slope measure.

120. COST The cost C of producing n computer laptop bags is given by

$$C = 1.25n + 15,750, \quad 0 < n.$$

Explain what the C-intercept and the slope measure.

121. DEPRECIATION A sub shop purchases a used pizza oven for $875. After 5 years, the oven will have to be replaced. Write a linear equation giving the value V of the equipment during the 5 years it will be in use.

122. DEPRECIATION A school district purchases a high-volume printer, copier, and scanner for $25,000. After 10 years, the equipment will have to be replaced. Its value at that time is expected to be $2000. Write a linear equation giving the value V of the equipment during the 10 years it will be in use.

123. SALES A discount outlet is offering a 20% discount on all items. Write a linear equation giving the sale price S for an item with a list price L.

124. HOURLY WAGE A microchip manufacturer pays its assembly line workers $12.25 per hour. In addition, workers receive a piecework rate of $0.75 per unit produced. Write a linear equation for the hourly wage W in terms of the number of units x produced per hour.

125. MONTHLY SALARY A pharmaceutical salesperson receives a monthly salary of $2500 plus a commission of 7% of sales. Write a linear equation for the salesperson's monthly wage W in terms of monthly sales S.

126. BUSINESS COSTS A sales representative of a company using a personal car receives $120 per day for lodging and meals plus $0.55 per mile driven. Write a linear equation giving the daily cost C to the company in terms of x, the number of miles driven.

127. CASH FLOW PER SHARE The cash flow per share for the Timberland Co. was $1.21 in 1999 and $1.46 in 2007. Write a linear equation that gives the cash flow per share in terms of the year. Let $t = 9$ represent 1999. Then predict the cash flows for the years 2012 and 2014.   (Source: The Timberland Co.)

128. NUMBER OF STORES In 2003 there were 1078 J.C. Penney stores and in 2007 there were 1067 stores. Write a linear equation that gives the number of stores in terms of the year. Let $t = 3$ represent 2003. Then predict the numbers of stores for the years 2012 and 2014. Are your answers reasonable? Explain. (Source: J.C. Penney Co.)

129. COLLEGE ENROLLMENT The Pennsylvania State University had enrollments of 40,571 students in 2000 and 44,112 students in 2008 at its main campus in University Park, Pennsylvania. (Source: *Penn State Fact Book*)

(a) Assuming the enrollment growth is linear, find a linear model that gives the enrollment in terms of the year t, where $t = 0$ corresponds to 2000.

(b) Use your model from part (a) to predict the enrollments in 2010 and 2015.

(c) What is the slope of your model? Explain its meaning in the context of the situation.

130. COLLEGE ENROLLMENT The University of Florida had enrollments of 46,107 students in 2000 and 51,413 students in 2008.   (Source: University of Florida)

(a) What was the average annual change in enrollment from 2000 to 2008?

(b) Use the average annual change in enrollment to estimate the enrollments in 2002, 2004, and 2006.

(c) Write the equation of a line that represents the given data in terms of the year t, where $t = 0$ corresponds to 2000. What is its slope? Interpret the slope in the context of the problem.

(d) Using the results of parts (a)–(c), write a short paragraph discussing the concepts of *slope* and *average rate of change*.

131. COST, REVENUE, AND PROFIT A roofing contractor purchases a shingle delivery truck with a shingle elevator for $42,000. The vehicle requires an average expenditure of $6.50 per hour for fuel and maintenance, and the operator is paid $11.50 per hour.

(a) Write a linear equation giving the total cost C of operating this equipment for t hours. (Include the purchase cost of the equipment.)

(b) Assuming that customers are charged $30 per hour of machine use, write an equation for the revenue R derived from t hours of use.

(c) Use the formula for profit

$$P = R - C$$

to write an equation for the profit derived from t hours of use.

(d) Use the result of part (c) to find the break-even point—that is, the number of hours this equipment must be used to yield a profit of 0 dollars.

132. RENTAL DEMAND A real estate office handles an apartment complex with 50 units. When the rent per unit is $580 per month, all 50 units are occupied. However, when the rent is $625 per month, the average number of occupied units drops to 47. Assume that the relationship between the monthly rent p and the demand x is linear.

(a) Write the equation of the line giving the demand x in terms of the rent p.

(b) Use this equation to predict the number of units occupied when the rent is $655.

(c) Predict the number of units occupied when the rent is $595.

133. GEOMETRY The length and width of a rectangular garden are 15 meters and 10 meters, respectively. A walkway of width x surrounds the garden.

(a) Draw a diagram that gives a visual representation of the problem.

(b) Write the equation for the perimeter y of the walkway in terms of x.

(c) Use a graphing utility to graph the equation for the perimeter.

(d) Determine the slope of the graph in part (c). For each additional one-meter increase in the width of the walkway, determine the increase in its perimeter.

134. AVERAGE ANNUAL SALARY The average salaries (in millions of dollars) of Major League Baseball players from 2000 through 2007 are shown in the scatter plot. Find the equation of the line that you think best fits these data. (Let y represent the average salary and let t represent the year, with $t = 0$ corresponding to 2000.) (Source: Major League Baseball Players Association)

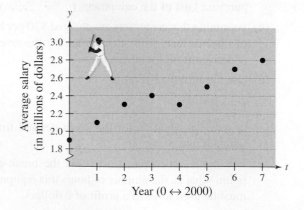

135. DATA ANALYSIS: NUMBER OF DOCTORS The numbers of doctors of osteopathic medicine y (in thousands) in the United States from 2000 through 2008, where x is the year, are shown as data points (x, y). (Source: American Osteopathic Association)

(2000, 44.9), (2001, 47.0), (2002, 49.2), (2003, 51.7), (2004, 54.1), (2005, 56.5), (2006, 58.9), (2007, 61.4), (2008, 64.0)

(a) Sketch a scatter plot of the data. Let $x = 0$ correspond to 2000.

(b) Use a straightedge to sketch the line that you think best fits the data.

(c) Find the equation of the line from part (b). Explain the procedure you used.

(d) Write a short paragraph explaining the meanings of the slope and y-intercept of the line in terms of the data.

(e) Compare the values obtained using your model with the actual values.

(f) Use your model to estimate the number of doctors of osteopathic medicine in 2012.

136. DATA ANALYSIS: AVERAGE SCORES An instructor gives regular 20-point quizzes and 100-point exams in an algebra course. Average scores for six students, given as data points (x, y), where x is the average quiz score and y is the average test score, are (18, 87), (10, 55), (19, 96), (16, 79), (13, 76), and (15, 82). [*Note:* There are many correct answers for parts (b)–(d).]

(a) Sketch a scatter plot of the data.

(b) Use a straightedge to sketch the line that you think best fits the data.

(c) Find an equation for the line you sketched in part (b).

(d) Use the equation in part (c) to estimate the average test score for a person with an average quiz score of 17.

(e) The instructor adds 4 points to the average test score of each student in the class. Describe the changes in the positions of the plotted points and the change in the equation of the line.

EXPLORATION

TRUE OR FALSE? In Exercises 137 and 138, determine whether the statement is true or false. Justify your answer.

137. A line with a slope of $-\frac{5}{7}$ is steeper than a line with a slope of $-\frac{6}{7}$.

138. The line through $(-8, 2)$ and $(-1, 4)$ and the line through $(0, -4)$ and $(-7, 7)$ are parallel.

139. Explain how you could show that the points $A(2, 3)$, $B(2, 9)$, and $C(4, 3)$ are the vertices of a right triangle.

140. Explain why the slope of a vertical line is said to be undefined.

141. With the information shown in the graphs, is it possible to determine the slope of each line? Is it possible that the lines could have the same slope? Explain.

(a)   (b)

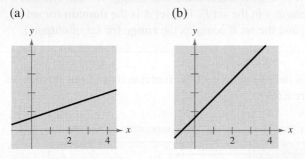

142. The slopes of two lines are -4 and $\frac{5}{2}$. Which is steeper? Explain.

143. Use a graphing utility to compare the slopes of the lines $y = mx$, where $m = 0.5$, 1, 2, and 4. Which line rises most quickly? Now, let $m = -0.5$, -1, -2, and -4. Which line falls most quickly? Use a square setting to obtain a true geometric perspective. What can you conclude about the slope and the "rate" at which the line rises or falls?

144. Find d_1 and d_2 in terms of m_1 and m_2, respectively (see figure). Then use the Pythagorean Theorem to find a relationship between m_1 and m_2.

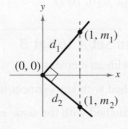

145. THINK ABOUT IT Is it possible for two lines with positive slopes to be perpendicular? Explain.

146. CAPSTONE Match the description of the situation with its graph. Also determine the slope and y-intercept of each graph and interpret the slope and y-intercept in the context of the situation. [The graphs are labeled (i), (ii), (iii), and (iv).]

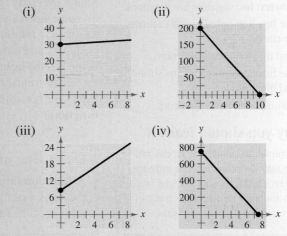

(a) A person is paying \$20 per week to a friend to repay a \$200 loan.

(b) An employee is paid \$8.50 per hour plus \$2 for each unit produced per hour.

(c) A sales representative receives \$30 per day for food plus \$0.32 for each mile traveled.

(d) A computer that was purchased for \$750 depreciates \$100 per year.

PROJECT: BACHELOR'S DEGREES To work an extended application analyzing the numbers of bachelor's degrees earned by women in the United States from 1996 through 2007, visit this text's website at *academic.cengage.com*. (Data Source: U.S. National Center for Education Statistics)

P.5 FUNCTIONS

What you should learn

- Determine whether relations between two variables are functions.
- Use function notation and evaluate functions.
- Find the domains of functions.
- Use functions to model and solve real-life problems.
- Evaluate difference quotients.

Why you should learn it

Functions can be used to model and solve real-life problems. For instance, in Exercise 100 on page 71, you will use a function to model the force of water against the face of a dam.

Introduction to Functions

Many everyday phenomena involve two quantities that are related to each other by some rule of correspondence. The mathematical term for such a rule of correspondence is a **relation.** In mathematics, relations are often represented by mathematical equations and formulas. For instance, the simple interest I earned on $1000 for 1 year is related to the annual interest rate r by the formula $I = 1000r$.

The formula $I = 1000r$ represents a special kind of relation that matches each item from one set with *exactly one* item from a different set. Such a relation is called a **function.**

Definition of Function

A **function** f from a set A to a set B is a relation that assigns to each element x in the set A exactly one element y in the set B. The set A is the **domain** (or set of inputs) of the function f, and the set B contains the **range** (or set of outputs).

To help understand this definition, look at the function that relates the time of day to the temperature in Figure P.53.

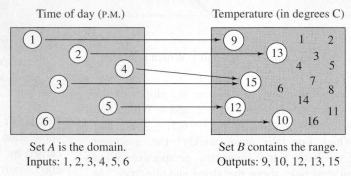

Set A is the domain.
Inputs: 1, 2, 3, 4, 5, 6

Set B contains the range.
Outputs: 9, 10, 12, 13, 15

FIGURE P.53

This function can be represented by the following ordered pairs, in which the first coordinate (x-value) is the input and the second coordinate (y-value) is the output.

$$\{(1, 9°), (2, 13°), (3, 15°), (4, 15°), (5, 12°), (6, 10°)\}$$

Characteristics of a Function from Set A to Set B

1. Each element in A must be matched with an element in B.

2. Some elements in B may not be matched with any element in A.

3. Two or more elements in A may be matched with the same element in B.

4. An element in A (the domain) cannot be matched with two different elements in B.

Functions are commonly represented in four ways.

Four Ways to Represent a Function

1. *Verbally* by a sentence that describes how the input variable is related to the output variable

2. *Numerically* by a table or a list of ordered pairs that matches input values with output values

3. *Graphically* by points on a graph in a coordinate plane in which the input values are represented by the horizontal axis and the output values are represented by the vertical axis

4. *Algebraically* by an equation in two variables

To determine whether or not a relation is a function, you must decide whether each input value is matched with exactly one output value. If any input value is matched with two or more output values, the relation is not a function.

Example 1   Testing for Functions

Determine whether the relation represents y as a function of x.

a. The input value x is the number of representatives from a state, and the output value y is the number of senators.

b.

Input, x	Output, y
2	11
2	10
3	8
4	5
5	1

c.

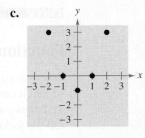

FIGURE P.54

Solution

a. This verbal description *does* describe y as a function of x. Regardless of the value of x, the value of y is always 2. Such functions are called *constant functions*.

b. This table *does not* describe y as a function of x. The input value 2 is matched with two different y-values.

c. The graph in Figure P.54 *does* describe y as a function of x. Each input value is matched with exactly one output value.

CHECK *Point*   Now try Exercise 11.

Representing functions by sets of ordered pairs is common in *discrete mathematics*. In algebra, however, it is more common to represent functions by equations or formulas involving two variables. For instance, the equation

$$y = x^2 \qquad \text{\small{y is a function of x.}}$$

represents the variable y as a function of the variable x. In this equation, x is

HISTORICAL NOTE

Leonhard Euler (1707–1783), a Swiss mathematician, is considered to have been the most prolific and productive mathematician in history. One of his greatest influences on mathematics was his use of symbols, or notation. The function notation $y = f(x)$ was introduced by Euler.

the **independent variable** and y is the **dependent variable.** The domain of the function is the set of all values taken on by the independent variable x, and the range of the function is the set of all values taken on by the dependent variable y.

Example 2   Testing for Functions Represented Algebraically

Which of the equations represent(s) y as a function of x?

a. $x^2 + y = 1$ **b.** $-x + y^2 = 1$

Solution

To determine whether y is a function of x, try to solve for y in terms of x.

a. Solving for y yields

$$x^2 + y = 1 \qquad \text{Write original equation.}$$
$$y = 1 - x^2. \qquad \text{Solve for } y.$$

To each value of x there corresponds exactly one value of y. So, y is a function of x.

b. Solving for y yields

$$-x + y^2 = 1 \qquad \text{Write original equation.}$$
$$y^2 = 1 + x \qquad \text{Add } x \text{ to each side.}$$
$$y = \pm\sqrt{1 + x}. \qquad \text{Solve for } y.$$

The $\pm$ indicates that to a given value of x there correspond two values of y. So, y is not a function of x.

CHECK *Point*   Now try Exercise 21.

Function Notation

When an equation is used to represent a function, it is convenient to name the function so that it can be referenced easily. For example, you know that the equation $y = 1 - x^2$ describes y as a function of x. Suppose you give this function the name "f." Then you can use the following **function notation.**

Input	Output	Equation
x	$f(x)$	$f(x) = 1 - x^2$

The symbol $f(x)$ is read as *the value of f at x* or simply *f of x.* The symbol $f(x)$ corresponds to the y-value for a given x. So, you can write $y = f(x)$. Keep in mind that f is the *name* of the function, whereas $f(x)$ is the *value* of the function at x. For instance, the function given by

$$f(x) = 3 - 2x$$

has *function values* denoted by $f(-1)$, $f(0)$, $f(2)$, and so on. To find these values, substitute the specified input values into the given equation.

For $x = -1$, $f(-1) = 3 - 2(-1) = 3 + 2 = 5$.

For $x = 0$, $f(0) = 3 - 2(0) = 3 - 0 = 3$.

For $x = 2$, $f(2) = 3 - 2(2) = 3 - 4 = -1$.

Although f is often used as a convenient function name and x is often used as the independent variable, you can use other letters. For instance,

$$f(x) = x^2 - 4x + 7, \quad f(t) = t^2 - 4t + 7, \quad \text{and} \quad g(s) = s^2 - 4s + 7$$

all define the same function. In fact, the role of the independent variable is that of a "placeholder." Consequently, the function could be described by

$$f\left(\,\rule{1.5em}{0.8em}\,\right) = \left(\,\rule{1.5em}{0.8em}\,\right)^2 - 4\left(\,\rule{1.5em}{0.8em}\,\right) + 7.$$

> **WARNING / CAUTION**
>
> In Example 3, note that $g(x + 2)$ is not equal to $g(x) + g(2)$. In general, $g(u + v) \neq g(u) + g(v)$.

Example 3   Evaluating a Function

Let $g(x) = -x^2 + 4x + 1$. Find each function value.

a. $g(2)$ **b.** $g(t)$ **c.** $g(x + 2)$

Solution

a. Replacing x with 2 in $g(x) = -x^2 + 4x + 1$ yields the following.

$$g(2) = -(2)^2 + 4(2) + 1 = -4 + 8 + 1 = 5$$

b. Replacing x with t yields the following.

$$g(t) = -(t)^2 + 4(t) + 1 = -t^2 + 4t + 1$$

c. Replacing x with $x + 2$ yields the following.

$$\begin{aligned} g(x + 2) &= -(x + 2)^2 + 4(x + 2) + 1 \\ &= -(x^2 + 4x + 4) + 4x + 8 + 1 \\ &= -x^2 - 4x - 4 + 4x + 8 + 1 \\ &= -x^2 + 5 \end{aligned}$$

CHECK *Point*   Now try Exercise 41.

A function defined by two or more equations over a specified domain is called a **piecewise-defined function.**

Example 4   A Piecewise-Defined Function

Evaluate the function when $x = -1, 0,$ and 1.

$$f(x) = \begin{cases} x^2 + 1, & x < 0 \\ x - 1, & x \geq 0 \end{cases}$$

Solution

Because $x = -1$ is less than 0, use $f(x) = x^2 + 1$ to obtain

$$f(-1) = (-1)^2 + 1 = 2.$$

For $x = 0$, use $f(x) = x - 1$ to obtain

$$f(0) = (0) - 1 = -1.$$

For $x = 1$, use $f(x) = x - 1$ to obtain

$$f(1) = (1) - 1 = 0.$$

CHECK *Point*   Now try Exercise 49.

Example 5 Finding Values for Which $f(x) = 0$

Find all real values of x such that $f(x) = 0$.

a. $f(x) = -2x + 10$

b. $f(x) = x^2 - 5x + 6$

Solution

For each function, set $f(x) = 0$ and solve for x.

a. $-2x + 10 = 0$ Set $f(x)$ equal to 0.

$\qquad -2x = -10$ Subtract 10 from each side.

$\qquad\quad x = 5$ Divide each side by -2.

So, $f(x) = 0$ when $x = 5$.

b. $x^2 - 5x + 6 = 0$ Set $f(x)$ equal to 0.

$(x - 2)(x - 3) = 0$ Factor.

$x - 2 = 0$ ⟹ $x = 2$ Set 1st factor equal to 0.

$x - 3 = 0$ ⟹ $x = 3$ Set 2nd factor equal to 0.

So, $f(x) = 0$ when $x = 2$ or $x = 3$.

CHECK Point Now try Exercise 59.

Example 6 Finding Values for Which $f(x) = g(x)$

Find the values of x for which $f(x) = g(x)$.

a. $f(x) = x^2 + 1$ and $g(x) = 3x - x^2$

b. $f(x) = x^2 - 1$ and $g(x) = -x^2 + x + 2$

Solution

a. $x^2 + 1 = 3x - x^2$ Set $f(x)$ equal to $g(x)$.

$2x^2 - 3x + 1 = 0$ Write in general form.

$(2x - 1)(x - 1) = 0$ Factor.

$2x - 1 = 0$ ⟹ $x = \frac{1}{2}$ Set 1st factor equal to 0.

$x - 1 = 0$ ⟹ $x = 1$ Set 2nd factor equal to 0.

So, $f(x) = g(x)$ when $x = \frac{1}{2}$ or $x = 1$.

b. $x^2 - 1 = -x^2 + x + 2$ Set $f(x)$ equal to $g(x)$.

$2x^2 - x - 3 = 0$ Write in general form.

$(2x - 3)(x + 1) = 0$ Factor.

$2x - 3 = 0$ ⟹ $x = \frac{3}{2}$ Set 1st factor equal to 0.

$x + 1 = 0$ ⟹ $x = -1$ Set 2nd factor equal to 0.

So, $f(x) = g(x)$ when $x = \frac{3}{2}$ or $x = -1$.

CHECK Point Now try Exercise 67.

The Domain of a Function

The domain of a function can be described explicitly or it can be *implied* by the expression used to define the function. The **implied domain** is the set of all real numbers for which the expression is defined. For instance, the function given by

$$f(x) = \frac{1}{x^2 - 4}$$ Domain excludes *x*-values that result in division by zero.

has an implied domain that consists of all real x other than $x = \pm 2$. These two values are excluded from the domain because division by zero is undefined. Another common type of implied domain is that used to avoid even roots of negative numbers. For example, the function given by

$$f(x) = \sqrt{x}$$ Domain excludes *x*-values that result in even roots of negative numbers.

is defined only for $x \geq 0$. So, its implied domain is the interval $[0, \infty)$. In general, the domain of a function *excludes* values that would cause division by zero *or* that would result in the even root of a negative number.

TECHNOLOGY

Use a graphing utility to graph the functions given by $y = \sqrt{4 - x^2}$ and $y = \sqrt{x^2 - 4}$. What is the domain of each function? Do the domains of these two functions overlap? If so, for what values do the domains overlap?

| **Example 7**   **Finding the Domain of a Function**

Find the domain of each function.

a. $f: \{(-3, 0), (-1, 4), (0, 2), (2, 2), (4, -1)\}$ **b.** $g(x) = \dfrac{1}{x + 5}$

c. Volume of a sphere: $V = \frac{4}{3}\pi r^3$ **d.** $h(x) = \sqrt{4 - 3x}$

Solution

a. The domain of f consists of all first coordinates in the set of ordered pairs.

   Domain $= \{-3, -1, 0, 2, 4\}$

b. Excluding *x*-values that yield zero in the denominator, the domain of g is the set of all real numbers x except $x = -5$.

c. Because this function represents the volume of a sphere, the values of the radius r must be positive. So, the domain is the set of all real numbers r such that $r > 0$.

d. This function is defined only for *x*-values for which

   $$4 - 3x \geq 0.$$

By solving this inequality, you can conclude that $x \leq \frac{4}{3}$. So, the domain is the interval $\left(-\infty, \frac{4}{3}\right]$.

CHECK *Point* ▸ Now try Exercise 73.   ▮

In Example 7(c), note that the domain of a function may be implied by the physical context. For instance, from the equation

$$V = \frac{4}{3}\pi r^3$$

you would have no reason to restrict r to positive values, but the physical context implies that a sphere cannot have a negative or zero radius.

$\dfrac{h}{r} = 4$

FIGURE **P.55**

Applications

Example 8 The Dimensions of a Container

You work in the marketing department of a soft-drink company and are experimenting with a new can for iced tea that is slightly narrower and taller than a standard can. For your experimental can, the ratio of the height to the radius is 4, as shown in Figure P.55.

a. Write the volume of the can as a function of the radius r.

b. Write the volume of the can as a function of the height h.

Solution

a. $V(r) = \pi r^2 h = \pi r^2 (4r) = 4\pi r^3$ Write V as a function of r.

b. $V(h) = \pi \left(\dfrac{h}{4}\right)^2 h = \dfrac{\pi h^3}{16}$ Write V as a function of h.

CHECK *Point*   Now try Exercise 87.

Example 9 The Path of a Baseball

A baseball is hit at a point 3 feet above ground at a velocity of 100 feet per second and an angle of 45°. The path of the baseball is given by the function

$$f(x) = -0.0032x^2 + x + 3$$

where x and $f(x)$ are measured in feet. Will the baseball clear a 10-foot fence located 300 feet from home plate?

Algebraic Solution

When $x = 300$, you can find the height of the baseball as follows.

$f(x) = -0.0032x^2 + x + 3$ Write original function.

$f(300) = -0.0032(300)^2 + 300 + 3$ Substitute 300 for x.

$= 15$ Simplify.

When $x = 300$, the height of the baseball is 15 feet, so the baseball will clear a 10-foot fence.

CHECK *Point*   Now try Exercise 93.

Graphical Solution

Use a graphing utility to graph the function $y = -0.0032x^2 + x + 3$. Use the *value* feature or the *zoom* and *trace* features of the graphing utility to estimate that $y = 15$ when $x = 300$, as shown in Figure P.56. So, the ball will clear a 10-foot fence.

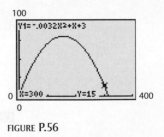

FIGURE **P.56**

In the equation in Example 9, the height of the baseball is a function of the distance from home plate.

Example 10   Alternative-Fueled Vehicles

Number of Alternative-Fueled
Vehicles in the U.S.

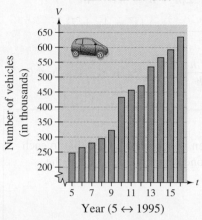

FIGURE P.57

The number V (in thousands) of alternative-fueled vehicles in the United States increased in a linear pattern from 1995 to 1999, as shown in Figure P.57. Then, in 2000, the number of vehicles took a jump and, until 2006, increased in a different linear pattern. These two patterns can be approximated by the function

$$V(t) = \begin{cases} 18.08t + 155.3, & 5 \le t \le 9 \\ 34.75t + 74.9, & 10 \le t \le 16 \end{cases}$$

where t represents the year, with $t = 5$ corresponding to 1995. Use this function to approximate the number of alternative-fueled vehicles for each year from 1995 to 2006. (Source: Science Applications International Corporation; Energy Information Administration)

Solution

From 1995 to 1999, use $V(t) = 18.08t + 155.3$.

245.7	263.8	281.9	299.9	318.0
1995	1996	1997	1998	1999

From 2000 to 2006, use $V(t) = 34.75t + 74.9$.

422.4	457.2	491.9	526.7	561.4	596.2	630.9
2000	2001	2002	2003	2004	2005	2006

CHECK Point   Now try Exercise 95.

Difference Quotients

One of the basic definitions in calculus employs the ratio

$$\frac{f(x + h) - f(x)}{h}, \quad h \ne 0.$$

This ratio is called a **difference quotient,** as illustrated in Example 11.

Example 11   Evaluating a Difference Quotient

For $f(x) = x^2 - 4x + 7$, find $\dfrac{f(x + h) - f(x)}{h}$.

Solution

$$\frac{f(x + h) - f(x)}{h} = \frac{[(x + h)^2 - 4(x + h) + 7] - (x^2 - 4x + 7)}{h}$$

$$= \frac{x^2 + 2xh + h^2 - 4x - 4h + 7 - x^2 + 4x - 7}{h}$$

$$= \frac{2xh + h^2 - 4h}{h} = \frac{h(2x + h - 4)}{h} = 2x + h - 4, \quad h \ne 0$$

CHECK Point   Now try Exercise 103.

The symbol ∫ indicates an example or exercise that highlights algebraic techniques specifically used in calculus.

You may find it easier to calculate the difference quotient in Example 11 by first finding $f(x + h)$, and then substituting the resulting expression into the difference quotient, as follows.

$$f(x + h) = (x + h)^2 - 4(x + h) + 7 = x^2 + 2xh + h^2 - 4x - 4h + 7$$

$$\frac{f(x + h) - f(x)}{h} = \frac{(x^2 + 2xh + h^2 - 4x - 4h + 7) - (x^2 - 4x + 7)}{h}$$

$$= \frac{2xh + h^2 - 4h}{h} = \frac{h(2x + h - 4)}{h} = 2x + h - 4, \quad h \neq 0$$

Summary of Function Terminology

Function: A **function** is a relationship between two variables such that to each value of the independent variable there corresponds exactly one value of the dependent variable.

Function Notation: $y = f(x)$

  f is the *name* of the function.

  y is the **dependent variable.**

  x is the **independent variable.**

  $f(x)$ is the *value of the function at x.*

Domain: The **domain** of a function is the set of all values (inputs) of the independent variable for which the function is defined. If x is in the domain of f, f is said to be *defined* at x. If x is not in the domain of f, f is said to be *undefined* at x.

Range: The **range** of a function is the set of all values (outputs) assumed by the dependent variable (that is, the set of all function values).

Implied Domain: If f is defined by an algebraic expression and the domain is not specified, the **implied domain** consists of all real numbers for which the expression is defined.

CLASSROOM DISCUSSION

Everyday Functions   In groups of two or three, identify common real-life functions. Consider everyday activities, events, and expenses, such as long distance telephone calls and car insurance. Here are two examples.

a. The statement, "Your happiness is a function of the grade you receive in this course" is *not* a correct mathematical use of the word "function." The word "happiness" is ambiguous.

b. The statement, "Your federal income tax is a function of your adjusted gross income" *is* a correct mathematical use of the word "function." Once you have determined your adjusted gross income, your income tax can be determined.

Describe your functions in words. Avoid using ambiguous words. Can you find an example of a piecewise-defined function?

P.5   EXERCISES

See www.CalcChat.com for worked-out solutions to odd-numbered exercises.

VOCABULARY: Fill in the blanks.

1. A relation that assigns to each element x from a set of inputs, or _____, exactly one element y in a set of outputs, or _____, is called a _____.

2. Functions are commonly represented in four different ways, _____, _____, _____, and _____.

3. For an equation that represents y as a function of x, the set of all values taken on by the _____ variable x is the domain, and the set of all values taken on by the _____ variable y is the range.

4. The function given by

$$f(x) = \begin{cases} 2x - 1, & x < 0 \\ x^2 + 4, & x \geq 0 \end{cases}$$

is an example of a _____ function.

5. If the domain of the function f is not given, then the set of values of the independent variable for which the expression is defined is called the _____ _____.

6. In calculus, one of the basic definitions is that of a _____ _____, given by $\dfrac{f(x + h) - f(x)}{h}$, $h \neq 0$.

SKILLS AND APPLICATIONS

In Exercises 7–10, is the relationship a function?

7. *Domain   Range*   8. *Domain   Range*

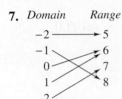   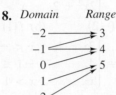

9. *Domain   Range*   10. *Domain   Range*
                              (Year)   (Number of North Atlantic tropical storms and hurricanes)

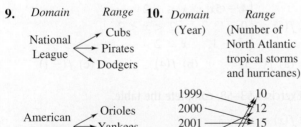

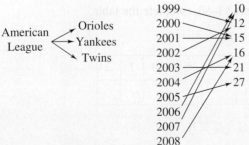

In Exercises 11–14, determine whether the relation represents y as a function of x.

11.

Input, x	-2	-1	0	1	2
Output, y	-8	-1	0	1	8

12.

Input, x	0	1	2	1	0
Output, y	-4	-2	0	2	4

13.

Input, x	10	7	4	7	10
Output, y	3	6	9	12	15

14.

Input, x	0	3	9	12	15
Output, y	3	3	3	3	3

In Exercises 15 and 16, which sets of ordered pairs represent functions from A to B? Explain.

15. $A = \{0, 1, 2, 3\}$ and $B = \{-2, -1, 0, 1, 2\}$

   (a) $\{(0, 1), (1, -2), (2, 0), (3, 2)\}$

   (b) $\{(0, -1), (2, 2), (1, -2), (3, 0), (1, 1)\}$

   (c) $\{(0, 0), (1, 0), (2, 0), (3, 0)\}$

   (d) $\{(0, 2), (3, 0), (1, 1)\}$

16. $A = \{a, b, c\}$ and $B = \{0, 1, 2, 3\}$

   (a) $\{(a, 1), (c, 2), (c, 3), (b, 3)\}$

   (b) $\{(a, 1), (b, 2), (c, 3)\}$

   (c) $\{(1, a), (0, a), (2, c), (3, b)\}$

   (d) $\{(c, 0), (b, 0), (a, 3)\}$

CIRCULATION OF NEWSPAPERS In Exercises 17 and 18, use the graph, which shows the circulation (in millions) of daily newspapers in the United States. (Source: Editor & Publisher Company)

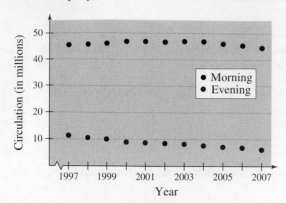

17. Is the circulation of morning newspapers a function of the year? Is the circulation of evening newspapers a function of the year? Explain.

18. Let $f(x)$ represent the circulation of evening newspapers in year x. Find $f(2002)$.

In Exercises 19–36, determine whether the equation represents y as a function of x.

19. $x^2 + y^2 = 4$ **20.** $x^2 - y^2 = 16$

21. $x^2 + y = 4$ **22.** $y - 4x^2 = 36$

23. $2x + 3y = 4$ **24.** $2x + 5y = 10$

25. $(x + 2)^2 + (y - 1)^2 = 25$

26. $(x - 2)^2 + y^2 = 4$

27. $y^2 = x^2 - 1$ **28.** $x + y^2 = 4$

29. $y = \sqrt{16 - x^2}$ **30.** $y = \sqrt{x + 5}$

31. $y = |4 - x|$ **32.** $|y| = 4 - x$

33. $x = 14$ **34.** $y = -75$

35. $y + 5 = 0$ **36.** $x - 1 = 0$

In Exercises 37–52, evaluate the function at each specified value of the independent variable and simplify.

37. $f(x) = 2x - 3$
  (a) $f(1)$ (b) $f(-3)$ (c) $f(x - 1)$

38. $g(y) = 7 - 3y$
  (a) $g(0)$ (b) $g\left(\frac{7}{3}\right)$ (c) $g(s + 2)$

39. $V(r) = \frac{4}{3}\pi r^3$
  (a) $V(3)$ (b) $V\left(\frac{3}{2}\right)$ (c) $V(2r)$

40. $S(r) = 4\pi r^2$
  (a) $S(2)$ (b) $S\left(\frac{1}{2}\right)$ (c) $S(3r)$

41. $g(t) = 4t^2 - 3t + 5$
  (a) $g(2)$ (b) $g(t - 2)$ (c) $g(t) - g(2)$

42. $h(t) = t^2 - 2t$
  (a) $h(2)$ (b) $h(1.5)$ (c) $h(x + 2)$

43. $f(y) = 3 - \sqrt{y}$
  (a) $f(4)$ (b) $f(0.25)$ (c) $f(4x^2)$

44. $f(x) = \sqrt{x + 8} + 2$
  (a) $f(-8)$ (b) $f(1)$ (c) $f(x - 8)$

45. $q(x) = 1/(x^2 - 9)$
  (a) $q(0)$ (b) $q(3)$ (c) $q(y + 3)$

46. $q(t) = (2t^2 + 3)/t^2$
  (a) $q(2)$ (b) $q(0)$ (c) $q(-x)$

47. $f(x) = |x|/x$
  (a) $f(2)$ (b) $f(-2)$ (c) $f(x - 1)$

48. $f(x) = |x| + 4$
  (a) $f(2)$ (b) $f(-2)$ (c) $f(x^2)$

49. $f(x) = \begin{cases} 2x + 1, & x < 0 \\ 2x + 2, & x \geq 0 \end{cases}$
  (a) $f(-1)$ (b) $f(0)$ (c) $f(2)$

50. $f(x) = \begin{cases} x^2 + 2, & x \leq 1 \\ 2x^2 + 2, & x > 1 \end{cases}$
  (a) $f(-2)$ (b) $f(1)$ (c) $f(2)$

51. $f(x) = \begin{cases} 3x - 1, & x < -1 \\ 4, & -1 \leq x \leq 1 \\ x^2, & x > 1 \end{cases}$
  (a) $f(-2)$ (b) $f\left(-\frac{1}{2}\right)$ (c) $f(3)$

52. $f(x) = \begin{cases} 4 - 5x, & x \leq -2 \\ 0, & -2 < x < 2 \\ x^2 + 1, & x \geq 2 \end{cases}$
  (a) $f(-3)$ (b) $f(4)$ (c) $f(-1)$

In Exercises 53–58, complete the table.

53. $f(x) = x^2 - 3$

x	-2	-1	0	1	2
$f(x)$					

54. $g(x) = \sqrt{x - 3}$

x	3	4	5	6	7
$g(x)$					

55. $h(t) = \frac{1}{2}|t + 3|$

t	-5	-4	-3	-2	-1
$h(t)$					

56. $f(s) = \dfrac{|s - 2|}{s - 2}$

s	0	1	$\frac{3}{2}$	$\frac{5}{2}$	4
$f(s)$					

57. $f(x) = \begin{cases} -\frac{1}{2}x + 4, & x \le 0 \\ (x - 2)^2, & x > 0 \end{cases}$

x	-2	-1	0	1	2
$f(x)$					

58. $f(x) = \begin{cases} 9 - x^2, & x < 3 \\ x - 3, & x \ge 3 \end{cases}$

x	1	2	3	4	5
$f(x)$					

In Exercises 59–66, find all real values of x such that $f(x) = 0$.

59. $f(x) = 15 - 3x$ **60.** $f(x) = 5x + 1$

61. $f(x) = \dfrac{3x - 4}{5}$ **62.** $f(x) = \dfrac{12 - x^2}{5}$

63. $f(x) = x^2 - 9$ **64.** $f(x) = x^2 - 8x + 15$

65. $f(x) = x^3 - x$

66. $f(x) = x^3 - x^2 - 4x + 4$

In Exercises 67–70, find the value(s) of x for which $f(x) = g(x)$.

67. $f(x) = x^2, \quad g(x) = x + 2$

68. $f(x) = x^2 + 2x + 1, \quad g(x) = 7x - 5$

69. $f(x) = x^4 - 2x^2, \quad g(x) = 2x^2$

70. $f(x) = \sqrt{x} - 4, \quad g(x) = 2 - x$

In Exercises 71–82, find the domain of the function.

71. $f(x) = 5x^2 + 2x - 1$ **72.** $g(x) = 1 - 2x^2$

73. $h(t) = \dfrac{4}{t}$ **74.** $s(y) = \dfrac{3y}{y + 5}$

75. $g(y) = \sqrt{y - 10}$ **76.** $f(t) = \sqrt[3]{t + 4}$

77. $g(x) = \dfrac{1}{x} - \dfrac{3}{x + 2}$ **78.** $h(x) = \dfrac{10}{x^2 - 2x}$

79. $f(s) = \dfrac{\sqrt{s - 1}}{s - 4}$ **80.** $f(x) = \dfrac{\sqrt{x + 6}}{6 + x}$

81. $f(x) = \dfrac{x - 4}{\sqrt{x}}$ **82.** $f(x) = \dfrac{x + 2}{\sqrt{x - 10}}$

In Exercises 83–86, assume that the domain of f is the set $A = \{-2, -1, 0, 1, 2\}$. Determine the set of ordered pairs that represents the function f.

83. $f(x) = x^2$ **84.** $f(x) = (x - 3)^2$

85. $f(x) = |x| + 2$ **86.** $f(x) = |x + 1|$

87. GEOMETRY Write the area A of a square as a function of its perimeter P.

88. GEOMETRY Write the area A of a circle as a function of its circumference C.

89. MAXIMUM VOLUME An open box of maximum volume is to be made from a square piece of material 24 centimeters on a side by cutting equal squares from the corners and turning up the sides (see figure).

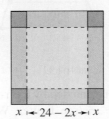
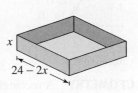

(a) The table shows the volumes V (in cubic centimeters) of the box for various heights x (in centimeters). Use the table to estimate the maximum volume.

Height, x	1	2	3	4	5	6
Volume, V	484	800	972	1024	980	864

(b) Plot the points (x, V) from the table in part (a). Does the relation defined by the ordered pairs represent V as a function of x?

(c) If V is a function of x, write the function and determine its domain.

90. MAXIMUM PROFIT The cost per unit in the production of an MP3 player is $60. The manufacturer charges $90 per unit for orders of 100 or less. To encourage large orders, the manufacturer reduces the charge by $0.15 per MP3 player for each unit ordered in excess of 100 (for example, there would be a charge of $87 per MP3 player for an order size of 120).

(a) The table shows the profits P (in dollars) for various numbers of units ordered, x. Use the table to estimate the maximum profit.

Units, x	110	120	130	140
Profit, P	3135	3240	3315	3360

Units, x	150	160	170
Profit, P	3375	3360	3315

(b) Plot the points (x, P) from the table in part (a). Does the relation defined by the ordered pairs represent P as a function of x?

(c) If P is a function of x, write the function and determine its domain.

91. GEOMETRY A right triangle is formed in the first quadrant by the x- and y-axes and a line through the point $(2, 1)$ (see figure). Write the area A of the triangle as a function of x, and determine the domain of the function.

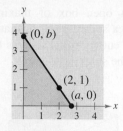

FIGURE FOR **91**

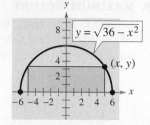

FIGURE FOR **92**

92. GEOMETRY A rectangle is bounded by the x-axis and the semicircle $y = \sqrt{36 - x^2}$ (see figure). Write the area A of the rectangle as a function of x, and graphically determine the domain of the function.

93. PATH OF A BALL The height y (in feet) of a baseball thrown by a child is

$$y = -\frac{1}{10}x^2 + 3x + 6$$

where x is the horizontal distance (in feet) from where the ball was thrown. Will the ball fly over the head of another child 30 feet away trying to catch the ball? (Assume that the child who is trying to catch the ball holds a baseball glove at a height of 5 feet.)

94. PRESCRIPTION DRUGS The numbers d (in millions) of drug prescriptions filled by independent outlets in the United States from 2000 through 2007 (see figure) can be approximated by the model

$$d(t) = \begin{cases} 10.6t + 699, & 0 \le t \le 4 \\ 15.5t + 637, & 5 \le t \le 7 \end{cases}$$

where t represents the year, with $t = 0$ corresponding to 2000. Use this model to find the number of drug prescriptions filled by independent outlets in each year from 2000 through 2007. (Source: National Association of Chain Drug Stores)

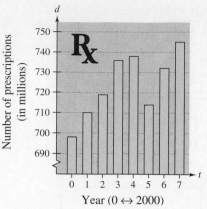

FIGURE FOR **94**

95. MEDIAN SALES PRICE The median sale prices p (in thousands of dollars) of an existing one-family home in the United States from 1998 through 2007 (see figure) can be approximated by the model

$$p(t) = \begin{cases} 1.011t^2 - 12.38t + 170.5, & 8 \le t \le 13 \\ -6.950t^2 + 222.55t - 1557.6, & 14 \le t \le 17 \end{cases}$$

where t represents the year, with $t = 8$ corresponding to 1998. Use this model to find the median sale price of an existing one-family home in each year from 1998 through 2007. (Source: National Association of Realtors)

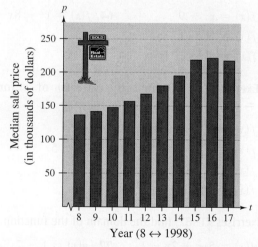

96. POSTAL REGULATIONS A rectangular package to be sent by the U.S. Postal Service can have a maximum combined length and girth (perimeter of a cross section) of 108 inches (see figure).

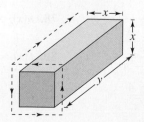

(a) Write the volume V of the package as a function of x. What is the domain of the function?

(b) Use a graphing utility to graph your function. Be sure to use an appropriate window setting.

(c) What dimensions will maximize the volume of the package? Explain your answer.

97. COST, REVENUE, AND PROFIT A company produces a product for which the variable cost is $12.30 per unit and the fixed costs are $98,000. The product sells for $17.98. Let x be the number of units produced and sold.

(a) The total cost for a business is the sum of the variable cost and the fixed costs. Write the total cost C as a function of the number of units produced.

(b) Write the revenue R as a function of the number of units sold.

(c) Write the profit P as a function of the number of units sold. (*Note: $P = R - C$*)

98. AVERAGE COST The inventor of a new game believes that the variable cost for producing the game is $0.95 per unit and the fixed costs are $6000. The inventor sells each game for $1.69. Let x be the number of games sold.

(a) The total cost for a business is the sum of the variable cost and the fixed costs. Write the total cost C as a function of the number of games sold.

(b) Write the average cost per unit $\overline{C} = C/x$ as a function of x.

99. TRANSPORTATION For groups of 80 or more people, a charter bus company determines the rate per person according to the formula

Rate $= 8 - 0.05(n - 80), \quad n \geq 80$

where the rate is given in dollars and n is the number of people.

(a) Write the revenue R for the bus company as a function of n.

(b) Use the function in part (a) to complete the table. What can you conclude?

n	90	100	110	120	130	140	150
$R(n)$							

100. PHYSICS The force F (in tons) of water against the face of a dam is estimated by the function $F(y) = 149.76\sqrt{10}y^{5/2}$, where y is the depth of the water (in feet).

(a) Complete the table. What can you conclude from the table?

y	5	10	20	30	40
$F(y)$					

(b) Use the table to approximate the depth at which the force against the dam is 1,000,000 tons.

(c) Find the depth at which the force against the dam is 1,000,000 tons algebraically.

101. HEIGHT OF A BALLOON A balloon carrying a transmitter ascends vertically from a point 3000 feet from the receiving station.

(a) Draw a diagram that gives a visual representation of the problem. Let h represent the height of the balloon and let d represent the distance between the balloon and the receiving station.

(b) Write the height of the balloon as a function of d. What is the domain of the function?

102. E-FILING The table shows the numbers of tax returns (in millions) made through e-file from 2000 through 2007. Let $f(t)$ represent the number of tax returns made through e-file in the year t.   (Source: Internal Revenue Service)

Year	Number of tax returns made through e-file
2000	35.4
2001	40.2
2002	46.9
2003	52.9
2004	61.5
2005	68.5
2006	73.3
2007	80.0

(a) Find $\dfrac{f(2007) - f(2000)}{2007 - 2000}$ and interpret the result in the context of the problem.

(b) Make a scatter plot of the data.

(c) Find a linear model for the data algebraically. Let N represent the number of tax returns made through e-file and let $t = 0$ correspond to 2000.

(d) Use the model found in part (c) to complete the table.

t	0	1	2	3	4	5	6	7
N								

(e) Compare your results from part (d) with the actual data.

 (f) Use a graphing utility to find a linear model for the data. Let $x = 0$ correspond to 2000. How does the model you found in part (c) compare with the model given by the graphing utility?

⌐∫ In Exercises 103–110, find the difference quotient and simplify your answer.

103. $f(x) = x^2 - x + 1$, $\dfrac{f(2 + h) - f(2)}{h}$, $h \neq 0$

104. $f(x) = 5x - x^2$, $\dfrac{f(5 + h) - f(5)}{h}$, $h \neq 0$

105. $f(x) = x^3 + 3x$, $\dfrac{f(x + h) - f(x)}{h}$, $h \neq 0$

106. $f(x) = 4x^2 - 2x$, $\dfrac{f(x + h) - f(x)}{h}$, $h \neq 0$

107. $g(x) = \dfrac{1}{x^2}$, $\dfrac{g(x) - g(3)}{x - 3}$, $x \neq 3$

108. $f(t) = \dfrac{1}{t - 2}$, $\dfrac{f(t) - f(1)}{t - 1}$, $t \neq 1$

109. $f(x) = \sqrt{5x}$, $\dfrac{f(x) - f(5)}{x - 5}$, $x \neq 5$

110. $f(x) = x^{2/3} + 1$, $\dfrac{f(x) - f(8)}{x - 8}$, $x \neq 8$

In Exercises 111–114, match the data with one of the following functions

$$f(x) = cx, \quad g(x) = cx^2, \quad h(x) = c\sqrt{|x|}, \quad and \quad r(x) = \dfrac{c}{x}$$

and determine the value of the constant c that will make the function fit the data in the table.

111.

x	-4	-1	0	1	4
y	-32	-2	0	-2	-32

112.

x	-4	-1	0	1	4
y	-1	$-\frac{1}{4}$	0	$\frac{1}{4}$	1

113.

x	-4	-1	0	1	4
y	-8	-32	Undefined	32	8

114.

x	-4	-1	0	1	4
y	6	3	0	3	6

EXPLORATION

TRUE OR FALSE? In Exercises 115–118, determine whether the statement is true or false. Justify your answer.

115. Every relation is a function.

116. Every function is a relation.

117. The domain of the function given by $f(x) = x^4 - 1$ is $(-\infty, \infty)$, and the range of $f(x)$ is $(0, \infty)$.

118. The set of ordered pairs $\{(-8, -2), (-6, 0), (-4, 0), (-2, 2), (0, 4), (2, -2)\}$ represents a function.

119. THINK ABOUT IT Consider

$$f(x) = \sqrt{x - 1} \quad and \quad g(x) = \dfrac{1}{\sqrt{x - 1}}.$$

Why are the domains of f and g different?

120. THINK ABOUT IT Consider $f(x) = \sqrt{x - 2}$ and $g(x) = \sqrt[3]{x - 2}$. Why are the domains of f and g different?

121. THINK ABOUT IT Given $f(x) = x^2$, is f the independent variable? Why or why not?

122. CAPSTONE

(a) Describe any differences between a *relation* and a *function*.

(b) In your own words, explain the meanings of *domain* and *range*.

In Exercises 123 and 124, determine whether the statements use the word *function* in ways that are mathematically correct. Explain your reasoning.

123. (a) The sales tax on a purchased item is a function of the selling price.

(b) Your score on the next algebra exam is a function of the number of hours you study the night before the exam.

124. (a) The amount in your savings account is a function of your salary.

(b) The speed at which a free-falling baseball strikes the ground is a function of the height from which it was dropped.

The symbol **⌐∫** indicates an example or exercise that highlights algebraic techniques specifically used in calculus.

P.6   ANALYZING GRAPHS OF FUNCTIONS

What you should learn

- Use the Vertical Line Test for functions.
- Find the zeros of functions.
- Determine intervals on which functions are increasing or decreasing and determine relative maximum and relative minimum values of functions.
- Determine the average rate of change of a function.
- Identify even and odd functions.

Why you should learn it

Graphs of functions can help you visualize relationships between variables in real life. For instance, in Exercise 110 on page 83, you will use the graph of a function to represent visually the temperature of a city over a 24-hour period.

The Graph of a Function

In Section P.5, you studied functions from an algebraic point of view. In this section, you will study functions from a graphical perspective.

The **graph of a function** f is the collection of ordered pairs $(x, f(x))$ such that x is in the domain of f. As you study this section, remember that

$x = $ the directed distance from the y-axis

$y = f(x) = $ the directed distance from the x-axis

as shown in Figure P.58.

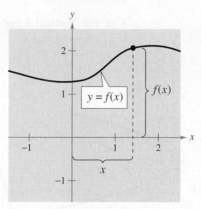

FIGURE P.58

Example 1   Finding the Domain and Range of a Function

Use the graph of the function f, shown in Figure P.59, to find (a) the domain of f, (b) the function values $f(-1)$ and $f(2)$, and (c) the range of f.

Solution

a. The closed dot at $(-1, 1)$ indicates that $x = -1$ is in the domain of f, whereas the open dot at $(5, 2)$ indicates that $x = 5$ is not in the domain. So, the domain of f is all x in the interval $[-1, 5)$.

b. Because $(-1, 1)$ is a point on the graph of f, it follows that $f(-1) = 1$. Similarly, because $(2, -3)$ is a point on the graph of f, it follows that $f(2) = -3$.

c. Because the graph does not extend below $f(2) = -3$ or above $f(0) = 3$, the range of f is the interval $[-3, 3]$.

CHECKPoint   Now try Exercise 9.

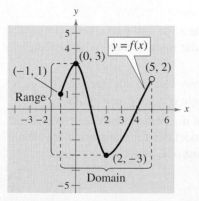

FIGURE P.59

The use of dots (open or closed) at the extreme left and right points of a graph indicates that the graph does not extend beyond these points. If no such dots are shown, assume that the graph extends beyond these points.

By the definition of a function, at most one y-value corresponds to a given x-value. This means that the graph of a function cannot have two or more different points with the same x-coordinate, and no two points on the graph of a function can be vertically above or below each other. It follows, then, that a vertical line can intersect the graph of a function at most once. This observation provides a convenient visual test called the **Vertical Line Test** for functions.

Vertical Line Test for Functions

A set of points in a coordinate plane is the graph of y as a function of x if and only if no *vertical* line intersects the graph at more than one point.

Example 2   Vertical Line Test for Functions

Use the Vertical Line Test to decide whether the graphs in Figure P.60 represent y as a function of x.

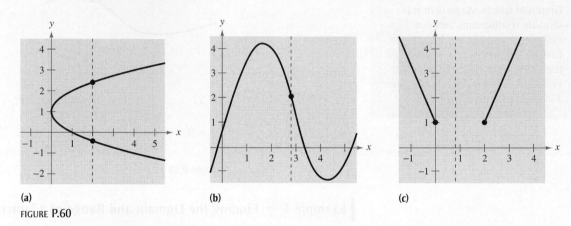

(a)

(b)

(c)

FIGURE P.60

Solution

a. This *is not* a graph of y as a function of x, because you can find a vertical line that intersects the graph twice. That is, for a particular input x, there is more than one output y.

b. This *is* a graph of y as a function of x, because every vertical line intersects the graph at most once. That is, for a particular input x, there is at most one output y.

c. This *is* a graph of y as a function of x. (Note that if a vertical line does not intersect the graph, it simply means that the function is undefined for that particular value of x.) That is, for a particular input x, there is at most one output y.

CHECK *Point* ► Now try Exercise 17.

TECHNOLOGY

Most graphing utilities are designed to graph functions of x more easily than other types of equations. For instance, the graph shown in Figure P.60(a) represents the equation $x - (y - 1)^2 = 0$. To use a graphing utility to duplicate this graph, you must first solve the equation for y to obtain $y = 1 \pm \sqrt{x}$, and then graph the two equations $y_1 = 1 + \sqrt{x}$ and $y_2 = 1 - \sqrt{x}$ in the same viewing window.

Zeros of a Function

If the graph of a function of x has an x-intercept at $(a, 0)$, then a is a **zero** of the function.

> ### Zeros of a Function
>
> The **zeros of a function** f of x are the x-values for which $f(x) = 0$.

Example 3   Finding the Zeros of a Function

Find the zeros of each function.

a. $f(x) = 3x^2 + x - 10$ **b.** $g(x) = \sqrt{10 - x^2}$ **c.** $h(t) = \dfrac{2t - 3}{t + 5}$

Solution

To find the zeros of a function, set the function equal to zero and solve for the independent variable.

a. $3x^2 + x - 10 = 0$ Set $f(x)$ equal to 0.

$(3x - 5)(x + 2) = 0$ Factor.

$3x - 5 = 0$ ⟹ $x = \dfrac{5}{3}$ Set 1st factor equal to 0.

$x + 2 = 0$ ⟹ $x = -2$ Set 2nd factor equal to 0.

The zeros of f are $x = \dfrac{5}{3}$ and $x = -2$. In Figure P.61, note that the graph of f has $\left(\dfrac{5}{3}, 0\right)$ and $(-2, 0)$ as its x-intercepts.

b. $\sqrt{10 - x^2} = 0$ Set $g(x)$ equal to 0.

$10 - x^2 = 0$ Square each side.

$10 = x^2$ Add x^2 to each side.

$\pm\sqrt{10} = x$ Extract square roots.

The zeros of g are $x = -\sqrt{10}$ and $x = \sqrt{10}$. In Figure P.62, note that the graph of g has $\left(-\sqrt{10}, 0\right)$ and $\left(\sqrt{10}, 0\right)$ as its x-intercepts.

c. $\dfrac{2t - 3}{t + 5} = 0$ Set $h(t)$ equal to 0.

$2t - 3 = 0$ Multiply each side by $t + 5$.

$2t = 3$ Add 3 to each side.

$t = \dfrac{3}{2}$ Divide each side by 2.

The zero of h is $t = \dfrac{3}{2}$. In Figure P.63, note that the graph of h has $\left(\dfrac{3}{2}, 0\right)$ as its t-intercept.

CHECK*Point* Now try Exercise 23.

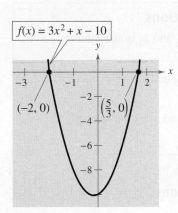

$f(x) = 3x^2 + x - 10$

$(-2, 0)$ $\left(\dfrac{5}{3}, 0\right)$

Zeros of f: $x = -2$, $x = \dfrac{5}{3}$

FIGURE P.61

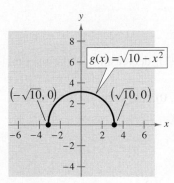

$g(x) = \sqrt{10 - x^2}$

$\left(-\sqrt{10}, 0\right)$ $\left(\sqrt{10}, 0\right)$

Zeros of g: $x = \pm\sqrt{10}$

FIGURE P.62

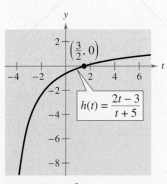

$\left(\dfrac{3}{2}, 0\right)$

$h(t) = \dfrac{2t - 3}{t + 5}$

Zero of h: $t = \dfrac{3}{2}$

FIGURE P.63

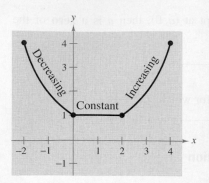

FIGURE **P.64**

Increasing and Decreasing Functions

The more you know about the graph of a function, the more you know about the function itself. Consider the graph shown in Figure P.64. As you move from *left to right*, this graph falls from $x = -2$ to $x = 0$, is constant from $x = 0$ to $x = 2$, and rises from $x = 2$ to $x = 4$.

Increasing, Decreasing, and Constant Functions

A function f is **increasing** on an interval if, for any x_1 and x_2 in the interval, $x_1 < x_2$ implies $f(x_1) < f(x_2)$.

A function f is **decreasing** on an interval if, for any x_1 and x_2 in the interval, $x_1 < x_2$ implies $f(x_1) > f(x_2)$.

A function f is **constant** on an interval if, for any x_1 and x_2 in the interval, $f(x_1) = f(x_2)$.

Example 4 Increasing and Decreasing Functions

Use the graphs in Figure P.65 to describe the increasing or decreasing behavior of each function.

Solution

a. This function is increasing over the entire real line.

b. This function is increasing on the interval $(-\infty, -1)$, decreasing on the interval $(-1, 1)$, and increasing on the interval $(1, \infty)$.

c. This function is increasing on the interval $(-\infty, 0)$, constant on the interval $(0, 2)$, and decreasing on the interval $(2, \infty)$.

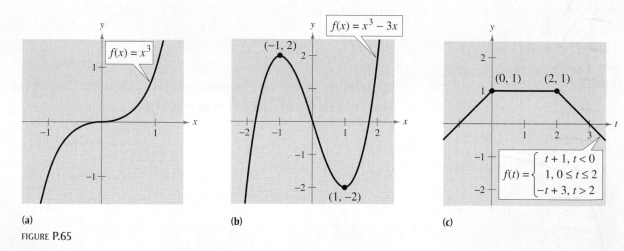

(a)   (b)   (c)

FIGURE **P.65**

CHECK*Point* Now try Exercise 41.

To help you decide whether a function is increasing, decreasing, or constant on an interval, you can evaluate the function for several values of x. However, calculus is needed to determine, for certain, all intervals on which a function is increasing, decreasing, or constant.

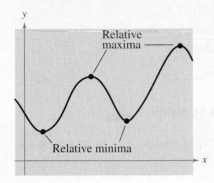

FIGURE **P.66**

The points at which a function changes its increasing, decreasing, or constant behavior are helpful in determining the **relative minimum** or **relative maximum** values of the function.

Definitions of Relative Minimum and Relative Maximum

A function value $f(a)$ is called a **relative minimum** of f if there exists an interval (x_1, x_2) that contains a such that

$$x_1 < x < x_2 \quad \text{implies} \quad f(a) \leq f(x).$$

A function value $f(a)$ is called a **relative maximum** of f if there exists an interval (x_1, x_2) that contains a such that

$$x_1 < x < x_2 \quad \text{implies} \quad f(a) \geq f(x).$$

Figure P.66 shows several different examples of relative minima and relative maxima. By writing a second-degree equation in standard form, $y = a(k - h)^2 + k$, you can find the *exact point* (h, k) at which it has a relative minimum or relative maximum. For the time being, however, you can use a graphing utility to find reasonable approximations of these points.

Example 5   Approximating a Relative Minimum

Use a graphing utility to approximate the relative minimum of the function given by $f(x) = 3x^2 - 4x - 2$.

Solution

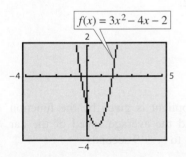

FIGURE **P.67**

The graph of f is shown in Figure P.67. By using the *zoom* and *trace* features or the *minimum* feature of a graphing utility, you can estimate that the function has a relative minimum at the point

$$(0.67, -3.33). \qquad \text{Relative minimum}$$

By writing this second-degree equation in standard form, $f(x) = 3\left(x - \frac{2}{3}\right)^2 - \frac{10}{3}$, you can determine that the exact point at which the relative minimum occurs is $\left(\frac{2}{3}, -\frac{10}{3}\right)$.

CHECK *Point*   Now try Exercise 57.

You can also use the *table* feature of a graphing utility to approximate numerically the relative minimum of the function in Example 5. Using a table that begins at 0.6 and increments the value of x by 0.01, you can approximate that the minimum of $f(x) = 3x^2 - 4x - 2$ occurs at the point $(0.67, -3.33)$.

TECHNOLOGY

If you use a graphing utility to estimate the x- and y-values of a relative minimum or relative maximum, the *zoom* feature will often produce graphs that are nearly flat. To overcome this problem, you can manually change the vertical setting of the viewing window. The graph will stretch vertically if the values of Ymin and Ymax are closer together.

Average Rate of Change

In Section P.4, you learned that the slope of a line can be interpreted as a *rate of change*. For a nonlinear graph whose slope changes at each point, the **average rate of change** between any two points $(x_1, f(x_1))$ and $(x_2, f(x_2))$ is the slope of the line through the two points (see Figure P.68). The line through the two points is called the **secant line**, and the slope of this line is denoted as m_{sec}.

$$\text{Average rate of change of } f \text{ from } x_1 \text{ to } x_2 = \frac{f(x_2) - f(x_1)}{x_2 - x_1}$$

$$= \frac{\text{change in } y}{\text{change in } x}$$

$$= m_{\text{sec}}$$

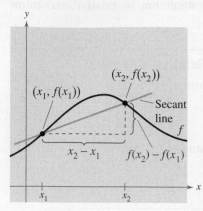

FIGURE P.68

Example 6   Average Rate of Change of a Function

Find the average rates of change of $f(x) = x^3 - 3x$ (a) from $x_1 = -2$ to $x_2 = 0$ and (b) from $x_1 = 0$ to $x_2 = 1$ (see Figure P.69).

Solution

a. The average rate of change of f from $x_1 = -2$ to $x_2 = 0$ is

$$\frac{f(x_2) - f(x_1)}{x_2 - x_1} = \frac{f(0) - f(-2)}{0 - (-2)} = \frac{0 - (-2)}{2} = 1. \qquad \text{Secant line has positive slope.}$$

b. The average rate of change of f from $x_1 = 0$ to $x_2 = 1$ is

$$\frac{f(x_2) - f(x_1)}{x_2 - x_1} = \frac{f(1) - f(0)}{1 - 0} = \frac{-2 - 0}{1} = -2. \qquad \text{Secant line has negative slope.}$$

CHECK*Point* ▶ Now try Exercise 75.

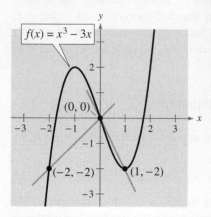

FIGURE P.69

Example 7   Finding Average Speed

The distance s (in feet) a moving car is from a stoplight is given by the function $s(t) = 20t^{3/2}$, where t is the time (in seconds). Find the average speed of the car (a) from $t_1 = 0$ to $t_2 = 4$ seconds and (b) from $t_1 = 4$ to $t_2 = 9$ seconds.

Solution

a. The average speed of the car from $t_1 = 0$ to $t_2 = 4$ seconds is

$$\frac{s(t_2) - s(t_1)}{t_2 - t_1} = \frac{s(4) - s(0)}{4 - (0)} = \frac{160 - 0}{4} = 40 \text{ feet per second.}$$

b. The average speed of the car from $t_1 = 4$ to $t_2 = 9$ seconds is

$$\frac{s(t_2) - s(t_1)}{t_2 - t_1} = \frac{s(9) - s(4)}{9 - 4} = \frac{540 - 160}{5} = 76 \text{ feet per second.}$$

CHECK*Point* ▶ Now try Exercise 113.

Even and Odd Functions

In Section P.3, you studied different types of symmetry of a graph. In the terminology of functions, a function is said to be **even** if its graph is symmetric with respect to the y-axis and to be **odd** if its graph is symmetric with respect to the origin. The symmetry tests in Section P.3 yield the following tests for even and odd functions.

Tests for Even and Odd Functions

A function $y = f(x)$ is **even** if, for each x in the domain of f,

$$f(-x) = f(x).$$

A function $y = f(x)$ is **odd** if, for each x in the domain of f,

$$f(-x) = -f(x).$$

Example 8   Even and Odd Functions

a. The function $g(x) = x^3 - x$ is odd because $g(-x) = -g(x)$, as follows.

$$g(-x) = (-x)^3 - (-x) \qquad \text{Substitute } -x \text{ for } x.$$

$$= -x^3 + x \qquad \text{Simplify.}$$

$$= -(x^3 - x) \qquad \text{Distributive Property}$$

$$= -g(x) \qquad \text{Test for odd function}$$

b. The function $h(x) = x^2 + 1$ is even because $h(-x) = h(x)$, as follows.

$$h(-x) = (-x)^2 + 1 \qquad \text{Substitute } -x \text{ for } x.$$

$$= x^2 + 1 \qquad \text{Simplify.}$$

$$= h(x) \qquad \text{Test for even function}$$

The graphs and symmetry of these two functions are shown in Figure P.70.

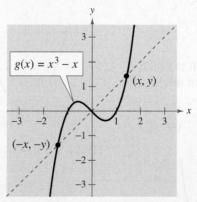

(a) **Symmetric to origin: Odd Function**

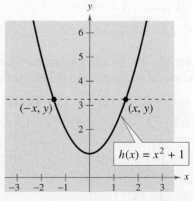

(b) **Symmetric to y-axis: Even Function**

FIGURE **P.70**

CHECK *Point*   Now try Exercise 83.

P.6 EXERCISES

See www.CalcChat.com for worked-out solutions to odd-numbered exercises.

VOCABULARY: Fill in the blanks.

1. The graph of a function f is the collection of _____ _____ $(x, f(x))$ such that x is in the domain of f.

2. The _____ _____ _____ is used to determine whether the graph of an equation is a function of y in terms of x.

3. The _____ of a function f are the values of x for which $f(x) = 0$.

4. A function f is _____ on an interval if, for any x_1 and x_2 in the interval, $x_1 < x_2$ implies $f(x_1) > f(x_2)$.

5. A function value $f(a)$ is a relative _____ of f if there exists an interval (x_1, x_2) containing a such that $x_1 < x < x_2$ implies $f(a) \geq f(x)$.

6. The _____ _____ _____ _____ between any two points $(x_1, f(x_1))$ and $(x_2, f(x_2))$ is the slope of the line through the two points, and this line is called the _____ line.

7. A function f is _____ if, for each x in the domain of f, $f(-x) = -f(x)$.

8. A function f is _____ if its graph is symmetric with respect to the y-axis.

SKILLS AND APPLICATIONS

In Exercises 9–12, use the graph of the function to find the domain and range of f.

9.
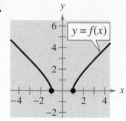

10.
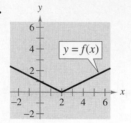

11.
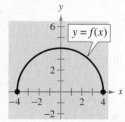

12.
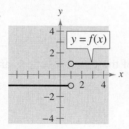

In Exercises 13–16, use the graph of the function to find the domain and range of f and the indicated function values.

13. (a) $f(-2)$ (b) $f(-1)$
    (c) $f\left(\frac{1}{2}\right)$ (d) $f(1)$

14. (a) $f(-1)$ (b) $f(2)$
    (c) $f(0)$ (d) $f(1)$

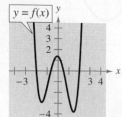

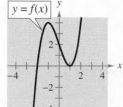

15. (a) $f(2)$ (b) $f(1)$
    (c) $f(3)$ (d) $f(-1)$

16. (a) $f(-2)$ (b) $f(1)$
    (c) $f(0)$ (d) $f(2)$

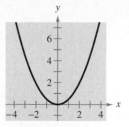

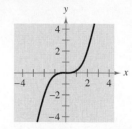

In Exercises 17–22, use the Vertical Line Test to determine whether y is a function of x. To print an enlarged copy of the graph, go to the website *www.mathgraphs.com*.

17. $y = \frac{1}{2}x^2$

18. $y = \frac{1}{4}x^3$

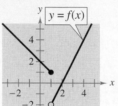

19. $x - y^2 = 1$

20. $x^2 + y^2 = 25$

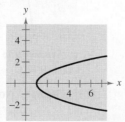

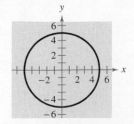

21. $x^2 = 2xy - 1$

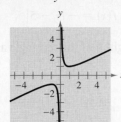

22. $x = |y + 2|$

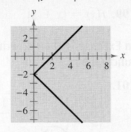

In Exercises 23–32, find the zeros of the function algebraically.

23. $f(x) = 2x^2 - 7x - 30$

24. $f(x) = 3x^2 + 22x - 16$

25. $f(x) = \dfrac{x}{9x^2 - 4}$ **26.** $f(x) = \dfrac{x^2 - 9x + 14}{4x}$

27. $f(x) = \frac{1}{2}x^3 - x$

28. $f(x) = x^3 - 4x^2 - 9x + 36$

29. $f(x) = 4x^3 - 24x^2 - x + 6$

30. $f(x) = 9x^4 - 25x^2$

31. $f(x) = \sqrt{2x} - 1$ **32.** $f(x) = \sqrt{3x + 2}$

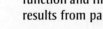 In Exercises 33–38, (a) use a graphing utility to graph the function and find the zeros of the function and (b) verify your results from part (a) algebraically.

33. $f(x) = 3 + \dfrac{5}{x}$ **34.** $f(x) = x(x - 7)$

35. $f(x) = \sqrt{2x + 11}$ **36.** $f(x) = \sqrt{3x - 14} - 8$

37. $f(x) = \dfrac{3x - 1}{x - 6}$ **38.** $f(x) = \dfrac{2x^2 - 9}{3 - x}$

In Exercises 39–46, determine the intervals over which the function is increasing, decreasing, or constant.

39. $f(x) = \frac{3}{2}x$ **40.** $f(x) = x^2 - 4x$

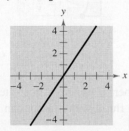

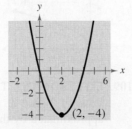

41. $f(x) = x^3 - 3x^2 + 2$ **42.** $f(x) = \sqrt{x^2 - 1}$

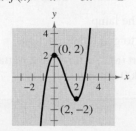

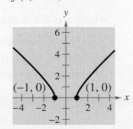

43. $f(x) = |x + 1| + |x - 1|$ **44.** $f(x) = \dfrac{x^2 + x + 1}{x + 1}$

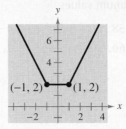

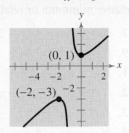

45. $f(x) = \begin{cases} x + 3, & x \le 0 \\ 3, & 0 < x \le 2 \\ 2x + 1, & x > 2 \end{cases}$

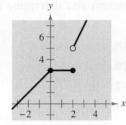

46. $f(x) = \begin{cases} 2x + 1, & x \le -1 \\ x^2 - 2, & x > -1 \end{cases}$

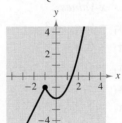

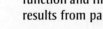 In Exercises 47–56, (a) use a graphing utility to graph the function and visually determine the intervals over which the function is increasing, decreasing, or constant, and (b) make a table of values to verify whether the function is increasing, decreasing, or constant over the intervals you identified in part (a).

47. $f(x) = 3$ **48.** $g(x) = x$

49. $g(s) = \dfrac{s^2}{4}$ **50.** $h(x) = x^2 - 4$

51. $f(t) = -t^4$ **52.** $f(x) = 3x^4 - 6x^2$

53. $f(x) = \sqrt{1 - x}$ **54.** $f(x) = x\sqrt{x + 3}$

55. $f(x) = x^{3/2}$ **56.** $f(x) = x^{2/3}$

In Exercises 57–66, use a graphing utility to graph the function and approximate (to two decimal places) any relative minimum or relative maximum values.

57. $f(x) = (x - 4)(x + 2)$ **58.** $f(x) = 3x^2 - 2x - 5$

59. $f(x) = -x^2 + 3x - 2$ **60.** $f(x) = -2x^2 + 9x$

61. $f(x) = x(x - 2)(x + 3)$

62. $f(x) = x^3 - 3x^2 - x + 1$

63. $g(x) = 2x^3 + 3x^2 - 12x$

64. $h(x) = x^3 - 6x^2 + 15$

65. $h(x) = (x - 1)\sqrt{x}$

66. $g(x) = x\sqrt{4 - x}$

In Exercises 67–74, graph the function and determine the interval(s) for which $f(x) \geq 0$.

67. $f(x) = 4 - x$ **68.** $f(x) = 4x + 2$

69. $f(x) = 9 - x^2$ **70.** $f(x) = x^2 - 4x$

71. $f(x) = \sqrt{x - 1}$ **72.** $f(x) = \sqrt{x + 2}$

73. $f(x) = -(1 + |x|)$ **74.** $f(x) = \frac{1}{2}(2 + |x|)$

In Exercises 75–82, find the average rate of change of the function from x_1 to x_2.

Function	*x-Values*
75. $f(x) = -2x + 15$	$x_1 = 0, x_2 = 3$
76. $f(x) = 3x + 8$	$x_1 = 0, x_2 = 3$
77. $f(x) = x^2 + 12x - 4$	$x_1 = 1, x_2 = 5$
78. $f(x) = x^2 - 2x + 8$	$x_1 = 1, x_2 = 5$
79. $f(x) = x^3 - 3x^2 - x$	$x_1 = 1, x_2 = 3$
80. $f(x) = -x^3 + 6x^2 + x$	$x_1 = 1, x_2 = 6$
81. $f(x) = -\sqrt{x - 2} + 5$	$x_1 = 3, x_2 = 11$
82. $f(x) = -\sqrt{x + 1} + 3$	$x_1 = 3, x_2 = 8$

In Exercises 83–90, determine whether the function is even, odd, or neither. Then describe the symmetry.

83. $f(x) = x^6 - 2x^2 + 3$ **84.** $h(x) = x^3 - 5$

85. $g(x) = x^3 - 5x$ **86.** $f(t) = t^2 + 2t - 3$

87. $h(x) = x\sqrt{x + 5}$ **88.** $f(x) = x\sqrt{1 - x^2}$

89. $f(s) = 4s^{3/2}$ **90.** $g(s) = 4s^{2/3}$

In Exercises 91–100, sketch a graph of the function and determine whether it is even, odd, or neither. Verify your answers algebraically.

91. $f(x) = 5$ **92.** $f(x) = -9$

93. $f(x) = 3x - 2$ **94.** $f(x) = 5 - 3x$

95. $h(x) = x^2 - 4$ **96.** $f(x) = -x^2 - 8$

97. $f(x) = \sqrt{1 - x}$ **98.** $g(t) = \sqrt[3]{t - 1}$

99. $f(x) = |x + 2|$ **100.** $f(x) = -|x - 5|$

In Exercises 101–104, write the height h of the rectangle as a function of x.

101.
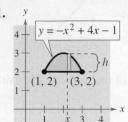

102.
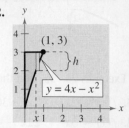

103.
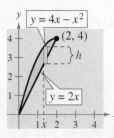

104.
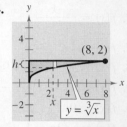

In Exercises 105–108, write the length L of the rectangle as a function of y.

105.
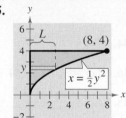

106.
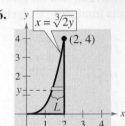

107.
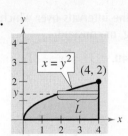

108.
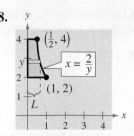

109. ELECTRONICS The number of lumens (time rate of flow of light) L from a fluorescent lamp can be approximated by the model

$$L = -0.294x^2 + 97.744x - 664.875, \quad 20 \leq x \leq 90$$

where x is the wattage of the lamp.

(a) Use a graphing utility to graph the function.

(b) Use the graph from part (a) to estimate the wattage necessary to obtain 2000 lumens.

110. DATA ANALYSIS: TEMPERATURE   The table shows the temperatures y (in degrees Fahrenheit) in a certain city over a 24-hour period. Let x represent the time of day, where $x = 0$ corresponds to 6 A.M.

Time, x	Temperature, y
0	34
2	50
4	60
6	64
8	63
10	59
12	53
14	46
16	40
18	36
20	34
22	37
24	45

A model that represents these data is given by

$$y = 0.026x^3 - 1.03x^2 + 10.2x + 34, \quad 0 \le x \le 24.$$

(a) Use a graphing utility to create a scatter plot of the data. Then graph the model in the same viewing window.

(b) How well does the model fit the data?

(c) Use the graph to approximate the times when the temperature was increasing and decreasing.

(d) Use the graph to approximate the maximum and minimum temperatures during this 24-hour period.

(e) Could this model be used to predict the temperatures in the city during the next 24-hour period? Why or why not?

111. COORDINATE AXIS SCALE   Each function described below models the specified data for the years 1998 through 2008, with $t = 8$ corresponding to 1998. Estimate a reasonable scale for the vertical axis (e.g., hundreds, thousands, millions, etc.) of the graph and justify your answer. (There are many correct answers.)

(a) $f(t)$ represents the average salary of college professors.

(b) $f(t)$ represents the U.S. population.

(c) $f(t)$ represents the percent of the civilian work force that is unemployed.

112. GEOMETRY   Corners of equal size are cut from a square with sides of length 8 meters (see figure).

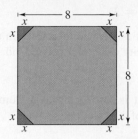

(a) Write the area A of the resulting figure as a function of x. Determine the domain of the function.

 (b) Use a graphing utility to graph the area function over its domain. Use the graph to find the range of the function.

(c) Identify the figure that would result if x were chosen to be the maximum value in the domain of the function. What would be the length of each side of the figure?

113. ENROLLMENT RATE   The enrollment rates r of children in preschool in the United States from 1970 through 2005 can be approximated by the model

$$r = -0.021t^2 + 1.44t + 39.3, \quad 0 \le t \le 35$$

where t represents the year, with $t = 0$ corresponding to 1970.   (Source: U.S. Census Bureau)

(a) Use a graphing utility to graph the model.

(b) Find the average rate of change of the model from 1970 through 2005. Interpret your answer in the context of the problem.

114. VEHICLE TECHNOLOGY SALES   The estimated revenues r (in millions of dollars) from sales of in-vehicle technologies in the United States from 2003 through 2008 can be approximated by the model

$$r = 157.30t^2 - 397.4t + 6114, \quad 3 \le t \le 8$$

where t represents the year, with $t = 3$ corresponding to 2003.   (Source: Consumer Electronics Association)

(a) Use a graphing utility to graph the model.

(b) Find the average rate of change of the model from 2003 through 2008. Interpret your answer in the context of the problem.

 PHYSICS   In Exercises 115–120, (a) use the position equation $s = -16t^2 + v_0t + s_0$ to write a function that represents the situation, (b) use a graphing utility to graph the function, (c) find the average rate of change of the function from t_1 to t_2, (d) describe the slope of the secant line through t_1 and t_2, (e) find the equation of the secant line through t_1 and t_2, and (f) graph the secant line in the same viewing window as your position function.

115. An object is thrown upward from a height of 6 feet at a velocity of 64 feet per second.

$t_1 = 0, t_2 = 3$

116. An object is thrown upward from a height of 6.5 feet at a velocity of 72 feet per second.

$t_1 = 0, t_2 = 4$

117. An object is thrown upward from ground level at a velocity of 120 feet per second.

$t_1 = 3, t_2 = 5$

118. An object is thrown upward from ground level at a velocity of 96 feet per second.

$t_1 = 2, t_2 = 5$

119. An object is dropped from a height of 120 feet.

$t_1 = 0, t_2 = 2$

120. An object is dropped from a height of 80 feet.

$t_1 = 1, t_2 = 2$

EXPLORATION

TRUE OR FALSE? In Exercises 121 and 122, determine whether the statement is true or false. Justify your answer.

121. A function with a square root cannot have a domain that is the set of real numbers.

122. It is possible for an odd function to have the interval $[0, \infty)$ as its domain.

123. If f is an even function, determine whether g is even, odd, or neither. Explain.

(a) $g(x) = -f(x)$ (b) $g(x) = f(-x)$

(c) $g(x) = f(x) - 2$ (d) $g(x) = f(x - 2)$

124. THINK ABOUT IT Does the graph in Exercise 19 represent x as a function of y? Explain.

THINK ABOUT IT In Exercises 125–130, find the coordinates of a second point on the graph of a function f if the given point is on the graph and the function is (a) even and (b) odd.

125. $\left(-\frac{3}{2}, 4\right)$ **126.** $\left(-\frac{5}{3}, -7\right)$

127. $(4, 9)$ **128.** $(5, -1)$

129. $(x, -y)$ **130.** $(2a, 2c)$

 131. WRITING Use a graphing utility to graph each function. Write a paragraph describing any similarities and differences you observe among the graphs.

(a) $y = x$ (b) $y = x^2$ (c) $y = x^3$

(d) $y = x^4$ (e) $y = x^5$ (f) $y = x^6$

132. CONJECTURE Use the results of Exercise 131 to make a conjecture about the graphs of $y = x^7$ and $y = x^8$. Use a graphing utility to graph the functions and compare the results with your conjecture.

133. Use the information in Example 7 to find the average speed of the car from $t_1 = 0$ to $t_2 = 9$ seconds. Explain why the result is less than the value obtained in part (b) of Example 7.

134. Graph each of the functions with a graphing utility. Determine whether the function is *even*, *odd*, or *neither*.

$f(x) = x^2 - x^4$

$g(x) = 2x^3 + 1$

$h(x) = x^5 - 2x^3 + x$

$j(x) = 2 - x^6 - x^8$

$k(x) = x^5 - 2x^4 + x - 2$

$p(x) = x^9 + 3x^5 - x^3 + x$

What do you notice about the equations of functions that are odd? What do you notice about the equations of functions that are even? Can you describe a way to identify a function as odd or even by inspecting the equation? Can you describe a way to identify a function as neither odd nor even by inspecting the equation?

135. WRITING Write a short paragraph describing three different functions that represent the behaviors of quantities between 1998 and 2009. Describe one quantity that decreased during this time, one that increased, and one that was constant. Present your results graphically.

136. CAPSTONE Use the graph of the function to answer (a)–(e).

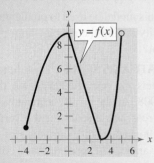

(a) Find the domain and range of f.

(b) Find the zero(s) of f.

(c) Determine the intervals over which f is increasing, decreasing, or constant.

(d) Approximate any relative minimum or relative maximum values of f.

(e) Is f even, odd, or neither?

A LIBRARY OF PARENT FUNCTIONS

What you should learn

- Identify and graph linear and squaring functions.
- Identify and graph cubic, square root, and reciprocal functions.
- Identify and graph step and other piecewise-defined functions.
- Recognize graphs of parent functions.

Why you should learn it

Step functions can be used to model real-life situations. For instance, in Exercise 69 on page 91, you will use a step function to model the cost of sending an overnight package from Los Angeles to Miami.

Linear and Squaring Functions

One of the goals of this text is to enable you to recognize the basic shapes of the graphs of different types of functions. For instance, you know that the graph of the **linear function** $f(x) = ax + b$ is a line with slope $m = a$ and y-intercept at $(0, b)$. The graph of the linear function has the following characteristics.

- The domain of the function is the set of all real numbers.
- The range of the function is the set of all real numbers.
- The graph has an x-intercept of $(-b/m, 0)$ and a y-intercept of $(0, b)$.
- The graph is increasing if $m > 0$, decreasing if $m < 0$, and constant if $m = 0$.

Example 1   Writing a Linear Function

Write the linear function f for which $f(1) = 3$ and $f(4) = 0$.

Solution

To find the equation of the line that passes through $(x_1, y_1) = (1, 3)$ and $(x_2, y_2) = (4, 0)$, first find the slope of the line.

$$m = \frac{y_2 - y_1}{x_2 - x_1} = \frac{0 - 3}{4 - 1} = \frac{-3}{3} = -1$$

Next, use the point-slope form of the equation of a line.

$$y - y_1 = m(x - x_1) \qquad \text{Point-slope form}$$
$$y - 3 = -1(x - 1) \qquad \text{Substitute for } x_1, y_1, \text{ and } m.$$
$$y = -x + 4 \qquad \text{Simplify.}$$
$$f(x) = -x + 4 \qquad \text{Function notation}$$

The graph of this function is shown in Figure P.71.

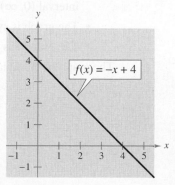

FIGURE P.71

CHECK*Point* Now try Exercise 11.

There are two special types of linear functions, the **constant function** and the **identity function.** A constant function has the form

$$f(x) = c$$

and has the domain of all real numbers with a range consisting of a single real number c. The graph of a constant function is a horizontal line, as shown in Figure P.72. The identity function has the form

$$f(x) = x.$$

Its domain and range are the set of all real numbers. The identity function has a slope of $m = 1$ and a y-intercept at $(0, 0)$. The graph of the identity function is a line for which each x-coordinate equals the corresponding y-coordinate. The graph is always increasing, as shown in Figure P.73.

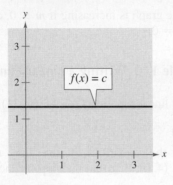

FIGURE P.72   FIGURE P.73

The graph of the **squaring function**

$$f(x) = x^2$$

is a U-shaped curve with the following characteristics.

- The domain of the function is the set of all real numbers.
- The range of the function is the set of all nonnegative real numbers.
- The function is even.
- The graph has an intercept at $(0, 0)$.
- The graph is decreasing on the interval $(-\infty, 0)$ and increasing on the interval $(0, \infty)$.
- The graph is symmetric with respect to the y-axis.
- The graph has a relative minimum at $(0, 0)$.

The graph of the squaring function is shown in Figure P.74.

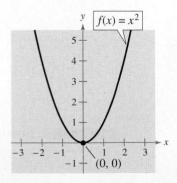

FIGURE P.74

Cubic, Square Root, and Reciprocal Functions

The basic characteristics of the graphs of the **cubic, square root,** and **reciprocal functions** are summarized below.

1. The graph of the *cubic* function $f(x) = x^3$ has the following characteristics.

- The domain of the function is the set of all real numbers.
- The range of the function is the set of all real numbers.
- The function is odd.
- The graph has an intercept at $(0, 0)$.
- The graph is increasing on the interval $(-\infty, \infty)$.
- The graph is symmetric with respect to the origin.

The graph of the cubic function is shown in Figure P.75.

2. The graph of the *square root* function $f(x) = \sqrt{x}$ has the following characteristics.

- The domain of the function is the set of all nonnegative real numbers.
- The range of the function is the set of all nonnegative real numbers.
- The graph has an intercept at $(0, 0)$.
- The graph is increasing on the interval $(0, \infty)$.

The graph of the square root function is shown in Figure P.76.

3. The graph of the *reciprocal* function $f(x) = \dfrac{1}{x}$ has the following characteristics.

- The domain of the function is $(-\infty, 0) \cup (0, \infty)$.
- The range of the function is $(-\infty, 0) \cup (0, \infty)$.
- The function is odd.
- The graph does not have any intercepts.
- The graph is decreasing on the intervals $(-\infty, 0)$ and $(0, \infty)$.
- The graph is symmetric with respect to the origin.

The graph of the reciprocal function is shown in Figure P.77.

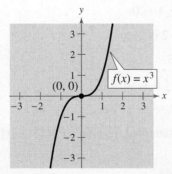

Cubic function
FIGURE P.75

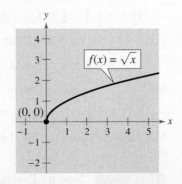

Square root function
FIGURE P.76

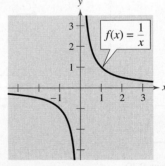

Reciprocal function
FIGURE P.77

Step and Piecewise-Defined Functions

Functions whose graphs resemble sets of stairsteps are known as **step functions.** The most famous of the step functions is the **greatest integer function,** which is denoted by $[\![x]\!]$ and defined as

$$f(x) = [\![x]\!] = \text{the greatest integer less than or equal to } x.$$

Some values of the greatest integer function are as follows.

$$[\![-1]\!] = (\text{greatest integer} \le -1) = -1$$

$$\left[\!\left[-\tfrac{1}{2}\right]\!\right] = \left(\text{greatest integer} \le -\tfrac{1}{2}\right) = -1$$

$$\left[\!\left[\tfrac{1}{10}\right]\!\right] = \left(\text{greatest integer} \le \tfrac{1}{10}\right) = 0$$

$$[\![1.5]\!] = (\text{greatest integer} \le 1.5) = 1$$

The graph of the greatest integer function

$$f(x) = [\![x]\!]$$

has the following characteristics, as shown in Figure P.78.

- The domain of the function is the set of all real numbers.
- The range of the function is the set of all integers.
- The graph has a y-intercept at $(0, 0)$ and x-intercepts in the interval $[0, 1)$.
- The graph is constant between each pair of consecutive integers.
- The graph jumps vertically one unit at each integer value.

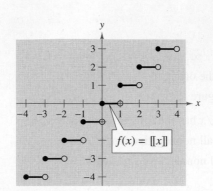

FIGURE P.78

TECHNOLOGY

When graphing a step function, you should set your graphing utility to *dot* mode.

Example 2 Evaluating a Step Function

Evaluate the function when $x = -1$, 2, and $\tfrac{3}{2}$.

$$f(x) = [\![x]\!] + 1$$

Solution

For $x = -1$, the greatest integer ≤ -1 is -1, so

$$f(-1) = [\![-1]\!] + 1 = -1 + 1 = 0.$$

For $x = 2$, the greatest integer ≤ 2 is 2, so

$$f(2) = [\![2]\!] + 1 = 2 + 1 = 3.$$

For $x = \tfrac{3}{2}$, the greatest integer $\le \tfrac{3}{2}$ is 1, so

$$f\left(\tfrac{3}{2}\right) = \left[\!\left[\tfrac{3}{2}\right]\!\right] + 1 = 1 + 1 = 2.$$

You can verify your answers by examining the graph of $f(x) = [\![x]\!] + 1$ shown in Figure P.79.

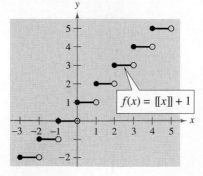

FIGURE P.79

CHECK*Point* Now try Exercise 43.

Recall from Section P.5 that a piecewise-defined function is defined by two or more equations over a specified domain. To graph a piecewise-defined function, graph each equation separately over the specified domain, as shown in Example 3.

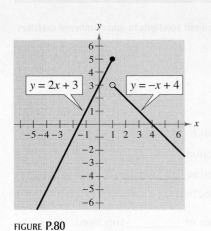

FIGURE P.80

Example 3   Graphing a Piecewise-Defined Function

Sketch the graph of

$$f(x) = \begin{cases} 2x + 3, & x \le 1 \\ -x + 4, & x > 1 \end{cases}.$$

Solution

This piecewise-defined function is composed of two linear functions. At $x = 1$ and to the left of $x = 1$ the graph is the line $y = 2x + 3$, and to the right of $x = 1$ the graph is the line $y = -x + 4$, as shown in Figure P.80. Notice that the point $(1, 5)$ is a solid dot and the point $(1, 3)$ is an open dot. This is because $f(1) = 2(1) + 3 = 5$.

CHECK**Point**   Now try Exercise 57.

Parent Functions

The eight graphs shown in Figure P.81 represent the most commonly used functions in algebra. Familiarity with the basic characteristics of these simple graphs will help you analyze the shapes of more complicated graphs—in particular, graphs obtained from these graphs by the rigid and nonrigid transformations studied in the next section.

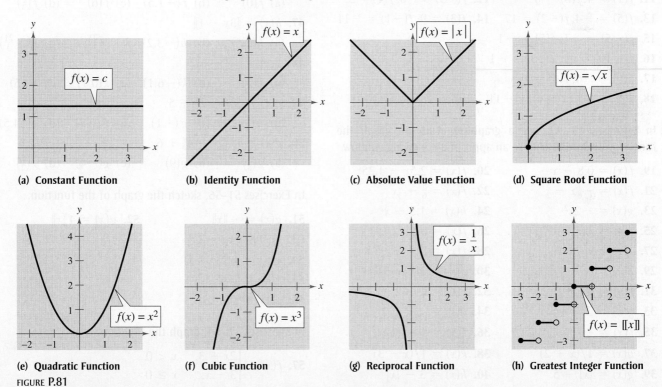

(a) Constant Function

(b) Identity Function

(c) Absolute Value Function

(d) Square Root Function

(e) Quadratic Function

(f) Cubic Function

(g) Reciprocal Function

(h) Greatest Integer Function

FIGURE P.81

P.7 EXERCISES

See www.CalcChat.com for worked-out solutions to odd-numbered exercises.

VOCABULARY

In Exercises 1–9, match each function with its name.

1. $f(x) = [\![x]\!]$ **2.** $f(x) = x$ **3.** $f(x) = 1/x$

4. $f(x) = x^2$ **5.** $f(x) = \sqrt{x}$ **6.** $f(x) = c$

7. $f(x) = |x|$ **8.** $f(x) = x^3$ **9.** $f(x) = ax + b$

   (a) squaring function   (b) square root function   (c) cubic function

   (d) linear function   (e) constant function   (f) absolute value function

   (g) greatest integer function   (h) reciprocal function   (i) identity function

10. Fill in the blank: The constant function and the identity function are two special types of _____ functions.

SKILLS AND APPLICATIONS

In Exercises 11–18, (a) write the linear function f such that it has the indicated function values and (b) sketch the graph of the function.

11. $f(1) = 4, f(0) = 6$ **12.** $f(-3) = -8, f(1) = 2$

13. $f(5) = -4, f(-2) = 17$ **14.** $f(3) = 9, f(-1) = -11$

15. $f(-5) = -1, f(5) = -1$

16. $f(-10) = 12, f(16) = -1$

17. $f\left(\frac{1}{2}\right) = -6, f(4) = -3$

18. $f\left(\frac{2}{3}\right) = -\frac{15}{2}, f(-4) = -11$

In Exercises 19–42, use a graphing utility to graph the function. Be sure to choose an appropriate viewing window.

19. $f(x) = 0.8 - x$ **20.** $f(x) = 2.5x - 4.25$

21. $f(x) = -\frac{1}{6}x - \frac{5}{2}$ **22.** $f(x) = \frac{5}{6} - \frac{2}{3}x$

23. $g(x) = -2x^2$ **24.** $h(x) = 1.5 - x^2$

25. $f(x) = 3x^2 - 1.75$ **26.** $f(x) = 0.5x^2 + 2$

27. $f(x) = x^3 - 1$ **28.** $f(x) = 8 - x^3$

29. $f(x) = (x - 1)^3 + 2$ **30.** $g(x) = 2(x + 3)^3 + 1$

31. $f(x) = 4\sqrt{x}$ **32.** $f(x) = 4 - 2\sqrt{x}$

33. $g(x) = 2 - \sqrt{x + 4}$ **34.** $h(x) = \sqrt{x + 2} + 3$

35. $f(x) = -1/x$ **36.** $f(x) = 4 + (1/x)$

37. $h(x) = 1/(x + 2)$ **38.** $k(x) = 1/(x - 3)$

39. $g(x) = |x| - 5$ **40.** $h(x) = 3 - |x|$

41. $f(x) = |x + 4|$ **42.** $f(x) = |x - 1|$

In Exercises 43–50, evaluate the function for the indicated values.

43. $f(x) = [\![x]\!]$

   (a) $f(2.1)$ (b) $f(2.9)$ (c) $f(-3.1)$ (d) $f\left(\frac{7}{2}\right)$

44. $g(x) = 2[\![x]\!]$

   (a) $g(-3)$ (b) $g(0.25)$ (c) $g(9.5)$ (d) $g\left(\frac{11}{3}\right)$

45. $h(x) = [\![x + 3]\!]$

   (a) $h(-2)$ (b) $h\left(\frac{1}{2}\right)$ (c) $h(4.2)$ (d) $h(-21.6)$

46. $f(x) = 4[\![x]\!] + 7$

   (a) $f(0)$ (b) $f(-1.5)$ (c) $f(6)$ (d) $f\left(\frac{5}{3}\right)$

47. $h(x) = [\![3x - 1]\!]$

   (a) $h(2.5)$ (b) $h(-3.2)$ (c) $h\left(\frac{7}{3}\right)$ (d) $h\left(-\frac{21}{3}\right)$

48. $k(x) = \left[\!\left[\frac{1}{2}x + 6\right]\!\right]$

   (a) $k(5)$ (b) $k(-6.1)$ (c) $k(0.1)$ (d) $k(15)$

49. $g(x) = 3[\![x - 2]\!] + 5$

   (a) $g(-2.7)$ (b) $g(-1)$ (c) $g(0.8)$ (d) $g(14.5)$

50. $g(x) = -7[\![x + 4]\!] + 6$

   (a) $g\left(\frac{1}{8}\right)$ (b) $g(9)$ (c) $g(-4)$ (d) $g\left(\frac{3}{2}\right)$

In Exercises 51–56, sketch the graph of the function.

51. $g(x) = -[\![x]\!]$ **52.** $g(x) = 4[\![x]\!]$

53. $g(x) = [\![x]\!] - 2$

54. $g(x) = [\![x]\!] - 1$

55. $g(x) = [\![x + 1]\!]$

56. $g(x) = [\![x - 3]\!]$

In Exercises 57–64, graph the function.

57. $f(x) = \begin{cases} 2x + 3, & x < 0 \\ 3 - x, & x \geq 0 \end{cases}$

58. $g(x) = \begin{cases} x + 6, & x \leq -4 \\ \frac{1}{2}x - 4, & x > -4 \end{cases}$

59. $f(x) = \begin{cases} \sqrt{4 + x}, & x < 0 \\ \sqrt{4 - x}, & x \geq 0 \end{cases}$

60. $f(x) = \begin{cases} 1 - (x - 1)^2, & x \leq 2 \\ \sqrt{x - 2}, & x > 2 \end{cases}$

61. $f(x) = \begin{cases} x^2 + 5, & x \leq 1 \\ -x^2 + 4x + 3, & x > 1 \end{cases}$

62. $h(x) = \begin{cases} 3 - x^2, & x < 0 \\ x^2 + 2, & x \geq 0 \end{cases}$

63. $h(x) = \begin{cases} 4 - x^2, & x < -2 \\ 3 + x, & -2 \leq x < 0 \\ x^2 + 1, & x \geq 0 \end{cases}$

64. $k(x) = \begin{cases} 2x + 1, & x \leq -1 \\ 2x^2 - 1, & -1 < x \leq 1 \\ 1 - x^2, & x > 1 \end{cases}$

In Exercises 65–68, (a) use a graphing utility to graph the function, (b) state the domain and range of the function, and (c) describe the pattern of the graph.

65. $s(x) = 2\left(\frac{1}{4}x - \left[\!\left[\frac{1}{4}x\right]\!\right]\right)$

66. $g(x) = 2\left(\frac{1}{4}x - \left[\!\left[\frac{1}{4}x\right]\!\right]\right)^2$

67. $h(x) = 4\left(\frac{1}{2}x - \left[\!\left[\frac{1}{2}x\right]\!\right]\right)$

68. $k(x) = 4\left(\frac{1}{2}x - \left[\!\left[\frac{1}{2}x\right]\!\right]\right)^2$

69. DELIVERY CHARGES The cost of sending an overnight package from Los Angeles to Miami is $23.40 for a package weighing up to but not including 1 pound and $3.75 for each additional pound or portion of a pound. A model for the total cost C (in dollars) of sending the package is $C = 23.40 + 3.75[\![x]\!]$, $x > 0$, where x is the weight in pounds.

(a) Sketch a graph of the model.

(b) Determine the cost of sending a package that weighs 9.25 pounds.

70. DELIVERY CHARGES The cost of sending an overnight package from New York to Atlanta is $22.65 for a package weighing up to but not including 1 pound and $3.70 for each additional pound or portion of a pound.

(a) Use the greatest integer function to create a model for the cost C of overnight delivery of a package weighing x pounds, $x > 0$.

(b) Sketch the graph of the function.

71. WAGES A mechanic is paid $14.00 per hour for regular time and time-and-a-half for overtime. The weekly wage function is given by

$$W(h) = \begin{cases} 14h, & 0 < h \leq 40 \\ 21(h - 40) + 560, & h > 40 \end{cases}$$

where h is the number of hours worked in a week.

(a) Evaluate $W(30)$, $W(40)$, $W(45)$, and $W(50)$.

(b) The company increased the regular work week to 45 hours. What is the new weekly wage function?

72. SNOWSTORM During a nine-hour snowstorm, it snows at a rate of 1 inch per hour for the first 2 hours, at a rate of 2 inches per hour for the next 6 hours, and at a rate of 0.5 inch per hour for the final hour. Write and graph a piecewise-defined function that gives the depth of the snow during the snowstorm. How many inches of snow accumulated from the storm?

73. REVENUE The table shows the monthly revenue y (in thousands of dollars) of a landscaping business for each month of the year 2008, with $x = 1$ representing January.

Month, x	Revenue, y
1	5.2
2	5.6
3	6.6
4	8.3
5	11.5
6	15.8
7	12.8
8	10.1
9	8.6
10	6.9
11	4.5
12	2.7

A mathematical model that represents these data is

$$f(x) = \begin{cases} -1.97x + 26.3 \\ 0.505x^2 - 1.47x + 6.3 \end{cases}$$

(a) Use a graphing utility to graph the model. What is the domain of each part of the piecewise-defined function? How can you tell? Explain your reasoning.

(b) Find $f(5)$ and $f(11)$, and interpret your results in the context of the problem.

(c) How do the values obtained from the model in part (a) compare with the actual data values?

EXPLORATION

TRUE OR FALSE? In Exercises 74 and 75, determine whether the statement is true or false. Justify your answer.

74. A piecewise-defined function will always have at least one x-intercept or at least one y-intercept.

75. A linear equation will always have an x-intercept and a y-intercept.

76. CAPSTONE For each graph of f shown in Figure P.81, do the following.

(a) Find the domain and range of f.

(b) Find the x- and y-intercepts of the graph of f.

(c) Determine the intervals over which f is increasing, decreasing, or constant.

(d) Determine whether f is even, odd, or neither. Then describe the symmetry.

What you should learn

- Use vertical and horizontal shifts to sketch graphs of functions.
- Use reflections to sketch graphs of functions.
- Use nonrigid transformations to sketch graphs of functions.

Why you should learn it

Transformations of functions can be used to model real-life applications. For instance, Exercise 79 on page 100 shows how a transformation of a function can be used to model the total numbers of miles driven by vans, pickups, and sport utility vehicles in the United States.

P.8 **TRANSFORMATIONS OF FUNCTIONS**

Shifting Graphs

Many functions have graphs that are simple transformations of the parent graphs summarized in Section P.7. For example, you can obtain the graph of

$$h(x) = x^2 + 2$$

by shifting the graph of $f(x) = x^2$ *upward* two units, as shown in Figure P.82. In function notation, h and f are related as follows.

$$h(x) = x^2 + 2 = f(x) + 2 \qquad \text{Upward shift of two units}$$

Similarly, you can obtain the graph of

$$g(x) = (x - 2)^2$$

by shifting the graph of $f(x) = x^2$ to the *right* two units, as shown in Figure P.83. In this case, the functions g and f have the following relationship.

$$g(x) = (x - 2)^2 = f(x - 2) \qquad \text{Right shift of two units}$$

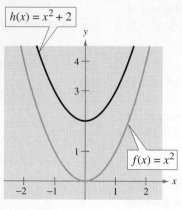

FIGURE P.82

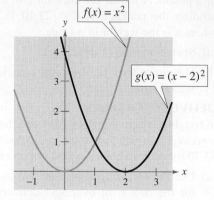

FIGURE P.83

The following list summarizes this discussion about horizontal and vertical shifts.

Vertical and Horizontal Shifts

Let c be a positive real number. **Vertical and horizontal shifts** in the graph of $y = f(x)$ are represented as follows.

1. Vertical shift c units *upward:* $\qquad h(x) = f(x) + c$

2. Vertical shift c units *downward:* $\qquad h(x) = f(x) - c$

3. Horizontal shift c units to the *right:* $h(x) = f(x - c)$

4. Horizontal shift c units to the *left:* $\quad h(x) = f(x + c)$

WARNING/CAUTION

In items 3 and 4, be sure you see that $h(x) = f(x - c)$ corresponds to a *right* shift and $h(x) = f(x + c)$ corresponds to a *left* shift for $c > 0$.

Some graphs can be obtained from combinations of vertical and horizontal shifts, as demonstrated in Example 1(b). Vertical and horizontal shifts generate a *family of functions*, each with the same shape but at different locations in the plane.

Example 1   Shifts in the Graphs of a Function

Use the graph of $f(x) = x^3$ to sketch the graph of each function.

a. $g(x) = x^3 - 1$ **b.** $h(x) = (x + 2)^3 + 1$

Solution

a. Relative to the graph of $f(x) = x^3$, the graph of

$$g(x) = x^3 - 1$$

is a downward shift of one unit, as shown in Figure P.84.

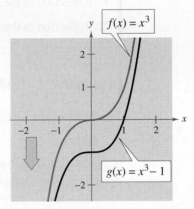

FIGURE **P.84**

b. Relative to the graph of $f(x) = x^3$, the graph of

$$h(x) = (x + 2)^3 + 1$$

involves a left shift of two units and an upward shift of one unit, as shown in Figure P.85.

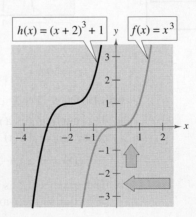

FIGURE **P.85**

CHECK Point   Now try Exercise 7.

In Figure P.85, notice that the same result is obtained if the vertical shift precedes the horizontal shift *or* if the horizontal shift precedes the vertical shift.

Study Tip

In Example 1(a), note that $g(x) = f(x) - 1$ and that in Example 1(b), $h(x) = f(x + 2) + 1$.

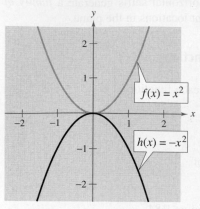

FIGURE P.86

Reflecting Graphs

The second common type of transformation is a **reflection.** For instance, if you consider the *x*-axis to be a mirror, the graph of

$$h(x) = -x^2$$

is the mirror image (or reflection) of the graph of

$$f(x) = x^2,$$

as shown in Figure P.86.

Reflections in the Coordinate Axes

Reflections in the coordinate axes of the graph of $y = f(x)$ are represented as follows.

1. Reflection in the *x*-axis: $h(x) = -f(x)$

2. Reflection in the *y*-axis: $h(x) = f(-x)$

Example 2 Finding Equations from Graphs

The graph of the function given by

$$f(x) = x^4$$

is shown in Figure P.87. Each of the graphs in Figure P.88 is a transformation of the graph of *f*. Find an equation for each of these functions.

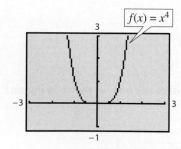

FIGURE P.87

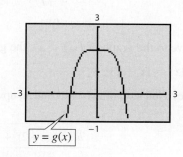

(a)

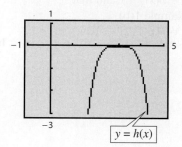

(b)

FIGURE P.88

Solution

a. The graph of *g* is a reflection in the *x*-axis *followed by* an upward shift of two units of the graph of $f(x) = x^4$. So, the equation for *g* is

$$g(x) = -x^4 + 2.$$

b. The graph of *h* is a horizontal shift of three units to the right *followed by* a reflection in the *x*-axis of the graph of $f(x) = x^4$. So, the equation for *h* is

$$h(x) = -(x - 3)^4.$$

CHECK Point ▶ Now try Exercise 15.

Example 3   Reflections and Shifts

Compare the graph of each function with the graph of $f(x) = \sqrt{x}$.

a. $g(x) = -\sqrt{x}$ **b.** $h(x) = \sqrt{-x}$ **c.** $k(x) = -\sqrt{x+2}$

Algebraic Solution

a. The graph of g is a reflection of the graph of f in the x-axis because

$$g(x) = -\sqrt{x}$$

$$= -f(x).$$

b. The graph of h is a reflection of the graph of f in the y-axis because

$$h(x) = \sqrt{-x}$$

$$= f(-x).$$

c. The graph of k is a left shift of two units followed by a reflection in the x-axis because

$$k(x) = -\sqrt{x+2}$$

$$= -f(x+2).$$

Graphical Solution

a. Graph f and g on the same set of coordinate axes. From the graph in Figure P.89, you can see that the graph of g is a reflection of the graph of f in the x-axis.

b. Graph f and h on the same set of coordinate axes. From the graph in Figure P.90, you can see that the graph of h is a reflection of the graph of f in the y-axis.

c. Graph f and k on the same set of coordinate axes. From the graph in Figure P.91, you can see that the graph of k is a left shift of two units of the graph of f, followed by a reflection in the x-axis.

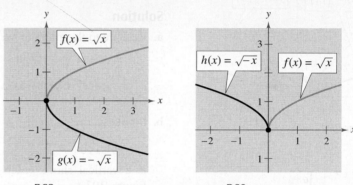

FIGURE **P.89**   FIGURE **P.90**

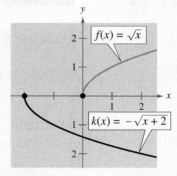

FIGURE **P.91**

CHECK*Point*   Now try Exercise 25.

When sketching the graphs of functions involving square roots, remember that the domain must be restricted to exclude negative numbers inside the radical. For instance, here are the domains of the functions in Example 3.

Domain of $g(x) = -\sqrt{x}$: $x \geq 0$

Domain of $h(x) = \sqrt{-x}$: $x \leq 0$

Domain of $k(x) = -\sqrt{x+2}$: $x \geq -2$

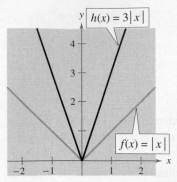

FIGURE **P.92**

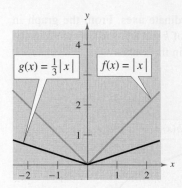

FIGURE **P.93**

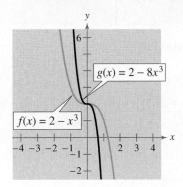

FIGURE **P.94**

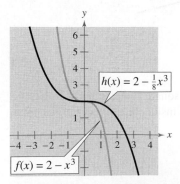

FIGURE **P.95**

Nonrigid Transformations

Horizontal shifts, vertical shifts, and reflections are **rigid transformations** because the basic shape of the graph is unchanged. These transformations change only the *position* of the graph in the coordinate plane. **Nonrigid transformations** are those that cause a *distortion*—a change in the shape of the original graph. For instance, a nonrigid transformation of the graph of $y = f(x)$ is represented by $g(x) = cf(x)$, where the transformation is a **vertical stretch** if $c > 1$ and a **vertical shrink** if $0 < c < 1$. Another nonrigid transformation of the graph of $y = f(x)$ is represented by $h(x) = f(cx)$, where the transformation is a **horizontal shrink** if $c > 1$ and a **horizontal stretch** if $0 < c < 1$.

Example 4   Nonrigid Transformations

Compare the graph of each function with the graph of $f(x) = |x|$.

a. $h(x) = 3|x|$ **b.** $g(x) = \frac{1}{3}|x|$

Solution

a. Relative to the graph of $f(x) = |x|$, the graph of

$$h(x) = 3|x| = 3f(x)$$

is a vertical stretch (each y-value is multiplied by 3) of the graph of f. (See Figure P.92.)

b. Similarly, the graph of

$$g(x) = \frac{1}{3}|x| = \frac{1}{3}f(x)$$

is a vertical shrink $\left(\text{each } y\text{-value is multiplied by } \frac{1}{3}\right)$ of the graph of f. (See Figure P.93.)

CHECK*Point*   Now try Exercise 29.

Example 5   Nonrigid Transformations

Compare the graph of each function with the graph of $f(x) = 2 - x^3$.

a. $g(x) = f(2x)$ **b.** $h(x) = f\left(\frac{1}{2}x\right)$

Solution

a. Relative to the graph of $f(x) = 2 - x^3$, the graph of

$$g(x) = f(2x) = 2 - (2x)^3 = 2 - 8x^3$$

is a horizontal shrink $(c > 1)$ of the graph of f. (See Figure P.94.)

b. Similarly, the graph of

$$h(x) = f\left(\frac{1}{2}x\right) = 2 - \left(\frac{1}{2}x\right)^3 = 2 - \frac{1}{8}x^3$$

is a horizontal stretch $(0 < c < 1)$ of the graph of f. (See Figure P.95.)

CHECK*Point*   Now try Exercise 35.

P.8   EXERCISES

See www.CalcChat.com for worked-out solutions to odd-numbered exercises.

VOCABULARY

In Exercises 1–5, fill in the blanks.

1. Horizontal shifts, vertical shifts, and reflections are called _____ transformations.

2. A reflection in the x-axis of $y = f(x)$ is represented by $h(x) =$ _____, while a reflection in the y-axis of $y = f(x)$ is represented by $h(x) =$ _____.

3. Transformations that cause a distortion in the shape of the graph of $y = f(x)$ are called _____ transformations.

4. A nonrigid transformation of $y = f(x)$ represented by $h(x) = f(cx)$ is a _____ _____ if $c > 1$ and a _____ _____ if $0 < c < 1$.

5. A nonrigid transformation of $y = f(x)$ represented by $g(x) = cf(x)$ is a _____ _____ if $c > 1$ and a _____ _____ if $0 < c < 1$.

6. Match the rigid transformation of $y = f(x)$ with the correct representation of the graph of h, where $c > 0$.
   (a) $h(x) = f(x) + c$ (i) A horizontal shift of f, c units to the right
   (b) $h(x) = f(x) - c$ (ii) A vertical shift of f, c units downward
   (c) $h(x) = f(x + c)$ (iii) A horizontal shift of f, c units to the left
   (d) $h(x) = f(x - c)$ (iv) A vertical shift of f, c units upward

SKILLS AND APPLICATIONS

7. For each function, sketch (on the same set of coordinate axes) a graph of each function for $c = -1, 1,$ and 3.
   (a) $f(x) = |x| + c$
   (b) $f(x) = |x - c|$
   (c) $f(x) = |x + 4| + c$

8. For each function, sketch (on the same set of coordinate axes) a graph of each function for $c = -3, -1, 1,$ and 3.
   (a) $f(x) = \sqrt{x} + c$
   (b) $f(x) = \sqrt{x - c}$
   (c) $f(x) = \sqrt{x - 3} + c$

9. For each function, sketch (on the same set of coordinate axes) a graph of each function for $c = -2, 0,$ and 2.
   (a) $f(x) = [\![x]\!] + c$
   (b) $f(x) = [\![x + c]\!]$
   (c) $f(x) = [\![x - 1]\!] + c$

10. For each function, sketch (on the same set of coordinate axes) a graph of each function for $c = -3, -1, 1,$ and 3.
    (a) $f(x) = \begin{cases} x^2 + c, & x < 0 \\ -x^2 + c, & x \geq 0 \end{cases}$
    (b) $f(x) = \begin{cases} (x + c)^2, & x < 0 \\ -(x + c)^2, & x \geq 0 \end{cases}$

In Exercises 11–14, use the graph of f to sketch each graph. To print an enlarged copy of the graph, go to the website *www.mathgraphs.com.*

11. (a) $y = f(x) + 2$
    (b) $y = f(x - 2)$
    (c) $y = 2f(x)$
    (d) $y = -f(x)$
    (e) $y = f(x + 3)$
    (f) $y = f(-x)$
    (g) $y = f\left(\frac{1}{2}x\right)$

12. (a) $y = f(-x)$
    (b) $y = f(x) + 4$
    (c) $y = 2f(x)$
    (d) $y = -f(x - 4)$
    (e) $y = f(x) - 3$
    (f) $y = -f(x) - 1$
    (g) $y = f(2x)$

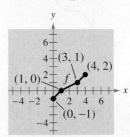

FIGURE FOR **11**

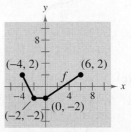

FIGURE FOR **12**

13. (a) $y = f(x) - 1$
    (b) $y = f(x - 1)$
    (c) $y = f(-x)$
    (d) $y = f(x + 1)$
    (e) $y = -f(x - 2)$
    (f) $y = \frac{1}{2}f(x)$
    (g) $y = f(2x)$

14. (a) $y = f(x - 5)$
    (b) $y = -f(x) + 3$
    (c) $y = \frac{1}{3}f(x)$
    (d) $y = -f(x + 1)$
    (e) $y = f(-x)$
    (f) $y = f(x) - 10$
    (g) $y = f\left(\frac{1}{3}x\right)$

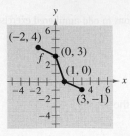

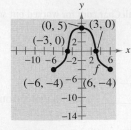

FIGURE FOR **13**   FIGURE FOR **14**

15. Use the graph of $f(x) = x^2$ to write an equation for each function whose graph is shown.

(a)

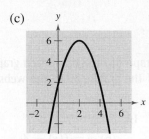

(b)

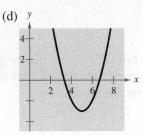

(c)

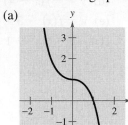

(d)

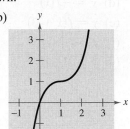

16. Use the graph of $f(x) = x^3$ to write an equation for each function whose graph is shown.

(a)

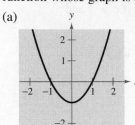

(b)

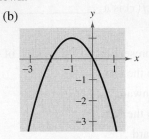

(c)

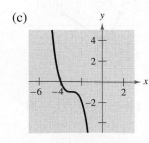

(d)

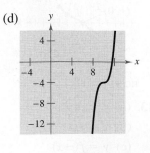

17. Use the graph of $f(x) = |x|$ to write an equation for each function whose graph is shown.

(a)

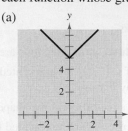

(b)

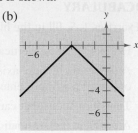

(c)

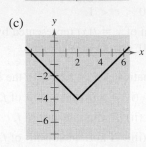

(d)

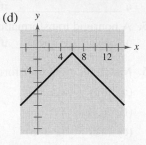

18. Use the graph of $f(x) = \sqrt{x}$ to write an equation for each function whose graph is shown.

(a)

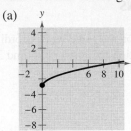

(b)

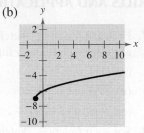

(c)

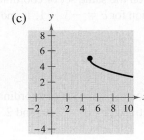

(d)

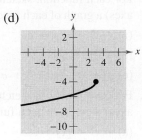

In Exercises 19–24, identify the parent function and the transformation shown in the graph. Write an equation for the function shown in the graph.

19.

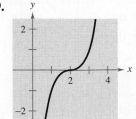

20.

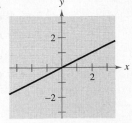

21.

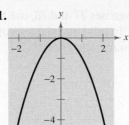

22.

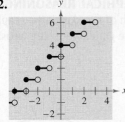

23.

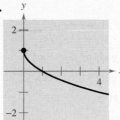

24.

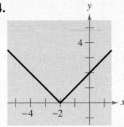

In Exercises 25–54, g is related to one of the parent functions described in Section P.7. (a) Identify the parent function f. (b) Describe the sequence of transformations from f to g. (c) Sketch the graph of g. (d) Use function notation to write g in terms of f.

25. $g(x) = 12 - x^2$ **26.** $g(x) = (x - 8)^2$

27. $g(x) = x^3 + 7$ **28.** $g(x) = -x^3 - 1$

29. $g(x) = \frac{2}{3}x^2 + 4$ **30.** $g(x) = 2(x - 7)^2$

31. $g(x) = 2 - (x + 5)^2$ **32.** $g(x) = -(x + 10)^2 + 5$

33. $g(x) = 3 + 2(x - 4)^2$ **34.** $g(x) = -\frac{1}{4}(x + 2)^2 - 2$

35. $g(x) = \sqrt{3x}$ **36.** $g(x) = \sqrt{\frac{1}{4}x}$

37. $g(x) = (x - 1)^3 + 2$ **38.** $g(x) = (x + 3)^3 - 10$

39. $g(x) = 3(x - 2)^3$ **40.** $g(x) = -\frac{1}{2}(x + 1)^3$

41. $g(x) = -|x| - 2$ **42.** $g(x) = 6 - |x + 5|$

43. $g(x) = -|x + 4| + 8$ **44.** $g(x) = |-x + 3| + 9$

45. $g(x) = -2|x - 1| - 4$ **46.** $g(x) = \frac{1}{2}|x - 2| - 3$

47. $g(x) = 3 - [\![x]\!]$ **48.** $g(x) = 2[\![x + 5]\!]$

49. $g(x) = \sqrt{x - 9}$ **50.** $g(x) = \sqrt{x + 4} + 8$

51. $g(x) = \sqrt{7 - x} - 2$ **52.** $g(x) = -\frac{1}{2}\sqrt{x + 3} - 1$

53. $g(x) = \sqrt{\frac{1}{2}x} - 4$ **54.** $g(x) = \sqrt{3x} + 1$

In Exercises 55–62, write an equation for the function that is described by the given characteristics.

55. The shape of $f(x) = x^2$, but shifted three units to the right and seven units downward

56. The shape of $f(x) = x^2$, but shifted two units to the left, nine units upward, and reflected in the x-axis

57. The shape of $f(x) = x^3$, but shifted 13 units to the right

58. The shape of $f(x) = x^3$, but shifted six units to the left, six units downward, and reflected in the y-axis

59. The shape of $f(x) = |x|$, but shifted 12 units upward and reflected in the x-axis

60. The shape of $f(x) = |x|$, but shifted four units to the left and eight units downward

61. The shape of $f(x) = \sqrt{x}$, but shifted six units to the left and reflected in both the x-axis and the y-axis

62. The shape of $f(x) = \sqrt{x}$, but shifted nine units downward and reflected in both the x-axis and the y-axis

63. Use the graph of $f(x) = x^2$ to write an equation for each function whose graph is shown.

(a)

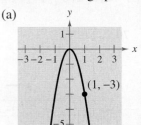

(b)

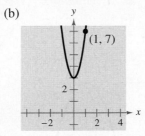

64. Use the graph of $f(x) = x^3$ to write an equation for each function whose graph is shown.

(a)

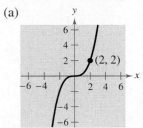

(b)

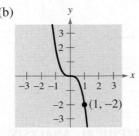

65. Use the graph of $f(x) = |x|$ to write an equation for each function whose graph is shown.

(a)

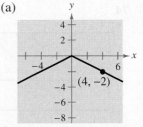

(b)

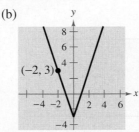

66. Use the graph of $f(x) = \sqrt{x}$ to write an equation for each function whose graph is shown.

(a)   (b)

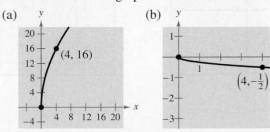

In Exercises 67–72, identify the parent function and the transformation shown in the graph. Write an equation for the function shown in the graph. Then use a graphing utility to verify your answer.

67.
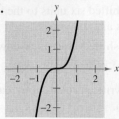

68.
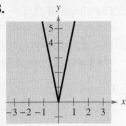

69.
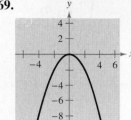

70.
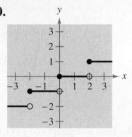

71.
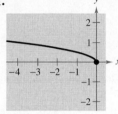

72.
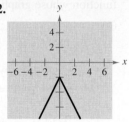

⚓ **GRAPHICAL ANALYSIS** In Exercises 73–76, use the viewing window shown to write a possible equation for the transformation of the parent function.

73.
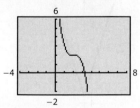

74.
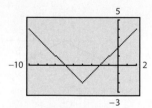

75.
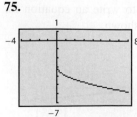

76.
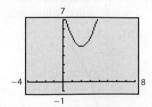

GRAPHICAL REASONING In Exercises 77 and 78, use the graph of f to sketch the graph of g. To print an enlarged copy of the graph, go to the website *www.mathgraphs.com*.

77.
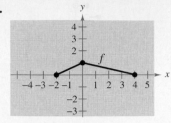

(a) $g(x) = f(x) + 2$ (b) $g(x) = f(x) - 1$
(c) $g(x) = f(-x)$ (d) $g(x) = -2f(x)$
(e) $g(x) = f(4x)$ (f) $g(x) = f\left(\frac{1}{2}x\right)$

78.
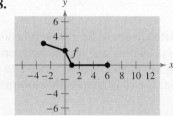

(a) $g(x) = f(x) - 5$ (b) $g(x) = f(x) + \frac{1}{2}$
(c) $g(x) = f(-x)$ (d) $g(x) = -4f(x)$
(e) $g(x) = f(2x) + 1$ (f) $g(x) = f\left(\frac{1}{4}x\right) - 2$

79. MILES DRIVEN The total numbers of miles M (in billions) driven by vans, pickups, and SUVs (sport utility vehicles) in the United States from 1990 through 2006 can be approximated by the function

$$M = 527 + 128.0\sqrt{t}, \quad 0 \le t \le 16$$

where t represents the year, with $t = 0$ corresponding to 1990. (Source: U.S. Federal Highway Administration)

⚓ (a) Describe the transformation of the parent function $f(x) = \sqrt{x}$. Then use a graphing utility to graph the function over the specified domain.

∫ (b) Find the average rate of change of the function from 1990 to 2006. Interpret your answer in the context of the problem.

(c) Rewrite the function so that $t = 0$ represents 2000. Explain how you got your answer.

(d) Use the model from part (c) to predict the number of miles driven by vans, pickups, and SUVs in 2012. Does your answer seem reasonable? Explain.

80. MARRIED COUPLES   The numbers N (in thousands) of married couples with stay-at-home mothers from 2000 through 2007 can be approximated by the function

$$N = -24.70(t - 5.99)^2 + 5617, \quad 0 \le t \le 7$$

where t represents the year, with $t = 0$ corresponding to 2000.   (Source: U.S. Census Bureau)

(a) Describe the transformation of the parent function $f(x) = x^2$. Then use a graphing utility to graph the function over the specified domain.

(b) Find the average rate of the change of the function from 2000 to 2007. Interpret your answer in the context of the problem.

(c) Use the model to predict the number of married couples with stay-at-home mothers in 2015. Does your answer seem reasonable? Explain.

EXPLORATION

TRUE OR FALSE?   In Exercises 81–84, determine whether the statement is true or false. Justify your answer.

81. The graph of $y = f(-x)$ is a reflection of the graph of $y = f(x)$ in the x-axis.

82. The graph of $y = -f(x)$ is a reflection of the graph of $y = f(x)$ in the y-axis.

83. The graphs of

$$f(x) = |x| + 6$$

and

$$f(x) = |-x| + 6$$

are identical.

84. If the graph of the parent function $f(x) = x^2$ is shifted six units to the right, three units upward, and reflected in the x-axis, then the point $(-2, 19)$ will lie on the graph of the transformation.

85. DESCRIBING PROFITS   Management originally predicted that the profits from the sales of a new product would be approximated by the graph of the function f shown. The actual profits are shown by the function g along with a verbal description. Use the concepts of transformations of graphs to write g in terms of f.

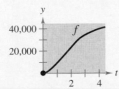

(a) The profits were only three-fourths as large as expected.

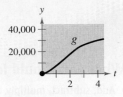

(b) The profits were consistently $10,000 greater than predicted.

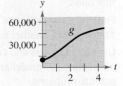

(c) There was a two-year delay in the introduction of the product. After sales began, profits grew as expected.

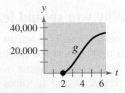

86. THINK ABOUT IT   You can use either of two methods to graph a function: plotting points or translating a parent function as shown in this section. Which method of graphing do you prefer to use for each function? Explain.

(a) $f(x) = 3x^2 - 4x + 1$

(b) $f(x) = 2(x - 1)^2 - 6$

87. The graph of $y = f(x)$ passes through the points $(0, 1)$, $(1, 2)$, and $(2, 3)$. Find the corresponding points on the graph of $y = f(x + 2) - 1$.

88. Use a graphing utility to graph f, g, and h in the same viewing window. Before looking at the graphs, try to predict how the graphs of g and h relate to the graph of f.

(a) $f(x) = x^2$, $g(x) = (x - 4)^2$,
    $h(x) = (x - 4)^2 + 3$

(b) $f(x) = x^2$, $g(x) = (x + 1)^2$,
    $h(x) = (x + 1)^2 - 2$

(c) $f(x) = x^2$, $g(x) = (x + 4)^2$,
    $h(x) = (x + 4)^2 + 2$

89. Reverse the order of transformations in Example 2(a). Do you obtain the same graph? Do the same for Example 2(b). Do you obtain the same graph? Explain.

90. CAPSTONE   Use the fact that the graph of $y = f(x)$ is increasing on the intervals $(-\infty, -1)$ and $(2, \infty)$ and decreasing on the interval $(-1, 2)$ to find the intervals on which the graph is increasing and decreasing. If not possible, state the reason.

(a) $y = f(-x)$ (b) $y = -f(x)$ (c) $y = \frac{1}{2}f(x)$

(d) $y = -f(x - 1)$ (e) $y = f(x - 2) + 1$

P.9   COMBINATIONS OF FUNCTIONS: COMPOSITE FUNCTIONS

What you should learn

- Add, subtract, multiply, and divide functions.
- Find the composition of one function with another function.
- Use combinations and compositions of functions to model and solve real-life problems.

Why you should learn it

Compositions of functions can be used to model and solve real-life problems. For instance, in Exercise 76 on page 110, compositions of functions are used to determine the price of a new hybrid car.

© Jim West/The Image Works

Arithmetic Combinations of Functions

Just as two real numbers can be combined by the operations of addition, subtraction, multiplication, and division to form other real numbers, two *functions* can be combined to create new functions. For example, the functions given by $f(x) = 2x - 3$ and $g(x) = x^2 - 1$ can be combined to form the sum, difference, product, and quotient of f and g.

$$f(x) + g(x) = (2x - 3) + (x^2 - 1)$$
$$= x^2 + 2x - 4 \qquad \text{Sum}$$
$$f(x) - g(x) = (2x - 3) - (x^2 - 1)$$
$$= -x^2 + 2x - 2 \qquad \text{Difference}$$
$$f(x)g(x) = (2x - 3)(x^2 - 1)$$
$$= 2x^3 - 3x^2 - 2x + 3 \qquad \text{Product}$$
$$\frac{f(x)}{g(x)} = \frac{2x - 3}{x^2 - 1}, \qquad x \neq \pm 1 \qquad \text{Quotient}$$

The domain of an **arithmetic combination** of functions f and g consists of all real numbers that are common to the domains of f and g. In the case of the quotient $f(x)/g(x)$, there is the further restriction that $g(x) \neq 0$.

Sum, Difference, Product, and Quotient of Functions

Let f and g be two functions with overlapping domains. Then, for all x common to both domains, the *sum, difference, product,* and *quotient* of f and g are defined as follows.

1. *Sum:* $(f + g)(x) = f(x) + g(x)$

2. *Difference:* $(f - g)(x) = f(x) - g(x)$

3. *Product:* $(fg)(x) = f(x) \cdot g(x)$

4. *Quotient:* $\left(\dfrac{f}{g}\right)(x) = \dfrac{f(x)}{g(x)}, \quad g(x) \neq 0$

Example 1   Finding the Sum of Two Functions

Given $f(x) = 2x + 1$ and $g(x) = x^2 + 2x - 1$, find $(f + g)(x)$. Then evaluate the sum when $x = 3$.

Solution

$$(f + g)(x) = f(x) + g(x) = (2x + 1) + (x^2 + 2x - 1) = x^2 + 4x$$

When $x = 3$, the value of this sum is

$$(f + g)(3) = 3^2 + 4(3) = 21.$$

CHECK*Point*   Now try Exercise 9(a).

| **Example 2** | **Finding the Difference of Two Functions** |

Given $f(x) = 2x + 1$ and $g(x) = x^2 + 2x - 1$, find $(f - g)(x)$. Then evaluate the difference when $x = 2$.

Solution

The difference of f and g is

$$(f - g)(x) = f(x) - g(x) = (2x + 1) - (x^2 + 2x - 1) = -x^2 + 2.$$

When $x = 2$, the value of this difference is

$$(f - g)(2) = -(2)^2 + 2 = -2.$$

CHECK *Point*   Now try Exercise 9(b).

| **Example 3** | **Finding the Product of Two Functions** |

Given $f(x) = x^2$ and $g(x) = x - 3$, find $(fg)(x)$. Then evaluate the product when $x = 4$.

Solution

$$(fg)(x) = f(x)g(x) = (x^2)(x - 3) = x^3 - 3x^2$$

When $x = 4$, the value of this product is

$$(fg)(4) = 4^3 - 3(4)^2 = 16.$$

CHECK *Point*   Now try Exercise 9(c).

In Examples 1–3, both f and g have domains that consist of all real numbers. So, the domains of $f + g, f - g$, and fg are also the set of all real numbers. Remember that any restrictions on the domains of f and g must be considered when forming the sum, difference, product, or quotient of f and g.

| **Example 4** | **Finding the Quotients of Two Functions** |

Find $(f/g)(x)$ and $(g/f)(x)$ for the functions given by $f(x) = \sqrt{x}$ and $g(x) = \sqrt{4 - x^2}$. Then find the domains of f/g and g/f.

Solution

The quotient of f and g is

$$\left(\frac{f}{g}\right)(x) = \frac{f(x)}{g(x)} = \frac{\sqrt{x}}{\sqrt{4 - x^2}}$$

and the quotient of g and f is

$$\left(\frac{g}{f}\right)(x) = \frac{g(x)}{f(x)} = \frac{\sqrt{4 - x^2}}{\sqrt{x}}.$$

The domain of f is $[0, \infty)$ and the domain of g is $[-2, 2]$. The intersection of these domains is $[0, 2]$. So, the domains of f/g and g/f are as follows.

Domain of f/g: $[0, 2)$ Domain of g/f: $(0, 2]$

CHECK *Point*   Now try Exercise 9(d).

Study Tip

Note that the domain of f/g includes $x = 0$, but not $x = 2$, because $x = 2$ yields a zero in the denominator, whereas the domain of g/f includes $x = 2$, but not $x = 0$, because $x = 0$ yields a zero in the denominator.

Composition of Functions

Another way of combining two functions is to form the **composition** of one with the other. For instance, if $f(x) = x^2$ and $g(x) = x + 1$, the composition of f with g is

$$f(g(x)) = f(x + 1)$$
$$= (x + 1)^2.$$

This composition is denoted as $f \circ g$ and reads as "f composed with g."

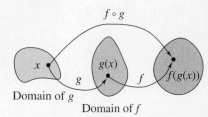

$f \circ g$

Domain of g

Domain of f

FIGURE P.96

Definition of Composition of Two Functions

The **composition** of the function f with the function g is

$$(f \circ g)(x) = f(g(x)).$$

The domain of $f \circ g$ is the set of all x in the domain of g such that $g(x)$ is in the domain of f. (See Figure P.96.)

Example 5   Composition of Functions

Given $f(x) = x + 2$ and $g(x) = 4 - x^2$, find the following.

a. $(f \circ g)(x)$ **b.** $(g \circ f)(x)$ **c.** $(g \circ f)(-2)$

Solution

a. The composition of f with g is as follows.

$$\begin{aligned}(f \circ g)(x) &= f(g(x)) && \text{Definition of } f \circ g\\ &= f(4 - x^2) && \text{Definition of } g(x)\\ &= (4 - x^2) + 2 && \text{Definition of } f(x)\\ &= -x^2 + 6 && \text{Simplify.}\end{aligned}$$

b. The composition of g with f is as follows.

$$\begin{aligned}(g \circ f)(x) &= g(f(x)) && \text{Definition of } g \circ f\\ &= g(x + 2) && \text{Definition of } f(x)\\ &= 4 - (x + 2)^2 && \text{Definition of } g(x)\\ &= 4 - (x^2 + 4x + 4) && \text{Expand.}\\ &= -x^2 - 4x && \text{Simplify.}\end{aligned}$$

Note that, in this case, $(f \circ g)(x) \neq (g \circ f)(x)$.

c. Using the result of part (b), you can write the following.

$$\begin{aligned}(g \circ f)(-2) &= -(-2)^2 - 4(-2) && \text{Substitute.}\\ &= -4 + 8 && \text{Simplify.}\\ &= 4 && \text{Simplify.}\end{aligned}$$

CHECK Point   Now try Exercise 37.

Study Tip

The following tables of values help illustrate the composition $(f \circ g)(x)$ given in Example 5.

x	0	1	2	3
$g(x)$	4	3	0	-5

$g(x)$	4	3	0	-5
$f(g(x))$	6	5	2	-3

x	0	1	2	3
$f(g(x))$	6	5	2	-3

Note that the first two tables can be combined (or "composed") to produce the values given in the third table.

Example 6   Finding the Domain of a Composite Function

Find the domain of $(f \circ g)(x)$ for the functions given by

$$f(x) = x^2 - 9 \quad \text{and} \quad g(x) = \sqrt{9 - x^2}.$$

Algebraic Solution

The composition of the functions is as follows.

$$(f \circ g)(x) = f(g(x))$$

$$= f\left(\sqrt{9 - x^2}\right)$$

$$= \left(\sqrt{9 - x^2}\right)^2 - 9$$

$$= 9 - x^2 - 9$$

$$= -x^2$$

From this, it might appear that the domain of the composition is the set of all real numbers. This, however, is not true. Because the domain of f is the set of all real numbers and the domain of g is $[-3, 3]$, the domain of $f \circ g$ is $[-3, 3]$.

CHECK Point   Now try Exercise 41.

Graphical Solution

You can use a graphing utility to graph the composition of the functions $(f \circ g)(x)$ as $y = \left(\sqrt{9 - x^2}\right)^2 - 9$. Enter the functions as follows.

$$y_1 = \sqrt{9 - x^2} \qquad y_2 = y_1^2 - 9$$

Graph y_2, as shown in Figure P.97. Use the *trace* feature to determine that the x-coordinates of points on the graph extend from -3 to 3. So, you can graphically estimate the domain of $f \circ g$ to be $[-3, 3]$.

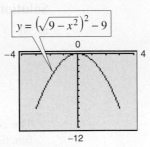

FIGURE **P.97**

In Examples 5 and 6, you formed the composition of two given functions. In calculus, it is also important to be able to identify two functions that make up a given composite function. For instance, the function h given by $h(x) = (3x - 5)^3$ is the composition of f with g, where $f(x) = x^3$ and $g(x) = 3x - 5$. That is,

$$h(x) = (3x - 5)^3 = [g(x)]^3 = f(g(x)).$$

Basically, to "decompose" a composite function, look for an "inner" function and an "outer" function. In the function h above, $g(x) = 3x - 5$ is the inner function and $f(x) = x^3$ is the outer function.

Example 7   Decomposing a Composite Function

Write the function given by $h(x) = \dfrac{1}{(x - 2)^2}$ as a composition of two functions.

Solution

One way to write h as a composition of two functions is to take the inner function to be $g(x) = x - 2$ and the outer function to be

$$f(x) = \frac{1}{x^2} = x^{-2}.$$

Then you can write

$$h(x) = \frac{1}{(x - 2)^2} = (x - 2)^{-2} = f(x - 2) = f(g(x)).$$

CHECK Point   Now try Exercise 53.

Application

Example 8 Bacteria Count

The number N of bacteria in a refrigerated food is given by

$$N(T) = 20T^2 - 80T + 500, \quad 2 \le T \le 14$$

where T is the temperature of the food in degrees Celsius. When the food is removed from refrigeration, the temperature of the food is given by

$$T(t) = 4t + 2, \quad 0 \le t \le 3$$

where t is the time in hours. (a) Find the composition $N(T(t))$ and interpret its meaning in context. (b) Find the time when the bacteria count reaches 2000.

Solution

a. $N(T(t)) = 20(4t + 2)^2 - 80(4t + 2) + 500$

$\qquad = 20(16t^2 + 16t + 4) - 320t - 160 + 500$

$\qquad = 320t^2 + 320t + 80 - 320t - 160 + 500$

$\qquad = 320t^2 + 420$

The composite function $N(T(t))$ represents the number of bacteria in the food as a function of the amount of time the food has been out of refrigeration.

b. The bacteria count will reach 2000 when $320t^2 + 420 = 2000$. Solve this equation to find that the count will reach 2000 when $t \approx 2.2$ hours. When you solve this equation, note that the negative value is rejected because it is not in the domain of the composite function.

CHECK *Point* ▶ Now try Exercise 73.

CLASSROOM DISCUSSION

Analyzing Arithmetic Combinations of Functions

a. Use the graphs of f and $(f + g)$ in Figure P.98 to make a table showing the values of $g(x)$ when $x = 1, 2, 3, 4, 5,$ and 6. Explain your reasoning.

b. Use the graphs of f and $(f - h)$ in Figure P.98 to make a table showing the values of $h(x)$ when $x = 1, 2, 3, 4, 5,$ and 6. Explain your reasoning.

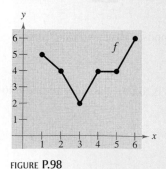
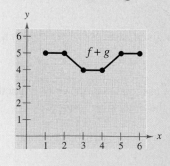
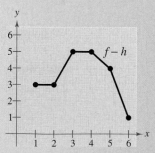

FIGURE P.98

P.9   EXERCISES

See www.CalcChat.com for worked-out solutions to odd-numbered exercises.

VOCABULARY: Fill in the blanks.

1. Two functions f and g can be combined by the arithmetic operations of _____, _____, _____, and _____ to create new functions.

2. The _____ of the function f with g is $(f \circ g)(x) = f(g(x))$.

3. The domain of $(f \circ g)$ is all x in the domain of g such that _____ is in the domain of f.

4. To decompose a composite function, look for an _____ function and an _____ function.

SKILLS AND APPLICATIONS

In Exercises 5–8, use the graphs of f and g to graph $h(x) = (f + g)(x)$. To print an enlarged copy of the graph, go to the website www.mathgraphs.com.

5.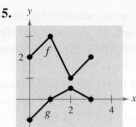

6.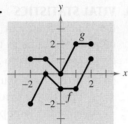

7.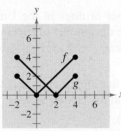

8.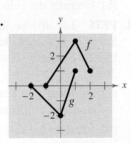

In Exercises 9–16, find (a) $(f + g)(x)$, (b) $(f - g)(x)$, (c) $(fg)(x)$, and (d) $(f/g)(x)$. What is the domain of f/g?

9. $f(x) = x + 2$, $g(x) = x - 2$

10. $f(x) = 2x - 5$, $g(x) = 2 - x$

11. $f(x) = x^2$, $g(x) = 4x - 5$

12. $f(x) = 3x + 1$, $g(x) = 5x - 4$

13. $f(x) = x^2 + 6$, $g(x) = \sqrt{1 - x}$

14. $f(x) = \sqrt{x^2 - 4}$, $g(x) = \dfrac{x^2}{x^2 + 1}$

15. $f(x) = \dfrac{1}{x}$, $g(x) = \dfrac{1}{x^2}$

16. $f(x) = \dfrac{x}{x + 1}$, $g(x) = x^3$

In Exercises 17–28, evaluate the indicated function for $f(x) = x^2 + 1$ and $g(x) = x - 4$.

17. $(f + g)(2)$

18. $(f - g)(-1)$

19. $(f - g)(0)$

20. $(f + g)(1)$

21. $(f - g)(3t)$

22. $(f + g)(t - 2)$

23. $(fg)(6)$

24. $(fg)(-6)$

25. $(f/g)(5)$

26. $(f/g)(0)$

27. $(f/g)(-1) - g(3)$

28. $(fg)(5) + f(4)$

In Exercises 29–32, graph the functions f, g, and $f + g$ on the same set of coordinate axes.

29. $f(x) = \frac{1}{2}x$, $g(x) = x - 1$

30. $f(x) = \frac{1}{3}x$, $g(x) = -x + 4$

31. $f(x) = x^2$, $g(x) = -2x$

32. $f(x) = 4 - x^2$, $g(x) = x$

GRAPHICAL REASONING In Exercises 33–36, use a graphing utility to graph f, g, and $f + g$ in the same viewing window. Which function contributes most to the magnitude of the sum when $0 \le x \le 2$? Which function contributes most to the magnitude of the sum when $x > 6$?

33. $f(x) = 3x$, $g(x) = -\dfrac{x^3}{10}$

34. $f(x) = \dfrac{x}{2}$, $g(x) = \sqrt{x}$

35. $f(x) = 3x + 2$, $g(x) = -\sqrt{x + 5}$

36. $f(x) = x^2 - \frac{1}{2}$, $g(x) = -3x^2 - 1$

In Exercises 37–40, find (a) $f \circ g$, (b) $g \circ f$, and (c) $g \circ g$.

37. $f(x) = x^2$, $g(x) = x - 1$

38. $f(x) = 3x + 5$, $g(x) = 5 - x$

39. $f(x) = \sqrt[3]{x - 1}$, $g(x) = x^3 + 1$

40. $f(x) = x^3$, $g(x) = \dfrac{1}{x}$

In Exercises 41–48, find (a) $f \circ g$ and (b) $g \circ f$. Find the domain of each function and each composite function.

41. $f(x) = \sqrt{x + 4}$, $g(x) = x^2$

42. $f(x) = \sqrt[3]{x - 5}$, $g(x) = x^3 + 1$

43. $f(x) = x^2 + 1$, $g(x) = \sqrt{x}$

44. $f(x) = x^{2/3}$, $g(x) = x^6$

45. $f(x) = |x|$, $g(x) = x + 6$

46. $f(x) = |x - 4|$, $g(x) = 3 - x$

47. $f(x) = \dfrac{1}{x}$, $g(x) = x + 3$

48. $f(x) = \dfrac{3}{x^2 - 1}$, $g(x) = x + 1$

In Exercises 49–52, use the graphs of f and g to evaluate the functions.

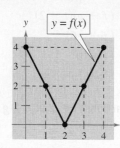
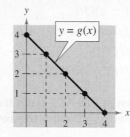

49. (a) $(f + g)(3)$ (b) $(f/g)(2)$

50. (a) $(f - g)(1)$ (b) $(fg)(4)$

51. (a) $(f \circ g)(2)$ (b) $(g \circ f)(2)$

52. (a) $(f \circ g)(1)$ (b) $(g \circ f)(3)$

In Exercises 53–60, find two functions f and g such that $(f \circ g)(x) = h(x)$. (There are many correct answers.)

53. $h(x) = (2x + 1)^2$ **54.** $h(x) = (1 - x)^3$

55. $h(x) = \sqrt[3]{x^2 - 4}$ **56.** $h(x) = \sqrt{9 - x}$

57. $h(x) = \dfrac{1}{x + 2}$ **58.** $h(x) = \dfrac{4}{(5x + 2)^2}$

59. $h(x) = \dfrac{-x^2 + 3}{4 - x^2}$ **60.** $h(x) = \dfrac{27x^3 + 6x}{10 - 27x^3}$

61. STOPPING DISTANCE The research and development department of an automobile manufacturer has determined that when a driver is required to stop quickly to avoid an accident, the distance (in feet) the car travels during the driver's reaction time is given by $R(x) = \frac{3}{4}x$, where x is the speed of the car in miles per hour. The distance (in feet) traveled while the driver is braking is given by $B(x) = \frac{1}{15}x^2$.

(a) Find the function that represents the total stopping distance T.

(b) Graph the functions R, B, and T on the same set of coordinate axes for $0 \le x \le 60$.

(c) Which function contributes most to the magnitude of the sum at higher speeds? Explain.

62. SALES From 2003 through 2008, the sales R_1 (in thousands of dollars) for one of two restaurants owned by the same parent company can be modeled by

$$R_1 = 480 - 8t - 0.8t^2, \quad t = 3, 4, 5, 6, 7, 8$$

where $t = 3$ represents 2003. During the same six-year period, the sales R_2 (in thousands of dollars) for the second restaurant can be modeled by

$$R_2 = 254 + 0.78t, \quad t = 3, 4, 5, 6, 7, 8.$$

(a) Write a function R_3 that represents the total sales of the two restaurants owned by the same parent company.

(b) Use a graphing utility to graph R_1, R_2, and R_3 in the same viewing window.

63. VITAL STATISTICS Let $b(t)$ be the number of births in the United States in year t, and let $d(t)$ represent the number of deaths in the United States in year t, where $t = 0$ corresponds to 2000.

(a) If $p(t)$ is the population of the United States in year t, find the function $c(t)$ that represents the percent change in the population of the United States.

(b) Interpret the value of $c(5)$.

64. PETS Let $d(t)$ be the number of dogs in the United States in year t, and let $c(t)$ be the number of cats in the United States in year t, where $t = 0$ corresponds to 2000.

(a) Find the function $p(t)$ that represents the total number of dogs and cats in the United States.

(b) Interpret the value of $p(5)$.

(c) Let $n(t)$ represent the population of the United States in year t, where $t = 0$ corresponds to 2000. Find and interpret

$$h(t) = \frac{p(t)}{n(t)}.$$

65. MILITARY PERSONNEL The total numbers of Navy personnel N (in thousands) and Marines personnel M (in thousands) from 2000 through 2007 can be approximated by the models

$$N(t) = 0.192t^3 - 3.88t^2 + 12.9t + 372$$

and

$$M(t) = 0.035t^3 - 0.23t^2 + 1.7t + 172$$

where t represents the year, with $t = 0$ corresponding to 2000. (Source: Department of Defense)

(a) Find and interpret $(N + M)(t)$. Evaluate this function for $t = 0$, 6, and 12.

(b) Find and interpret $(N - M)(t)$. Evaluate this function for $t = 0$, 6, and 12.

66. SPORTS The numbers of people playing tennis T (in millions) in the United States from 2000 through 2007 can be approximated by the function

$$T(t) = 0.0233t^4 - 0.3408t^3 + 1.556t^2 - 1.86t + 22.8$$

and the U.S. population P (in millions) from 2000 through 2007 can be approximated by the function $P(t) = 2.78t + 282.5$, where t represents the year, with $t = 0$ corresponding to 2000. (Source: Tennis Industry Association, U.S. Census Bureau)

(a) Find and interpret $h(t) = \dfrac{T(t)}{P(t)}$.

(b) Evaluate the function in part (a) for $t = 0$, 3, and 6.

BIRTHS AND DEATHS In Exercises 67 and 68, use the table, which shows the total numbers of births B (in thousands) and deaths D (in thousands) in the United States from 1990 through 2006. (Source: U.S. Census Bureau)

Year, t	Births, B	Deaths, D
1990	4158	2148
1991	4111	2170
1992	4065	2176
1993	4000	2269
1994	3953	2279
1995	3900	2312
1996	3891	2315
1997	3881	2314
1998	3942	2337
1999	3959	2391
2000	4059	2403
2001	4026	2416
2002	4022	2443
2003	4090	2448
2004	4112	2398
2005	4138	2448
2006	4266	2426

The models for these data are

$$B(t) = -0.197t^3 + 8.96t^2 - 90.0t + 4180$$

and

$$D(t) = -1.21t^2 + 38.0t + 2137$$

where t represents the year, with $t = 0$ corresponding to 1990.

67. Find and interpret $(B - D)(t)$.

68. Evaluate $B(t)$, $D(t)$, and $(B - D)(t)$ for the years 2010 and 2012. What does each function value represent?

69. GRAPHICAL REASONING An electronically controlled thermostat in a home is programmed to lower the temperature automatically during the night. The temperature in the house T (in degrees Fahrenheit) is given in terms of t, the time in hours on a 24-hour clock (see figure).

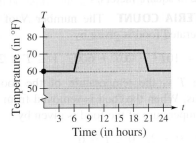

(a) Explain why T is a function of t.

(b) Approximate $T(4)$ and $T(15)$.

(c) The thermostat is reprogrammed to produce a temperature H for which $H(t) = T(t - 1)$. How does this change the temperature?

(d) The thermostat is reprogrammed to produce a temperature H for which $H(t) = T(t) - 1$. How does this change the temperature?

(e) Write a piecewise-defined function that represents the graph.

70. GEOMETRY A square concrete foundation is prepared as a base for a cylindrical tank (see figure).

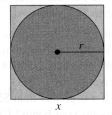

(a) Write the radius r of the tank as a function of the length x of the sides of the square.

(b) Write the area A of the circular base of the tank as a function of the radius r.

(c) Find and interpret $(A \circ r)(x)$.

71. RIPPLES A pebble is dropped into a calm pond, causing ripples in the form of concentric circles. The radius r (in feet) of the outer ripple is $r(t) = 0.6t$, where t is the time in seconds after the pebble strikes the water. The area A of the circle is given by the function $A(r) = \pi r^2$. Find and interpret $(A \circ r)(t)$.

72. POLLUTION The spread of a contaminant is increasing in a circular pattern on the surface of a lake. The radius of the contaminant can be modeled by $r(t) = 5.25\sqrt{t}$, where r is the radius in meters and t is the time in hours since contamination.

Definition of Inverse Function

Let f and g be two functions such that

$$f(g(x)) = x \qquad \text{for every } x \text{ in the domain of } g$$

and

$$g(f(x)) = x \qquad \text{for every } x \text{ in the domain of } f.$$

Under these conditions, the function g is the **inverse function** of the function f. The function g is denoted by f^{-1} (read "f-inverse"). So,

$$f(f^{-1}(x)) = x \qquad \text{and} \qquad f^{-1}(f(x)) = x.$$

The domain of f must be equal to the range of f^{-1}, and the range of f must be equal to the domain of f^{-1}.

Do not be confused by the use of -1 to denote the inverse function f^{-1}. In this text, whenever f^{-1} is written, it *always* refers to the inverse function of the function f and *not* to the reciprocal of $f(x)$.

If the function g is the inverse function of the function f, it must also be true that the function f is the inverse function of the function g. For this reason, you can say that the functions f and g are *inverse functions of each other*.

Example 2   Verifying Inverse Functions

Which of the functions is the inverse function of $f(x) = \dfrac{5}{x-2}$?

$$g(x) = \frac{x-2}{5} \qquad\qquad h(x) = \frac{5}{x} + 2$$

Solution

By forming the composition of f with g, you have

$$f(g(x)) = f\left(\frac{x-2}{5}\right) = \frac{5}{\left(\dfrac{x-2}{5}\right) - 2} = \frac{25}{x-12} \neq x.$$

Because this composition is not equal to the identity function x, it follows that g *is not* the inverse function of f. By forming the composition of f with h, you have

$$f(h(x)) = f\left(\frac{5}{x} + 2\right) = \frac{5}{\left(\dfrac{5}{x} + 2\right) - 2} = \frac{5}{\left(\dfrac{5}{x}\right)} = x.$$

So, it appears that h *is* the inverse function of f. You can confirm this by showing that the composition of h with f is also equal to the identity function, as shown below.

$$h(f(x)) = h\left(\frac{5}{x-2}\right) = \frac{5}{\left(\dfrac{5}{x-2}\right)} + 2 = x - 2 + 2 = x$$

CHECK*Point* ➤ Now try Exercise 19.

The Graph of an Inverse Function

The graphs of a function f and its inverse function f^{-1} are related to each other in the following way. If the point (a, b) lies on the graph of f, then the point (b, a) must lie on the graph of f^{-1}, and vice versa. This means that the graph of f^{-1} is a *reflection* of the graph of f in the line $y = x$, as shown in Figure P.100.

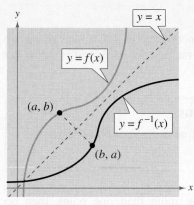
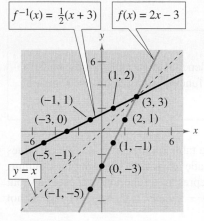

FIGURE **P.100**

FIGURE **P.101**

Example 3   Finding Inverse Functions Graphically

Sketch the graphs of the inverse functions $f(x) = 2x - 3$ and $f^{-1}(x) = \frac{1}{2}(x + 3)$ on the same rectangular coordinate system and show that the graphs are reflections of each other in the line $y = x$.

Solution

The graphs of f and f^{-1} are shown in Figure P.101. It appears that the graphs are reflections of each other in the line $y = x$. You can further verify this reflective property by testing a few points on each graph. Note in the following list that if the point (a, b) is on the graph of f, the point (b, a) is on the graph of f^{-1}.

Graph of $f(x) = 2x - 3$	*Graph of* $f^{-1}(x) = \frac{1}{2}(x + 3)$
$(-1, -5)$	$(-5, -1)$
$(0, -3)$	$(-3, 0)$
$(1, -1)$	$(-1, 1)$
$(2, 1)$	$(1, 2)$
$(3, 3)$	$(3, 3)$

CHECK *Point*   Now try Exercise 25.

Example 4   Finding Inverse Functions Graphically

Sketch the graphs of the inverse functions $f(x) = x^2$ $(x \geq 0)$ and $f^{-1}(x) = \sqrt{x}$ on the same rectangular coordinate system and show that the graphs are reflections of each other in the line $y = x$.

Solution

The graphs of f and f^{-1} are shown in Figure P.102. It appears that the graphs are reflections of each other in the line $y = x$. You can further verify this reflective property by testing a few points on each graph. Note in the following list that if the point (a, b) is on the graph of f, the point (b, a) is on the graph of f^{-1}.

Graph of $f(x) = x^2$, $x \geq 0$	*Graph of* $f^{-1}(x) = \sqrt{x}$
$(0, 0)$	$(0, 0)$
$(1, 1)$	$(1, 1)$
$(2, 4)$	$(4, 2)$
$(3, 9)$	$(9, 3)$

Try showing that $f(f^{-1}(x)) = x$ and $f^{-1}(f(x)) = x$.

CHECK *Point*   Now try Exercise 27.

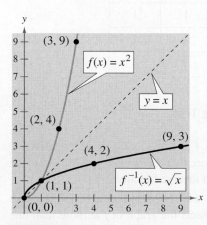

FIGURE **P.102**

One-to-One Functions

The reflective property of the graphs of inverse functions gives you a nice *geometric* test for determining whether a function has an inverse function. This test is called the **Horizontal Line Test** for inverse functions.

Horizontal Line Test for Inverse Functions

A function f has an inverse function if and only if no *horizontal* line intersects the graph of f at more than one point.

If no horizontal line intersects the graph of f at more than one point, then no y-value is matched with more than one x-value. This is the essential characteristic of what are called **one-to-one functions.**

One-to-One Functions

A function f is **one-to-one** if each value of the dependent variable corresponds to exactly one value of the independent variable. A function f has an inverse function if and only if f is one-to-one.

Consider the function given by $f(x) = x^2$. The table on the left is a table of values for $f(x) = x^2$. The table of values on the right is made up by interchanging the columns of the first table. The table on the right does not represent a function because the input $x = 4$ is matched with two different outputs: $y = -2$ and $y = 2$. So, $f(x) = x^2$ is not one-to-one and does not have an inverse function.

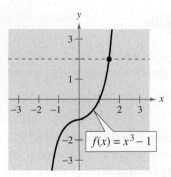

FIGURE P.103

x	$f(x) = x^2$
-2	4
-1	1
0	0
1	1
2	4
3	9

x	y
4	-2
1	-1
0	0
1	1
4	2
9	3

Example 5   Applying the Horizontal Line Test

a. The graph of the function given by $f(x) = x^3 - 1$ is shown in Figure P.103. Because no horizontal line intersects the graph of f at more than one point, you can conclude that f is a one-to-one function and *does* have an inverse function.

b. The graph of the function given by $f(x) = x^2 - 1$ is shown in Figure P.104. Because it is possible to find a horizontal line that intersects the graph of f at more than one point, you can conclude that f *is not* a one-to-one function and *does not* have an inverse function.

FIGURE P.104

CHECK Point   Now try Exercise 39.

Finding Inverse Functions Algebraically

For simple functions (such as the one in Example 1), you can find inverse functions by inspection. For more complicated functions, however, it is best to use the following guidelines. The key step in these guidelines is Step 3—interchanging the roles of x and y. This step corresponds to the fact that inverse functions have ordered pairs with the coordinates reversed.

Note what happens when you try to find the inverse function of a function that is not one-to-one.

$f(x) = x^2 + 1$	Original function
$y = x^2 + 1$	Replace $f(x)$ by y.
$x = y^2 + 1$	Interchange x and y.
$x - 1 = y^2$	Isolate y-term.
$y = \pm\sqrt{x - 1}$	Solve for y.

You obtain two y-values for each x.

Finding an Inverse Function

1. Use the Horizontal Line Test to decide whether f has an inverse function.

2. In the equation for $f(x)$, replace $f(x)$ by y.

3. Interchange the roles of x and y, and solve for y.

4. Replace y by $f^{-1}(x)$ in the new equation.

5. Verify that f and f^{-1} are inverse functions of each other by showing that the domain of f is equal to the range of f^{-1}, the range of f is equal to the domain of f^{-1}, and $f(f^{-1}(x)) = x$ and $f^{-1}(f(x)) = x$.

Example 6   Finding an Inverse Function Algebraically

Find the inverse function of

$$f(x) = \frac{5 - 3x}{2}.$$

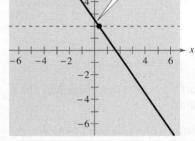

FIGURE **P.105**

Solution

The graph of f is a line, as shown in Figure P.105. This graph passes the Horizontal Line Test. So, you know that f is one-to-one and has an inverse function.

$f(x) = \dfrac{5 - 3x}{2}$	Write original function.
$y = \dfrac{5 - 3x}{2}$	Replace $f(x)$ by y.
$x = \dfrac{5 - 3y}{2}$	Interchange x and y.
$2x = 5 - 3y$	Multiply each side by 2.
$3y = 5 - 2x$	Isolate the y-term.
$y = \dfrac{5 - 2x}{3}$	Solve for y.
$f^{-1}(x) = \dfrac{5 - 2x}{3}$	Replace y by $f^{-1}(x)$.

Note that both f and f^{-1} have domains and ranges that consist of the entire set of real numbers. Check that $f(f^{-1}(x)) = x$ and $f^{-1}(f(x)) = x$.

CHECK*Point* ▶ Now try Exercise 63.

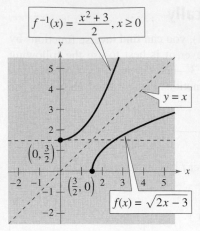

$f^{-1}(x) = \dfrac{x^2 + 3}{2}, \; x \geq 0$

$\left(0, \dfrac{3}{2}\right)$

$y = x$

$\left(\dfrac{3}{2}, 0\right)$

$f(x) = \sqrt{2x - 3}$

FIGURE P.106

Example 7 Finding an Inverse Function

Find the inverse function of

$$f(x) = \sqrt{2x - 3}.$$

Solution

The graph of f is a curve, as shown in Figure P.106. Because this graph passes the Horizontal Line Test, you know that f is one-to-one and has an inverse function.

$f(x) = \sqrt{2x - 3}$	Write original function.
$y = \sqrt{2x - 3}$	Replace $f(x)$ by y.
$x = \sqrt{2y - 3}$	Interchange x and y.
$x^2 = 2y - 3$	Square each side.
$2y = x^2 + 3$	Isolate y.
$y = \dfrac{x^2 + 3}{2}$	Solve for y.
$f^{-1}(x) = \dfrac{x^2 + 3}{2}, \quad x \geq 0$	Replace y by $f^{-1}(x)$.

The graph of f^{-1} in Figure P.106 is the reflection of the graph of f in the line $y = x$. Note that the range of f is the interval $[0, \infty)$, which implies that the domain of f^{-1} is the interval $[0, \infty)$. Moreover, the domain of f is the interval $\left[\frac{3}{2}, \infty\right)$, which implies that the range of f^{-1} is the interval $\left[\frac{3}{2}, \infty\right)$. Verify that $f(f^{-1}(x)) = x$ and $f^{-1}(f(x)) = x$.

CHECK*Point* Now try Exercise 69.

CLASSROOM DISCUSSION

The Existence of an Inverse Function   Write a short paragraph describing why the following functions do or do not have inverse functions.

a. Let x represent the retail price of an item (in dollars), and let $f(x)$ represent the sales tax on the item. Assume that the sales tax is 6% of the retail price *and* that the sales tax is rounded to the nearest cent. Does this function have an inverse function? (*Hint:* Can you undo this function? For instance, if you know that the sales tax is $0.12, can you determine exactly what the retail price is?)

b. Let x represent the temperature in degrees Celsius, and let $f(x)$ represent the temperature in degrees Fahrenheit. Does this function have an inverse function? (*Hint:* The formula for converting from degrees Celsius to degrees Fahrenheit is $F = \frac{9}{5}C + 32$.)

P.10   EXERCISES

See www.CalcChat.com for worked-out solutions to odd-numbered exercises.

VOCABULARY: Fill in the blanks.

1. If the composite functions $f(g(x))$ and $g(f(x))$ both equal x, then the function g is the _____ function of f.

2. The inverse function of f is denoted by _____.

3. The domain of f is the _____ of f^{-1}, and the _____ of f^{-1} is the range of f.

4. The graphs of f and f^{-1} are reflections of each other in the line _____.

5. A function f is _____ if each value of the dependent variable corresponds to exactly one value of the independent variable.

6. A graphical test for the existence of an inverse function of f is called the _____ Line Test.

SKILLS AND APPLICATIONS

In Exercises 7–14, find the inverse function of f informally. Verify that $f(f^{-1}(x)) = x$ and $f^{-1}(f(x)) = x$.

7. $f(x) = 6x$

8. $f(x) = \frac{1}{3}x$

9. $f(x) = x + 9$

10. $f(x) = x - 4$

11. $f(x) = 3x + 1$

12. $f(x) = \dfrac{x - 1}{5}$

13. $f(x) = \sqrt[3]{x}$

14. $f(x) = x^5$

In Exercises 15–18, match the graph of the function with the graph of its inverse function. [The graphs of the inverse functions are labeled (a), (b), (c), and (d).]

(a)
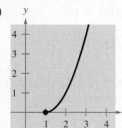

(b)
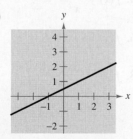

(c)
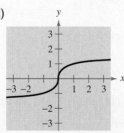

(d)
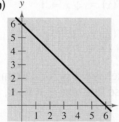

15.
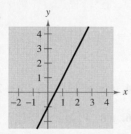

16.
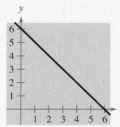

17.
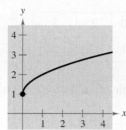

18.
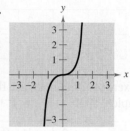

In Exercises 19–22, verify that f and g are inverse functions.

19. $f(x) = -\dfrac{7}{2}x - 3$, $g(x) = -\dfrac{2x + 6}{7}$

20. $f(x) = \dfrac{x - 9}{4}$, $g(x) = 4x + 9$

21. $f(x) = x^3 + 5$, $g(x) = \sqrt[3]{x - 5}$

22. $f(x) = \dfrac{x^3}{2}$, $g(x) = \sqrt[3]{2x}$

In Exercises 23–34, show that f and g are inverse functions (a) algebraically and (b) graphically.

23. $f(x) = 2x$, $g(x) = \dfrac{x}{2}$

24. $f(x) = x - 5$, $g(x) = x + 5$

25. $f(x) = 7x + 1$, $g(x) = \dfrac{x - 1}{7}$

26. $f(x) = 3 - 4x$, $g(x) = \dfrac{3 - x}{4}$

27. $f(x) = \dfrac{x^3}{8}$, $g(x) = \sqrt[3]{8x}$

28. $f(x) = \dfrac{1}{x}$, $g(x) = \dfrac{1}{x}$

29. $f(x) = \sqrt{x - 4}$, $g(x) = x^2 + 4$, $x \geq 0$

30. $f(x) = 1 - x^3$, $g(x) = \sqrt[3]{1 - x}$

31. $f(x) = 9 - x^2$, $x \geq 0$, $g(x) = \sqrt{9 - x}$, $x \leq 9$

32. $f(x) = \dfrac{1}{1 + x}, \quad x \geq 0, \quad g(x) = \dfrac{1 - x}{x}, \quad 0 < x \leq 1$

33. $f(x) = \dfrac{x - 1}{x + 5}, \quad g(x) = -\dfrac{5x + 1}{x - 1}$

34. $f(x) = \dfrac{x + 3}{x - 2}, \quad g(x) = \dfrac{2x + 3}{x - 1}$

In Exercises 35 and 36, does the function have an inverse function?

35.

x	-1	0	1	2	3	4
$f(x)$	-2	1	2	1	-2	-6

36.

x	-3	-2	-1	0	2	3
$f(x)$	10	6	4	1	-3	-10

In Exercises 37 and 38, use the table of values for $y = f(x)$ to complete a table for $y = f^{-1}(x)$.

37.

x	-2	-1	0	1	2	3
$f(x)$	-2	0	2	4	6	8

38.

x	-3	-2	-1	0	1	2
$f(x)$	-10	-7	-4	-1	2	5

In Exercises 39–42, does the function have an inverse function?

39.

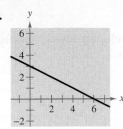

40.

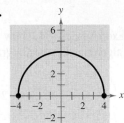

41.

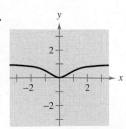

42.

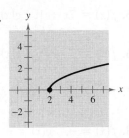

In Exercises 43–48, use a graphing utility to graph the function, and use the Horizontal Line Test to determine whether the function is one-to-one and so has an inverse function.

43. $g(x) = \dfrac{4 - x}{6}$

44. $f(x) = 10$

45. $h(x) = |x + 4| - |x - 4|$

46. $g(x) = (x + 5)^3$

47. $f(x) = -2x\sqrt{16 - x^2}$

48. $f(x) = \frac{1}{8}(x + 2)^2 - 1$

In Exercises 49–62, (a) find the inverse function of f, (b) graph both f and f^{-1} on the same set of coordinate axes, (c) describe the relationship between the graphs of f and f^{-1}, and (d) state the domain and range of f and f^{-1}.

49. $f(x) = 2x - 3$ **50.** $f(x) = 3x + 1$

51. $f(x) = x^5 - 2$ **52.** $f(x) = x^3 + 1$

53. $f(x) = \sqrt{4 - x^2}, \quad 0 \leq x \leq 2$

54. $f(x) = x^2 - 2, \quad x \leq 0$

55. $f(x) = \dfrac{4}{x}$ **56.** $f(x) = -\dfrac{2}{x}$

57. $f(x) = \dfrac{x + 1}{x - 2}$ **58.** $f(x) = \dfrac{x - 3}{x + 2}$

59. $f(x) = \sqrt[3]{x - 1}$ **60.** $f(x) = x^{3/5}$

61. $f(x) = \dfrac{6x + 4}{4x + 5}$ **62.** $f(x) = \dfrac{8x - 4}{2x + 6}$

In Exercises 63–76, determine whether the function has an inverse function. If it does, find the inverse function.

63. $f(x) = x^4$ **64.** $f(x) = \dfrac{1}{x^2}$

65. $g(x) = \dfrac{x}{8}$ **66.** $f(x) = 3x + 5$

67. $p(x) = -4$ **68.** $f(x) = \dfrac{3x + 4}{5}$

69. $f(x) = (x + 3)^2, \quad x \geq -3$

70. $q(x) = (x - 5)^2$

71. $f(x) = \begin{cases} x + 3, & x < 0 \\ 6 - x, & x \geq 0 \end{cases}$

72. $f(x) = \begin{cases} -x, & x \leq 0 \\ x^2 - 3x, & x > 0 \end{cases}$

73. $h(x) = -\dfrac{4}{x^2}$ **74.** $f(x) = |x - 2|, \quad x \leq 2$

75. $f(x) = \sqrt{2x + 3}$ **76.** $f(x) = \sqrt{x - 2}$

THINK ABOUT IT   In Exercises 77–86, restrict the domain of the function f so that the function is one-to-one and has an inverse function. Then find the inverse function f^{-1}. State the domains and ranges of f and f^{-1}. Explain your results. (There are many correct answers.)

77. $f(x) = (x - 2)^2$
78. $f(x) = 1 - x^4$
79. $f(x) = |x + 2|$
80. $f(x) = |x - 5|$
81. $f(x) = (x + 6)^2$
82. $f(x) = (x - 4)^2$
83. $f(x) = -2x^2 + 5$
84. $f(x) = \frac{1}{2}x^2 - 1$
85. $f(x) = |x - 4| + 1$
86. $f(x) = -|x - 1| - 2$

In Exercises 87–92, use the functions given by $f(x) = \frac{1}{8}x - 3$ and $g(x) = x^3$ to find the indicated value or function.

87. $(f^{-1} \circ g^{-1})(1)$
88. $(g^{-1} \circ f^{-1})(-3)$
89. $(f^{-1} \circ f^{-1})(6)$
90. $(g^{-1} \circ g^{-1})(-4)$
91. $(f \circ g)^{-1}$
92. $g^{-1} \circ f^{-1}$

In Exercises 93–96, use the functions given by $f(x) = x + 4$ and $g(x) = 2x - 5$ to find the specified function.

93. $g^{-1} \circ f^{-1}$
94. $f^{-1} \circ g^{-1}$
95. $(f \circ g)^{-1}$
96. $(g \circ f)^{-1}$

97. SHOE SIZES   The table shows men's shoe sizes in the United States and the corresponding European shoe sizes. Let $y = f(x)$ represent the function that gives the men's European shoe size in terms of x, the men's U.S. size.

Men's U.S. shoe size	Men's European shoe size
8	41
9	42
10	43
11	45
12	46
13	47

(a) Is f one-to-one? Explain.
(b) Find $f(11)$.
(c) Find $f^{-1}(43)$, if possible.
(d) Find $f(f^{-1}(41))$.
(e) Find $f^{-1}(f(13))$.

98. SHOE SIZES   The table shows women's shoe sizes in the United States and the corresponding European shoe sizes. Let $y = g(x)$ represent the function that gives the women's European shoe size in terms of x, the women's U.S. size.

Women's U.S. shoe size	Women's European shoe size
4	35
5	37
6	38
7	39
8	40
9	42

(a) Is g one-to-one? Explain.
(b) Find $g(6)$.
(c) Find $g^{-1}(42)$.
(d) Find $g(g^{-1}(39))$.
(e) Find $g^{-1}(g(5))$.

99. LCD TVS   The sales S (in millions of dollars) of LCD televisions in the United States from 2001 through 2007 are shown in the table. The time (in years) is given by t, with $t = 1$ corresponding to 2001. (Source: Consumer Electronics Association)

Year, t	Sales, $S(t)$
1	62
2	246
3	664
4	1579
5	3258
6	8430
7	14,532

(a) Does S^{-1} exist?
(b) If S^{-1} exists, what does it represent in the context of the problem?
(c) If S^{-1} exists, find $S^{-1}(8430)$.
(d) If the table was extended to 2009 and if the sales of LCD televisions for that year was $14,532 million, would S^{-1} exist? Explain.

100. POPULATION The projected populations P (in millions of people) in the United States for 2015 through 2040 are shown in the table. The time (in years) is given by t, with $t = 15$ corresponding to 2015. (Source: U.S. Census Bureau)

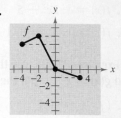

Year, t	Population, $P(t)$
15	325.5
20	341.4
25	357.5
30	373.5
35	389.5
40	405.7

(a) Does P^{-1} exist?

(b) If P^{-1} exists, what does it represent in the context of the problem?

(c) If P^{-1} exists, find $P^{-1}(357.5)$.

(d) If the table was extended to 2050 and if the projected population of the U.S. for that year was 373.5 million, would P^{-1} exist? Explain.

101. HOURLY WAGE Your wage is $10.00 per hour plus $0.75 for each unit produced per hour. So, your hourly wage y in terms of the number of units produced x is $y = 10 + 0.75x$.

(a) Find the inverse function. What does each variable represent in the inverse function?

(b) Determine the number of units produced when your hourly wage is $24.25.

102. DIESEL MECHANICS The function given by

$$y = 0.03x^2 + 245.50, \quad 0 < x < 100$$

approximates the exhaust temperature y in degrees Fahrenheit, where x is the percent load for a diesel engine.

(a) Find the inverse function. What does each variable represent in the inverse function?

(b) Use a graphing utility to graph the inverse function.

(c) The exhaust temperature of the engine must not exceed 500 degrees Fahrenheit. What is the percent load interval?

EXPLORATION

TRUE OR FALSE? In Exercises 103 and 104, determine whether the statement is true or false. Justify your answer.

103. If f is an even function, then f^{-1} exists.

104. If the inverse function of f exists and the graph of f has a y-intercept, then the y-intercept of f is an x-intercept of f^{-1}.

105. PROOF Prove that if f and g are one-to-one functions, then $(f \circ g)^{-1}(x) = (g^{-1} \circ f^{-1})(x)$.

106. PROOF Prove that if f is a one-to-one odd function, then f^{-1} is an odd function.

In Exercises 107 and 108, use the graph of the function f to create a table of values for the given points. Then create a second table that can be used to find f^{-1}, and sketch the graph of f^{-1} if possible.

107.

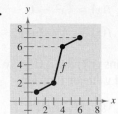

108.

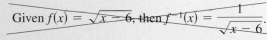

In Exercises 109–112, determine if the situation could be represented by a one-to-one function. If so, write a statement that describes the inverse function.

109. The number of miles n a marathon runner has completed in terms of the time t in hours

110. The population p of South Carolina in terms of the year t from 1960 through 2008

111. The depth of the tide d at a beach in terms of the time t over a 24-hour period

112. The height h in inches of a human born in the year 2000 in terms of his or her age n in years.

113. THINK ABOUT IT The function given by $f(x) = k(2 - x - x^3)$ has an inverse function, and $f^{-1}(3) = -2$. Find k.

114. THINK ABOUT IT Consider the functions given by $f(x) = x + 2$ and $f^{-1}(x) = x - 2$. Evaluate $f(f^{-1}(x))$ and $f^{-1}(f(x))$ for the indicated values of x. What can you conclude about the functions?

x	-10	0	7	45
$f(f^{-1}(x))$				
$f^{-1}(f(x))$				

115. THINK ABOUT IT Restrict the domain of $f(x) = x^2 + 1$ to $x \geq 0$. Use a graphing utility to graph the function. Does the restricted function have an inverse function? Explain.

116. CAPSTONE Describe and correct the error.

~~Given $f(x) = \sqrt{x - 6}$, then $f^{-1}(x) = \dfrac{1}{\sqrt{x - 6}}$.~~

P CHAPTER SUMMARY

	What Did You Learn?	Explanation/Examples	Review Exercises
Section P.1	Represent and classify real numbers *(p. 2)*.	Real numbers include both rational and irrational numbers. Real numbers are represented graphically by a real number line.	1, 2
	Order real numbers and use inequalities *(p. 4)*.	$a < b$: a is less than b. $\quad$ $a > b$: a is greater than b. $a \le b$: a is less than or equal to b. $a \ge b$: a is greater than or equal to b.	3–6
	Find the absolute values of real numbers and find the distance between two real numbers *(p. 6)*.	**Absolute value of a:** $\|a\| = \begin{cases} a, & \text{if } a \ge 0 \\ -a, & \text{if } a < 0 \end{cases}$ **Distance between a and b:** $d(a, b) = \|b - a\| = \|a - b\|$	7–12
	Evaluate algebraic expressions *(p. 8)*.	To evaluate an algebraic expression, substitute numerical values for each of the variables in the expression.	13–16
	Use the basic rules and properties of algebra *(p. 9)*.	The basic rules of algebra, the properties of negation and equality, the properties of zero, and the properties and operations of fractions can be used to perform operations.	17–28
Section P.2	Identify different types of equations *(p. 15)*.	**Identity:** true for *every* real number in the domain. **Conditional equation:** true for just some (or even none) of the real numbers in the domain.	29, 30
	Solve linear equations in one variable and equations that lead to linear equations *(p. 15)*.	**Linear equation in one variable:** An equation that can be written in the standard form $ax + b = 0$, where a and b are real numbers with $a \ne 0$.	31–38
	Solve quadratic equations *(p. 18)*, polynomial equations of degree three or greater *(p. 22)*, equations involving radicals *(p. 23)*, and equations with absolute values *(p. 24)*.	Four methods for solving quadratic equations are factoring, extracting square roots, completing the square, and the Quadratic Formula. These methods can sometimes be extended to solve polynomial equations of higher degree. When solving equations involving radicals and absolute values, be sure to check for extraneous solutions.	39–62
Section P.3	Plot points in the Cartesian plane *(p. 29)*.	For an ordered pair (x, y), the x-coordinate is the directed distance from the y-axis to the point, and the y-coordinate is the directed distance from the x-axis to the point.	63–66
	Use the Distance Formula *(p. 31)* and the Midpoint Formula *(p. 32)*.	**Distance Formula:** $d = \sqrt{(x_2 - x_1)^2 + (y_2 - y_1)^2}$ **Midpoint Formula:** $\left(\dfrac{x_1 + x_2}{2}, \dfrac{y_1 + y_2}{2} \right)$	67–70
	Use a coordinate plane to model and solve real-life problems *(p. 33)*.	The coordinate plane can be used to find the length of a football pass (See Example 6).	71, 72
	Sketch graphs of equations *(p. 34)*, and find x- and y-intercepts of graphs of equations *(p. 35)*.	To graph an equation, make a table of values, plot the points, and connect the points with a smooth curve or line. To find x-intercepts, set y equal to zero and solve for x. To find y-intercepts, set x equal to zero and solve for y.	73–78
	Use symmetry to sketch graphs of equations *(p. 36)*.	Graphs can have symmetry with respect to one of the coordinate axes or with respect to the origin. You can test for symmetry algebraically and graphically.	79–86

What Did You Learn?	**Explanation/Examples**	**Review Exercises**
Section P.3 Find equations of and sketch graphs of circles *(p. 38)*.	The point (x, y) lies on the circle of radius r and center (h, k) if and only if $(x - h)^2 + (y - k)^2 = r^2$.	87–92
Section P.4 Use slope to graph linear equations in two variables *(p. 43)*.	**The Slope-Intercept Form of the Equation of a Line** The graph of the equation $y = mx + b$ is a line whose slope is m and whose y-intercept is $(0, b)$.	93–100
Find the slope of a line given two points on the line *(p. 45)*.	The slope m of the nonvertical line through (x_1, y_1) and (x_2, y_2) is $m = (y_2 - y_1)/(x_2 - x_1)$, where $x_1 \neq x_2$.	101–104
Write linear equations in two variables *(p. 47)*.	**Point-Slope Form of the Equation of a Line** The equation of the line with slope m passing through the point (x_1, y_1) is $y - y_1 = m(x - x_1)$.	105–112
Use slope to identify parallel and perpendicular lines *(p. 48)*.	**Parallel lines:** Slopes are equal. **Perpendicular lines:** Slopes are negative reciprocals of each other.	113, 114
Use slope and linear equations in two variables to model and solve real-life problems *(p. 49)*.	A linear equation in two variables can be used to describe the book value of exercise equipment in a given year. (See Example 7.)	115, 116
Section P.5 Determine whether relations between two variables are functions *(p. 58)*.	A function f from a set A (domain) to a set B (range) is a relation that assigns to each element x in the set A exactly one element y in the set B.	117–122
Use function notation, evaluate functions *(p. 60)*, and find domains *(p. 63)*.	**Equation:** $f(x) = 5 - x^2$ $f(2)$: $f(2) = 5 - 2^2 = 1$ **Domain of $f(x) = 5 - x^2$:** All real numbers	123–130
Use functions to model and solve real-life problems *(p. 64)*.	A function can be used to model the number of alternative-fueled vehicles in the United States. (See Example 10.)	131, 132
Evaluate difference quotients *(p. 65)*.	**Difference quotient:** $[f(x + h) - f(x)]/h, h \neq 0$	133, 134
Section P.6 Use the Vertical Line Test for functions *(p. 74)*.	A graph represents a function if and only if no *vertical* line intersects the graph at more than one point.	135–138
Find the zeros of functions *(p. 75)*.	**Zeros of $f(x)$:** x-values for which $f(x) = 0$	139–142
Determine intervals on which functions are increasing or decreasing *(p. 76)*, find relative minimum and maximum values *(p. 77)*, and find the average rate of change of a function *(p. 78)*.	To determine whether a function is increasing, decreasing, or constant on an interval, evaluate the function for several values of x. The points at which the behavior of a function changes can help determine the relative minimum or relative maximum. The average rate of change between any two points is the slope of the line (secant line) through the two points.	143–152
Identify even and odd functions *(p. 79)*.	**Even:** For each x in the domain of f, $f(-x) = f(x)$. **Odd:** For each x in the domain of f, $f(-x) = -f(x)$.	153–156
Section P.7 Identify and graph linear *(p. 85)* and squaring functions *(p. 86)*.	**Linear:** $f(x) = ax + b$ **Squaring:** $f(x) = x^2$	157–160

What Did You Learn?	Explanation/Examples	Review Exercises
Section P.7 Identify and graph cubic, square root, reciprocal *(p. 87)*, step, and other piecewise-defined functions *(p. 88)*.	**Cubic:** $f(x) = x^3$ **Square Root:** $f(x) = \sqrt{x}$ **Reciprocal:** $f(x) = 1/x$ **Step:** $f(x) = [\![x]\!]$	161–170
Recognize graphs of parent functions *(p. 89)*.	Eight of the most commonly used functions in algebra are shown in Figure P.81.	171, 172
Section P.8 Use vertical and horizontal shifts *(p. 92)*, reflections *(p. 94)*, and nonrigid transformations *(p. 96)* to sketch graphs of functions.	**Vertical shifts:** $h(x) = f(x) + c$ or $h(x) = f(x) - c$ **Horizontal shifts:** $h(x) = f(x - c)$ or $h(x) = f(x + c)$ **Reflection in *x*-axis:** $h(x) = -f(x)$ **Reflection in *y*-axis:** $h(x) = f(-x)$ **Nonrigid transformations:** $h(x) = cf(x)$ or $h(x) = f(cx)$	173–186
Section P.9 Add, subtract, multiply, and divide functions *(p. 102)*.	$(f + g)(x) = f(x) + g(x)$ $(f - g)(x) = f(x) - g(x)$ $(fg)(x) = f(x) \cdot g(x)$ $(f/g)(x) = f(x)/g(x),\ g(x) \neq 0$	187, 188
Find the composition of one function with another function *(p. 104)*.	The composition of the function f with the function g is $(f \circ g)(x) = f(g(x))$.	189–192
Use combinations and compositions of functions to model and solve real-life problems *(p. 106)*.	A composite function can be used to represent the number of bacteria in food as a function of the amount of time the food has been out of refrigeration. (See Example 8.)	193, 194
Section P.10 Find inverse functions informally and verify that two functions are inverse functions of each other *(p. 111)*.	Let f and g be two functions such that $f(g(x)) = x$ for every x in the domain of g and $g(f(x)) = x$ for every x in the domain of f. Under these conditions, the function g is the inverse function of the function f.	195, 196
Use graphs of functions to determine whether functions have inverse functions *(p. 113)*.	If the point (a, b) lies on the graph of f, then the point (b, a) must lie on the graph of f^{-1}, and vice versa. In short, f^{-1} is a reflection of f in the line $y = x$.	197, 198
Use the Horizontal Line Test to determine if functions are one-to-one *(p. 114)*.	**Horizontal Line Test for Inverse Functions** A function f has an inverse function if and only if no *horizontal* line intersects f at more than one point.	199–202
Find inverse functions algebraically *(p. 115)*.	To find inverse functions, replace $f(x)$ by y, interchange the roles of x and y, and solve for y. Replace y by $f^{-1}(x)$.	203–208

P REVIEW EXERCISES

See www.CalcChat.com for worked-out solutions to odd-numbered exercises.

P.1 In Exercises 1 and 2, determine which numbers in the set are (a) natural numbers, (b) whole numbers, (c) integers, (d) rational numbers, and (e) irrational numbers.

1. $\left\{11, -14, -\frac{8}{9}, \frac{5}{2}, \sqrt{6}, 0.4\right\}$

2. $\left\{\sqrt{15}, -22, -\frac{10}{3}, 0, 5.2, \frac{3}{7}\right\}$

In Exercises 3 and 4, use a calculator to find the decimal form of each rational number. If it is a nonterminating decimal, write the repeating pattern. Then plot the numbers on the real number line and place the appropriate inequality sign (< or >) between them.

3. (a) $\frac{5}{6}$ (b) $\frac{7}{8}$ **4.** (a) $\frac{9}{25}$ (b) $\frac{5}{7}$

In Exercises 5 and 6, give a verbal description of the subset of real numbers represented by the inequality, and sketch the subset on the real number line.

5. $x \le 7$ **6.** $x > 1$

In Exercises 7 and 8, find the distance between a and b.

7. $a = -74$, $b = 48$ **8.** $a = -112$, $b = -6$

In Exercises 9–12, use absolute value notation to describe the situation.

9. The distance between x and 7 is at least 4.

10. The distance between x and 25 is no more than 10.

11. The distance between y and -30 is less than 5.

12. The distance between z and -16 is greater than 8.

In Exercises 13–16, evaluate the expression for each value of x.

	Expression	Values
13.	$12x - 7$	(a) $x = 0$ (b) $x = -1$
14.	$x^2 - 6x + 5$	(a) $x = -2$ (b) $x = 2$
15.	$-x^2 + x - 1$	(a) $x = 1$ (b) $x = -1$
16.	$\dfrac{x}{x - 3}$	(a) $x = -3$ (b) $x = 2$

In Exercises 17–22, identify the rule of algebra illustrated by the statement.

17. $2x + (3x - 10) = (2x + 3x) - 10$

18. $4(t + 2) = 4 \cdot t + 4 \cdot 2$

19. $0 + (a - 5) = a - 5$

20. $\dfrac{2}{y + 4} \cdot \dfrac{y + 4}{2} = 1$, $y \ne -4$

21. $(t^2 + 1) + 3 = 3 + (t^2 + 1)$

22. $1 \cdot (3x + 4) = (3x + 4)$

In Exercises 23–28, perform the operation(s). (Write fractional answers in simplest form.)

23. $|-3| + 4(-2) - 6$ **24.** $\dfrac{|-10|}{-10}$

25. $\frac{5}{18} \div \frac{10}{3}$ **26.** $(16 - 8) \div 4$

27. $6[4 - 2(6 + 8)]$ **28.** $-4[16 - 3(7 - 10)]$

P.2 In Exercises 29 and 30, determine whether the equation is an identity or a conditional equation.

29. $2(x - 1) = 2x - 2$ **30.** $3(x + 2) = 5x + 4$

In Exercises 31–38, solve the equation and check your solution. (If not possible, explain why.)

31. $3x - 2(x + 5) = 10$ **32.** $4x + 2(7 - x) = 5$

33. $4(x + 3) - 3 = 2(4 - 3x) - 4$

34. $\frac{1}{2}(x - 3) - 2(x + 1) = 5$

35. $\dfrac{x}{5} - 3 = \dfrac{x}{3} + 1$ **36.** $\dfrac{4x - 3}{6} + \dfrac{x}{4} = x - 2$

37. $\dfrac{18}{x} = \dfrac{10}{x - 4}$ **38.** $\dfrac{5}{x - 2} = \dfrac{13}{2x - 3}$

In Exercises 39–48, use any method to solve the quadratic equation.

39. $2x^2 + 5x + 3 = 0$ **40.** $3x^2 + 7x + 4 = 0$

41. $6 = 3x^2$ **42.** $16x^2 = 25$

43. $(x + 4)^2 = 18$ **44.** $(x - 8)^2 = 15$

45. $x^2 - 12x + 30 = 0$ **46.** $x^2 + 6x - 3 = 0$

47. $-2x^2 - 5x + 27 = 0$ **48.** $-20 - 3x + 3x^2 = 0$

In Exercises 49–62, find all solutions of the equation. Check your solutions in the original equation.

49. $5x^4 - 12x^3 = 0$ **50.** $4x^3 - 6x^2 = 0$

51. $x^4 - 5x^2 + 6 = 0$

52. $9x^4 + 27x^3 - 4x^2 - 12x = 0$

53. $\sqrt{x + 4} = 3$ **54.** $\sqrt{x - 2} - 8 = 0$

55. $\sqrt{2x + 3} + \sqrt{x - 2} = 2$

56. $5\sqrt{x} - \sqrt{x - 1} = 6$

57. $(x - 1)^{2/3} - 25 = 0$

58. $(x + 2)^{3/4} = 27$

59. $|x - 5| = 10$

60. $|2x + 3| = 7$

61. $|x^2 - 3| = 2x$

62. $|x^2 - 6| = x$

P.3 In Exercises 63 and 64, plot the points in the Cartesian plane.

63. $(2, 2), (0, -4), (-3, 6), (-1, -7)$

64. $(5, 0), (8, 1), (4, -2), (-3, -3)$

In Exercises 65 and 66, determine the quadrant(s) in which (x, y) is located so that the condition(s) is (are) satisfied.

65. $x > 0$ and $y = -2$ **66.** $y > 0$

In Exercises 67–70, (a) plot the points, (b) find the distance between the points, and (c) find the midpoint of the line segment joining the points.

67. $(5, 1), (1, 4)$ **68.** $(6, -2), (5, 3)$

69. $(5.6, 0), (0, 8.2)$ **70.** $(0, -1.2), (-3.6, 0)$

71. SALES The Cheesecake Factory had annual sales of $1315.3 million in 2006 and $1606.4 million in 2008. Use the Midpoint Formula to estimate the sales in 2007. (Source: The Cheesecake Factory, Inc.)

72. METEOROLOGY The apparent temperature is a measure of relative discomfort to a person from heat and high humidity. The table shows the actual temperatures x (in degrees Fahrenheit) versus the apparent temperatures y (in degrees Fahrenheit) for a relative humidity of 75%.

x	70	75	80	85	90	95	100
y	70	77	85	95	109	130	150

(a) Sketch a scatter plot of the data shown in the table.

(b) Find the change in the apparent temperature when the actual temperature changes from 70°F to 100°F.

In Exercises 73–76, complete a table of values. Use the solution points to sketch the graph of the equation.

73. $y = 2x - 6$ **74.** $y = -\frac{1}{2}x + 2$

75. $y = x^2 - 3x$ **76.** $y = 2x^2 - x - 9$

In Exercises 77 and 78, find the x- and y-intercepts of the graph of the equation.

77. $y = (x - 3)^2 - 4$ **78.** $y = |x + 1| - 3$

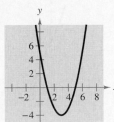

In Exercises 79–86, use the algebraic tests to check for symmetry with respect to both axes and the origin. Then sketch the graph of the equation.

79. $y = -4x + 1$ **80.** $y = 5x - 6$

81. $y = 5 - x^2$ **82.** $y = x^2 - 10$

83. $y = x^3 + 3$ **84.** $y = -6 - x^3$

85. $y = \sqrt{x + 5}$ **86.** $y = |x| + 9$

In Exercises 87–90, find the center and radius of the circle and sketch its graph.

87. $x^2 + y^2 = 9$ **88.** $x^2 + y^2 = 4$

89. $\left(x - \frac{1}{2}\right)^2 + (y + 1)^2 = 36$

90. $(x + 4)^2 + \left(y - \frac{3}{2}\right)^2 = 100$

91. Find the standard form of the equation of the circle for which the endpoints of a diameter are $(0, 0)$ and $(4, -6)$.

92. Find the standard form of the equation of the circle for which the endpoints of a diameter are $(-2, -3)$ and $(4, -10)$.

P.4 In Exercises 93–100, find the slope and y-intercept (if possible) of the equation of the line. Sketch the line.

93. $y = -2x - 7$ **94.** $y = 4x - 3$

95. $y = 6$ **96.** $x = -3$

97. $y = 3x + 13$ **98.** $y = -10x + 9$

99. $y = -\frac{5}{2}x - 1$ **100.** $y = \frac{5}{6}x + 5$

In Exercises 101–104, plot the points and find the slope of the line passing through the pair of points.

101. $(6, 4), (-3, -4)$ **102.** $\left(\frac{3}{2}, 1\right), \left(5, \frac{5}{2}\right)$

103. $(-4.5, 6), (2.1, 3)$ **104.** $(-3, 2), (8, 2)$

In Exercises 105–108, find the slope-intercept form of the equation of the line that passes through the given point and has the indicated slope. Sketch the line.

	Point	Slope
105.	$(3, 0)$	$m = \frac{2}{3}$
106.	$(-8, 5)$	$m = 0$
107.	$(10, -3)$	$m = -\frac{1}{2}$
108.	$(12, -6)$	m is undefined.

In Exercises 109–112, find the slope-intercept form of the equation of the line passing through the points.

109. $(0, 0), (0, 10)$ **110.** $(2, -1), (4, -1)$

111. $(-1, 0), (6, 2)$ **112.** $(11, -2), (6, -1)$

In Exercises 113 and 114, write the slope-intercept forms of the equations of the lines through the given point (a) parallel to the given line and (b) perpendicular to the given line.

Point	Line
113. $(3, -2)$	$5x - 4y = 8$
114. $(-8, 3)$	$2x + 3y = 5$

115. SALES During the second and third quarters of the year, a salvage yard had sales of $160,000 and $185,000, respectively. The growth of sales follows a linear pattern. Estimate sales during the fourth quarter.

116. INFLATION The dollar value of a product in 2010 is $85, and the product is expected to increase in value at a rate of $3.75 per year.

(a) Write a linear equation that gives the dollar value V of the product in terms of the year t. (Let $t = 10$ represent 2010.)

(b) Use a graphing utility to graph the equation found in part (a).

(c) Move the cursor along the graph of the sales model to estimate the dollar value of the product in 2015.

P.5 In Exercises 117 and 118, which sets of ordered pairs represent functions from A to B? Explain.

117. $A = \{10, 20, 30, 40\}$ and $B = \{0, 2, 4, 6\}$

(a) $\{(20, 4), (40, 0), (20, 6), (30, 2)\}$

(b) $\{(10, 4), (20, 4), (30, 4), (40, 4)\}$

(c) $\{(40, 0), (30, 2), (20, 4), (10, 6)\}$

(d) $\{(20, 2), (10, 0), (40, 4)\}$

118. $A = \{u, v, w\}$ and $B = \{-2, -1, 0, 1, 2\}$

(a) $\{(v, -1), (u, 2), (w, 0), (u, -2)\}$

(b) $\{(u, -2), (v, 2), (w, 1)\}$

(c) $\{(u, 2), (v, 2), (w, 1), (w, 1)\}$

(d) $\{(w, -2), (v, 0), (w, 2)\}$

In Exercises 119–122, determine whether the equation represents y as a function of x.

119. $16x - y^4 = 0$ **120.** $2x - y - 3 = 0$

121. $y = \sqrt{1 - x}$ **122.** $|y| = x + 2$

In Exercises 123 and 124, evaluate the function at each specified value of the independent variable and simplify.

123. $h(x) = \begin{cases} 2x + 1, & x \le -1 \\ x^2 + 2, & x > -1 \end{cases}$

(a) $h(-2)$ (b) $h(-1)$ (c) $h(0)$ (d) $h(2)$

124. $f(x) = x^2 + 1$

(a) $f(2)$ (b) $f(-4)$ (c) $f(t^2)$ (d) $f(t + 1)$

In Exercises 125–130, find the domain of the function. Verify your result with a graph.

125. $f(x) = \sqrt{25 - x^2}$

126. $f(x) = 3x + 4$

127. $g(s) = \dfrac{5s + 5}{3s - 9}$

128. $f(x) = \sqrt{x^2 + 8x}$

129. $h(x) = \dfrac{x}{x^2 - x - 6}$

130. $h(t) = |t + 1|$

131. PHYSICS The velocity of a ball projected upward from ground level is given by $v(t) = -32t + 48$, where t is the time in seconds and v is the velocity in feet per second.

(a) Find the velocity when $t = 1$.

(b) Find the time when the ball reaches its maximum height. [Hint: Find the time when $v(t) = 0$.]

(c) Find the velocity when $t = 2$.

132. MIXTURE PROBLEM From a full 50-liter container of a 40% concentration of acid, x liters is removed and replaced with 100% acid.

(a) Write the amount of acid in the final mixture as a function of x.

(b) Determine the domain and range of the function.

(c) Determine x if the final mixture is 50% acid.

In Exercises 133 and 134, find the difference quotient and simplify your answer.

133. $f(x) = 2x^2 + 3x - 1$, $\dfrac{f(x + h) - f(x)}{h}$, $h \ne 0$

134. $f(x) = x^3 - 5x^2 + x$, $\dfrac{f(x + h) - f(x)}{h}$, $h \ne 0$

P.6 In Exercises 135–138, use the Vertical Line Test to determine whether y is a function of x. To print an enlarged copy of the graph, go to the website *www.mathgraphs.com*.

135. $y = (x - 3)^2$ **136.** $y = -\frac{3}{5}x^3 - 2x + 1$

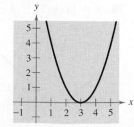
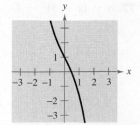

137. $x - 4 = y^2$

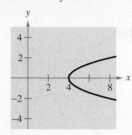

138. $x = -|4 - y|$

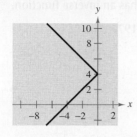

In Exercises 139–142, find the zeros of the function algebraically.

139. $f(x) = 3x^2 - 16x + 21$

140. $f(x) = 5x^2 + 4x - 1$

141. $f(x) = \dfrac{8x + 3}{11 - x}$

142. $f(x) = x^3 - x^2 - 25x + 25$

In Exercises 143 and 144, determine the intervals over which the function is increasing, decreasing, or constant.

143. $f(x) = |x| + |x + 1|$

144. $f(x) = (x^2 - 4)^2$

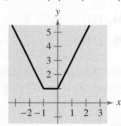
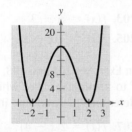

 In Exercises 145–148, use a graphing utility to graph the function and approximate (to two decimal places) any relative minimum or relative maximum values.

145. $f(x) = -x^2 + 2x + 1$

146. $f(x) = x^4 - 4x^2 - 2$

147. $f(x) = x^3 - 6x^4$

148. $f(x) = x^3 - 4x^2 - 1$

In Exercises 149–152, find the average rate of change of the function from x_1 to x_2.

Function	x-Values
149. $f(x) = -x^2 + 6x - 2$	$x_1 = 0, x_2 = 4$
150. $f(x) = x^3 + 12x - 2$	$x_1 = 0, x_2 = 4$
151. $f(x) = 2 - \sqrt{x + 1}$	$x_1 = 3, x_2 = 7$
152. $f(x) = 1 - \sqrt{x + 3}$	$x_1 = 1, x_2 = 6$

In Exercises 153–156, determine whether the function is even, odd, or neither.

153. $f(x) = x^5 + 4x - 7$ **154.** $f(x) = x^4 - 20x^2$

155. $f(x) = 2x\sqrt{x^2 + 3}$ **156.** $f(x) = \sqrt[5]{6x^2}$

P.7 In Exercises 157–160, write the linear function f such that it has the indicated function values. Then sketch the graph of the function.

157. $f(2) = -8, f(-1) = 4$

158. $f(0) = -5, f(4) = -8$

159. $f\left(-\frac{4}{5}\right) = 2, f\left(\frac{11}{5}\right) = 7$

160. $f(3.3) = 5.6, f(-4.7) = -1.4$

In Exercises 161–170, graph the function.

161. $f(x) = 3 - x^2$ **162.** $h(x) = x^3 - 2$

163. $f(x) = -\sqrt{x}$ **164.** $f(x) = \sqrt{x + 1}$

165. $g(x) = \dfrac{3}{x}$ **166.** $g(x) = \dfrac{1}{x + 5}$

167. $f(x) = [\![x]\!] - 2$ **168.** $g(x) = [\![x + 4]\!]$

169. $f(x) = \begin{cases} 5x - 3, & x \geq -1 \\ -4x + 5, & x < -1 \end{cases}$

170. $f(x) = \begin{cases} x^2 - 2, & x < -2 \\ 5, & -2 \leq x \leq 0 \\ 8x - 5, & x > 0 \end{cases}$

In Exercises 171 and 172, the figure shows the graph of a transformed parent function. Identify the parent function.

171.

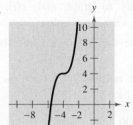

172.

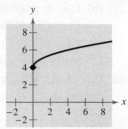

P.8 In Exercises 173–186, h is related to one of the parent functions described in this chapter. (a) Identify the parent function f. (b) Describe the sequence of transformations from f to h. (c) Sketch the graph of h. (d) Use function notation to write h in terms of f.

173. $h(x) = x^2 - 9$ **174.** $h(x) = (x - 2)^3 + 2$

175. $h(x) = -\sqrt{x} + 4$ **176.** $h(x) = |x + 3| - 5$

177. $h(x) = -(x + 2)^2 + 3$

178. $h(x) = \frac{1}{2}(x - 1)^2 - 2$

179. $h(x) = -[\![x]\!] + 6$

180. $h(x) = -\sqrt{x + 1} + 9$

181. $h(x) = -|-x + 4| + 6$

182. $h(x) = -(x + 1)^2 - 3$

183. $h(x) = 5[\![x - 9]\!]$ **184.** $h(x) = -\frac{1}{3}x^3$

185. $h(x) = -2\sqrt{x - 4}$ **186.** $h(x) = \frac{1}{2}|x| - 1$

P.9 In Exercises 187 and 188, find (a) $(f + g)(x)$, (b) $(f - g)(x)$, (c) $(fg)(x)$, and (d) $(f/g)(x)$. What is the domain of f/g?

187. $f(x) = x^2 + 3$, $g(x) = 2x - 1$

188. $f(x) = x^2 - 4$, $g(x) = \sqrt{3 - x}$

In Exercises 189 and 190, find (a) $f \circ g$ and (b) $g \circ f$. Find the domain of each function and each composite function.

189. $f(x) = \frac{1}{3}x - 3$, $g(x) = 3x + 1$

190. $f(x) = x^3 - 4$, $g(x) = \sqrt[3]{x + 7}$

In Exercises 191 and 192, find two functions f and g such that $(f \circ g)(x) = h(x)$. (There are many correct answers.)

191. $h(x) = (1 - 2x)^3$

192. $h(x) = \sqrt[3]{x + 2}$

193. PHONE EXPENDITURES The average annual expenditures (in dollars) for residential $r(t)$ and cellular $c(t)$ phone services from 2001 through 2006 can be approximated by the functions

$$r(t) = 27.5t + 705 \quad \text{and} \quad c(t) = 151.3t + 151$$

where t represents the year, with $t = 1$ corresponding to 2001.   (Source: Bureau of Labor Statistics)

(a) Find and interpret $(r + c)(t)$.

(b) Use a graphing utility to graph $r(t)$, $c(t)$, and $(r + c)(t)$ in the same viewing window.

(c) Find $(r + c)(13)$. Use the graph in part (b) to verify your result.

194. BACTERIA COUNT The number N of bacteria in a refrigerated food is given by

$$N(T) = 25T^2 - 50T + 300, \quad 2 \le T \le 20$$

where T is the temperature of the food in degrees Celsius. When the food is removed from refrigeration, the temperature of the food is given by

$$T(t) = 2t + 1, \quad 0 \le t \le 9$$

where t is the time in hours. (a) Find the composition $N(T(t))$ and interpret its meaning in context and (b) find the time when the bacteria count reaches 750.

P.10 In Exercises 195 and 196, find the inverse function of f informally. Verify that $f(f^{-1}(x)) = x$ and $f^{-1}(f(x)) = x$.

195. $f(x) = 3x + 8$

196. $f(x) = \dfrac{x - 4}{5}$

In Exercises 197 and 198, determine whether the function has an inverse function.

197.

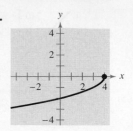

198.

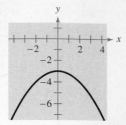

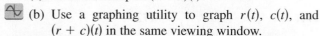 In Exercises 199–202, use a graphing utility to graph the function, and use the Horizontal Line Test to determine whether the function is one-to-one and so has an inverse function.

199. $f(x) = 4 - \frac{1}{3}x$

200. $f(x) = (x - 1)^2$

201. $h(t) = \dfrac{2}{t - 3}$

202. $g(x) = \sqrt{x + 6}$

In Exercises 203–206, (a) find the inverse function of f, (b) graph both f and f^{-1} on the same set of coordinate axes, (c) describe the relationship between the graphs of f and f^{-1}, and (d) state the domain and range of f and f^{-1}.

203. $f(x) = \frac{1}{2}x - 3$

204. $f(x) = 5x - 7$

205. $f(x) = \sqrt{x + 1}$

206. $f(x) = x^3 + 2$

In Exercises 207 and 208, restrict the domain of the function f to an interval over which the function is increasing and determine f^{-1} over that interval.

207. $f(x) = 2(x - 4)^2$

208. $f(x) = |x - 2|$

EXPLORATION

TRUE OR FALSE? In Exercises 209 and 210, determine whether the statement is true or false. Justify your answer.

209. Relative to the graph of $f(x) = \sqrt{x}$, the function given by

$$h(x) = -\sqrt{x + 9} - 13$$

is shifted 9 units to the left and 13 units downward, then reflected in the x-axis.

210. If f and g are two inverse functions, then the domain of g is equal to the range of f.

211. WRITING Explain why it is essential to check your solutions to radical, absolute value, and rational equations.

212. WRITING Explain how to tell whether a relation between two variables is a function.

213. WRITING Explain the difference between the Vertical Line Test and the Horizontal Line Test.

P CHAPTER TEST   See www.CalcChat.com for worked-out solutions to odd-numbered exercises.

Take this test as you would take a test in class. When you are finished, check your work against the answers given in the back of the book.

1. Place < or > between the real numbers $-\frac{10}{3}$ and $-|-4|$.

2. Find the distance between the real numbers -5.4 and $3\frac{3}{4}$.

3. Identify the rule of algebra illustrated by $(5 - x) + 0 = 5 - x$.

In Exercises 4–7, solve the equation (if possible).

4. $\frac{2}{3}(x - 1) + \frac{1}{4}x = 10$ 5. $(x - 3)(x + 2) = 14$

6. $\dfrac{x - 2}{x + 2} + \dfrac{4}{x + 2} + 4 = 0$ 7. $x^4 + x^2 - 6 = 0$

8. Plot the points $(-2, 5)$ and $(6, 0)$. Find the coordinates of the midpoint of the line segment joining the points and the distance between the points.

In Exercises 9–11, check for symmetry with respect to both axes and the origin. Then sketch the graph of the equation. Identify any x- and y-intercepts.

9. $y = 4 - \frac{3}{4}x$ 10. $y = 4 - |x|$ 11. $y = x - x^3$

12. Find the center and radius of the circle given by $(x - 3)^2 + y^2 = 9$. Then sketch its graph.

In Exercises 13 and 14, find the slope-intercept form of the equation of the line passing through the points.

13. $(2, -3), (-4, 9)$ 14. $(3, 0.8), (7, -6)$

15. Find equations of the lines that pass through the point $(0, 4)$ and are (a) parallel to and (b) perpendicular to the line $5x + 2y = 3$.

16. Evaluate the functions given by $f(x) = |x + 2| - 15$ at each specified value of the independent variable and simplify.

   (a) $f(-8)$ (b) $f(14)$ (c) $f(x - 6)$

In Exercises 17–19, (a) use a graphing utility to graph the function, (b) determine the domain of the function, (c) approximate the intervals over which the function is increasing, decreasing, or constant, and (d) determine whether the function is even, odd, or neither.

17. $f(x) = 2x^6 + 5x^4 - x^2$ 18. $f(x) = 4x\sqrt{3 - x}$ 19. $f(x) = |x + 5|$

In Exercises 20–22, (a) identify the parent function in the transformation, (b) describe the sequence of transformations from f to h, and (c) sketch the graph of h.

20. $h(x) = -[\![x]\!]$ 21. $h(x) = -\sqrt{x + 5} + 8$ 22. $h(x) = -2(x - 5)^3 + 3$

In Exercises 23 and 24, find (a) $(f + g)(x)$, (b) $(f - g)(x)$, (c) $(fg)(x)$, (d) $(f/g)(x)$, (e) $(f \circ g)(x)$, and (f) $(g \circ f)(x)$.

23. $f(x) = 3x^2 - 7, \quad g(x) = -x^2 - 4x + 5$ 24. $f(x) = 1/x, \quad g(x) = 2\sqrt{x}$

In Exercises 25–27, determine whether the function has an inverse function, and if so, find the inverse function.

25. $f(x) = x^3 + 8$ 26. $f(x) = |x^2 - 3| + 6$ 27. $f(x) = 3x\sqrt{x}$

PROOFS IN MATHEMATICS

What does the word *proof* mean to you? In mathematics, the word *proof* is used to mean simply a valid argument. When you are proving a statement or theorem, you must use facts, definitions, and accepted properties in a logical order. You can also use previously proved theorems in your proof. For instance, the Distance Formula is used in the proof of the Midpoint Formula below. There are several different proof methods, which you will see in later chapters.

The Midpoint Formula *(p. 32)*

The midpoint of the line segment joining the points (x_1, y_1) and (x_2, y_2) is given by the Midpoint Formula

$$\text{Midpoint} = \left(\frac{x_1 + x_2}{2}, \frac{y_1 + y_2}{2} \right).$$

The Cartesian Plane

The Cartesian plane was named after the French mathematician René Descartes (1596–1650). While Descartes was lying in bed, he noticed a fly buzzing around on the square ceiling tiles. He discovered that the position of the fly could be described by which ceiling tile the fly landed on. This led to the development of the Cartesian plane. Descartes felt that a coordinate plane could be used to facilitate description of the positions of objects.

Proof

Using the figure, you must show that $d_1 = d_2$ and $d_1 + d_2 = d_3$.

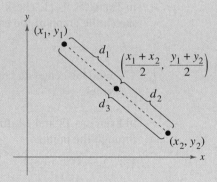

By the Distance Formula, you obtain

$$d_1 = \sqrt{\left(\frac{x_1 + x_2}{2} - x_1 \right)^2 + \left(\frac{y_1 + y_2}{2} - y_1 \right)^2}$$

$$= \frac{1}{2}\sqrt{(x_2 - x_1)^2 + (y_2 - y_1)^2}$$

$$d_2 = \sqrt{\left(x_2 - \frac{x_1 + x_2}{2} \right)^2 + \left(y_2 - \frac{y_1 + y_2}{2} \right)^2}$$

$$= \frac{1}{2}\sqrt{(x_2 - x_1)^2 + (y_2 - y_1)^2}$$

$$d_3 = \sqrt{(x_2 - x_1)^2 + (y_2 - y_1)^2}$$

So, it follows that $d_1 = d_2$ and $d_1 + d_2 = d_3$.

PROBLEM SOLVING

This collection of thought-provoking and challenging exercises further explores and expands upon concepts learned in this chapter.

1. As a salesperson, you receive a monthly salary of $2000, plus a commission of 7% of sales. You are offered a new job at $2300 per month, plus a commission of 5% of sales.

    (a) Write a linear equation for your current monthly wage W_1 in terms of your monthly sales S.

    (b) Write a linear equation for the monthly wage W_2 of your new job offer in terms of the monthly sales S.

    (c) Use a graphing utility to graph both equations in the same viewing window. Find the point of intersection. What does it signify?

    (d) You think you can sell $20,000 per month. Should you change jobs? Explain.

2. For the numbers 2 through 9 on a telephone keypad (see figure), create two relations: one mapping numbers onto letters, and the other mapping letters onto numbers. Are both relations functions? Explain.

3. What can be said about the sum and difference of each of the following?

    (a) Two even functions   (b) Two odd functions

    (c) An odd function and an even function

4. The two functions given by

$$f(x) = x \quad \text{and} \quad g(x) = -x$$

are their own inverse functions. Graph each function and explain why this is true. Graph other linear functions that are their own inverse functions. Find a general formula for a family of linear functions that are their own inverse functions.

5. Prove that a function of the following form is even.

$$y = a_{2n}x^{2n} + a_{2n-2}x^{2n-2} + \cdots + a_2x^2 + a_0$$

6. A miniature golf professional is trying to make a hole-in-one on the miniature golf green shown. A coordinate plane is placed over the golf green. The golf ball is at the point $(2.5, 2)$ and the hole is at the point $(9.5, 2)$. The professional wants to bank the ball off the side wall of the green at the point (x, y). Find the coordinates of the point (x, y). Then write an equation for the path of the ball.

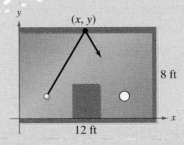

FIGURE FOR 6

7. At 2:00 P.M. on April 11, 1912, the *Titanic* left Cobh, Ireland, on her voyage to New York City. At 11:40 P.M. on April 14, the *Titanic* struck an iceberg and sank, having covered only about 2100 miles of the approximately 3400-mile trip.

    (a) What was the total duration of the voyage in hours?

    (b) What was the average speed in miles per hour?

    (c) Write a function relating the distance of the *Titanic* from New York City and the number of hours traveled. Find the domain and range of the function.

    (d) Graph the function from part (c).

8. Consider the function given by $f(x) = -x^2 + 4x - 3$. Find the average rate of change of the function from x_1 to x_2.

    (a) $x_1 = 1, x_2 = 2$ (b) $x_1 = 1, x_2 = 1.5$

    (c) $x_1 = 1, x_2 = 1.25$ (d) $x_1 = 1, x_2 = 1.125$

    (e) $x_1 = 1, x_2 = 1.0625$

    (f) Does the average rate of change seem to be approaching one value? If so, what value?

    (g) Find the equations of the secant lines through the points $(x_1, f(x_1))$ and $(x_2, f(x_2))$ for parts (a)–(e).

    (h) Find the equation of the line through the point $(1, f(1))$ using your answer from part (f) as the slope of the line.

9. Consider the functions given by $f(x) = 4x$ and $g(x) = x + 6$.

    (a) Find $(f \circ g)(x)$.   (b) Find $(f \circ g)^{-1}(x)$.

    (c) Find $f^{-1}(x)$ and $g^{-1}(x)$.

    (d) Find $(g^{-1} \circ f^{-1})(x)$ and compare the result with that of part (b).

    (e) Repeat parts (a) through (d) for $f(x) = x^3 + 1$ and $g(x) = 2x$.

    (f) Write two one-to-one functions f and g, and repeat parts (a) through (d) for these functions.

    (g) Make a conjecture about $(f \circ g)^{-1}(x)$ and $(g^{-1} \circ f^{-1})(x)$.

10. You are in a boat 2 miles from the nearest point on the coast. You are to travel to a point Q, 3 miles down the coast and 1 mile inland (see figure). You can row at 2 miles per hour and you can walk at 4 miles per hour.

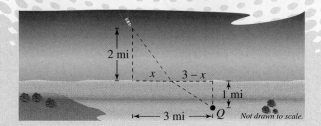

Not drawn to scale.

(a) Write the total time T of the trip as a function of x.

(b) Determine the domain of the function.

(c) Use a graphing utility to graph the function. Be sure to choose an appropriate viewing window.

(d) Use the *zoom* and *trace* features to find the value of x that minimizes T.

(e) Write a brief paragraph interpreting these values.

11. The **Heaviside function** $H(x)$ is widely used in engineering applications. (See figure.) To print an enlarged copy of the graph, go to the website *www.mathgraphs.com*.

$$H(x) = \begin{cases} 1, & x \geq 0 \\ 0, & x < 0 \end{cases}$$

Sketch the graph of each function by hand.

(a) $H(x) - 2$ (b) $H(x - 2)$ (c) $-H(x)$

(d) $H(-x)$ (e) $\frac{1}{2}H(x)$ (f) $-H(x - 2) + 2$

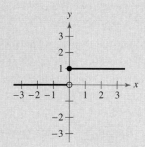

12. Let $f(x) = \dfrac{1}{1 - x}$.

(a) What are the domain and range of f?

(b) Find $f(f(x))$. What is the domain of this function?

(c) Find $f(f(f(x)))$. Is the graph a line? Why or why not?

13. Show that the Associative Property holds for compositions of functions—that is,

$$(f \circ (g \circ h))(x) = ((f \circ g) \circ h)(x).$$

14. Consider the graph of the function f shown in the figure. Use this graph to sketch the graph of each function. To print an enlarged copy of the graph, go to the website *www.mathgraphs.com*.

(a) $f(x + 1)$ (b) $f(x) + 1$ (c) $2f(x)$ (d) $f(-x)$

(e) $-f(x)$ (f) $|f(x)|$ (g) $f(|x|)$

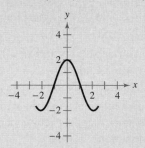

15. Use the graphs of f and f^{-1} to complete each table of function values.

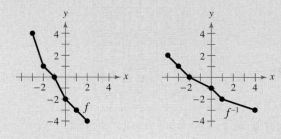

(a)

x	-4	-2	0	4
$(f(f^{-1}(x))$				

(b)

x	-3	-2	0	1
$(f + f^{-1})(x)$				

(c)

x	-3	-2	0	1
$(f \cdot f^{-1})(x)$				

(d)

x	-4	-3	0	4
$\lvert f^{-1}(x) \rvert$				

Trigonometry

1

In Mathematics

Trigonometry is used to find relationships between the sides and angles of triangles, and to write trigonometric functions as models of real-life quantities.

In Real Life

Trigonometric functions are used to model quantities that are periodic. For instance, throughout the day, the depth of water at the end of a dock in Bar Harbor, Maine varies with the tides. The depth can be modeled by a trigonometric function. (See Example 7, page 179.)

Andre Jenny/Alamy

IN CAREERS

There are many careers that use trigonometry. Several are listed below.

1.1 RADIAN AND DEGREE MEASURE

What you should learn
- Describe angles.
- Use radian measure.
- Use degree measure.
- Use angles to model and solve real-life problems.

Why you should learn it
You can use angles to model and solve real-life problems. For instance, in Exercise 119 on page 145, you are asked to use angles to find the speed of a bicycle.

Angles

As derived from the Greek language, the word **trigonometry** means "measurement of triangles." Initially, trigonometry dealt with relationships among the sides and angles of triangles and was used in the development of astronomy, navigation, and surveying. With the development of calculus and the physical sciences in the 17th century, a different perspective arose—one that viewed the classic trigonometric relationships as *functions* with the set of real numbers as their domains. Consequently, the applications of trigonometry expanded to include a vast number of physical phenomena involving rotations and vibrations. These phenomena include sound waves, light rays, planetary orbits, vibrating strings, pendulums, and orbits of atomic particles.

The approach in this text incorporates *both* perspectives, starting with angles and their measure.

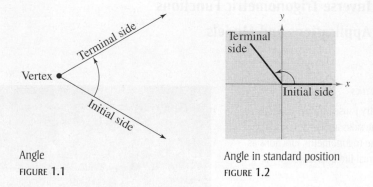

Angle
FIGURE **1.1**

Angle in standard position
FIGURE **1.2**

An **angle** is determined by rotating a ray (half-line) about its endpoint. The starting position of the ray is the **initial side** of the angle, and the position after rotation is the **terminal side,** as shown in Figure 1.1. The endpoint of the ray is the **vertex** of the angle. This perception of an angle fits a coordinate system in which the origin is the vertex and the initial side coincides with the positive x-axis. Such an angle is in **standard position,** as shown in Figure 1.2. **Positive angles** are generated by counterclockwise rotation, and **negative angles** by clockwise rotation, as shown in Figure 1.3. Angles are labeled with Greek letters α (alpha), β (beta), and θ (theta), as well as uppercase letters $A, B,$ and $C.$ In Figure 1.4, note that angles α and β have the same initial and terminal sides. Such angles are **coterminal.**

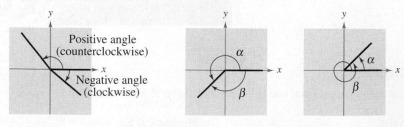

FIGURE **1.3**

FIGURE **1.4** Coterminal angles

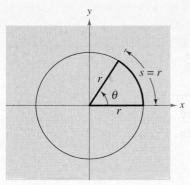

Arc length = radius when $\theta = 1$ radian

FIGURE **1.5**

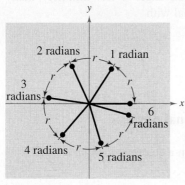

FIGURE **1.6**

Radian Measure

The **measure of an angle** is determined by the amount of rotation from the initial side to the terminal side. One way to measure angles is in *radians*. This type of measure is especially useful in calculus. To define a radian, you can use a **central angle** of a circle, one whose vertex is the center of the circle, as shown in Figure 1.5.

Definition of Radian

One **radian** is the measure of a central angle θ that intercepts an arc s equal in length to the radius r of the circle. See Figure 1.5. Algebraically, this means that

$$\theta = \frac{s}{r}$$

where θ is measured in radians.

Because the circumference of a circle is $2\pi r$ units, it follows that a central angle of one full revolution (counterclockwise) corresponds to an arc length of

$$s = 2\pi r.$$

Moreover, because $2\pi \approx 6.28$, there are just over six radius lengths in a full circle, as shown in Figure 1.6. Because the units of measure for s and r are the same, the ratio s/r has no units—it is simply a real number.

Because the radian measure of an angle of one full revolution is 2π, you can obtain the following.

$$\frac{1}{2} \text{ revolution} = \frac{2\pi}{2} = \pi \text{ radians}$$

$$\frac{1}{4} \text{ revolution} = \frac{2\pi}{4} = \frac{\pi}{2} \text{ radians}$$

$$\frac{1}{6} \text{ revolution} = \frac{2\pi}{6} = \frac{\pi}{3} \text{ radians}$$

These and other common angles are shown in Figure 1.7.

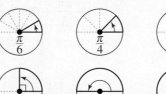

FIGURE **1.7**

Recall that the four quadrants in a coordinate system are numbered I, II, III, and IV. Figure 1.8 on page 136 shows which angles between 0 and 2π lie in each of the four quadrants. Note that angles between 0 and $\pi/2$ are **acute** angles and angles between $\pi/2$ and π are **obtuse** angles.

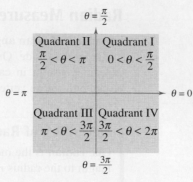

$$\theta = \frac{\pi}{2}$$

Quadrant II $\frac{\pi}{2} < \theta < \pi$	Quadrant I $0 < \theta < \frac{\pi}{2}$
Quadrant III $\pi < \theta < \frac{3\pi}{2}$	Quadrant IV $\frac{3\pi}{2} < \theta < 2\pi$

$\theta = \pi$ $\theta = 0$

$$\theta = \frac{3\pi}{2}$$

FIGURE **1.8**

Study Tip

The phrase "the terminal side of θ lies in a quadrant" is often abbreviated by simply saying that "θ lies in a quadrant." The terminal sides of the "quadrant angles" 0, $\pi/2$, π, and $3\pi/2$ do not lie within quadrants.

Algebra Help

You can review operations involving fractions in Section P.1.

Two angles are coterminal if they have the same initial and terminal sides. For instance, the angles 0 and 2π are coterminal, as are the angles $\pi/6$ and $13\pi/6$. You can find an angle that is coterminal to a given angle θ by adding or subtracting 2π (one revolution), as demonstrated in Example 1. A given angle θ has infinitely many coterminal angles. For instance, $\theta = \pi/6$ is coterminal with

$$\frac{\pi}{6} + 2n\pi$$

where n is an integer.

Example 1   Sketching and Finding Coterminal Angles

a. For the positive angle $13\pi/6$, subtract 2π to obtain a coterminal angle

$$\frac{13\pi}{6} - 2\pi = \frac{\pi}{6}. \qquad \text{See Figure 1.9.}$$

b. For the positive angle $3\pi/4$, subtract 2π to obtain a coterminal angle

$$\frac{3\pi}{4} - 2\pi = -\frac{5\pi}{4}. \qquad \text{See Figure 1.10.}$$

c. For the negative angle $-2\pi/3$, add 2π to obtain a coterminal angle

$$-\frac{2\pi}{3} + 2\pi = \frac{4\pi}{3}. \qquad \text{See Figure 1.11.}$$

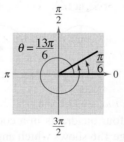

FIGURE **1.9**

FIGURE **1.10**

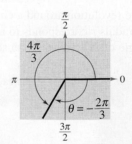

FIGURE **1.11**

CHECK *Point*   Now try Exercise 27.

Two positive angles α and β are **complementary** (complements of each other) if their sum is $\pi/2$. Two positive angles are **supplementary** (supplements of each other) if their sum is π. See Figure 1.12.

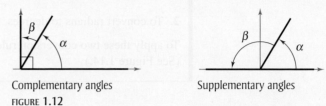

Complementary angles   Supplementary angles

FIGURE **1.12**

Example 2   Complementary and Supplementary Angles

If possible, find the complement and the supplement of (a) $2\pi/5$ and (b) $4\pi/5$.

Solution

a. The complement of $2\pi/5$ is

$$\frac{\pi}{2} - \frac{2\pi}{5} = \frac{5\pi}{10} - \frac{4\pi}{10} = \frac{\pi}{10}.$$

The supplement of $2\pi/5$ is

$$\pi - \frac{2\pi}{5} = \frac{5\pi}{5} - \frac{2\pi}{5} = \frac{3\pi}{5}.$$

b. Because $4\pi/5$ is greater than $\pi/2$, it has no complement. (Remember that complements are *positive* angles.) The supplement is

$$\pi - \frac{4\pi}{5} = \frac{5\pi}{5} - \frac{4\pi}{5} = \frac{\pi}{5}.$$

CHECK *Point*   Now try Exercise 31.

Degree Measure

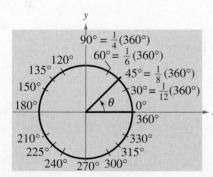

FIGURE **1.13**

A second way to measure angles is in terms of **degrees,** denoted by the symbol $°$. A measure of one degree ($1°$) is equivalent to a rotation of $\frac{1}{360}$ of a complete revolution about the vertex. To measure angles, it is convenient to mark degrees on the circumference of a circle, as shown in Figure 1.13. So, a full revolution (counterclockwise) corresponds to $360°$, a half revolution to $180°$, a quarter revolution to $90°$, and so on.

Because 2π radians corresponds to one complete revolution, degrees and radians are related by the equations

$$360° = 2\pi \text{ rad} \qquad \text{and} \qquad 180° = \pi \text{ rad}.$$

From the latter equation, you obtain

$$1° = \frac{\pi}{180} \text{ rad} \qquad \text{and} \qquad 1 \text{ rad} = \left(\frac{180°}{\pi}\right)$$

which lead to the conversion rules at the top of the next page.

> ## Conversions Between Degrees and Radians
>
> **1.** To convert degrees to radians, multiply degrees by $\dfrac{\pi \text{ rad}}{180°}$.
>
> **2.** To convert radians to degrees, multiply radians by $\dfrac{180°}{\pi \text{ rad}}$.
>
> To apply these two conversion rules, use the basic relationship $\pi \text{ rad} = 180°$. (See Figure 1.14.)

$\dfrac{\pi}{6}$
30°

$\dfrac{\pi}{2}$
90°

$\dfrac{\pi}{4}$
45°

π
180°

$\dfrac{\pi}{3}$
60°

$\dfrac{2\pi}{}$
360°

FIGURE 1.14

When no units of angle measure are specified, *radian measure is implied*. For instance, if you write $\theta = 2$, you imply that $\theta = 2$ radians.

Example 3   Converting from Degrees to Radians

a. $135° = (135 \text{ deg})\left(\dfrac{\pi \text{ rad}}{180 \text{ deg}}\right) = \dfrac{3\pi}{4}$ radians   Multiply by $\pi/180$.

b. $540° = (540 \text{ deg})\left(\dfrac{\pi \text{ rad}}{180 \text{ deg}}\right) = 3\pi$ radians   Multiply by $\pi/180$.

c. $-270° = (-270 \text{ deg})\left(\dfrac{\pi \text{ rad}}{180 \text{ deg}}\right) = -\dfrac{3\pi}{2}$ radians   Multiply by $\pi/180$.

CHECK*Point* ▶ Now try Exercise 57.

Example 4   Converting from Radians to Degrees

a. $-\dfrac{\pi}{2} \text{ rad} = \left(-\dfrac{\pi}{2} \text{ rad}\right)\left(\dfrac{180 \text{ deg}}{\pi \text{ rad}}\right) = -90°$ Multiply by $180/\pi$.

b. $\dfrac{9\pi}{2} \text{ rad} = \left(\dfrac{9\pi}{2} \text{ rad}\right)\left(\dfrac{180 \text{ deg}}{\pi \text{ rad}}\right) = 810°$ Multiply by $180/\pi$.

c. $2 \text{ rad} = (2 \text{ rad})\left(\dfrac{180 \text{ deg}}{\pi \text{ rad}}\right) = \dfrac{360°}{\pi} \approx 114.59°$ Multiply by $180/\pi$.

CHECK*Point* ▶ Now try Exercise 61.

If you have a calculator with a "radian-to-degree" conversion key, try using it to verify the result shown in part (b) of Example 4.

TECHNOLOGY

With calculators it is convenient to use *decimal* degrees to denote fractional parts of degrees. Historically, however, fractional parts of degrees were expressed in *minutes* and *seconds*, using the prime (′) and double prime (″) notations, respectively. That is,

$1' = \text{one minute} = \frac{1}{60}(1°)$

$1'' = \text{one second} = \frac{1}{3600}(1°)$.

Consequently, an angle of 64 degrees, 32 minutes, and 47 seconds is represented by $\theta = 64° \, 32' \, 47''$. Many calculators have special keys for converting an angle in degrees, minutes, and seconds (D° M′ S″) to decimal degree form, and vice versa.

Applications

The *radian measure* formula, $\theta = s/r$, can be used to measure arc length along a circle.

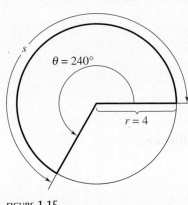

$\theta = 240°$

$r = 4$

FIGURE **1.15**

Arc Length

For a circle of radius r, a central angle θ intercepts an arc of length s given by

$$s = r\theta \qquad \text{Length of circular arc}$$

where θ is measured in radians. Note that if $r = 1$, then $s = \theta$, and the radian measure of θ equals the arc length.

Example 5   Finding Arc Length

A circle has a radius of 4 inches. Find the length of the arc intercepted by a central angle of 240°, as shown in Figure 1.15.

Solution

To use the formula $s = r\theta$, first convert 240° to radian measure.

$$240° = (240 \text{ deg})\left(\frac{\pi \text{ rad}}{180 \text{ deg}}\right)$$

$$= \frac{4\pi}{3} \text{ radians}$$

Then, using a radius of $r = 4$ inches, you can find the arc length to be

$$s = r\theta$$

$$= 4\left(\frac{4\pi}{3}\right)$$

$$= \frac{16\pi}{3} \approx 16.76 \text{ inches.}$$

Note that the units for $r\theta$ are determined by the units for r because θ is given in radian measure, which has no units.

CHECK Point   Now try Exercise 89.   ∎

The formula for the length of a circular arc can be used to analyze the motion of a particle moving at a *constant speed* along a circular path.

Linear and Angular Speeds

Consider a particle moving at a constant speed along a circular arc of radius r. If s is the length of the arc traveled in time t, then the **linear speed** v of the particle is

$$\text{Linear speed } v = \frac{\text{arc length}}{\text{time}} = \frac{s}{t}.$$

Moreover, if θ is the angle (in radian measure) corresponding to the arc length s, then the **angular speed** ω (the lowercase Greek letter omega) of the particle is

$$\text{Angular speed } \omega = \frac{\text{central angle}}{\text{time}} = \frac{\theta}{t}.$$

Study Tip

Linear speed measures how fast the particle moves, and angular speed measures how fast the angle changes. By dividing the formula for arc length by t, you can establish a relationship between linear speed v and angular speed ω, as shown.

$$s = r\theta$$

$$\frac{s}{t} = \frac{r\theta}{t}$$

$$v = r\omega$$

FIGURE 1.16

FIGURE 1.17

Example 6   Finding Linear Speed

The second hand of a clock is 10.2 centimeters long, as shown in Figure 1.16. Find the linear speed of the tip of this second hand as it passes around the clock face.

Solution

In one revolution, the arc length traveled is

$$s = 2\pi r$$

$$= 2\pi(10.2) \qquad \text{Substitute for } r.$$

$$= 20.4\pi \text{ centimeters.}$$

The time required for the second hand to travel this distance is

$$t = 1 \text{ minute} = 60 \text{ seconds.}$$

So, the linear speed of the tip of the second hand is

$$\text{Linear speed} = \frac{s}{t}$$

$$= \frac{20.4\pi \text{ centimeters}}{60 \text{ seconds}}$$

$$\approx 1.068 \text{ centimeters per second.}$$

CHECK Point   Now try Exercise 111.

Example 7   Finding Angular and Linear Speeds

The blades of a wind turbine are 116 feet long (see Figure 1.17). The propeller rotates at 15 revolutions per minute.

a. Find the angular speed of the propeller in radians per minute.

b. Find the linear speed of the tips of the blades.

Solution

a. Because each revolution generates 2π radians, it follows that the propeller turns $(15)(2\pi) = 30\pi$ radians per minute. In other words, the angular speed is

$$\text{Angular speed} = \frac{\theta}{t}$$

$$= \frac{30\pi \text{ radians}}{1 \text{ minute}} = 30\pi \text{ radians per minute.}$$

b. The linear speed is

$$\text{Linear speed} = \frac{s}{t}$$

$$= \frac{r\theta}{t}$$

$$= \frac{(116)(30\pi) \text{ feet}}{1 \text{ minute}} \approx 10{,}933 \text{ feet per minute.}$$

CHECK Point   Now try Exercise 113.

A **sector** of a circle is the region bounded by two radii of the circle and their intercepted arc (see Figure 1.18).

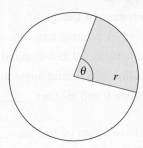

FIGURE **1.18**

Area of a Sector of a Circle

For a circle of radius r, the area A of a sector of the circle with central angle θ is given by

$$A = \frac{1}{2}r^2\theta$$

where θ is measured in radians.

Example 8   Area of a Sector of a Circle

A sprinkler on a golf course fairway sprays water over a distance of 70 feet and rotates through an angle of 120° (see Figure 1.19). Find the area of the fairway watered by the sprinkler.

Solution

First convert 120° to radian measure as follows.

$$\theta = 120°$$

$$= (120 \text{ deg})\left(\frac{\pi \text{ rad}}{180 \text{ deg}}\right) \qquad \text{Multiply by } \pi/180.$$

$$= \frac{2\pi}{3} \text{ radians}$$

Then, using $\theta = 2\pi/3$ and $r = 70$, the area is

$$A = \frac{1}{2}r^2\theta \qquad \text{Formula for the area of a sector of a circle}$$

$$= \frac{1}{2}(70)^2\left(\frac{2\pi}{3}\right) \qquad \text{Substitute for } r \text{ and } \theta.$$

$$= \frac{4900\pi}{3} \qquad \text{Simplify.}$$

$$\approx 5131 \text{ square feet.} \qquad \text{Simplify.}$$

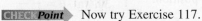 **Point**   Now try Exercise 117.

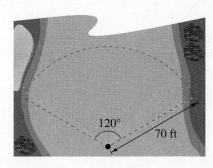

FIGURE **1.19**

120°

70 ft

1.1 EXERCISES

See www.CalcChat.com for worked-out solutions to odd-numbered exercises.

VOCABULARY: Fill in the blanks.

1. _____ means "measurement of triangles."

2. An _____ is determined by rotating a ray about its endpoint.

3. Two angles that have the same initial and terminal sides are _____.

4. One _____ is the measure of a central angle that intercepts an arc equal to the radius of the circle.

5. Angles that measure between 0 and $\pi/2$ are _____ angles, and angles that measure between $\pi/2$ and π are _____ angles.

6. Two positive angles that have a sum of $\pi/2$ are _____ angles, whereas two positive angles that have a sum of π are _____ angles.

7. The angle measure that is equivalent to a rotation of $\frac{1}{360}$ of a complete revolution about an angle's vertex is one _____.

8. 180 degrees = _____ radians.

9. The _____ speed of a particle is the ratio of arc length to time traveled, and the _____ speed of a particle is the ratio of central angle to time traveled.

10. The area A of a sector of a circle with radius r and central angle θ, where θ is measured in radians, is given by the formula _____.

SKILLS AND APPLICATIONS

In Exercises 11–16, estimate the angle to the nearest one-half radian.

11.

12.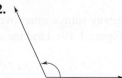

13.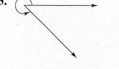

14.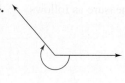

15.

16.

In Exercises 17–22, determine the quadrant in which each angle lies. (The angle measure is given in radians.)

17. (a) $\dfrac{\pi}{4}$ (b) $\dfrac{5\pi}{4}$ **18.** (a) $\dfrac{11\pi}{8}$ (b) $\dfrac{9\pi}{8}$

19. (a) $-\dfrac{\pi}{6}$ (b) $-\dfrac{\pi}{3}$ **20.** (a) $-\dfrac{5\pi}{6}$ (b) $-\dfrac{11\pi}{9}$

21. (a) 3.5 (b) 2.25 **22.** (a) 6.02 (b) −4.25

In Exercises 23–26, sketch each angle in standard position.

23. (a) $\dfrac{\pi}{3}$ (b) $-\dfrac{2\pi}{3}$ **24.** (a) $-\dfrac{7\pi}{4}$ (b) $\dfrac{5\pi}{2}$

25. (a) $\dfrac{11\pi}{6}$ (b) −3

26. (a) 4 (b) 7π

In Exercises 27–30, determine two coterminal angles (one positive and one negative) for each angle. Give your answers in radians.

27. (a) 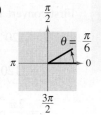 (b)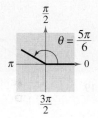

28. (a) 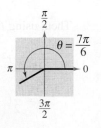 (b)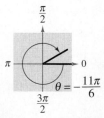

29. (a) $\theta = \dfrac{2\pi}{3}$ (b) $\theta = \dfrac{\pi}{12}$

30. (a) $\theta = -\dfrac{9\pi}{4}$ (b) $\theta = -\dfrac{2\pi}{15}$

In Exercises 31–34, find (if possible) the complement and supplement of each angle.

31. (a) $\pi/3$ (b) $\pi/4$ **32.** (a) $\pi/12$ (b) $11\pi/12$

33. (a) 1 (b) 2   **34.** (a) 3 (b) 1.5

In Exercises 35–40, estimate the number of degrees in the angle. Use a protractor to check your answer.

35.    **36.**

37. 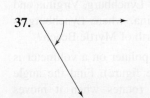   **38.**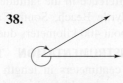

39.    **40.**

In Exercises 41–44, determine the quadrant in which each angle lies.

41. (a) $130°$ (b) $285°$

42. (a) $8.3°$ (b) $257° \, 30'$

43. (a) $-132° \, 50'$ (b) $-336°$

44. (a) $-260°$ (b) $-3.4°$

In Exercises 45–48, sketch each angle in standard position.

45. (a) $90°$ (b) $180°$ **46.** (a) $270°$ (b) $120°$

47. (a) $-30°$ (b) $-135°$

48. (a) $-750°$ (b) $-600°$

In Exercises 49–52, determine two coterminal angles (one positive and one negative) for each angle. Give your answers in degrees.

49. (a)   (b)

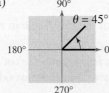

50. (a)   (b)

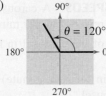

51. (a) $\theta = 240°$ (b) $\theta = -180°$

52. (a) $\theta = -390°$ (b) $\theta = 230°$

In Exercises 53–56, find (if possible) the complement and supplement of each angle.

53. (a) $18°$ (b) $85°$ **54.** (a) $46°$ (b) $93°$

55. (a) $150°$ (b) $79°$ **56.** (a) $130°$ (b) $170°$

In Exercises 57–60, rewrite each angle in radian measure as a multiple of π. (Do not use a calculator.)

57. (a) $30°$ (b) $45°$ **58.** (a) $315°$ (b) $120°$

59. (a) $-20°$ (b) $-60°$ **60.** (a) $-270°$ (b) $144°$

In Exercises 61–64, rewrite each angle in degree measure. (Do not use a calculator.)

61. (a) $\dfrac{3\pi}{2}$ (b) $\dfrac{7\pi}{6}$ **62.** (a) $-\dfrac{7\pi}{12}$ (b) $\dfrac{\pi}{9}$

63. (a) $\dfrac{5\pi}{4}$ (b) $-\dfrac{7\pi}{3}$ **64.** (a) $\dfrac{11\pi}{6}$ (b) $\dfrac{34\pi}{15}$

In Exercises 65–72, convert the angle measure from degrees to radians. Round to three decimal places.

65. $45°$ **66.** $87.4°$

67. $-216.35°$ **68.** $-48.27°$

69. $532°$ **70.** $345°$

71. $-0.83°$ **72.** $0.54°$

In Exercises 73–80, convert the angle measure from radians to degrees. Round to three decimal places.

73. $\pi/7$ **74.** $5\pi/11$

75. $15\pi/8$ **76.** $13\pi/2$

77. -4.2π **78.** 4.8π

79. -2 **80.** -0.57

In Exercises 81–84, convert each angle measure to decimal degree form without using a calculator. Then check your answers using a calculator.

81. (a) $54° \, 45'$ (b) $-128° \, 30'$

82. (a) $245° \, 10'$ (b) $2° \, 12'$

83. (a) $85° \, 18' \, 30''$ (b) $330° \, 25''$

84. (a) $-135° \, 36''$ (b) $-408° \, 16' \, 20''$

In Exercises 85–88, convert each angle measure to degrees, minutes, and seconds without using a calculator. Then check your answers using a calculator.

85. (a) $240.6°$ (b) $-145.8°$

86. (a) $-345.12°$ (b) $0.45°$

87. (a) $2.5°$ (b) $-3.58°$

88. (a) $-0.36°$ (b) $0.79°$

In Exercises 89–92, find the length of the arc on a circle of radius r intercepted by a central angle θ.

Radius r	Central Angle θ
89. 15 inches	120°
90. 9 feet	60°
91. 3 meters	150°
92. 20 centimeters	45°

In Exercises 93–96, find the radian measure of the central angle of a circle of radius r that intercepts an arc of length s.

Radius r	Arc Length s
93. 4 inches	18 inches
94. 14 feet	8 feet
95. 25 centimeters	10.5 centimeters
96. 80 kilometers	150 kilometers

In Exercises 97–100, use the given arc length and radius to find the angle θ (in radians).

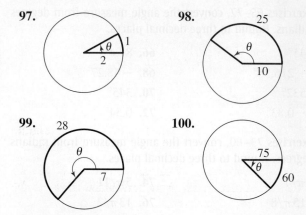

In Exercises 101–104, find the area of the sector of the circle with radius r and central angle θ.

Radius r	Central Angle θ
101. 6 inches	$\pi/3$
102. 12 millimeters	$\pi/4$
103. 2.5 feet	225°
104. 1.4 miles	330°

DISTANCE BETWEEN CITIES In Exercises 105 and 106, find the distance between the cities. Assume that Earth is a sphere of radius 4000 miles and that the cities are on the same longitude (one city is due north of the other).

City	Latitude
105. Dallas, Texas	32° 47′ 39″ N
Omaha, Nebraska	41° 15′ 50″ N

City	Latitude
106. San Francisco, California	37° 47′ 36″ N
Seattle, Washington	47° 37′ 18″ N

107. DIFFERENCE IN LATITUDES Assuming that Earth is a sphere of radius 6378 kilometers, what is the difference in the latitudes of Syracuse, New York and Annapolis, Maryland, where Syracuse is about 450 kilometers due north of Annapolis?

108. DIFFERENCE IN LATITUDES Assuming that Earth is a sphere of radius 6378 kilometers, what is the difference in the latitudes of Lynchburg, Virginia and Myrtle Beach, South Carolina, where Lynchburg is about 400 kilometers due north of Myrtle Beach?

109. INSTRUMENTATION The pointer on a voltmeter is 6 centimeters in length (see figure). Find the angle through which the pointer rotates when it moves 2.5 centimeters on the scale.

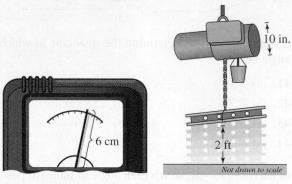

FIGURE FOR **109**   FIGURE FOR **110**

110. ELECTRIC HOIST An electric hoist is being used to lift a beam (see figure). The diameter of the drum on the hoist is 10 inches, and the beam must be raised 2 feet. Find the number of degrees through which the drum must rotate.

111. LINEAR AND ANGULAR SPEEDS A circular power saw has a $7\frac{1}{4}$-inch-diameter blade that rotates at 5000 revolutions per minute.

(a) Find the angular speed of the saw blade in radians per minute.

(b) Find the linear speed (in feet per minute) of one of the 24 cutting teeth as they contact the wood being cut.

112. LINEAR AND ANGULAR SPEEDS A carousel with a 50-foot diameter makes 4 revolutions per minute.

(a) Find the angular speed of the carousel in radians per minute.

(b) Find the linear speed (in feet per minute) of the platform rim of the carousel.

113. LINEAR AND ANGULAR SPEEDS The diameter of a DVD is approximately 12 centimeters. The drive motor of the DVD player is controlled to rotate precisely between 200 and 500 revolutions per minute, depending on what track is being read.

(a) Find an interval for the angular speed of a DVD as it rotates.

(b) Find an interval for the linear speed of a point on the outermost track as the DVD rotates.

114. ANGULAR SPEED A two-inch-diameter pulley on an electric motor that runs at 1700 revolutions per minute is connected by a belt to a four-inch-diameter pulley on a saw arbor.

(a) Find the angular speed (in radians per minute) of each pulley.

(b) Find the revolutions per minute of the saw.

115. ANGULAR SPEED A car is moving at a rate of 65 miles per hour, and the diameter of its wheels is 2 feet.

(a) Find the number of revolutions per minute the wheels are rotating.

(b) Find the angular speed of the wheels in radians per minute.

116. ANGULAR SPEED A computerized spin balance machine rotates a 25-inch-diameter tire at 480 revolutions per minute.

(a) Find the road speed (in miles per hour) at which the tire is being balanced.

(b) At what rate should the spin balance machine be set so that the tire is being tested for 55 miles per hour?

117. AREA A sprinkler on a golf green is set to spray water over a distance of 15 meters and to rotate through an angle of 140°. Draw a diagram that shows the region that can be irrigated with the sprinkler. Find the area of the region.

118. AREA A car's rear windshield wiper rotates 125°. The total length of the wiper mechanism is 25 inches and wipes the windshield over a distance of 14 inches. Find the area covered by the wiper.

119. SPEED OF A BICYCLE The radii of the pedal sprocket, the wheel sprocket, and the wheel of the bicycle in the figure are 4 inches, 2 inches, and 14 inches, respectively. A cyclist is pedaling at a rate of 1 revolution per second.

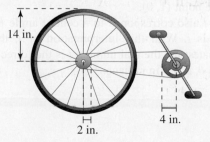

(a) Find the speed of the bicycle in feet per second and miles per hour.

(b) Use your result from part (a) to write a function for the distance d (in miles) a cyclist travels in terms of the number n of revolutions of the pedal sprocket.

(c) Write a function for the distance d (in miles) a cyclist travels in terms of the time t (in seconds). Compare this function with the function from part (b).

(d) Classify the types of functions you found in parts (b) and (c). Explain your reasoning.

120. CAPSTONE Write a short paper in your own words explaining the meaning of each of the following concepts to a classmate.

(a) an angle in standard position

(b) positive and negative angles

(c) coterminal angles

(d) angle measure in degrees and radians

(e) obtuse and acute angles

(f) complementary and supplementary angles

EXPLORATION

TRUE OR FALSE? In Exercises 121–123, determine whether the statement is true or false. Justify your answer.

121. A measurement of 4 radians corresponds to two complete revolutions from the initial side to the terminal side of an angle.

122. The difference between the measures of two coterminal angles is always a multiple of 360° if expressed in degrees and is always a multiple of 2π radians if expressed in radians.

123. An angle that measures $-1260°$ lies in Quadrant III.

124. THINK ABOUT IT A fan motor turns at a given angular speed. How does the speed of the tips of the blades change if a fan of greater diameter is installed on the motor? Explain.

125. THINK ABOUT IT Is a degree or a radian the larger unit of measure? Explain.

126. WRITING If the radius of a circle is increasing and the magnitude of a central angle is held constant, how is the length of the intercepted arc changing? Explain your reasoning.

127. PROOF Prove that the area of a circular sector of radius r with central angle θ is $A = \frac{1}{2}\theta r^2$, where θ is measured in radians.

1.2 TRIGONOMETRIC FUNCTIONS: THE UNIT CIRCLE

What you should learn

- Identify a unit circle and describe its relationship to real numbers.
- Evaluate trigonometric functions using the unit circle.
- Use the domain and period to evaluate sine and cosine functions.
- Use a calculator to evaluate trigonometric functions.

Why you should learn it

Trigonometric functions are used to model the movement of an oscillating weight. For instance, in Exercise 60 on page 152, the displacement from equilibrium of an oscillating weight suspended by a spring is modeled as a function of time.

The Unit Circle

The two historical perspectives of trigonometry incorporate different methods for introducing the trigonometric functions. Our first introduction to these functions is based on the unit circle.

Consider the **unit circle** given by

$$x^2 + y^2 = 1 \qquad \text{Unit circle}$$

as shown in Figure 1.20.

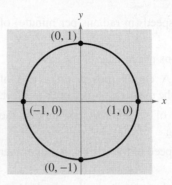

FIGURE **1.20**

Imagine that the real number line is wrapped around this circle, with positive numbers corresponding to a counterclockwise wrapping and negative numbers corresponding to a clockwise wrapping, as shown in Figure 1.21.

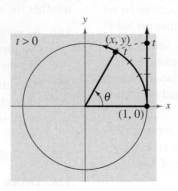
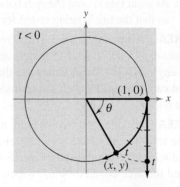

FIGURE **1.21**

As the real number line is wrapped around the unit circle, each real number t corresponds to a point (x, y) on the circle. For example, the real number 0 corresponds to the point $(1, 0)$. Moreover, because the unit circle has a circumference of 2π, the real number 2π also corresponds to the point $(1, 0)$.

In general, each real number t also corresponds to a central angle θ (in standard position) whose radian measure is t. With this interpretation of t, the arc length formula $s = r\theta$ (with $r = 1$) indicates that the real number t is the (directional) length of the arc intercepted by the angle θ, given in radians.

The Trigonometric Functions

From the preceding discussion, it follows that the coordinates x and y are two functions of the real variable t. You can use these coordinates to define the six trigonometric functions of t.

sine   cosecant   cosine   secant   tangent   cotangent

These six functions are normally abbreviated sin, csc, cos, sec, tan, and cot, respectively.

> ### Study Tip
>
> Note in the definition at the right that the functions in the second row are the *reciprocals* of the corresponding functions in the first row.

Definitions of Trigonometric Functions

Let t be a real number and let (x, y) be the point on the unit circle corresponding to t.

$$\sin t = y \qquad\qquad \cos t = x \qquad\qquad \tan t = \frac{y}{x}, \quad x \neq 0$$

$$\csc t = \frac{1}{y}, \quad y \neq 0 \qquad \sec t = \frac{1}{x}, \quad x \neq 0 \qquad \cot t = \frac{x}{y}, \quad y \neq 0$$

In the definitions of the trigonometric functions, note that the tangent and secant are not defined when $x = 0$. For instance, because $t = \pi/2$ corresponds to $(x, y) = (0, 1)$, it follows that $\tan(\pi/2)$ and $\sec(\pi/2)$ are *undefined*. Similarly, the cotangent and cosecant are not defined when $y = 0$. For instance, because $t = 0$ corresponds to $(x, y) = (1, 0)$, cot 0 and csc 0 are *undefined*.

In Figure 1.22, the unit circle has been divided into eight equal arcs, corresponding to t-values of

$$0, \frac{\pi}{4}, \frac{\pi}{2}, \frac{3\pi}{4}, \pi, \frac{5\pi}{4}, \frac{3\pi}{2}, \frac{7\pi}{4}, \text{ and } 2\pi.$$

Similarly, in Figure 1.23, the unit circle has been divided into 12 equal arcs, corresponding to t-values of

$$0, \frac{\pi}{6}, \frac{\pi}{3}, \frac{\pi}{2}, \frac{2\pi}{3}, \frac{5\pi}{6}, \pi, \frac{7\pi}{6}, \frac{4\pi}{3}, \frac{3\pi}{2}, \frac{5\pi}{3}, \frac{11\pi}{6}, \text{ and } 2\pi.$$

To verify the points on the unit circle in Figure 1.22, note that $\left(\frac{\sqrt{2}}{2}, \frac{\sqrt{2}}{2}\right)$ also lies on the line $y = x$. So, substituting x for y in the equation of the unit circle produces the following.

$$x^2 + x^2 = 1 \implies 2x^2 = 1 \implies x^2 = \frac{1}{2} \implies x = \pm\frac{\sqrt{2}}{2}$$

Because the point is in the first quadrant, $x = \frac{\sqrt{2}}{2}$ and because $y = x$, you also have $y = \frac{\sqrt{2}}{2}$. You can use similar reasoning to verify the rest of the points in Figure 1.22 and the points in Figure 1.23.

Using the (x, y) coordinates in Figures 1.22 and 1.23, you can evaluate the trigonometric functions for common t-values. This procedure is demonstrated in Examples 1, 2, and 3. You should study and learn these exact function values for common t-values because they will help you in later sections to perform calculations.

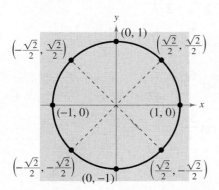

FIGURE **1.22**

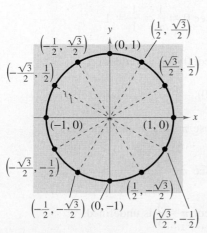

FIGURE **1.23**

Algebra Help

You can review dividing fractions in Section P.1.

Example 1 Evaluating Trigonometric Functions

Evaluate the six trigonometric functions at each real number.

a. $t = \dfrac{\pi}{6}$ **b.** $t = \dfrac{5\pi}{4}$ **c.** $t = 0$ **d.** $t = \pi$

Solution

For each t-value, begin by finding the corresponding point (x, y) on the unit circle. Then use the definitions of trigonometric functions listed on page 147.

a. $t = \dfrac{\pi}{6}$ corresponds to the point $(x, y) = \left(\dfrac{\sqrt{3}}{2}, \dfrac{1}{2}\right)$.

$$\sin \frac{\pi}{6} = y = \frac{1}{2} \qquad\qquad \csc \frac{\pi}{6} = \frac{1}{y} = \frac{1}{1/2} = 2$$

$$\cos \frac{\pi}{6} = x = \frac{\sqrt{3}}{2} \qquad\qquad \sec \frac{\pi}{6} = \frac{1}{x} = \frac{2}{\sqrt{3}} = \frac{2\sqrt{3}}{3}$$

$$\tan \frac{\pi}{6} = \frac{y}{x} = \frac{1/2}{\sqrt{3}/2} = \frac{1}{\sqrt{3}} = \frac{\sqrt{3}}{3} \qquad\qquad \cot \frac{\pi}{6} = \frac{x}{y} = \frac{\sqrt{3}/2}{1/2} = \sqrt{3}$$

b. $t = \dfrac{5\pi}{4}$ corresponds to the point $(x, y) = \left(-\dfrac{\sqrt{2}}{2}, -\dfrac{\sqrt{2}}{2}\right)$.

$$\sin \frac{5\pi}{4} = y = -\frac{\sqrt{2}}{2} \qquad\qquad \csc \frac{5\pi}{4} = \frac{1}{y} = -\frac{2}{\sqrt{2}} = -\sqrt{2}$$

$$\cos \frac{5\pi}{4} = x = -\frac{\sqrt{2}}{2} \qquad\qquad \sec \frac{5\pi}{4} = \frac{1}{x} = -\frac{2}{\sqrt{2}} = -\sqrt{2}$$

$$\tan \frac{5\pi}{4} = \frac{y}{x} = \frac{-\sqrt{2}/2}{-\sqrt{2}/2} = 1 \qquad\qquad \cot \frac{5\pi}{4} = \frac{x}{y} = \frac{-\sqrt{2}/2}{-\sqrt{2}/2} = 1$$

c. $t = 0$ corresponds to the point $(x, y) = (1, 0)$.

$$\sin 0 = y = 0 \qquad\qquad \csc 0 = \frac{1}{y} \text{ is undefined.}$$

$$\cos 0 = x = 1 \qquad\qquad \sec 0 = \frac{1}{x} = \frac{1}{1} = 1$$

$$\tan 0 = \frac{y}{x} = \frac{0}{1} = 0 \qquad\qquad \cot 0 = \frac{x}{y} \text{ is undefined.}$$

d. $t = \pi$ corresponds to the point $(x, y) = (-1, 0)$.

$$\sin \pi = y = 0 \qquad\qquad \csc \pi = \frac{1}{y} \text{ is undefined.}$$

$$\cos \pi = x = -1 \qquad\qquad \sec \pi = \frac{1}{x} = \frac{1}{-1} = -1$$

$$\tan \pi = \frac{y}{x} = \frac{0}{-1} = 0 \qquad\qquad \cot \pi = \frac{x}{y} \text{ is undefined.}$$

CHECK*Point*   Now try Exercise 23.

Example 2   Evaluating Trigonometric Functions

Evaluate the six trigonometric functions at $t = -\dfrac{\pi}{3}$.

Solution

Moving *clockwise* around the unit circle, it follows that $t = -\pi/3$ corresponds to the point $(x, y) = \left(1/2, -\sqrt{3}/2\right)$.

$$\sin\left(-\frac{\pi}{3}\right) = -\frac{\sqrt{3}}{2} \qquad\qquad \csc\left(-\frac{\pi}{3}\right) = -\frac{2}{\sqrt{3}} = -\frac{2\sqrt{3}}{3}$$

$$\cos\left(-\frac{\pi}{3}\right) = \frac{1}{2} \qquad\qquad \sec\left(-\frac{\pi}{3}\right) = 2$$

$$\tan\left(-\frac{\pi}{3}\right) = \frac{-\sqrt{3}/2}{1/2} = -\sqrt{3} \qquad\qquad \cot\left(-\frac{\pi}{3}\right) = \frac{1/2}{-\sqrt{3}/2} = -\frac{1}{\sqrt{3}} = -\frac{\sqrt{3}}{3}$$

CHECK *Point*   Now try Exercise 33.

Domain and Period of Sine and Cosine

The *domain* of the sine and cosine functions is the set of all real numbers. To determine the *range* of these two functions, consider the unit circle shown in Figure 1.24. By definition, $\sin t = y$ and $\cos t = x$. Because (x, y) is on the unit circle, you know that $-1 \le y \le 1$ and $-1 \le x \le 1$. So, the values of sine and cosine also range between -1 and 1.

$$
\begin{array}{ccc}
-1 \le \; y \; \le 1 & & -1 \le \; x \; \le 1 \\
 & \text{and} & \\
-1 \le \sin t \le 1 & & -1 \le \cos t \le 1
\end{array}
$$

Adding 2π to each value of t in the interval $[0, 2\pi]$ completes a second revolution around the unit circle, as shown in Figure 1.25. The values of $\sin(t + 2\pi)$ and $\cos(t + 2\pi)$ correspond to those of $\sin t$ and $\cos t$. Similar results can be obtained for repeated revolutions (positive or negative) on the unit circle. This leads to the general result

$$\sin(t + 2\pi n) = \sin t$$

and

$$\cos(t + 2\pi n) = \cos t$$

for any integer n and real number t. Functions that behave in such a repetitive (or cyclic) manner are called **periodic.**

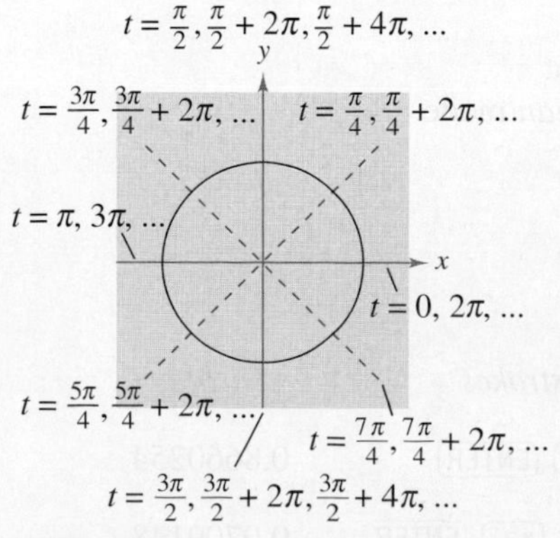

> ### Definition of Periodic Function
>
> A function f is **periodic** if there exists a positive real number c such that
>
> $$f(t + c) = f(t)$$
>
> for all t in the domain of f. The smallest number c for which f is periodic is called the **period** of f.

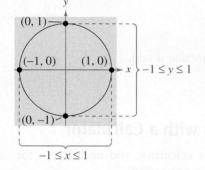

(0, 1)   (−1, 0)   (1, 0)   (0, −1) $-1 \le y \le 1$ $-1 \le x \le 1$

FIGURE 1.24

$t = \dfrac{\pi}{2}, \dfrac{\pi}{2} + 2\pi, \dfrac{\pi}{2} + 4\pi, \dots$

$t = \dfrac{3\pi}{4}, \dfrac{3\pi}{4} + 2\pi, \dots$ $t = \dfrac{\pi}{4}, \dfrac{\pi}{4} + 2\pi, \dots$

$t = \pi, 3\pi, \dots$

$t = 0, 2\pi, \dots$

$t = \dfrac{5\pi}{4}, \dfrac{5\pi}{4} + 2\pi, \dots$ $t = \dfrac{7\pi}{4}, \dfrac{7\pi}{4} + 2\pi, \dots$

$t = \dfrac{3\pi}{2}, \dfrac{3\pi}{2} + 2\pi, \dfrac{3\pi}{2} + 4\pi, \dots$

FIGURE 1.25

Recall from Section P.6 that a function f is *even* if $f(-t) = f(t)$, and is *odd* if $f(-t) = -f(t)$.

Even and Odd Trigonometric Functions

The cosine and secant functions are *even*.

$$\cos(-t) = \cos t \qquad \sec(-t) = \sec t$$

The sine, cosecant, tangent, and cotangent functions are *odd*.

$$\sin(-t) = -\sin t \qquad \csc(-t) = -\csc t$$

$$\tan(-t) = -\tan t \qquad \cot(-t) = -\cot t$$

Example 3   Using the Period to Evaluate the Sine and Cosine

a. Because $\dfrac{13\pi}{6} = 2\pi + \dfrac{\pi}{6}$, you have $\sin \dfrac{13\pi}{6} = \sin\left(2\pi + \dfrac{\pi}{6}\right) = \sin \dfrac{\pi}{6} = \dfrac{1}{2}$.

b. Because $-\dfrac{7\pi}{2} = -4\pi + \dfrac{\pi}{2}$, you have

$$\cos\left(-\dfrac{7\pi}{2}\right) = \cos\left(-4\pi + \dfrac{\pi}{2}\right) = \cos \dfrac{\pi}{2} = 0.$$

c. For $\sin t = \dfrac{4}{5}$, $\sin(-t) = -\dfrac{4}{5}$ because the sine function is odd.

CHECK*Point*   Now try Exercise 37.

Evaluating Trigonometric Functions with a Calculator

When evaluating a trigonometric function with a calculator, you need to set the calculator to the desired *mode* of measurement (*degree* or *radian*).

Most calculators do not have keys for the cosecant, secant, and cotangent functions. To evaluate these functions, you can use the $\boxed{x^{-1}}$ key with their respective reciprocal functions sine, cosine, and tangent. For instance, to evaluate $\csc(\pi/8)$, use the fact that

$$\csc \dfrac{\pi}{8} = \dfrac{1}{\sin(\pi/8)}$$

and enter the following keystroke sequence in *radian* mode.

$\boxed{(}\ \boxed{\text{SIN}}\ \boxed{(}\ \boxed{\pi}\ \boxed{\div}\, 8\ \boxed{)}\ \boxed{)}\ \boxed{x^{-1}}\ \boxed{\text{ENTER}}$ Display 2.6131259

Example 4   Using a Calculator

Function	Mode	Calculator Keystrokes	Display
a. $\sin \dfrac{2\pi}{3}$	Radian	$\boxed{\text{SIN}}\ \boxed{(}\, 2\ \boxed{\pi}\ \boxed{\div}\, 3\ \boxed{)}\ \boxed{\text{ENTER}}$	0.8660254
b. $\cot 1.5$	Radian	$\boxed{(}\ \boxed{\text{TAN}}\ \boxed{(}\, 1.5\ \boxed{)}\ \boxed{)}\ \boxed{x^{-1}}\ \boxed{\text{ENTER}}$	0.0709148

CHECK*Point*   Now try Exercise 55.

Study Tip

From the definition of periodic function, it follows that the sine and cosine functions are periodic and have a period of 2π. The other four trigonometric functions are also periodic, and will be discussed further in Section 1.6.

TECHNOLOGY

When evaluating trigonometric functions with a calculator, remember to enclose all fractional angle measures in parentheses. For instance, if you want to evaluate *sin t* for $t = \pi/6$, you should enter

$\boxed{\text{SIN}}\ \boxed{(}\ \boxed{\pi}\ \boxed{\div}\, 6\ \boxed{)}\ \boxed{\text{ENTER}}$.

These keystrokes yield the correct value of 0.5. Note that some calculators automatically place a left parenthesis after trigonometric functions. Check the user's guide for your calculator for specific keystrokes on how to evaluate trigonometric functions.

1.2   EXERCISES

See www.CalcChat.com for worked-out solutions to odd-numbered exercises.

VOCABULARY: Fill in the blanks.

1. Each real number t corresponds to a point (x, y) on the _____ _____.

2. A function f is _____ if there exists a positive real number c such that $f(t + c) = f(t)$ for all t in the domain of f.

3. The smallest number c for which a function f is periodic is called the _____ of f.

4. A function f is _____ if $f(-t) = -f(t)$ and _____ if $f(-t) = f(t)$.

SKILLS AND APPLICATIONS

In Exercises 5–8, determine the exact values of the six trigonometric functions of the real number t.

5.
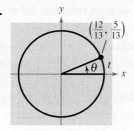

6.
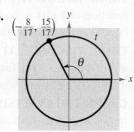

7.
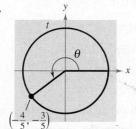

8.
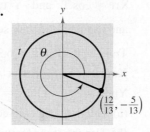

In Exercises 9–16, find the point (x, y) on the unit circle that corresponds to the real number t.

9. $t = \dfrac{\pi}{2}$

10. $t = \pi$

11. $t = \dfrac{\pi}{4}$

12. $t = \dfrac{\pi}{3}$

13. $t = \dfrac{5\pi}{6}$

14. $t = \dfrac{3\pi}{4}$

15. $t = \dfrac{4\pi}{3}$

16. $t = \dfrac{5\pi}{3}$

In Exercises 17–26, evaluate (if possible) the sine, cosine, and tangent of the real number.

17. $t = \dfrac{\pi}{4}$

18. $t = \dfrac{\pi}{3}$

19. $t = -\dfrac{\pi}{6}$

20. $t = -\dfrac{\pi}{4}$

21. $t = -\dfrac{7\pi}{4}$

22. $t = -\dfrac{4\pi}{3}$

23. $t = \dfrac{11\pi}{6}$

24. $t = \dfrac{5\pi}{3}$

25. $t = -\dfrac{3\pi}{2}$

26. $t = -2\pi$

In Exercises 27–34, evaluate (if possible) the six trigonometric functions of the real number.

27. $t = \dfrac{2\pi}{3}$

28. $t = \dfrac{5\pi}{6}$

29. $t = \dfrac{4\pi}{3}$

30. $t = \dfrac{7\pi}{4}$

31. $t = \dfrac{3\pi}{4}$

32. $t = \dfrac{3\pi}{2}$

33. $t = -\dfrac{\pi}{2}$

34. $t = -\pi$

In Exercises 35–42, evaluate the trigonometric function using its period as an aid.

35. $\sin 4\pi$

36. $\cos 3\pi$

37. $\cos \dfrac{7\pi}{3}$

38. $\sin \dfrac{9\pi}{4}$

39. $\cos \dfrac{17\pi}{4}$

40. $\sin \dfrac{19\pi}{6}$

41. $\sin\left(-\dfrac{8\pi}{3}\right)$

42. $\cos\left(-\dfrac{9\pi}{4}\right)$

In Exercises 43–48, use the value of the trigonometric function to evaluate the indicated functions.

43. $\sin t = \dfrac{1}{2}$
   (a) $\sin(-t)$
   (b) $\csc(-t)$

44. $\sin(-t) = \dfrac{3}{8}$
   (a) $\sin t$
   (b) $\csc t$

45. $\cos(-t) = -\dfrac{1}{5}$
   (a) $\cos t$
   (b) $\sec(-t)$

46. $\cos t = -\dfrac{3}{4}$
   (a) $\cos(-t)$
   (b) $\sec(-t)$

47. $\sin t = \dfrac{4}{5}$
   (a) $\sin(\pi - t)$
   (b) $\sin(t + \pi)$

48. $\cos t = \dfrac{4}{5}$
   (a) $\cos(\pi - t)$
   (b) $\cos(t + \pi)$

In Exercises 49–58, use a calculator to evaluate the trigonometric function. Round your answer to four decimal places. (Be sure the calculator is set in the correct angle mode.)

49. $\sin \dfrac{\pi}{4}$ **50.** $\tan \dfrac{\pi}{3}$

51. $\cot \dfrac{\pi}{4}$ **52.** $\csc \dfrac{2\pi}{3}$

53. $\cos(-1.7)$ **54.** $\cos(-2.5)$

55. $\csc 0.8$ **56.** $\sec 1.8$

57. $\sec(-22.8)$ **58.** $\cot(-0.9)$

59. HARMONIC MOTION The displacement from equilibrium of an oscillating weight suspended by a spring is given by $y(t) = \frac{1}{4}\cos 6t$, where y is the displacement (in feet) and t is the time (in seconds). Find the displacements when (a) $t = 0$, (b) $t = \frac{1}{4}$, and (c) $t = \frac{1}{2}$.

60. HARMONIC MOTION The displacement from equilibrium of an oscillating weight suspended by a spring is given by $y(t) = 3\sin(\pi t/4)$, where y is the displacement (in feet) and t is the time (in seconds).

(a) Complete the table.

t	0	$\frac{1}{2}$	1	$\frac{3}{2}$	2
y					

(b) Use the *table* feature of a graphing utility to determine when the displacement is maximum.

(c) Use the *table* feature of a graphing utility to approximate the time t ($0 < t < 8$) when the weight reaches equilibrium.

EXPLORATION

TRUE OR FALSE? In Exercises 61–64, determine whether the statement is true or false. Justify your answer.

61. Because $\sin(-t) = -\sin t$, it can be said that the sine of a negative angle is a negative number.

62. $\tan a = \tan(a - 6\pi)$

63. The real number 0 corresponds to the point $(0, 1)$ on the unit circle.

64. $\cos\left(-\dfrac{7\pi}{2}\right) = \cos\left(\pi + \dfrac{\pi}{2}\right)$

65. Let (x_1, y_1) and (x_2, y_2) be points on the unit circle corresponding to $t = t_1$ and $t = \pi - t_1$, respectively.

(a) Identify the symmetry of the points (x_1, y_1) and (x_2, y_2).

(b) Make a conjecture about any relationship between $\sin t_1$ and $\sin(\pi - t_1)$.

(c) Make a conjecture about any relationship between $\cos t_1$ and $\cos(\pi - t_1)$.

66. Use the unit circle to verify that the cosine and secant functions are even and that the sine, cosecant, tangent, and cotangent functions are odd.

67. Verify that $\cos 2t \neq 2\cos t$ by approximating $\cos 1.5$ and $2\cos 0.75$.

68. Verify that $\sin(t_1 + t_2) \neq \sin t_1 + \sin t_2$ by approximating $\sin 0.25$, $\sin 0.75$, and $\sin 1$.

69. THINK ABOUT IT Because $f(t) = \sin t$ is an odd function and $g(t) = \cos t$ is an even function, what can be said about the function $h(t) = f(t)g(t)$?

70. THINK ABOUT IT Because $f(t) = \sin t$ and $g(t) = \tan t$ are odd functions, what can be said about the function $h(t) = f(t)g(t)$?

71. GRAPHICAL ANALYSIS With your graphing utility in *radian* and *parametric* modes, enter the equations

$$X_{1T} = \cos T \quad \text{and} \quad Y_{1T} = \sin T$$

and use the following settings.

Tmin = 0, Tmax = 6.3, Tstep = 0.1

Xmin = −1.5, Xmax = 1.5, Xscl = 1

Ymin = −1, Ymax = 1, Yscl = 1

(a) Graph the entered equations and describe the graph.

(b) Use the *trace* feature to move the cursor around the graph. What do the t-values represent? What do the x- and y-values represent?

(c) What are the least and greatest values of x and y?

72. CAPSTONE A student you are tutoring has used a unit circle divided into 8 equal parts to complete the table for selected values of t. What is wrong?

t	0	$\dfrac{\pi}{4}$	$\dfrac{\pi}{2}$	$\dfrac{3\pi}{4}$	π
x	1	$\dfrac{\sqrt{2}}{2}$	0	$-\dfrac{\sqrt{2}}{2}$	−1
y	0	$\dfrac{\sqrt{2}}{2}$	1	$\dfrac{\sqrt{2}}{2}$	0
$\sin t$	1	$\dfrac{\sqrt{2}}{2}$	0	$-\dfrac{\sqrt{2}}{2}$	−1
$\cos t$	0	$\dfrac{\sqrt{2}}{2}$	1	$\dfrac{\sqrt{2}}{2}$	0
$\tan t$	Undef.	1	0	−1	Undef.

1.3 RIGHT TRIANGLE TRIGONOMETRY

What you should learn

- Evaluate trigonometric functions of acute angles.
- Use fundamental trigonometric identities.
- Use a calculator to evaluate trigonometric functions.
- Use trigonometric functions to model and solve real-life problems.

Why you should learn it

Trigonometric functions are often used to analyze real-life situations. For instance, in Exercise 76 on page 163, you can use trigonometric functions to find the height of a helium-filled balloon.

The Six Trigonometric Functions

Our second look at the trigonometric functions is from a *right triangle* perspective. Consider a right triangle, with one acute angle labeled θ, as shown in Figure 1.26. Relative to the angle θ, the three sides of the triangle are the **hypotenuse,** the **opposite side** (the side opposite the angle θ), and the **adjacent side** (the side adjacent to the angle θ).

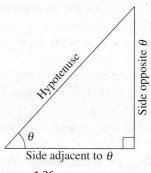

Side adjacent to θ

FIGURE **1.26**

Using the lengths of these three sides, you can form six ratios that define the six trigonometric functions of the acute angle θ.

> **sine   cosecant   cosine   secant   tangent   cotangent**

In the following definitions, it is important to see that $0° < \theta < 90°$ (θ lies in the first quadrant) and that for such angles the value of each trigonometric function is *positive*.

Right Triangle Definitions of Trigonometric Functions

Let θ be an *acute* angle of a right triangle. The six trigonometric functions of the angle θ are defined as follows. (Note that the functions in the second row are the *reciprocals* of the corresponding functions in the first row.)

$$\sin \theta = \frac{\text{opp}}{\text{hyp}} \qquad \cos \theta = \frac{\text{adj}}{\text{hyp}} \qquad \tan \theta = \frac{\text{opp}}{\text{adj}}$$

$$\csc \theta = \frac{\text{hyp}}{\text{opp}} \qquad \sec \theta = \frac{\text{hyp}}{\text{adj}} \qquad \cot \theta = \frac{\text{adj}}{\text{opp}}$$

The abbreviations opp, adj, and hyp represent the lengths of the three sides of a right triangle.

opp = the length of the side *opposite* θ

adj = the length of the side *adjacent to* θ

hyp = the length of the *hypotenuse*

FIGURE **1.27**

Example 1 Evaluating Trigonometric Functions

Use the triangle in Figure 1.27 to find the values of the six trigonometric functions of θ.

Solution

By the Pythagorean Theorem, $(\text{hyp})^2 = (\text{opp})^2 + (\text{adj})^2$, it follows that

$$\text{hyp} = \sqrt{4^2 + 3^2}$$
$$= \sqrt{25}$$
$$= 5.$$

So, the six trigonometric functions of θ are

$$\sin \theta = \frac{\text{opp}}{\text{hyp}} = \frac{4}{5} \qquad \csc \theta = \frac{\text{hyp}}{\text{opp}} = \frac{5}{4}$$

$$\cos \theta = \frac{\text{adj}}{\text{hyp}} = \frac{3}{5} \qquad \sec \theta = \frac{\text{hyp}}{\text{adj}} = \frac{5}{3}$$

$$\tan \theta = \frac{\text{opp}}{\text{adj}} = \frac{4}{3} \qquad \cot \theta = \frac{\text{adj}}{\text{opp}} = \frac{3}{4}.$$

CHECK*Point* Now try Exercise 7.

In Example 1, you were given the lengths of two sides of the right triangle, but not the angle θ. Often, you will be asked to find the trigonometric functions of a *given* acute angle θ. To do this, construct a right triangle having θ as one of its angles.

Example 2 Evaluating Trigonometric Functions of 45°

Find the values of sin 45°, cos 45°, and tan 45°.

Solution

Construct a right triangle having 45° as one of its acute angles, as shown in Figure 1.28. Choose the length of the adjacent side to be 1. From geometry, you know that the other acute angle is also 45°. So, the triangle is isosceles and the length of the opposite side is also 1. Using the Pythagorean Theorem, you find the length of the hypotenuse to be $\sqrt{2}$.

$$\sin 45° = \frac{\text{opp}}{\text{hyp}} = \frac{1}{\sqrt{2}} = \frac{\sqrt{2}}{2}$$

$$\cos 45° = \frac{\text{adj}}{\text{hyp}} = \frac{1}{\sqrt{2}} = \frac{\sqrt{2}}{2}$$

$$\tan 45° = \frac{\text{opp}}{\text{adj}} = \frac{1}{1} = 1$$

CHECK*Point* Now try Exercise 23.

HISTORICAL NOTE

Georg Joachim Rhaeticus (1514–1574) was the leading Teutonic mathematical astronomer of the 16th century. He was the first to define the trigonometric functions as ratios of the sides of a right triangle.

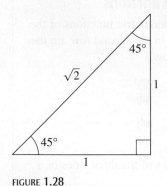

FIGURE **1.28**

Study Tip

Because the angles 30°, 45°, and 60° ($\pi/6$, $\pi/4$, and $\pi/3$) occur frequently in trigonometry, you should learn to construct the triangles shown in Figures 1.28 and 1.29.

TECHNOLOGY

You can use a calculator to convert the answers in Example 3 to decimals. However, the radical form is the exact value and in most cases, the exact value is preferred.

| Example 3 | Evaluating Trigonometric Functions of 30° and 60° |

Use the equilateral triangle shown in Figure 1.29 to find the values of sin 60°, cos 60°, sin 30°, and cos 30°.

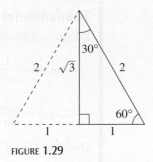

FIGURE 1.29

Solution

Use the Pythagorean Theorem and the equilateral triangle in Figure 1.29 to verify the lengths of the sides shown in the figure. For $\theta = 60°$, you have adj $= 1$, opp $= \sqrt{3}$, and hyp $= 2$. So,

$$\sin 60° = \frac{\text{opp}}{\text{hyp}} = \frac{\sqrt{3}}{2} \quad \text{and} \quad \cos 60° = \frac{\text{adj}}{\text{hyp}} = \frac{1}{2}.$$

For $\theta = 30°$, adj $= \sqrt{3}$, opp $= 1$, and hyp $= 2$. So,

$$\sin 30° = \frac{\text{opp}}{\text{hyp}} = \frac{1}{2} \quad \text{and} \quad \cos 30° = \frac{\text{adj}}{\text{hyp}} = \frac{\sqrt{3}}{2}.$$

CHECK*Point*   Now try Exercise 27.

Sines, Cosines, and Tangents of Special Angles

$$\sin 30° = \sin \frac{\pi}{6} = \frac{1}{2} \qquad \cos 30° = \cos \frac{\pi}{6} = \frac{\sqrt{3}}{2} \qquad \tan 30° = \tan \frac{\pi}{6} = \frac{\sqrt{3}}{3}$$

$$\sin 45° = \sin \frac{\pi}{4} = \frac{\sqrt{2}}{2} \qquad \cos 45° = \cos \frac{\pi}{4} = \frac{\sqrt{2}}{2} \qquad \tan 45° = \tan \frac{\pi}{4} = 1$$

$$\sin 60° = \sin \frac{\pi}{3} = \frac{\sqrt{3}}{2} \qquad \cos 60° = \cos \frac{\pi}{3} = \frac{1}{2} \qquad \tan 60° = \tan \frac{\pi}{3} = \sqrt{3}$$

In the box, note that $\sin 30° = \frac{1}{2} = \cos 60°$. This occurs because 30° and 60° are complementary angles. In general, it can be shown from the right triangle definitions that *cofunctions of complementary angles are equal*. That is, if θ is an acute angle, the following relationships are true.

$$\sin(90° - \theta) = \cos \theta \qquad \cos(90° - \theta) = \sin \theta$$

$$\tan(90° - \theta) = \cot \theta \qquad \cot(90° - \theta) = \tan \theta$$

$$\sec(90° - \theta) = \csc \theta \qquad \csc(90° - \theta) = \sec \theta$$

Trigonometric Identities

In trigonometry, a great deal of time is spent studying relationships between trigonometric functions (identities).

Fundamental Trigonometric Identities

Reciprocal Identities

$$\sin \theta = \frac{1}{\csc \theta} \qquad \cos \theta = \frac{1}{\sec \theta} \qquad \tan \theta = \frac{1}{\cot \theta}$$

$$\csc \theta = \frac{1}{\sin \theta} \qquad \sec \theta = \frac{1}{\cos \theta} \qquad \cot \theta = \frac{1}{\tan \theta}$$

Quotient Identities

$$\tan \theta = \frac{\sin \theta}{\cos \theta} \qquad \cot \theta = \frac{\cos \theta}{\sin \theta}$$

Pythagorean Identities

$$\sin^2 \theta + \cos^2 \theta = 1 \qquad 1 + \tan^2 \theta = \sec^2 \theta$$

$$1 + \cot^2 \theta = \csc^2 \theta$$

Note that $\sin^2 \theta$ represents $(\sin \theta)^2$, $\cos^2 \theta$ represents $(\cos \theta)^2$, and so on.

Example 4   Applying Trigonometric Identities

Let θ be an acute angle such that $\sin \theta = 0.6$. Find the values of (a) $\cos \theta$ and (b) $\tan \theta$ using trigonometric identities.

Solution

a. To find the value of $\cos \theta$, use the Pythagorean identity

$$\sin^2 \theta + \cos^2 \theta = 1.$$

So, you have

$(0.6)^2 + \cos^2 \theta = 1$	Substitute 0.6 for $\sin \theta$.
$\cos^2 \theta = 1 - (0.6)^2 = 0.64$	Subtract $(0.6)^2$ from each side.
$\cos \theta = \sqrt{0.64} = 0.8.$	Extract the positive square root.

b. Now, knowing the sine and cosine of θ, you can find the tangent of θ to be

$$\tan \theta = \frac{\sin \theta}{\cos \theta}$$

$$= \frac{0.6}{0.8}$$

$$= 0.75.$$

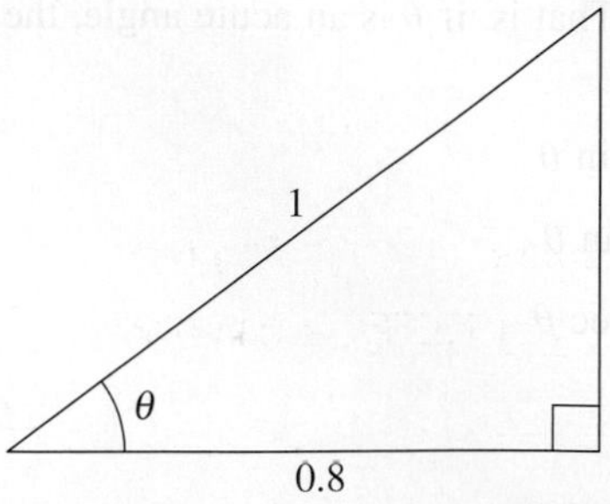

Use the definitions of $\cos \theta$ and $\tan \theta$, and the triangle shown in Figure 1.30, to check these results.

FIGURE **1.30**

CHECK *Point* Now try Exercise 33.

Example 5   Applying Trigonometric Identities

Let θ be an acute angle such that $\tan \theta = 3$. Find the values of (a) $\cot \theta$ and (b) $\sec \theta$ using trigonometric identities.

Solution

a. $\cot \theta = \dfrac{1}{\tan \theta}$ Reciprocal identity

$\cot \theta = \dfrac{1}{3}$

b. $\sec^2 \theta = 1 + \tan^2 \theta$ Pythagorean identity

$\sec^2 \theta = 1 + 3^2$

$\sec^2 \theta = 10$

$\sec \theta = \sqrt{10}$

Use the definitions of $\cot \theta$ and $\sec \theta$, and the triangle shown in Figure 1.31, to check these results.

CHECK*Point*   Now try Exercise 35.

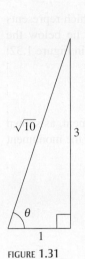

FIGURE 1.31

Evaluating Trigonometric Functions with a Calculator

To use a calculator to evaluate trigonometric functions of angles measured in degrees, first set the calculator to *degree* mode and then proceed as demonstrated in Section 1.2. For instance, you can find values of cos 28° and sec 28° as follows.

Function	Mode	Calculator Keystrokes	Display
a. cos 28°	Degree	(COS) 28 (ENTER)	0.8829476
b. sec 28°	Degree	((COS) (28)) (x⁻¹) (ENTER)	1.1325701

Throughout this text, angles are assumed to be measured in radians unless noted otherwise. For example, sin 1 means the sine of 1 radian and sin 1° means the sine of 1 degree.

Example 6   Using a Calculator

Use a calculator to evaluate $\sec(5°\, 40'\, 12'')$.

Solution

Begin by converting to decimal degree form. [Recall that $1' = \frac{1}{60}(1°)$ and $1'' = \frac{1}{3600}(1°)$].

$$5°\, 40'\, 12'' = 5° + \left(\frac{40}{60}\right)° + \left(\frac{12}{3600}\right)° = 5.67°$$

Then, use a calculator to evaluate sec 5.67°.

Function	Calculator Keystrokes	Display
$\sec(5°\,40'\,12'') = \sec 5.67°$	((COS) (5.67)) (x⁻¹) (ENTER)	1.0049166

CHECK*Point*   Now try Exercise 51.

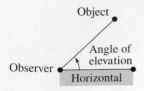

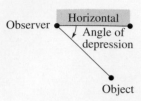

FIGURE **1.32**

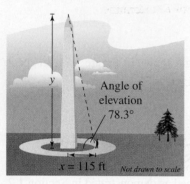

FIGURE **1.33**

Applications Involving Right Triangles

Many applications of trigonometry involve a process called **solving right triangles.** In this type of application, you are usually given one side of a right triangle and one of the acute angles and are asked to find one of the other sides, *or* you are given two sides and are asked to find one of the acute angles.

In Example 7, the angle you are given is the **angle of elevation,** which represents the angle from the horizontal upward to an object. For objects that lie below the horizontal, it is common to use the term **angle of depression,** as shown in Figure 1.32.

Example 7   Using Trigonometry to Solve a Right Triangle

A surveyor is standing 115 feet from the base of the Washington Monument, as shown in Figure 1.33. The surveyor measures the angle of elevation to the top of the monument as 78.3°. How tall is the Washington Monument?

Solution

From Figure 1.33, you can see that

$$\tan 78.3° = \frac{\text{opp}}{\text{adj}} = \frac{y}{x}$$

where $x = 115$ and y is the height of the monument. So, the height of the Washington Monument is

$$y = x \tan 78.3° \approx 115(4.82882) \approx 555 \text{ feet}.$$

CHECK Point   Now try Exercise 67.

Example 8   Using Trigonometry to Solve a Right Triangle

A historic lighthouse is 200 yards from a bike path along the edge of a lake. A walkway to the lighthouse is 400 yards long. Find the acute angle θ between the bike path and the walkway, as illustrated in Figure 1.34.

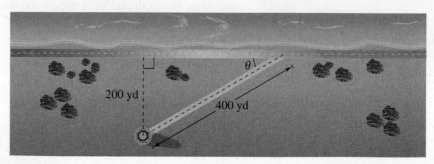

FIGURE **1.34**

Solution

From Figure 1.34, you can see that the sine of the angle θ is

$$\sin \theta = \frac{\text{opp}}{\text{hyp}} = \frac{200}{400} = \frac{1}{2}.$$

Now you should recognize that $\theta = 30°$.

CHECK Point   Now try Exercise 69.

By now you are able to recognize that $\theta = 30°$ is the acute angle that satisfies the equation $\sin \theta = \frac{1}{2}$. Suppose, however, that you were given the equation $\sin \theta = 0.6$ and were asked to find the acute angle θ. Because

$$\sin 30° = \frac{1}{2}$$

$$= 0.5000$$

and

$$\sin 45° = \frac{1}{\sqrt{2}}$$

$$\approx 0.7071$$

you might guess that θ lies somewhere between 30° and 45°. In a later section, you will study a method by which a more precise value of θ can be determined.

Example 9   Solving a Right Triangle

Find the length c of the skateboard ramp shown in Figure 1.35.

FIGURE **1.35**

Solution

From Figure 1.35, you can see that

$$\sin 18.4° = \frac{\text{opp}}{\text{hyp}}$$

$$= \frac{4}{c}.$$

So, the length of the skateboard ramp is

$$c = \frac{4}{\sin 18.4°}$$

$$\approx \frac{4}{0.3156}$$

$$\approx 12.7 \text{ feet.}$$

CHECK*Point*   Now try Exercise 71.

1.3 EXERCISES

See www.CalcChat.com for worked-out solutions to odd-numbered exercises.

VOCABULARY

1. Match the trigonometric function with its right triangle definition.

   (a) Sine   (b) Cosine   (c) Tangent   (d) Cosecant   (e) Secant   (f) Cotangent

   (i) $\dfrac{\text{hypotenuse}}{\text{adjacent}}$ (ii) $\dfrac{\text{adjacent}}{\text{opposite}}$ (iii) $\dfrac{\text{hypotenuse}}{\text{opposite}}$ (iv) $\dfrac{\text{adjacent}}{\text{hypotenuse}}$ (v) $\dfrac{\text{opposite}}{\text{hypotenuse}}$ (vi) $\dfrac{\text{opposite}}{\text{adjacent}}$

In Exercises 2–4, fill in the blanks.

2. Relative to the angle θ, the three sides of a right triangle are the _____ side, the _____ side, and the _____.

3. Cofunctions of _____ angles are equal.

4. An angle that measures from the horizontal upward to an object is called the angle of _____, whereas an angle that measures from the horizontal downward to an object is called the angle of _____.

SKILLS AND APPLICATIONS

In Exercises 5–8, find the exact values of the six trigonometric functions of the angle θ shown in the figure. (Use the Pythagorean Theorem to find the third side of the triangle.)

5.

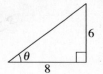

6.

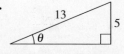

7.

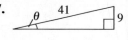

8.

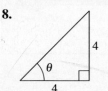

In Exercises 9–12, find the exact values of the six trigonometric functions of the angle θ for each of the two triangles. Explain why the function values are the same.

9.

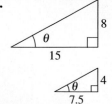

10. 1.25

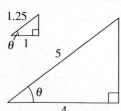

11.

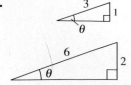

12.

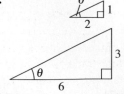

In Exercises 13–20, sketch a right triangle corresponding to the trigonometric function of the acute angle θ. Use the Pythagorean Theorem to determine the third side and then find the other five trigonometric functions of θ.

13. $\tan \theta = \frac{3}{4}$ **14.** $\cos \theta = \frac{5}{6}$

15. $\sec \theta = \frac{3}{2}$ **16.** $\tan \theta = \frac{4}{5}$

17. $\sin \theta = \frac{1}{5}$ **18.** $\sec \theta = \frac{17}{7}$

19. $\cot \theta = 3$ **20.** $\csc \theta = 9$

In Exercises 21–30, construct an appropriate triangle to complete the table. ($0° \le \theta \le 90°, 0 \le \theta \le \pi/2$)

	Function	θ (deg)	θ (rad)	Function Value
21.	sin	30°		
22.	cos	45°		
23.	sec		$\dfrac{\pi}{4}$	
24.	tan		$\dfrac{\pi}{3}$	
25.	cot			$\dfrac{\sqrt{3}}{3}$
26.	csc			$\sqrt{2}$
27.	csc		$\dfrac{\pi}{6}$	
28.	sin		$\dfrac{\pi}{4}$	
29.	cot			1
30.	tan			$\dfrac{\sqrt{3}}{3}$

In Exercises 31–36, use the given function value(s), and trigonometric identities (including the cofunction identities), to find the indicated trigonometric functions.

31. $\sin 60° = \dfrac{\sqrt{3}}{2}$, $\cos 60° = \dfrac{1}{2}$

    (a) $\sin 30°$ (b) $\cos 30°$

    (c) $\tan 60°$ (d) $\cot 60°$

32. $\sin 30° = \dfrac{1}{2}$, $\tan 30° = \dfrac{\sqrt{3}}{3}$

    (a) $\csc 30°$ (b) $\cot 60°$

    (c) $\cos 30°$ (d) $\cot 30°$

33. $\cos \theta = \dfrac{1}{3}$

    (a) $\sin \theta$ (b) $\tan \theta$

    (c) $\sec \theta$ (d) $\csc(90° - \theta)$

34. $\sec \theta = 5$

    (a) $\cos \theta$ (b) $\cot \theta$

    (c) $\cot(90° - \theta)$ (d) $\sin \theta$

35. $\cot \alpha = 5$

    (a) $\tan \alpha$ (b) $\csc \alpha$

    (c) $\cot(90° - \alpha)$ (d) $\cos \alpha$

36. $\cos \beta = \dfrac{\sqrt{7}}{4}$

    (a) $\sec \beta$ (b) $\sin \beta$

    (c) $\cot \beta$ (d) $\sin(90° - \beta)$

In Exercises 37–46, use trigonometric identities to transform the left side of the equation into the right side $(0 < \theta < \pi/2)$.

37. $\tan \theta \cot \theta = 1$

38. $\cos \theta \sec \theta = 1$

39. $\tan \alpha \cos \alpha = \sin \alpha$

40. $\cot \alpha \sin \alpha = \cos \alpha$

41. $(1 + \sin \theta)(1 - \sin \theta) = \cos^2 \theta$

42. $(1 + \cos \theta)(1 - \cos \theta) = \sin^2 \theta$

43. $(\sec \theta + \tan \theta)(\sec \theta - \tan \theta) = 1$

44. $\sin^2 \theta - \cos^2 \theta = 2 \sin^2 \theta - 1$

45. $\dfrac{\sin \theta}{\cos \theta} + \dfrac{\cos \theta}{\sin \theta} = \csc \theta \sec \theta$

46. $\dfrac{\tan \beta + \cot \beta}{\tan \beta} = \csc^2 \beta$

In Exercises 47–56, use a calculator to evaluate each function. Round your answers to four decimal places. (Be sure the calculator is in the correct angle mode.)

47. (a) $\sin 10°$ (b) $\cos 80°$

48. (a) $\tan 23.5°$ (b) $\cot 66.5°$

49. (a) $\sin 16.35°$ (b) $\csc 16.35°$

50. (a) $\cot 79.56°$ (b) $\sec 79.56°$

51. (a) $\cos 4° 50' 15''$ (b) $\sec 4° 50' 15''$

52. (a) $\sec 42° 12'$ (b) $\csc 48° 7'$

53. (a) $\cot 11° 15'$ (b) $\tan 11° 15'$

54. (a) $\sec 56° 8' 10''$ (b) $\cos 56° 8' 10''$

55. (a) $\csc 32° 40' 3''$ (b) $\tan 44° 28' 16''$

56. (a) $\sec\left(\dfrac{9}{5} \cdot 20 + 32\right)°$ (b) $\cot\left(\dfrac{9}{5} \cdot 30 + 32\right)°$

In Exercises 57–62, find the values of θ in degrees $(0° < \theta < 90°)$ and radians $(0 < \theta < \pi/2)$ without the aid of a calculator.

57. (a) $\sin \theta = \dfrac{1}{2}$ (b) $\csc \theta = 2$

58. (a) $\cos \theta = \dfrac{\sqrt{2}}{2}$ (b) $\tan \theta = 1$

59. (a) $\sec \theta = 2$ (b) $\cot \theta = 1$

60. (a) $\tan \theta = \sqrt{3}$ (b) $\cos \theta = \dfrac{1}{2}$

61. (a) $\csc \theta = \dfrac{2\sqrt{3}}{3}$ (b) $\sin \theta = \dfrac{\sqrt{2}}{2}$

62. (a) $\cot \theta = \dfrac{\sqrt{3}}{3}$ (b) $\sec \theta = \sqrt{2}$

In Exercises 63–66, solve for x, y, or r as indicated.

63. Solve for y.   **64.** Solve for x.

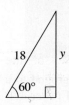   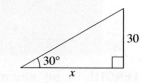

65. Solve for x.   **66.** Solve for r.

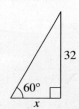   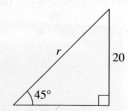

67. EMPIRE STATE BUILDING   You are standing 45 meters from the base of the Empire State Building. You estimate that the angle of elevation to the top of the 86th floor (the observatory) is 82°. If the total height of the building is another 123 meters above the 86th floor, what is the approximate height of the building? One of your friends is on the 86th floor. What is the distance between you and your friend?

68. HEIGHT A six-foot person walks from the base of a broadcasting tower directly toward the tip of the shadow cast by the tower. When the person is 132 feet from the tower and 3 feet from the tip of the shadow, the person's shadow starts to appear beyond the tower's shadow.

(a) Draw a right triangle that gives a visual representation of the problem. Show the known quantities of the triangle and use a variable to indicate the height of the tower.

(b) Use a trigonometric function to write an equation involving the unknown quantity.

(c) What is the height of the tower?

69. ANGLE OF ELEVATION You are skiing down a mountain with a vertical height of 1500 feet. The distance from the top of the mountain to the base is 3000 feet. What is the angle of elevation from the base to the top of the mountain?

70. WIDTH OF A RIVER A biologist wants to know the width w of a river so that instruments for studying the pollutants in the water can be set properly. From point A, the biologist walks downstream 100 feet and sights to point C (see figure). From this sighting, it is determined that $\theta = 54°$. How wide is the river?

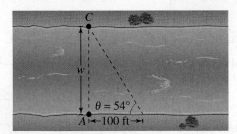

71. LENGTH A guy wire runs from the ground to a cell tower. The wire is attached to the cell tower 150 feet above the ground. The angle formed between the wire and the ground is 43° (see figure).

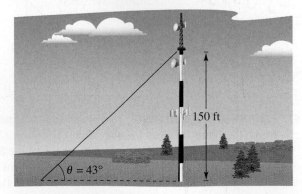

(a) How long is the guy wire?

(b) How far from the base of the tower is the guy wire anchored to the ground?

72. HEIGHT OF A MOUNTAIN In traveling across flat land, you notice a mountain directly in front of you. Its angle of elevation (to the peak) is 3.5°. After you drive 13 miles closer to the mountain, the angle of elevation is 9°. Approximate the height of the mountain.

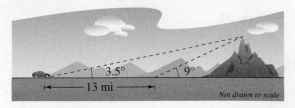

73. MACHINE SHOP CALCULATIONS A steel plate has the form of one-fourth of a circle with a radius of 60 centimeters. Two two-centimeter holes are to be drilled in the plate positioned as shown in the figure. Find the coordinates of the center of each hole.

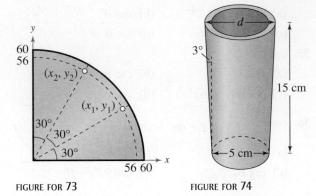

FIGURE FOR 73   FIGURE FOR 74

74. MACHINE SHOP CALCULATIONS A tapered shaft has a diameter of 5 centimeters at the small end and is 15 centimeters long (see figure). The taper is 3°. Find the diameter d of the large end of the shaft.

75. GEOMETRY Use a compass to sketch a quarter of a circle of radius 10 centimeters. Using a protractor, construct an angle of 20° in standard position (see figure). Drop a perpendicular line from the point of intersection of the terminal side of the angle and the arc of the circle. By actual measurement, calculate the coordinates (x, y) of the point of intersection and use these measurements to approximate the six trigonometric functions of a 20° angle.

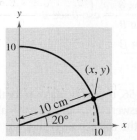

76. HEIGHT A 20-meter line is used to tether a helium-filled balloon. Because of a breeze, the line makes an angle of approximately 85° with the ground.

(a) Draw a right triangle that gives a visual representation of the problem. Show the known quantities of the triangle and use a variable to indicate the height of the balloon.

(b) Use a trigonometric function to write an equation involving the unknown quantity.

(c) What is the height of the balloon?

(d) The breeze becomes stronger and the angle the balloon makes with the ground decreases. How does this affect the triangle you drew in part (a)?

(e) Complete the table, which shows the heights (in meters) of the balloon for decreasing angle measures θ.

Angle, θ	80°	70°	60°	50°
Height				

Angle, θ	40°	30°	20°	10°
Height				

(f) As the angle the balloon makes with the ground approaches 0°, how does this affect the height of the balloon? Draw a right triangle to explain your reasoning.

EXPLORATION

TRUE OR FALSE? In Exercises 77–82, determine whether the statement is true or false. Justify your answer.

77. $\sin 60° \csc 60° = 1$

78. $\sec 30° = \csc 60°$

79. $\sin 45° + \cos 45° = 1$

80. $\cot^2 10° - \csc^2 10° = -1$

81. $\dfrac{\sin 60°}{\sin 30°} = \sin 2°$

82. $\tan[(5°)^2] = \tan^2 5°$

83. THINK ABOUT IT

(a) Complete the table.

θ	0.1	0.2	0.3	0.4	0.5
$\sin \theta$					

(b) Is θ or $\sin \theta$ greater for θ in the interval $(0, 0.5]$?

(c) As θ approaches 0, how do θ and $\sin \theta$ compare? Explain.

84. THINK ABOUT IT

(a) Complete the table.

θ	0°	18°	36°	54°	72°	90°
$\sin \theta$						
$\cos \theta$						

(b) Discuss the behavior of the sine function for θ in the range from 0° to 90°.

(c) Discuss the behavior of the cosine function for θ in the range from 0° to 90°.

(d) Use the definitions of the sine and cosine functions to explain the results of parts (b) and (c).

85. WRITING In right triangle trigonometry, explain why $\sin 30° = \frac{1}{2}$ regardless of the size of the triangle.

86. GEOMETRY Use the equilateral triangle shown in Figure 1.29 and similar triangles to verify the points in Figure 1.23 (in Section 1.2) that do not lie on the axes.

87. THINK ABOUT IT You are given only the value $\tan \theta$. Is it possible to find the value of $\sec \theta$ without finding the measure of θ? Explain.

88. CAPSTONE The Johnstown Inclined Plane in Pennsylvania is one of the longest and steepest hoists in the world. The railway cars travel a distance of 896.5 feet at an angle of approximately 35.4°, rising to a height of 1693.5 feet above sea level.

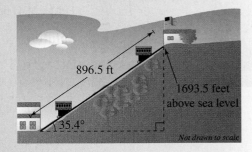

(a) Find the vertical rise of the inclined plane.

(b) Find the elevation of the lower end of the inclined plane.

(c) The cars move up the mountain at a rate of 300 feet per minute. Find the rate at which they rise vertically.

1.4 TRIGONOMETRIC FUNCTIONS OF ANY ANGLE

What you should learn

- Evaluate trigonometric functions of any angle.
- Find reference angles.
- Evaluate trigonometric functions of real numbers.

Why you should learn it

You can use trigonometric functions to model and solve real-life problems. For instance, in Exercise 99 on page 172, you can use trigonometric functions to model the monthly normal temperatures in New York City and Fairbanks, Alaska.

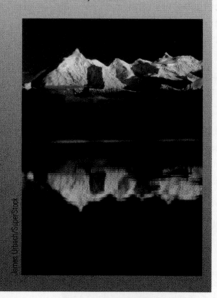

James Urbach/SuperStock

Introduction

In Section 1.3, the definitions of trigonometric functions were restricted to acute angles. In this section, the definitions are extended to cover *any* angle. If θ is an *acute* angle, these definitions coincide with those given in the preceding section.

Definitions of Trigonometric Functions of Any Angle

Let θ be an angle in standard position with (x, y) a point on the terminal side of θ and $r = \sqrt{x^2 + y^2} \neq 0$.

$$\sin \theta = \frac{y}{r} \qquad \cos \theta = \frac{x}{r}$$

$$\tan \theta = \frac{y}{x}, \quad x \neq 0 \qquad \cot \theta = \frac{x}{y}, \quad y \neq 0$$

$$\sec \theta = \frac{r}{x}, \quad x \neq 0 \qquad \csc \theta = \frac{r}{y}, \quad y \neq 0$$

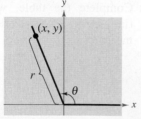

Because $r = \sqrt{x^2 + y^2}$ *cannot* be zero, it follows that the sine and cosine functions are defined for any real value of θ. However, if $x = 0$, the tangent and secant of θ are undefined. For example, the tangent of $90°$ is undefined. Similarly, if $y = 0$, the cotangent and cosecant of θ are undefined.

Example 1   Evaluating Trigonometric Functions

Let $(-3, 4)$ be a point on the terminal side of θ. Find the sine, cosine, and tangent of θ.

Solution

Referring to Figure 1.36, you can see that $x = -3$, $y = 4$, and

$$r = \sqrt{x^2 + y^2} = \sqrt{(-3)^2 + 4^2} = \sqrt{25} = 5.$$

So, you have the following.

$$\sin \theta = \frac{y}{r} = \frac{4}{5} \qquad \cos \theta = \frac{x}{r} = -\frac{3}{5} \qquad \tan \theta = \frac{y}{x} = -\frac{4}{3}$$

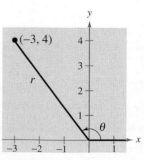

FIGURE 1.36

CHECK*Point* Now try Exercise 9.

Algebra Help

The formula $r = \sqrt{x^2 + y^2}$ is a result of the Distance Formula. You can review the Distance Formula in Section P.3.

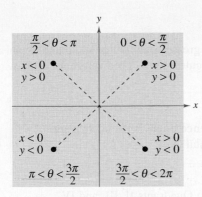

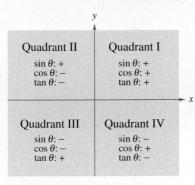

FIGURE **1.37**

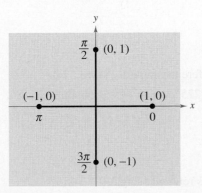

FIGURE **1.38**

The *signs* of the trigonometric functions in the four quadrants can be determined from the definitions of the functions. For instance, because $\cos \theta = x/r$, it follows that $\cos \theta$ is positive wherever $x > 0$, which is in Quadrants I and IV. (Remember, r is always positive.) In a similar manner, you can verify the results shown in Figure 1.37.

Example 2   Evaluating Trigonometric Functions

Given $\tan \theta = -\frac{5}{4}$ and $\cos \theta > 0$, find $\sin \theta$ and $\sec \theta$.

Solution

Note that θ lies in Quadrant IV because that is the only quadrant in which the tangent is negative and the cosine is positive. Moreover, using

$$\tan \theta = \frac{y}{x}$$

$$= -\frac{5}{4}$$

and the fact that y is negative in Quadrant IV, you can let $y = -5$ and $x = 4$. So, $r = \sqrt{16 + 25} = \sqrt{41}$ and you have

$$\sin \theta = \frac{y}{r} = \frac{-5}{\sqrt{41}}$$

$$\approx -0.7809$$

$$\sec \theta = \frac{r}{x} = \frac{\sqrt{41}}{4}$$

$$\approx 1.6008.$$

CHECK Point   Now try Exercise 23.

Example 3   Trigonometric Functions of Quadrant Angles

Evaluate the cosine and tangent functions at the four quadrant angles 0, $\dfrac{\pi}{2}$, π, and $\dfrac{3\pi}{2}$.

Solution

To begin, choose a point on the terminal side of each angle, as shown in Figure 1.38. For each of the four points, $r = 1$, and you have the following.

$$\cos 0 = \frac{x}{r} = \frac{1}{1} = 1 \qquad \tan 0 = \frac{y}{x} = \frac{0}{1} = 0 \qquad\qquad (x, y) = (1, 0)$$

$$\cos \frac{\pi}{2} = \frac{x}{r} = \frac{0}{1} = 0 \qquad \tan \frac{\pi}{2} = \frac{y}{x} = \frac{1}{0} \Rightarrow \text{undefined} \qquad (x, y) = (0, 1)$$

$$\cos \pi = \frac{x}{r} = \frac{-1}{1} = -1 \qquad \tan \pi = \frac{y}{x} = \frac{0}{-1} = 0 \qquad\qquad (x, y) = (-1, 0)$$

$$\cos \frac{3\pi}{2} = \frac{x}{r} = \frac{0}{1} = 0 \qquad \tan \frac{3\pi}{2} = \frac{y}{x} = \frac{-1}{0} \Rightarrow \text{undefined} \qquad (x, y) = (0, -1)$$

CHECK Point   Now try Exercise 37.

Reference Angles

The values of the trigonometric functions of angles greater than 90° (or less than 0°) can be determined from their values at corresponding acute angles called **reference angles.**

> ### Definition of Reference Angle
> Let θ be an angle in standard position. Its **reference angle** is the acute angle θ' formed by the terminal side of θ and the horizontal axis.

Figure 1.39 shows the reference angles for θ in Quadrants II, III, and IV.

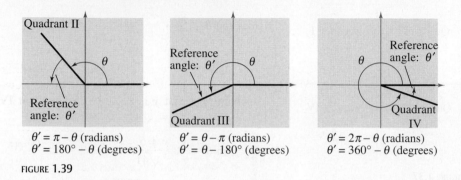

$\theta' = \pi - \theta$ (radians)
$\theta' = 180° - \theta$ (degrees)

$\theta' = \theta - \pi$ (radians)
$\theta' = \theta - 180°$ (degrees)

$\theta' = 2\pi - \theta$ (radians)
$\theta' = 360° - \theta$ (degrees)

FIGURE 1.39

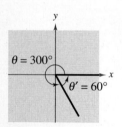

FIGURE 1.40

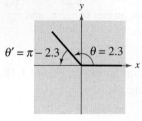

FIGURE 1.41

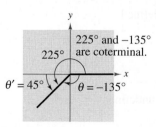

FIGURE 1.42

Example 4 Finding Reference Angles

Find the reference angle θ'.

a. $\theta = 300°$ **b.** $\theta = 2.3$ **c.** $\theta = -135°$

Solution

a. Because 300° lies in Quadrant IV, the angle it makes with the x-axis is

$$\theta' = 360° - 300°$$

$$= 60°. \qquad \text{Degrees}$$

Figure 1.40 shows the angle $\theta = 300°$ and its reference angle $\theta' = 60°$.

b. Because 2.3 lies between $\pi/2 \approx 1.5708$ and $\pi \approx 3.1416$, it follows that it is in Quadrant II and its reference angle is

$$\theta' = \pi - 2.3$$

$$\approx 0.8416. \qquad \text{Radians}$$

Figure 1.41 shows the angle $\theta = 2.3$ and its reference angle $\theta' = \pi - 2.3$.

c. First, determine that $-135°$ is coterminal with 225°, which lies in Quadrant III. So, the reference angle is

$$\theta' = 225° - 180°$$

$$= 45°. \qquad \text{Degrees}$$

Figure 1.42 shows the angle $\theta = -135°$ and its reference angle $\theta' = 45°$.

CHECK Point Now try Exercise 45.

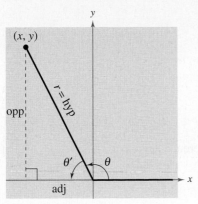

$$\text{opp} = |y|, \text{adj} = |x|$$

FIGURE 1.43

Trigonometric Functions of Real Numbers

To see how a reference angle is used to evaluate a trigonometric function, consider the point (x, y) on the terminal side of θ, as shown in Figure 1.43. By definition, you know that

$$\sin \theta = \frac{y}{r} \quad \text{and} \quad \tan \theta = \frac{y}{x}.$$

For the right triangle with acute angle θ' and sides of lengths $|x|$ and $|y|$, you have

$$\sin \theta' = \frac{\text{opp}}{\text{hyp}} = \frac{|y|}{r}$$

and

$$\tan \theta' = \frac{\text{opp}}{\text{adj}} = \frac{|y|}{|x|}.$$

So, it follows that $\sin \theta$ and $\sin \theta'$ are equal, *except possibly in sign*. The same is true for $\tan \theta$ and $\tan \theta'$ *and* for the other four trigonometric functions. In all cases, the sign of the function value can be determined by the quadrant in which θ lies.

Evaluating Trigonometric Functions of Any Angle

To find the value of a trigonometric function of any angle θ:

1. Determine the function value for the associated reference angle θ'.

2. Depending on the quadrant in which θ lies, affix the appropriate sign to the function value.

By using reference angles and the special angles discussed in the preceding section, you can greatly extend the scope of *exact* trigonometric values. For instance, knowing the function values of 30° means that you know the function values of all angles for which 30° is a reference angle. For convenience, the table below shows the exact values of the trigonometric functions of special angles and quadrant angles.

Trigonometric Values of Common Angles

θ (degrees)	0°	30°	45°	60°	90°	180°	270°
θ (radians)	0	$\frac{\pi}{6}$	$\frac{\pi}{4}$	$\frac{\pi}{3}$	$\frac{\pi}{2}$	π	$\frac{3\pi}{2}$
$\sin \theta$	0	$\frac{1}{2}$	$\frac{\sqrt{2}}{2}$	$\frac{\sqrt{3}}{2}$	1	0	-1
$\cos \theta$	1	$\frac{\sqrt{3}}{2}$	$\frac{\sqrt{2}}{2}$	$\frac{1}{2}$	0	-1	0
$\tan \theta$	0	$\frac{\sqrt{3}}{3}$	1	$\sqrt{3}$	Undef.	0	Undef.

| Example 5 | Using Reference Angles |

Evaluate each trigonometric function.

a. $\cos \dfrac{4\pi}{3}$ **b.** $\tan(-210°)$ **c.** $\csc \dfrac{11\pi}{4}$

Solution

a. Because $\theta = 4\pi/3$ lies in Quadrant III, the reference angle is

$$\theta' = \frac{4\pi}{3} - \pi = \frac{\pi}{3}$$

as shown in Figure 1.44. Moreover, the cosine is negative in Quadrant III, so

$$\cos \frac{4\pi}{3} = (-) \cos \frac{\pi}{3}$$

$$= -\frac{1}{2}.$$

b. Because $-210° + 360° = 150°$, it follows that $-210°$ is coterminal with the second-quadrant angle 150°. So, the reference angle is $\theta' = 180° - 150° = 30°$, as shown in Figure 1.45. Finally, because the tangent is negative in Quadrant II, you have

$$\tan(-210°) = (-) \tan 30°$$

$$= -\frac{\sqrt{3}}{3}.$$

c. Because $(11\pi/4) - 2\pi = 3\pi/4$, it follows that $11\pi/4$ is coterminal with the second-quadrant angle $3\pi/4$. So, the reference angle is $\theta' = \pi - (3\pi/4) = \pi/4$, as shown in Figure 1.46. Because the cosecant is positive in Quadrant II, you have

$$\csc \frac{11\pi}{4} = (+) \csc \frac{\pi}{4}$$

$$= \frac{1}{\sin(\pi/4)}$$

$$= \sqrt{2}.$$

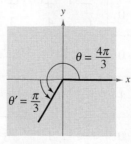

FIGURE **1.44**

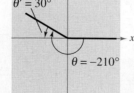

FIGURE **1.45**

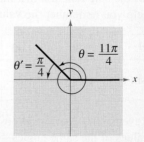

FIGURE **1.46**

CHECK*Point* Now try Exercise 59.

Example 6   Using Trigonometric Identities

Let θ be an angle in Quadrant II such that $\sin \theta = \frac{1}{3}$. Find (a) $\cos \theta$ and (b) $\tan \theta$ by using trigonometric identities.

Solution

a. Using the Pythagorean identity $\sin^2 \theta + \cos^2 \theta = 1$, you obtain

$$\left(\frac{1}{3}\right)^2 + \cos^2 \theta = 1 \qquad\qquad \text{Substitute } \tfrac{1}{3} \text{ for } \sin \theta.$$

$$\cos^2 \theta = 1 - \frac{1}{9} = \frac{8}{9}.$$

Because $\cos \theta < 0$ in Quadrant II, you can use the negative root to obtain

$$\cos \theta = -\frac{\sqrt{8}}{\sqrt{9}}$$

$$= -\frac{2\sqrt{2}}{3}.$$

b. Using the trigonometric identity $\tan \theta = \dfrac{\sin \theta}{\cos \theta}$, you obtain

$$\tan \theta = \frac{1/3}{-2\sqrt{2}/3} \qquad\qquad \text{Substitute for } \sin \theta \text{ and } \cos \theta.$$

$$= -\frac{1}{2\sqrt{2}}$$

$$= -\frac{\sqrt{2}}{4}.$$

CHECK Point   Now try Exercise 69.

You can use a calculator to evaluate trigonometric functions, as shown in the next example.

Example 7   Using a Calculator

Use a calculator to evaluate each trigonometric function.

a. $\cot 410°$ **b.** $\sin(-7)$ **c.** $\sec \dfrac{\pi}{9}$

Solution

Function	Mode	Calculator Keystrokes	Display
a. $\cot 410°$	Degree	(TAN) (410))) x⁻¹ ENTER	0.8390996
b. $\sin(-7)$	Radian	SIN ((−) 7) ENTER	−0.6569866
c. $\sec \dfrac{\pi}{9}$	Radian	(COS) (π ÷ 9))) x⁻¹ ENTER	1.0641778

CHECK Point   Now try Exercise 79.

1.4 EXERCISES

See www.CalcChat.com for worked-out solutions to odd-numbered exercises.

VOCABULARY: Fill in the blanks.

In Exercises 1–6, let θ be an angle in standard position, with (x, y) a point on the terminal side of θ and $r = \sqrt{x^2 + y^2} \neq 0$.

1. $\sin \theta =$ _____

2. $\dfrac{r}{y} =$ _____

3. $\tan \theta =$ _____

4. $\sec \theta =$ _____

5. $\dfrac{x}{r} =$ _____

6. $\dfrac{x}{y} =$ _____

7. Because $r = \sqrt{x^2 + y^2}$ cannot be _____, the sine and cosine functions are _____ for any real value of θ.

8. The acute positive angle that is formed by the terminal side of the angle θ and the horizontal axis is called the _____ angle of θ and is denoted by θ'.

SKILLS AND APPLICATIONS

In Exercises 9–12, determine the exact values of the six trigonometric functions of the angle θ.

9. (a) 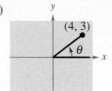   (b)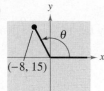

10. (a) 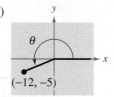   (b)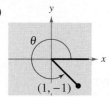

11. (a) 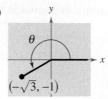   (b)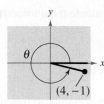

12. (a) 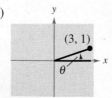   (b)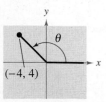

In Exercises 13–18, the point is on the terminal side of an angle in standard position. Determine the exact values of the six trigonometric functions of the angle.

13. $(5, 12)$

14. $(8, 15)$

15. $(-5, -2)$

16. $(-4, 10)$

17. $(-5.4, 7.2)$

18. $\left(3\frac{1}{2}, -7\frac{3}{4}\right)$

In Exercises 19–22, state the quadrant in which θ lies.

19. $\sin \theta > 0$ and $\cos \theta > 0$

20. $\sin \theta < 0$ and $\cos \theta < 0$

21. $\sin \theta > 0$ and $\cos \theta < 0$

22. $\sec \theta > 0$ and $\cot \theta < 0$

In Exercises 23–32, find the values of the six trigonometric functions of θ with the given constraint.

	Function Value	*Constraint*
23.	$\tan \theta = -\dfrac{15}{8}$	$\sin \theta > 0$
24.	$\cos \theta = \dfrac{8}{17}$	$\tan \theta < 0$
25.	$\sin \theta = \dfrac{3}{5}$	θ lies in Quadrant II.
26.	$\cos \theta = -\dfrac{4}{5}$	θ lies in Quadrant III.
27.	$\cot \theta = -3$	$\cos \theta > 0$
28.	$\csc \theta = 4$	$\cot \theta < 0$
29.	$\sec \theta = -2$	$\sin \theta < 0$
30.	$\sin \theta = 0$	$\sec \theta = -1$
31.	$\cot \theta$ is undefined.	$\pi/2 \leq \theta \leq 3\pi/2$
32.	$\tan \theta$ is undefined.	$\pi \leq \theta \leq 2\pi$

In Exercises 33–36, the terminal side of θ lies on the given line in the specified quadrant. Find the values of the six trigonometric functions of θ by finding a point on the line.

	Line	*Quadrant*
33.	$y = -x$	II
34.	$y = \frac{1}{3}x$	III
35.	$2x - y = 0$	III
36.	$4x + 3y = 0$	IV

In Exercises 37–44, evaluate the trigonometric function of the quadrant angle.

37. $\sin \pi$

38. $\csc \dfrac{3\pi}{2}$

39. $\sec \dfrac{3\pi}{2}$

40. $\sec \pi$

41. $\sin \dfrac{\pi}{2}$

42. $\cot \pi$

43. $\csc \pi$

44. $\cot \dfrac{\pi}{2}$

In Exercises 45–52, find the reference angle θ', and sketch θ and θ' in standard position.

45. $\theta = 160°$

46. $\theta = 309°$

47. $\theta = -125°$

48. $\theta = -215°$

49. $\theta = \dfrac{2\pi}{3}$

50. $\theta = \dfrac{7\pi}{6}$

51. $\theta = 4.8$

52. $\theta = 11.6$

In Exercises 53–68, evaluate the sine, cosine, and tangent of the angle without using a calculator.

53. $225°$

54. $300°$

55. $750°$

56. $-405°$

57. $-150°$

58. $-840°$

59. $\dfrac{2\pi}{3}$

60. $\dfrac{3\pi}{4}$

61. $\dfrac{5\pi}{4}$

62. $\dfrac{7\pi}{6}$

63. $-\dfrac{\pi}{6}$

64. $-\dfrac{\pi}{2}$

65. $\dfrac{9\pi}{4}$

66. $\dfrac{10\pi}{3}$

67. $-\dfrac{3\pi}{2}$

68. $-\dfrac{23\pi}{4}$

In Exercises 69–74, find the indicated trigonometric value in the specified quadrant.

Function	Quadrant	Trigonometric Value
69. $\sin \theta = -\dfrac{3}{5}$	IV	$\cos \theta$
70. $\cot \theta = -3$	II	$\sin \theta$
71. $\tan \theta = \dfrac{3}{2}$	III	$\sec \theta$
72. $\csc \theta = -2$	IV	$\cot \theta$
73. $\cos \theta = \dfrac{5}{8}$	I	$\sec \theta$
74. $\sec \theta = -\dfrac{9}{4}$	III	$\tan \theta$

In Exercises 75–90, use a calculator to evaluate the trigonometric function. Round your answer to four decimal places. (Be sure the calculator is set in the correct angle mode.)

75. $\sin 10°$

76. $\sec 225°$

77. $\cos(-110°)$

78. $\csc(-330°)$

79. $\tan 304°$

80. $\cot 178°$

81. $\sec 72°$

82. $\tan(-188°)$

83. $\tan 4.5$

84. $\cot 1.35$

85. $\tan \dfrac{\pi}{9}$

86. $\tan\left(-\dfrac{\pi}{9}\right)$

87. $\sin(-0.65)$

88. $\sec 0.29$

89. $\cot\left(-\dfrac{11\pi}{8}\right)$

90. $\csc\left(-\dfrac{15\pi}{14}\right)$

In Exercises 91–96, find two solutions of the equation. Give your answers in degrees ($0° \le \theta < 360°$) and in radians ($0 \le \theta < 2\pi$). Do not use a calculator.

91. (a) $\sin \theta = \dfrac{1}{2}$ (b) $\sin \theta = -\dfrac{1}{2}$

92. (a) $\cos \theta = \dfrac{\sqrt{2}}{2}$ (b) $\cos \theta = -\dfrac{\sqrt{2}}{2}$

93. (a) $\csc \theta = \dfrac{2\sqrt{3}}{3}$ (b) $\cot \theta = -1$

94. (a) $\sec \theta = 2$ (b) $\sec \theta = -2$

95. (a) $\tan \theta = 1$ (b) $\cot \theta = -\sqrt{3}$

96. (a) $\sin \theta = \dfrac{\sqrt{3}}{2}$ (b) $\sin \theta = -\dfrac{\sqrt{3}}{2}$

97. DISTANCE   An airplane, flying at an altitude of 6 miles, is on a flight path that passes directly over an observer (see figure). If θ is the angle of elevation from the observer to the plane, find the distance d from the observer to the plane when (a) $\theta = 30°$, (b) $\theta = 90°$, and (c) $\theta = 120°$.

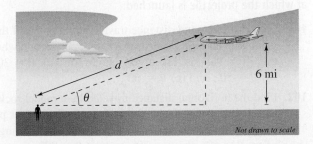

Not drawn to scale

98. HARMONIC MOTION   The displacement from equilibrium of an oscillating weight suspended by a spring is given by $y(t) = 2 \cos 6t$, where y is the displacement (in centimeters) and t is the time (in seconds). Find the displacement when (a) $t = 0$, (b) $t = \dfrac{1}{4}$, and (c) $t = \dfrac{1}{2}$.

99. DATA ANALYSIS: METEOROLOGY The table shows the monthly normal temperatures (in degrees Fahrenheit) for selected months in New York City (N) and Fairbanks, Alaska (F). (Source: National Climatic Data Center)

Month	New York City, N	Fairbanks, F
January	33	−10
April	52	32
July	77	62
October	58	24
December	38	−6

(a) Use the *regression* feature of a graphing utility to find a model of the form $y = a \sin(bt + c) + d$ for each city. Let t represent the month, with $t = 1$ corresponding to January.

(b) Use the models from part (a) to find the monthly normal temperatures for the two cities in February, March, May, June, August, September, and November.

(c) Compare the models for the two cities.

100. SALES A company that produces snowboards, which are seasonal products, forecasts monthly sales over the next 2 years to be $S = 23.1 + 0.442t + 4.3 \cos(\pi t/6)$, where S is measured in thousands of units and t is the time in months, with $t = 1$ representing January 2010. Predict sales for each of the following months.

(a) February 2010   (b) February 2011

(c) June 2010   (d) June 2011

PATH OF A PROJECTILE In Exercises 101 and 102, use the following information. The horizontal distance d (in feet) traveled by a projectile with an initial speed of v feet per second is modeled by $d = v^2/32 \sin 2\theta$, where θ is the angle at which the projectile is launched.

101. Find the horizontal distance traveled by a golf ball that is hit with an initial speed of 100 feet per second when the golf ball is hit at an angle of (a) $\theta = 30°$, (b) $\theta = 50°$, and (c) $\theta = 60°$.

102. Find the horizontal distance traveled by a model rocket that is launched with an initial speed of 120 feet per second when the model rocket is launched at an angle of (a) $\theta = 60°$, (b) $\theta = 70°$, and (c) $\theta = 80°$.

EXPLORATION

TRUE OR FALSE? In Exercises 103 and 104, determine whether the statement is true or false. Justify your answer.

103. In each of the four quadrants, the signs of the secant function and sine function will be the same.

104. To find the reference angle for an angle θ (given in degrees), find the integer n such that $0 \le 360°n - \theta \le 360°$. The difference $360°n - \theta$ is the reference angle.

105. WRITING Consider an angle in standard position with $r = 12$ centimeters, as shown in the figure. Write a short paragraph describing the changes in the values of x, y, $\sin \theta$, $\cos \theta$, and $\tan \theta$ as θ increases continuously from $0°$ to $90°$.

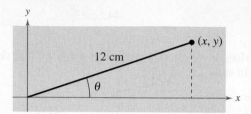

106. CAPSTONE Write a short paper in your own words explaining to a classmate how to evaluate the six trigonometric functions of any angle θ in standard position. Include an explanation of reference angles and how to use them, the signs of the functions in each of the four quadrants, and the trigonometric values of common angles. Be sure to include figures or diagrams in your paper.

107. THINK ABOUT IT The figure shows point $P(x, y)$ on a unit circle and right triangle OAP.

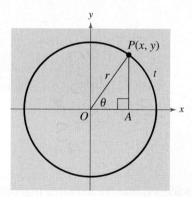

(a) Find $\sin t$ and $\cos t$ using the unit circle definitions of sine and cosine (from Section 1.2).

(b) What is the value of r? Explain.

(c) Use the definitions of sine and cosine given in this section to find $\sin \theta$ and $\cos \theta$. Write your answers in terms of x and y.

(d) Based on your answers to parts (a) and (c), what can you conclude?

What you should learn

- Sketch the graphs of basic sine and cosine functions.
- Use amplitude and period to help sketch the graphs of sine and cosine functions.
- Sketch translations of the graphs of sine and cosine functions.
- Use sine and cosine functions to model real-life data.

Why you should learn it

Sine and cosine functions are often used in scientific calculations. For instance, in Exercise 87 on page 182, you can use a trigonometric function to model the airflow of your respiratory cycle.

© Karl Weatherly/Corbis

1.5 GRAPHS OF SINE AND COSINE FUNCTIONS

Basic Sine and Cosine Curves

In this section, you will study techniques for sketching the graphs of the sine and cosine functions. The graph of the sine function is a **sine curve.** In Figure 1.47, the black portion of the graph represents one period of the function and is called **one cycle** of the sine curve. The gray portion of the graph indicates that the basic sine curve repeats indefinitely in the positive and negative directions. The graph of the cosine function is shown in Figure 1.48.

Recall from Section 1.2 that the domain of the sine and cosine functions is the set of all real numbers. Moreover, the range of each function is the interval $[-1, 1]$, and each function has a period of 2π. Do you see how this information is consistent with the basic graphs shown in Figures 1.47 and 1.48?

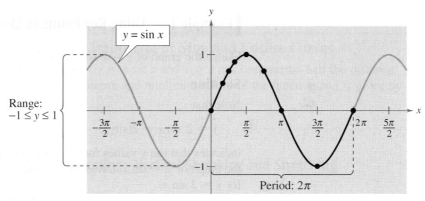

FIGURE 1.47

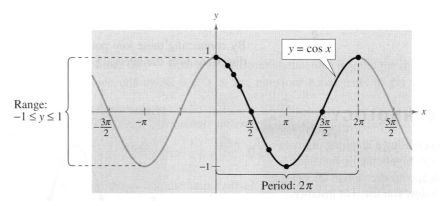

FIGURE 1.48

Note in Figures 1.47 and 1.48 that the sine curve is symmetric with respect to the *origin*, whereas the cosine curve is symmetric with respect to the *y-axis*. These properties of symmetry follow from the fact that the sine function is odd and the cosine function is even.

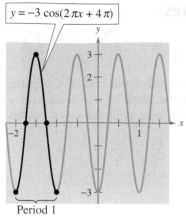

$y = -3 \cos(2\pi x + 4\pi)$

Period 1

FIGURE **1.55**

Example 5 Horizontal Translation

Sketch the graph of

$$y = -3 \cos(2\pi x + 4\pi).$$

Solution

The amplitude is 3 and the period is $2\pi/2\pi = 1$. By solving the equations

$$2\pi x + 4\pi = 0$$
$$2\pi x = -4\pi$$
$$x = -2$$

and

$$2\pi x + 4\pi = 2\pi$$
$$2\pi x = -2\pi$$
$$x = -1$$

you see that the interval $[-2, -1]$ corresponds to one cycle of the graph. Dividing this interval into four equal parts produces the key points

Minimum	Intercept	Maximum	Intercept		Minimum
$(-2, -3),$	$\left(-\dfrac{7}{4}, 0\right),$	$\left(-\dfrac{3}{2}, 3\right),$	$\left(-\dfrac{5}{4}, 0\right),$	and	$(-1, -3).$

The graph is shown in Figure 1.55.

CHECK Point ▶ Now try Exercise 51.

The final type of transformation is the *vertical translation* caused by the constant d in the equations

$$y = d + a \sin(bx - c)$$

and

$$y = d + a \cos(bx - c).$$

The shift is d units upward for $d > 0$ and d units downward for $d < 0$. In other words, the graph oscillates about the horizontal line $y = d$ instead of about the x-axis.

Example 6 Vertical Translation

Sketch the graph of

$$y = 2 + 3 \cos 2x.$$

Solution

The amplitude is 3 and the period is π. The key points over the interval $[0, \pi]$ are

$$(0, 5), \qquad \left(\frac{\pi}{4}, 2\right), \qquad \left(\frac{\pi}{2}, -1\right), \qquad \left(\frac{3\pi}{4}, 2\right), \qquad \text{and} \qquad (\pi, 5).$$

The graph is shown in Figure 1.56. Compared with the graph of $f(x) = 3 \cos 2x$, the graph of $y = 2 + 3 \cos 2x$ is shifted upward two units.

CHECK Point ▶ Now try Exercise 57.

$y = 2 + 3 \cos 2x$

Period π

FIGURE **1.56**

Mathematical Modeling

Sine and cosine functions can be used to model many real-life situations, including electric currents, musical tones, radio waves, tides, and weather patterns.

Time, t	Depth, y
Midnight	3.4
2 A.M.	8.7
4 A.M.	11.3
6 A.M.	9.1
8 A.M.	3.8
10 A.M.	0.1
Noon	1.2

Example 7   Finding a Trigonometric Model

Throughout the day, the depth of water at the end of a dock in Bar Harbor, Maine varies with the tides. The table shows the depths (in feet) at various times during the morning.   (Source: Nautical Software, Inc.)

a. Use a trigonometric function to model the data.

b. Find the depths at 9 A.M. and 3 P.M.

c. A boat needs at least 10 feet of water to moor at the dock. During what times in the afternoon can it safely dock?

Solution

a. Begin by graphing the data, as shown in Figure 1.57. You can use either a sine or a cosine model. Suppose you use a cosine model of the form

$$y = a \cos(bt - c) + d.$$

The difference between the maximum height and the minimum height of the graph is twice the amplitude of the function. So, the amplitude is

$$a = \frac{1}{2}[(\text{maximum depth}) - (\text{minimum depth})] = \frac{1}{2}(11.3 - 0.1) = 5.6.$$

The cosine function completes one half of a cycle between the times at which the maximum and minimum depths occur. So, the period is

$$p = 2[(\text{time of min. depth}) - (\text{time of max. depth})] = 2(10 - 4) = 12$$

which implies that $b = 2\pi/p \approx 0.524$. Because high tide occurs 4 hours after midnight, consider the left endpoint to be $c/b = 4$, so $c \approx 2.094$. Moreover, because the average depth is $\frac{1}{2}(11.3 + 0.1) = 5.7$, it follows that $d = 5.7$. So, you can model the depth with the function given by

$$y = 5.6 \cos(0.524t - 2.094) + 5.7.$$

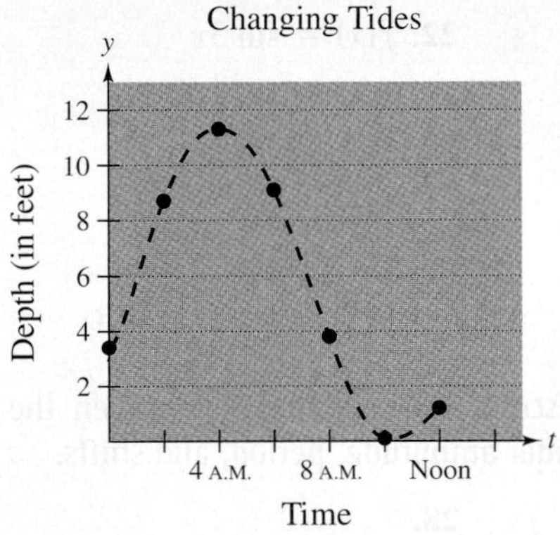

b. The depths at 9 A.M. and 3 P.M. are as follows.

$$y = 5.6 \cos(0.524 \cdot 9 - 2.094) + 5.7$$

$$\approx 0.84 \text{ foot} \qquad\qquad\qquad 9 \text{ A.M.}$$

$$y = 5.6 \cos(0.524 \cdot 15 - 2.094) + 5.7$$

$$\approx 10.57 \text{ feet} \qquad\qquad\qquad 3 \text{ P.M.}$$

c. To find out when the depth y is at least 10 feet, you can graph the model with the line $y = 10$ using a graphing utility, as shown in Figure 1.58. Using the *intersect* feature, you can determine that the depth is at least 10 feet between 2:42 P.M. ($t \approx 14.7$) and 5:18 P.M. ($t \approx 17.3$).

CHECK*Point* ▸   Now try Exercise 91.   ▮

Changing Tides

Depth (in feet)

4 A.M.   8 A.M.   Noon

Time

FIGURE **1.57**

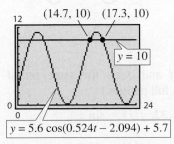

$$y = 5.6 \cos(0.524t - 2.094) + 5.7$$

FIGURE **1.58**

1.5 EXERCISES

See www.CalcChat.com for worked-out solutions to odd-numbered exercises.

VOCABULARY: Fill in the blanks.

1. One period of a sine or cosine function is called one _____ of the sine or cosine curve.

2. The _____ of a sine or cosine curve represents half the distance between the maximum and minimum values of the function.

3. For the function given by $y = a \sin(bx - c)$, $\dfrac{c}{b}$ represents the _____ _____ of the graph of the function.

4. For the function given by $y = d + a \cos(bx - c)$, d represents a _____ _____ of the graph of the function.

SKILLS AND APPLICATIONS

In Exercises 5–18, find the period and amplitude.

5. $y = 2 \sin 5x$

6. $y = 3 \cos 2x$

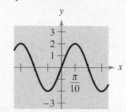
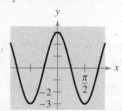

7. $y = \dfrac{3}{4} \cos \dfrac{x}{2}$

8. $y = -3 \sin \dfrac{x}{3}$

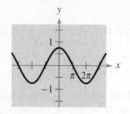
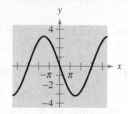

9. $y = \dfrac{1}{2} \sin \dfrac{\pi x}{3}$

10. $y = \dfrac{3}{2} \cos \dfrac{\pi x}{2}$

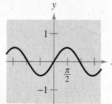
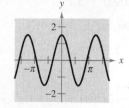

11. $y = -4 \sin x$

12. $y = -\cos \dfrac{2x}{3}$

13. $y = 3 \sin 10x$

14. $y = \dfrac{1}{5} \sin 6x$

15. $y = \dfrac{5}{3} \cos \dfrac{4x}{5}$

16. $y = \dfrac{5}{2} \cos \dfrac{x}{4}$

17. $y = \dfrac{1}{4} \sin 2\pi x$

18. $y = \dfrac{2}{3} \cos \dfrac{\pi x}{10}$

In Exercises 19–26, describe the relationship between the graphs of f and g. Consider amplitude, period, and shifts.

19. $f(x) = \sin x$
    $g(x) = \sin(x - \pi)$

20. $f(x) = \cos x$
    $g(x) = \cos(x + \pi)$

21. $f(x) = \cos 2x$
    $g(x) = -\cos 2x$

22. $f(x) = \sin 3x$
    $g(x) = \sin(-3x)$

23. $f(x) = \cos x$
    $g(x) = \cos 2x$

24. $f(x) = \sin x$
    $g(x) = \sin 3x$

25. $f(x) = \sin 2x$
    $g(x) = 3 + \sin 2x$

26. $f(x) = \cos 4x$
    $g(x) = -2 + \cos 4x$

In Exercises 27–30, describe the relationship between the graphs of f and g. Consider amplitude, period, and shifts.

27.
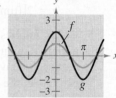

28.
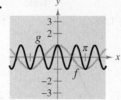

29.
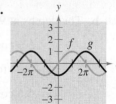

30.
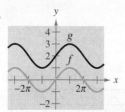

In Exercises 31–38, graph f and g on the same set of coordinate axes. (Include two full periods.)

31. $f(x) = -2 \sin x$
    $g(x) = 4 \sin x$

32. $f(x) = \sin x$
    $g(x) = \sin \dfrac{x}{3}$

33. $f(x) = \cos x$
    $g(x) = 2 + \cos x$

34. $f(x) = 2 \cos 2x$
    $g(x) = -\cos 4x$

35. $f(x) = -\dfrac{1}{2}\sin\dfrac{x}{2}$ **36.** $f(x) = 4\sin\pi x$

$g(x) = 3 - \dfrac{1}{2}\sin\dfrac{x}{2}$ $g(x) = 4\sin\pi x - 3$

37. $f(x) = 2\cos x$ **38.** $f(x) = -\cos x$

$g(x) = 2\cos(x + \pi)$ $g(x) = -\cos(x - \pi)$

In Exercises 39–60, sketch the graph of the function. (Include two full periods.)

39. $y = 5\sin x$ **40.** $y = \dfrac{1}{4}\sin x$

41. $y = \dfrac{1}{3}\cos x$ **42.** $y = 4\cos x$

43. $y = \cos\dfrac{x}{2}$ **44.** $y = \sin 4x$

45. $y = \cos 2\pi x$ **46.** $y = \sin\dfrac{\pi x}{4}$

47. $y = -\sin\dfrac{2\pi x}{3}$ **48.** $y = -10\cos\dfrac{\pi x}{6}$

49. $y = \sin\left(x - \dfrac{\pi}{2}\right)$ **50.** $y = \sin(x - 2\pi)$

51. $y = 3\cos(x + \pi)$ **52.** $y = 4\cos\left(x + \dfrac{\pi}{4}\right)$

53. $y = 2 - \sin\dfrac{2\pi x}{3}$ **54.** $y = -3 + 5\cos\dfrac{\pi t}{12}$

55. $y = 2 + \dfrac{1}{10}\cos 60\pi x$ **56.** $y = 2\cos x - 3$

57. $y = 3\cos(x + \pi) - 3$ **58.** $y = 4\cos\left(x + \dfrac{\pi}{4}\right) + 4$

59. $y = \dfrac{2}{3}\cos\left(\dfrac{x}{2} - \dfrac{\pi}{4}\right)$ **60.** $y = -3\cos(6x + \pi)$

In Exercises 61–66, g is related to a parent function $f(x) = \sin(x)$ or $f(x) = \cos(x)$. (a) Describe the sequence of transformations from f to g. (b) Sketch the graph of g. (c) Use function notation to write g in terms of f.

61. $g(x) = \sin(4x - \pi)$ **62.** $g(x) = \sin(2x + \pi)$

63. $g(x) = \cos(x - \pi) + 2$ **64.** $g(x) = 1 + \cos(x + \pi)$

65. $g(x) = 2\sin(4x - \pi) - 3$ **66.** $g(x) = 4 - \sin(2x + \pi)$

In Exercises 67–72, use a graphing utility to graph the function. Include two full periods. Be sure to choose an appropriate viewing window.

67. $y = -2\sin(4x + \pi)$ **68.** $y = -4\sin\left(\dfrac{2}{3}x - \dfrac{\pi}{3}\right)$

69. $y = \cos\left(2\pi x - \dfrac{\pi}{2}\right) + 1$

70. $y = 3\cos\left(\dfrac{\pi x}{2} + \dfrac{\pi}{2}\right) - 2$

71. $y = -0.1\sin\left(\dfrac{\pi x}{10} + \pi\right)$ **72.** $y = \dfrac{1}{100}\sin 120\pi t$

GRAPHICAL REASONING In Exercises 73–76, find a and d for the function $f(x) = a\cos x + d$ such that the graph of f matches the figure.

73.
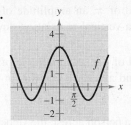

74.
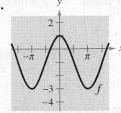

75.
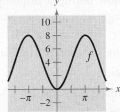

76.
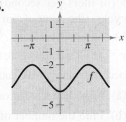

GRAPHICAL REASONING In Exercises 77–80, find a, b, and c for the function $f(x) = a\sin(bx - c)$ such that the graph of f matches the figure.

77.
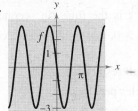

78.
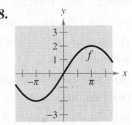

79.
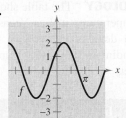

80.
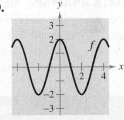

In Exercises 81 and 82, use a graphing utility to graph y_1 and y_2 in the interval $[-2\pi, 2\pi]$. Use the graphs to find real numbers x such that $y_1 = y_2$.

81. $y_1 = \sin x$ **82.** $y_1 = \cos x$

$y_2 = -\dfrac{1}{2}$ $y_2 = -1$

In Exercises 83–86, write an equation for the function that is described by the given characteristics.

83. A sine curve with a period of π, an amplitude of 2, a right phase shift of $\pi/2$, and a vertical translation up 1 unit

84. A sine curve with a period of 4π, an amplitude of 3, a left phase shift of $\pi/4$, and a vertical translation down 1 unit

85. A cosine curve with a period of π, an amplitude of 1, a left phase shift of π, and a vertical translation down $\frac{3}{2}$ units

86. A cosine curve with a period of 4π, an amplitude of 3, a right phase shift of $\pi/2$, and a vertical translation up 2 units

87. RESPIRATORY CYCLE For a person at rest, the velocity v (in liters per second) of airflow during a respiratory cycle (the time from the beginning of one breath to the beginning of the next) is given by $v = 0.85 \sin \dfrac{\pi t}{3}$, where t is the time (in seconds). (Inhalation occurs when $v > 0$, and exhalation occurs when $v < 0$.)

(a) Find the time for one full respiratory cycle.

(b) Find the number of cycles per minute.

(c) Sketch the graph of the velocity function.

88. RESPIRATORY CYCLE After exercising for a few minutes, a person has a respiratory cycle for which the velocity of airflow is approximated by $v = 1.75 \sin \dfrac{\pi t}{2}$, where t is the time (in seconds). (Inhalation occurs when $v > 0$, and exhalation occurs when $v < 0$.)

(a) Find the time for one full respiratory cycle.

(b) Find the number of cycles per minute.

(c) Sketch the graph of the velocity function.

89. DATA ANALYSIS: METEOROLOGY The table shows the maximum daily high temperatures in Las Vegas L and International Falls I (in degrees Fahrenheit) for month t, with $t = 1$ corresponding to January. (Source: National Climatic Data Center)

Month, t	Las Vegas, L	International Falls, I
1	57.1	13.8
2	63.0	22.4
3	69.5	34.9
4	78.1	51.5
5	87.8	66.6
6	98.9	74.2
7	104.1	78.6
8	101.8	76.3
9	93.8	64.7
10	80.8	51.7
11	66.0	32.5
12	57.3	18.1

(a) A model for the temperature in Las Vegas is given by

$$L(t) = 80.60 + 23.50 \cos\left(\frac{\pi t}{6} - 3.67\right).$$

Find a trigonometric model for International Falls.

(b) Use a graphing utility to graph the data points and the model for the temperatures in Las Vegas. How well does the model fit the data?

(c) Use a graphing utility to graph the data points and the model for the temperatures in International Falls. How well does the model fit the data?

(d) Use the models to estimate the average maximum temperature in each city. Which term of the models did you use? Explain.

(e) What is the period of each model? Are the periods what you expected? Explain.

(f) Which city has the greater variability in temperature throughout the year? Which factor of the models determines this variability? Explain.

90. HEALTH The function given by

$$P = 100 - 20 \cos \frac{5\pi t}{3}$$

approximates the blood pressure P (in millimeters of mercury) at time t (in seconds) for a person at rest.

(a) Find the period of the function.

(b) Find the number of heartbeats per minute.

91. PIANO TUNING When tuning a piano, a technician strikes a tuning fork for the A above middle C and sets up a wave motion that can be approximated by $y = 0.001 \sin 880\pi t$, where t is the time (in seconds).

(a) What is the period of the function?

(b) The frequency f is given by $f = 1/p$. What is the frequency of the note?

92. DATA ANALYSIS: ASTRONOMY The percents y (in decimal form) of the moon's face that was illuminated on day x in the year 2009, where $x = 1$ represents January 1, are shown in the table. (Source: U.S. Naval Observatory)

x	y
4	0.5
11	1.0
18	0.5
26	0.0
33	0.5
40	1.0

(a) Create a scatter plot of the data.

(b) Find a trigonometric model that fits the data.

(c) Add the graph of your model in part (b) to the scatter plot. How well does the model fit the data?

(d) What is the period of the model?

(e) Estimate the moon's percent illumination for March 12, 2009.

93. FUEL CONSUMPTION The daily consumption C (in gallons) of diesel fuel on a farm is modeled by

$$C = 30.3 + 21.6 \sin\left(\frac{2\pi t}{365} + 10.9\right)$$

where t is the time (in days), with $t = 1$ corresponding to January 1.

(a) What is the period of the model? Is it what you expected? Explain.

(b) What is the average daily fuel consumption? Which term of the model did you use? Explain.

(c) Use a graphing utility to graph the model. Use the graph to approximate the time of the year when consumption exceeds 40 gallons per day.

94. FERRIS WHEEL A Ferris wheel is built such that the height h (in feet) above ground of a seat on the wheel at time t (in seconds) can be modeled by

$$h(t) = 53 + 50 \sin\left(\frac{\pi}{10}t - \frac{\pi}{2}\right).$$

(a) Find the period of the model. What does the period tell you about the ride?

(b) Find the amplitude of the model. What does the amplitude tell you about the ride?

(c) Use a graphing utility to graph one cycle of the model.

EXPLORATION

TRUE OR FALSE? In Exercises 95–97, determine whether the statement is true or false. Justify your answer.

95. The graph of the function given by $f(x) = \sin(x + 2\pi)$ translates the graph of $f(x) = \sin x$ exactly one period to the right so that the two graphs look identical.

96. The function given by $y = \frac{1}{2}\cos 2x$ has an amplitude that is twice that of the function given by $y = \cos x$.

97. The graph of $y = -\cos x$ is a reflection of the graph of $y = \sin(x + \pi/2)$ in the x-axis.

98. WRITING Sketch the graph of $y = \cos bx$ for $b = \frac{1}{2}$, 2, and 3. How does the value of b affect the graph? How many complete cycles occur between 0 and 2π for each value of b?

99. WRITING Sketch the graph of $y = \sin(x - c)$ for $c = -\pi/4, 0$, and $\pi/4$. How does the value of c affect the graph?

100. CAPSTONE Use a graphing utility to graph the function given by $y = d + a\sin(bx - c)$, for several different values of a, b, c, and d. Write a paragraph describing the changes in the graph corresponding to changes in each constant.

CONJECTURE In Exercises 101 and 102, graph f and g on the same set of coordinate axes. Include two full periods. Make a conjecture about the functions.

101. $f(x) = \sin x, \quad g(x) = \cos\left(x - \frac{\pi}{2}\right)$

102. $f(x) = \sin x, \quad g(x) = -\cos\left(x + \frac{\pi}{2}\right)$

103. Using calculus, it can be shown that the sine and cosine functions can be approximated by the polynomials

$$\sin x \approx x - \frac{x^3}{3!} + \frac{x^5}{5!} \quad \text{and} \quad \cos x \approx 1 - \frac{x^2}{2!} + \frac{x^4}{4!}$$

where x is in radians.

(a) Use a graphing utility to graph the sine function and its polynomial approximation in the same viewing window. How do the graphs compare?

(b) Use a graphing utility to graph the cosine function and its polynomial approximation in the same viewing window. How do the graphs compare?

(c) Study the patterns in the polynomial approximations of the sine and cosine functions and predict the next term in each. Then repeat parts (a) and (b). How did the accuracy of the approximations change when an additional term was added?

104. Use the polynomial approximations of the sine and cosine functions in Exercise 103 to approximate the following function values. Compare the results with those given by a calculator. Is the error in the approximation the same in each case? Explain.

(a) $\sin \frac{1}{2}$ (b) $\sin 1$ (c) $\sin \frac{\pi}{6}$

(d) $\cos(-0.5)$ (e) $\cos 1$ (f) $\cos \frac{\pi}{4}$

PROJECT: METEOROLOGY To work an extended application analyzing the mean monthly temperature and mean monthly precipitation in Honolulu, Hawaii, visit this text's website at *academic.cengage.com*. (Data Source: National Climatic Data Center)

1.6 GRAPHS OF OTHER TRIGONOMETRIC FUNCTIONS

What you should learn

- Sketch the graphs of tangent functions.
- Sketch the graphs of cotangent functions.
- Sketch the graphs of secant and cosecant functions.
- Sketch the graphs of damped trigonometric functions.

Why you should learn it

Graphs of trigonometric functions can be used to model real-life situations such as the distance from a television camera to a unit in a parade, as in Exercise 92 on page 193.

Photodisc/Getty Images

Graph of the Tangent Function

Recall that the tangent function is odd. That is, $\tan(-x) = -\tan x$. Consequently, the graph of $y = \tan x$ is symmetric with respect to the origin. You also know from the identity $\tan x = \sin x/\cos x$ that the tangent is undefined for values at which $\cos x = 0$. Two such values are $x = \pm\pi/2 \approx \pm 1.5708$.

x	$-\dfrac{\pi}{2}$	-1.57	-1.5	$-\dfrac{\pi}{4}$	0	$\dfrac{\pi}{4}$	1.5	1.57	$\dfrac{\pi}{2}$
$\tan x$	Undef.	-1255.8	-14.1	-1	0	1	14.1	1255.8	Undef.

As indicated in the table, $\tan x$ increases without bound as x approaches $\pi/2$ from the left, and decreases without bound as x approaches $-\pi/2$ from the right. So, the graph of $y = \tan x$ has *vertical asymptotes* at $x = \pi/2$ and $x = -\pi/2$, as shown in Figure 1.59. Moreover, because the period of the tangent function is π, vertical asymptotes also occur when $x = \pi/2 + n\pi$, where n is an integer. The domain of the tangent function is the set of all real numbers other than $x = \pi/2 + n\pi$, and the range is the set of all real numbers.

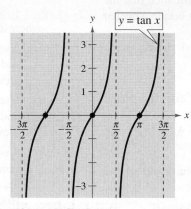

$y = \tan x$

PERIOD: π
DOMAIN: ALL $x \neq \dfrac{\pi}{2} + n\pi$
RANGE: $(-\infty, \infty)$
VERTICAL ASYMPTOTES: $x = \dfrac{\pi}{2} + n\pi$
SYMMETRY: ORIGIN

FIGURE 1.59

Algebra Help

- You can review odd and even functions in Section P.6.
- You can review symmetry of a graph in Section P.3.
- You can review trigonometric identities in Section 1.3.
- You can review domain and range of a function in Section P.5.
- You can review intercepts of a graph in Section P.3.

Sketching the graph of $y = a\tan(bx - c)$ is similar to sketching the graph of $y = a\sin(bx - c)$ in that you locate key points that identify the intercepts and asymptotes. Two consecutive vertical asymptotes can be found by solving the equations

$$bx - c = -\frac{\pi}{2} \quad \text{and} \quad bx - c = \frac{\pi}{2}.$$

The midpoint between two consecutive vertical asymptotes is an x-intercept of the graph. The period of the function $y = a\tan(bx - c)$ is the distance between two consecutive vertical asymptotes. The amplitude of a tangent function is not defined. After plotting the asymptotes and the x-intercept, plot a few additional points between the two asymptotes and sketch one cycle. Finally, sketch one or two additional cycles to the left and right.

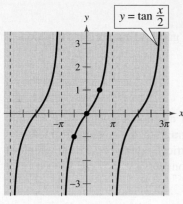

FIGURE **1.60**

Example 1   Sketching the Graph of a Tangent Function

Sketch the graph of $y = \tan(x/2)$.

Solution

By solving the equations

$$\frac{x}{2} = -\frac{\pi}{2} \quad \text{and} \quad \frac{x}{2} = \frac{\pi}{2}$$

$$x = -\pi \qquad\qquad x = \pi$$

you can see that two consecutive vertical asymptotes occur at $x = -\pi$ and $x = \pi$. Between these two asymptotes, plot a few points, including the x-intercept, as shown in the table. Three cycles of the graph are shown in Figure 1.60.

x	$-\pi$	$-\dfrac{\pi}{2}$	0	$\dfrac{\pi}{2}$	π
$\tan\dfrac{x}{2}$	Undef.	-1	0	1	Undef.

CHECK*Point* Now try Exercise 15.

Example 2   Sketching the Graph of a Tangent Function

Sketch the graph of $y = -3\tan 2x$.

Solution

By solving the equations

$$2x = -\frac{\pi}{2} \quad \text{and} \quad 2x = \frac{\pi}{2}$$

$$x = -\frac{\pi}{4} \qquad\qquad x = \frac{\pi}{4}$$

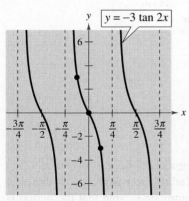

FIGURE **1.61**

you can see that two consecutive vertical asymptotes occur at $x = -\pi/4$ and $x = \pi/4$. Between these two asymptotes, plot a few points, including the x-intercept, as shown in the table. Three cycles of the graph are shown in Figure 1.61.

x	$-\dfrac{\pi}{4}$	$-\dfrac{\pi}{8}$	0	$\dfrac{\pi}{8}$	$\dfrac{\pi}{4}$
$-3\tan 2x$	Undef.	3	0	-3	Undef.

By comparing the graphs in Examples 1 and 2, you can see that the graph of $y = a\tan(bx - c)$ increases between consecutive vertical asymptotes when $a > 0$, and decreases between consecutive vertical asymptotes when $a < 0$. In other words, the graph for $a < 0$ is a reflection in the x-axis of the graph for $a > 0$.

CHECK*Point* Now try Exercise 17.

Graph of the Cotangent Function

The graph of the cotangent function is similar to the graph of the tangent function. It also has a period of π. However, from the identity

$$y = \cot x = \frac{\cos x}{\sin x}$$

you can see that the cotangent function has vertical asymptotes when $\sin x$ is zero, which occurs at $x = n\pi$, where n is an integer. The graph of the cotangent function is shown in Figure 1.62. Note that two consecutive vertical asymptotes of the graph of $y = a\cot(bx - c)$ can be found by solving the equations $bx - c = 0$ and $bx - c = \pi$.

TECHNOLOGY

Some graphing utilities have difficulty graphing trigonometric functions that have vertical asymptotes. Your graphing utility may connect parts of the graphs of tangent, cotangent, secant, and cosecant functions that are not supposed to be connected. To eliminate this problem, change the mode of the graphing utility to *dot* mode.

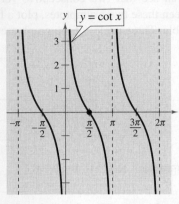

PERIOD: π
DOMAIN: ALL $x \neq n\pi$
RANGE: $(-\infty, \infty)$
VERTICAL ASYMPTOTES: $x = n\pi$
SYMMETRY: ORIGIN

FIGURE **1.62**

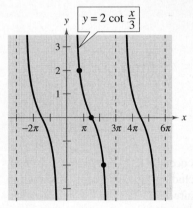

FIGURE **1.63**

Example 3 Sketching the Graph of a Cotangent Function

Sketch the graph of $y = 2\cot\dfrac{x}{3}$.

Solution

By solving the equations

$$\frac{x}{3} = 0 \qquad \text{and} \qquad \frac{x}{3} = \pi$$

$$x = 0 \qquad\qquad\qquad x = 3\pi$$

you can see that two consecutive vertical asymptotes occur at $x = 0$ and $x = 3\pi$. Between these two asymptotes, plot a few points, including the x-intercept, as shown in the table. Three cycles of the graph are shown in Figure 1.63. Note that the period is 3π, the distance between consecutive asymptotes.

x	0	$\dfrac{3\pi}{4}$	$\dfrac{3\pi}{2}$	$\dfrac{9\pi}{4}$	3π
$2\cot\dfrac{x}{3}$	Undef.	2	0	-2	Undef.

CHECKPoint ► Now try Exercise 27.

Graphs of the Reciprocal Functions

The graphs of the two remaining trigonometric functions can be obtained from the graphs of the sine and cosine functions using the reciprocal identities

$$\csc x = \frac{1}{\sin x} \quad \text{and} \quad \sec x = \frac{1}{\cos x}.$$

For instance, at a given value of x, the y-coordinate of $\sec x$ is the reciprocal of the y-coordinate of $\cos x$. Of course, when $\cos x = 0$, the reciprocal does not exist. Near such values of x, the behavior of the secant function is similar to that of the tangent function. In other words, the graphs of

$$\tan x = \frac{\sin x}{\cos x} \quad \text{and} \quad \sec x = \frac{1}{\cos x}$$

have vertical asymptotes at $x = \pi/2 + n\pi$, where n is an integer, and the cosine is zero at these x-values. Similarly,

$$\cot x = \frac{\cos x}{\sin x} \quad \text{and} \quad \csc x = \frac{1}{\sin x}$$

have vertical asymptotes where $\sin x = 0$—that is, at $x = n\pi$.

To sketch the graph of a secant or cosecant function, you should first make a sketch of its reciprocal function. For instance, to sketch the graph of $y = \csc x$, first sketch the graph of $y = \sin x$. Then take reciprocals of the y-coordinates to obtain points on the graph of $y = \csc x$. This procedure is used to obtain the graphs shown in Figure 1.64.

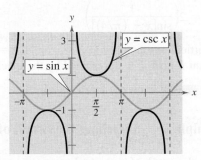

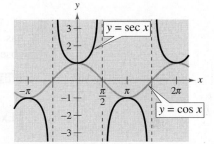

PERIOD: 2π

DOMAIN: ALL $x \neq n\pi$

RANGE: $(-\infty, -1] \cup [1, \infty)$

VERTICAL ASYMPTOTES: $x = n\pi$

SYMMETRY: ORIGIN

FIGURE **1.64**

PERIOD: 2π

DOMAIN: ALL $x \neq \frac{\pi}{2} + n\pi$

RANGE: $(-\infty, -1] \cup [1, \infty)$

VERTICAL ASYMPTOTES: $x = \frac{\pi}{2} + n\pi$

SYMMETRY: y-AXIS

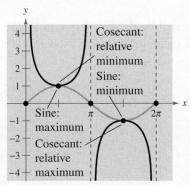

FIGURE **1.65**

In comparing the graphs of the cosecant and secant functions with those of the sine and cosine functions, note that the "hills" and "valleys" are interchanged. For example, a hill (or maximum point) on the sine curve corresponds to a valley (a relative minimum) on the cosecant curve, and a valley (or minimum point) on the sine curve corresponds to a hill (a relative maximum) on the cosecant curve, as shown in Figure 1.65. Additionally, x-intercepts of the sine and cosine functions become vertical asymptotes of the cosecant and secant functions, respectively (see Figure 1.65).

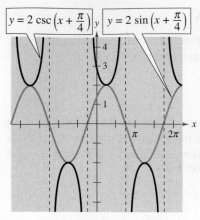

FIGURE **1.66**

Example 4   Sketching the Graph of a Cosecant Function

Sketch the graph of $y = 2 \csc\left(x + \dfrac{\pi}{4}\right)$.

Solution

Begin by sketching the graph of

$$y = 2 \sin\left(x + \frac{\pi}{4}\right).$$

For this function, the amplitude is 2 and the period is 2π. By solving the equations

$$x + \frac{\pi}{4} = 0 \qquad \text{and} \qquad x + \frac{\pi}{4} = 2\pi$$

$$x = -\frac{\pi}{4} \qquad\qquad\qquad x = \frac{7\pi}{4}$$

you can see that one cycle of the sine function corresponds to the interval from $x = -\pi/4$ to $x = 7\pi/4$. The graph of this sine function is represented by the gray curve in Figure 1.66. Because the sine function is zero at the midpoint and endpoints of this interval, the corresponding cosecant function

$$y = 2 \csc\left(x + \frac{\pi}{4}\right)$$

$$= 2\left(\frac{1}{\sin[x + (\pi/4)]}\right)$$

has vertical asymptotes at $x = -\pi/4$, $x = 3\pi/4$, $x = 7\pi/4$, etc. The graph of the cosecant function is represented by the black curve in Figure 1.66.

CHECK*Point* Now try Exercise 33.

Example 5   Sketching the Graph of a Secant Function

Sketch the graph of $y = \sec 2x$.

Solution

Begin by sketching the graph of $y = \cos 2x$, as indicated by the gray curve in Figure 1.67. Then, form the graph of $y = \sec 2x$ as the black curve in the figure. Note that the x-intercepts of $y = \cos 2x$

$$\left(-\frac{\pi}{4}, 0\right), \qquad \left(\frac{\pi}{4}, 0\right), \qquad \left(\frac{3\pi}{4}, 0\right), \dots$$

correspond to the vertical asymptotes

$$x = -\frac{\pi}{4}, \qquad x = \frac{\pi}{4}, \qquad x = \frac{3\pi}{4}, \dots$$

of the graph of $y = \sec 2x$. Moreover, notice that the period of $y = \cos 2x$ and $y = \sec 2x$ is π.

CHECK*Point* Now try Exercise 35.

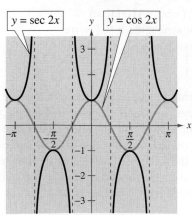

FIGURE **1.67**

Damped Trigonometric Graphs

A *product* of two functions can be graphed using properties of the individual functions. For instance, consider the function

$$f(x) = x \sin x$$

as the product of the functions $y = x$ and $y = \sin x$. Using properties of absolute value and the fact that $|\sin x| \leq 1$, you have $0 \leq |x||\sin x| \leq |x|$. Consequently,

$$-|x| \leq x \sin x \leq |x|$$

which means that the graph of $f(x) = x \sin x$ lies between the lines $y = -x$ and $y = x$. Furthermore, because

$$f(x) = x \sin x = \pm x \qquad \text{at} \qquad x = \frac{\pi}{2} + n\pi$$

and

$$f(x) = x \sin x = 0 \qquad \text{at} \qquad x = n\pi$$

the graph of f touches the line $y = -x$ or the line $y = x$ at $x = \pi/2 + n\pi$ and has x-intercepts at $x = n\pi$. A sketch of f is shown in Figure 1.68. In the function $f(x) = x \sin x$, the factor x is called the **damping factor.**

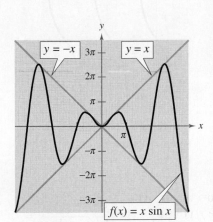

FIGURE **1.68**

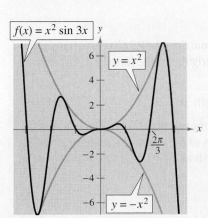

FIGURE **1.69**

Study Tip

Do you see why the graph of $f(x) = x \sin x$ touches the lines $y = \pm x$ at $x = \pi/2 + n\pi$ and why the graph has x-intercepts at $x = n\pi$? Recall that the sine function is equal to 1 at $\pi/2$, $3\pi/2, 5\pi/2, \ldots$ (odd multiples of $\pi/2$) and is equal to 0 at π, $2\pi, 3\pi, \ldots$ (multiples of π).

Example 6   Damped Sine Wave

Sketch the graph of

$$f(x) = x^2 \sin 3x.$$

Solution

Consider $f(x)$ as the product of the two functions

$$y = x^2 \qquad \text{and} \qquad y = \sin 3x$$

each of which has the set of real numbers as its domain. For any real number x, you know that $x^2 \geq 0$ and $|\sin 3x| \leq 1$. So, $x^2 |\sin 3x| \leq x^2$, which means that

$$-x^2 \leq x^2 \sin 3x \leq x^2.$$

Furthermore, because

$$f(x) = x^2 \sin 3x = \pm x^2 \qquad \text{at} \qquad x = \frac{\pi}{6} + \frac{n\pi}{3}$$

and

$$f(x) = x^2 \sin 3x = 0 \qquad \text{at} \qquad x = \frac{n\pi}{3}$$

the graph of f touches the curves $y = -x^2$ and $y = x^2$ at $x = \pi/6 + n\pi/3$ and has intercepts at $x = n\pi/3$. A sketch is shown in Figure 1.69.

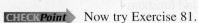 Now try Exercise 81.

Figure 1.70 summarizes the characteristics of the six basic trigonometric functions.

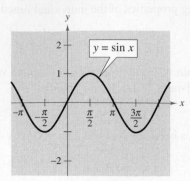

DOMAIN: $(-\infty, \infty)$
RANGE: $[-1, 1]$
PERIOD: 2π

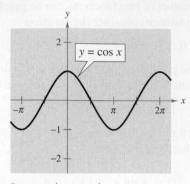

DOMAIN: $(-\infty, \infty)$
RANGE: $[-1, 1]$
PERIOD: 2π

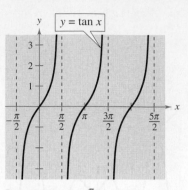

DOMAIN: ALL $x \neq \frac{\pi}{2} + n\pi$
RANGE: $(-\infty, \infty)$
PERIOD: π

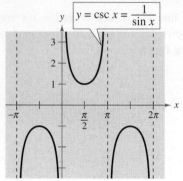

DOMAIN: ALL $x \neq n\pi$
RANGE: $(-\infty, -1] \cup [1, \infty)$
PERIOD: 2π

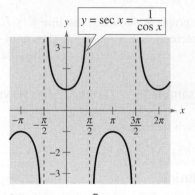

DOMAIN: ALL $x \neq \frac{\pi}{2} + n\pi$
RANGE: $(-\infty, -1] \cup [1, \infty)$
PERIOD: 2π

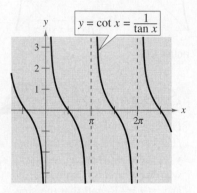

DOMAIN: ALL $x \neq n\pi$
RANGE: $(-\infty, \infty)$
PERIOD: π

FIGURE 1.70

CLASSROOM DISCUSSION

Combining Trigonometric Functions Recall from Section P.9 that functions can be combined arithmetically. This also applies to trigonometric functions. For each of the functions

$$h(x) = x + \sin x \quad \text{and} \quad h(x) = \cos x - \sin 3x$$

(a) identify two simpler functions f and g that comprise the combination, (b) use a table to show how to obtain the numerical values of $h(x)$ from the numerical values of $f(x)$ and $g(x)$, and (c) use graphs of f and g to show how the graph of h may be formed.

Can you find functions

$$f(x) = d + a\sin(bx + c) \quad \text{and} \quad g(x) = d + a\cos(bx + c)$$

such that $f(x) + g(x) = 0$ for all x?

1.6   EXERCISES

See www.CalcChat.com for worked-out solutions to odd-numbered exercises.

VOCABULARY: Fill in the blanks.

1. The tangent, cotangent, and cosecant functions are _____ , so the graphs of these functions have symmetry with respect to the _____.
2. The graphs of the tangent, cotangent, secant, and cosecant functions all have _____ asymptotes.
3. To sketch the graph of a secant or cosecant function, first make a sketch of its corresponding _____ function.
4. For the functions given by $f(x) = g(x) \cdot \sin x$, $g(x)$ is called the _____ factor of the function $f(x)$.
5. The period of $y = \tan x$ is _____.
6. The domain of $y = \cot x$ is all real numbers such that _____.
7. The range of $y = \sec x$ is _____.
8. The period of $y = \csc x$ is _____.

SKILLS AND APPLICATIONS

In Exercises 9–14, match the function with its graph. State the period of the function. [The graphs are labeled (a), (b), (c), (d), (e), and (f).]

(a)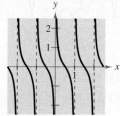

(b)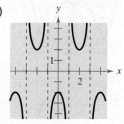

(c)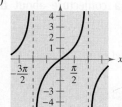

(d)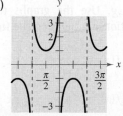

(e)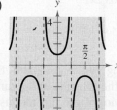

(f)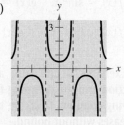

9. $y = \sec 2x$

10. $y = \tan \dfrac{x}{2}$

11. $y = \dfrac{1}{2} \cot \pi x$

12. $y = -\csc x$

13. $y = \dfrac{1}{2} \sec \dfrac{\pi x}{2}$

14. $y = -2 \sec \dfrac{\pi x}{2}$

In Exercises 15–38, sketch the graph of the function. Include two full periods.

15. $y = \dfrac{1}{3} \tan x$

16. $y = \tan 4x$

17. $y = -2 \tan 3x$

18. $y = -3 \tan \pi x$

19. $y = -\dfrac{1}{2} \sec x$

20. $y = \dfrac{1}{4} \sec x$

21. $y = \csc \pi x$

22. $y = 3 \csc 4x$

23. $y = \dfrac{1}{2} \sec \pi x$

24. $y = -2 \sec 4x + 2$

25. $y = \csc \dfrac{x}{2}$

26. $y = \csc \dfrac{x}{3}$

27. $y = 3 \cot 2x$

28. $y = 3 \cot \dfrac{\pi x}{2}$

29. $y = 2 \sec 3x$

30. $y = -\dfrac{1}{2} \tan x$

31. $y = \tan \dfrac{\pi x}{4}$

32. $y = \tan(x + \pi)$

33. $y = 2 \csc(x - \pi)$

34. $y = \csc(2x - \pi)$

35. $y = 2 \sec(x + \pi)$

36. $y = -\sec \pi x + 1$

37. $y = \dfrac{1}{4} \csc\left(x + \dfrac{\pi}{4}\right)$

38. $y = 2 \cot\left(x + \dfrac{\pi}{2}\right)$

In Exercises 39–48, use a graphing utility to graph the function. Include two full periods.

39. $y = \tan \dfrac{x}{3}$

40. $y = -\tan 2x$

41. $y = -2 \sec 4x$

42. $y = \sec \pi x$

43. $y = \tan\left(x - \dfrac{\pi}{4}\right)$

44. $y = \dfrac{1}{4} \cot\left(x - \dfrac{\pi}{2}\right)$

45. $y = -\csc(4x - \pi)$

46. $y = 2 \sec(2x - \pi)$

47. $y = 0.1 \tan\left(\dfrac{\pi x}{4} + \dfrac{\pi}{4}\right)$

48. $y = \dfrac{1}{3} \sec\left(\dfrac{\pi x}{2} + \dfrac{\pi}{2}\right)$

In Exercises 49–56, use a graph to solve the equation on the interval $[-2\pi, 2\pi]$.

49. $\tan x = 1$

50. $\tan x = \sqrt{3}$

51. $\cot x = -\dfrac{\sqrt{3}}{3}$

52. $\cot x = 1$

53. $\sec x = -2$

54. $\sec x = 2$

55. $\csc x = \sqrt{2}$

56. $\csc x = -\dfrac{2\sqrt{3}}{3}$

In Exercises 57–64, use the graph of the function to determine whether the function is even, odd, or neither. Verify your answer algebraically.

57. $f(x) = \sec x$

58. $f(x) = \tan x$

59. $g(x) = \cot x$

60. $g(x) = \csc x$

61. $f(x) = x + \tan x$

62. $f(x) = x^2 - \sec x$

63. $g(x) = x \csc x$

64. $g(x) = x^2 \cot x$

65. GRAPHICAL REASONING Consider the functions given by

$$f(x) = 2 \sin x \quad \text{and} \quad g(x) = \frac{1}{2} \csc x$$

on the interval $(0, \pi)$.

(a) Graph f and g in the same coordinate plane.

(b) Approximate the interval in which $f > g$.

(c) Describe the behavior of each of the functions as x approaches π. How is the behavior of g related to the behavior of f as x approaches π?

66. GRAPHICAL REASONING Consider the functions given by

$$f(x) = \tan \frac{\pi x}{2} \quad \text{and} \quad g(x) = \frac{1}{2} \sec \frac{\pi x}{2}$$

on the interval $(-1, 1)$.

(a) Use a graphing utility to graph f and g in the same viewing window.

(b) Approximate the interval in which $f < g$.

(c) Approximate the interval in which $2f < 2g$. How does the result compare with that of part (b)? Explain.

In Exercises 67–72, use a graphing utility to graph the two equations in the same viewing window. Use the graphs to determine whether the expressions are equivalent. Verify the results algebraically.

67. $y_1 = \sin x \csc x$, $\quad y_2 = 1$

68. $y_1 = \sin x \sec x$, $\quad y_2 = \tan x$

69. $y_1 = \dfrac{\cos x}{\sin x}$, $\quad y_2 = \cot x$

70. $y_1 = \tan x \cot^2 x$, $\quad y_2 = \cot x$

71. $y_1 = 1 + \cot^2 x$, $\quad y_2 = \csc^2 x$

72. $y_1 = \sec^2 x - 1$, $\quad y_2 = \tan^2 x$

In Exercises 73–76, match the function with its graph. Describe the behavior of the function as x approaches zero. [The graphs are labeled (a), (b), (c), and (d).]

(a)
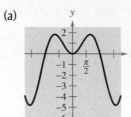

(b)
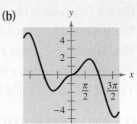

(c)
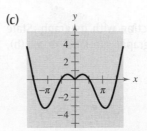

(d)
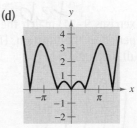

73. $f(x) = |x \cos x|$

74. $f(x) = x \sin x$

75. $g(x) = |x| \sin x$

76. $g(x) = |x| \cos x$

CONJECTURE In Exercises 77–80, graph the functions f and g. Use the graphs to make a conjecture about the relationship between the functions.

77. $f(x) = \sin x + \cos\left(x + \dfrac{\pi}{2}\right)$, $\quad g(x) = 0$

78. $f(x) = \sin x - \cos\left(x + \dfrac{\pi}{2}\right)$, $\quad g(x) = 2 \sin x$

79. $f(x) = \sin^2 x$, $\quad g(x) = \frac{1}{2}(1 - \cos 2x)$

80. $f(x) = \cos^2 \dfrac{\pi x}{2}$, $\quad g(x) = \frac{1}{2}(1 + \cos \pi x)$

In Exercises 81–84, use a graphing utility to graph the function and the damping factor of the function in the same viewing window. Describe the behavior of the function as x increases without bound.

81. $g(x) = x \cos \pi x$

82. $f(x) = x^2 \cos x$

83. $f(x) = x^3 \sin x$

84. $h(x) = x^3 \cos x$

In Exercises 85–90, use a graphing utility to graph the function. Describe the behavior of the function as x approaches zero.

85. $y = \dfrac{6}{x} + \cos x$, $\quad x > 0$ **86.** $y = \dfrac{4}{x} + \sin 2x$, $\quad x > 0$

87. $g(x) = \dfrac{\sin x}{x}$ **88.** $f(x) = \dfrac{1 - \cos x}{x}$

89. $f(x) = \sin \dfrac{1}{x}$ **90.** $h(x) = x \sin \dfrac{1}{x}$

91. DISTANCE A plane flying at an altitude of 7 miles above a radar antenna will pass directly over the radar antenna (see figure). Let d be the ground distance from the antenna to the point directly under the plane and let x be the angle of elevation to the plane from the antenna. (d is positive as the plane approaches the antenna.) Write d as a function of x and graph the function over the interval $0 < x < \pi$.

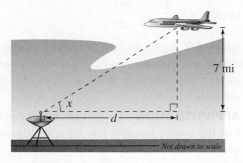

Not drawn to scale

92. TELEVISION COVERAGE A television camera is on a reviewing platform 27 meters from the street on which a parade will be passing from left to right (see figure). Write the distance d from the camera to a particular unit in the parade as a function of the angle x, and graph the function over the interval $-\pi/2 < x < \pi/2$. (Consider x as negative when a unit in the parade approaches from the left.)

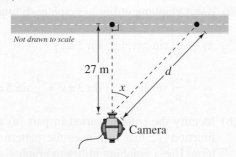

Not drawn to scale

93. METEOROLOGY The normal monthly high temperatures H (in degrees Fahrenheit) in Erie, Pennsylvania are approximated by

$$H(t) = 56.94 - 20.86 \cos(\pi t/6) - 11.58 \sin(\pi t/6)$$

and the normal monthly low temperatures L are approximated by

$$L(t) = 41.80 - 17.13 \cos(\pi t/6) - 13.39 \sin(\pi t/6)$$

where t is the time (in months), with $t = 1$ corresponding to January (see figure). (Source: National Climatic Data Center)

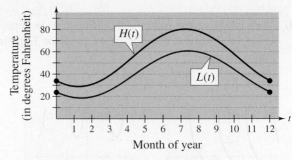

(a) What is the period of each function?

(b) During what part of the year is the difference between the normal high and normal low temperatures greatest? When is it smallest?

(c) The sun is northernmost in the sky around June 21, but the graph shows the warmest temperatures at a later date. Approximate the lag time of the temperatures relative to the position of the sun.

94. SALES The projected monthly sales S (in thousands of units) of lawn mowers (a seasonal product) are modeled by $S = 74 + 3t - 40 \cos(\pi t/6)$, where t is the time (in months), with $t = 1$ corresponding to January. Graph the sales function over 1 year.

95. HARMONIC MOTION An object weighing W pounds is suspended from the ceiling by a steel spring (see figure). The weight is pulled downward (positive direction) from its equilibrium position and released. The resulting motion of the weight is described by the function $y = (4/t)\cos 4t,\ t > 0$, where y is the distance (in feet) and t is the time (in seconds).

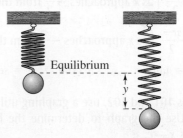

(a) Use a graphing utility to graph the function.

(b) Describe the behavior of the displacement function for increasing values of time t.

EXPLORATION

TRUE OR FALSE? In Exercises 96 and 97, determine whether the statement is true or false. Justify your answer.

96. The graph of $y = \csc x$ can be obtained on a calculator by graphing the reciprocal of $y = \sin x$.

97. The graph of $y = \sec x$ can be obtained on a calculator by graphing a translation of the reciprocal of $y = \sin x$.

98. CAPSTONE Determine which function is represented by the graph. Do not use a calculator. Explain your reasoning.

(a)

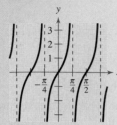

(b)

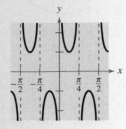

(i) $f(x) = \tan 2x$ (i) $f(x) = \sec 4x$

(ii) $f(x) = \tan(x/2)$ (ii) $f(x) = \csc 4x$

(iii) $f(x) = 2 \tan x$ (iii) $f(x) = \csc(x/4)$

(iv) $f(x) = -\tan 2x$ (iv) $f(x) = \sec(x/4)$

(v) $f(x) = -\tan(x/2)$ (v) $f(x) = \csc(4x - \pi)$

In Exercises 99 and 100, use a graphing utility to graph the function. Use the graph to determine the behavior of the function as $x \to c$.

(a) $x \to \dfrac{\pi^+}{2}$ $\left(\text{as } x \text{ approaches } \dfrac{\pi}{2} \text{ from the right}\right)$

(b) $x \to \dfrac{\pi^-}{2}$ $\left(\text{as } x \text{ approaches } \dfrac{\pi}{2} \text{ from the left}\right)$

(c) $x \to -\dfrac{\pi^+}{2}$ $\left(\text{as } x \text{ approaches } -\dfrac{\pi}{2} \text{ from the right}\right)$

(d) $x \to -\dfrac{\pi^-}{2}$ $\left(\text{as } x \text{ approaches } -\dfrac{\pi}{2} \text{ from the left}\right)$

99. $f(x) = \tan x$ **100.** $f(x) = \sec x$

In Exercises 101 and 102, use a graphing utility to graph the function. Use the graph to determine the behavior of the function as $x \to c$.

(a) As $x \to 0^+$, the value of $f(x) \to$ ▮ .

(b) As $x \to 0^-$, the value of $f(x) \to$ ▮ .

(c) As $x \to \pi^+$, the value of $f(x) \to$ ▮ .

(d) As $x \to \pi^-$, the value of $f(x) \to$ ▮ .

101. $f(x) = \cot x$ **102.** $f(x) = \csc x$

103. THINK ABOUT IT Consider the function given by $f(x) = x - \cos x$.

 (a) Use a graphing utility to graph the function and verify that there exists a zero between 0 and 1. Use the graph to approximate the zero.

(b) Starting with $x_0 = 1$, generate a sequence $x_1, x_2, x_3, \ldots$, where $x_n = \cos(x_{n-1})$. For example,

$$x_0 = 1$$
$$x_1 = \cos(x_0)$$
$$x_2 = \cos(x_1)$$
$$x_3 = \cos(x_2)$$
$$\vdots$$

What value does the sequence approach?

104. APPROXIMATION Using calculus, it can be shown that the tangent function can be approximated by the polynomial

$$\tan x \approx x + \frac{2x^3}{3!} + \frac{16x^5}{5!}$$

where x is in radians. Use a graphing utility to graph the tangent function and its polynomial approximation in the same viewing window. How do the graphs compare?

105. APPROXIMATION Using calculus, it can be shown that the secant function can be approximated by the polynomial

$$\sec x \approx 1 + \frac{x^2}{2!} + \frac{5x^4}{4!}$$

where x is in radians. Use a graphing utility to graph the secant function and its polynomial approximation in the same viewing window. How do the graphs compare?

106. PATTERN RECOGNITION

(a) Use a graphing utility to graph each function.

$$y_1 = \frac{4}{\pi}\left(\sin \pi x + \frac{1}{3} \sin 3\pi x\right)$$

$$y_2 = \frac{4}{\pi}\left(\sin \pi x + \frac{1}{3} \sin 3\pi x + \frac{1}{5} \sin 5\pi x\right)$$

(b) Identify the pattern started in part (a) and find a function y_3 that continues the pattern one more term. Use a graphing utility to graph y_3.

(c) The graphs in parts (a) and (b) approximate the periodic function in the figure. Find a function y_4 that is a better approximation.

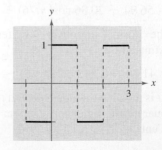

INVERSE TRIGONOMETRIC FUNCTIONS

What you should learn

- Evaluate and graph the inverse sine function.
- Evaluate and graph the other inverse trigonometric functions.
- Evaluate and graph the compositions of trigonometric functions.

Why you should learn it

You can use inverse trigonometric functions to model and solve real-life problems. For instance, in Exercise 106 on page 203, an inverse trigonometric function can be used to model the angle of elevation from a television camera to a space shuttle launch.

NASA

Inverse Sine Function

Recall from Section P.10 that, for a function to have an inverse function, it must be one-to-one—that is, it must pass the Horizontal Line Test. From Figure 1.71, you can see that $y = \sin x$ does not pass the test because different values of x yield the same y-value.

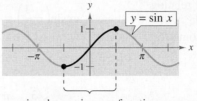

$\sin x$ has an inverse function on this interval.

FIGURE **1.71**

However, if you restrict the domain to the interval $-\pi/2 \le x \le \pi/2$ (corresponding to the black portion of the graph in Figure 1.71), the following properties hold.

1. On the interval $[-\pi/2, \pi/2]$, the function $y = \sin x$ is increasing.

2. On the interval $[-\pi/2, \pi/2]$, $y = \sin x$ takes on its full range of values, $-1 \le \sin x \le 1$.

3. On the interval $[-\pi/2, \pi/2]$, $y = \sin x$ is one-to-one.

So, on the restricted domain $-\pi/2 \le x \le \pi/2$, $y = \sin x$ has a unique inverse function called the **inverse sine function.** It is denoted by

$$y = \arcsin x \qquad \text{or} \qquad y = \sin^{-1} x.$$

The notation $\sin^{-1} x$ is consistent with the inverse function notation $f^{-1}(x)$. The arcsin x notation (read as "the arcsine of x") comes from the association of a central angle with its intercepted *arc length* on a unit circle. So, arcsin x means the angle (or arc) whose sine is x. Both notations, arcsin x and $\sin^{-1} x$, are commonly used in mathematics, so remember that $\sin^{-1} x$ denotes the *inverse* sine function rather than $1/\sin x$. The values of arcsin x lie in the interval $-\pi/2 \le \arcsin x \le \pi/2$. The graph of $y = \arcsin x$ is shown in Example 2.

Definition of Inverse Sine Function

The **inverse sine function** is defined by

$$y = \arcsin x \qquad \text{if and only if} \qquad \sin y = x$$

where $-1 \le x \le 1$ and $-\pi/2 \le y \le \pi/2$. The domain of $y = \arcsin x$ is $[-1, 1]$, and the range is $[-\pi/2, \pi/2]$.

Study Tip

As with the trigonometric functions, much of the work with the inverse trigonometric functions can be done by *exact* calculations rather than by calculator approximations. Exact calculations help to increase your understanding of the inverse functions by relating them to the right triangle definitions of the trigonometric functions.

Example 1   Evaluating the Inverse Sine Function

If possible, find the exact value.

a. $\arcsin\left(-\dfrac{1}{2}\right)$ **b.** $\sin^{-1}\dfrac{\sqrt{3}}{2}$ **c.** $\sin^{-1}2$

Solution

a. Because $\sin\left(-\dfrac{\pi}{6}\right) = -\dfrac{1}{2}$ for $-\dfrac{\pi}{2} \le y \le \dfrac{\pi}{2}$, it follows that

$$\arcsin\left(-\dfrac{1}{2}\right) = -\dfrac{\pi}{6}. \qquad \text{Angle whose sine is } -\tfrac{1}{2}$$

b. Because $\sin\dfrac{\pi}{3} = \dfrac{\sqrt{3}}{2}$ for $-\dfrac{\pi}{2} \le y \le \dfrac{\pi}{2}$, it follows that

$$\sin^{-1}\dfrac{\sqrt{3}}{2} = \dfrac{\pi}{3}. \qquad \text{Angle whose sine is } \sqrt{3}/2$$

c. It is not possible to evaluate $y = \sin^{-1}x$ when $x = 2$ because there is no angle whose sine is 2. Remember that the domain of the inverse sine function is $[-1, 1]$.

CHECK *Point*   Now try Exercise 5.

Example 2   Graphing the Arcsine Function

Sketch a graph of

$$y = \arcsin x.$$

Solution

By definition, the equations $y = \arcsin x$ and $\sin y = x$ are equivalent for $-\pi/2 \le y \le \pi/2$. So, their graphs are the same. From the interval $[-\pi/2, \pi/2]$, you can assign values to y in the second equation to make a table of values. Then plot the points and draw a smooth curve through the points.

y	$-\dfrac{\pi}{2}$	$-\dfrac{\pi}{4}$	$-\dfrac{\pi}{6}$	0	$\dfrac{\pi}{6}$	$\dfrac{\pi}{4}$	$\dfrac{\pi}{2}$
$x = \sin y$	-1	$-\dfrac{\sqrt{2}}{2}$	$-\dfrac{1}{2}$	0	$\dfrac{1}{2}$	$\dfrac{\sqrt{2}}{2}$	1

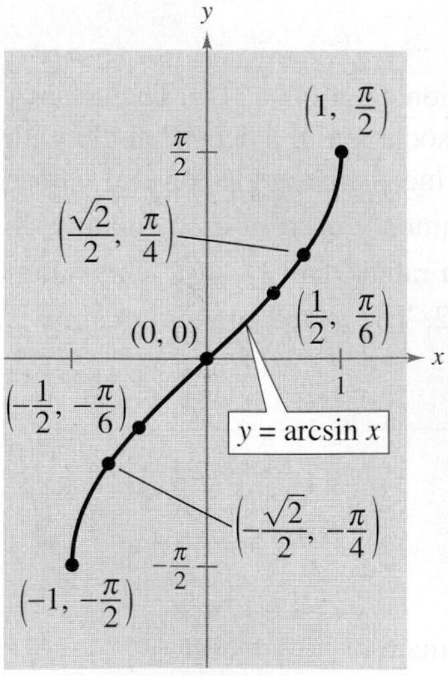

The resulting graph for $y = \arcsin x$ is shown in Figure 1.72. Note that it is the reflection (in the line $y = x$) of the black portion of the graph in Figure 1.71. Be sure you see that Figure 1.72 shows the *entire* graph of the inverse sine function. Remember that the domain of $y = \arcsin x$ is the closed interval $[-1, 1]$ and the range is the closed interval $[-\pi/2, \pi/2]$.

CHECK *Point*   Now try Exercise 21.

FIGURE **1.72**

Other Inverse Trigonometric Functions

The cosine function is decreasing and one-to-one on the interval $0 \leq x \leq \pi$, as shown in Figure 1.73.

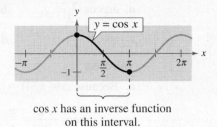

cos x has an inverse function
on this interval.

FIGURE **1.73**

Consequently, on this interval the cosine function has an inverse function—the **inverse cosine function**—denoted by

$$y = \arccos x \qquad \text{or} \qquad y = \cos^{-1} x.$$

Similarly, you can define an **inverse tangent function** by restricting the domain of $y = \tan x$ to the interval $(-\pi/2, \pi/2)$. The following list summarizes the definitions of the three most common inverse trigonometric functions. The remaining three are defined in Exercises 115–117.

Definitions of the Inverse Trigonometric Functions

Function	Domain	Range
$y = \arcsin x$ if and only if $\sin y = x$	$-1 \leq x \leq 1$	$-\dfrac{\pi}{2} \leq y \leq \dfrac{\pi}{2}$
$y = \arccos x$ if and only if $\cos y = x$	$-1 \leq x \leq 1$	$0 \leq y \leq \pi$
$y = \arctan x$ if and only if $\tan y = x$	$-\infty < x < \infty$	$-\dfrac{\pi}{2} < y < \dfrac{\pi}{2}$

The graphs of these three inverse trigonometric functions are shown in Figure 1.74.

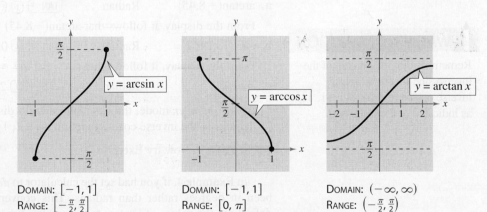

DOMAIN: $[-1, 1]$
RANGE: $\left[-\frac{\pi}{2}, \frac{\pi}{2}\right]$

DOMAIN: $[-1, 1]$
RANGE: $[0, \pi]$

DOMAIN: $(-\infty, \infty)$
RANGE: $\left(-\frac{\pi}{2}, \frac{\pi}{2}\right)$

FIGURE **1.74**

| **Example 3** | **Evaluating Inverse Trigonometric Functions** |

Find the exact value.

a. $\arccos \dfrac{\sqrt{2}}{2}$ **b.** $\cos^{-1}(-1)$

c. $\arctan 0$ **d.** $\tan^{-1}(-1)$

Solution

a. Because $\cos(\pi/4) = \sqrt{2}/2$, and $\pi/4$ lies in $[0, \pi]$, it follows that

$$\arccos \dfrac{\sqrt{2}}{2} = \dfrac{\pi}{4}. \qquad \text{Angle whose cosine is } \sqrt{2}/2$$

b. Because $\cos \pi = -1$, and π lies in $[0, \pi]$, it follows that

$$\cos^{-1}(-1) = \pi. \qquad \text{Angle whose cosine is } -1$$

c. Because $\tan 0 = 0$, and 0 lies in $(-\pi/2, \pi/2)$, it follows that

$$\arctan 0 = 0. \qquad \text{Angle whose tangent is } 0$$

d. Because $\tan(-\pi/4) = -1$, and $-\pi/4$ lies in $(-\pi/2, \pi/2)$, it follows that

$$\tan^{-1}(-1) = -\dfrac{\pi}{4}. \qquad \text{Angle whose tangent is } -1$$

CHECK *Point* Now try Exercise 15.

| **Example 4** | **Calculators and Inverse Trigonometric Functions** |

Use a calculator to approximate the value (if possible).

a. $\arctan(-8.45)$

b. $\sin^{-1} 0.2447$

c. $\arccos 2$

Solution

Function	Mode	Calculator Keystrokes
a. $\arctan(-8.45)$	Radian	TAN⁻¹ ((–) 8.45) ENTER

From the display, it follows that $\arctan(-8.45) \approx -1.453001$.

| **b.** $\sin^{-1} 0.2447$ | Radian | SIN⁻¹ (0.2447) ENTER |

From the display, it follows that $\sin^{-1} 0.2447 \approx 0.2472103$.

| **c.** $\arccos 2$ | Radian | COS⁻¹ (2) ENTER |

In *real number* mode, the calculator should display an *error message* because the domain of the inverse cosine function is $[-1, 1]$.

CHECK *Point* Now try Exercise 29.

WARNING/CAUTION

Remember that the domain of the inverse sine function and the inverse cosine function is $[-1, 1]$, as indicated in Example 4(c).

In Example 4, if you had set the calculator to *degree* mode, the displays would have been in degrees rather than radians. This convention is peculiar to calculators. By definition, the values of inverse trigonometric functions are *always in radians*.

Compositions of Functions

Algebra Help

You can review the composition of functions in Section P.9.

Recall from Section P.10 that for all x in the domains of f and f^{-1}, inverse functions have the properties

$$f(f^{-1}(x)) = x \qquad \text{and} \qquad f^{-1}(f(x)) = x.$$

Inverse Properties of Trigonometric Functions

If $-1 \leq x \leq 1$ and $-\pi/2 \leq y \leq \pi/2$, then

$$\sin(\arcsin x) = x \qquad \text{and} \qquad \arcsin(\sin y) = y.$$

If $-1 \leq x \leq 1$ and $0 \leq y \leq \pi$, then

$$\cos(\arccos x) = x \qquad \text{and} \qquad \arccos(\cos y) = y.$$

If x is a real number and $-\pi/2 < y < \pi/2$, then

$$\tan(\arctan x) = x \qquad \text{and} \qquad \arctan(\tan y) = y.$$

Keep in mind that these inverse properties do not apply for arbitrary values of x and y. For instance,

$$\arcsin\left(\sin \frac{3\pi}{2}\right) = \arcsin(-1) = -\frac{\pi}{2} \neq \frac{3\pi}{2}.$$

In other words, the property

$$\arcsin(\sin y) = y$$

is not valid for values of y outside the interval $[-\pi/2, \pi/2]$.

Example 5   Using Inverse Properties

If possible, find the exact value.

a. $\tan[\arctan(-5)]$ **b.** $\arcsin\left(\sin \dfrac{5\pi}{3}\right)$ **c.** $\cos(\cos^{-1} \pi)$

Solution

a. Because -5 lies in the domain of the arctan function, the inverse property applies, and you have

$$\tan[\arctan(-5)] = -5.$$

b. In this case, $5\pi/3$ does not lie within the range of the arcsine function, $-\pi/2 \leq y \leq \pi/2$. However, $5\pi/3$ is coterminal with

$$\frac{5\pi}{3} - 2\pi = -\frac{\pi}{3}$$

which does lie in the range of the arcsine function, and you have

$$\arcsin\left(\sin \frac{5\pi}{3}\right) = \arcsin\left[\sin\left(-\frac{\pi}{3}\right)\right] = -\frac{\pi}{3}.$$

c. The expression $\cos(\cos^{-1} \pi)$ is not defined because $\cos^{-1} \pi$ is not defined. Remember that the domain of the inverse cosine function is $[-1, 1]$.

CHECK*Point*   Now try Exercise 49.

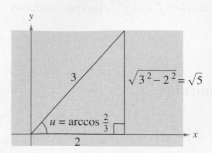

Angle whose cosine is $\frac{2}{3}$

FIGURE **1.75**

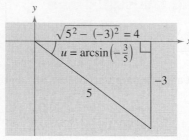

Angle whose sine is $-\frac{3}{5}$

FIGURE **1.76**

Example 6 shows how to use right triangles to find exact values of compositions of inverse functions. Then, Example 7 shows how to use right triangles to convert a trigonometric expression into an algebraic expression. This conversion technique is used frequently in calculus.

Example 6   Evaluating Compositions of Functions

Find the exact value.

a. $\tan\left(\arccos \dfrac{2}{3}\right)$ **b.** $\cos\left[\arcsin\left(-\dfrac{3}{5}\right)\right]$

Solution

a. If you let $u = \arccos \frac{2}{3}$, then $\cos u = \frac{2}{3}$. Because $\cos u$ is positive, u is a *first*-quadrant angle. You can sketch and label angle u as shown in Figure 1.75. Consequently,

$$\tan\left(\arccos \frac{2}{3}\right) = \tan u = \frac{\text{opp}}{\text{adj}} = \frac{\sqrt{5}}{2}.$$

b. If you let $u = \arcsin\left(-\frac{3}{5}\right)$, then $\sin u = -\frac{3}{5}$. Because $\sin u$ is negative, u is a *fourth*-quadrant angle. You can sketch and label angle u as shown in Figure 1.76. Consequently,

$$\cos\left[\arcsin\left(-\frac{3}{5}\right)\right] = \cos u = \frac{\text{adj}}{\text{hyp}} = \frac{4}{5}.$$

CHECK *Point*   Now try Exercise 57.

Example 7   Some Problems from Calculus

Write each of the following as an algebraic expression in x.

a. $\sin(\arccos 3x)$, $0 \le x \le \dfrac{1}{3}$ **b.** $\cot(\arccos 3x)$, $0 \le x < \dfrac{1}{3}$

Solution

If you let $u = \arccos 3x$, then $\cos u = 3x$, where $-1 \le 3x \le 1$. Because

$$\cos u = \frac{\text{adj}}{\text{hyp}} = \frac{3x}{1}$$

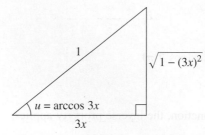

Angle whose cosine is $3x$

FIGURE **1.77**

you can sketch a right triangle with acute angle u, as shown in Figure 1.77. From this triangle, you can easily convert each expression to algebraic form.

a. $\sin(\arccos 3x) = \sin u = \dfrac{\text{opp}}{\text{hyp}} = \sqrt{1 - 9x^2}$, $0 \le x \le \dfrac{1}{3}$

b. $\cot(\arccos 3x) = \cot u = \dfrac{\text{adj}}{\text{opp}} = \dfrac{3x}{\sqrt{1 - 9x^2}}$, $0 \le x < \dfrac{1}{3}$

CHECK *Point*   Now try Exercise 67.

In Example 7, similar arguments can be made for x-values lying in the interval $\left[-\frac{1}{3}, 0\right]$.

1.7   EXERCISES

See www.CalcChat.com for worked-out solutions to odd-numbered exercises.

VOCABULARY: Fill in the blanks.

Function	Alternative Notation	Domain	Range
1. $y = \arcsin x$	_____	_____	$-\dfrac{\pi}{2} \le y \le \dfrac{\pi}{2}$
2. _____	$y = \cos^{-1} x$	$-1 \le x \le 1$	_____
3. $y = \arctan x$	_____	_____	_____

4. Without restrictions, no trigonometric function has a(n) _____ function.

SKILLS AND APPLICATIONS

In Exercises 5–20, evaluate the expression without using a calculator.

5. $\arcsin \dfrac{1}{2}$

6. $\arcsin 0$

7. $\arccos \dfrac{1}{2}$

8. $\arccos 0$

9. $\arctan \dfrac{\sqrt{3}}{3}$

10. $\arctan(1)$

11. $\cos^{-1}\left(-\dfrac{\sqrt{3}}{2}\right)$

12. $\sin^{-1}\left(-\dfrac{\sqrt{2}}{2}\right)$

13. $\arctan(-\sqrt{3})$

14. $\arctan \sqrt{3}$

15. $\arccos\left(-\dfrac{1}{2}\right)$

16. $\arcsin \dfrac{\sqrt{2}}{2}$

17. $\sin^{-1}\left(-\dfrac{\sqrt{3}}{2}\right)$

18. $\tan^{-1}\left(-\dfrac{\sqrt{3}}{3}\right)$

19. $\tan^{-1} 0$

20. $\cos^{-1} 1$

In Exercises 21 and 22, use a graphing utility to graph f, g, and $y = x$ in the same viewing window to verify geometrically that g is the inverse function of f. (Be sure to restrict the domain of f properly.)

21. $f(x) = \sin x, \quad g(x) = \arcsin x$

22. $f(x) = \tan x, \quad g(x) = \arctan x$

In Exercises 23–40, use a calculator to evaluate the expression. Round your result to two decimal places.

23. $\arccos 0.37$

24. $\arcsin 0.65$

25. $\arcsin(-0.75)$

26. $\arccos(-0.7)$

27. $\arctan(-3)$

28. $\arctan 25$

29. $\sin^{-1} 0.31$

30. $\cos^{-1} 0.26$

31. $\arccos(-0.41)$

32. $\arcsin(-0.125)$

33. $\arctan 0.92$

34. $\arctan 2.8$

35. $\arcsin \dfrac{7}{8}$

36. $\arccos\left(-\dfrac{1}{3}\right)$

37. $\tan^{-1} \dfrac{19}{4}$

38. $\tan^{-1}\left(-\dfrac{95}{7}\right)$

39. $\tan^{-1}(-\sqrt{372})$

40. $\tan^{-1}(-\sqrt{2165})$

In Exercises 41 and 42, determine the missing coordinates of the points on the graph of the function.

41.
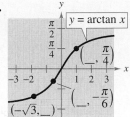

42.
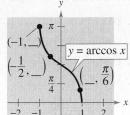

In Exercises 43–48, use an inverse trigonometric function to write θ as a function of x.

43.
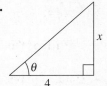

44.
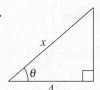

45.
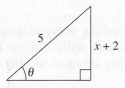

46.
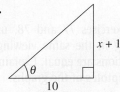

47.
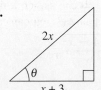

48.
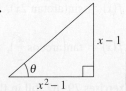

In Exercises 49–54, use the properties of inverse trigonometric functions to evaluate the expression.

49. $\sin(\arcsin 0.3)$

50. $\tan(\arctan 45)$

51. $\cos[\arccos(-0.1)]$

52. $\sin[\arcsin(-0.2)]$

53. $\arcsin(\sin 3\pi)$

54. $\arccos\left(\cos \dfrac{7\pi}{2}\right)$

In Exercises 55–66, find the exact value of the expression. (*Hint:* Sketch a right triangle.)

55. $\sin\left(\arctan\frac{3}{4}\right)$

56. $\sec\left(\arcsin\frac{4}{5}\right)$

57. $\cos(\tan^{-1}2)$

58. $\sin\left(\cos^{-1}\frac{\sqrt{5}}{5}\right)$

59. $\cos\left(\arcsin\frac{5}{13}\right)$

60. $\csc\left[\arctan\left(-\frac{5}{12}\right)\right]$

61. $\sec\left[\arctan\left(-\frac{3}{5}\right)\right]$

62. $\tan\left[\arcsin\left(-\frac{3}{4}\right)\right]$

63. $\sin\left[\arccos\left(-\frac{2}{3}\right)\right]$

64. $\cot\left(\arctan\frac{5}{8}\right)$

65. $\csc\left[\cos^{-1}\left(\frac{\sqrt{3}}{2}\right)\right]$

66. $\sec\left[\sin^{-1}\left(-\frac{\sqrt{2}}{2}\right)\right]$

In Exercises 67–76, write an algebraic expression that is equivalent to the expression. (*Hint:* Sketch a right triangle, as demonstrated in Example 7.)

67. $\cot(\arctan x)$

68. $\sin(\arctan x)$

69. $\cos(\arcsin 2x)$

70. $\sec(\arctan 3x)$

71. $\sin(\arccos x)$

72. $\sec[\arcsin(x-1)]$

73. $\tan\left(\arccos\frac{x}{3}\right)$

74. $\cot\left(\arctan\frac{1}{x}\right)$

75. $\csc\left(\arctan\frac{x}{\sqrt{2}}\right)$

76. $\cos\left(\arcsin\frac{x-h}{r}\right)$

In Exercises 77 and 78, use a graphing utility to graph f and g in the same viewing window to verify that the two functions are equal. Explain why they are equal. Identify any asymptotes of the graphs.

77. $f(x) = \sin(\arctan 2x), \quad g(x) = \dfrac{2x}{\sqrt{1+4x^2}}$

78. $f(x) = \tan\left(\arccos\dfrac{x}{2}\right), \quad g(x) = \dfrac{\sqrt{4-x^2}}{x}$

In Exercises 79–82, fill in the blank.

79. $\arctan\dfrac{9}{x} = \arcsin(\;\rule{1.5em}{0.8em}\;), \quad x \neq 0$

80. $\arcsin\dfrac{\sqrt{36-x^2}}{6} = \arccos(\;\rule{1.5em}{0.8em}\;), \quad 0 \leq x \leq 6$

81. $\arccos\dfrac{3}{\sqrt{x^2-2x+10}} = \arcsin(\;\rule{1.5em}{0.8em}\;)$

82. $\arccos\dfrac{x-2}{2} = \arctan(\;\rule{1.5em}{0.8em}\;), \quad |x-2| \leq 2$

In Exercises 83 and 84, sketch a graph of the function and compare the graph of g with the graph of $f(x) = \arcsin x$.

83. $g(x) = \arcsin(x-1)$

84. $g(x) = \arcsin\dfrac{x}{2}$

In Exercises 85–90, sketch a graph of the function.

85. $y = 2\arccos x$

86. $g(t) = \arccos(t+2)$

87. $f(x) = \arctan 2x$

88. $f(x) = \dfrac{\pi}{2} + \arctan x$

89. $h(v) = \tan(\arccos v)$

90. $f(x) = \arccos\dfrac{x}{4}$

In Exercises 91–96, use a graphing utility to graph the function.

91. $f(x) = 2\arccos(2x)$

92. $f(x) = \pi\arcsin(4x)$

93. $f(x) = \arctan(2x-3)$

94. $f(x) = -3 + \arctan(\pi x)$

95. $f(x) = \pi - \sin^{-1}\left(\dfrac{2}{3}\right)$

96. $f(x) = \dfrac{\pi}{2} + \cos^{-1}\left(\dfrac{1}{\pi}\right)$

In Exercises 97 and 98, write the function in terms of the sine function by using the identity

$$A\cos\omega t + B\sin\omega t = \sqrt{A^2+B^2}\,\sin\left(\omega t + \arctan\dfrac{A}{B}\right).$$

Use a graphing utility to graph both forms of the function. What does the graph imply?

97. $f(t) = 3\cos 2t + 3\sin 2t$

98. $f(t) = 4\cos \pi t + 3\sin \pi t$

In Exercises 99–104, fill in the blank. If not possible, state the reason. (*Note:* The notation $x \to c^+$ indicates that x approaches c from the right and $x \to c^-$ indicates that x approaches c from the left.)

99. As $x \to 1^-$, the value of $\arcsin x \to \;\rule{1.5em}{0.8em}\;$.

100. As $x \to 1^-$, the value of $\arccos x \to \;\rule{1.5em}{0.8em}\;$.

101. As $x \to \infty$, the value of arctan $x \to$ ▓▓ .

102. As $x \to -1^+$, the value of arcsin $x \to$ ▓▓ .

103. As $x \to -1^+$, the value of arccos $x \to$ ▓▓ .

104. As $x \to -\infty$, the value of arctan $x \to$ ▓▓ .

105. DOCKING A BOAT A boat is pulled in by means of a winch located on a dock 5 feet above the deck of the boat (see figure). Let θ be the angle of elevation from the boat to the winch and let s be the length of the rope from the winch to the boat.

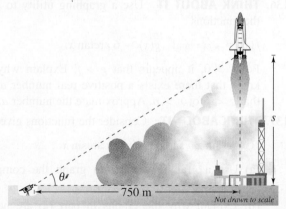

(a) Write θ as a function of s.

(b) Find θ when $s = 40$ feet and $s = 20$ feet.

106. PHOTOGRAPHY A television camera at ground level is filming the lift-off of a space shuttle at a point 750 meters from the launch pad (see figure). Let θ be the angle of elevation to the shuttle and let s be the height of the shuttle.

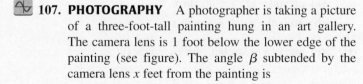

Not drawn to scale

(a) Write θ as a function of s.

(b) Find θ when $s = 300$ meters and $s = 1200$ meters.

107. PHOTOGRAPHY A photographer is taking a picture of a three-foot-tall painting hung in an art gallery. The camera lens is 1 foot below the lower edge of the painting (see figure). The angle β subtended by the camera lens x feet from the painting is

$$\beta = \arctan \frac{3x}{x^2 + 4}, \quad x > 0.$$

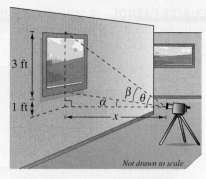

Not drawn to scale

(a) Use a graphing utility to graph β as a function of x.

(b) Move the cursor along the graph to approximate the distance from the picture when β is maximum.

(c) Identify the asymptote of the graph and discuss its meaning in the context of the problem.

108. GRANULAR ANGLE OF REPOSE Different types of granular substances naturally settle at different angles when stored in cone-shaped piles. This angle θ is called the *angle of repose* (see figure). When rock salt is stored in a cone-shaped pile 11 feet high, the diameter of the pile's base is about 34 feet. (Source: Bulk-Store Structures, Inc.)

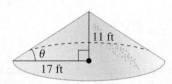

(a) Find the angle of repose for rock salt.

(b) How tall is a pile of rock salt that has a base diameter of 40 feet?

109. GRANULAR ANGLE OF REPOSE When whole corn is stored in a cone-shaped pile 20 feet high, the diameter of the pile's base is about 82 feet.

(a) Find the angle of repose for whole corn.

(b) How tall is a pile of corn that has a base diameter of 100 feet?

110. ANGLE OF ELEVATION An airplane flies at an altitude of 6 miles toward a point directly over an observer. Consider θ and x as shown in the figure.

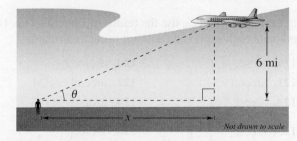

Not drawn to scale

(a) Write θ as a function of x.

(b) Find θ when $x = 7$ miles and $x = 1$ mile.

111. SECURITY PATROL A security car with its spotlight on is parked 20 meters from a warehouse. Consider θ and x as shown in the figure.

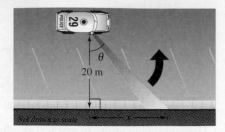

Not drawn to scale

(a) Write θ as a function of x.

(b) Find θ when $x = 5$ meters and $x = 12$ meters.

EXPLORATION

TRUE OR FALSE? In Exercises 112–114, determine whether the statement is true or false. Justify your answer.

112. $\sin \dfrac{5\pi}{6} = \dfrac{1}{2} \implies \arcsin \dfrac{1}{2} = \dfrac{5\pi}{6}$

113. $\tan \dfrac{5\pi}{4} = 1 \implies \arctan 1 = \dfrac{5\pi}{4}$

114. $\arctan x = \dfrac{\arcsin x}{\arccos x}$

115. Define the inverse cotangent function by restricting the domain of the cotangent function to the interval $(0, \pi)$, and sketch its graph.

116. Define the inverse secant function by restricting the domain of the secant function to the intervals $[0, \pi/2)$ and $(\pi/2, \pi]$, and sketch its graph.

117. Define the inverse cosecant function by restricting the domain of the cosecant function to the intervals $[-\pi/2, 0)$ and $(0, \pi/2]$, and sketch its graph.

118. CAPSTONE Use the results of Exercises 115–117 to explain how to graph (a) the inverse cotangent function, (b) the inverse secant function, and (c) the inverse cosecant function on a graphing utility.

In Exercises 119–126, use the results of Exercises 115–117 to evaluate each expression without using a calculator.

119. $\operatorname{arcsec} \sqrt{2}$

120. $\operatorname{arcsec} 1$

121. $\operatorname{arccot}(-1)$

122. $\operatorname{arccot}\left(-\sqrt{3}\right)$

123. $\operatorname{arccsc} 2$

124. $\operatorname{arccsc}(-1)$

125. $\operatorname{arccsc}\left(\dfrac{2\sqrt{3}}{3}\right)$

126. $\operatorname{arcsec}\left(-\dfrac{2\sqrt{3}}{3}\right)$

In Exercises 127–134, use the results of Exercises 115–117 and a calculator to approximate the value of the expression. Round your result to two decimal places.

127. $\operatorname{arcsec} 2.54$

128. $\operatorname{arcsec}(-1.52)$

129. $\operatorname{arccot} 5.25$

130. $\operatorname{arccot}(-10)$

131. $\operatorname{arccot} \dfrac{5}{3}$

132. $\operatorname{arccot}\left(-\dfrac{16}{7}\right)$

133. $\operatorname{arccsc}\left(-\dfrac{25}{3}\right)$

134. $\operatorname{arccsc}(-12)$

135. AREA In calculus, it is shown that the area of the region bounded by the graphs of $y = 0$, $y = 1/(x^2 + 1)$, $x = a$, and $x = b$ is given by

Area = $\arctan b - \arctan a$

(see figure). Find the area for the following values of a and b.

(a) $a = 0, b = 1$ (b) $a = -1, b = 1$

(c) $a = 0, b = 3$ (d) $a = -1, b = 3$

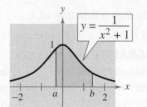

136. THINK ABOUT IT Use a graphing utility to graph the functions

$$f(x) = \sqrt{x} \quad \text{and} \quad g(x) = 6 \arctan x.$$

For $x > 0$, it appears that $g > f$. Explain why you know that there exists a positive real number a such that $g < f$ for $x > a$. Approximate the number a.

137. THINK ABOUT IT Consider the functions given by

$$f(x) = \sin x \quad \text{and} \quad f^{-1}(x) = \arcsin x.$$

(a) Use a graphing utility to graph the composite functions $f \circ f^{-1}$ and $f^{-1} \circ f$.

(b) Explain why the graphs in part (a) are not the graph of the line $y = x$. Why do the graphs of $f \circ f^{-1}$ and $f^{-1} \circ f$ differ?

138. PROOF Prove each identity.

(a) $\arcsin(-x) = -\arcsin x$

(b) $\arctan(-x) = -\arctan x$

(c) $\arctan x + \arctan \dfrac{1}{x} = \dfrac{\pi}{2}, \quad x > 0$

(d) $\arcsin x + \arccos x = \dfrac{\pi}{2}$

(e) $\arcsin x = \arctan \dfrac{x}{\sqrt{1 - x^2}}$

1.8 APPLICATIONS AND MODELS

What you should learn

- Solve real-life problems involving right triangles.
- Solve real-life problems involving directional bearings.
- Solve real-life problems involving harmonic motion.

Why you should learn it

Right triangles often occur in real-life situations. For instance, in Exercise 65 on page 215, right triangles are used to determine the shortest grain elevator for a grain storage bin on a farm.

Applications Involving Right Triangles

In this section, the three angles of a right triangle are denoted by the letters A, B, and C (where C is the right angle), and the lengths of the sides opposite these angles by the letters a, b, and c (where c is the hypotenuse).

Example 1   Solving a Right Triangle

Solve the right triangle shown in Figure 1.78 for all unknown sides and angles.

FIGURE **1.78**

Solution

Because $C = 90°$, it follows that $A + B = 90°$ and $B = 90° - 34.2° = 55.8°$. To solve for a, use the fact that

$$\tan A = \frac{\text{opp}}{\text{adj}} = \frac{a}{b} \qquad\Longrightarrow\qquad a = b \tan A.$$

So, $a = 19.4 \tan 34.2° \approx 13.18$. Similarly, to solve for c, use the fact that

$$\cos A = \frac{\text{adj}}{\text{hyp}} = \frac{b}{c} \qquad\Longrightarrow\qquad c = \frac{b}{\cos A}.$$

So, $c = \dfrac{19.4}{\cos 34.2°} \approx 23.46.$

CHECK*Point*   Now try Exercise 5.

Example 2   Finding a Side of a Right Triangle

A safety regulation states that the maximum angle of elevation for a rescue ladder is 72°. A fire department's longest ladder is 110 feet. What is the maximum safe rescue height?

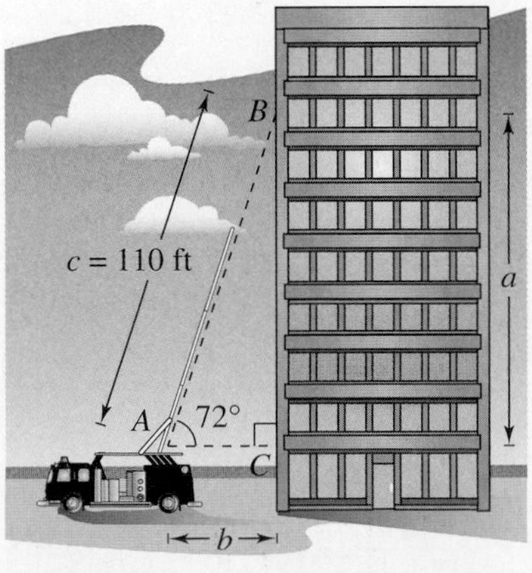

Solution

A sketch is shown in Figure 1.79. From the equation $\sin A = a/c$, it follows that

$$a = c \sin A = 110 \sin 72° \approx 104.6.$$

So, the maximum safe rescue height is about 104.6 feet above the height of the fire truck.

CHECK*Point*   Now try Exercise 19.

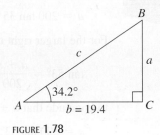

FIGURE **1.79**

Example 3 Finding a Side of a Right Triangle

At a point 200 feet from the base of a building, the angle of elevation to the *bottom* of a smokestack is 35°, whereas the angle of elevation to the *top* is 53°, as shown in Figure 1.80. Find the height s of the smokestack alone.

Solution

Note from Figure 1.80 that this problem involves two right triangles. For the smaller right triangle, use the fact that

$$\tan 35° = \frac{a}{200}$$

to conclude that the height of the building is

$$a = 200 \tan 35°.$$

For the larger right triangle, use the equation

$$\tan 53° = \frac{a + s}{200}$$

to conclude that $a + s = 200 \tan 53°$. So, the height of the smokestack is

$$s = 200 \tan 53° - a$$

$$= 200 \tan 53° - 200 \tan 35°$$

$$\approx 125.4 \text{ feet.}$$

CHECK Point Now try Exercise 23.

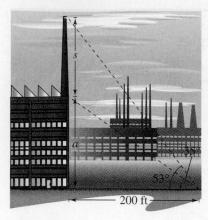

FIGURE **1.80**

Example 4 Finding an Acute Angle of a Right Triangle

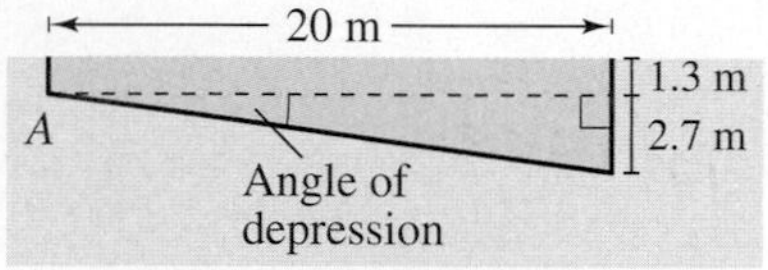

A swimming pool is 20 meters long and 12 meters wide. The bottom of the pool is slanted so that the water depth is 1.3 meters at the shallow end and 4 meters at the deep end, as shown in Figure 1.81. Find the angle of depression of the bottom of the pool.

Solution

Using the tangent function, you can see that

$$\tan A = \frac{\text{opp}}{\text{adj}}$$

$$= \frac{2.7}{20}$$

$$= 0.135.$$

So, the angle of depression is

$$A = \arctan 0.135$$

$$\approx 0.13419 \text{ radian}$$

$$\approx 7.69°.$$

CHECK Point Now try Exercise 29.

FIGURE **1.81**

Trigonometry and Bearings

In surveying and navigation, directions can be given in terms of **bearings.** A bearing measures the acute angle that a path or line of sight makes with a fixed north-south line, as shown in Figure 1.82. For instance, the bearing S 35° E in Figure 1.82 means 35 degrees east of south.

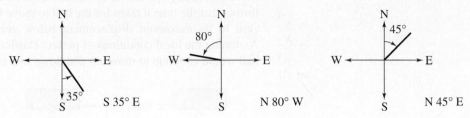

FIGURE **1.82**

| Example 5 | **Finding Directions in Terms of Bearings** |

A ship leaves port at noon and heads due west at 20 knots, or 20 nautical miles (nm) per hour. At 2 P.M. the ship changes course to N 54° W, as shown in Figure 1.83. Find the ship's bearing and distance from the port of departure at 3 P.M.

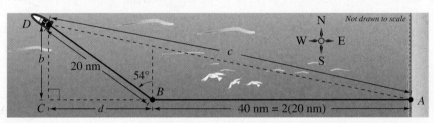

FIGURE **1.83**

Study Tip

In *air navigation*, bearings are measured in degrees *clockwise* from north. Examples of air navigation bearings are shown below.

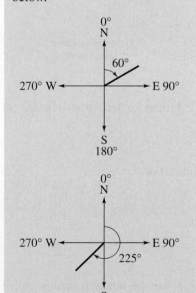

Solution

For triangle *BCD*, you have $B = 90° - 54° = 36°$. The two sides of this triangle can be determined to be

$$b = 20 \sin 36° \quad \text{and} \quad d = 20 \cos 36°.$$

For triangle *ACD*, you can find angle *A* as follows.

$$\tan A = \frac{b}{d + 40} = \frac{20 \sin 36°}{20 \cos 36° + 40} \approx 0.2092494$$

$$A \approx \arctan 0.2092494 \approx 11.82°$$

The angle with the north-south line is $90° - 11.82° = 78.18°$. So, the bearing of the ship is N 78.18° W. Finally, from triangle *ACD*, you have $\sin A = b/c$, which yields

$$c = \frac{b}{\sin A} = \frac{20 \sin 36°}{\sin 11.82°}$$

$$\approx 57.4 \text{ nautical miles.} \qquad \text{Distance from port}$$

CHECK**Point**   Now try Exercise 37.

Harmonic Motion

The periodic nature of the trigonometric functions is useful for describing the motion of a point on an object that vibrates, oscillates, rotates, or is moved by wave motion.

For example, consider a ball that is bobbing up and down on the end of a spring, as shown in Figure 1.84. Suppose that 10 centimeters is the maximum distance the ball moves vertically upward or downward from its equilibrium (at rest) position. Suppose further that the time it takes for the ball to move from its maximum displacement above zero to its maximum displacement below zero and back again is $t = 4$ seconds. Assuming the ideal conditions of perfect elasticity and no friction or air resistance, the ball would continue to move up and down in a uniform and regular manner.

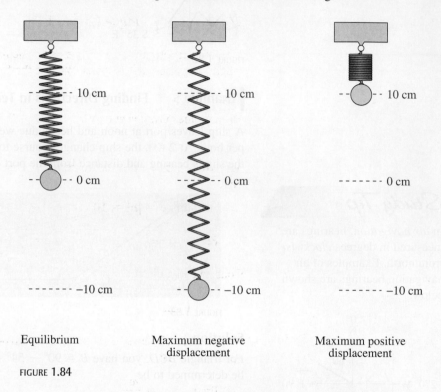

| Equilibrium | Maximum negative displacement | Maximum positive displacement |

FIGURE **1.84**

From this spring you can conclude that the period (time for one complete cycle) of the motion is

Period = 4 seconds

its amplitude (maximum displacement from equilibrium) is

Amplitude = 10 centimeters

and its **frequency** (number of cycles per second) is

Frequency $= \dfrac{1}{4}$ cycle per second.

Motion of this nature can be described by a sine or cosine function, and is called **simple harmonic motion.**

Definition of Simple Harmonic Motion

A point that moves on a coordinate line is said to be in **simple harmonic motion** if its distance d from the origin at time t is given by either

$$d = a \sin \omega t \quad \text{or} \quad d = a \cos \omega t$$

where a and ω are real numbers such that $\omega > 0$. The motion has amplitude $|a|$, period $\dfrac{2\pi}{\omega}$, and frequency $\dfrac{\omega}{2\pi}$.

Example 6   Simple Harmonic Motion

Write the equation for the simple harmonic motion of the ball described in Figure 1.84, where the period is 4 seconds. What is the frequency of this harmonic motion?

Solution

Because the spring is at equilibrium $(d = 0)$ when $t = 0$, you use the equation

$$d = a \sin \omega t.$$

Moreover, because the maximum displacement from zero is 10 and the period is 4, you have

$$\text{Amplitude} = |a| = 10$$

$$\text{Period} = \frac{2\pi}{\omega} = 4 \quad \implies \quad \omega = \frac{\pi}{2}.$$

Consequently, the equation of motion is

$$d = 10 \sin \frac{\pi}{2} t.$$

Note that the choice of $a = 10$ or $a = -10$ depends on whether the ball initially moves up or down. The frequency is

$$\text{Frequency} = \frac{\omega}{2\pi}$$

$$= \frac{\pi/2}{2\pi}$$

$$= \frac{1}{4} \text{ cycle per second.}$$

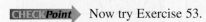 **CHECKPoint**   Now try Exercise 53.

One illustration of the relationship between sine waves and harmonic motion can be seen in the wave motion resulting when a stone is dropped into a calm pool of water. The waves move outward in roughly the shape of sine (or cosine) waves, as shown in Figure 1.85. As an example, suppose you are fishing and your fishing bob is attached so that it does not move horizontally. As the waves move outward from the dropped stone, your fishing bob will move up and down in simple harmonic motion, as shown in Figure 1.86.

FIGURE **1.85**

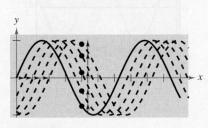

FIGURE **1.86**

Example 7 Simple Harmonic Motion

Given the equation for simple harmonic motion

$$d = 6 \cos \frac{3\pi}{4}t$$

find (a) the maximum displacement, (b) the frequency, (c) the value of d when $t = 4$, and (d) the least positive value of t for which $d = 0$.

Algebraic Solution

The given equation has the form $d = a \cos \omega t$, with $a = 6$ and $\omega = 3\pi/4$.

a. The maximum displacement (from the point of equilibrium) is given by the amplitude. So, the maximum displacement is 6.

b. Frequency $= \dfrac{\omega}{2\pi}$

$$= \frac{3\pi/4}{2\pi}$$

$$= \frac{3}{8} \text{ cycle per unit of time}$$

c. $d = 6 \cos \left[\dfrac{3\pi}{4}(4) \right]$

$$= 6 \cos 3\pi$$

$$= 6(-1)$$

$$= -6$$

d. To find the least positive value of t for which $d = 0$, solve the equation

$$d = 6 \cos \frac{3\pi}{4}t = 0.$$

First divide each side by 6 to obtain

$$\cos \frac{3\pi}{4}t = 0.$$

This equation is satisfied when

$$\frac{3\pi}{4}t = \frac{\pi}{2}, \frac{3\pi}{2}, \frac{5\pi}{2}, \dots$$

Multiply these values by $4/(3\pi)$ to obtain

$$t = \frac{2}{3}, 2, \frac{10}{3}, \dots$$

So, the least positive value of t is $t = \frac{2}{3}$.

CHECK**Point** Now try Exercise 57.

Graphical Solution

Use a graphing utility set in *radian* mode to graph

$$y = 6 \cos \frac{3\pi}{4}x.$$

a. Use the *maximum* feature of the graphing utility to estimate that the maximum displacement from the point of equilibrium $y = 0$ is 6, as shown in Figure 1.87.

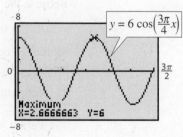

FIGURE 1.87

b. The period is the time for the graph to complete one cycle, which is $x \approx 2.667$. You can estimate the frequency as follows.

$$\text{Frequency} \approx \frac{1}{2.667} \approx 0.375 \text{ cycle per unit of time}$$

c. Use the *trace* or *value* feature to estimate that the value of y when $x = 4$ is $y = -6$, as shown in Figure 1.88.

d. Use the *zero* or *root* feature to estimate that the least positive value of x for which $y = 0$ is $x \approx 0.6667$, as shown in Figure 1.89.

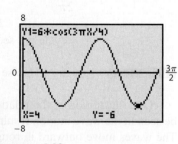

FIGURE 1.88

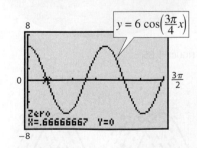

FIGURE 1.89

1.8 ▸ EXERCISES

See www.CalcChat.com for worked-out solutions to odd-numbered exercises.

VOCABULARY: Fill in the blanks.

1. A _____ measures the acute angle a path or line of sight makes with a fixed north-south line.

2. A point that moves on a coordinate line is said to be in simple _____ _____ if its distance d from the origin at time t is given by either $d = a \sin \omega t$ or $d = a \cos \omega t$.

3. The time for one complete cycle of a point in simple harmonic motion is its _____.

4. The number of cycles per second of a point in simple harmonic motion is its _____.

SKILLS AND APPLICATIONS

In Exercises 5–14, solve the right triangle shown in the figure for all unknown sides and angles. Round your answers to two decimal places.

5. $A = 30°$, $b = 3$ **6.** $B = 54°$, $c = 15$

7. $B = 71°$, $b = 24$ **8.** $A = 8.4°$, $a = 40.5$

9. $a = 3$, $b = 4$ **10.** $a = 25$, $c = 35$

11. $b = 16$, $c = 52$ **12.** $b = 1.32$, $c = 9.45$

13. $A = 12°15'$, $c = 430.5$

14. $B = 65°12'$, $a = 14.2$

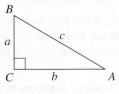

FIGURE FOR 5–14

FIGURE FOR 15–18

In Exercises 15–18, find the altitude of the isosceles triangle shown in the figure. Round your answers to two decimal places.

15. $\theta = 45°$, $b = 6$ **16.** $\theta = 18°$, $b = 10$

17. $\theta = 32°$, $b = 8$ **18.** $\theta = 27°$, $b = 11$

19. LENGTH The sun is 25° above the horizon. Find the length of a shadow cast by a building that is 100 feet tall (see figure).

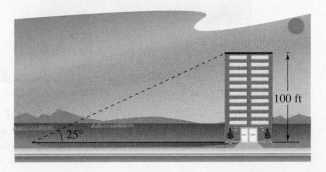

20. LENGTH The sun is 20° above the horizon. Find the length of a shadow cast by a park statue that is 12 feet tall.

21. HEIGHT A ladder 20 feet long leans against the side of a house. Find the height from the top of the ladder to the ground if the angle of elevation of the ladder is 80°.

22. HEIGHT The length of a shadow of a tree is 125 feet when the angle of elevation of the sun is 33°. Approximate the height of the tree.

23. HEIGHT From a point 50 feet in front of a church, the angles of elevation to the base of the steeple and the top of the steeple are 35° and 47° 40′, respectively. Find the height of the steeple.

24. DISTANCE An observer in a lighthouse 350 feet above sea level observes two ships directly offshore. The angles of depression to the ships are 4° and 6.5° (see figure). How far apart are the ships?

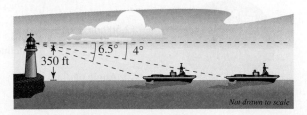

Not drawn to scale

25. DISTANCE A passenger in an airplane at an altitude of 10 kilometers sees two towns directly to the east of the plane. The angles of depression to the towns are 28° and 55° (see figure). How far apart are the towns?

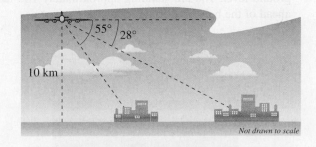

Not drawn to scale

26. ALTITUDE You observe a plane approaching overhead and assume that its speed is 550 miles per hour. The angle of elevation of the plane is 16° at one time and 57° one minute later. Approximate the altitude of the plane.

27. ANGLE OF ELEVATION An engineer erects a 75-foot cellular telephone tower. Find the angle of elevation to the top of the tower at a point on level ground 50 feet from its base.

28. ANGLE OF ELEVATION The height of an outdoor basketball backboard is $12\frac{1}{2}$ feet, and the backboard casts a shadow $17\frac{1}{3}$ feet long.

(a) Draw a right triangle that gives a visual representation of the problem. Label the known and unknown quantities.

(b) Use a trigonometric function to write an equation involving the unknown quantity.

(c) Find the angle of elevation of the sun.

29. ANGLE OF DEPRESSION A cellular telephone tower that is 150 feet tall is placed on top of a mountain that is 1200 feet above sea level. What is the angle of depression from the top of the tower to a cell phone user who is 5 horizontal miles away and 400 feet above sea level?

30. ANGLE OF DEPRESSION A Global Positioning System satellite orbits 12,500 miles above Earth's surface (see figure). Find the angle of depression from the satellite to the horizon. Assume the radius of Earth is 4000 miles.

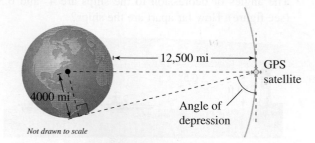

Not drawn to scale

31. HEIGHT You are holding one of the tethers attached to the top of a giant character balloon in a parade. Before the start of the parade the balloon is upright and the bottom is floating approximately 20 feet above ground level. You are standing approximately 100 feet ahead of the balloon (see figure).

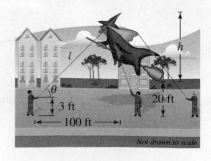

Not drawn to scale

(a) Find the length l of the tether you are holding in terms of h, the height of the balloon from top to bottom.

(b) Find an expression for the angle of elevation θ from you to the top of the balloon.

(c) Find the height h of the balloon if the angle of elevation to the top of the balloon is 35°.

32. HEIGHT The designers of a water park are creating a new slide and have sketched some preliminary drawings. The length of the ladder is 30 feet, and its angle of elevation is 60° (see figure).

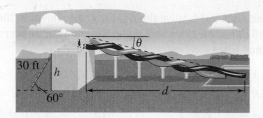

(a) Find the height h of the slide.

(b) Find the angle of depression θ from the top of the slide to the end of the slide at the ground in terms of the horizontal distance d the rider travels.

(c) The angle of depression of the ride is bounded by safety restrictions to be no less than 25° and not more than 30°. Find an interval for how far the rider travels horizontally.

33. SPEED ENFORCEMENT A police department has set up a speed enforcement zone on a straight length of highway. A patrol car is parked parallel to the zone, 200 feet from one end and 150 feet from the other end (see figure).

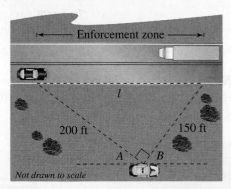

Not drawn to scale

(a) Find the length l of the zone and the measures of the angles A and B (in degrees).

(b) Find the minimum amount of time (in seconds) it takes for a vehicle to pass through the zone without exceeding the posted speed limit of 35 miles per hour.

34. AIRPLANE ASCENT During takeoff, an airplane's angle of ascent is 18° and its speed is 275 feet per second.

(a) Find the plane's altitude after 1 minute.

(b) How long will it take the plane to climb to an altitude of 10,000 feet?

35. NAVIGATION An airplane flying at 600 miles per hour has a bearing of 52°. After flying for 1.5 hours, how far north and how far east will the plane have traveled from its point of departure?

36. NAVIGATION A jet leaves Reno, Nevada and is headed toward Miami, Florida at a bearing of 100°. The distance between the two cities is approximately 2472 miles.

(a) How far north and how far west is Reno relative to Miami?

(b) If the jet is to return directly to Reno from Miami, at what bearing should it travel?

37. NAVIGATION A ship leaves port at noon and has a bearing of S 29° W. The ship sails at 20 knots.

(a) How many nautical miles south and how many nautical miles west will the ship have traveled by 6:00 P.M.?

(b) At 6:00 P.M., the ship changes course to due west. Find the ship's bearing and distance from the port of departure at 7:00 P.M.

38. NAVIGATION A privately owned yacht leaves a dock in Myrtle Beach, South Carolina and heads toward Freeport in the Bahamas at a bearing of S 1.4° E. The yacht averages a speed of 20 knots over the 428 nautical-mile trip.

(a) How long will it take the yacht to make the trip?

(b) How far east and south is the yacht after 12 hours?

(c) If a plane leaves Myrtle Beach to fly to Freeport, what bearing should be taken?

39. NAVIGATION A ship is 45 miles east and 30 miles south of port. The captain wants to sail directly to port. What bearing should be taken?

40. NAVIGATION An airplane is 160 miles north and 85 miles east of an airport. The pilot wants to fly directly to the airport. What bearing should be taken?

41. SURVEYING A surveyor wants to find the distance across a swamp (see figure). The bearing from A to B is N 32° W. The surveyor walks 50 meters from A, and at the point C the bearing to B is N 68° W. Find (a) the bearing from A to C and (b) the distance from A to B.

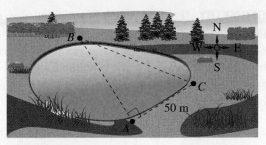

FIGURE FOR **41**

42. LOCATION OF A FIRE Two fire towers are 30 kilometers apart, where tower A is due west of tower B. A fire is spotted from the towers, and the bearings from A and B are N 76° E and N 56° W, respectively (see figure). Find the distance d of the fire from the line segment AB.

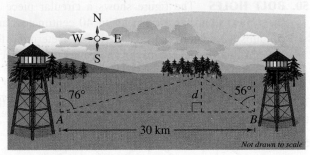

GEOMETRY In Exercises 43 and 44, find the angle α between two nonvertical lines L_1 and L_2. The angle α satisfies the equation

$$\tan \alpha = \left| \frac{m_2 - m_1}{1 + m_2 m_1} \right|$$

where m_1 and m_2 are the slopes of L_1 and L_2, respectively. (Assume that $m_1 m_2 \neq -1$.)

43. L_1: $3x - 2y = 5$ **44.** L_1: $2x - y = 8$

      L_2: $x + y = 1$ L_2: $x - 5y = -4$

45. GEOMETRY Determine the angle between the diagonal of a cube and the diagonal of its base, as shown in the figure.

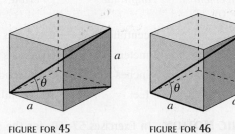

FIGURE FOR **45**   FIGURE FOR **46**

46. GEOMETRY Determine the angle between the diagonal of a cube and its edge, as shown in the figure.

47. GEOMETRY Find the length of the sides of a regular pentagon inscribed in a circle of radius 25 inches.

48. GEOMETRY Find the length of the sides of a regular hexagon inscribed in a circle of radius 25 inches.

49. HARDWARE Write the distance y across the flat sides of a hexagonal nut as a function of r (see figure).

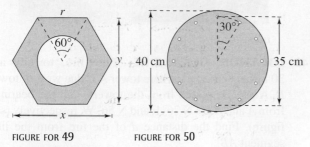

FIGURE FOR **49**   FIGURE FOR **50**

50. BOLT HOLES The figure shows a circular piece of sheet metal that has a diameter of 40 centimeters and contains 12 equally-spaced bolt holes. Determine the straight-line distance between the centers of consecutive bolt holes.

TRUSSES In Exercises 51 and 52, find the lengths of all the unknown members of the truss.

51.

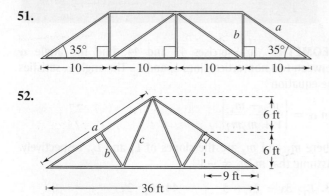

52.

HARMONIC MOTION In Exercises 53–56, find a model for simple harmonic motion satisfying the specified conditions.

Displacement ($t = 0$)	Amplitude	Period
53. 0	4 centimeters	2 seconds
54. 0	3 meters	6 seconds
55. 3 inches	3 inches	1.5 seconds
56. 2 feet	2 feet	10 seconds

HARMONIC MOTION In Exercises 57–60, for the simple harmonic motion described by the trigonometric function, find (a) the maximum displacement, (b) the frequency, (c) the value of d when $t = 5$, and (d) the least positive value of t for which $d = 0$. Use a graphing utility to verify your results.

57. $d = 9 \cos \dfrac{6\pi}{5} t$ **58.** $d = \dfrac{1}{2} \cos 20\pi t$

59. $d = \dfrac{1}{4} \sin 6\pi t$ **60.** $d = \dfrac{1}{64} \sin 792\pi t$

61. TUNING FORK A point on the end of a tuning fork moves in simple harmonic motion described by $d = a \sin \omega t$. Find ω given that the tuning fork for middle C has a frequency of 264 vibrations per second.

62. WAVE MOTION A buoy oscillates in simple harmonic motion as waves go past. It is noted that the buoy moves a total of 3.5 feet from its low point to its high point (see figure), and that it returns to its high point every 10 seconds. Write an equation that describes the motion of the buoy if its high point is at $t = 0$.

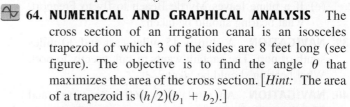

63. OSCILLATION OF A SPRING A ball that is bobbing up and down on the end of a spring has a maximum displacement of 3 inches. Its motion (in ideal conditions) is modeled by $y = \dfrac{1}{4} \cos 16t$ ($t > 0$), where y is measured in feet and t is the time in seconds.

(a) Graph the function.

(b) What is the period of the oscillations?

(c) Determine the first time the weight passes the point of equilibrium ($y = 0$).

64. NUMERICAL AND GRAPHICAL ANALYSIS The cross section of an irrigation canal is an isosceles trapezoid of which 3 of the sides are 8 feet long (see figure). The objective is to find the angle θ that maximizes the area of the cross section. [*Hint:* The area of a trapezoid is $(h/2)(b_1 + b_2)$.]

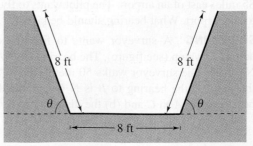

(a) Complete seven additional rows of the table.

Base 1	Base 2	Altitude	Area
8	$8 + 16 \cos 10°$	$8 \sin 10°$	22.1
8	$8 + 16 \cos 20°$	$8 \sin 20°$	42.5

(b) Use a graphing utility to generate additional rows of the table. Use the table to estimate the maximum cross-sectional area.

(c) Write the area A as a function of θ.

(d) Use a graphing utility to graph the function. Use the graph to estimate the maximum cross-sectional area. How does your estimate compare with that of part (b)?

 65. NUMERICAL AND GRAPHICAL ANALYSIS A 2-meter-high fence is 3 meters from the side of a grain storage bin. A grain elevator must reach from ground level outside the fence to the storage bin (see figure). The objective is to determine the shortest elevator that meets the constraints.

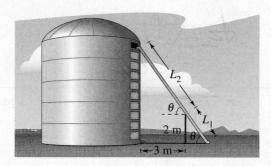

(a) Complete four rows of the table.

θ	L_1	L_2	$L_1 + L_2$
0.1	$\dfrac{2}{\sin 0.1}$	$\dfrac{3}{\cos 0.1}$	23.0
0.2	$\dfrac{2}{\sin 0.2}$	$\dfrac{3}{\cos 0.2}$	13.1

(b) Use a graphing utility to generate additional rows of the table. Use the table to estimate the minimum length of the elevator.

(c) Write the length $L_1 + L_2$ as a function of θ.

(d) Use a graphing utility to graph the function. Use the graph to estimate the minimum length. How does your estimate compare with that of part (b)?

66. DATA ANALYSIS The table shows the average sales S (in millions of dollars) of an outerwear manufacturer for each month t, where $t = 1$ represents January.

Time, t	1	2	3	4	5	6
Sales, S	13.46	11.15	8.00	4.85	2.54	1.70

Time, t	7	8	9	10	11	12
Sales, S	2.54	4.85	8.00	11.15	13.46	14.30

(a) Create a scatter plot of the data.

(b) Find a trigonometric model that fits the data. Graph the model with your scatter plot. How well does the model fit the data?

(c) What is the period of the model? Do you think it is reasonable given the context? Explain your reasoning.

(d) Interpret the meaning of the model's amplitude in the context of the problem.

67. DATA ANALYSIS The number of hours H of daylight in Denver, Colorado on the 15th of each month are: 1(9.67), 2(10.72), 3(11.92), 4(13.25), 5(14.37), 6(14.97), 7(14.72), 8(13.77), 9(12.48), 10(11.18), 11(10.00), 12(9.38). The month is represented by t, with $t = 1$ corresponding to January. A model for the data is given by

$$H(t) = 12.13 + 2.77 \sin[(\pi t/6) - 1.60].$$

 (a) Use a graphing utility to graph the data points and the model in the same viewing window.

(b) What is the period of the model? Is it what you expected? Explain.

(c) What is the amplitude of the model? What does it represent in the context of the problem? Explain.

EXPLORATION

68. CAPSTONE While walking across flat land, you notice a wind turbine tower of height h feet directly in front of you. The angle of elevation to the top of the tower is A degrees. After you walk d feet closer to the tower, the angle of elevation increases to B degrees.

(a) Draw a diagram to represent the situation.

(b) Write an expression for the height h of the tower in terms of the angles A and B and the distance d.

TRUE OR FALSE? In Exercises 69 and 70, determine whether the statement is true or false. Justify your answer.

69. The Leaning Tower of Pisa is not vertical, but if you know the angle of elevation θ to the top of the tower when you stand d feet away from it, you can find its height h using the formula $h = d \tan \theta$.

70. N 24° E means 24 degrees north of east.

1 CHAPTER SUMMARY

	What Did You Learn?	Explanation/Examples	Review Exercises
Section 1.1	Describe angles *(p. 134)*.		1–8
	Convert between degrees and radians *(p. 138)*.	To convert degrees to radians, multiply degrees by $(\pi \text{ rad})/180°$. To convert radians to degrees, multiply radians by $180°/(\pi \text{ rad})$.	9–20
	Use angles to model and solve real-life problems *(p. 139)*.	Angles can be used to find the length of a circular arc and the area of a sector of a circle. (See Examples 5 and 8.)	21–24
Section 1.2	Identify a unit circle and describe its relationship to real numbers *(p. 146)*.		25–28
	Evaluate trigonometric functions using the unit circle *(p. 147)*.	$t = \dfrac{2\pi}{3}$ corresponds to $(x, y) = \left(-\dfrac{1}{2}, \dfrac{\sqrt{3}}{2}\right)$. So $\cos \dfrac{2\pi}{3} = -\dfrac{1}{2}$, $\sin \dfrac{2\pi}{3} = \dfrac{\sqrt{3}}{2}$, and $\tan \dfrac{2\pi}{3} = -\sqrt{3}$.	29–32
	Use domain and period to evaluate sine and cosine functions *(p. 149)*.	Because $\dfrac{9\pi}{4} = 2\pi + \dfrac{\pi}{4}$, $\sin \dfrac{9\pi}{4} = \sin \dfrac{\pi}{4} = \dfrac{\sqrt{2}}{2}$.	33–36
	Use a calculator to evaluate trigonometric functions *(p. 150)*.	$\sin \dfrac{3\pi}{8} \approx 0.9239$, $\cot(-1.2) \approx -0.3888$	37–40
Section 1.3	Evaluate trigonometric functions of acute angles *(p. 153)*.	$\sin \theta = \dfrac{\text{opp}}{\text{hyp}}$, $\cos \theta = \dfrac{\text{adj}}{\text{hyp}}$, $\tan \theta = \dfrac{\text{opp}}{\text{adj}}$ $\csc \theta = \dfrac{\text{hyp}}{\text{opp}}$, $\sec \theta = \dfrac{\text{hyp}}{\text{adj}}$, $\cot \theta = \dfrac{\text{adj}}{\text{opp}}$	41, 42
	Use fundamental trigonometric identities *(p. 156)*.	$\sin \theta = \dfrac{1}{\csc \theta}$, $\tan \theta = \dfrac{\sin \theta}{\cos \theta}$, $\sin^2 \theta + \cos^2 \theta = 1$	43–46
	Use a calculator to evaluate trigonometric functions *(p. 157)*.	$\tan 34.7° \approx 0.6924$, $\csc 29° \, 15' \approx 2.0466$	47–54
	Use trigonometric functions to model and solve real-life problems *(p. 158)*.	Trigonometric functions can be used to find the height of a monument, the angle between two paths, and the length of a ramp. (See Examples 7–9.)	55, 56

What Did You Learn?	Explanation/Examples	Review Exercises
Section 1.4 Evaluate trigonometric functions of any angle (p. 164).	Let (3, 4) be a point on the terminal side of θ. Then $\sin\theta = \frac{4}{5}$, $\cos\theta = \frac{3}{5}$, and $\tan\theta = \frac{4}{3}$.	57–70
Find reference angles (p. 166).	Let θ be an angle in standard position. Its reference angle is the acute angle θ' formed by the terminal side of θ and the horizontal axis.	71–74
Evaluate trigonometric functions of real numbers (p. 167).	$\cos\frac{7\pi}{3} = \frac{1}{2}$ because $\theta' = \frac{7\pi}{3} - 2\pi = \frac{\pi}{3}$. So, $\cos\frac{7\pi}{3} = \cos\frac{\pi}{3} = \frac{1}{2}$.	75–84
Section 1.5 Sketch the graphs of sine and cosine functions using amplitude and period (p. 173).		85–88
Sketch translations of the graphs of sine and cosine functions (p. 177).	For $y = d + a\sin(bx - c)$ and $y = d + a\cos(bx - c)$, the constant c creates a horizontal translation. The constant d creates a vertical translation. (See Examples 4–6.)	89–92
Use sine and cosine functions to model real-life data (p. 179).	A cosine function can be used to model the depth of the water at the end of a dock at various times. (See Example 7.)	93, 94
Section 1.6 Sketch the graphs of tangent (p. 184), cotangent (p. 186), secant (p. 187), and cosecant (p. 187) functions.		95–102
Sketch the graphs of damped trigonometric functions (p. 189).	For $f(x) = x\cos 2x$ and $g(x) = x^2\sin 4x$, the factors x and x^2 are called damping factors.	103, 104
Section 1.7 Evaluate and graph inverse trigonometric functions (p. 195).	$\sin^{-1}\frac{1}{2} = \frac{\pi}{6}$, $\cos^{-1}\left(-\frac{\sqrt{2}}{2}\right) = \frac{3\pi}{4}$, $\tan^{-1}\sqrt{3} = \frac{\pi}{3}$	105–122, 131–138
Evaluate and graph the compositions of trigonometric functions (p. 199).	$\cos[\arctan(5/12)] = 12/13$, $\sin(\sin^{-1} 0.4) = 0.4$	123–130
Section 1.8 Solve real-life problems involving right triangles (p. 205).	A trigonometric function can be used to find the height of a smokestack on top of a building. (See Example 3.)	139, 140
Solve real-life problems involving directional bearings (p. 207).	Trigonometric functions can be used to find a ship's bearing and distance from a port at a given time. (See Example 5.)	141
Solve real-life problems involving harmonic motion (p. 208).	Sine or cosine functions can be used to describe the motion of an object that vibrates, oscillates, rotates, or is moved by wave motion. (See Examples 6 and 7.)	142

1 REVIEW EXERCISES

See www.CalcChat.com for worked-out solutions to odd-numbered exercises.

1.1 In Exercises 1–8, (a) sketch the angle in standard position, (b) determine the quadrant in which the angle lies, and (c) determine one positive and one negative coterminal angle.

1. $15\pi/4$
2. $2\pi/9$
3. $-4\pi/3$
4. $-23\pi/3$
5. $70°$
6. $280°$
7. $-110°$
8. $-405°$

In Exercises 9–12, convert the angle measure from degrees to radians. Round your answer to three decimal places.

9. $450°$
10. $-112.5°$
11. $-33°\,45'$
12. $197°\,17'$

In Exercises 13–16, convert the angle measure from radians to degrees. Round your answer to three decimal places.

13. $3\pi/10$
14. $-11\pi/6$
15. -3.5
16. 5.7

In Exercises 17–20, convert each angle measure to degrees, minutes, and seconds without using a calculator.

17. $198.4°$
18. $-70.2°$
19. $0.65°$
20. $-5.96°$

21. **ARC LENGTH** Find the length of the arc on a circle with a radius of 20 inches intercepted by a central angle of $138°$.

22. **PHONOGRAPH** Phonograph records are vinyl discs that rotate on a turntable. A typical record album is 12 inches in diameter and plays at $33\frac{1}{3}$ revolutions per minute.

    (a) What is the angular speed of a record album?

    (b) What is the linear speed of the outer edge of a record album?

23. **CIRCULAR SECTOR** Find the area of the sector of a circle with a radius of 18 inches and central angle $\theta = 120°$.

24. **CIRCULAR SECTOR** Find the area of the sector of a circle with a radius of 6.5 millimeters and central angle $\theta = 5\pi/6$.

1.2 In Exercises 25–28, find the point (x, y) on the unit circle that corresponds to the real number t.

25. $t = 2\pi/3$
26. $t = 7\pi/4$
27. $t = 7\pi/6$
28. $t = -4\pi/3$

In Exercises 29–32, evaluate (if possible) the six trigonometric functions of the real number.

29. $t = 7\pi/6$
30. $t = 3\pi/4$
31. $t = -2\pi/3$
32. $t = 2\pi$

In Exercises 33–36, evaluate the trigonometric function using its period as an aid.

33. $\sin(11\pi/4)$
34. $\cos 4\pi$
35. $\sin(-17\pi/6)$
36. $\cos(-13\pi/3)$

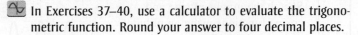 In Exercises 37–40, use a calculator to evaluate the trigonometric function. Round your answer to four decimal places.

37. $\tan 33$
38. $\csc 10.5$
39. $\sec(12\pi/5)$
40. $\sin(-\pi/9)$

1.3 In Exercises 41 and 42, find the exact values of the six trigonometric functions of the angle θ shown in the figure.

41.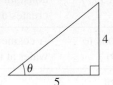
42.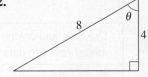

In Exercises 43–46, use the given function value and trigonometric identities (including the cofunction identities) to find the indicated trigonometric functions.

43. $\sin \theta = \frac{1}{3}$ (a) $\csc \theta$ (b) $\cos \theta$
               (c) $\sec \theta$ (d) $\tan \theta$

44. $\tan \theta = 4$ (a) $\cot \theta$ (b) $\sec \theta$
               (c) $\cos \theta$ (d) $\csc \theta$

45. $\csc \theta = 4$ (a) $\sin \theta$ (b) $\cos \theta$
               (c) $\sec \theta$ (d) $\tan \theta$

46. $\csc \theta = 5$ (a) $\sin \theta$ (b) $\cot \theta$
               (c) $\tan \theta$ (d) $\sec(90° - \theta)$

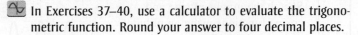 In Exercises 47–54, use a calculator to evaluate the trigonometric function. Round your answer to four decimal places.

47. $\tan 33°$
48. $\csc 11°$
49. $\sin 34.2°$
50. $\sec 79.3°$
51. $\cot 15°\,14'$
52. $\csc 44°\,35'$
53. $\tan 31°\,24'\,5''$
54. $\cos 78°\,11'\,58''$

55. **RAILROAD GRADE** A train travels 3.5 kilometers on a straight track with a grade of $1°\,10'$ (see figure on the next page). What is the vertical rise of the train in that distance?

FIGURE FOR 55

56. GUY WIRE A guy wire runs from the ground to the top of a 25-foot telephone pole. The angle formed between the wire and the ground is 52°. How far from the base of the pole is the wire attached to the ground?

1.4 In Exercises 57–64, the point is on the terminal side of an angle θ in standard position. Determine the exact values of the six trigonometric functions of the angle θ.

57. $(12, 16)$ **58.** $(3, -4)$
59. $\left(\frac{2}{3}, \frac{5}{2}\right)$ **60.** $\left(-\frac{10}{3}, -\frac{2}{3}\right)$
61. $(-0.5, 4.5)$ **62.** $(0.3, 0.4)$
63. $(x, 4x), \quad x > 0$ **64.** $(-2x, -3x), \quad x > 0$

In Exercises 65–70, find the values of the remaining five trigonometric functions of θ.

Function Value	*Constraint*
65. $\sec \theta = \frac{6}{5}$	$\tan \theta < 0$
66. $\csc \theta = \frac{3}{2}$	$\cos \theta < 0$
67. $\sin \theta = \frac{3}{8}$	$\cos \theta < 0$
68. $\tan \theta = \frac{5}{4}$	$\cos \theta < 0$
69. $\cos \theta = -\frac{2}{5}$	$\sin \theta > 0$
70. $\sin \theta = -\frac{1}{2}$	$\cos \theta > 0$

In Exercises 71–74, find the reference angle θ' and sketch θ and θ' in standard position.

71. $\theta = 264°$ **72.** $\theta = 635°$
73. $\theta = -6\pi/5$ **74.** $\theta = 17\pi/3$

In Exercises 75–80, evaluate the sine, cosine, and tangent of the angle without using a calculator.

75. $\pi/3$ **76.** $\pi/4$
77. $-7\pi/3$ **78.** $-5\pi/4$
79. $495°$ **80.** $-150°$

In Exercises 81–84, use a calculator to evaluate the trigonometric function. Round your answer to four decimal places.

81. $\sin 4$ **82.** $\cot(-4.8)$
83. $\sin(12\pi/5)$ **84.** $\tan(-25\pi/7)$

1.5 In Exercises 85–92, sketch the graph of the function. Include two full periods.

85. $y = \sin 6x$ **86.** $y = -\cos 3x$
87. $f(x) = 5 \sin(2x/5)$ **88.** $f(x) = 8 \cos(-x/4)$
89. $y = 5 + \sin x$ **90.** $y = -4 - \cos \pi x$
91. $g(t) = \frac{5}{2} \sin(t - \pi)$ **92.** $g(t) = 3 \cos(t + \pi)$

93. SOUND WAVES Sound waves can be modeled by sine functions of the form $y = a \sin bx$, where x is measured in seconds.

(a) Write an equation of a sound wave whose amplitude is 2 and whose period is $\frac{1}{264}$ second.

(b) What is the frequency of the sound wave described in part (a)?

94. DATA ANALYSIS: METEOROLOGY The times S of sunset (Greenwich Mean Time) at 40° north latitude on the 15th of each month are: 1(16:59), 2(17:35), 3(18:06), 4(18:38), 5(19:08), 6(19:30), 7(19:28), 8(18:57), 9(18:09), 10(17:21), 11(16:44), 12(16:36). The month is represented by t, with $t = 1$ corresponding to January. A model (in which minutes have been converted to the decimal parts of an hour) for the data is $S(t) = 18.09 + 1.41 \sin[(\pi t/6) + 4.60]$.

(a) Use a graphing utility to graph the data points and the model in the same viewing window.

(b) What is the period of the model? Is it what you expected? Explain.

(c) What is the amplitude of the model? What does it represent in the model? Explain.

1.6 In Exercises 95–102, sketch a graph of the function. Include two full periods.

95. $f(x) = 3 \tan 2x$ **96.** $f(t) = \tan\left(t + \frac{\pi}{2}\right)$
97. $f(x) = \frac{1}{2} \cot x$ **98.** $g(t) = 2 \cot 2t$
99. $f(x) = 3 \sec x$ **100.** $h(t) = \sec\left(t - \frac{\pi}{4}\right)$
101. $f(x) = \frac{1}{2} \csc \frac{x}{2}$ **102.** $f(t) = 3 \csc\left(2t + \frac{\pi}{4}\right)$

In Exercises 103 and 104, use a graphing utility to graph the function and the damping factor of the function in the same viewing window. Describe the behavior of the function as x increases without bound.

103. $f(x) = x \cos x$ **104.** $g(x) = x^4 \cos x$

1.7 In Exercises 105–110, evaluate the expression. If necessary, round your answer to two decimal places.

105. $\arcsin\left(-\frac{1}{2}\right)$ **106.** $\arcsin(-1)$
107. $\arcsin 0.4$ **108.** $\arcsin 0.213$
109. $\sin^{-1}(-0.44)$ **110.** $\sin^{-1} 0.89$

In Exercises 111–114, evaluate the expression without using a calculator.

111. $\arccos(-\sqrt{2}/2)$ **112.** $\arccos(\sqrt{2}/2)$
113. $\cos^{-1}(-1)$ **114.** $\cos^{-1}(\sqrt{3}/2)$

 In Exercises 115–118, use a calculator to evaluate the expression. Round your answer to two decimal places.

115. $\arccos 0.324$ **116.** $\arccos(-0.888)$
117. $\tan^{-1}(-1.5)$ **118.** $\tan^{-1} 8.2$

 In Exercises 119–122, use a graphing utility to graph the function.

119. $f(x) = 2\arcsin x$ **120.** $f(x) = 3\arccos x$
121. $f(x) = \arctan(x/2)$ **122.** $f(x) = -\arcsin 2x$

In Exercises 123–128, find the exact value of the expression.

123. $\cos\left(\arctan \frac{3}{4}\right)$ **124.** $\tan\left(\arccos \frac{3}{5}\right)$
125. $\sec\left(\tan^{-1} \frac{12}{5}\right)$ **126.** $\sec\left[\sin^{-1}\left(-\frac{1}{4}\right)\right]$
127. $\cot\left(\arctan \frac{7}{10}\right)$ **128.** $\cot\left[\arcsin\left(-\frac{12}{13}\right)\right]$

In Exercises 129 and 130, write an algebraic expression that is equivalent to the expression.

129. $\tan[\arccos(x/2)]$ **130.** $\sec[\arcsin(x-1)]$

In Exercises 131–134, evaluate each expression without using a calculator.

131. $\operatorname{arccot} \sqrt{3}$ **132.** $\operatorname{arcsec}(-1)$
133. $\operatorname{arcsec}(-\sqrt{2})$ **134.** $\operatorname{arccsc} 1$

 In Exercises 135–138, use a calculator to approximate the value of the expression. Round your result to two decimal places.

135. $\operatorname{arccot}(10.5)$ **136.** $\operatorname{arcsec}(-7.5)$
137. $\operatorname{arcsec}\left(-\frac{5}{2}\right)$ **138.** $\operatorname{arccsc}(-2.01)$

1.8 **139. ANGLE OF ELEVATION** The height of a radio transmission tower is 70 meters, and it casts a shadow of length 30 meters. Draw a diagram and find the angle of elevation of the sun.

140. HEIGHT Your football has landed at the edge of the roof of your school building. When you are 25 feet from the base of the building, the angle of elevation to your football is 21°. How high off the ground is your football?

141. DISTANCE From city A to city B, a plane flies 650 miles at a bearing of 48°. From city B to city C, the plane flies 810 miles at a bearing of 115°. Find the distance from city A to city C and the bearing from city A to city C.

142. WAVE MOTION Your fishing bobber oscillates in simple harmonic motion from the waves in the lake where you fish. Your bobber moves a total of 1.5 inches from its high point to its low point and returns to its high point every 3 seconds. Write an equation modeling the motion of your bobber if it is at its high point at time $t = 0$.

EXPLORATION

TRUE OR FALSE? In Exercises 143 and 144, determine whether the statement is true or false. Justify your answer.

143. $y = \sin\theta$ is not a function because $\sin 30° = \sin 150°$.

144. Because $\tan 3\pi/4 = -1$, $\arctan(-1) = 3\pi/4$.

145. WRITING Describe the behavior of $f(\theta) = \sec\theta$ at the zeros of $g(\theta) = \cos\theta$. Explain your reasoning.

146. CONJECTURE

 (a) Use a graphing utility to complete the table.

θ	0.1	0.4	0.7	1.0	1.3
$\tan\left(\theta - \dfrac{\pi}{2}\right)$					
$-\cot\theta$					

(b) Make a conjecture about the relationship between $\tan[\theta - (\pi/2)]$ and $-\cot\theta$.

147. WRITING When graphing the sine and cosine functions, determining the amplitude is part of the analysis. Explain why this is not true for the other four trigonometric functions.

148. GRAPHICAL REASONING The formulas for the area of a circular sector and arc length are $A = \frac{1}{2}r^2\theta$ and $s = r\theta$, respectively. (r is the radius and θ is the angle measured in radians.)

(a) For $\theta = 0.8$, write the area and arc length as functions of r. What is the domain of each function? Use a graphing utility to graph the functions. Use the graphs to determine which function changes more rapidly as r increases. Explain.

(b) For $r = 10$ centimeters, write the area and arc length as functions of θ. What is the domain of each function? Use a graphing utility to graph and identify the functions.

149. WRITING Describe a real-life application that can be represented by a simple harmonic motion model and is different from any that you've seen in this chapter. Explain which function you would use to model your application and why. Explain how you would determine the amplitude, period, and frequency of the model for your application.

1 CHAPTER TEST

See www.CalcChat.com for worked-out solutions to odd-numbered exercises.

Take this test as you would take a test in class. When you are finished, check your work against the answers given in the back of the book.

1. Consider an angle that measures $\dfrac{5\pi}{4}$ radians.

    (a) Sketch the angle in standard position.
    (b) Determine two coterminal angles (one positive and one negative).
    (c) Convert the angle to degree measure.

2. A truck is moving at a rate of 105 kilometers per hour, and the diameter of its wheels is 1 meter. Find the angular speed of the wheels in radians per minute.

3. A water sprinkler sprays water on a lawn over a distance of 25 feet and rotates through an angle of 130°. Find the area of the lawn watered by the sprinkler.

4. Find the exact values of the six trigonometric functions of the angle θ shown in the figure.

5. Given that $\tan \theta = \frac{3}{2}$, find the other five trigonometric functions of θ.

6. Determine the reference angle θ' for the angle $\theta = 205°$ and sketch θ and θ' in standard position.

7. Determine the quadrant in which θ lies if $\sec \theta < 0$ and $\tan \theta > 0$.

8. Find two exact values of θ in degrees ($0 \le \theta < 360°$) if $\cos \theta = -\sqrt{3}/2$. (Do not use a calculator.)

9. Use a calculator to approximate two values of θ in radians ($0 \le \theta < 2\pi$) if $\csc \theta = 1.030$. Round the results to two decimal places.

In Exercises 10 and 11, find the remaining five trigonometric functions of θ satisfying the conditions.

10. $\cos \theta = \frac{3}{5}$, $\tan \theta < 0$

11. $\sec \theta = -\frac{29}{20}$, $\sin \theta > 0$

In Exercises 12 and 13, sketch the graph of the function. (Include two full periods.)

12. $g(x) = -2 \sin\left(x - \dfrac{\pi}{4}\right)$

13. $f(\alpha) = \dfrac{1}{2} \tan 2\alpha$

In Exercises 14 and 15, use a graphing utility to graph the function. If the function is periodic, find its period.

14. $y = \sin 2\pi x + 2 \cos \pi x$

15. $y = 6t \cos(0.25t)$, $0 \le t \le 32$

16. Find a, b, and c for the function $f(x) = a \sin(bx + c)$ such that the graph of f matches the figure.

17. Find the exact value of $\cot\left(\arcsin \frac{3}{8}\right)$ without the aid of a calculator.

18. Graph the function $f(x) = 2 \arcsin\left(\frac{1}{2}x\right)$.

19. A plane is 90 miles south and 110 miles east of London Heathrow Airport. What bearing should be taken to fly directly to the airport?

20. Write the equation for the simple harmonic motion of a ball on a spring that starts at its lowest point of 6 inches below equilibrium, bounces to its maximum height of 6 inches above equilibrium, and returns to its lowest point in a total of 2 seconds.

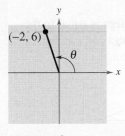

FIGURE FOR 4

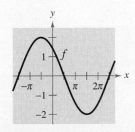

FIGURE FOR 16

PROOFS IN MATHEMATICS

The Pythagorean Theorem

The Pythagorean Theorem is one of the most famous theorems in mathematics. More than 100 different proofs now exist. James A. Garfield, the twentieth president of the United States, developed a proof of the Pythagorean Theorem in 1876. His proof, shown below, involved the fact that a trapezoid can be formed from two congruent right triangles and an isosceles right triangle.

The Pythagorean Theorem

In a right triangle, the sum of the squares of the lengths of the legs is equal to the square of the length of the hypotenuse, where a and b are the legs and c is the hypotenuse.

$$a^2 + b^2 = c^2$$

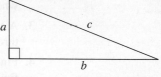

Proof

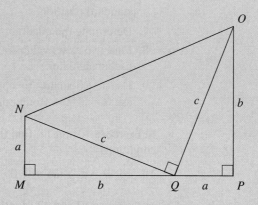

$$
\begin{array}{c}
\text{Area of} \\
\text{trapezoid } MNOP
\end{array}
=
\begin{array}{c}
\text{Area of} \\
\triangle MNQ
\end{array}
+
\begin{array}{c}
\text{Area of} \\
\triangle PQO
\end{array}
+
\begin{array}{c}
\text{Area of} \\
\triangle NOQ
\end{array}
$$

$$\frac{1}{2}(a+b)(a+b) = \frac{1}{2}ab + \frac{1}{2}ab + \frac{1}{2}c^2$$

$$\frac{1}{2}(a+b)(a+b) = ab + \frac{1}{2}c^2$$

$$(a+b)(a+b) = 2ab + c^2$$

$$a^2 + 2ab + b^2 = 2ab + c^2$$

$$a^2 + b^2 = c^2$$

PROBLEM SOLVING

This collection of thought-provoking and challenging exercises further explores and expands upon concepts learned in this chapter.

1. The restaurant at the top of the Space Needle in Seattle, Washington is circular and has a radius of 47.25 feet. The dining part of the restaurant revolves, making about one complete revolution every 48 minutes. A dinner party was seated at the edge of the revolving restaurant at 6:45 P.M. and was finished at 8:57 P.M.

   (a) Find the angle through which the dinner party rotated.

   (b) Find the distance the party traveled during dinner.

2. A bicycle's gear ratio is the number of times the freewheel turns for every one turn of the chainwheel (see figure). The table shows the numbers of teeth in the freewheel and chainwheel for the first five gears of an 18-speed touring bicycle. The chainwheel completes one rotation for each gear. Find the angle through which the freewheel turns for each gear. Give your answers in both degrees and radians.

Gear number	Number of teeth in freewheel	Number of teeth in chainwheel
1	32	24
2	26	24
3	22	24
4	32	40
5	19	24

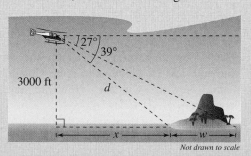

Freewheel

Chainwheel

3. A surveyor in a helicopter is trying to determine the width of an island, as shown in the figure.

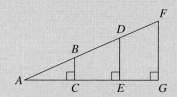

27°

39°

3000 ft

d

x

w

Not drawn to scale

(a) What is the shortest distance d the helicopter would have to travel to land on the island?

(b) What is the horizontal distance x that the helicopter would have to travel before it would be directly over the nearer end of the island?

(c) Find the width w of the island. Explain how you obtained your answer.

4. Use the figure below.

```
                                    F
                              D
                         B
              A _____
                      C    E    G
```

(a) Explain why $\triangle ABC$, $\triangle ADE$, and $\triangle AFG$ are similar triangles.

(b) What does similarity imply about the ratios

$$\frac{BC}{AB}, \frac{DE}{AD}, \text{ and } \frac{FG}{AF}?$$

(c) Does the value of sin A depend on which triangle from part (a) is used to calculate it? Would the value of sin A change if it were found using a different right triangle that was similar to the three given triangles?

(d) Do your conclusions from part (c) apply to the other five trigonometric functions? Explain.

5. Use a graphing utility to graph h, and use the graph to decide whether h is even, odd, or neither.

   (a) $h(x) = \cos^2 x$

   (b) $h(x) = \sin^2 x$

6. If f is an even function and g is an odd function, use the results of Exercise 5 to make a conjecture about h, where

   (a) $h(x) = [f(x)]^2$

   (b) $h(x) = [g(x)]^2$.

7. The model for the height h (in feet) of a Ferris wheel car is

$$h = 50 + 50 \sin 8\pi t$$

where t is the time (in minutes). (The Ferris wheel has a radius of 50 feet.) This model yields a height of 50 feet when $t = 0$. Alter the model so that the height of the car is 1 foot when $t = 0$.

8. The pressure P (in millimeters of mercury) against the walls of the blood vessels of a patient is modeled by

$$P = 100 - 20 \cos\left(\frac{8\pi}{3}t\right)$$

where t is time (in seconds).

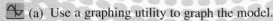 (a) Use a graphing utility to graph the model.

(b) What is the period of the model? What does the period tell you about this situation?

(c) What is the amplitude of the model? What does it tell you about this situation?

(d) If one cycle of this model is equivalent to one heartbeat, what is the pulse of this patient?

(e) If a physician wants this patient's pulse rate to be 64 beats per minute or less, what should the period be? What should the coefficient of t be?

9. A popular theory that attempts to explain the ups and downs of everyday life states that each of us has three cycles, called biorhythms, which begin at birth. These three cycles can be modeled by sine waves.

Physical (23 days): $P = \sin\dfrac{2\pi t}{23}, \quad t \geq 0$

Emotional (28 days): $E = \sin\dfrac{2\pi t}{28}, \quad t \geq 0$

Intellectual (33 days): $I = \sin\dfrac{2\pi t}{33}, \quad t \geq 0$

where t is the number of days since birth. Consider a person who was born on July 20, 1988.

(a) Use a graphing utility to graph the three models in the same viewing window for $7300 \leq t \leq 7380$.

(b) Describe the person's biorhythms during the month of September 2008.

(c) Calculate the person's three energy levels on September 22, 2008.

10. (a) Use a graphing utility to graph the functions given by

$$f(x) = 2 \cos 2x + 3 \sin 3x \quad \text{and}$$

$$g(x) = 2 \cos 2x + 3 \sin 4x.$$

(b) Use the graphs from part (a) to find the period of each function.

(c) If α and β are positive integers, is the function given by $h(x) = A \cos \alpha x + B \sin \beta x$ periodic? Explain your reasoning.

11. Two trigonometric functions f and g have periods of 2, and their graphs intersect at $x = 5.35$.

(a) Give one smaller and one larger positive value of x at which the functions have the same value.

(b) Determine one negative value of x at which the graphs intersect.

(c) Is it true that $f(13.35) = g(-4.65)$? Explain your reasoning.

12. The function f is periodic, with period c. So, $f(t + c) = f(t)$. Are the following equal? Explain.

(a) $f(t - 2c) = f(t)$

(b) $f\left(t + \frac{1}{2}c\right) = f\left(\frac{1}{2}t\right)$

(c) $f\left(\frac{1}{2}(t + c)\right) = f\left(\frac{1}{2}t\right)$

13. If you stand in shallow water and look at an object below the surface of the water, the object will look farther away from you than it really is. This is because when light rays pass between air and water, the water refracts, or bends, the light rays. The index of refraction for water is 1.333. This is the ratio of the sine of θ_1 and the sine of θ_2 (see figure).

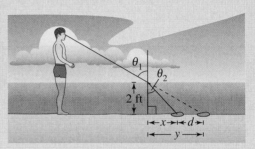

(a) You are standing in water that is 2 feet deep and are looking at a rock at angle $\theta_1 = 60°$ (measured from a line perpendicular to the surface of the water). Find θ_2.

(b) Find the distances x and y.

(c) Find the distance d between where the rock is and where it appears to be.

(d) What happens to d as you move closer to the rock? Explain your reasoning.

14. In calculus, it can be shown that the arctangent function can be approximated by the polynomial

$$\arctan x \approx x - \frac{x^3}{3} + \frac{x^5}{5} - \frac{x^7}{7}$$

where x is in radians.

(a) Use a graphing utility to graph the arctangent function and its polynomial approximation in the same viewing window. How do the graphs compare?

(b) Study the pattern in the polynomial approximation of the arctangent function and guess the next term. Then repeat part (a). How does the accuracy of the approximation change when additional terms are added?

Analytic Trigonometry

2

2.1   Using Fundamental Identities

2.2   Verifying Trigonometric Identities

2.3   Solving Trigonometric Equations

2.4   Sum and Difference Formulas

2.5   Multiple-Angle and Product-to-Sum Formulas

In Mathematics

Analytic trigonometry is used to simplify trigonometric expressions and solve trigonometric equations.

In Real Life

Analytic trigonometry is used to model real-life phenomena. For instance, when an airplane travels faster than the speed of sound, the sound waves form a cone behind the airplane. Concepts of trigonometry can be used to describe the apex angle of the cone. (See Exercise 137, page 269.)

Christopher Pasatier/Reuters/Landov

IN CAREERS

There are many careers that use analytic trigonometry. Several are listed below.

- Mechanical Engineer
  Exercise 89, page 250

- Physicist
  Exercise 90, page 257

- Athletic Trainer
  Exercise 135, page 269

- Physical Therapist
  Exercise 8, page 279

2.1 USING FUNDAMENTAL IDENTITIES

What you should learn

- Recognize and write the fundamental trigonometric identities.
- Use the fundamental trigonometric identities to evaluate trigonometric functions, simplify trigonometric expressions, and rewrite trigonometric expressions.

Why you should learn it

Fundamental trigonometric identities can be used to simplify trigonometric expressions. For instance, in Exercise 117 on page 233, you can use trigonometric identities to simplify an expression for the coefficient of friction.

Introduction

In Chapter 1, you studied the basic definitions, properties, graphs, and applications of the individual trigonometric functions. In this chapter, you will learn how to use the fundamental identities to do the following.

1. Evaluate trigonometric functions.
2. Simplify trigonometric expressions.
3. Develop additional trigonometric identities.
4. Solve trigonometric equations.

Fundamental Trigonometric Identities

Reciprocal Identities

$$\sin u = \frac{1}{\csc u} \qquad \cos u = \frac{1}{\sec u} \qquad \tan u = \frac{1}{\cot u}$$

$$\csc u = \frac{1}{\sin u} \qquad \sec u = \frac{1}{\cos u} \qquad \cot u = \frac{1}{\tan u}$$

Quotient Identities

$$\tan u = \frac{\sin u}{\cos u} \qquad \cot u = \frac{\cos u}{\sin u}$$

Pythagorean Identities

$$\sin^2 u + \cos^2 u = 1 \qquad 1 + \tan^2 u = \sec^2 u \qquad 1 + \cot^2 u = \csc^2 u$$

Cofunction Identities

$$\sin\left(\frac{\pi}{2} - u\right) = \cos u \qquad \cos\left(\frac{\pi}{2} - u\right) = \sin u$$

$$\tan\left(\frac{\pi}{2} - u\right) = \cot u \qquad \cot\left(\frac{\pi}{2} - u\right) = \tan u$$

$$\sec\left(\frac{\pi}{2} - u\right) = \csc u \qquad \csc\left(\frac{\pi}{2} - u\right) = \sec u$$

Even/Odd Identities

$$\sin(-u) = -\sin u \qquad \cos(-u) = \cos u \qquad \tan(-u) = -\tan u$$

$$\csc(-u) = -\csc u \qquad \sec(-u) = \sec u \qquad \cot(-u) = -\cot u$$

Study Tip

You should learn the fundamental trigonometric identities well, because they are used frequently in trigonometry and they will also appear later in calculus. Note that u can be an angle, a real number, or a variable.

Pythagorean identities are sometimes used in radical form such as

$$\sin u = \pm\sqrt{1 - \cos^2 u}$$

or

$$\tan u = \pm\sqrt{\sec^2 u - 1}$$

where the sign depends on the choice of u.

Using the Fundamental Identities

One common application of trigonometric identities is to use given values of trigonometric functions to evaluate other trigonometric functions.

Example 1   Using Identities to Evaluate a Function

Use the values $\sec u = -\frac{3}{2}$ and $\tan u > 0$ to find the values of all six trigonometric functions.

Solution

Using a reciprocal identity, you have

$$\cos u = \frac{1}{\sec u} = \frac{1}{-3/2} = -\frac{2}{3}.$$

Using a Pythagorean identity, you have

$\sin^2 u = 1 - \cos^2 u$	Pythagorean identity
$\quad = 1 - \left(-\frac{2}{3}\right)^2$	Substitute $-\frac{2}{3}$ for $\cos u$.
$\quad = 1 - \frac{4}{9} = \frac{5}{9}.$	Simplify.

Because $\sec u < 0$ and $\tan u > 0$, it follows that u lies in Quadrant III. Moreover, because $\sin u$ is negative when u is in Quadrant III, you can choose the negative root and obtain $\sin u = -\sqrt{5}/3$. Now, knowing the values of the sine and cosine, you can find the values of all six trigonometric functions.

$$\sin u = -\frac{\sqrt{5}}{3} \qquad\qquad \csc u = \frac{1}{\sin u} = -\frac{3}{\sqrt{5}} = -\frac{3\sqrt{5}}{5}$$

$$\cos u = -\frac{2}{3} \qquad\qquad \sec u = \frac{1}{\cos u} = -\frac{3}{2}$$

$$\tan u = \frac{\sin u}{\cos u} = \frac{-\sqrt{5}/3}{-2/3} = \frac{\sqrt{5}}{2} \qquad\qquad \cot u = \frac{1}{\tan u} = \frac{2}{\sqrt{5}} = \frac{2\sqrt{5}}{5}$$

CHECK*Point*   Now try Exercise 21.

Example 2   Simplifying a Trigonometric Expression

Simplify $\sin x \cos^2 x - \sin x$.

Solution

First factor out a common monomial factor and then use a fundamental identity.

$\sin x \cos^2 x - \sin x = \sin x(\cos^2 x - 1)$	Factor out common monomial factor.
$\quad = -\sin x(1 - \cos^2 x)$	Factor out -1.
$\quad = -\sin x(\sin^2 x)$	Pythagorean identity
$\quad = -\sin^3 x$	Multiply.

CHECK*Point*   Now try Exercise 59.

TECHNOLOGY

You can use a graphing utility to check the result of Example 2. To do this, graph

$$y_1 = \sin x \cos^2 x - \sin x$$

and

$$y_2 = -\sin^3 x$$

in the same viewing window, as shown below. Because Example 2 shows the equivalence algebraically and the two graphs appear to coincide, you can conclude that the expressions are equivalent.

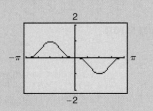

When factoring trigonometric expressions, it is helpful to find a special polynomial factoring form that fits the expression, as shown in Example 3.

Example 3 Factoring Trigonometric Expressions

Factor each expression.

a. $\sec^2 \theta - 1$ **b.** $4 \tan^2 \theta + \tan \theta - 3$

Solution

a. This expression has the form $u^2 - v^2$, which is the difference of two squares. It factors as

$$\sec^2 \theta - 1 = (\sec \theta - 1)(\sec \theta + 1).$$

b. This expression has the polynomial form $ax^2 + bx + c$, and it factors as

$$4 \tan^2 \theta + \tan \theta - 3 = (4 \tan \theta - 3)(\tan \theta + 1).$$

CHECK **Point**   Now try Exercise 61.

On occasion, factoring or simplifying can best be done by first rewriting the expression in terms of just *one* trigonometric function or in terms of *sine and cosine only*. These strategies are shown in Examples 4 and 5, respectively.

Example 4 Factoring a Trigonometric Expression

Factor $\csc^2 x - \cot x - 3$.

Solution

Use the identity $\csc^2 x = 1 + \cot^2 x$ to rewrite the expression in terms of the cotangent.

$$\csc^2 x - \cot x - 3 = (1 + \cot^2 x) - \cot x - 3 \qquad \text{Pythagorean identity}$$

$$= \cot^2 x - \cot x - 2 \qquad \text{Combine like terms.}$$

$$= (\cot x - 2)(\cot x + 1) \qquad \text{Factor.}$$

CHECK **Point**   Now try Exercise 65.

Example 5 Simplifying a Trigonometric Expression

Simplify $\sin t + \cot t \cos t$.

Solution

Begin by rewriting $\cot t$ in terms of sine and cosine.

$$\sin t + \cot t \cos t = \sin t + \left(\frac{\cos t}{\sin t}\right) \cos t \qquad \text{Quotient identity}$$

$$= \frac{\sin^2 t + \cos^2 t}{\sin t} \qquad \text{Add fractions.}$$

$$= \frac{1}{\sin t} \qquad \text{Pythagorean identity}$$

$$= \csc t \qquad \text{Reciprocal identity}$$

CHECK **Point**   Now try Exercise 71.

Study Tip

Remember that when adding rational expressions, you must first find the least common denominator (LCD). In Example 5, the LCD is $\sin t$.

Example 6   Adding Trigonometric Expressions

Perform the addition and simplify.

$$\frac{\sin \theta}{1 + \cos \theta} + \frac{\cos \theta}{\sin \theta}$$

Solution

$$\frac{\sin \theta}{1 + \cos \theta} + \frac{\cos \theta}{\sin \theta} = \frac{(\sin \theta)(\sin \theta) + (\cos \theta)(1 + \cos \theta)}{(1 + \cos \theta)(\sin \theta)}$$

$$= \frac{\sin^2 \theta + \cos^2 \theta + \cos \theta}{(1 + \cos \theta)(\sin \theta)} \qquad \text{Multiply.}$$

$$= \frac{1 + \cos \theta}{(1 + \cos \theta)(\sin \theta)} \qquad \begin{array}{l}\text{Pythagorean identity:} \\ \sin^2 \theta + \cos^2 \theta = 1\end{array}$$

$$= \frac{1}{\sin \theta} \qquad \text{Divide out common factor.}$$

$$= \csc \theta \qquad \text{Reciprocal identity}$$

CHECK*Point*   Now try Exercise 75.

The last two examples in this section involve techniques for rewriting expressions in forms that are used in calculus.

Example 7   Rewriting a Trigonometric Expression

Rewrite $\dfrac{1}{1 + \sin x}$ so that it is *not* in fractional form.

Solution

From the Pythagorean identity $\cos^2 x = 1 - \sin^2 x = (1 - \sin x)(1 + \sin x)$, you can see that multiplying both the numerator and the denominator by $(1 - \sin x)$ will produce a monomial denominator.

$$\frac{1}{1 + \sin x} = \frac{1}{1 + \sin x} \cdot \frac{1 - \sin x}{1 - \sin x} \qquad \begin{array}{l}\text{Multiply numerator and} \\ \text{denominator by } (1 - \sin x).\end{array}$$

$$= \frac{1 - \sin x}{1 - \sin^2 x} \qquad \text{Multiply.}$$

$$= \frac{1 - \sin x}{\cos^2 x} \qquad \text{Pythagorean identity}$$

$$= \frac{1}{\cos^2 x} - \frac{\sin x}{\cos^2 x} \qquad \text{Write as separate fractions.}$$

$$= \frac{1}{\cos^2 x} - \frac{\sin x}{\cos x} \cdot \frac{1}{\cos x} \qquad \text{Product of fractions}$$

$$= \sec^2 x - \tan x \sec x \qquad \text{Reciprocal and quotient identities}$$

CHECK*Point*   Now try Exercise 81.

Example 8   Trigonometric Substitution

Use the substitution $x = 2 \tan \theta$, $0 < \theta < \pi/2$, to write

$$\sqrt{4 + x^2}$$

as a trigonometric function of θ.

Solution

Begin by letting $x = 2 \tan \theta$. Then, you can obtain

$$\sqrt{4 + x^2} = \sqrt{4 + (2 \tan \theta)^2}$$ Substitute $2 \tan \theta$ for x.

$$= \sqrt{4 + 4 \tan^2 \theta}$$ Rule of exponents

$$= \sqrt{4(1 + \tan^2 \theta)}$$ Factor.

$$= \sqrt{4 \sec^2 \theta}$$ Pythagorean identity

$$= 2 \sec \theta.$$ $\sec \theta > 0$ for $0 < \theta < \pi/2$

CHECK Point   Now try Exercise 93.

Figure 2.1 shows the right triangle illustration of the trigonometric substitution $x = 2 \tan \theta$ in Example 8. You can use this triangle to check the solution of Example 8. For $0 < \theta < \pi/2$, you have

$$\text{opp} = x, \quad \text{adj} = 2, \quad \text{and} \quad \text{hyp} = \sqrt{4 + x^2}.$$

With these expressions, you can write the following.

$$\sec \theta = \frac{\text{hyp}}{\text{adj}}$$

$$\sec \theta = \frac{\sqrt{4 + x^2}}{2}$$

$$2 \sec \theta = \sqrt{4 + x^2}.$$

So, the solution checks.

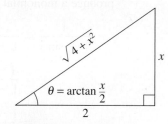

Angle whose tangent is $x/2$.

FIGURE **2.1**

2.1   EXERCISES

See www.CalcChat.com for worked-out solutions to odd-numbered exercises.

VOCABULARY: Fill in the blank to complete the trigonometric identity.

1. $\dfrac{\sin u}{\cos u} = $ _____

2. $\dfrac{1}{\csc u} = $ _____

3. $\dfrac{1}{\tan u} = $ _____

4. $\dfrac{1}{\cos u} = $ _____

5. $1 + $ _____ $= \csc^2 u$

6. $1 + \tan^2 u = $ _____

7. $\sin\left(\dfrac{\pi}{2} - u\right) = $ _____

8. $\sec\left(\dfrac{\pi}{2} - u\right) = $ _____

9. $\cos(-u) = $ _____

10. $\tan(-u) = $ _____

SKILLS AND APPLICATIONS

In Exercises 11–24, use the given values to evaluate (if possible) all six trigonometric functions.

11. $\sin x = \dfrac{1}{2}, \quad \cos x = \dfrac{\sqrt{3}}{2}$

12. $\tan x = \dfrac{\sqrt{3}}{3}, \quad \cos x = -\dfrac{\sqrt{3}}{2}$

13. $\sec \theta = \sqrt{2}, \quad \sin \theta = -\dfrac{\sqrt{2}}{2}$

14. $\csc \theta = \dfrac{25}{7}, \quad \tan \theta = \dfrac{7}{24}$

15. $\tan x = \dfrac{8}{15}, \quad \sec x = -\dfrac{17}{15}$

16. $\cot \phi = -3, \quad \sin \phi = \dfrac{\sqrt{10}}{10}$

17. $\sec \phi = \dfrac{3}{2}, \quad \csc \phi = -\dfrac{3\sqrt{5}}{5}$

18. $\cos\left(\dfrac{\pi}{2} - x\right) = \dfrac{3}{5}, \quad \cos x = \dfrac{4}{5}$

19. $\sin(-x) = -\dfrac{1}{3}, \quad \tan x = -\dfrac{\sqrt{2}}{4}$

20. $\sec x = 4, \quad \sin x > 0$

21. $\tan \theta = 2, \quad \sin \theta < 0$

22. $\csc \theta = -5, \quad \cos \theta < 0$

23. $\sin \theta = -1, \quad \cot \theta = 0$

24. $\tan \theta$ is undefined, $\quad \sin \theta > 0$

In Exercises 25–30, match the trigonometric expression with one of the following.

(a) $\sec x$ (b) -1 (c) $\cot x$

(d) 1 (e) $-\tan x$ (f) $\sin x$

25. $\sec x \cos x$

26. $\tan x \csc x$

27. $\cot^2 x - \csc^2 x$

28. $(1 - \cos^2 x)(\csc x)$

29. $\dfrac{\sin(-x)}{\cos(-x)}$

30. $\dfrac{\sin[(\pi/2) - x]}{\cos[(\pi/2) - x]}$

In Exercises 31–36, match the trigonometric expression with one of the following.

(a) $\csc x$ (b) $\tan x$ (c) $\sin^2 x$

(d) $\sin x \tan x$ (e) $\sec^2 x$ (f) $\sec^2 x + \tan^2 x$

31. $\sin x \sec x$

32. $\cos^2 x(\sec^2 x - 1)$

33. $\sec^4 x - \tan^4 x$

34. $\cot x \sec x$

35. $\dfrac{\sec^2 x - 1}{\sin^2 x}$

36. $\dfrac{\cos^2[(\pi/2) - x]}{\cos x}$

In Exercises 37–58, use the fundamental identities to simplify the expression. There is more than one correct form of each answer.

37. $\cot \theta \sec \theta$

38. $\cos \beta \tan \beta$

39. $\tan(-x) \cos x$

40. $\sin x \cot(-x)$

41. $\sin \phi(\csc \phi - \sin \phi)$

42. $\sec^2 x(1 - \sin^2 x)$

43. $\dfrac{\cot x}{\csc x}$

44. $\dfrac{\csc \theta}{\sec \theta}$

45. $\dfrac{1 - \sin^2 x}{\csc^2 x - 1}$

46. $\dfrac{1}{\tan^2 x + 1}$

47. $\dfrac{\tan \theta \cot \theta}{\sec \theta}$

48. $\dfrac{\sin \theta \csc \theta}{\tan \theta}$

49. $\sec \alpha \cdot \dfrac{\sin \alpha}{\tan \alpha}$

50. $\dfrac{\tan^2 \theta}{\sec^2 \theta}$

51. $\cos\left(\dfrac{\pi}{2} - x\right) \sec x$

52. $\cot\left(\dfrac{\pi}{2} - x\right) \cos x$

53. $\dfrac{\cos^2 y}{1 - \sin y}$

54. $\cos t(1 + \tan^2 t)$

55. $\sin \beta \tan \beta + \cos \beta$

56. $\csc \phi \tan \phi + \sec \phi$

57. $\cot u \sin u + \tan u \cos u$

58. $\sin \theta \sec \theta + \cos \theta \csc \theta$

In Exercises 59–70, factor the expression and use the fundamental identities to simplify. There is more than one correct form of each answer.

59. $\tan^2 x - \tan^2 x \sin^2 x$ **60.** $\sin^2 x \csc^2 x - \sin^2 x$

61. $\sin^2 x \sec^2 x - \sin^2 x$ **62.** $\cos^2 x + \cos^2 x \tan^2 x$

63. $\dfrac{\sec^2 x - 1}{\sec x - 1}$ **64.** $\dfrac{\cos^2 x - 4}{\cos x - 2}$

65. $\tan^4 x + 2\tan^2 x + 1$ **66.** $1 - 2\cos^2 x + \cos^4 x$

67. $\sin^4 x - \cos^4 x$ **68.** $\sec^4 x - \tan^4 x$

69. $\csc^3 x - \csc^2 x - \csc x + 1$

70. $\sec^3 x - \sec^2 x - \sec x + 1$

In Exercises 71–74, perform the multiplication and use the fundamental identities to simplify. There is more than one correct form of each answer.

71. $(\sin x + \cos x)^2$

72. $(\cot x + \csc x)(\cot x - \csc x)$

73. $(2\csc x + 2)(2\csc x - 2)$

74. $(3 - 3\sin x)(3 + 3\sin x)$

In Exercises 75–80, perform the addition or subtraction and use the fundamental identities to simplify. There is more than one correct form of each answer.

75. $\dfrac{1}{1 + \cos x} + \dfrac{1}{1 - \cos x}$ **76.** $\dfrac{1}{\sec x + 1} - \dfrac{1}{\sec x - 1}$

77. $\dfrac{\cos x}{1 + \sin x} + \dfrac{1 + \sin x}{\cos x}$ **78.** $\dfrac{\tan x}{1 + \sec x} + \dfrac{1 + \sec x}{\tan x}$

79. $\tan x + \dfrac{\cos x}{1 + \sin x}$ **80.** $\tan x - \dfrac{\sec^2 x}{\tan x}$

In Exercises 81–84, rewrite the expression so that it is not in fractional form. There is more than one correct form of each answer.

81. $\dfrac{\sin^2 y}{1 - \cos y}$ **82.** $\dfrac{5}{\tan x + \sec x}$

83. $\dfrac{3}{\sec x - \tan x}$ **84.** $\dfrac{\tan^2 x}{\csc x + 1}$

NUMERICAL AND GRAPHICAL ANALYSIS In Exercises 85–88, use a graphing utility to complete the table and graph the functions. Make a conjecture about y_1 and y_2.

x	0.2	0.4	0.6	0.8	1.0	1.2	1.4
y_1							
y_2							

85. $y_1 = \cos\left(\dfrac{\pi}{2} - x\right)$, $y_2 = \sin x$

86. $y_1 = \sec x - \cos x$, $y_2 = \sin x \tan x$

87. $y_1 = \dfrac{\cos x}{1 - \sin x}$, $y_2 = \dfrac{1 + \sin x}{\cos x}$

88. $y_1 = \sec^4 x - \sec^2 x$, $y_2 = \tan^2 x + \tan^4 x$

In Exercises 89–92, use a graphing utility to determine which of the six trigonometric functions is equal to the expression. Verify your answer algebraically.

89. $\cos x \cot x + \sin x$ **90.** $\sec x \csc x - \tan x$

91. $\dfrac{1}{\sin x}\left(\dfrac{1}{\cos x} - \cos x\right)$

92. $\dfrac{1}{2}\left(\dfrac{1 + \sin \theta}{\cos \theta} + \dfrac{\cos \theta}{1 + \sin \theta}\right)$

In Exercises 93–104, use the trigonometric substitution to write the algebraic expression as a trigonometric function of θ, where $0 < \theta < \pi/2$.

93. $\sqrt{9 - x^2}$, $x = 3\cos \theta$

94. $\sqrt{64 - 16x^2}$, $x = 2\cos \theta$

95. $\sqrt{16 - x^2}$, $x = 4\sin \theta$

96. $\sqrt{49 - x^2}$, $x = 7\sin \theta$

97. $\sqrt{x^2 - 9}$, $x = 3\sec \theta$

98. $\sqrt{x^2 - 4}$, $x = 2\sec \theta$

99. $\sqrt{x^2 + 25}$, $x = 5\tan \theta$

100. $\sqrt{x^2 + 100}$, $x = 10\tan \theta$

101. $\sqrt{4x^2 + 9}$, $2x = 3\tan \theta$

102. $\sqrt{9x^2 + 25}$, $3x = 5\tan \theta$

103. $\sqrt{2 - x^2}$, $x = \sqrt{2}\sin \theta$

104. $\sqrt{10 - x^2}$, $x = \sqrt{10}\sin \theta$

In Exercises 105–108, use the trigonometric substitution to write the algebraic equation as a trigonometric equation of θ, where $-\pi/2 < \theta < \pi/2$. Then find $\sin \theta$ and $\cos \theta$.

105. $3 = \sqrt{9 - x^2}$, $x = 3\sin \theta$

106. $3 = \sqrt{36 - x^2}$, $x = 6\sin \theta$

107. $2\sqrt{2} = \sqrt{16 - 4x^2}$, $x = 2\cos \theta$

108. $-5\sqrt{3} = \sqrt{100 - x^2}$, $x = 10\cos \theta$

In Exercises 109–112, use a graphing utility to solve the equation for θ, where $0 \le \theta < 2\pi$.

109. $\sin \theta = \sqrt{1 - \cos^2 \theta}$

110. $\cos \theta = -\sqrt{1 - \sin^2 \theta}$

111. $\sec \theta = \sqrt{1 + \tan^2 \theta}$

112. $\csc \theta = \sqrt{1 + \cot^2 \theta}$

In Exercises 113–116, use a calculator to demonstrate the identity for each value of θ.

113. $\csc^2 \theta - \cot^2 \theta = 1$

(a) $\theta = 132°$ (b) $\theta = \dfrac{2\pi}{7}$

114. $\tan^2 \theta + 1 = \sec^2 \theta$

(a) $\theta = 346°$ (b) $\theta = 3.1$

115. $\cos\left(\dfrac{\pi}{2} - \theta\right) = \sin \theta$

(a) $\theta = 80°$ (b) $\theta = 0.8$

116. $\sin(-\theta) = -\sin \theta$

(a) $\theta = 250°$ (b) $\theta = \frac{1}{2}$

117. FRICTION   The forces acting on an object weighing W units on an inclined plane positioned at an angle of θ with the horizontal (see figure) are modeled by

$$\mu W \cos \theta = W \sin \theta$$

where μ is the coefficient of friction. Solve the equation for μ and simplify the result.

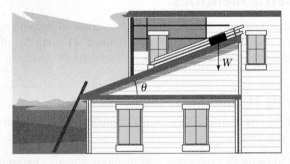

118. RATE OF CHANGE   The rate of change of the function $f(x) = -x + \tan x$ is given by the expression $-1 + \sec^2 x$. Show that this expression can also be written as $\tan^2 x$.

119. RATE OF CHANGE   The rate of change of the function

$$f(x) = \sec x + \cos x$$

is given by the expression

$$\sec x \tan x - \sin x.$$

Show that this expression can also be written as $\sin x \tan^2 x$.

120. RATE OF CHANGE   The rate of change of the function

$$f(x) = -\csc x - \sin x$$

is given by the expression

$$\csc x \cot x - \cos x.$$

Show that this expression can also be written as $\cos x \cot^2 x$.

EXPLORATION

TRUE OR FALSE?   In Exercises 121 and 122, determine whether the statement is true or false. Justify your answer.

121. The even and odd trigonometric identities are helpful for determining whether the value of a trigonometric function is positive or negative.

122. A cofunction identity can be used to transform a tangent function so that it can be represented by a cosecant function.

In Exercises 123–126, fill in the blanks. (*Note:* The notation $x \to c^+$ indicates that x approaches c from the right and $x \to c^-$ indicates that x approaches c from the left.)

123. As $x \to \dfrac{\pi^-}{2}$, $\sin x \to$ ▮▮ and $\csc x \to$ ▮▮ .

124. As $x \to 0^+$, $\cos x \to$ ▮▮ and $\sec x \to$ ▮▮ .

125. As $x \to \dfrac{\pi^-}{2}$, $\tan x \to$ ▮▮ and $\cot x \to$ ▮▮ .

126. As $x \to \pi^+$, $\sin x \to$ ▮▮ and $\csc x \to$ ▮▮ .

In Exercises 127–132, determine whether or not the equation is an identity, and give a reason for your answer.

127. $\cos \theta = \sqrt{1 - \sin^2 \theta}$ **128.** $\cot \theta = \sqrt{\csc^2 \theta + 1}$

129. $\dfrac{(\sin k\theta)}{(\cos k\theta)} = \tan \theta$, k is a constant.

130. $\dfrac{1}{(5 \cos \theta)} = 5 \sec \theta$

131. $\sin \theta \csc \theta = 1$ **132.** $\csc^2 \theta = 1$

133. Use the trigonometric substitution $u = a \sin \theta$, where $-\pi/2 < \theta < \pi/2$ and $a > 0$, to simplify the expression $\sqrt{a^2 - u^2}$.

134. Use the trigonometric substitution $u = a \tan \theta$, where $-\pi/2 < \theta < \pi/2$ and $a > 0$, to simplify the expression $\sqrt{a^2 + u^2}$.

135. Use the trigonometric substitution $u = a \sec \theta$, where $0 < \theta < \pi/2$ and $a > 0$, to simplify the expression $\sqrt{u^2 - a^2}$.

136. CAPSTONE

(a) Use the definitions of sine and cosine to derive the Pythagorean identity $\sin^2 \theta + \cos^2 \theta = 1$.

(b) Use the Pythagorean identity $\sin^2 \theta + \cos^2 \theta = 1$ to derive the other Pythagorean identities, $1 + \tan^2 \theta = \sec^2 \theta$ and $1 + \cot^2 \theta = \csc^2 \theta$. Discuss how to remember these identities and other fundamental identities.

2.2 ## VERIFYING TRIGONOMETRIC IDENTITIES

What you should learn

• Verify trigonometric identities.

Why you should learn it

You can use trigonometric identities to rewrite trigonometric equations that model real-life situations. For instance, in Exercise 70 on page 240, you can use trigonometric identities to simplify the equation that models the length of a shadow cast by a gnomon (a device used to tell time).

Robert W. Ginn/PhotoEdit

Introduction

In this section, you will study techniques for verifying trigonometric identities. In the next section, you will study techniques for solving trigonometric equations. The key to verifying identities *and* solving equations is the ability to use the fundamental identities and the rules of algebra to rewrite trigonometric expressions.

Remember that a *conditional equation* is an equation that is true for only some of the values in its domain. For example, the conditional equation

$$\sin x = 0 \qquad \text{Conditional equation}$$

is true only for $x = n\pi$, where n is an integer. When you find these values, you are *solving* the equation.

On the other hand, an equation that is true for all real values in the domain of the variable is an *identity*. For example, the familiar equation

$$\sin^2 x = 1 - \cos^2 x \qquad \text{Identity}$$

is true for all real numbers x. So, it is an identity.

Verifying Trigonometric Identities

Although there are similarities, verifying that a trigonometric equation is an identity is quite different from solving an equation. There is no well-defined set of rules to follow in verifying trigonometric identities, and the process is best learned by practice.

Guidelines for Verifying Trigonometric Identities

1. Work with one side of the equation at a time. It is often better to work with the more complicated side first.

2. Look for opportunities to factor an expression, add fractions, square a binomial, or create a monomial denominator.

3. Look for opportunities to use the fundamental identities. Note which functions are in the final expression you want. Sines and cosines pair up well, as do secants and tangents, and cosecants and cotangents.

4. If the preceding guidelines do not help, try converting all terms to sines and cosines.

5. Always try *something*. Even paths that lead to dead ends provide insights.

Verifying trigonometric identities is a useful process if you need to convert a trigonometric expression into a form that is more useful algebraically. When you verify an identity, you cannot *assume* that the two sides of the equation are equal because you are trying to verify that they *are* equal. As a result, when verifying identities, you cannot use operations such as adding the same quantity to each side of the equation or cross multiplication.

| **Example 1** | **Verifying a Trigonometric Identity** |

Verify the identity $(\sec^2 \theta - 1)/\sec^2 \theta = \sin^2 \theta$.

Solution

The left side is more complicated, so start with it.

$$\frac{\sec^2 \theta - 1}{\sec^2 \theta} = \frac{(\tan^2 \theta + 1) - 1}{\sec^2 \theta} \qquad \text{Pythagorean identity}$$

$$= \frac{\tan^2 \theta}{\sec^2 \theta} \qquad \text{Simplify.}$$

$$= \tan^2 \theta (\cos^2 \theta) \qquad \text{Reciprocal identity}$$

$$= \frac{\sin^2 \theta}{(\cos^2 \theta)} (\cos^2 \theta) \qquad \text{Quotient identity}$$

$$= \sin^2 \theta \qquad \text{Simplify.}$$

Notice how the identity is verified. You start with the left side of the equation (the more complicated side) and use the fundamental trigonometric identities to simplify it until you obtain the right side.

CHECK**Point**   Now try Exercise 15.

There can be more than one way to verify an identity. Here is another way to verify the identity in Example 1.

$$\frac{\sec^2 \theta - 1}{\sec^2 \theta} = \frac{\sec^2 \theta}{\sec^2 \theta} - \frac{1}{\sec^2 \theta} \qquad \text{Rewrite as the difference of fractions.}$$

$$= 1 - \cos^2 \theta \qquad \text{Reciprocal identity}$$

$$= \sin^2 \theta \qquad \text{Pythagorean identity}$$

> ⚠️ **WARNING / CAUTION**
>
> Remember that an identity is only true for all real values in the domain of the variable. For instance, in Example 1 the identity is not true when $\theta = \pi/2$ because $\sec^2 \theta$ is not defined when $\theta = \pi/2$.

| **Example 2** | **Verifying a Trigonometric Identity** |

Verify the identity $2 \sec^2 \alpha = \dfrac{1}{1 - \sin \alpha} + \dfrac{1}{1 + \sin \alpha}$.

Algebraic Solution

The right side is more complicated, so start with it.

$$\frac{1}{1 - \sin \alpha} + \frac{1}{1 + \sin \alpha} = \frac{1 + \sin \alpha + 1 - \sin \alpha}{(1 - \sin \alpha)(1 + \sin \alpha)} \qquad \text{Add fractions.}$$

$$= \frac{2}{1 - \sin^2 \alpha} \qquad \text{Simplify.}$$

$$= \frac{2}{\cos^2 \alpha} \qquad \text{Pythagorean identity}$$

$$= 2 \sec^2 \alpha \qquad \text{Reciprocal identity}$$

Numerical Solution

Use the *table* feature of a graphing utility set in *radian* mode to create a table that shows the values of $y_1 = 2/\cos^2 x$ and $y_2 = 1/(1 - \sin x) + 1/(1 + \sin x)$ for different values of x, as shown in Figure 2.2. From the table, you can see that the values appear to be identical, so $2 \sec^2 x = 1/(1 - \sin x) + 1/(1 + \sin x)$ appears to be an identity.

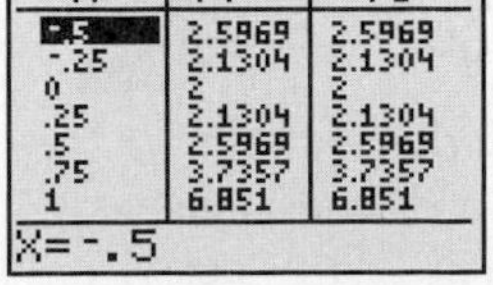

X	Y1	Y2
-.5	2.5969	2.5969
-.25	2.1304	2.1304
0	2	2
.25	2.1304	2.1304
.5	2.5969	2.5969
.75	3.7357	3.7357
1	6.851	6.851

X=-.5

FIGURE 2.2

CHECK**Point**   Now try Exercise 31.

Example 3 Verifying a Trigonometric Identity

Verify the identity $(\tan^2 x + 1)(\cos^2 x - 1) = -\tan^2 x$.

Algebraic Solution

By applying identities before multiplying, you obtain the following.

$$(\tan^2 x + 1)(\cos^2 x - 1) = (\sec^2 x)(-\sin^2 x) \qquad \text{Pythagorean identities}$$

$$= -\frac{\sin^2 x}{\cos^2 x} \qquad \text{Reciprocal identity}$$

$$= -\left(\frac{\sin x}{\cos x}\right)^2 \qquad \text{Rule of exponents}$$

$$= -\tan^2 x \qquad \text{Quotient identity}$$

Graphical Solution

Use a graphing utility set in *radian* mode to graph the left side of the identity $y_1 = (\tan^2 x + 1)(\cos^2 x - 1)$ and the right side of the identity $y_2 = -\tan^2 x$ in the same viewing window, as shown in Figure 2.3. (Select the *line* style for y_1 and the *path* style for y_2.) Because the graphs appear to coincide, $(\tan^2 x + 1)(\cos^2 x - 1) = -\tan^2 x$ appears to be an identity.

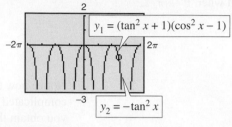

FIGURE 2.3

CHECK Point Now try Exercise 53.

Example 4 Converting to Sines and Cosines

Verify the identity $\tan x + \cot x = \sec x \csc x$.

Solution

Try converting the left side into sines and cosines.

$$\tan x + \cot x = \frac{\sin x}{\cos x} + \frac{\cos x}{\sin x} \qquad \text{Quotient identities}$$

$$= \frac{\sin^2 x + \cos^2 x}{\cos x \sin x} \qquad \text{Add fractions.}$$

$$= \frac{1}{\cos x \sin x} \qquad \text{Pythagorean identity}$$

$$= \frac{1}{\cos x} \cdot \frac{1}{\sin x} \qquad \text{Product of fractions.}$$

$$= \sec x \csc x \qquad \text{Reciprocal identities}$$

CHECK Point Now try Exercise 25.

> **WARNING / CAUTION**
>
> Although a graphing utility can be useful in helping to verify an identity, you must use algebraic techniques to produce a *valid* proof.

> *Study Tip*
>
> As shown at the right, $\csc^2 x(1 + \cos x)$ is considered a simplified form of $1/(1 - \cos x)$ because the expression does not contain any fractions.

Recall from algebra that *rationalizing the denominator* using conjugates is, on occasion, a powerful simplification technique. A related form of this technique, shown below, works for simplifying trigonometric expressions as well.

$$\frac{1}{1 - \cos x} = \frac{1}{1 - \cos x}\left(\frac{1 + \cos x}{1 + \cos x}\right) = \frac{1 + \cos x}{1 - \cos^2 x} = \frac{1 + \cos x}{\sin^2 x}$$

$$= \csc^2 x(1 + \cos x)$$

This technique is demonstrated in the next example.

Example 5   Verifying a Trigonometric Identity

Verify the identity $\sec x + \tan x = \dfrac{\cos x}{1 - \sin x}$.

Algebraic Solution

Begin with the *right* side because you can create a monomial denominator by multiplying the numerator and denominator by $1 + \sin x$.

$$\frac{\cos x}{1 - \sin x} = \frac{\cos x}{1 - \sin x}\left(\frac{1 + \sin x}{1 + \sin x}\right) \quad\quad \begin{array}{l}\text{Multiply numerator and}\\\text{denominator by } 1 + \sin x.\end{array}$$

$$= \frac{\cos x + \cos x \sin x}{1 - \sin^2 x} \quad\quad \text{Multiply.}$$

$$= \frac{\cos x + \cos x \sin x}{\cos^2 x} \quad\quad \text{Pythagorean identity}$$

$$= \frac{\cos x}{\cos^2 x} + \frac{\cos x \sin x}{\cos^2 x} \quad\quad \text{Write as separate fractions.}$$

$$= \frac{1}{\cos x} + \frac{\sin x}{\cos x} \quad\quad \text{Simplify.}$$

$$= \sec x + \tan x \quad\quad \text{Identities}$$

CHECK Point   Now try Exercise 59.

Graphical Solution

Use a graphing utility set in the *radian* and *dot* modes to graph $y_1 = \sec x + \tan x$ and $y_2 = \cos x/(1 - \sin x)$ in the same viewing window, as shown in Figure 2.4. Because the graphs appear to coincide, $\sec x + \tan x = \cos x/(1 - \sin x)$ appears to be an identity.

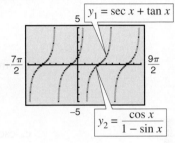

FIGURE 2.4

In Examples 1 through 5, you have been verifying trigonometric identities by working with one side of the equation and converting to the form given on the other side. On occasion, it is practical to work with each side *separately*, to obtain one common form equivalent to both sides. This is illustrated in Example 6.

Example 6   Working with Each Side Separately

Verify the identity $\dfrac{\cot^2 \theta}{1 + \csc \theta} = \dfrac{1 - \sin \theta}{\sin \theta}$.

Algebraic Solution

Working with the left side, you have

$$\frac{\cot^2 \theta}{1 + \csc \theta} = \frac{\csc^2 \theta - 1}{1 + \csc \theta} \quad\quad \text{Pythagorean identity}$$

$$= \frac{(\csc \theta - 1)(\csc \theta + 1)}{1 + \csc \theta} \quad\quad \text{Factor.}$$

$$= \csc \theta - 1. \quad\quad \text{Simplify.}$$

Now, simplifying the right side, you have

$$\frac{1 - \sin \theta}{\sin \theta} = \frac{1}{\sin \theta} - \frac{\sin \theta}{\sin \theta} \quad\quad \text{Write as separate fractions.}$$

$$= \csc \theta - 1. \quad\quad \text{Reciprocal identity}$$

The identity is verified because both sides are equal to $\csc \theta - 1$.

CHECK Point   Now try Exercise 19.

Numerical Solution

Use the *table* feature of a graphing utility set in *radian* mode to create a table that shows the values of $y_1 = \cot^2 x/(1 + \csc x)$ and $y_2 = (1 - \sin x)/\sin x$ for different values of x, as shown in Figure 2.5. From the table you can see that the values appear to be identical, so $\cot^2 x/(1 + \csc x) = (1 - \sin x)/\sin x$ appears to be an identity.

FIGURE 2.5

In Example 7, powers of trigonometric functions are rewritten as more complicated sums of products of trigonometric functions. This is a common procedure used in calculus.

Example 7 Three Examples from Calculus

Verify each identity.

a. $\tan^4 x = \tan^2 x \sec^2 x - \tan^2 x$

b. $\sin^3 x \cos^4 x = (\cos^4 x - \cos^6 x) \sin x$

c. $\csc^4 x \cot x = \csc^2 x(\cot x + \cot^3 x)$

Solution

a. $\tan^4 x = (\tan^2 x)(\tan^2 x)$ Write as separate factors.

$\qquad = \tan^2 x(\sec^2 x - 1)$ Pythagorean identity

$\qquad = \tan^2 x \sec^2 x - \tan^2 x$ Multiply.

b. $\sin^3 x \cos^4 x = \sin^2 x \cos^4 x \sin x$ Write as separate factors.

$\qquad = (1 - \cos^2 x) \cos^4 x \sin x$ Pythagorean identity

$\qquad = (\cos^4 x - \cos^6 x) \sin x$ Multiply.

c. $\csc^4 x \cot x = \csc^2 x \csc^2 x \cot x$ Write as separate factors.

$\qquad = \csc^2 x(1 + \cot^2 x) \cot x$ Pythagorean identity

$\qquad = \csc^2 x(\cot x + \cot^3 x)$ Multiply.

CHECK*Point* ▶ Now try Exercise 63.

CLASSROOM DISCUSSION

Error Analysis   You are tutoring a student in trigonometry. One of the homework problems your student encounters asks whether the following statement is an identity.

$$\tan^2 x \sin^2 x \stackrel{?}{=} \frac{5}{6}\tan^2 x$$

Your student does not attempt to verify the equivalence algebraically, but mistakenly uses only a graphical approach. Using range settings of

Xmin $= -3\pi$	Ymin $= -20$
Xmax $= 3\pi$	Ymax $= 20$
Xscl $= \pi/2$	Yscl $= 1$

your student graphs both sides of the expression on a graphing utility and concludes that the statement is an identity.

What is wrong with your student's reasoning? Explain. Discuss the limitations of verifying identities graphically.

2.2   **EXERCISES**

See www.CalcChat.com for worked-out solutions to odd-numbered exercises.

VOCABULARY

In Exercises 1 and 2, fill in the blanks.

1. An equation that is true for all real values in its domain is called an _____.

2. An equation that is true for only some values in its domain is called a _____ _____.

In Exercises 3–8, fill in the blank to complete the trigonometric identity.

3. $\dfrac{1}{\cot u} =$ _____

4. $\dfrac{\cos u}{\sin u} =$ _____

5. $\sin^2 u +$ _____ $= 1$

6. $\cos\left(\dfrac{\pi}{2} - u\right) =$ _____

7. $\csc(-u) =$ _____

8. $\sec(-u) =$ _____

SKILLS AND APPLICATIONS

In Exercises 9–50, verify the identity.

9. $\tan t \cot t = 1$

10. $\sec y \cos y = 1$

11. $\cot^2 y(\sec^2 y - 1) = 1$

12. $\cos x + \sin x \tan x = \sec x$

13. $(1 + \sin \alpha)(1 - \sin \alpha) = \cos^2 \alpha$

14. $\cos^2 \beta - \sin^2 \beta = 2\cos^2 \beta - 1$

15. $\cos^2 \beta - \sin^2 \beta = 1 - 2\sin^2 \beta$

16. $\sin^2 \alpha - \sin^4 \alpha = \cos^2 \alpha - \cos^4 \alpha$

17. $\dfrac{\tan^2 \theta}{\sec \theta} = \sin \theta \tan \theta$

18. $\dfrac{\cot^3 t}{\csc t} = \cos t(\csc^2 t - 1)$

19. $\dfrac{\cot^2 t}{\csc t} = \dfrac{1 - \sin^2 t}{\sin t}$

20. $\dfrac{1}{\tan \beta} + \tan \beta = \dfrac{\sec^2 \beta}{\tan \beta}$

21. $\sin^{1/2} x \cos x - \sin^{5/2} x \cos x = \cos^3 x \sqrt{\sin x}$

22. $\sec^6 x(\sec x \tan x) - \sec^4 x(\sec x \tan x) = \sec^5 x \tan^3 x$

23. $\dfrac{\cot x}{\sec x} = \csc x - \sin x$

24. $\dfrac{\sec \theta - 1}{1 - \cos \theta} = \sec \theta$

25. $\csc x - \sin x = \cos x \cot x$

26. $\sec x - \cos x = \sin x \tan x$

27. $\dfrac{1}{\tan x} + \dfrac{1}{\cot x} = \tan x + \cot x$

28. $\dfrac{1}{\sin x} - \dfrac{1}{\csc x} = \csc x - \sin x$

29. $\dfrac{1 + \sin \theta}{\cos \theta} + \dfrac{\cos \theta}{1 + \sin \theta} = 2 \sec \theta$

30. $\dfrac{\cos \theta \cot \theta}{1 - \sin \theta} - 1 = \csc \theta$

31. $\dfrac{1}{\cos x + 1} + \dfrac{1}{\cos x - 1} = -2 \csc x \cot x$

32. $\cos x - \dfrac{\cos x}{1 - \tan x} = \dfrac{\sin x \cos x}{\sin x - \cos x}$

33. $\tan\left(\dfrac{\pi}{2} - \theta\right)\tan \theta = 1$

34. $\dfrac{\cos[(\pi/2) - x]}{\sin[(\pi/2) - x]} = \tan x$

35. $\dfrac{\tan x \cot x}{\cos x} = \sec x$

36. $\dfrac{\csc(-x)}{\sec(-x)} = -\cot x$

37. $(1 + \sin y)[1 + \sin(-y)] = \cos^2 y$

38. $\dfrac{\tan x + \tan y}{1 - \tan x \tan y} = \dfrac{\cot x + \cot y}{\cot x \cot y - 1}$

39. $\dfrac{\tan x + \cot y}{\tan x \cot y} = \tan y + \cot x$

40. $\dfrac{\cos x - \cos y}{\sin x + \sin y} + \dfrac{\sin x - \sin y}{\cos x + \cos y} = 0$

41. $\sqrt{\dfrac{1 + \sin \theta}{1 - \sin \theta}} = \dfrac{1 + \sin \theta}{|\cos \theta|}$

42. $\sqrt{\dfrac{1 - \cos \theta}{1 + \cos \theta}} = \dfrac{1 - \cos \theta}{|\sin \theta|}$

43. $\cos^2 \beta + \cos^2\left(\dfrac{\pi}{2} - \beta\right) = 1$

44. $\sec^2 y - \cot^2\left(\dfrac{\pi}{2} - y\right) = 1$

45. $\sin t \csc\left(\dfrac{\pi}{2} - t\right) = \tan t$

46. $\sec^2\left(\dfrac{\pi}{2} - x\right) - 1 = \cot^2 x$

47. $\tan(\sin^{-1} x) = \dfrac{x}{\sqrt{1 - x^2}}$

48. $\cos(\sin^{-1} x) = \sqrt{1 - x^2}$

49. $\tan\left(\sin^{-1}\dfrac{x - 1}{4}\right) = \dfrac{x - 1}{\sqrt{16 - (x - 1)^2}}$

50. $\tan\left(\cos^{-1}\dfrac{x + 1}{2}\right) = \dfrac{\sqrt{4 - (x + 1)^2}}{x + 1}$

Example 1 Collecting Like Terms

Solve $\sin x + \sqrt{2} = -\sin x$.

Solution

Begin by rewriting the equation so that $\sin x$ is isolated on one side of the equation.

$\sin x + \sqrt{2} = -\sin x$	Write original equation.
$\sin x + \sin x + \sqrt{2} = 0$	Add $\sin x$ to each side.
$\sin x + \sin x = -\sqrt{2}$	Subtract $\sqrt{2}$ from each side.
$2 \sin x = -\sqrt{2}$	Combine like terms.
$\sin x = -\dfrac{\sqrt{2}}{2}$	Divide each side by 2.

Because $\sin x$ has a period of 2π, first find all solutions in the interval $[0, 2\pi)$. These solutions are $x = 5\pi/4$ and $x = 7\pi/4$. Finally, add multiples of 2π to each of these solutions to get the general form

$$x = \frac{5\pi}{4} + 2n\pi \quad \text{and} \quad x = \frac{7\pi}{4} + 2n\pi \qquad \text{General solution}$$

where n is an integer.

CHECK Point Now try Exercise 11.

Example 2 Extracting Square Roots

Solve $3 \tan^2 x - 1 = 0$.

Solution

Begin by rewriting the equation so that $\tan x$ is isolated on one side of the equation.

$3 \tan^2 x - 1 = 0$	Write original equation.
$3 \tan^2 x = 1$	Add 1 to each side.
$\tan^2 x = \dfrac{1}{3}$	Divide each side by 3.
$\tan x = \pm \dfrac{1}{\sqrt{3}} = \pm \dfrac{\sqrt{3}}{3}$	Extract square roots.

Because $\tan x$ has a period of π, first find all solutions in the interval $[0, \pi)$. These solutions are $x = \pi/6$ and $x = 5\pi/6$. Finally, add multiples of π to each of these solutions to get the general form

$$x = \frac{\pi}{6} + n\pi \quad \text{and} \quad x = \frac{5\pi}{6} + n\pi \qquad \text{General solution}$$

where n is an integer.

CHECK Point Now try Exercise 15.

WARNING / CAUTION

When you extract square roots, make sure you account for both the positive and negative solutions.

The equations in Examples 1 and 2 involved only one trigonometric function. When two or more functions occur in the same equation, collect all terms on one side and try to separate the functions by factoring or by using appropriate identities. This may produce factors that yield no solutions, as illustrated in Example 3.

Example 3   Factoring

Solve $\cot x \cos^2 x = 2 \cot x$.

Solution

Begin by rewriting the equation so that all terms are collected on one side of the equation.

$$\cot x \cos^2 x = 2 \cot x \qquad \text{Write original equation.}$$

$$\cot x \cos^2 x - 2 \cot x = 0 \qquad \text{Subtract } 2 \cot x \text{ from each side.}$$

$$\cot x(\cos^2 x - 2) = 0 \qquad \text{Factor.}$$

By setting each of these factors equal to zero, you obtain

$$\cot x = 0 \qquad \text{and} \qquad \cos^2 x - 2 = 0$$

$$x = \frac{\pi}{2} \qquad\qquad\qquad \cos^2 x = 2$$

$$\cos x = \pm\sqrt{2}.$$

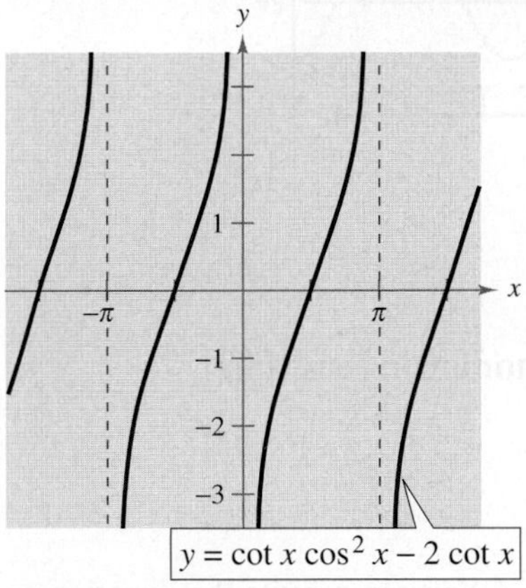

The equation $\cot x = 0$ has the solution $x = \pi/2$ [in the interval $(0, \pi)$]. No solution is obtained for $\cos x = \pm\sqrt{2}$ because $\pm\sqrt{2}$ are outside the range of the cosine function. Because $\cot x$ has a period of π, the general form of the solution is obtained by adding multiples of π to $x = \pi/2$, to get

$$x = \frac{\pi}{2} + n\pi \qquad \text{General solution}$$

where n is an integer. You can confirm this graphically by sketching the graph of $y = \cot x \cos^2 x - 2 \cot x$, as shown in Figure 2.8. From the graph you can see that the x-intercepts occur at $-3\pi/2$, $-\pi/2$, $\pi/2$, $3\pi/2$, and so on. These x-intercepts correspond to the solutions of $\cot x \cos^2 x - 2 \cot x = 0$.

CHECK *Point*   Now try Exercise 19.

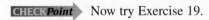

FIGURE **2.8**

$y = \cot x \cos^2 x - 2 \cot x$

Equations of Quadratic Type

Many trigonometric equations are of quadratic type $ax^2 + bx + c = 0$. Here are a couple of examples.

Quadratic in sin x	*Quadratic in sec x*
$2 \sin^2 x - \sin x - 1 = 0$	$\sec^2 x - 3 \sec x - 2 = 0$
$2(\sin x)^2 - \sin x - 1 = 0$	$(\sec x)^2 - 3(\sec x) - 2 = 0$

To solve equations of this type, factor the quadratic or, if this is not possible, use the Quadratic Formula.

Algebra Help

You can review the techniques for solving quadratic equations in Section P.2.

Example 4   Factoring an Equation of Quadratic Type

Find all solutions of $2 \sin^2 x - \sin x - 1 = 0$ in the interval $[0, 2\pi)$.

Algebraic Solution

Begin by treating the equation as a quadratic in $\sin x$ and factoring.

$$2 \sin^2 x - \sin x - 1 = 0 \qquad \text{Write original equation.}$$

$$(2 \sin x + 1)(\sin x - 1) = 0 \qquad \text{Factor.}$$

Setting each factor equal to zero, you obtain the following solutions in the interval $[0, 2\pi)$.

$$2 \sin x + 1 = 0 \qquad \text{and} \qquad \sin x - 1 = 0$$

$$\sin x = -\frac{1}{2} \qquad\qquad \sin x = 1$$

$$x = \frac{7\pi}{6}, \frac{11\pi}{6} \qquad\qquad x = \frac{\pi}{2}$$

Graphical Solution

Use a graphing utility set in *radian* mode to graph $y = 2 \sin^2 x - \sin x - 1$ for $0 \leq x < 2\pi$, as shown in Figure 2.9. Use the *zero* or *root* feature or the *zoom* and *trace* features to approximate the x-intercepts to be

$$x \approx 1.571 \approx \frac{\pi}{2}, \quad x \approx 3.665 \approx \frac{7\pi}{6}, \quad \text{and} \quad x \approx 5.760 \approx \frac{11\pi}{6}.$$

These values are the approximate solutions of $2 \sin^2 x - \sin x - 1 = 0$ in the interval $[0, 2\pi)$.

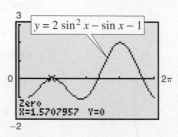

FIGURE 2.9

CHECK Point   Now try Exercise 33.

Example 5   Rewriting with a Single Trigonometric Function

Solve $2 \sin^2 x + 3 \cos x - 3 = 0$.

Solution

This equation contains both sine and cosine functions. You can rewrite the equation so that it has only cosine functions by using the identity $\sin^2 x = 1 - \cos^2 x$.

$$2 \sin^2 x + 3 \cos x - 3 = 0 \qquad \text{Write original equation.}$$

$$2(1 - \cos^2 x) + 3 \cos x - 3 = 0 \qquad \text{Pythagorean identity}$$

$$2 \cos^2 x - 3 \cos x + 1 = 0 \qquad \text{Multiply each side by } -1.$$

$$(2 \cos x - 1)(\cos x - 1) = 0 \qquad \text{Factor.}$$

Set each factor equal to zero to find the solutions in the interval $[0, 2\pi)$.

$$2 \cos x - 1 = 0 \implies \cos x = \frac{1}{2} \implies x = \frac{\pi}{3}, \frac{5\pi}{3}$$

$$\cos x - 1 = 0 \implies \cos x = 1 \implies x = 0$$

Because $\cos x$ has a period of 2π, the general form of the solution is obtained by adding multiples of 2π to get

$$x = 2n\pi, \quad x = \frac{\pi}{3} + 2n\pi, \quad x = \frac{5\pi}{3} + 2n\pi \qquad \text{General solution}$$

where n is an integer.

CHECK Point   Now try Exercise 35.

Sometimes you must square each side of an equation to obtain a quadratic, as demonstrated in the next example. Because this procedure can introduce extraneous solutions, you should check any solutions in the original equation to see whether they are valid or extraneous.

Example 6   Squaring and Converting to Quadratic Type

Find all solutions of $\cos x + 1 = \sin x$ in the interval $[0, 2\pi)$.

Solution

It is not clear how to rewrite this equation in terms of a single trigonometric function. Notice what happens when you square each side of the equation.

$\cos x + 1 = \sin x$	Write original equation.
$\cos^2 x + 2\cos x + 1 = \sin^2 x$	Square each side.
$\cos^2 x + 2\cos x + 1 = 1 - \cos^2 x$	Pythagorean identity
$\cos^2 x + \cos^2 x + 2\cos x + 1 - 1 = 0$	Rewrite equation.
$2\cos^2 x + 2\cos x = 0$	Combine like terms.
$2\cos x(\cos x + 1) = 0$	Factor.

Setting each factor equal to zero produces

$$2\cos x = 0 \qquad \text{and} \qquad \cos x + 1 = 0$$

$$\cos x = 0 \qquad\qquad\qquad \cos x = -1$$

$$x = \frac{\pi}{2}, \frac{3\pi}{2} \qquad\qquad\qquad x = \pi.$$

Because you squared the original equation, check for extraneous solutions.

Check $x = \pi/2$

$$\cos\frac{\pi}{2} + 1 \overset{?}{=} \sin\frac{\pi}{2} \qquad \text{Substitute } \pi/2 \text{ for } x.$$

$$0 + 1 = 1 \qquad \text{Solution checks. } \checkmark$$

Check $x = 3\pi/2$

$$\cos\frac{3\pi}{2} + 1 \overset{?}{=} \sin\frac{3\pi}{2} \qquad \text{Substitute } 3\pi/2 \text{ for } x.$$

$$0 + 1 \neq -1 \qquad \text{Solution does not check.}$$

Check $x = \pi$

$$\cos\pi + 1 \overset{?}{=} \sin\pi \qquad \text{Substitute } \pi \text{ for } x.$$

$$-1 + 1 = 0 \qquad \text{Solution checks. } \checkmark$$

Of the three possible solutions, $x = 3\pi/2$ is extraneous. So, in the interval $[0, 2\pi)$, the only two solutions are $x = \pi/2$ and $x = \pi$.

CHECK*Point*   Now try Exercise 37.

Study Tip

You square each side of the equation in Example 6 because the squares of the sine and cosine functions are related by a Pythagorean identity. The same is true for the squares of the secant and tangent functions and for the squares of the cosecant and cotangent functions.

Functions Involving Multiple Angles

The next two examples involve trigonometric functions of multiple angles of the forms $\sin ku$ and $\cos ku$. To solve equations of these forms, first solve the equation for ku, then divide your result by k.

Example 7 **Functions of Multiple Angles**

Solve $2 \cos 3t - 1 = 0$.

Solution

$2 \cos 3t - 1 = 0$	Write original equation.
$2 \cos 3t = 1$	Add 1 to each side.
$\cos 3t = \dfrac{1}{2}$	Divide each side by 2.

In the interval $[0, 2\pi)$, you know that $3t = \pi/3$ and $3t = 5\pi/3$ are the only solutions, so, in general, you have

$$3t = \frac{\pi}{3} + 2n\pi \quad \text{and} \quad 3t = \frac{5\pi}{3} + 2n\pi.$$

Dividing these results by 3, you obtain the general solution

$$t = \frac{\pi}{9} + \frac{2n\pi}{3} \quad \text{and} \quad t = \frac{5\pi}{9} + \frac{2n\pi}{3} \qquad \text{General solution}$$

where n is an integer.

CHECK*Point* Now try Exercise 39.

Example 8 **Functions of Multiple Angles**

Solve $3 \tan \dfrac{x}{2} + 3 = 0$.

Solution

$3 \tan \dfrac{x}{2} + 3 = 0$	Write original equation.
$3 \tan \dfrac{x}{2} = -3$	Subtract 3 from each side.
$\tan \dfrac{x}{2} = -1$	Divide each side by 3.

In the interval $[0, \pi)$, you know that $x/2 = 3\pi/4$ is the only solution, so, in general, you have

$$\frac{x}{2} = \frac{3\pi}{4} + n\pi.$$

Multiplying this result by 2, you obtain the general solution

$$x = \frac{3\pi}{2} + 2n\pi \qquad \text{General solution}$$

where n is an integer.

CHECK*Point* Now try Exercise 43.

Using Inverse Functions

In the next example, you will see how inverse trigonometric functions can be used to solve an equation.

Example 9   Using Inverse Functions

Solve $\sec^2 x - 2 \tan x = 4$.

Solution

$$\sec^2 x - 2 \tan x = 4 \qquad \text{Write original equation.}$$

$$1 + \tan^2 x - 2 \tan x - 4 = 0 \qquad \text{Pythagorean identity}$$

$$\tan^2 x - 2 \tan x - 3 = 0 \qquad \text{Combine like terms.}$$

$$(\tan x - 3)(\tan x + 1) = 0 \qquad \text{Factor.}$$

Setting each factor equal to zero, you obtain two solutions in the interval $(-\pi/2, \pi/2)$. [Recall that the range of the inverse tangent function is $(-\pi/2, \pi/2)$.]

$$\tan x - 3 = 0 \qquad \text{and} \qquad \tan x + 1 = 0$$

$$\tan x = 3 \qquad\qquad\qquad \tan x = -1$$

$$x = \arctan 3 \qquad\qquad\qquad x = -\frac{\pi}{4}$$

Finally, because $\tan x$ has a period of π, you obtain the general solution by adding multiples of π

$$x = \arctan 3 + n\pi \qquad \text{and} \qquad x = -\frac{\pi}{4} + n\pi \qquad \text{General solution}$$

where n is an integer. You can use a calculator to approximate the value of $\arctan 3$.

CHECK*Point*   Now try Exercise 63.

CLASSROOM DISCUSSION

Equations with No Solutions   One of the following equations has solutions and the other two do not. Which two equations do not have solutions?

a. $\sin^2 x - 5 \sin x + 6 = 0$

b. $\sin^2 x - 4 \sin x + 6 = 0$

c. $\sin^2 x - 5 \sin x - 6 = 0$

Find conditions involving the constants b and c that will guarantee that the equation

$$\sin^2 x + b \sin x + c = 0$$

has at least one solution on some interval of length 2π.

2.3 EXERCISES

See www.CalcChat.com for worked-out solutions to odd-numbered exercises.

VOCABULARY: Fill in the blanks.

1. When solving a trigonometric equation, the preliminary goal is to _____ the trigonometric function involved in the equation.

2. The equation $2 \sin \theta + 1 = 0$ has the solutions $\theta = \dfrac{7\pi}{6} + 2n\pi$ and $\theta = \dfrac{11\pi}{6} + 2n\pi$, which are called _____ solutions.

3. The equation $2 \tan^2 x - 3 \tan x + 1 = 0$ is a trigonometric equation that is of _____ type.

4. A solution of an equation that does not satisfy the original equation is called an _____ solution.

SKILLS AND APPLICATIONS

In Exercises 5–10, verify that the x-values are solutions of the equation.

5. $2 \cos x - 1 = 0$

  (a) $x = \dfrac{\pi}{3}$ (b) $x = \dfrac{5\pi}{3}$

6. $\sec x - 2 = 0$

  (a) $x = \dfrac{\pi}{3}$ (b) $x = \dfrac{5\pi}{3}$

7. $3 \tan^2 2x - 1 = 0$

  (a) $x = \dfrac{\pi}{12}$ (b) $x = \dfrac{5\pi}{12}$

8. $2 \cos^2 4x - 1 = 0$

  (a) $x = \dfrac{\pi}{16}$ (b) $x = \dfrac{3\pi}{16}$

9. $2 \sin^2 x - \sin x - 1 = 0$

  (a) $x = \dfrac{\pi}{2}$ (b) $x = \dfrac{7\pi}{6}$

10. $\csc^4 x - 4 \csc^2 x = 0$

  (a) $x = \dfrac{\pi}{6}$ (b) $x = \dfrac{5\pi}{6}$

In Exercises 11–24, solve the equation.

11. $2 \cos x + 1 = 0$

12. $2 \sin x + 1 = 0$

13. $\sqrt{3} \csc x - 2 = 0$

14. $\tan x + \sqrt{3} = 0$

15. $3 \sec^2 x - 4 = 0$

16. $3 \cot^2 x - 1 = 0$

17. $\sin x(\sin x + 1) = 0$

18. $(3 \tan^2 x - 1)(\tan^2 x - 3) = 0$

19. $4 \cos^2 x - 1 = 0$

20. $\sin^2 x = 3 \cos^2 x$

21. $2 \sin^2 2x = 1$

22. $\tan^2 3x = 3$

23. $\tan 3x(\tan x - 1) = 0$

24. $\cos 2x(2 \cos x + 1) = 0$

In Exercises 25–38, find all solutions of the equation in the interval $[0, 2\pi)$.

25. $\cos^3 x = \cos x$

26. $\sec^2 x - 1 = 0$

27. $3 \tan^3 x = \tan x$

28. $2 \sin^2 x = 2 + \cos x$

29. $\sec^2 x - \sec x = 2$

30. $\sec x \csc x = 2 \csc x$

31. $2 \sin x + \csc x = 0$

32. $\sec x + \tan x = 1$

33. $2 \cos^2 x + \cos x - 1 = 0$

34. $2 \sin^2 x + 3 \sin x + 1 = 0$

35. $2 \sec^2 x + \tan^2 x - 3 = 0$

36. $\cos x + \sin x \tan x = 2$

37. $\csc x + \cot x = 1$

38. $\sin x - 2 = \cos x - 2$

In Exercises 39–44, solve the multiple-angle equation.

39. $\cos 2x = \dfrac{1}{2}$

40. $\sin 2x = -\dfrac{\sqrt{3}}{2}$

41. $\tan 3x = 1$

42. $\sec 4x = 2$

43. $\cos \dfrac{x}{2} = \dfrac{\sqrt{2}}{2}$

44. $\sin \dfrac{x}{2} = -\dfrac{\sqrt{3}}{2}$

In Exercises 45–48, find the x-intercepts of the graph.

45. $y = \sin \dfrac{\pi x}{2} + 1$

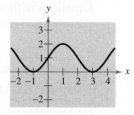

46. $y = \sin \pi x + \cos \pi x$

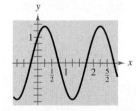

47. $y = \tan^2\left(\dfrac{\pi x}{6}\right) - 3$

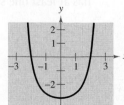

48. $y = \sec^4\left(\dfrac{\pi x}{8}\right) - 4$

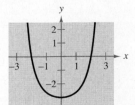

In Exercises 49–58, use a graphing utility to approximate the solutions (to three decimal places) of the equation in the interval $[0, 2\pi)$.

49. $2 \sin x + \cos x = 0$

50. $4 \sin^3 x + 2 \sin^2 x - 2 \sin x - 1 = 0$

51. $\dfrac{1 + \sin x}{\cos x} + \dfrac{\cos x}{1 + \sin x} = 4$ **52.** $\dfrac{\cos x \cot x}{1 - \sin x} = 3$

53. $x \tan x - 1 = 0$ **54.** $x \cos x - 1 = 0$

55. $\sec^2 x + 0.5 \tan x - 1 = 0$

56. $\csc^2 x + 0.5 \cot x - 5 = 0$

57. $2 \tan^2 x + 7 \tan x - 15 = 0$

58. $6 \sin^2 x - 7 \sin x + 2 = 0$

In Exercises 59–62, use the Quadratic Formula to solve the equation in the interval $[0, 2\pi)$. Then use a graphing utility to approximate the angle x.

59. $12 \sin^2 x - 13 \sin x + 3 = 0$

60. $3 \tan^2 x + 4 \tan x - 4 = 0$

61. $\tan^2 x + 3 \tan x + 1 = 0$

62. $4 \cos^2 x - 4 \cos x - 1 = 0$

In Exercises 63–74, use inverse functions where needed to find all solutions of the equation in the interval $[0, 2\pi)$.

63. $\tan^2 x + \tan x - 12 = 0$

64. $\tan^2 x - \tan x - 2 = 0$

65. $\tan^2 x - 6 \tan x + 5 = 0$

66. $\sec^2 x + \tan x - 3 = 0$

67. $2 \cos^2 x - 5 \cos x + 2 = 0$

68. $2 \sin^2 x - 7 \sin x + 3 = 0$

69. $\cot^2 x - 9 = 0$

70. $\cot^2 x - 6 \cot x + 5 = 0$

71. $\sec^2 x - 4 \sec x = 0$

72. $\sec^2 x + 2 \sec x - 8 = 0$

73. $\csc^2 x + 3 \csc x - 4 = 0$

74. $\csc^2 x - 5 \csc x = 0$

In Exercises 75–78, use a graphing utility to approximate the solutions (to three decimal places) of the equation in the given interval.

75. $3 \tan^2 x + 5 \tan x - 4 = 0$, $\left[-\dfrac{\pi}{2}, \dfrac{\pi}{2} \right]$

76. $\cos^2 x - 2 \cos x - 1 = 0$, $[0, \pi]$

77. $4 \cos^2 x - 2 \sin x + 1 = 0$, $\left[-\dfrac{\pi}{2}, \dfrac{\pi}{2} \right]$

78. $2 \sec^2 x + \tan x - 6 = 0$, $\left[-\dfrac{\pi}{2}, \dfrac{\pi}{2} \right]$

In Exercises 79–84, (a) use a graphing utility to graph the function and approximate the maximum and minimum points on the graph in the interval $[0, 2\pi)$, and (b) solve the trigonometric equation and demonstrate that its solutions are the x-coordinates of the maximum and minimum points of f. (Calculus is required to find the trigonometric equation.)

	Function	*Trigonometric Equation*
79.	$f(x) = \sin^2 x + \cos x$	$2 \sin x \cos x - \sin x = 0$
80.	$f(x) = \cos^2 x - \sin x$	$-2 \sin x \cos x - \cos x = 0$
81.	$f(x) = \sin x + \cos x$	$\cos x - \sin x = 0$
82.	$f(x) = 2 \sin x + \cos 2x$	$2 \cos x - 4 \sin x \cos x = 0$
83.	$f(x) = \sin x \cos x$	$-\sin^2 x + \cos^2 x = 0$
84.	$f(x) = \sec x + \tan x - x$	

$$\sec x \tan x + \sec^2 x - 1 = 0$$

FIXED POINT   In Exercises 85 and 86, find the smallest positive fixed point of the function f. [A *fixed point* of a function f is a real number c such that $f(c) = c$.]

85. $f(x) = \tan \dfrac{\pi x}{4}$ **86.** $f(x) = \cos x$

87. GRAPHICAL REASONING   Consider the function given by

$$f(x) = \cos \frac{1}{x}$$

and its graph shown in the figure.

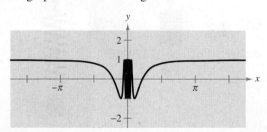

(a) What is the domain of the function?

(b) Identify any symmetry and any asymptotes of the graph.

(c) Describe the behavior of the function as $x \to 0$.

(d) How many solutions does the equation

$$\cos \frac{1}{x} = 0$$

have in the interval $[-1, 1]$? Find the solutions.

(e) Does the equation $\cos(1/x) = 0$ have a greatest solution? If so, approximate the solution. If not, explain why.

88. GRAPHICAL REASONING Consider the function given by $f(x) = (\sin x)/x$ and its graph shown in the figure.

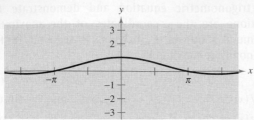

(a) What is the domain of the function?

(b) Identify any symmetry and any asymptotes of the graph.

(c) Describe the behavior of the function as $x \to 0$.

(d) How many solutions does the equation

$$\frac{\sin x}{x} = 0$$

have in the interval $[-8, 8]$? Find the solutions.

89. HARMONIC MOTION A weight is oscillating on the end of a spring (see figure). The position of the weight relative to the point of equilibrium is given by $y = \frac{1}{12}(\cos 8t - 3 \sin 8t)$, where y is the displacement (in meters) and t is the time (in seconds). Find the times when the weight is at the point of equilibrium ($y = 0$) for $0 \le t \le 1$.

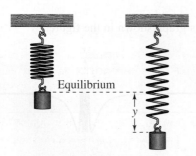

Equilibrium

90. DAMPED HARMONIC MOTION The displacement from equilibrium of a weight oscillating on the end of a spring is given by $y = 1.56t^{-1/2} \cos 1.9t$, where y is the displacement (in feet) and t is the time (in seconds). Use a graphing utility to graph the displacement function for $0 \le t \le 10$. Find the time beyond which the displacement does not exceed 1 foot from equilibrium.

91. SALES The monthly sales S (in thousands of units) of a seasonal product are approximated by

$$S = 74.50 + 43.75 \sin \frac{\pi t}{6}$$

where t is the time (in months), with $t = 1$ corresponding to January. Determine the months in which sales exceed 100,000 units.

92. SALES The monthly sales S (in hundreds of units) of skiing equipment at a sports store are approximated by

$$S = 58.3 + 32.5 \cos \frac{\pi t}{6}$$

where t is the time (in months), with $t = 1$ corresponding to January. Determine the months in which sales exceed 7500 units.

93. PROJECTILE MOTION A batted baseball leaves the bat at an angle of θ with the horizontal and an initial velocity of $v_0 = 100$ feet per second. The ball is caught by an outfielder 300 feet from home plate (see figure). Find θ if the range r of a projectile is given by $r = \frac{1}{32}v_0^2 \sin 2\theta$.

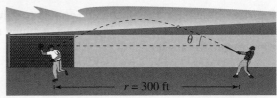

$r = 300$ ft

Not drawn to scale

94. PROJECTILE MOTION A sharpshooter intends to hit a target at a distance of 1000 yards with a gun that has a muzzle velocity of 1200 feet per second (see figure). Neglecting air resistance, determine the gun's minimum angle of elevation θ if the range r is given by

$$r = \frac{1}{32}v_0^2 \sin 2\theta.$$

$r = 1000$ yd

Not drawn to scale

95. FERRIS WHEEL A Ferris wheel is built such that the height h (in feet) above ground of a seat on the wheel at time t (in minutes) can be modeled by

$$h(t) = 53 + 50 \sin\left(\frac{\pi}{16}t - \frac{\pi}{2}\right).$$

The wheel makes one revolution every 32 seconds. The ride begins when $t = 0$.

(a) During the first 32 seconds of the ride, when will a person on the Ferris wheel be 53 feet above ground?

(b) When will a person be at the top of the Ferris wheel for the first time during the ride? If the ride lasts 160 seconds, how many times will a person be at the top of the ride, and at what times?

96. **DATA ANALYSIS: METEOROLOGY**   The table shows the average daily high temperatures in Houston H (in degrees Fahrenheit) for month t, with $t = 1$ corresponding to January.   (Source: National Climatic Data Center)

Month, t	Houston, H
1	62.3
2	66.5
3	73.3
4	79.1
5	85.5
6	90.7
7	93.6
8	93.5
9	89.3
10	82.0
11	72.0
12	64.6

(a) Create a scatter plot of the data.

(b) Find a cosine model for the temperatures in Houston.

(c) Use a graphing utility to graph the data points and the model for the temperatures in Houston. How well does the model fit the data?

(d) What is the overall average daily high temperature in Houston?

(e) Use a graphing utility to describe the months during which the average daily high temperature is above 86°F and below 86°F.

97. **GEOMETRY**   The area of a rectangle (see figure) inscribed in one arc of the graph of $y = \cos x$ is given by $A = 2x \cos x$, $0 < x < \pi/2$.

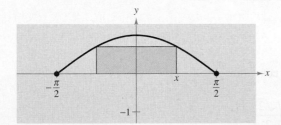

(a) Use a graphing utility to graph the area function, and approximate the area of the largest inscribed rectangle.

(b) Determine the values of x for which $A \geq 1$.

98. **QUADRATIC APPROXIMATION**   Consider the function given by $f(x) = 3 \sin(0.6x - 2)$.

(a) Approximate the zero of the function in the interval $[0, 6]$.

(b) A quadratic approximation agreeing with f at $x = 5$ is $g(x) = -0.45x^2 + 5.52x - 13.70$. Use a graphing utility to graph f and g in the same viewing window. Describe the result.

(c) Use the Quadratic Formula to find the zeros of g. Compare the zero in the interval $[0, 6]$ with the result of part (a).

EXPLORATION

TRUE OR FALSE? In Exercises 99 and 100, determine whether the statement is true or false. Justify your answer.

99. The equation $2 \sin 4t - 1 = 0$ has four times the number of solutions in the interval $[0, 2\pi)$ as the equation $2 \sin t - 1 = 0$.

100. If you correctly solve a trigonometric equation to the statement $\sin x = 3.4$, then you can finish solving the equation by using an inverse function.

101. **THINK ABOUT IT**   Explain what would happen if you divided each side of the equation $\cot x \cos^2 x = 2 \cot x$ by $\cot x$. Is this a correct method to use when solving equations?

102. **GRAPHICAL REASONING**   Use a graphing utility to confirm the solutions found in Example 6 in two different ways.

(a) Graph both sides of the equation and find the x-coordinates of the points at which the graphs intersect.

Left side: $y = \cos x + 1$

Right side: $y = \sin x$

(b) Graph the equation $y = \cos x + 1 - \sin x$ and find the x-intercepts of the graph. Do both methods produce the same x-values? Which method do you prefer? Explain.

103. Explain in your own words how knowledge of algebra is important when solving trigonometric equations.

104. **CAPSTONE**   Consider the equation $2 \sin x - 1 = 0$. Explain the similarities and differences between finding all solutions in the interval $\left[0, \dfrac{\pi}{2}\right)$, finding all solutions in the interval $[0, 2\pi)$, and finding the general solution.

PROJECT: METEOROLOGY   To work an extended application analyzing the normal daily high temperatures in Phoenix and in Seattle, visit this text's website at *academic.cengage.com*.   (Data Source: NOAA)

2.4 SUM AND DIFFERENCE FORMULAS

What you should learn

• Use sum and difference formulas to evaluate trigonometric functions, verify identities, and solve trigonometric equations.

Why you should learn it

You can use identities to rewrite trigonometric expressions. For instance, in Exercise 89 on page 257, you can use an identity to rewrite a trigonometric expression in a form that helps you analyze a harmonic motion equation.

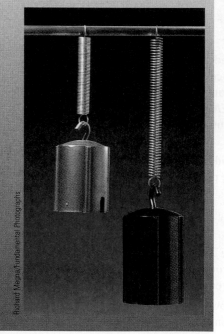

Using Sum and Difference Formulas

In this and the following section, you will study the uses of several trigonometric identities and formulas.

Sum and Difference Formulas

$$\sin(u + v) = \sin u \cos v + \cos u \sin v$$

$$\sin(u - v) = \sin u \cos v - \cos u \sin v$$

$$\cos(u + v) = \cos u \cos v - \sin u \sin v$$

$$\cos(u - v) = \cos u \cos v + \sin u \sin v$$

$$\tan(u + v) = \frac{\tan u + \tan v}{1 - \tan u \tan v} \qquad \tan(u - v) = \frac{\tan u - \tan v}{1 + \tan u \tan v}$$

For a proof of the sum and difference formulas, see Proofs in Mathematics on page 276.

Examples 1 and 2 show how **sum and difference formulas** can be used to find exact values of trigonometric functions involving sums or differences of special angles.

Example 1   Evaluating a Trigonometric Function

Find the exact value of $\sin \dfrac{\pi}{12}$.

Solution

To find the *exact* value of $\sin \dfrac{\pi}{12}$, use the fact that

$$\frac{\pi}{12} = \frac{\pi}{3} - \frac{\pi}{4}.$$

Consequently, the formula for $\sin(u - v)$ yields

$$\sin \frac{\pi}{12} = \sin\left(\frac{\pi}{3} - \frac{\pi}{4}\right)$$

$$= \sin \frac{\pi}{3} \cos \frac{\pi}{4} - \cos \frac{\pi}{3} \sin \frac{\pi}{4}$$

$$= \frac{\sqrt{3}}{2}\left(\frac{\sqrt{2}}{2}\right) - \frac{1}{2}\left(\frac{\sqrt{2}}{2}\right)$$

$$= \frac{\sqrt{6} - \sqrt{2}}{4}.$$

Try checking this result on your calculator. You will find that $\sin \dfrac{\pi}{12} \approx 0.259$.

CHECK*Point*▶   Now try Exercise 7.

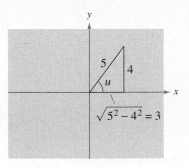

FIGURE **2.10**

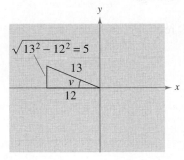

FIGURE **2.11**

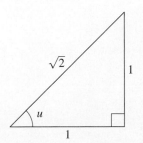

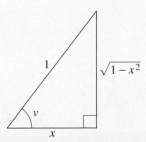

FIGURE **2.12**

Example 2   Evaluating a Trigonometric Function

Find the exact value of $\cos 75°$.

Solution

Using the fact that $75° = 30° + 45°$, together with the formula for $\cos(u + v)$, you obtain

$$\cos 75° = \cos(30° + 45°)$$

$$= \cos 30° \cos 45° - \sin 30° \sin 45°$$

$$= \frac{\sqrt{3}}{2}\left(\frac{\sqrt{2}}{2}\right) - \frac{1}{2}\left(\frac{\sqrt{2}}{2}\right) = \frac{\sqrt{6} - \sqrt{2}}{4}.$$

CHECK*Point*   Now try Exercise 11.

Example 3   Evaluating a Trigonometric Expression

Find the exact value of $\sin(u + v)$ given

$$\sin u = \frac{4}{5}, \text{ where } 0 < u < \frac{\pi}{2}, \quad \text{and} \quad \cos v = -\frac{12}{13}, \text{ where } \frac{\pi}{2} < v < \pi.$$

Solution

Because $\sin u = 4/5$ and u is in Quadrant I, $\cos u = 3/5$, as shown in Figure 2.10. Because $\cos v = -12/13$ and v is in Quadrant II, $\sin v = 5/13$, as shown in Figure 2.11. You can find $\sin(u + v)$ as follows.

$$\sin(u + v) = \sin u \cos v + \cos u \sin v$$

$$= \left(\frac{4}{5}\right)\left(-\frac{12}{13}\right) + \left(\frac{3}{5}\right)\left(\frac{5}{13}\right)$$

$$= -\frac{48}{65} + \frac{15}{65}$$

$$= -\frac{33}{65}$$

CHECK*Point*   Now try Exercise 43.

Example 4   An Application of a Sum Formula

Write $\cos(\arctan 1 + \arccos x)$ as an algebraic expression.

Solution

This expression fits the formula for $\cos(u + v)$. Angles $u = \arctan 1$ and $v = \arccos x$ are shown in Figure 2.12. So

$$\cos(u + v) = \cos(\arctan 1) \cos(\arccos x) - \sin(\arctan 1) \sin(\arccos x)$$

$$= \frac{1}{\sqrt{2}} \cdot x - \frac{1}{\sqrt{2}} \cdot \sqrt{1 - x^2}$$

$$= \frac{x - \sqrt{1 - x^2}}{\sqrt{2}}.$$

CHECK*Point*   Now try Exercise 57.

HISTORICAL NOTE

The Granger Collection, New York

Hipparchus, considered the most eminent of Greek astronomers, was born about 190 B.C. in Nicaea. He was credited with the invention of trigonometry. He also derived the sum and difference formulas for $\sin(A \pm B)$ and $\cos(A \pm B)$.

Example 5 shows how to use a difference formula to prove the cofunction identity

$$\cos\left(\frac{\pi}{2} - x\right) = \sin x.$$

Example 5 Proving a Cofunction Identity

Prove the cofunction identity $\cos\left(\frac{\pi}{2} - x\right) = \sin x$.

Solution
Using the formula for $\cos(u - v)$, you have

$$\cos\left(\frac{\pi}{2} - x\right) = \cos\frac{\pi}{2}\cos x + \sin\frac{\pi}{2}\sin x$$

$$= (0)(\cos x) + (1)(\sin x)$$

$$= \sin x.$$

CHECK Point Now try Exercise 61.

Sum and difference formulas can be used to rewrite expressions such as

$$\sin\left(\theta + \frac{n\pi}{2}\right) \quad \text{and} \quad \cos\left(\theta + \frac{n\pi}{2}\right), \quad \text{where } n \text{ is an integer}$$

as expressions involving only $\sin\theta$ or $\cos\theta$. The resulting formulas are called **reduction formulas.**

Example 6 Deriving Reduction Formulas

Simplify each expression.

a. $\cos\left(\theta - \frac{3\pi}{2}\right)$ **b.** $\tan(\theta + 3\pi)$

Solution
a. Using the formula for $\cos(u - v)$, you have

$$\cos\left(\theta - \frac{3\pi}{2}\right) = \cos\theta\cos\frac{3\pi}{2} + \sin\theta\sin\frac{3\pi}{2}$$

$$= (\cos\theta)(0) + (\sin\theta)(-1)$$

$$= -\sin\theta.$$

b. Using the formula for $\tan(u + v)$, you have

$$\tan(\theta + 3\pi) = \frac{\tan\theta + \tan 3\pi}{1 - \tan\theta\tan 3\pi}$$

$$= \frac{\tan\theta + 0}{1 - (\tan\theta)(0)}$$

$$= \tan\theta.$$

CHECK Point Now try Exercise 73.

| **Example 7** | **Solving a Trigonometric Equation** |

Find all solutions of $\sin\left(x + \dfrac{\pi}{4}\right) + \sin\left(x - \dfrac{\pi}{4}\right) = -1$ in the interval $[0, 2\pi)$.

Algebraic Solution

Using sum and difference formulas, rewrite the equation as

$$\sin x \cos \frac{\pi}{4} + \cos x \sin \frac{\pi}{4} + \sin x \cos \frac{\pi}{4} - \cos x \sin \frac{\pi}{4} = -1$$

$$2 \sin x \cos \frac{\pi}{4} = -1$$

$$2(\sin x)\left(\frac{\sqrt{2}}{2}\right) = -1$$

$$\sin x = -\frac{1}{\sqrt{2}}$$

$$\sin x = -\frac{\sqrt{2}}{2}.$$

So, the only solutions in the interval $[0, 2\pi)$ are

$$x = \frac{5\pi}{4} \quad \text{and} \quad x = \frac{7\pi}{4}.$$

Graphical Solution

Sketch the graph of

$$y = \sin\left(x + \frac{\pi}{4}\right) + \sin\left(x - \frac{\pi}{4}\right) + 1 \text{ for } 0 \le x < 2\pi.$$

as shown in Figure 2.13. From the graph you can see that the x-intercepts are $5\pi/4$ and $7\pi/4$. So, the solutions in the interval $[0, 2\pi)$ are

$$x = \frac{5\pi}{4} \quad \text{and} \quad x = \frac{7\pi}{4}.$$

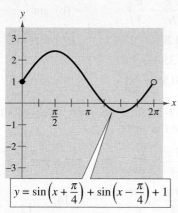

$$y = \sin\left(x + \frac{\pi}{4}\right) + \sin\left(x - \frac{\pi}{4}\right) + 1$$

FIGURE 2.13

CHECK*Point*▸   Now try Exercise 79.

The next example was taken from calculus. It is used to derive the derivative of the sine function.

| **Example 8** | **An Application from Calculus** | |

Verify that $\dfrac{\sin(x + h) - \sin x}{h} = (\cos x)\left(\dfrac{\sin h}{h}\right) - (\sin x)\left(\dfrac{1 - \cos h}{h}\right)$ where $h \ne 0$.

Solution

Using the formula for $\sin(u + v)$, you have

$$\frac{\sin(x + h) - \sin x}{h} = \frac{\sin x \cos h + \cos x \sin h - \sin x}{h}$$

$$= \frac{\cos x \sin h - \sin x(1 - \cos h)}{h}$$

$$= (\cos x)\left(\frac{\sin h}{h}\right) - (\sin x)\left(\frac{1 - \cos h}{h}\right).$$

CHECK*Point*▸   Now try Exercise 105.

2.4 EXERCISES

See www.CalcChat.com for worked-out solutions to odd-numbered exercises.

VOCABULARY: Fill in the blank.

1. $\sin(u - v) = $ _____
2. $\cos(u + v) = $ _____
3. $\tan(u + v) = $ _____
4. $\sin(u + v) = $ _____
5. $\cos(u - v) = $ _____
6. $\tan(u - v) = $ _____

SKILLS AND APPLICATIONS

In Exercises 7–12, find the exact value of each expression.

7. (a) $\cos\left(\dfrac{\pi}{4} + \dfrac{\pi}{3}\right)$ (b) $\cos\dfrac{\pi}{4} + \cos\dfrac{\pi}{3}$

8. (a) $\sin\left(\dfrac{3\pi}{4} + \dfrac{5\pi}{6}\right)$ (b) $\sin\dfrac{3\pi}{4} + \sin\dfrac{5\pi}{6}$

9. (a) $\sin\left(\dfrac{7\pi}{6} - \dfrac{\pi}{3}\right)$ (b) $\sin\dfrac{7\pi}{6} - \sin\dfrac{\pi}{3}$

10. (a) $\cos(120° + 45°)$ (b) $\cos 120° + \cos 45°$

11. (a) $\sin(135° - 30°)$ (b) $\sin 135° - \cos 30°$

12. (a) $\sin(315° - 60°)$ (b) $\sin 315° - \sin 60°$

In Exercises 13–28, find the exact values of the sine, cosine, and tangent of the angle.

13. $\dfrac{11\pi}{12} = \dfrac{3\pi}{4} + \dfrac{\pi}{6}$

14. $\dfrac{7\pi}{12} = \dfrac{\pi}{3} + \dfrac{\pi}{4}$

15. $\dfrac{17\pi}{12} = \dfrac{9\pi}{4} - \dfrac{5\pi}{6}$

16. $-\dfrac{\pi}{12} = \dfrac{\pi}{6} - \dfrac{\pi}{4}$

17. $105° = 60° + 45°$

18. $165° = 135° + 30°$

19. $195° = 225° - 30°$

20. $255° = 300° - 45°$

21. $\dfrac{13\pi}{12}$

22. $-\dfrac{7\pi}{12}$

23. $-\dfrac{13\pi}{12}$

24. $\dfrac{5\pi}{12}$

25. $285°$

26. $-105°$

27. $-165°$

28. $15°$

In Exercises 29–36, write the expression as the sine, cosine, or tangent of an angle.

29. $\sin 3 \cos 1.2 - \cos 3 \sin 1.2$

30. $\cos\dfrac{\pi}{7} \cos\dfrac{\pi}{5} - \sin\dfrac{\pi}{7} \sin\dfrac{\pi}{5}$

31. $\sin 60° \cos 15° + \cos 60° \sin 15°$

32. $\cos 130° \cos 40° - \sin 130° \sin 40°$

33. $\dfrac{\tan 45° - \tan 30°}{1 + \tan 45° \tan 30°}$

34. $\dfrac{\tan 140° - \tan 60°}{1 + \tan 140° \tan 60°}$

35. $\dfrac{\tan 2x + \tan x}{1 - \tan 2x \tan x}$

36. $\cos 3x \cos 2y + \sin 3x \sin 2y$

In Exercises 37–42, find the exact value of the expression.

37. $\sin\dfrac{\pi}{12} \cos\dfrac{\pi}{4} + \cos\dfrac{\pi}{12} \sin\dfrac{\pi}{4}$

38. $\cos\dfrac{\pi}{16} \cos\dfrac{3\pi}{16} - \sin\dfrac{\pi}{16} \sin\dfrac{3\pi}{16}$

39. $\sin 120° \cos 60° - \cos 120° \sin 60°$

40. $\cos 120° \cos 30° + \sin 120° \sin 30°$

41. $\dfrac{\tan(5\pi/6) - \tan(\pi/6)}{1 + \tan(5\pi/6) \tan(\pi/6)}$

42. $\dfrac{\tan 25° + \tan 110°}{1 - \tan 25° \tan 110°}$

In Exercises 43–50, find the exact value of the trigonometric function given that $\sin u = \frac{5}{13}$ and $\cos v = -\frac{3}{5}$. (Both u and v are in Quadrant II.)

43. $\sin(u + v)$ 44. $\cos(u - v)$
45. $\cos(u + v)$ 46. $\sin(v - u)$
47. $\tan(u + v)$ 48. $\csc(u - v)$
49. $\sec(v - u)$ 50. $\cot(u + v)$

In Exercises 51–56, find the exact value of the trigonometric function given that $\sin u = -\frac{7}{25}$ and $\cos v = -\frac{4}{5}$. (Both u and v are in Quadrant III.)

51. $\cos(u + v)$ 52. $\sin(u + v)$
53. $\tan(u - v)$ 54. $\cot(v - u)$
55. $\csc(u - v)$ 56. $\sec(v - u)$

In Exercises 57–60, write the trigonometric expression as an algebraic expression.

57. $\sin(\arcsin x + \arccos x)$
58. $\sin(\arctan 2x - \arccos x)$
59. $\cos(\arccos x + \arcsin x)$
60. $\cos(\arccos x - \arctan x)$

In Exercises 61–70, prove the identity.

61. $\sin\left(\dfrac{\pi}{2} - x\right) = \cos x$ **62.** $\sin\left(\dfrac{\pi}{2} + x\right) = \cos x$

63. $\sin\left(\dfrac{\pi}{6} + x\right) = \dfrac{1}{2}(\cos x + \sqrt{3}\sin x)$

64. $\cos\left(\dfrac{5\pi}{4} - x\right) = -\dfrac{\sqrt{2}}{2}(\cos x + \sin x)$

65. $\cos(\pi - \theta) + \sin\left(\dfrac{\pi}{2} + \theta\right) = 0$

66. $\tan\left(\dfrac{\pi}{4} - \theta\right) = \dfrac{1 - \tan\theta}{1 + \tan\theta}$

67. $\cos(x + y)\cos(x - y) = \cos^2 x - \sin^2 y$

68. $\sin(x + y)\sin(x - y) = \sin^2 x - \sin^2 y$

69. $\sin(x + y) + \sin(x - y) = 2\sin x \cos y$

70. $\cos(x + y) + \cos(x - y) = 2\cos x \cos y$

In Exercises 71–74, simplify the expression algebraically and use a graphing utility to confirm your answer graphically.

71. $\cos\left(\dfrac{3\pi}{2} - x\right)$ **72.** $\cos(\pi + x)$

73. $\sin\left(\dfrac{3\pi}{2} + \theta\right)$ **74.** $\tan(\pi + \theta)$

In Exercises 75–84, find all solutions of the equation in the interval $[0, 2\pi)$.

75. $\sin(x + \pi) - \sin x + 1 = 0$

76. $\sin(x + \pi) - \sin x - 1 = 0$

77. $\cos(x + \pi) - \cos x - 1 = 0$

78. $\cos(x + \pi) - \cos x + 1 = 0$

79. $\sin\left(x + \dfrac{\pi}{6}\right) - \sin\left(x - \dfrac{\pi}{6}\right) = \dfrac{1}{2}$

80. $\sin\left(x + \dfrac{\pi}{3}\right) + \sin\left(x - \dfrac{\pi}{3}\right) = 1$

81. $\cos\left(x + \dfrac{\pi}{4}\right) - \cos\left(x - \dfrac{\pi}{4}\right) = 1$

82. $\tan(x + \pi) + 2\sin(x + \pi) = 0$

83. $\sin\left(x + \dfrac{\pi}{2}\right) - \cos^2 x = 0$

84. $\cos\left(x - \dfrac{\pi}{2}\right) + \sin^2 x = 0$

In Exercises 85–88, use a graphing utility to approximate the solutions in the interval $[0, 2\pi)$.

85. $\cos\left(x + \dfrac{\pi}{4}\right) + \cos\left(x - \dfrac{\pi}{4}\right) = 1$

86. $\tan(x + \pi) - \cos\left(x + \dfrac{\pi}{2}\right) = 0$

87. $\sin\left(x + \dfrac{\pi}{2}\right) + \cos^2 x = 0$

88. $\cos\left(x - \dfrac{\pi}{2}\right) - \sin^2 x = 0$

89. HARMONIC MOTION A weight is attached to a spring suspended vertically from a ceiling. When a driving force is applied to the system, the weight moves vertically from its equilibrium position, and this motion is modeled by

$$y = \dfrac{1}{3}\sin 2t + \dfrac{1}{4}\cos 2t$$

where y is the distance from equilibrium (in feet) and t is the time (in seconds).

(a) Use the identity

$$a\sin B\theta + b\cos B\theta = \sqrt{a^2 + b^2}\,\sin(B\theta + C)$$

where $C = \arctan(b/a)$, $a > 0$, to write the model in the form $y = \sqrt{a^2 + b^2}\,\sin(Bt + C)$.

(b) Find the amplitude of the oscillations of the weight.

(c) Find the frequency of the oscillations of the weight.

90. STANDING WAVES The equation of a standing wave is obtained by adding the displacements of two waves traveling in opposite directions (see figure). Assume that each of the waves has amplitude A, period T, and wavelength λ. If the models for these waves are

$$y_1 = A\cos 2\pi\left(\dfrac{t}{T} - \dfrac{x}{\lambda}\right) \quad \text{and} \quad y_2 = A\cos 2\pi\left(\dfrac{t}{T} + \dfrac{x}{\lambda}\right)$$

show that

$$y_1 + y_2 = 2A\cos\dfrac{2\pi t}{T}\cos\dfrac{2\pi x}{\lambda}.$$

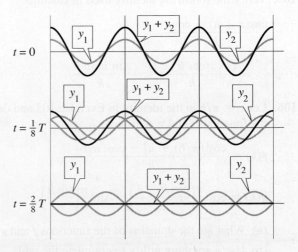

EXPLORATION

TRUE OR FALSE? In Exercises 91–94, determine whether the statement is true or false. Justify your answer.

91. $\sin(u \pm v) = \sin u \cos v \pm \cos u \sin v$

92. $\cos(u \pm v) = \cos u \cos v \pm \sin u \sin v$

93. $\tan\left(x - \dfrac{\pi}{4}\right) = \dfrac{\tan x + 1}{1 - \tan x}$

94. $\sin\left(x - \dfrac{\pi}{2}\right) = -\cos x$

In Exercises 95–98, verify the identity.

95. $\cos(n\pi + \theta) = (-1)^n \cos\theta$, n is an integer

96. $\sin(n\pi + \theta) = (-1)^n \sin\theta$, n is an integer

97. $a \sin B\theta + b \cos B\theta = \sqrt{a^2 + b^2}\, \sin(B\theta + C)$,

where $C = \arctan(b/a)$ and $a > 0$

98. $a \sin B\theta + b \cos B\theta = \sqrt{a^2 + b^2}\, \cos(B\theta - C)$,

where $C = \arctan(a/b)$ and $b > 0$

In Exercises 99–102, use the formulas given in Exercises 97 and 98 to write the trigonometric expression in the following forms.

(a) $\sqrt{a^2 + b^2}\, \sin(B\theta + C)$ (b) $\sqrt{a^2 + b^2}\, \cos(B\theta - C)$

99. $\sin\theta + \cos\theta$ **100.** $3 \sin 2\theta + 4 \cos 2\theta$

101. $12 \sin 3\theta + 5 \cos 3\theta$ **102.** $\sin 2\theta + \cos 2\theta$

In Exercises 103 and 104, use the formulas given in Exercises 97 and 98 to write the trigonometric expression in the form $a \sin B\theta + b \cos B\theta$.

103. $2 \sin\left(\theta + \dfrac{\pi}{4}\right)$ **104.** $5 \cos\left(\theta - \dfrac{\pi}{4}\right)$

105. Verify the following identity used in calculus.

$$\frac{\cos(x + h) - \cos x}{h}$$

$$= \frac{\cos x(\cos h - 1)}{h} - \frac{\sin x \sin h}{h}$$

 106. Let $x = \pi/6$ in the identity in Exercise 105 and define the functions f and g as follows.

$$f(h) = \frac{\cos[(\pi/6) + h] - \cos(\pi/6)}{h}$$

$$g(h) = \cos\frac{\pi}{6}\left(\frac{\cos h - 1}{h}\right) - \sin\frac{\pi}{6}\left(\frac{\sin h}{h}\right)$$

(a) What are the domains of the functions f and g?

(b) Use a graphing utility to complete the table.

h	0.5	0.2	0.1	0.05	0.02	0.01
$f(h)$						
$g(h)$						

(c) Use a graphing utility to graph the functions f and g.

(d) Use the table and the graphs to make a conjecture about the values of the functions f and g as $h \to 0$.

In Exercises 107 and 108, use the figure, which shows two lines whose equations are $y_1 = m_1 x + b_1$ and $y_2 = m_2 x + b_2$. Assume that both lines have positive slopes. Derive a formula for the angle between the two lines. Then use your formula to find the angle between the given pair of lines.

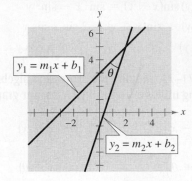

107. $y = x$ and $y = \sqrt{3}\,x$

108. $y = x$ and $y = \dfrac{1}{\sqrt{3}}\,x$

In Exercises 109 and 110, use a graphing utility to graph y_1 and y_2 in the same viewing window. Use the graphs to determine whether $y_1 = y_2$. Explain your reasoning.

109. $y_1 = \cos(x + 2)$, $y_2 = \cos x + \cos 2$

110. $y_1 = \sin(x + 4)$, $y_2 = \sin x + \sin 4$

111. PROOF

(a) Write a proof of the formula for $\sin(u + v)$.

(b) Write a proof of the formula for $\sin(u - v)$.

112. CAPSTONE Give an example to justify each statement.

(a) $\sin(u + v) \neq \sin u + \sin v$

(b) $\sin(u - v) \neq \sin u - \sin v$

(c) $\cos(u + v) \neq \cos u + \cos v$

(d) $\cos(u - v) \neq \cos u - \cos v$

(e) $\tan(u + v) \neq \tan u + \tan v$

(f) $\tan(u - v) \neq \tan u - \tan v$

MULTIPLE-ANGLE AND PRODUCT-TO-SUM FORMULAS

What you should learn

- Use multiple-angle formulas to rewrite and evaluate trigonometric functions.
- Use power-reducing formulas to rewrite and evaluate trigonometric functions.
- Use half-angle formulas to rewrite and evaluate trigonometric functions.
- Use product-to-sum and sum-to-product formulas to rewrite and evaluate trigonometric functions.
- Use trigonometric formulas to rewrite real-life models.

Why you should learn it

You can use a variety of trigonometric formulas to rewrite trigonometric functions in more convenient forms. For instance, in Exercise 135 on page 269, you can use a double-angle formula to determine at what angle an athlete must throw a javelin.

Mark Dadswell/Getty Images

Multiple-Angle Formulas

In this section, you will study four other categories of trigonometric identities.

1. The first category involves *functions of multiple angles* such as sin ku and cos ku.

2. The second category involves *squares of trigonometric functions* such as $\sin^2 u$.

3. The third category involves *functions of half-angles* such as $\sin(u/2)$.

4. The fourth category involves *products of trigonometric functions* such as sin u cos v.

You should learn the **double-angle formulas** because they are used often in trigonometry and calculus. For proofs of these formulas, see Proofs in Mathematics on page 277.

Double-Angle Formulas

$$\sin 2u = 2 \sin u \cos u$$

$$\cos 2u = \cos^2 u - \sin^2 u$$

$$= 2 \cos^2 u - 1$$

$$\tan 2u = \frac{2 \tan u}{1 - \tan^2 u}$$

$$= 1 - 2 \sin^2 u$$

Example 1   Solving a Multiple-Angle Equation

Solve $2 \cos x + \sin 2x = 0$.

Solution

Begin by rewriting the equation so that it involves functions of x (rather than $2x$). Then factor and solve.

$2 \cos x + \sin 2x = 0$	Write original equation.
$2 \cos x + 2 \sin x \cos x = 0$	Double-angle formula
$2 \cos x(1 + \sin x) = 0$	Factor.
$2 \cos x = 0$ and $1 + \sin x = 0$	Set factors equal to zero.
$x = \dfrac{\pi}{2}, \dfrac{3\pi}{2}$ $x = \dfrac{3\pi}{2}$	Solutions in $[0, 2\pi)$

So, the general solution is

$$x = \frac{\pi}{2} + 2n\pi \quad \text{and} \quad x = \frac{3\pi}{2} + 2n\pi$$

where n is an integer. Try verifying these solutions graphically.

CHECK*Point* ▸ Now try Exercise 19.

Example 2   Using Double-Angle Formulas to Analyze Graphs

Use a double-angle formula to rewrite the equation

$$y = 4 \cos^2 x - 2.$$

Then sketch the graph of the equation over the interval $[0, 2\pi]$.

Solution

Using the double-angle formula for $\cos 2u$, you can rewrite the original equation as

$y = 4 \cos^2 x - 2$ Write original equation.

$= 2(2 \cos^2 x - 1)$ Factor.

$= 2 \cos 2x.$ Use double-angle formula.

Using the techniques discussed in Section 1.5, you can recognize that the graph of this function has an amplitude of 2 and a period of π. The key points in the interval $[0, \pi]$ are as follows.

Maximum	Intercept	Minimum	Intercept	Maximum
$(0, 2)$	$\left(\dfrac{\pi}{4}, 0\right)$	$\left(\dfrac{\pi}{2}, -2\right)$	$\left(\dfrac{3\pi}{4}, 0\right)$	$(\pi, 2)$

Two cycles of the graph are shown in Figure 2.14.

CHECK *Point*   Now try Exercise 33.

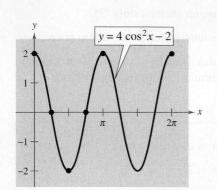

FIGURE **2.14**

Example 3   Evaluating Functions Involving Double Angles

Use the following to find $\sin 2\theta$, $\cos 2\theta$, and $\tan 2\theta$.

$$\cos \theta = \frac{5}{13}, \qquad \frac{3\pi}{2} < \theta < 2\pi$$

Solution

From Figure 2.15, you can see that $\sin \theta = y/r = -12/13$. Consequently, using each of the double-angle formulas, you can write

$$\sin 2\theta = 2 \sin \theta \cos \theta = 2\left(-\frac{12}{13}\right)\left(\frac{5}{13}\right) = -\frac{120}{169}$$

$$\cos 2\theta = 2 \cos^2 \theta - 1 = 2\left(\frac{25}{169}\right) - 1 = -\frac{119}{169}$$

$$\tan 2\theta = \frac{\sin 2\theta}{\cos 2\theta} = \frac{120}{119}.$$

CHECK *Point*   Now try Exercise 37.

The double-angle formulas are not restricted to angles 2θ and θ. Other *double* combinations, such as 4θ and 2θ or 6θ and 3θ, are also valid. Here are two examples.

$$\sin 4\theta = 2 \sin 2\theta \cos 2\theta \qquad \text{and} \qquad \cos 6\theta = \cos^2 3\theta - \sin^2 3\theta$$

By using double-angle formulas together with the sum formulas given in the preceding section, you can form other multiple-angle formulas.

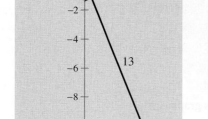

FIGURE **2.15**

| **Example 4** | **Deriving a Triple-Angle Formula** |

$$\sin 3x = \sin(2x + x)$$

$$= \sin 2x \cos x + \cos 2x \sin x$$

$$= 2 \sin x \cos x \cos x + (1 - 2 \sin^2 x) \sin x$$

$$= 2 \sin x \cos^2 x + \sin x - 2 \sin^3 x$$

$$= 2 \sin x(1 - \sin^2 x) + \sin x - 2 \sin^3 x$$

$$= 2 \sin x - 2 \sin^3 x + \sin x - 2 \sin^3 x$$

$$= 3 \sin x - 4 \sin^3 x$$

CHECK*Point*   Now try Exercise 117.

Power-Reducing Formulas

The double-angle formulas can be used to obtain the following **power-reducing formulas.** Example 5 shows a typical power reduction that is used in calculus.

Power-Reducing Formulas

$$\sin^2 u = \frac{1 - \cos 2u}{2} \qquad \cos^2 u = \frac{1 + \cos 2u}{2} \qquad \tan^2 u = \frac{1 - \cos 2u}{1 + \cos 2u}$$

For a proof of the power-reducing formulas, see Proofs in Mathematics on page 277.

| **Example 5** | **Reducing a Power**   ∫ |

Rewrite $\sin^4 x$ as a sum of first powers of the cosines of multiple angles.

Solution

Note the repeated use of power-reducing formulas.

$$\sin^4 x = (\sin^2 x)^2 \qquad\qquad \text{Property of exponents}$$

$$= \left(\frac{1 - \cos 2x}{2}\right)^2 \qquad \text{Power-reducing formula}$$

$$= \frac{1}{4}(1 - 2 \cos 2x + \cos^2 2x) \qquad \text{Expand.}$$

$$= \frac{1}{4}\left(1 - 2 \cos 2x + \frac{1 + \cos 4x}{2}\right) \qquad \text{Power-reducing formula}$$

$$= \frac{1}{4} - \frac{1}{2} \cos 2x + \frac{1}{8} + \frac{1}{8} \cos 4x \qquad \text{Distributive Property}$$

$$= \frac{1}{8}(3 - 4 \cos 2x + \cos 4x) \qquad \text{Factor out common factor.}$$

CHECK*Point*   Now try Exercise 43.

Half-Angle Formulas

You can derive some useful alternative forms of the power-reducing formulas by replacing u with $u/2$. The results are called **half-angle formulas.**

Half-Angle Formulas

$$\sin \frac{u}{2} = \pm \sqrt{\frac{1 - \cos u}{2}}$$

$$\cos \frac{u}{2} = \pm \sqrt{\frac{1 + \cos u}{2}}$$

$$\tan \frac{u}{2} = \frac{1 - \cos u}{\sin u} = \frac{\sin u}{1 + \cos u}$$

The signs of $\sin \dfrac{u}{2}$ and $\cos \dfrac{u}{2}$ depend on the quadrant in which $\dfrac{u}{2}$ lies.

| **Example 6** | **Using a Half-Angle Formula** |

Find the exact value of $\sin 105°$.

Solution

Begin by noting that $105°$ is half of $210°$. Then, using the half-angle formula for $\sin(u/2)$ and the fact that $105°$ lies in Quadrant II, you have

$$\sin 105° = \sqrt{\frac{1 - \cos 210°}{2}}$$

$$= \sqrt{\frac{1 - (-\cos 30°)}{2}}$$

$$= \sqrt{\frac{1 + \left(\sqrt{3}/2\right)}{2}}$$

$$= \frac{\sqrt{2 + \sqrt{3}}}{2}.$$

The positive square root is chosen because $\sin \theta$ is positive in Quadrant II.

CHECKPoint Now try Exercise 59.

Use your calculator to verify the result obtained in Example 6. That is, evaluate $\sin 105°$ and $\left(\sqrt{2 + \sqrt{3}}\right)/2$.

$$\sin 105° \approx 0.9659258$$

$$\frac{\sqrt{2 + \sqrt{3}}}{2} \approx 0.9659258$$

You can see that both values are approximately 0.9659258.

Study Tip

To find the exact value of a trigonometric function with an angle measure in D°M′S″ form using a half-angle formula, first convert the angle measure to decimal degree form. Then multiply the resulting angle measure by 2.

Example 7   Solving a Trigonometric Equation

Find all solutions of $2 - \sin^2 x = 2 \cos^2 \dfrac{x}{2}$ in the interval $[0, 2\pi)$.

Algebraic Solution

$$2 - \sin^2 x = 2 \cos^2 \frac{x}{2} \qquad \text{Write original equation.}$$

$$2 - \sin^2 x = 2\left(\pm\sqrt{\frac{1 + \cos x}{2}}\right)^2 \qquad \text{Half-angle formula}$$

$$2 - \sin^2 x = 2\left(\frac{1 + \cos x}{2}\right) \qquad \text{Simplify.}$$

$$2 - \sin^2 x = 1 + \cos x \qquad \text{Simplify.}$$

$$2 - (1 - \cos^2 x) = 1 + \cos x \qquad \text{Pythagorean identity}$$

$$\cos^2 x - \cos x = 0 \qquad \text{Simplify.}$$

$$\cos x(\cos x - 1) = 0 \qquad \text{Factor.}$$

By setting the factors $\cos x$ and $\cos x - 1$ equal to zero, you find that the solutions in the interval $[0, 2\pi)$ are

$$x = \frac{\pi}{2}, \quad x = \frac{3\pi}{2}, \quad \text{and} \quad x = 0.$$

CHECK*Point*   Now try Exercise 77.

Graphical Solution

Use a graphing utility set in *radian* mode to graph $y = 2 - \sin^2 x - 2 \cos^2(x/2)$, as shown in Figure 2.16. Use the *zero* or *root* feature or the *zoom* and *trace* features to approximate the x-intercepts in the interval $[0, 2\pi)$ to be

$$x = 0,\ x \approx 1.571 \approx \frac{\pi}{2},\text{ and } x \approx 4.712 \approx \frac{3\pi}{2}.$$

These values are the approximate solutions of $2 - \sin^2 x - 2 \cos^2(x/2) = 0$ in the interval $[0, 2\pi)$.

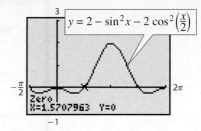

FIGURE 2.16

Product-to-Sum Formulas

Each of the following **product-to-sum formulas** can be verified using the sum and difference formulas discussed in the preceding section.

Product-to-Sum Formulas

$$\sin u \sin v = \frac{1}{2}[\cos(u - v) - \cos(u + v)]$$

$$\cos u \cos v = \frac{1}{2}[\cos(u - v) + \cos(u + v)]$$

$$\sin u \cos v = \frac{1}{2}[\sin(u + v) + \sin(u - v)]$$

$$\cos u \sin v = \frac{1}{2}[\sin(u + v) - \sin(u - v)]$$

Product-to-sum formulas are used in calculus to evaluate integrals involving the products of sines and cosines of two different angles.

| **Example 8** | **Writing Products as Sums** |

Rewrite the product $\cos 5x \sin 4x$ as a sum or difference.

Solution

Using the appropriate product-to-sum formula, you obtain

$$\cos 5x \sin 4x = \tfrac{1}{2}[\sin(5x + 4x) - \sin(5x - 4x)]$$

$$= \tfrac{1}{2}\sin 9x - \tfrac{1}{2}\sin x.$$

CHECK Point Now try Exercise 85.

Occasionally, it is useful to reverse the procedure and write a sum of trigonometric functions as a product. This can be accomplished with the following **sum-to-product formulas.**

Sum-to-Product Formulas

$$\sin u + \sin v = 2 \sin\left(\frac{u + v}{2}\right)\cos\left(\frac{u - v}{2}\right)$$

$$\sin u - \sin v = 2 \cos\left(\frac{u + v}{2}\right)\sin\left(\frac{u - v}{2}\right)$$

$$\cos u + \cos v = 2 \cos\left(\frac{u + v}{2}\right)\cos\left(\frac{u - v}{2}\right)$$

$$\cos u - \cos v = -2 \sin\left(\frac{u + v}{2}\right)\sin\left(\frac{u - v}{2}\right)$$

For a proof of the sum-to-product formulas, see Proofs in Mathematics on page 278.

| **Example 9** | **Using a Sum-to-Product Formula** |

Find the exact value of $\cos 195° + \cos 105°$.

Solution

Using the appropriate sum-to-product formula, you obtain

$$\cos 195° + \cos 105° = 2 \cos\left(\frac{195° + 105°}{2}\right)\cos\left(\frac{195° - 105°}{2}\right)$$

$$= 2 \cos 150° \cos 45°$$

$$= 2\left(-\frac{\sqrt{3}}{2}\right)\left(\frac{\sqrt{2}}{2}\right)$$

$$= -\frac{\sqrt{6}}{2}.$$

CHECK Point Now try Exercise 99.

Example 10   Solving a Trigonometric Equation

Solve $\sin 5x + \sin 3x = 0$.

Algebraic Solution

$$\sin 5x + \sin 3x = 0 \qquad \text{Write original equation.}$$

$$2 \sin\left(\frac{5x + 3x}{2}\right) \cos\left(\frac{5x - 3x}{2}\right) = 0 \qquad \text{Sum-to-product formula}$$

$$2 \sin 4x \cos x = 0 \qquad \text{Simplify.}$$

By setting the factor $2 \sin 4x$ equal to zero, you can find that the solutions in the interval $[0, 2\pi)$ are

$$x = 0, \frac{\pi}{4}, \frac{\pi}{2}, \frac{3\pi}{4}, \pi, \frac{5\pi}{4}, \frac{3\pi}{2}, \frac{7\pi}{4}.$$

The equation $\cos x = 0$ yields no additional solutions, so you can conclude that the solutions are of the form

$$x = \frac{n\pi}{4}$$

where n is an integer.

Graphical Solution

Sketch the graph of

$$y = \sin 5x + \sin 3x,$$

as shown in Figure 2.17. From the graph you can see that the x-intercepts occur at multiples of $\pi/4$. So, you can conclude that the solutions are of the form

$$x = \frac{n\pi}{4}$$

where n is an integer.

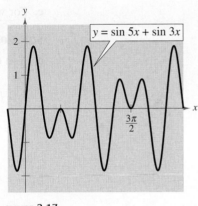

FIGURE 2.17

CHECK Point   Now try Exercise 103.

Example 11   Verifying a Trigonometric Identity

Verify the identity $\dfrac{\sin 3x - \sin x}{\cos x + \cos 3x} = \tan x$.

Solution

Using appropriate sum-to-product formulas, you have

$$\frac{\sin 3x - \sin x}{\cos x + \cos 3x} = \frac{2 \cos\left(\dfrac{3x + x}{2}\right) \sin\left(\dfrac{3x - x}{2}\right)}{2 \cos\left(\dfrac{x + 3x}{2}\right) \cos\left(\dfrac{x - 3x}{2}\right)}$$

$$= \frac{2 \cos(2x) \sin x}{2 \cos(2x) \cos(-x)}$$

$$= \frac{\sin x}{\cos(-x)}$$

$$= \frac{\sin x}{\cos x} = \tan x.$$

CHECK Point   Now try Exercise 121.

Application

Example 12 Projectile Motion

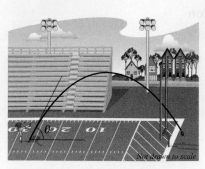

FIGURE 2.18

Ignoring air resistance, the range of a projectile fired at an angle θ with the horizontal and with an initial velocity of v_0 feet per second is given by

$$r = \frac{1}{16}v_0^2 \sin \theta \cos \theta$$

where r is the horizontal distance (in feet) that the projectile will travel. A place kicker for a football team can kick a football from ground level with an initial velocity of 80 feet per second (see Figure 2.18).

a. Write the projectile motion model in a simpler form.

b. At what angle must the player kick the football so that the football travels 200 feet?

c. For what angle is the horizontal distance the football travels a maximum?

Solution

a. You can use a double-angle formula to rewrite the projectile motion model as

$$r = \frac{1}{32}v_0^2(2 \sin \theta \cos \theta) \qquad \text{Rewrite original projectile motion model.}$$

$$= \frac{1}{32}v_0^2 \sin 2\theta. \qquad \text{Rewrite model using a double-angle formula.}$$

b.
$$r = \frac{1}{32}v_0^2 \sin 2\theta \qquad \text{Write projectile motion model.}$$

$$200 = \frac{1}{32}(80)^2 \sin 2\theta \qquad \text{Substitute 200 for } r \text{ and 80 for } v_0.$$

$$200 = 200 \sin 2\theta \qquad \text{Simplify.}$$

$$1 = \sin 2\theta \qquad \text{Divide each side by 200.}$$

You know that $2\theta = \pi/2$, so dividing this result by 2 produces $\theta = \pi/4$. Because $\pi/4 = 45°$, you can conclude that the player must kick the football at an angle of $45°$ so that the football will travel 200 feet.

c. From the model $r = 200 \sin 2\theta$ you can see that the amplitude is 200. So the maximum range is $r = 200$ feet. From part (b), you know that this corresponds to an angle of $45°$. Therefore, kicking the football at an angle of $45°$ will produce a maximum horizontal distance of 200 feet.

CHECK Point Now try Exercise 135.

CLASSROOM DISCUSSION

Deriving an Area Formula Describe how you can use a double-angle formula or a half-angle formula to derive a formula for the area of an isosceles triangle. Use a labeled sketch to illustrate your derivation. Then write two examples that show how your formula can be used.

2.5 **EXERCISES**

See www.CalcChat.com for worked-out solutions to odd-numbered exercises.

VOCABULARY: Fill in the blank to complete the trigonometric formula.

1. $\sin 2u = $ _____

2. $\dfrac{1 + \cos 2u}{2} = $ _____

3. $\cos 2u = $ _____

4. $\dfrac{1 - \cos 2u}{1 + \cos 2u} = $ _____

5. $\sin \dfrac{u}{2} = $ _____

6. $\tan \dfrac{u}{2} = $ _____

7. $\cos u \cos v = $ _____

8. $\sin u \cos v = $ _____

9. $\sin u + \sin v = $ _____

10. $\cos u - \cos v = $ _____

SKILLS AND APPLICATIONS

In Exercises 11–18, use the figure to find the exact value of the trigonometric function.

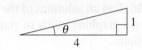

11. $\cos 2\theta$

12. $\sin 2\theta$

13. $\tan 2\theta$

14. $\sec 2\theta$

15. $\csc 2\theta$

16. $\cot 2\theta$

17. $\sin 4\theta$

18. $\tan 4\theta$

In Exercises 19–28, find the exact solutions of the equation in the interval $[0, 2\pi)$.

19. $\sin 2x - \sin x = 0$

20. $\sin 2x + \cos x = 0$

21. $4 \sin x \cos x = 1$

22. $\sin 2x \sin x = \cos x$

23. $\cos 2x - \cos x = 0$

24. $\cos 2x + \sin x = 0$

25. $\sin 4x = -2 \sin 2x$

26. $(\sin 2x + \cos 2x)^2 = 1$

27. $\tan 2x - \cot x = 0$

28. $\tan 2x - 2 \cos x = 0$

In Exercises 29–36, use a double-angle formula to rewrite the expression.

29. $6 \sin x \cos x$

30. $\sin x \cos x$

31. $6 \cos^2 x - 3$

32. $\cos^2 x - \dfrac{1}{2}$

33. $4 - 8 \sin^2 x$

34. $10 \sin^2 x - 5$

35. $(\cos x + \sin x)(\cos x - \sin x)$

36. $(\sin x - \cos x)(\sin x + \cos x)$

In Exercises 37–42, find the exact values of $\sin 2u$, $\cos 2u$, and $\tan 2u$ using the double-angle formulas.

37. $\sin u = -\dfrac{3}{5}, \quad \dfrac{3\pi}{2} < u < 2\pi$

38. $\cos u = -\dfrac{4}{5}, \quad \dfrac{\pi}{2} < u < \pi$

39. $\tan u = \dfrac{3}{5}, \quad 0 < u < \dfrac{\pi}{2}$

40. $\cot u = \sqrt{2}, \quad \pi < u < \dfrac{3\pi}{2}$

41. $\sec u = -2, \quad \dfrac{\pi}{2} < u < \pi$

42. $\csc u = 3, \quad \dfrac{\pi}{2} < u < \pi$

In Exercises 43–52, use the power-reducing formulas to rewrite the expression in terms of the first power of the cosine.

43. $\cos^4 x$

44. $\sin^4 2x$

45. $\cos^4 2x$

46. $\sin^8 x$

47. $\tan^4 2x$

48. $\sin^2 x \cos^4 x$

49. $\sin^2 2x \cos^2 2x$

50. $\tan^2 2x \cos^4 2x$

51. $\sin^4 x \cos^2 x$

52. $\sin^4 x \cos^4 x$

In Exercises 53–58, use the figure to find the exact value of the trigonometric function.

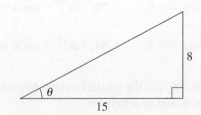

53. $\cos \dfrac{\theta}{2}$

54. $\sin \dfrac{\theta}{2}$

55. $\tan \dfrac{\theta}{2}$

56. $\sec \dfrac{\theta}{2}$

57. $\csc \dfrac{\theta}{2}$

58. $\cot \dfrac{\theta}{2}$

In Exercises 59–66, use the half-angle formulas to determine the exact values of the sine, cosine, and tangent of the angle.

59. $75°$ **60.** $165°$

61. $112° \, 30'$ **62.** $67° \, 30'$

63. $\pi/8$ **64.** $\pi/12$

65. $3\pi/8$ **66.** $7\pi/12$

In Exercises 67–72, (a) determine the quadrant in which $u/2$ lies, and (b) find the exact values of $\sin(u/2)$, $\cos(u/2)$, and $\tan(u/2)$ using the half-angle formulas.

67. $\cos u = \dfrac{7}{25}, \quad 0 < u < \dfrac{\pi}{2}$

68. $\sin u = \dfrac{5}{13}, \quad \dfrac{\pi}{2} < u < \pi$

69. $\tan u = -\dfrac{5}{12}, \quad \dfrac{3\pi}{2} < u < 2\pi$

70. $\cot u = 3, \quad \pi < u < \dfrac{3\pi}{2}$

71. $\csc u = -\dfrac{5}{3}, \quad \pi < u < \dfrac{3\pi}{2}$

72. $\sec u = \dfrac{7}{2}, \quad \dfrac{3\pi}{2} < u < 2\pi$

In Exercises 73–76, use the half-angle formulas to simplify the expression.

73. $\sqrt{\dfrac{1 - \cos 6x}{2}}$ **74.** $\sqrt{\dfrac{1 + \cos 4x}{2}}$

75. $-\sqrt{\dfrac{1 - \cos 8x}{1 + \cos 8x}}$ **76.** $-\sqrt{\dfrac{1 - \cos(x - 1)}{2}}$

 In Exercises 77–80, find all solutions of the equation in the interval $[0, 2\pi)$. Use a graphing utility to graph the equation and verify the solutions.

77. $\sin \dfrac{x}{2} + \cos x = 0$ **78.** $\sin \dfrac{x}{2} + \cos x - 1 = 0$

79. $\cos \dfrac{x}{2} - \sin x = 0$ **80.** $\tan \dfrac{x}{2} - \sin x = 0$

In Exercises 81–90, use the product-to-sum formulas to write the product as a sum or difference.

81. $\sin \dfrac{\pi}{3} \cos \dfrac{\pi}{6}$ **82.** $4 \cos \dfrac{\pi}{3} \sin \dfrac{5\pi}{6}$

83. $10 \cos 75° \cos 15°$ **84.** $6 \sin 45° \cos 15°$

85. $\sin 5\theta \sin 3\theta$ **86.** $3 \sin(-4\alpha) \sin 6\alpha$

87. $7 \cos(-5\beta) \sin 3\beta$ **88.** $\cos 2\theta \cos 4\theta$

89. $\sin(x + y) \sin(x - y)$ **90.** $\sin(x + y) \cos(x - y)$

In Exercises 91–98, use the sum-to-product formulas to write the sum or difference as a product.

91. $\sin 3\theta + \sin \theta$ **92.** $\sin 5\theta - \sin 3\theta$

93. $\cos 6x + \cos 2x$ **94.** $\cos x + \cos 4x$

95. $\sin(\alpha + \beta) - \sin(\alpha - \beta)$ **96.** $\cos(\phi + 2\pi) + \cos \phi$

97. $\cos\left(\theta + \dfrac{\pi}{2}\right) - \cos\left(\theta - \dfrac{\pi}{2}\right)$

98. $\sin\left(x + \dfrac{\pi}{2}\right) + \sin\left(x - \dfrac{\pi}{2}\right)$

In Exercises 99–102, use the sum-to-product formulas to find the exact value of the expression.

99. $\sin 75° + \sin 15°$ **100.** $\cos 120° + \cos 60°$

101. $\cos \dfrac{3\pi}{4} - \cos \dfrac{\pi}{4}$ **102.** $\sin \dfrac{5\pi}{4} - \sin \dfrac{3\pi}{4}$

 In Exercises 103–106, find all solutions of the equation in the interval $[0, 2\pi)$. Use a graphing utility to graph the equation and verify the solutions.

103. $\sin 6x + \sin 2x = 0$ **104.** $\cos 2x - \cos 6x = 0$

105. $\dfrac{\cos 2x}{\sin 3x - \sin x} - 1 = 0$ **106.** $\sin^2 3x - \sin^2 x = 0$

In Exercises 107–110, use the figure to find the exact value of the trigonometric function.

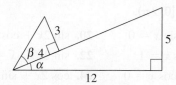

107. $\sin 2\alpha$ **108.** $\cos 2\beta$

109. $\cos(\beta/2)$ **110.** $\sin(\alpha + \beta)$

In Exercises 111–124, verify the identity.

111. $\csc 2\theta = \dfrac{\csc \theta}{2 \cos \theta}$ **112.** $\sec 2\theta = \dfrac{\sec^2 \theta}{2 - \sec^2 \theta}$

113. $\sin \dfrac{\alpha}{3} \cos \dfrac{\alpha}{3} = \dfrac{1}{2} \sin \dfrac{2\alpha}{3}$ **114.** $\dfrac{\cos 3\beta}{\cos \beta} = 1 - 4 \sin^2 \beta$

115. $1 + \cos 10y = 2 \cos^2 5y$

116. $\cos^4 x - \sin^4 x = \cos 2x$

117. $\cos 4\alpha = \cos^2 2\alpha - \sin^2 2\alpha$

118. $(\sin x + \cos x)^2 = 1 + \sin 2x$

119. $\tan \dfrac{u}{2} = \csc u - \cot u$

120. $\sec \dfrac{u}{2} = \pm\sqrt{\dfrac{2 \tan u}{\tan u + \sin u}}$

121. $\dfrac{\cos 4x + \cos 2x}{\sin 4x + \sin 2x} = \cot 3x$

122. $\dfrac{\sin x \pm \sin y}{\cos x + \cos y} = \tan \dfrac{x \pm y}{2}$

123. $\sin\left(\dfrac{\pi}{6} + x\right) + \sin\left(\dfrac{\pi}{6} - x\right) = \cos x$

124. $\cos\left(\dfrac{\pi}{3} + x\right) + \cos\left(\dfrac{\pi}{3} - x\right) = \cos x$

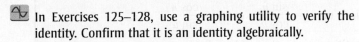 In Exercises 125–128, use a graphing utility to verify the identity. Confirm that it is an identity algebraically.

125. $\cos 3\beta = \cos^3 \beta - 3 \sin^2 \beta \cos \beta$

126. $\sin 4\beta = 4 \sin \beta \cos \beta(1 - 2 \sin^2 \beta)$

127. $(\cos 4x - \cos 2x)/(2 \sin 3x) = -\sin x$

128. $(\cos 3x - \cos x)/(\sin 3x - \sin x) = -\tan 2x$

In Exercises 129 and 130, graph the function by hand in the interval $[0, 2\pi]$ by using the power-reducing formulas.

129. $f(x) = \sin^2 x$ 130. $f(x) = \cos^2 x$

In Exercises 131–134, write the trigonometric expression as an algebraic expression.

131. $\sin(2 \arcsin x)$ 132. $\cos(2 \arccos x)$

133. $\cos(2 \arcsin x)$ 134. $\sin(2 \arccos x)$

135. **PROJECTILE MOTION** The range of a projectile fired at an angle θ with the horizontal and with an initial velocity of v_0 feet per second is

$$r = \dfrac{1}{32} v_0^2 \sin 2\theta$$

where r is measured in feet. An athlete throws a javelin at 75 feet per second. At what angle must the athlete throw the javelin so that the javelin travels 130 feet?

136. **RAILROAD TRACK** When two railroad tracks merge, the overlapping portions of the tracks are in the shapes of circular arcs (see figure). The radius of each arc r (in feet) and the angle θ are related by

$$\dfrac{x}{2} = 2r \sin^2 \dfrac{\theta}{2}.$$

Write a formula for x in terms of $\cos \theta$.

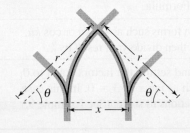

137. **MACH NUMBER** The mach number M of an airplane is the ratio of its speed to the speed of sound. When an airplane travels faster than the speed of sound, the sound waves form a cone behind the airplane (see figure). The mach number is related to the apex angle θ of the cone by $\sin(\theta/2) = 1/M$.

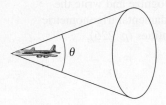

(a) Find the angle θ that corresponds to a mach number of 1.

(b) Find the angle θ that corresponds to a mach number of 4.5.

(c) The speed of sound is about 760 miles per hour. Determine the speed of an object with the mach numbers from parts (a) and (b).

(d) Rewrite the equation in terms of θ.

EXPLORATION

138. **CAPSTONE** Consider the function given by $f(x) = \sin^4 x + \cos^4 x$.

(a) Use the power-reducing formulas to write the function in terms of cosine to the first power.

(b) Determine another way of rewriting the function. Use a graphing utility to rule out incorrectly rewritten functions.

(c) Add a trigonometric term to the function so that it becomes a perfect square trinomial. Rewrite the function as a perfect square trinomial minus the term that you added. Use a graphing utility to rule out incorrectly rewritten functions.

(d) Rewrite the result of part (c) in terms of the sine of a double angle. Use a graphing utility to rule out incorrectly rewritten functions.

(e) When you rewrite a trigonometric expression, the result may not be the same as a friend's. Does this mean that one of you is wrong? Explain.

TRUE OR FALSE? In Exercises 139 and 140, determine whether the statement is true or false. Justify your answer.

139. Because the sine function is an odd function, for a negative number u, $\sin 2u = -2 \sin u \cos u$.

140. $\sin \dfrac{u}{2} = -\sqrt{\dfrac{1 - \cos u}{2}}$ when u is in the second quadrant.

2 CHAPTER SUMMARY

What Did You Learn?	Explanation/Examples	Review Exercises
Section 2.1 Recognize and write the fundamental trigonometric identities (p. 226).	**Reciprocal Identities** $\sin u = 1/\csc u \qquad \cos u = 1/\sec u \qquad \tan u = 1/\cot u$ $\csc u = 1/\sin u \qquad \sec u = 1/\cos u \qquad \cot u = 1/\tan u$ **Quotient Identities:** $\tan u = \dfrac{\sin u}{\cos u}, \quad \cot u = \dfrac{\cos u}{\sin u}$ **Pythagorean Identities:** $\sin^2 u + \cos^2 u = 1,$ $1 + \tan^2 u = \sec^2 u, \quad 1 + \cot^2 u = \csc^2 u$ **Cofunction Identities** $\sin[(\pi/2) - u] = \cos u \qquad \cos[(\pi/2) - u] = \sin u$ $\tan[(\pi/2) - u] = \cot u \qquad \cot[(\pi/2) - u] = \tan u$ $\sec[(\pi/2) - u] = \csc u \qquad \csc[(\pi/2) - u] = \sec u$ **Even/Odd Identities** $\sin(-u) = -\sin u \quad \cos(-u) = \cos u \quad \tan(-u) = -\tan u$ $\csc(-u) = -\csc u \quad \sec(-u) = \sec u \quad \cot(-u) = -\cot u$	1–6
Use the fundamental trigonometric identities to evaluate trigonometric functions, and simplify and rewrite trigonometric expressions (p. 227).	In some cases, when factoring or simplifying trigonometric expressions, it is helpful to rewrite the expression in terms of just *one* trigonometric function or in terms of *sine and cosine only.*	7–28
Section 2.2 Verify trigonometric identities (p. 234).	**Guidelines for Verifying Trigonometric Identities** **1.** Work with one side of the equation at a time. **2.** Look to factor an expression, add fractions, square a binomial, or create a monomial denominator. **3.** Look to use the fundamental identities. Note which functions are in the final expression you want. Sines and cosines pair up well, as do secants and tangents, and cosecants and cotangents. **4.** If the preceding guidelines do not help, try converting all terms to sines and cosines. **5.** Always try *something*.	29–36
Section 2.3 Use standard algebraic techniques to solve trigonometric equations (p. 241).	Use standard algebraic techniques such as collecting like terms, extracting square roots, and factoring to solve trigonometric equations.	37–42
Solve trigonometric equations of quadratic type (p. 243).	To solve trigonometric equations of quadratic type $ax^2 + bx + c = 0$, factor the quadratic or, if this is not possible, use the Quadratic Formula.	43–46
Solve trigonometric equations involving multiple angles (p. 246).	To solve equations that contain forms such as $\sin ku$ or $\cos ku$, first solve the equation for ku, then divide your result by k.	47–52
Use inverse trigonometric functions to solve trigonometric equations (p. 247).	After factoring an equation and setting the factors equal to 0, you may get an equation such as $\tan x - 3 = 0$. In this case, use inverse trigonometric functions to solve. (See Example 9.)	53–56

	What Did You Learn?	Explanation/Examples	Review Exercises
Section 2.4	Use sum and difference formulas to evaluate trigonometric functions, verify identities, and solve trigonometric equations *(p. 252)*.	**Sum and Difference Formulas** $\sin(u + v) = \sin u \cos v + \cos u \sin v$ $\sin(u - v) = \sin u \cos v - \cos u \sin v$ $\cos(u + v) = \cos u \cos v - \sin u \sin v$ $\cos(u - v) = \cos u \cos v + \sin u \sin v$ $\tan(u + v) = \dfrac{\tan u + \tan v}{1 - \tan u \tan v}$ $\tan(u - v) = \dfrac{\tan u - \tan v}{1 + \tan u \tan v}$	57–80
Section 2.5	Use multiple-angle formulas to rewrite and evaluate trigonometric functions *(p. 259)*.	**Double-Angle Formulas** $\sin 2u = 2 \sin u \cos u \qquad \cos 2u = \cos^2 u - \sin^2 u$ $\tan 2u = \dfrac{2 \tan u}{1 - \tan^2 u} \qquad \qquad = 2 \cos^2 u - 1$ $\qquad \qquad \qquad \qquad \qquad = 1 - 2 \sin^2 u$	81–86
	Use power-reducing formulas to rewrite and evaluate trigonometric functions *(p. 261)*.	**Power-Reducing Formulas** $\sin^2 u = \dfrac{1 - \cos 2u}{2}, \quad \cos^2 u = \dfrac{1 + \cos 2u}{2}$ $\tan^2 u = \dfrac{1 - \cos 2u}{1 + \cos 2u}$	87–90
	Use half-angle formulas to rewrite and evaluate trigonometric functions *(p. 262)*.	**Half-Angle Formulas** $\sin \dfrac{u}{2} = \pm \sqrt{\dfrac{1 - \cos u}{2}}, \quad \cos \dfrac{u}{2} = \pm \sqrt{\dfrac{1 + \cos u}{2}}$ $\tan \dfrac{u}{2} = \dfrac{1 - \cos u}{\sin u} = \dfrac{\sin u}{1 + \cos u}$ The signs of $\sin(u/2)$ and $\cos(u/2)$ depend on the quadrant in which $u/2$ lies.	91–100
	Use product-to-sum formulas *(p. 263)* and sum-to-product formulas *(p. 264)* to rewrite and evaluate trigonometric functions.	**Product-to-Sum Formulas** $\sin u \sin v = (1/2)[\cos(u - v) - \cos(u + v)]$ $\cos u \cos v = (1/2)[\cos(u - v) + \cos(u + v)]$ $\sin u \cos v = (1/2)[\sin(u + v) + \sin(u - v)]$ $\cos u \sin v = (1/2)[\sin(u + v) - \sin(u - v)]$ **Sum-to-Product Formulas** $\sin u + \sin v = 2 \sin\left(\dfrac{u + v}{2}\right)\cos\left(\dfrac{u - v}{2}\right)$ $\sin u - \sin v = 2 \cos\left(\dfrac{u + v}{2}\right)\sin\left(\dfrac{u - v}{2}\right)$ $\cos u + \cos v = 2 \cos\left(\dfrac{u + v}{2}\right)\cos\left(\dfrac{u - v}{2}\right)$ $\cos u - \cos v = -2 \sin\left(\dfrac{u + v}{2}\right)\sin\left(\dfrac{u - v}{2}\right)$	101–108
	Use trigonometric formulas to rewrite real-life models *(p. 266)*.	A trigonometric formula can be used to rewrite the projectile motion model $r = (1/16)v_0^2 \sin \theta \cos \theta$. (See Example 12.)	109–114

2 REVIEW EXERCISES

See www.CalcChat.com for worked-out solutions to odd-numbered exercises.

2.1 In Exercises 1–6, name the trigonometric function that is equivalent to the expression.

1. $\dfrac{\sin x}{\cos x}$

2. $\dfrac{1}{\sin x}$

3. $\dfrac{1}{\sec x}$

4. $\dfrac{1}{\tan x}$

5. $\sqrt{\cot^2 x + 1}$

6. $\sqrt{1 + \tan^2 x}$

In Exercises 7–10, use the given values and trigonometric identities to evaluate (if possible) all six trigonometric functions.

7. $\sin x = \frac{5}{13}, \quad \cos x = \frac{12}{13}$

8. $\tan \theta = \dfrac{2}{3}, \quad \sec \theta = \dfrac{\sqrt{13}}{3}$

9. $\sin\left(\dfrac{\pi}{2} - x\right) = \dfrac{\sqrt{2}}{2}, \quad \sin x = -\dfrac{\sqrt{2}}{2}$

10. $\csc\left(\dfrac{\pi}{2} - \theta\right) = 9, \quad \sin \theta = \dfrac{4\sqrt{5}}{9}$

In Exercises 11–24, use the fundamental trigonometric identities to simplify the expression.

11. $\dfrac{1}{\cot^2 x + 1}$

12. $\dfrac{\tan \theta}{1 - \cos^2 \theta}$

13. $\tan^2 x(\csc^2 x - 1)$

14. $\cot^2 x(\sin^2 x)$

15. $\dfrac{\sin\left(\dfrac{\pi}{2} - \theta\right)}{\sin \theta}$

16. $\dfrac{\cot\left(\dfrac{\pi}{2} - u\right)}{\cos u}$

17. $\dfrac{\sin^2 \theta + \cos^2 \theta}{\sin \theta}$

18. $\dfrac{\sec^2(-\theta)}{\csc^2 \theta}$

19. $\cos^2 x + \cos^2 x \cot^2 x$

20. $\tan^2 \theta \csc^2 \theta - \tan^2 \theta$

21. $(\tan x + 1)^2 \cos x$

22. $(\sec x - \tan x)^2$

23. $\dfrac{1}{\csc \theta + 1} - \dfrac{1}{\csc \theta - 1}$

24. $\dfrac{\tan^2 x}{1 + \sec x}$

In Exercises 25 and 26, use the trigonometric substitution to write the algebraic expression as a trigonometric function of θ, where $0 < \theta < \pi/2$.

25. $\sqrt{25 - x^2}, x = 5 \sin \theta$

26. $\sqrt{x^2 - 16}, x = 4 \sec \theta$

∮ **27. RATE OF CHANGE** The rate of change of the function $f(x) = \csc x - \cot x$ is given by the expression $\csc^2 x - \csc x \cot x$. Show that this expression can also be written as

$$\dfrac{1 - \cos x}{\sin^2 x}.$$

∮ **28. RATE OF CHANGE** The rate of change of the function $f(x) = 2\sqrt{\sin x}$ is given by the expression $\sin^{-1/2} x \cos x$. Show that this expression can also be written as $\cot x \sqrt{\sin x}$.

2.2 In Exercises 29–36, verify the identity.

29. $\cos x(\tan^2 x + 1) = \sec x$

30. $\sec^2 x \cot x - \cot x = \tan x$

31. $\sec\left(\dfrac{\pi}{2} - \theta\right) = \csc \theta$

32. $\cot\left(\dfrac{\pi}{2} - x\right) = \tan x$

33. $\dfrac{1}{\tan \theta \csc \theta} = \cos \theta$

34. $\dfrac{1}{\tan x \csc x \sin x} = \cot x$

35. $\sin^5 x \cos^2 x = (\cos^2 x - 2 \cos^4 x + \cos^6 x) \sin x$

36. $\cos^3 x \sin^2 x = (\sin^2 x - \sin^4 x) \cos x$

2.3 In Exercises 37–42, solve the equation.

37. $\sin x = \sqrt{3} - \sin x$

38. $4 \cos \theta = 1 + 2 \cos \theta$

39. $3\sqrt{3} \tan u = 3$

40. $\frac{1}{2} \sec x - 1 = 0$

41. $3 \csc^2 x = 4$

42. $4 \tan^2 u - 1 = \tan^2 u$

In Exercises 43–52, find all solutions of the equation in the interval $[0, 2\pi)$.

43. $2 \cos^2 x - \cos x = 1$

44. $2 \sin^2 x - 3 \sin x = -1$

45. $\cos^2 x + \sin x = 1$

46. $\sin^2 x + 2 \cos x = 2$

47. $2 \sin 2x - \sqrt{2} = 0$

48. $2 \cos \dfrac{x}{2} + 1 = 0$

49. $3 \tan^2\left(\dfrac{x}{3}\right) - 1 = 0$

50. $\sqrt{3} \tan 3x = 0$

51. $\cos 4x(\cos x - 1) = 0$

52. $3 \csc^2 5x = -4$

In Exercises 53–56, use inverse functions where needed to find all solutions of the equation in the interval $[0, 2\pi)$.

53. $\sin^2 x - 2 \sin x = 0$

54. $2 \cos^2 x + 3 \cos x = 0$

55. $\tan^2 \theta + \tan \theta - 6 = 0$

56. $\sec^2 x + 6 \tan x + 4 = 0$

2.4 In Exercises 57–60, find the exact values of the sine, cosine, and tangent of the angle.

57. $285° = 315° - 30°$

58. $345° = 300° + 45°$

59. $\dfrac{25\pi}{12} = \dfrac{11\pi}{6} + \dfrac{\pi}{4}$

60. $\dfrac{19\pi}{12} = \dfrac{11\pi}{6} - \dfrac{\pi}{4}$

In Exercises 61–64, write the expression as the sine, cosine, or tangent of an angle.

61. $\sin 60° \cos 45° - \cos 60° \sin 45°$

62. $\cos 45° \cos 120° - \sin 45° \sin 120°$

63. $\dfrac{\tan 25° + \tan 10°}{1 - \tan 25° \tan 10°}$

64. $\dfrac{\tan 68° - \tan 115°}{1 + \tan 68° \tan 115°}$

In Exercises 65–70, find the exact value of the trigonometric function given that $\tan u = \frac{3}{4}$ and $\cos v = -\frac{4}{5}$. (u is in Quadrant I and v is in Quadrant III.)

65. $\sin(u + v)$ **66.** $\tan(u + v)$

67. $\cos(u - v)$ **68.** $\sin(u - v)$

69. $\cos(u + v)$ **70.** $\tan(u - v)$

In Exercises 71–76, verify the identity.

71. $\cos\left(x + \dfrac{\pi}{2}\right) = -\sin x$ **72.** $\sin\left(x - \dfrac{3\pi}{2}\right) = \cos x$

73. $\tan\left(x - \dfrac{\pi}{2}\right) = -\cot x$ **74.** $\tan(\pi - x) = -\tan x$

75. $\cos 3x = 4 \cos^3 x - 3 \cos x$

76. $\dfrac{\sin(\alpha - \beta)}{\sin(\alpha + \beta)} = \dfrac{\tan \alpha - \tan \beta}{\tan \alpha + \tan \beta}$

In Exercises 77–80, find all solutions of the equation in the interval $[0, 2\pi)$.

77. $\sin\left(x + \dfrac{\pi}{4}\right) - \sin\left(x - \dfrac{\pi}{4}\right) = 1$

78. $\cos\left(x + \dfrac{\pi}{6}\right) - \cos\left(x - \dfrac{\pi}{6}\right) = 1$

79. $\sin\left(x + \dfrac{\pi}{2}\right) - \sin\left(x - \dfrac{\pi}{2}\right) = \sqrt{3}$

80. $\cos\left(x + \dfrac{3\pi}{4}\right) - \cos\left(x - \dfrac{3\pi}{4}\right) = 0$

2.5 In Exercises 81–84, find the exact values of $\sin 2u$, $\cos 2u$, and $\tan 2u$ using the double-angle formulas.

81. $\sin u = -\dfrac{4}{5}, \quad \pi < u < \dfrac{3\pi}{2}$

82. $\cos u = -\dfrac{2}{\sqrt{5}}, \quad \dfrac{\pi}{2} < u < \pi$

83. $\sec u = -3, \quad \dfrac{\pi}{2} < u < \pi$

84. $\cot u = 2, \quad \pi < u < \dfrac{3\pi}{2}$

In Exercises 85 and 86, use double-angle formulas to verify the identity algebraically and use a graphing utility to confirm your result graphically.

85. $\sin 4x = 8 \cos^3 x \sin x - 4 \cos x \sin x$

86. $\tan^2 x = \dfrac{1 - \cos 2x}{1 + \cos 2x}$

In Exercises 87–90, use the power-reducing formulas to rewrite the expression in terms of the first power of the cosine.

87. $\tan^2 2x$ **88.** $\cos^2 3x$

89. $\sin^2 x \tan^2 x$ **90.** $\cos^2 x \tan^2 x$

In Exercises 91–94, use the half-angle formulas to determine the exact values of the sine, cosine, and tangent of the angle.

91. $-75°$ **92.** $15°$

93. $\dfrac{19\pi}{12}$ **94.** $-\dfrac{17\pi}{12}$

In Exercises 95–98, (a) determine the quadrant in which $u/2$ lies, and (b) find the exact values of $\sin(u/2)$, $\cos(u/2)$, and $\tan(u/2)$ using the half-angle formulas.

95. $\sin u = \frac{7}{25}, \ 0 < u < \pi/2$

96. $\tan u = \frac{4}{3}, \ \pi < u < 3\pi/2$

97. $\cos u = -\frac{2}{7}, \ \pi/2 < u < \pi$

98. $\sec u = -6, \ \pi/2 < u < \pi$

In Exercises 99 and 100, use the half-angle formulas to simplify the expression.

99. $-\sqrt{\dfrac{1 + \cos 10x}{2}}$ **100.** $\dfrac{\sin 6x}{1 + \cos 6x}$

In Exercises 101–104, use the product-to-sum formulas to write the product as a sum or difference.

101. $\cos \dfrac{\pi}{6} \sin \dfrac{\pi}{6}$ **102.** $6 \sin 15° \sin 45°$

103. $\cos 4\theta \sin 6\theta$ **104.** $2 \sin 7\theta \cos 3\theta$

In Exercises 105–108, use the sum-to-product formulas to write the sum or difference as a product.

105. $\sin 4\theta - \sin 8\theta$

106. $\cos 6\theta + \cos 5\theta$

107. $\cos\left(x + \dfrac{\pi}{6}\right) - \cos\left(x - \dfrac{\pi}{6}\right)$

108. $\sin\left(x + \dfrac{\pi}{4}\right) - \sin\left(x - \dfrac{\pi}{4}\right)$

109. PROJECTILE MOTION A baseball leaves the hand of the player at first base at an angle of θ with the horizontal and at an initial velocity of $v_0 = 80$ feet per second. The ball is caught by the player at second base 100 feet away. Find θ if the range r of a projectile is

$$r = \frac{1}{32} v_0^2 \sin 2\theta.$$

110. GEOMETRY A trough for feeding cattle is 4 meters long and its cross sections are isosceles triangles with the two equal sides being $\frac{1}{2}$ meter (see figure). The angle between the two sides is θ.

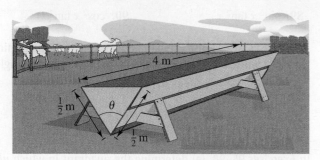

(a) Write the trough's volume as a function of $\theta/2$.

(b) Write the volume of the trough as a function of θ and determine the value of θ such that the volume is maximum.

HARMONIC MOTION In Exercises 111–114, use the following information. A weight is attached to a spring suspended vertically from a ceiling. When a driving force is applied to the system, the weight moves vertically from its equilibrium position, and this motion is described by the model $y = 1.5 \sin 8t - 0.5 \cos 8t$, where y is the distance from equilibrium (in feet) and t is the time (in seconds).

 111. Use a graphing utility to graph the model.

112. Write the model in the form

$$y = \sqrt{a^2 + b^2} \sin(Bt + C).$$

113. Find the amplitude of the oscillations of the weight.

114. Find the frequency of the oscillations of the weight.

EXPLORATION

TRUE OR FALSE? In Exercises 115–118, determine whether the statement is true or false. Justify your answer.

115. If $\dfrac{\pi}{2} < \theta < \pi$, then $\cos \dfrac{\theta}{2} < 0$.

116. $\sin(x + y) = \sin x + \sin y$

117. $4 \sin(-x) \cos(-x) = -2 \sin 2x$

118. $4 \sin 45° \cos 15° = 1 + \sqrt{3}$

119. List the reciprocal identities, quotient identities, and Pythagorean identities from memory.

120. THINK ABOUT IT If a trigonometric equation has an infinite number of solutions, is it true that the equation is an identity? Explain.

121. THINK ABOUT IT Explain why you know from observation that the equation $a \sin x - b = 0$ has no solution if $|a| < |b|$.

122. SURFACE AREA The surface area of a honeycomb is given by the equation

$$S = 6hs + \frac{3}{2}s^2\left(\frac{\sqrt{3} - \cos \theta}{\sin \theta}\right), \quad 0 < \theta \le 90°$$

where $h = 2.4$ inches, $s = 0.75$ inch, and θ is the angle shown in the figure.

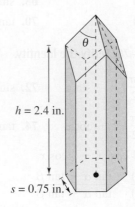

(a) For what value(s) of θ is the surface area 12 square inches?

(b) What value of θ gives the minimum surface area?

In Exercises 123 and 124, use the graphs of y_1 and y_2 to determine how to change one function to form the identity $y_1 = y_2$.

123. $y_1 = \sec^2\left(\dfrac{\pi}{2} - x\right)$

$y_2 = \cot^2 x$

124. $y_1 = \dfrac{\cos 3x}{\cos x}$

$y_2 = (2 \sin x)^2$

 In Exercises 125 and 126, use the *zero* or *root* feature of a graphing utility to approximate the zeros of the function.

125. $y = \sqrt{x + 3} + 4 \cos x$

126. $y = 2 - \dfrac{1}{2}x^2 + 3 \sin \dfrac{\pi x}{2}$

2 CHAPTER TEST

See www.CalcChat.com for worked-out solutions to odd-numbered exercises.

Take this test as you would take a test in class. When you are finished, check your work against the answers given in the back of the book.

1. If $\tan \theta = \frac{6}{5}$ and $\cos \theta < 0$, use the fundamental identities to evaluate all six trigonometric functions of θ.

2. Use the fundamental identities to simplify $\csc^2 \beta (1 - \cos^2 \beta)$.

3. Factor and simplify $\dfrac{\sec^4 x - \tan^4 x}{\sec^2 x + \tan^2 x}$.   **4.** Add and simplify $\dfrac{\cos \theta}{\sin \theta} + \dfrac{\sin \theta}{\cos \theta}$.

5. Determine the values of θ, $0 \le \theta < 2\pi$, for which $\tan \theta = -\sqrt{\sec^2 \theta - 1}$ is true.

6. Use a graphing utility to graph the functions $y_1 = \cos x + \sin x \tan x$ and $y_2 = \sec x$. Make a conjecture about y_1 and y_2. Verify the result algebraically.

In Exercises 7–12, verify the identity.

7. $\sin \theta \sec \theta = \tan \theta$

8. $\sec^2 x \tan^2 x + \sec^2 x = \sec^4 x$

9. $\dfrac{\csc \alpha + \sec \alpha}{\sin \alpha + \cos \alpha} = \cot \alpha + \tan \alpha$

10. $\tan\left(x + \dfrac{\pi}{2}\right) = -\cot x$

11. $\sin(n\pi + \theta) = (-1)^n \sin \theta$, n is an integer.

12. $(\sin x + \cos x)^2 = 1 + \sin 2x$

13. Rewrite $\sin^4 \dfrac{x}{2}$ in terms of the first power of the cosine.

14. Use a half-angle formula to simplify the expression $\sin 4\theta / (1 + \cos 4\theta)$.

15. Write $4 \sin 3\theta \cos 2\theta$ as a sum or difference.

16. Write $\cos 3\theta - \cos \theta$ as a product.

In Exercises 17–20, find all solutions of the equation in the interval $[0, 2\pi)$.

17. $\tan^2 x + \tan x = 0$

18. $\sin 2\alpha - \cos \alpha = 0$

19. $4 \cos^2 x - 3 = 0$

20. $\csc^2 x - \csc x - 2 = 0$

21. Use a graphing utility to approximate the solutions of the equation $5 \sin x - x = 0$ accurate to three decimal places.

22. Find the exact value of $\cos 105°$ using the fact that $105° = 135° - 30°$.

23. Use the figure to find the exact values of $\sin 2u$, $\cos 2u$, and $\tan 2u$.

24. Cheyenne, Wyoming has a latitude of $41°$N. At this latitude, the position of the sun at sunrise can be modeled by

$$D = 31 \sin\left(\frac{2\pi}{365}t - 1.4\right)$$

where t is the time (in days) and $t = 1$ represents January 1. In this model, D represents the number of degrees north or south of due east that the sun rises. Use a graphing utility to determine the days on which the sun is more than $20°$ north of due east at sunrise.

25. The heights h (in feet) of two people in different seats on a Ferris wheel can be modeled by

$$h_1 = 28 \cos 10t + 38 \quad \text{and} \quad h_2 = 28 \cos\left[10\left(t - \frac{\pi}{6}\right)\right] + 38, \ 0 \le t \le 2$$

where t is the time (in minutes). When are the two people at the same height?

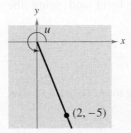

$(2, -5)$

FIGURE FOR 23

PROOFS IN MATHEMATICS

Sum and Difference Formulas *(p. 252)*

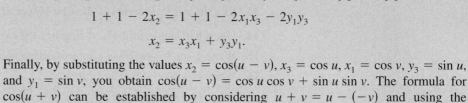

$$\sin(u + v) = \sin u \cos v + \cos u \sin v \qquad \tan(u + v) = \frac{\tan u + \tan v}{1 - \tan u \tan v}$$

$$\sin(u - v) = \sin u \cos v - \cos u \sin v$$

$$\cos(u + v) = \cos u \cos v - \sin u \sin v \qquad \tan(u - v) = \frac{\tan u - \tan v}{1 + \tan u \tan v}$$

$$\cos(u - v) = \cos u \cos v + \sin u \sin v$$

Proof

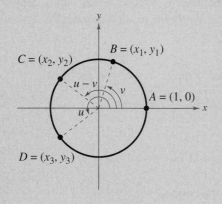

You can use the figures at the left for the proofs of the formulas for $\cos(u \pm v)$. In the top figure, let A be the point $(1, 0)$ and then use u and v to locate the points $B = (x_1, y_1)$, $C = (x_2, y_2)$, and $D = (x_3, y_3)$ on the unit circle. So, $x_i^2 + y_i^2 = 1$ for $i = 1, 2,$ and 3. For convenience, assume that $0 < v < u < 2\pi$. In the bottom figure, note that arcs AC and BD have the same length. So, line segments AC and BD are also equal in length, which implies that

$$\sqrt{(x_2 - 1)^2 + (y_2 - 0)^2} = \sqrt{(x_3 - x_1)^2 + (y_3 - y_1)^2}$$

$$x_2^2 - 2x_2 + 1 + y_2^2 = x_3^2 - 2x_1x_3 + x_1^2 + y_3^2 - 2y_1y_3 + y_1^2$$

$$(x_2^2 + y_2^2) + 1 - 2x_2 = (x_3^2 + y_3^2) + (x_1^2 + y_1^2) - 2x_1x_3 - 2y_1y_3$$

$$1 + 1 - 2x_2 = 1 + 1 - 2x_1x_3 - 2y_1y_3$$

$$x_2 = x_3x_1 + y_3y_1.$$

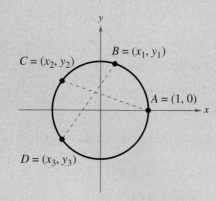

Finally, by substituting the values $x_2 = \cos(u - v)$, $x_3 = \cos u$, $x_1 = \cos v$, $y_3 = \sin u$, and $y_1 = \sin v$, you obtain $\cos(u - v) = \cos u \cos v + \sin u \sin v$. The formula for $\cos(u + v)$ can be established by considering $u + v = u - (-v)$ and using the formula just derived to obtain

$$\cos(u + v) = \cos[u - (-v)] = \cos u \cos(-v) + \sin u \sin(-v)$$

$$= \cos u \cos v - \sin u \sin v.$$

You can use the sum and difference formulas for sine and cosine to prove the formulas for $\tan(u \pm v)$.

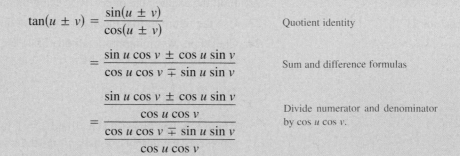

$$\tan(u \pm v) = \frac{\sin(u \pm v)}{\cos(u \pm v)} \qquad \text{Quotient identity}$$

$$= \frac{\sin u \cos v \pm \cos u \sin v}{\cos u \cos v \mp \sin u \sin v} \qquad \text{Sum and difference formulas}$$

$$= \frac{\dfrac{\sin u \cos v \pm \cos u \sin v}{\cos u \cos v}}{\dfrac{\cos u \cos v \mp \sin u \sin v}{\cos u \cos v}} \qquad \begin{array}{l}\text{Divide numerator and denominator} \\ \text{by } \cos u \cos v.\end{array}$$

$$= \frac{\dfrac{\sin u \cos v}{\cos u \cos v} \pm \dfrac{\cos u \sin v}{\cos u \cos v}}{\dfrac{\cos u \cos v}{\cos u \cos v} \mp \dfrac{\sin u \sin v}{\cos u \cos v}}$$

Write as separate fractions.

$$= \frac{\dfrac{\sin u}{\cos u} \pm \dfrac{\sin v}{\cos v}}{1 \mp \dfrac{\sin u}{\cos u} \cdot \dfrac{\sin v}{\cos v}}$$

Product of fractions

$$= \frac{\tan u \pm \tan v}{1 \mp \tan u \tan v}$$

Quotient identity

Trigonometry and Astronomy

Trigonometry was used by early astronomers to calculate measurements in the universe. Trigonometry was used to calculate the circumference of Earth and the distance from Earth to the moon. Another major accomplishment in astronomy using trigonometry was computing distances to stars.

Double-Angle Formulas *(p. 259)*

$$\sin 2u = 2 \sin u \cos u \qquad \cos 2u = \cos^2 u - \sin^2 u$$
$$\tan 2u = \frac{2 \tan u}{1 - \tan^2 u} \qquad\qquad = 2 \cos^2 u - 1 = 1 - 2 \sin^2 u$$

Proof

To prove all three formulas, let $v = u$ in the corresponding sum formulas.

$$\sin 2u = \sin(u + u) = \sin u \cos u + \cos u \sin u = 2 \sin u \cos u$$
$$\cos 2u = \cos(u + u) = \cos u \cos u - \sin u \sin u = \cos^2 u - \sin^2 u$$
$$\tan 2u = \tan(u + u) = \frac{\tan u + \tan u}{1 - \tan u \tan u} = \frac{2 \tan u}{1 - \tan^2 u}$$

Power-Reducing Formulas *(p. 261)*

$$\sin^2 u = \frac{1 - \cos 2u}{2} \qquad \cos^2 u = \frac{1 + \cos 2u}{2} \qquad \tan^2 u = \frac{1 - \cos 2u}{1 + \cos 2u}$$

Proof

To prove the first formula, solve for $\sin^2 u$ in the double-angle formula $\cos 2u = 1 - 2 \sin^2 u$, as follows.

$$\cos 2u = 1 - 2 \sin^2 u$$

Write double-angle formula.

$$2 \sin^2 u = 1 - \cos 2u$$

Subtract $\cos 2u$ from and add $2 \sin^2 u$ to each side.

$$\sin^2 u = \frac{1 - \cos 2u}{2}$$

Divide each side by 2.

In a similar way you can prove the second formula, by solving for $\cos^2 u$ in the double-angle formula

$$\cos 2u = 2\cos^2 u - 1.$$

To prove the third formula, use a quotient identity, as follows.

$$\tan^2 u = \frac{\sin^2 u}{\cos^2 u}$$

$$= \frac{\dfrac{1 - \cos 2u}{2}}{\dfrac{1 + \cos 2u}{2}}$$

$$= \frac{1 - \cos 2u}{1 + \cos 2u}$$

Sum-to-Product Formulas *(p. 264)*

$$\sin u + \sin v = 2 \sin\left(\frac{u + v}{2}\right) \cos\left(\frac{u - v}{2}\right)$$

$$\sin u - \sin v = 2 \cos\left(\frac{u + v}{2}\right) \sin\left(\frac{u - v}{2}\right)$$

$$\cos u + \cos v = 2 \cos\left(\frac{u + v}{2}\right) \cos\left(\frac{u - v}{2}\right)$$

$$\cos u - \cos v = -2 \sin\left(\frac{u + v}{2}\right) \sin\left(\frac{u - v}{2}\right)$$

Proof

To prove the first formula, let $x = u + v$ and $y = u - v$. Then substitute $u = (x + y)/2$ and $v = (x - y)/2$ in the product-to-sum formula.

$$\sin u \cos v = \frac{1}{2}[\sin(u + v) + \sin(u - v)]$$

$$\sin\left(\frac{x + y}{2}\right) \cos\left(\frac{x - y}{2}\right) = \frac{1}{2}(\sin x + \sin y)$$

$$2 \sin\left(\frac{x + y}{2}\right) \cos\left(\frac{x - y}{2}\right) = \sin x + \sin y$$

The other sum-to-product formulas can be proved in a similar manner.

PROBLEM SOLVING

This collection of thought-provoking and challenging exercises further explores and expands upon concepts learned in this chapter.

1. (a) Write each of the other trigonometric functions of θ in terms of $\sin \theta$.

   (b) Write each of the other trigonometric functions of θ in terms of $\cos \theta$.

2. Verify that for all integers n,

   $$\cos\left[\frac{(2n + 1)\pi}{2}\right] = 0.$$

3. Verify that for all integers n,

   $$\sin\left[\frac{(12n + 1)\pi}{6}\right] = \frac{1}{2}.$$

4. A particular sound wave is modeled by

   $$p(t) = \frac{1}{4\pi}\left(p_1(t) + 30p_2(t) + p_3(t) + p_5(t) + 30p_6(t)\right)$$

   where $p_n(t) = \frac{1}{n}\sin(524n\pi t)$, and t is the time (in seconds).

   (a) Find the sine components $p_n(t)$ and use a graphing utility to graph each component. Then verify the graph of p that is shown.

   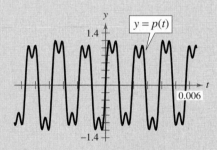

   (b) Find the period of each sine component of p. Is p periodic? If so, what is its period?

   (c) Use the *zero* or *root* feature or the *zoom* and *trace* features of a graphing utility to find the t-intercepts of the graph of p over one cycle.

   (d) Use the *maximum* and *minimum* features of a graphing utility to approximate the absolute maximum and absolute minimum values of p over one cycle.

5. Three squares of side s are placed side by side (see figure). Make a conjecture about the relationship between the sum $u + v$ and w. Prove your conjecture by using the identity for the tangent of the sum of two angles.

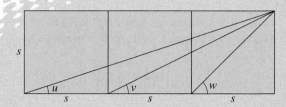

FIGURE FOR 5

6. The path traveled by an object (neglecting air resistance) that is projected at an initial height of h_0 feet, an initial velocity of v_0 feet per second, and an initial angle θ is given by

   $$y = -\frac{16}{v_0^2 \cos^2 \theta}x^2 + (\tan \theta)x + h_0$$

   where x and y are measured in feet. Find a formula for the maximum height of an object projected from ground level at velocity v_0 and angle θ. To do this, find half of the horizontal distance

   $$\frac{1}{32}v_0^2 \sin 2\theta$$

   and then substitute it for x in the general model for the path of a projectile (where $h_0 = 0$).

7. Use the figure to derive the formulas for

   $$\sin \frac{\theta}{2}, \cos \frac{\theta}{2}, \text{ and } \tan \frac{\theta}{2}$$

   where θ is an acute angle.

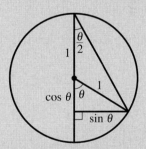

8. The force F (in pounds) on a person's back when he or she bends over at an angle θ is modeled by

   $$F = \frac{0.6W \sin(\theta + 90°)}{\sin 12°}$$

   where W is the person's weight (in pounds).

   (a) Simplify the model.

   (b) Use a graphing utility to graph the model, where $W = 185$ and $0° < \theta < 90°$.

   (c) At what angle is the force a maximum? At what angle is the force a minimum?

279

9. The number of hours of daylight that occur at any location on Earth depends on the time of year and the latitude of the location. The following equations model the numbers of hours of daylight in Seward, Alaska (60° latitude) and New Orleans, Louisiana (30° latitude).

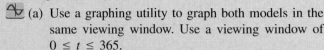

$$D = 12.2 - 6.4 \cos\left[\frac{\pi(t + 0.2)}{182.6}\right] \quad \text{Seward}$$

$$D = 12.2 - 1.9 \cos\left[\frac{\pi(t + 0.2)}{182.6}\right] \quad \text{New Orleans}$$

In these models, D represents the number of hours of daylight and t represents the day, with $t = 0$ corresponding to January 1.

(a) Use a graphing utility to graph both models in the same viewing window. Use a viewing window of $0 \le t \le 365$.

(b) Find the days of the year on which both cities receive the same amount of daylight.

(c) Which city has the greater variation in the number of daylight hours? Which constant in each model would you use to determine the difference between the greatest and least numbers of hours of daylight?

(d) Determine the period of each model.

10. The tide, or depth of the ocean near the shore, changes throughout the day. The water depth d (in feet) of a bay can be modeled by

$$d = 35 - 28 \cos\frac{\pi}{6.2}t$$

where t is the time in hours, with $t = 0$ corresponding to 12:00 A.M.

(a) Algebraically find the times at which the high and low tides occur.

(b) Algebraically find the time(s) at which the water depth is 3.5 feet.

(c) Use a graphing utility to verify your results from parts (a) and (b).

11. Find the solution of each inequality in the interval $[0, 2\pi]$.

(a) $\sin x \ge 0.5$ (b) $\cos x \le -0.5$

(c) $\tan x < \sin x$ (d) $\cos x \ge \sin x$

12. The index of refraction n of a transparent material is the ratio of the speed of light in a vacuum to the speed of light in the material. Some common materials and their indices are air (1.00), water (1.33), and glass (1.50). Triangular prisms are often used to measure the index of refraction based on the formula

$$n = \frac{\sin\left(\frac{\theta}{2} + \frac{\alpha}{2}\right)}{\sin\frac{\theta}{2}}.$$

For the prism shown in the figure, $\alpha = 60°$.

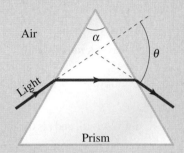

(a) Write the index of refraction as a function of $\cot(\theta/2)$.

(b) Find θ for a prism made of glass.

13. (a) Write a sum formula for $\sin(u + v + w)$.

(b) Write a sum formula for $\tan(u + v + w)$.

14. (a) Derive a formula for $\cos 3\theta$.

(b) Derive a formula for $\cos 4\theta$.

15. The heights h (in inches) of pistons 1 and 2 in an automobile engine can be modeled by

$$h_1 = 3.75 \sin 733t + 7.5$$

and

$$h_2 = 3.75 \sin 733\left(t + \frac{4\pi}{3}\right) + 7.5$$

where t is measured in seconds.

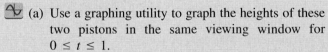

(a) Use a graphing utility to graph the heights of these two pistons in the same viewing window for $0 \le t \le 1$.

(b) How often are the pistons at the same height?

Additional Topics in Trigonometry

In Mathematics

Trigonometry is used to solve triangles, represent vectors, and to write trigonometric forms of complex numbers.

In Real Life

Trigonometry is used to find areas, estimate heights, and represent vectors involving force, velocity, and other quantities. For instance, trigonometry and vectors can be used to find the tension in the tow lines as a loaded barge is being towed by two tugboats. (See Exercise 93, page 310.)

Luca Tettoni/Terra/Corbis

IN CAREERS

There are many careers that use trigonometry. Several are listed below.

3.1 # LAW OF SINES

What you should learn

- Use the Law of Sines to solve oblique triangles (AAS or ASA).
- Use the Law of Sines to solve oblique triangles (SSA).
- Find the areas of oblique triangles.
- Use the Law of Sines to model and solve real-life problems.

Why you should learn it

You can use the Law of Sines to solve real-life problems involving oblique triangles. For instance, in Exercise 53 on page 290, you can use the Law of Sines to determine the distance from a boat to the shoreline.

© Owen Franken/Corbis

Introduction

In Chapter 1, you studied techniques for solving right triangles. In this section and the next, you will solve **oblique triangles**—triangles that have no right angles. As standard notation, the angles of a triangle are labeled A, B, and C, and their opposite sides are labeled a, b, and c, as shown in Figure 3.1.

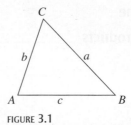

FIGURE **3.1**

To solve an oblique triangle, you need to know the measure of at least one side and any two other measures of the triangle—either two sides, two angles, or one angle and one side. This breaks down into the following four cases.

1. Two angles and any side (AAS or ASA)
2. Two sides and an angle opposite one of them (SSA)
3. Three sides (SSS)
4. Two sides and their included angle (SAS)

The first two cases can be solved using the **Law of Sines,** whereas the last two cases require the Law of Cosines (see Section 3.2).

Law of Sines

If ABC is a triangle with sides a, b, and c, then

$$\frac{a}{\sin A} = \frac{b}{\sin B} = \frac{c}{\sin C}.$$

A is acute.

A is obtuse.

The Law of Sines can also be written in the reciprocal form

$$\frac{\sin A}{a} = \frac{\sin B}{b} = \frac{\sin C}{c}.$$

For a proof of the Law of Sines, see Proofs in Mathematics on page 331.

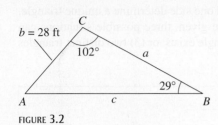

$b = 28$ ft

$102°$

a

$29°$

A c B

FIGURE **3.2**

Study Tip

When solving triangles, a careful sketch is useful as a quick test for the feasibility of an answer. Remember that the longest side lies opposite the largest angle, and the shortest side lies opposite the smallest angle.

Example 1   Given Two Angles and One Side—AAS

For the triangle in Figure 3.2, $C = 102°$, $B = 29°$, and $b = 28$ feet. Find the remaining angle and sides.

Solution

The third angle of the triangle is

$$A = 180° - B - C$$
$$= 180° - 29° - 102°$$
$$= 49°.$$

By the Law of Sines, you have

$$\frac{a}{\sin A} = \frac{b}{\sin B} = \frac{c}{\sin C}.$$

Using $b = 28$ produces

$$a = \frac{b}{\sin B}(\sin A) = \frac{28}{\sin 29°}(\sin 49°) \approx 43.59 \text{ feet}$$

and

$$c = \frac{b}{\sin B}(\sin C) = \frac{28}{\sin 29°}(\sin 102°) \approx 56.49 \text{ feet}.$$

CHECK*Point*   Now try Exercise 5.

Example 2   Given Two Angles and One Side—ASA

A pole tilts *toward* the sun at an 8° angle from the vertical, and it casts a 22-foot shadow. The angle of elevation from the tip of the shadow to the top of the pole is 43°. How tall is the pole?

Solution

From Figure 3.3, note that $A = 43°$ and $B = 90° + 8° = 98°$. So, the third angle is

$$C = 180° - A - B$$
$$= 180° - 43° - 98°$$
$$= 39°.$$

By the Law of Sines, you have

$$\frac{a}{\sin A} = \frac{c}{\sin C}.$$

Because $c = 22$ feet, the length of the pole is

$$a = \frac{c}{\sin C}(\sin A) = \frac{22}{\sin 39°}(\sin 43°) \approx 23.84 \text{ feet}.$$

CHECK*Point*   Now try Exercise 45.

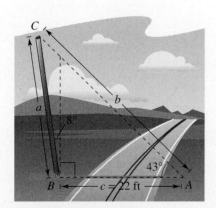

FIGURE **3.3**

For practice, try reworking Example 2 for a pole that tilts *away from* the sun under the same conditions.

The Ambiguous Case (SSA)

In Examples 1 and 2, you saw that two angles and one side determine a unique triangle. However, if two sides and one opposite angle are given, three possible situations can occur: (1) no such triangle exists, (2) one such triangle exists, or (3) two distinct triangles may satisfy the conditions.

The Ambiguous Case (SSA)

Consider a triangle in which you are given a, b, and A.   ($h = b \sin A$)

	A is acute.	A is acute.	A is acute.	A is acute.	A is obtuse.	A is obtuse.
Sketch	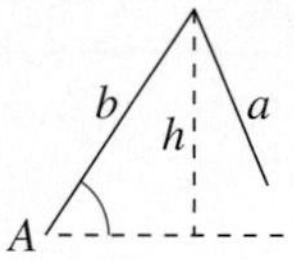	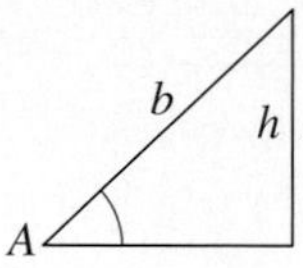	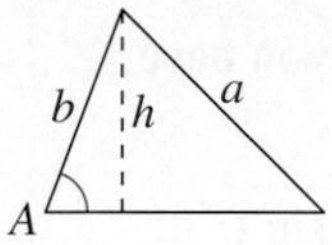	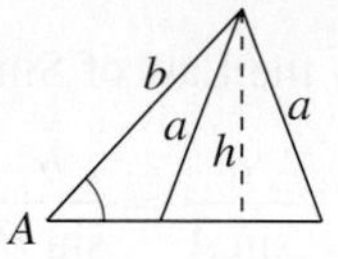	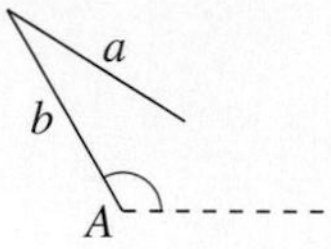	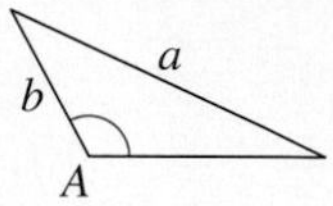
Necessary condition	$a < h$	$a = h$	$a \geq b$	$h < a < b$	$a \leq b$	$a > b$
Triangles possible	None	One	One	Two	None	One

Example 3   Single-Solution Case—SSA

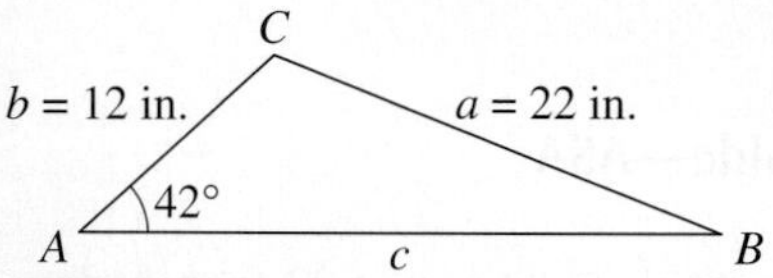

$b = 12$ in. C $a = 22$ in.

A $42°$ c B

One solution: $a \geq b$

FIGURE **3.4**

For the triangle in Figure 3.4, $a = 22$ inches, $b = 12$ inches, and $A = 42°$. Find the remaining side and angles.

Solution

By the Law of Sines, you have

$$\frac{\sin B}{b} = \frac{\sin A}{a} \qquad \text{Reciprocal form}$$

$$\sin B = b\left(\frac{\sin A}{a}\right) \qquad \text{Multiply each side by } b.$$

$$\sin B = 12\left(\frac{\sin 42°}{22}\right) \qquad \text{Substitute for } A, a, \text{ and } b.$$

$$B \approx 21.41°. \qquad B \text{ is acute.}$$

Now, you can determine that

$$C \approx 180° - 42° - 21.41° = 116.59°.$$

Then, the remaining side is

$$\frac{c}{\sin C} = \frac{a}{\sin A}$$

$$c = \frac{a}{\sin A}(\sin C) = \frac{22}{\sin 42°}(\sin 116.59°) \approx 29.40 \text{ inches.}$$

CHECK*Point* Now try Exercise 25.

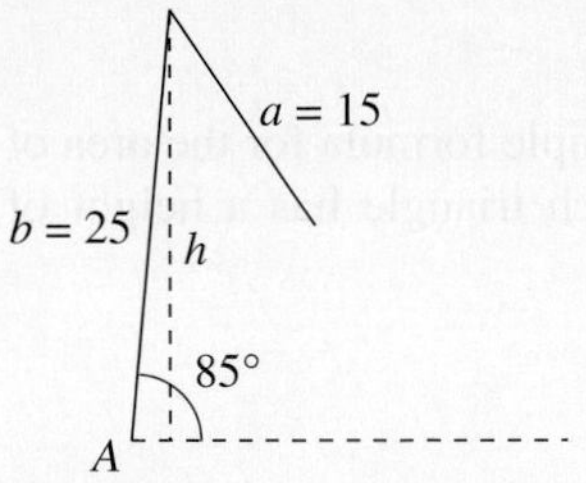

$a = 15$

$b = 25$ h

$85°$

A

No solution: $a < h$

FIGURE **3.5**

Example 4   No-Solution Case—SSA

Show that there is no triangle for which $a = 15$, $b = 25$, and $A = 85°$.

Solution

Begin by making the sketch shown in Figure 3.5. From this figure it appears that no triangle is formed. You can verify this using the Law of Sines.

$$\frac{\sin B}{b} = \frac{\sin A}{a} \qquad \text{Reciprocal form}$$

$$\sin B = b\left(\frac{\sin A}{a}\right) \qquad \text{Multiply each side by } b.$$

$$\sin B = 25\left(\frac{\sin 85°}{15}\right) \approx 1.660 > 1$$

This contradicts the fact that $|\sin B| \leq 1$. So, no triangle can be formed having sides $a = 15$ and $b = 25$ and an angle of $A = 85°$.

CHECK **Point**   Now try Exercise 27.

Example 5   Two-Solution Case—SSA

Find two triangles for which $a = 12$ meters, $b = 31$ meters, and $A = 20.5°$.

Solution

By the Law of Sines, you have

$$\frac{\sin B}{b} = \frac{\sin A}{a} \qquad \text{Reciprocal form}$$

$$\sin B = b\left(\frac{\sin A}{a}\right) = 31\left(\frac{\sin 20.5°}{12}\right) \approx 0.9047.$$

There are two angles, $B_1 \approx 64.8°$ and $B_2 \approx 180° - 64.8° = 115.2°$, between $0°$ and $180°$ whose sine is 0.9047. For $B_1 \approx 64.8°$, you obtain

$$C \approx 180° - 20.5° - 64.8° = 94.7°$$

$$c = \frac{a}{\sin A}(\sin C) = \frac{12}{\sin 20.5°}(\sin 94.7°) \approx 34.15 \text{ meters.}$$

For $B_2 \approx 115.2°$, you obtain

$$C \approx 180° - 20.5° - 115.2° = 44.3°$$

$$c = \frac{a}{\sin A}(\sin C) = \frac{12}{\sin 20.5°}(\sin 44.3°) \approx 23.93 \text{ meters.}$$

The resulting triangles are shown in Figure 3.6.

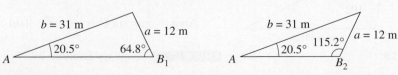

$b = 31$ m $a = 12$ m
A $20.5°$ $64.8°$ B_1

$b = 31$ m $a = 12$ m
A $20.5°$ $115.2°$ B_2

FIGURE **3.6**

CHECK **Point**   Now try Exercise 29.

Area of an Oblique Triangle

The procedure used to prove the Law of Sines leads to a simple formula for the area of an oblique triangle. Referring to Figure 3.7, note that each triangle has a height of $h = b \sin A$. Consequently, the area of each triangle is

$$\text{Area} = \frac{1}{2}(\text{base})(\text{height}) = \frac{1}{2}(c)(b \sin A) = \frac{1}{2}bc \sin A.$$

By similar arguments, you can develop the formulas

$$\text{Area} = \frac{1}{2}ab \sin C = \frac{1}{2}ac \sin B.$$

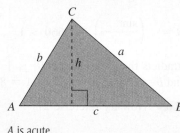
A is acute.

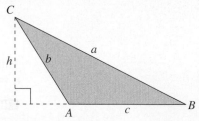
A is obtuse.

FIGURE 3.7

Area of an Oblique Triangle

The area of any triangle is one-half the product of the lengths of two sides times the sine of their included angle. That is,

$$\text{Area} = \frac{1}{2}bc \sin A = \frac{1}{2}ab \sin C = \frac{1}{2}ac \sin B.$$

Note that if angle A is 90°, the formula gives the area for a right triangle:

$$\text{Area} = \frac{1}{2}bc(\sin 90°) = \frac{1}{2}bc = \frac{1}{2}(\text{base})(\text{height}). \qquad \sin 90° = 1$$

Similar results are obtained for angles C and B equal to 90°.

Example 6   Finding the Area of a Triangular Lot

Find the area of a triangular lot having two sides of lengths 90 meters and 52 meters and an included angle of 102°.

Solution

Consider $a = 90$ meters, $b = 52$ meters, and angle $C = 102°$, as shown in Figure 3.8. Then, the area of the triangle is

$$\text{Area} = \frac{1}{2}ab \sin C = \frac{1}{2}(90)(52)(\sin 102°) \approx 2289 \text{ square meters.}$$

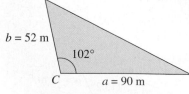

FIGURE 3.8

CHECK *Point* Now try Exercise 39.

Application

Example 7   An Application of the Law of Sines

The course for a boat race starts at point A in Figure 3.9 and proceeds in the direction S 52° W to point B, then in the direction S 40° E to point C, and finally back to A. Point C lies 8 kilometers directly south of point A. Approximate the total distance of the race course.

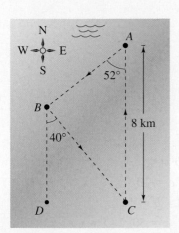

FIGURE **3.9**

Solution

Because lines BD and AC are parallel, it follows that $\angle BCA \cong \angle CBD$. Consequently, triangle ABC has the measures shown in Figure 3.10. The measure of angle B is $180° - 52° - 40° = 88°$. Using the Law of Sines,

$$\frac{a}{\sin 52°} = \frac{b}{\sin 88°} = \frac{c}{\sin 40°}.$$

Because $b = 8$,

$$a = \frac{8}{\sin 88°}(\sin 52°) \approx 6.308$$

and

$$c = \frac{8}{\sin 88°}(\sin 40°) \approx 5.145.$$

The total length of the course is approximately

$$\text{Length} \approx 8 + 6.308 + 5.145$$

$$= 19.453 \text{ kilometers.}$$

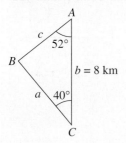

FIGURE **3.10**

CHECK **Point**   Now try Exercise 49.

CLASSROOM DISCUSSION

Using the Law of Sines   In this section, you have been using the Law of Sines to solve *oblique* triangles. Can the Law of Sines also be used to solve a right triangle? If so, write a short paragraph explaining how to use the Law of Sines to solve each triangle. Is there an easier way to solve these triangles?

a. (AAS)

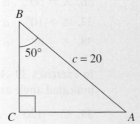

b. (ASA)

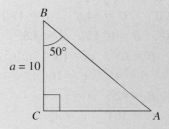

3.1   EXERCISES

See www.CalcChat.com for worked-out solutions to odd-numbered exercises.

VOCABULARY: Fill in the blanks.

1. An _____ triangle is a triangle that has no right angle.

2. For triangle ABC, the Law of Sines is given by $\dfrac{a}{\sin A} = $ _____ $= \dfrac{c}{\sin C}$.

3. Two _____ and one _____ determine a unique triangle.

4. The area of an oblique triangle is given by $\frac{1}{2}bc \sin A = \frac{1}{2}ab \sin C = $ _____ .

SKILLS AND APPLICATIONS

In Exercises 5–24, use the Law of Sines to solve the triangle. Round your answers to two decimal places.

5.

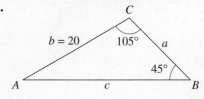

6.

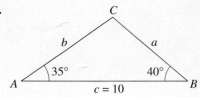

7.

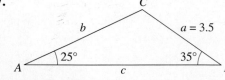

8.

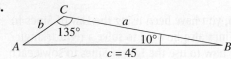

9. $A = 102.4°$, $C = 16.7°$, $a = 21.6$

10. $A = 24.3°$, $C = 54.6°$, $c = 2.68$

11. $A = 83° \, 20'$, $C = 54.6°$, $c = 18.1$

12. $A = 5° \, 40'$, $B = 8° \, 15'$, $b = 4.8$

13. $A = 35°$, $B = 65°$, $c = 10$

14. $A = 120°$, $B = 45°$, $c = 16$

15. $A = 55°$, $B = 42°$, $c = \frac{3}{4}$

16. $B = 28°$, $C = 104°$, $a = 3\frac{5}{8}$

17. $A = 36°$, $a = 8$, $b = 5$

18. $A = 60°$, $a = 9$, $c = 10$

19. $B = 15° \, 30'$, $a = 4.5$, $b = 6.8$

20. $B = 2° \, 45'$, $b = 6.2$, $c = 5.8$

21. $A = 145°$, $a = 14$, $b = 4$

22. $A = 100°$, $a = 125$, $c = 10$

23. $A = 110° \, 15'$, $a = 48$, $b = 16$

24. $C = 95.20°$, $a = 35$, $c = 50$

In Exercises 25–34, use the Law of Sines to solve (if possible) the triangle. If two solutions exist, find both. Round your answers to two decimal places.

25. $A = 110°$, $a = 125$, $b = 100$

26. $A = 110°$, $a = 125$, $b = 200$

27. $A = 76°$, $a = 18$, $b = 20$

28. $A = 76°$, $a = 34$, $b = 21$

29. $A = 58°$, $a = 11.4$, $b = 12.8$

30. $A = 58°$, $a = 4.5$, $b = 12.8$

31. $A = 120°$, $a = b = 25$

32. $A = 120°$, $a = 25$, $b = 24$

33. $A = 45°$, $a = b = 1$

34. $A = 25° \, 4'$, $a = 9.5$, $b = 22$

In Exercises 35–38, find values for b such that the triangle has (a) one solution, (b) two solutions, and (c) no solution.

35. $A = 36°$, $a = 5$

36. $A = 60°$, $a = 10$

37. $A = 10°$, $a = 10.8$

38. $A = 88°$, $a = 315.6$

In Exercises 39–44, find the area of the triangle having the indicated angle and sides.

39. $C = 120°$, $a = 4$, $b = 6$

40. $B = 130°$, $a = 62$, $c = 20$

41. $A = 43° \, 45'$, $b = 57$, $c = 85$

42. $A = 5° \, 15'$, $b = 4.5$, $c = 22$

43. $B = 72° \, 30'$, $a = 105$, $c = 64$

44. $C = 84° \, 30'$, $a = 16$, $b = 20$

45. HEIGHT Because of prevailing winds, a tree grew so that it was leaning 4° from the vertical. At a point 40 meters from the tree, the angle of elevation to the top of the tree is 30° (see figure). Find the height h of the tree.

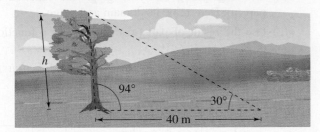

46. HEIGHT A flagpole at a right angle to the horizontal is located on a slope that makes an angle of 12° with the horizontal. The flagpole's shadow is 16 meters long and points directly up the slope. The angle of elevation from the tip of the shadow to the sun is 20°.

(a) Draw a triangle to represent the situation. Show the known quantities on the triangle and use a variable to indicate the height of the flagpole.

(b) Write an equation that can be used to find the height of the flagpole.

(c) Find the height of the flagpole.

47. ANGLE OF ELEVATION A 10-meter utility pole casts a 17-meter shadow directly down a slope when the angle of elevation of the sun is 42° (see figure). Find θ, the angle of elevation of the ground.

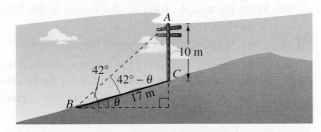

48. FLIGHT PATH A plane flies 500 kilometers with a bearing of 316° from Naples to Elgin (see figure). The plane then flies 720 kilometers from Elgin to Canton (Canton is due west of Naples). Find the bearing of the flight from Elgin to Canton.

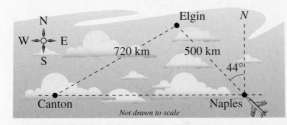

49. BRIDGE DESIGN A bridge is to be built across a small lake from a gazebo to a dock (see figure). The bearing from the gazebo to the dock is S 41° W. From a tree 100 meters from the gazebo, the bearings to the gazebo and the dock are S 74° E and S 28° E, respectively. Find the distance from the gazebo to the dock.

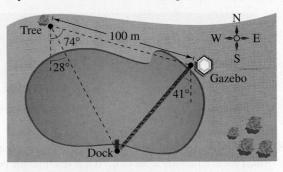

50. RAILROAD TRACK DESIGN The circular arc of a railroad curve has a chord of length 3000 feet corresponding to a central angle of 40°.

(a) Draw a diagram that visually represents the situation. Show the known quantities on the diagram and use the variables r and s to represent the radius of the arc and the length of the arc, respectively.

(b) Find the radius r of the circular arc.

(c) Find the length s of the circular arc.

51. GLIDE PATH A pilot has just started on the glide path for landing at an airport with a runway of length 9000 feet. The angles of depression from the plane to the ends of the runway are 17.5° and 18.8°.

(a) Draw a diagram that visually represents the situation.

(b) Find the air distance the plane must travel until touching down on the near end of the runway.

(c) Find the ground distance the plane must travel until touching down.

(d) Find the altitude of the plane when the pilot begins the descent.

52. LOCATING A FIRE The bearing from the Pine Knob fire tower to the Colt Station fire tower is N 65° E, and the two towers are 30 kilometers apart. A fire spotted by rangers in each tower has a bearing of N 80° E from Pine Knob and S 70° E from Colt Station (see figure). Find the distance of the fire from each tower.

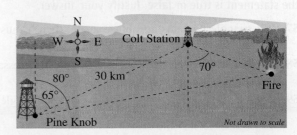

53. DISTANCE A boat is sailing due east parallel to the shoreline at a speed of 10 miles per hour. At a given time, the bearing to the lighthouse is S 70° E, and 15 minutes later the bearing is S 63° E (see figure). The lighthouse is located at the shoreline. What is the distance from the boat to the shoreline?

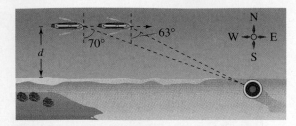

54. DISTANCE A family is traveling due west on a road that passes a famous landmark. At a given time the bearing to the landmark is N 62° W, and after the family travels 5 miles farther the bearing is N 38° W. What is the closest the family will come to the landmark while on the road?

55. ALTITUDE The angles of elevation to an airplane from two points A and B on level ground are 55° and 72°, respectively. The points A and B are 2.2 miles apart, and the airplane is east of both points in the same vertical plane. Find the altitude of the plane.

56. DISTANCE The angles of elevation θ and ϕ to an airplane from the airport control tower and from an observation post 2 miles away are being continuously monitored (see figure). Write an equation giving the distance d between the plane and observation post in terms of θ and ϕ.

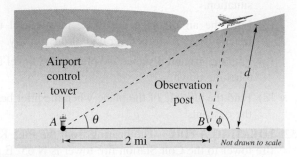

EXPLORATION

TRUE OR FALSE? In Exercises 57–59, determine whether the statement is true or false. Justify your answer.

57. If a triangle contains an obtuse angle, then it must be oblique.

58. Two angles and one side of a triangle do not necessarily determine a unique triangle.

59. If three sides or three angles of an oblique triangle are known, then the triangle can be solved.

60. GRAPHICAL AND NUMERICAL ANALYSIS In the figure, α and β are positive angles.

(a) Write α as a function of β.

(b) Use a graphing utility to graph the function in part (a). Determine its domain and range.

(c) Use the result of part (a) to write c as a function of β.

(d) Use a graphing utility to graph the function in part (c). Determine its domain and range.

(e) Complete the table. What can you infer?

β	0.4	0.8	1.2	1.6	2.0	2.4	2.8
α							
c							

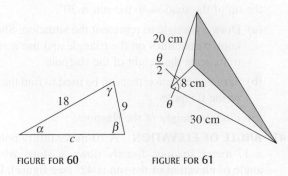

FIGURE FOR 60   FIGURE FOR 61

61. GRAPHICAL ANALYSIS

(a) Write the area A of the shaded region in the figure as a function of θ.

(b) Use a graphing utility to graph the function.

(c) Determine the domain of the function. Explain how the area of the region and the domain of the function would change if the eight-centimeter line segment were decreased in length.

62. CAPSTONE In the figure, a triangle is to be formed by drawing a line segment of length a from $(4, 3)$ to the positive x-axis. For what value(s) of a can you form (a) one triangle, (b) two triangles, and (c) no triangles? Explain your reasoning.

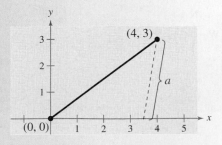

3.2   LAW OF COSINES

What you should learn

- Use the Law of Cosines to solve oblique triangles (SSS or SAS).
- Use the Law of Cosines to model and solve real-life problems.
- Use Heron's Area Formula to find the area of a triangle.

Why you should learn it

You can use the Law of Cosines to solve real-life problems involving oblique triangles. For instance, in Exercise 52 on page 297, you can use the Law of Cosines to approximate how far a baseball player has to run to make a catch.

Introduction

Two cases remain in the list of conditions needed to solve an oblique triangle—SSS and SAS. If you are given three sides (SSS), or two sides and their included angle (SAS), none of the ratios in the Law of Sines would be complete. In such cases, you can use the **Law of Cosines.**

Law of Cosines

Standard Form	*Alternative Form*
$a^2 = b^2 + c^2 - 2bc \cos A$	$\cos A = \dfrac{b^2 + c^2 - a^2}{2bc}$
$b^2 = a^2 + c^2 - 2ac \cos B$	$\cos B = \dfrac{a^2 + c^2 - b^2}{2ac}$
$c^2 = a^2 + b^2 - 2ab \cos C$	$\cos C = \dfrac{a^2 + b^2 - c^2}{2ab}$

For a proof of the Law of Cosines, see Proofs in Mathematics on page 332.

Example 1   **Three Sides of a Triangle—SSS**

Find the three angles of the triangle in Figure 3.11.

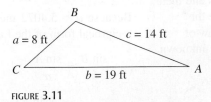

FIGURE **3.11**

Solution

It is a good idea first to find the angle opposite the longest side—side b in this case. Using the alternative form of the Law of Cosines, you find that

$$\cos B = \frac{a^2 + c^2 - b^2}{2ac} = \frac{8^2 + 14^2 - 19^2}{2(8)(14)} \approx -0.45089.$$

Because $\cos B$ is negative, you know that B is an *obtuse* angle given by $B \approx 116.80°$. At this point, it is simpler to use the Law of Sines to determine A.

$$\sin A = a\left(\frac{\sin B}{b}\right) \approx 8\left(\frac{\sin 116.80°}{19}\right) \approx 0.37583$$

You know that A must be acute because B is obtuse, and a triangle can have, at most, one obtuse angle. So, $A \approx 22.08°$ and $C \approx 180° - 22.08° - 116.80° = 41.12°$.

CHECKPoint   Now try Exercise 5.

Do you see why it was wise to find the largest angle *first* in Example 1? Knowing the cosine of an angle, you can determine whether the angle is acute or obtuse. That is,

$\cos \theta > 0$ for $0° < \theta < 90°$ Acute

$\cos \theta < 0$ for $90° < \theta < 180°$.   Obtuse

So, in Example 1, once you found that angle B was obtuse, you knew that angles A and C were both acute. If the largest angle is acute, the remaining two angles are acute also.

Example 2   Two Sides and the Included Angle—SAS

Find the remaining angles and side of the triangle in Figure 3.12.

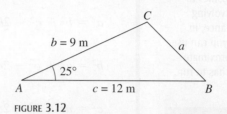

FIGURE 3.12

Solution

Use the Law of Cosines to find the unknown side a in the figure.

$a^2 = b^2 + c^2 - 2bc \cos A$

$a^2 = 9^2 + 12^2 - 2(9)(12) \cos 25°$

$a^2 \approx 29.2375$

$a \approx 5.4072$

Because $a \approx 5.4072$ meters, you now know the ratio $(\sin A)/a$ and you can use the reciprocal form of the Law of Sines to solve for B.

$\dfrac{\sin B}{b} = \dfrac{\sin A}{a}$

$\sin B = b\left(\dfrac{\sin A}{a}\right)$

$= 9\left(\dfrac{\sin 25°}{5.4072}\right)$

≈ 0.7034

There are two angles between $0°$ and $180°$ whose sine is 0.7034, $B_1 \approx 44.7°$ and $B_2 \approx 180° - 44.7° = 135.3°$.

For $B_1 \approx 44.7°$,

$C_1 \approx 180° - 25° - 44.7° = 110.3°$.

For $B_2 \approx 135.3°$,

$C_2 \approx 180° - 25° - 135.3° = 19.7°$.

Because side c is the longest side of the triangle, C must be the largest angle of the triangle. So, $B \approx 44.7°$ and $C \approx 110.3°$.

CHECK*Point*   Now try Exercise 7.

Applications

| Example 3 | **An Application of the Law of Cosines** |

The pitcher's mound on a women's softball field is 43 feet from home plate and the distance between the bases is 60 feet, as shown in Figure 3.13. (The pitcher's mound is not halfway between home plate and second base.) How far is the pitcher's mound from first base?

Solution

In triangle *HPF*, $H = 45°$ (line *HP* bisects the right angle at *H*), $f = 43$, and $p = 60$. Using the Law of Cosines for this SAS case, you have

$$h^2 = f^2 + p^2 - 2fp \cos H$$
$$= 43^2 + 60^2 - 2(43)(60) \cos 45° \approx 1800.3.$$

So, the approximate distance from the pitcher's mound to first base is

$$h \approx \sqrt{1800.3} \approx 42.43 \text{ feet.}$$

CHECK*Point*   Now try Exercise 43.

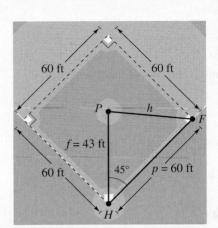

FIGURE **3.13**

| Example 4 | **An Application of the Law of Cosines** |

A ship travels 60 miles due east, then adjusts its course northward, as shown in Figure 3.14. After traveling 80 miles in that direction, the ship is 139 miles from its point of departure. Describe the bearing from point *B* to point *C*.

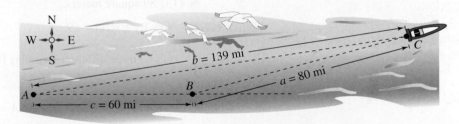

FIGURE **3.14**

Solution

You have $a = 80$, $b = 139$, and $c = 60$. So, using the alternative form of the Law of Cosines, you have

$$\cos B = \frac{a^2 + c^2 - b^2}{2ac}$$
$$= \frac{80^2 + 60^2 - 139^2}{2(80)(60)}$$
$$\approx -0.97094.$$

So, $B \approx \arccos(-0.97094) \approx 166.15°$, and thus the bearing measured from due north from point *B* to point *C* is

$$166.15° - 90° = 76.15°, \text{ or N } 76.15° \text{ E.}$$

CHECK*Point*   Now try Exercise 49.

Heron's Area Formula

The Law of Cosines can be used to establish the following formula for the area of a triangle. This formula is called **Heron's Area Formula** after the Greek mathematician Heron (c. 100 B.C.).

Heron's Area Formula

Given any triangle with sides of lengths a, b, and c, the area of the triangle is

$$\text{Area} = \sqrt{s(s-a)(s-b)(s-c)}$$

where $s = (a+b+c)/2$.

For a proof of Heron's Area Formula, see Proofs in Mathematics on page 333.

Example 5   Using Heron's Area Formula

Find the area of a triangle having sides of lengths $a = 43$ meters, $b = 53$ meters, and $c = 72$ meters.

Solution

Because $s = (a+b+c)/2 = 168/2 = 84$, Heron's Area Formula yields

$$\text{Area} = \sqrt{s(s-a)(s-b)(s-c)}$$
$$= \sqrt{84(41)(31)(12)}$$
$$\approx 1131.89 \text{ square meters.}$$

CHECK Point   Now try Exercise 59.

You have now studied three different formulas for the area of a triangle.

Standard Formula: $\text{Area} = \frac{1}{2}bh$

Oblique Triangle: $\text{Area} = \frac{1}{2}bc\sin A = \frac{1}{2}ab\sin C = \frac{1}{2}ac\sin B$

Heron's Area Formula: $\text{Area} = \sqrt{s(s-a)(s-b)(s-c)}$

CLASSROOM DISCUSSION

The Area of a Triangle   Use the most appropriate formula to find the area of each triangle below. Show your work and give your reasons for choosing each formula.

a.
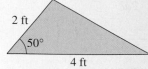

b.
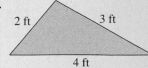

c.
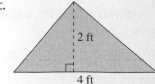

d.
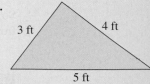

3.2   EXERCISES

See www.CalcChat.com for worked-out solutions to odd-numbered exercises.

VOCABULARY: Fill in the blanks.

1. If you are given three sides of a triangle, you would use the Law of _____ to find the three angles of the triangle.

2. If you are given two angles and any side of a triangle, you would use the Law of _____ to solve the triangle.

3. The standard form of the Law of Cosines for $\cos B = \dfrac{a^2 + c^2 - b^2}{2ac}$ is _____ .

4. The Law of Cosines can be used to establish a formula for finding the area of a triangle called _____ _____ Formula.

SKILLS AND APPLICATIONS

In Exercises 5–20, use the Law of Cosines to solve the triangle. Round your answers to two decimal places.

5.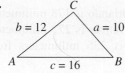

6.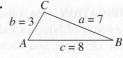

7.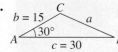

8.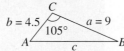

9. $a = 11$, $b = 15$, $c = 21$

10. $a = 55$, $b = 25$, $c = 72$

11. $a = 75.4$, $b = 52$, $c = 52$

12. $a = 1.42$, $b = 0.75$, $c = 1.25$

13. $A = 120°$, $b = 6$, $c = 7$

14. $A = 48°$, $b = 3$, $c = 14$

15. $B = 10° 35'$, $a = 40$, $c = 30$

16. $B = 75° 20'$, $a = 6.2$, $c = 9.5$

17. $B = 125° 40'$, $a = 37$, $c = 37$

18. $C = 15° 15'$, $a = 7.45$, $b = 2.15$

19. $C = 43°$, $a = \frac{4}{9}$, $b = \frac{7}{9}$

20. $C = 101°$, $a = \frac{3}{8}$, $b = \frac{3}{4}$

In Exercises 21–26, complete the table by solving the parallelogram shown in the figure. (The lengths of the diagonals are given by c and d.)

	a	b	c	d	θ	ϕ
21.	5	8			45°	
22.	25	35				120°
23.	10	14	20			
24.	40	60		80		
25.	15		25	20		
26.		25	50	35		

In Exercises 27–32, determine whether the Law of Sines or the Law of Cosines is needed to solve the triangle. Then solve the triangle.

27. $a = 8$, $c = 5$, $B = 40°$

28. $a = 10$, $b = 12$, $C = 70°$

29. $A = 24°$, $a = 4$, $b = 18$

30. $a = 11$, $b = 13$, $c = 7$

31. $A = 42°$, $B = 35°$, $c = 1.2$

32. $a = 160$, $B = 12°$, $C = 7°$

In Exercises 33–40, use Heron's Area Formula to find the area of the triangle.

33. $a = 8$, $b = 12$, $c = 17$

34. $a = 33$, $b = 36$, $c = 25$

35. $a = 2.5$, $b = 10.2$, $c = 9$

36. $a = 75.4$, $b = 52$, $c = 52$

37. $a = 12.32$, $b = 8.46$, $c = 15.05$

38. $a = 3.05$, $b = 0.75$, $c = 2.45$

39. $a = 1$, $b = \frac{1}{2}$, $c = \frac{3}{4}$

40. $a = \frac{3}{5}$, $b = \frac{5}{8}$, $c = \frac{3}{8}$

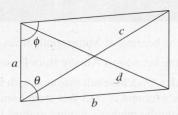

41. NAVIGATION A boat race runs along a triangular course marked by buoys *A*, *B*, and *C*. The race starts with the boats headed west for 3700 meters. The other two sides of the course lie to the north of the first side, and their lengths are 1700 meters and 3000 meters. Draw a figure that gives a visual representation of the situation, and find the bearings for the last two legs of the race.

42. NAVIGATION A plane flies 810 miles from Franklin to Centerville with a bearing of 75°. Then it flies 648 miles from Centerville to Rosemount with a bearing of 32°. Draw a figure that visually represents the situation, and find the straight-line distance and bearing from Franklin to Rosemount.

43. SURVEYING To approximate the length of a marsh, a surveyor walks 250 meters from point *A* to point *B*, then turns 75° and walks 220 meters to point *C* (see figure). Approximate the length *AC* of the marsh.

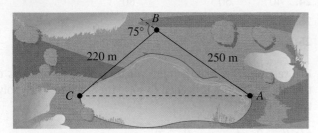

44. SURVEYING A triangular parcel of land has 115 meters of frontage, and the other boundaries have lengths of 76 meters and 92 meters. What angles does the frontage make with the two other boundaries?

45. SURVEYING A triangular parcel of ground has sides of lengths 725 feet, 650 feet, and 575 feet. Find the measure of the largest angle.

46. STREETLIGHT DESIGN Determine the angle *θ* in the design of the streetlight shown in the figure.

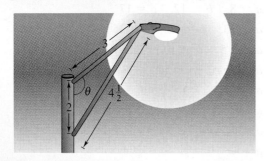

47. DISTANCE Two ships leave a port at 9 A.M. One travels at a bearing of N 53° W at 12 miles per hour, and the other travels at a bearing of S 67° W at 16 miles per hour. Approximate how far apart they are at noon that day.

48. LENGTH A 100-foot vertical tower is to be erected on the side of a hill that makes a 6° angle with the horizontal (see figure). Find the length of each of the two guy wires that will be anchored 75 feet uphill and downhill from the base of the tower.

49. NAVIGATION On a map, Orlando is 178 millimeters due south of Niagara Falls, Denver is 273 millimeters from Orlando, and Denver is 235 millimeters from Niagara Falls (see figure).

(a) Find the bearing of Denver from Orlando.

(b) Find the bearing of Denver from Niagara Falls.

50. NAVIGATION On a map, Minneapolis is 165 millimeters due west of Albany, Phoenix is 216 millimeters from Minneapolis, and Phoenix is 368 millimeters from Albany (see figure).

(a) Find the bearing of Minneapolis from Phoenix.

(b) Find the bearing of Albany from Phoenix.

51. BASEBALL On a baseball diamond with 90-foot sides, the pitcher's mound is 60.5 feet from home plate. How far is it from the pitcher's mound to third base?

52. BASEBALL   The baseball player in center field is playing approximately 330 feet from the television camera that is behind home plate. A batter hits a fly ball that goes to the wall 420 feet from the camera (see figure). The camera turns 8° to follow the play. Approximately how far does the center fielder have to run to make the catch?

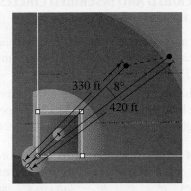

53. AIRCRAFT TRACKING   To determine the distance between two aircraft, a tracking station continuously determines the distance to each aircraft and the angle A between them (see figure). Determine the distance a between the planes when $A = 42°$, $b = 35$ miles, and $c = 20$ miles.

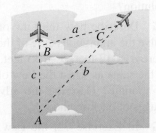

54. AIRCRAFT TRACKING   Use the figure for Exercise 53 to determine the distance a between the planes when $A = 11°$, $b = 20$ miles, and $c = 20$ miles.

55. TRUSSES Q is the midpoint of the line segment $\overline{PR}$ in the truss rafter shown in the figure. What are the lengths of the line segments $\overline{PQ}$, $\overline{QS}$, and $\overline{RS}$?

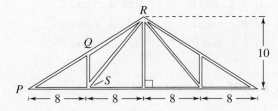

 56. ENGINE DESIGN   An engine has a seven-inch connecting rod fastened to a crank (see figure).

(a) Use the Law of Cosines to write an equation giving the relationship between x and θ.

(b) Write x as a function of θ. (Select the sign that yields positive values of x.)

(c) Use a graphing utility to graph the function in part (b).

(d) Use the graph in part (c) to determine the maximum distance the piston moves in one cycle.

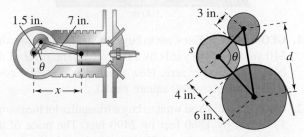

FIGURE FOR 56   FIGURE FOR 57

57. PAPER MANUFACTURING   In a process with continuous paper, the paper passes across three rollers of radii 3 inches, 4 inches, and 6 inches (see figure). The centers of the three-inch and six-inch rollers are d inches apart, and the length of the arc in contact with the paper on the four-inch roller is s inches. Complete the table.

d (inches)	9	10	12	13	14	15	16
θ (degrees)							
s (inches)							

58. AWNING DESIGN   A retractable awning above a patio door lowers at an angle of 50° from the exterior wall at a height of 10 feet above the ground (see figure). No direct sunlight is to enter the door when the angle of elevation of the sun is greater than 70°. What is the length x of the awning?

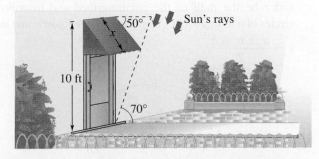

59. GEOMETRY   The lengths of the sides of a triangular parcel of land are approximately 200 feet, 500 feet, and 600 feet. Approximate the area of the parcel.

60. GEOMETRY A parking lot has the shape of a parallelogram (see figure). The lengths of two adjacent sides are 70 meters and 100 meters. The angle between the two sides is 70°. What is the area of the parking lot?

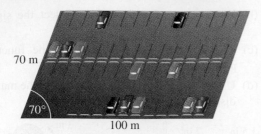

70 m

70°

100 m

61. GEOMETRY You want to buy a triangular lot measuring 510 yards by 840 yards by 1120 yards. The price of the land is $2000 per acre. How much does the land cost? (*Hint:* 1 acre = 4840 square yards)

62. GEOMETRY You want to buy a triangular lot measuring 1350 feet by 1860 feet by 2490 feet. The price of the land is $2200 per acre. How much does the land cost? (*Hint:* 1 acre = 43,560 square feet)

EXPLORATION

TRUE OR FALSE? In Exercises 63 and 64, determine whether the statement is true or false. Justify your answer.

63. In Heron's Area Formula, s is the average of the lengths of the three sides of the triangle.

64. In addition to SSS and SAS, the Law of Cosines can be used to solve triangles with SSA conditions.

65. WRITING A triangle has side lengths of 10 centimeters, 16 centimeters, and 5 centimeters. Can the Law of Cosines be used to solve the triangle? Explain.

66. WRITING Given a triangle with $b = 47$ meters, $A = 87°$, and $C = 110°$, can the Law of Cosines be used to solve the triangle? Explain.

67. CIRCUMSCRIBED AND INSCRIBED CIRCLES Let R and r be the radii of the circumscribed and inscribed circles of a triangle ABC, respectively (see figure), and let

$$s = \frac{a + b + c}{2}.$$

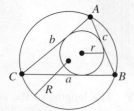

(a) Prove that $2R = \dfrac{a}{\sin A} = \dfrac{b}{\sin B} = \dfrac{c}{\sin C}$.

(b) Prove that $r = \sqrt{\dfrac{(s - a)(s - b)(s - c)}{s}}$.

CIRCUMSCRIBED AND INSCRIBED CIRCLES In Exercises 68 and 69, use the results of Exercise 67.

68. Given a triangle with $a = 25$, $b = 55$, and $c = 72$, find the areas of (a) the triangle, (b) the circumscribed circle, and (c) the inscribed circle.

69. Find the length of the largest circular running track that can be built on a triangular piece of property with sides of lengths 200 feet, 250 feet, and 325 feet.

70. THINK ABOUT IT What familiar formula do you obtain when you use the third form of the Law of Cosines $c^2 = a^2 + b^2 - 2ab \cos C$, and you let $C = 90°$? What is the relationship between the Law of Cosines and this formula?

71. THINK ABOUT IT In Example 2, suppose $A = 115°$. After solving for a, which angle would you solve for next, B or C? Are there two possible solutions for that angle? If so, how can you determine which angle is the correct solution?

72. WRITING Describe how the Law of Cosines can be used to solve the ambiguous case of the oblique triangle ABC, where $a = 12$ feet, $b = 30$ feet, and $A = 20°$. Is the result the same as when the Law of Sines is used to solve the triangle? Describe the advantages and the disadvantages of each method.

73. WRITING In Exercise 72, the Law of Cosines was used to solve a triangle in the two-solution case of SSA. Can the Law of Cosines be used to solve the no-solution and single-solution cases of SSA? Explain.

74. CAPSTONE Determine whether the Law of Sines or the Law of Cosines is needed to solve the triangle.

(a) A, C, and a (b) a, c, and C

(c) b, c, and A (d) A, B, and c

(e) b, c, and C (f) a, b, and c

75. PROOF Use the Law of Cosines to prove that

$$\frac{1}{2} bc(1 + \cos A) = \frac{a + b + c}{2} \cdot \frac{-a + b + c}{2}.$$

76. PROOF Use the Law of Cosines to prove that

$$\frac{1}{2} bc(1 - \cos A) = \frac{a - b + c}{2} \cdot \frac{a + b - c}{2}.$$

3.3 VECTORS IN THE PLANE

What you should learn

- Represent vectors as directed line segments.
- Write the component forms of vectors.
- Perform basic vector operations and represent them graphically.
- Write vectors as linear combinations of unit vectors.
- Find the direction angles of vectors.
- Use vectors to model and solve real-life problems.

Why you should learn it

You can use vectors to model and solve real-life problems involving magnitude and direction. For instance, in Exercise 102 on page 311, you can use vectors to determine the true direction of a commercial jet.

Introduction

Quantities such as force and velocity involve both *magnitude* and *direction* and cannot be completely characterized by a single real number. To represent such a quantity, you can use a **directed line segment,** as shown in Figure 3.15. The directed line segment $\overrightarrow{PQ}$ has **initial point** P and **terminal point** Q. Its **magnitude** (or length) is denoted by $\|\overrightarrow{PQ}\|$ and can be found using the Distance Formula.

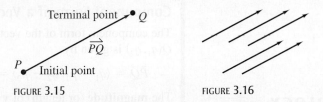

FIGURE 3.15   FIGURE 3.16

Two directed line segments that have the same magnitude and direction are equivalent. For example, the directed line segments in Figure 3.16 are all equivalent. The set of all directed line segments that are equivalent to the directed line segment $\overrightarrow{PQ}$ is a **vector v in the plane,** written $\mathbf{v} = \overrightarrow{PQ}$. Vectors are denoted by lowercase, boldface letters such as $\mathbf{u}$, $\mathbf{v}$, and $\mathbf{w}$.

Example 1   Vector Representation by Directed Line Segments

Let $\mathbf{u}$ be represented by the directed line segment from $P(0, 0)$ to $Q(3, 2)$, and let $\mathbf{v}$ be represented by the directed line segment from $R(1, 2)$ to $S(4, 4)$, as shown in Figure 3.17. Show that $\mathbf{u}$ and $\mathbf{v}$ are equivalent.

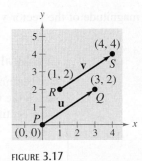

FIGURE 3.17

Solution

From the Distance Formula, it follows that $\overrightarrow{PQ}$ and $\overrightarrow{RS}$ have the *same magnitude*.

$$\|\overrightarrow{PQ}\| = \sqrt{(3-0)^2 + (2-0)^2} = \sqrt{13} \qquad \|\overrightarrow{RS}\| = \sqrt{(4-1)^2 + (4-2)^2} = \sqrt{13}$$

Moreover, both line segments have the *same direction* because they are both directed toward the upper right on lines having a slope of

$$\frac{4-2}{4-1} = \frac{2-0}{3-0} = \frac{2}{3}.$$

Because $\overrightarrow{PQ}$ and $\overrightarrow{RS}$ have the same magnitude and direction, $\mathbf{u}$ and $\mathbf{v}$ are equivalent.

CHECK Point   Now try Exercise 11.

Component Form of a Vector

The directed line segment whose initial point is the origin is often the most convenient representative of a set of equivalent directed line segments. This representative of the vector **v** is in **standard position.**

A vector whose initial point is the origin $(0, 0)$ can be uniquely represented by the coordinates of its terminal point (v_1, v_2). This is the **component form of a vector v,** written as $\mathbf{v} = \langle v_1, v_2 \rangle$. The coordinates v_1 and v_2 are the *components* of **v**. If both the initial point and the terminal point lie at the origin, **v** is the **zero vector** and is denoted by $\mathbf{0} = \langle 0, 0 \rangle$.

Component Form of a Vector

The component form of the vector with initial point $P(p_1, p_2)$ and terminal point $Q(q_1, q_2)$ is given by

$$\overrightarrow{PQ} = \langle q_1 - p_1, q_2 - p_2 \rangle = \langle v_1, v_2 \rangle = \mathbf{v}.$$

The **magnitude** (or length) of **v** is given by

$$\|\mathbf{v}\| = \sqrt{(q_1 - p_1)^2 + (q_2 - p_2)^2} = \sqrt{v_1^2 + v_2^2}.$$

If $\|\mathbf{v}\| = 1$, **v** is a **unit vector.** Moreover, $\|\mathbf{v}\| = 0$ if and only if **v** is the zero vector **0**.

TECHNOLOGY

You can graph vectors with a graphing utility by graphing directed line segments. Consult the user's guide for your graphing utility for specific instructions.

Two vectors $\mathbf{u} = \langle u_1, u_2 \rangle$ and $\mathbf{v} = \langle v_1, v_2 \rangle$ are *equal* if and only if $u_1 = v_1$ and $u_2 = v_2$. For instance, in Example 1, the vector **u** from $P(0, 0)$ to $Q(3, 2)$ is $\mathbf{u} = \overrightarrow{PQ} = \langle 3 - 0, 2 - 0 \rangle = \langle 3, 2 \rangle$, and the vector **v** from $R(1, 2)$ to $S(4, 4)$ is $\mathbf{v} = \overrightarrow{RS} = \langle 4 - 1, 4 - 2 \rangle = \langle 3, 2 \rangle$.

Example 2   Finding the Component Form of a Vector

Find the component form and magnitude of the vector **v** that has initial point $(4, -7)$ and terminal point $(-1, 5)$.

Algebraic Solution

Let

$$P(4, -7) = (p_1, p_2)$$

and

$$Q(-1, 5) = (q_1, q_2).$$

Then, the components of $\mathbf{v} = \langle v_1, v_2 \rangle$ are

$$v_1 = q_1 - p_1 = -1 - 4 = -5$$

$$v_2 = q_2 - p_2 = 5 - (-7) = 12.$$

So, $\mathbf{v} = \langle -5, 12 \rangle$ and the magnitude of **v** is

$$\|\mathbf{v}\| = \sqrt{(-5)^2 + 12^2}$$

$$= \sqrt{169} = 13.$$

Graphical Solution

Use centimeter graph paper to plot the points $P(4, -7)$ and $Q(-1, 5)$. Carefully sketch the vector **v**. Use the sketch to find the components of $\mathbf{v} = \langle v_1, v_2 \rangle$. Then use a centimeter ruler to find the magnitude of **v**.

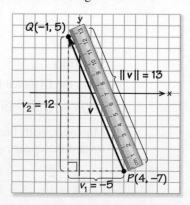

FIGURE 3.18

Figure 3.18 shows that the components of **v** are $v_1 = -5$ and $v_2 = 12$, so $\mathbf{v} = \langle -5, 12 \rangle$. Figure 3.18 also shows that the magnitude of **v** is $\|\mathbf{v}\| = 13$.

CHECK Point Now try Exercise 19.

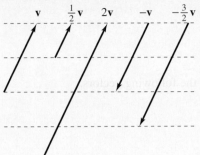

FIGURE **3.19**

Vector Operations

The two basic vector operations are **scalar multiplication** and **vector addition.** In operations with vectors, numbers are usually referred to as **scalars.** In this text, scalars will always be real numbers. Geometrically, the product of a vector **v** and a scalar k is the vector that is $|k|$ times as long as **v**. If k is positive, k**v** has the same direction as **v**, and if k is negative, k**v** has the direction opposite that of **v**, as shown in Figure 3.19.

To add two vectors **u** and **v** geometrically, first position them (without changing their lengths or directions) so that the initial point of the second vector **v** coincides with the terminal point of the first vector **u**. The sum **u** + **v** is the vector formed by joining the initial point of the first vector **u** with the terminal point of the second vector **v**, as shown in Figure 3.20. This technique is called the **parallelogram law** for vector addition because the vector **u** + **v**, often called the **resultant** of vector addition, is the diagonal of a parallelogram having adjacent sides **u** and **v**.

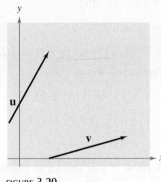
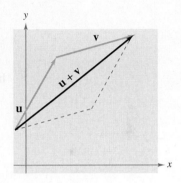

FIGURE **3.20**

Definitions of Vector Addition and Scalar Multiplication

Let $\mathbf{u} = \langle u_1, u_2 \rangle$ and $\mathbf{v} = \langle v_1, v_2 \rangle$ be vectors and let k be a scalar (a real number). Then the *sum* of **u** and **v** is the vector

$$\mathbf{u} + \mathbf{v} = \langle u_1 + v_1, u_2 + v_2 \rangle \qquad \text{Sum}$$

and the *scalar multiple* of k times **u** is the vector

$$k\mathbf{u} = k\langle u_1, u_2 \rangle = \langle ku_1, ku_2 \rangle. \qquad \text{Scalar multiple}$$

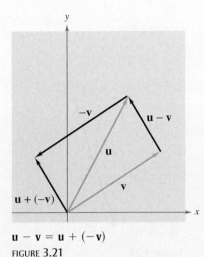

$\mathbf{u} - \mathbf{v} = \mathbf{u} + (-\mathbf{v})$

FIGURE **3.21**

The **negative** of $\mathbf{v} = \langle v_1, v_2 \rangle$ is

$$-\mathbf{v} = (-1)\mathbf{v}$$

$$= \langle -v_1, -v_2 \rangle \qquad \text{Negative}$$

and the **difference** of **u** and **v** is

$$\mathbf{u} - \mathbf{v} = \mathbf{u} + (-\mathbf{v}) \qquad \text{Add } (-\mathbf{v}). \text{ See Figure 6.21.}$$

$$= \langle u_1 - v_1, u_2 - v_2 \rangle. \qquad \text{Difference}$$

To represent $\mathbf{u} - \mathbf{v}$ geometrically, you can use directed line segments with the *same* initial point. The difference $\mathbf{u} - \mathbf{v}$ is the vector from the terminal point of **v** to the terminal point of **u**, which is equal to $\mathbf{u} + (-\mathbf{v})$, as shown in Figure 3.21.

The component definitions of vector addition and scalar multiplication are illustrated in Example 3. In this example, notice that each of the vector operations can be interpreted geometrically.

Example 3 Vector Operations

Let $\mathbf{v} = \langle -2, 5 \rangle$ and $\mathbf{w} = \langle 3, 4 \rangle$, and find each of the following vectors.

a. $2\mathbf{v}$ **b.** $\mathbf{w} - \mathbf{v}$ **c.** $\mathbf{v} + 2\mathbf{w}$

Solution

a. Because $\mathbf{v} = \langle -2, 5 \rangle$, you have

$$2\mathbf{v} = 2\langle -2, 5 \rangle$$
$$= \langle 2(-2), 2(5) \rangle$$
$$= \langle -4, 10 \rangle.$$

A sketch of $2\mathbf{v}$ is shown in Figure 3.22.

b. The difference of $\mathbf{w}$ and $\mathbf{v}$ is

$$\mathbf{w} - \mathbf{v} = \langle 3, 4 \rangle - \langle -2, 5 \rangle$$
$$= \langle 3 - (-2), 4 - 5 \rangle$$
$$= \langle 5, -1 \rangle.$$

A sketch of $\mathbf{w} - \mathbf{v}$ is shown in Figure 3.23. Note that the figure shows the vector difference $\mathbf{w} - \mathbf{v}$ as the sum $\mathbf{w} + (-\mathbf{v})$.

c. The sum of $\mathbf{v}$ and $2\mathbf{w}$ is

$$\mathbf{v} + 2\mathbf{w} = \langle -2, 5 \rangle + 2\langle 3, 4 \rangle$$
$$= \langle -2, 5 \rangle + \langle 2(3), 2(4) \rangle$$
$$= \langle -2, 5 \rangle + \langle 6, 8 \rangle$$
$$= \langle -2 + 6, 5 + 8 \rangle$$
$$= \langle 4, 13 \rangle.$$

A sketch of $\mathbf{v} + 2\mathbf{w}$ is shown in Figure 3.24.

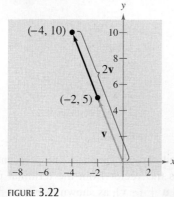

FIGURE 3.22

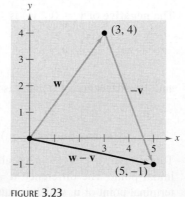

FIGURE 3.23

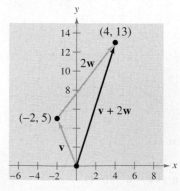

FIGURE 3.24

CHECK *Point* Now try Exercise 31.

Vector addition and scalar multiplication share many of the properties of ordinary arithmetic.

Properties of Vector Addition and Scalar Multiplication

Let **u**, **v**, and **w** be vectors and let c and d be scalars. Then the following properties are true.

1. $\mathbf{u} + \mathbf{v} = \mathbf{v} + \mathbf{u}$ 2. $(\mathbf{u} + \mathbf{v}) + \mathbf{w} = \mathbf{u} + (\mathbf{v} + \mathbf{w})$

3. $\mathbf{u} + \mathbf{0} = \mathbf{u}$ 4. $\mathbf{u} + (-\mathbf{u}) = \mathbf{0}$

5. $c(d\mathbf{u}) = (cd)\mathbf{u}$ 6. $(c + d)\mathbf{u} = c\mathbf{u} + d\mathbf{u}$

7. $c(\mathbf{u} + \mathbf{v}) = c\mathbf{u} + c\mathbf{v}$ 8. $1(\mathbf{u}) = \mathbf{u}, \quad 0(\mathbf{u}) = \mathbf{0}$

9. $\|c\mathbf{v}\| = |c|\,\|\mathbf{v}\|$

Property 9 can be stated as follows: the magnitude of the vector $c\mathbf{v}$ is the absolute value of c times the magnitude of **v**.

Unit Vectors

In many applications of vectors, it is useful to find a unit vector that has the same direction as a given nonzero vector **v**. To do this, you can divide **v** by its magnitude to obtain

$$\mathbf{u} = \text{unit vector} = \frac{\mathbf{v}}{\|\mathbf{v}\|} = \left(\frac{1}{\|\mathbf{v}\|}\right)\mathbf{v}. \qquad \text{Unit vector in direction of } \mathbf{v}$$

Note that **u** is a scalar multiple of **v**. The vector **u** has a magnitude of 1 and the same direction as **v**. The vector **u** is called a **unit vector in the direction of v**.

Example 4   Finding a Unit Vector

Find a unit vector in the direction of $\mathbf{v} = \langle -2, 5 \rangle$ and verify that the result has a magnitude of 1.

Solution

The unit vector in the direction of **v** is

$$\frac{\mathbf{v}}{\|\mathbf{v}\|} = \frac{\langle -2, 5 \rangle}{\sqrt{(-2)^2 + (5)^2}}$$

$$= \frac{1}{\sqrt{29}}\langle -2, 5 \rangle$$

$$= \left\langle \frac{-2}{\sqrt{29}}, \frac{5}{\sqrt{29}} \right\rangle.$$

This vector has a magnitude of 1 because

$$\sqrt{\left(\frac{-2}{\sqrt{29}}\right)^2 + \left(\frac{5}{\sqrt{29}}\right)^2} = \sqrt{\frac{4}{29} + \frac{25}{29}} = \sqrt{\frac{29}{29}} = 1.$$

CHECK*Point*   Now try Exercise 41.

HISTORICAL NOTE

The Granger Collection

William Rowan Hamilton (1805–1865), an Irish mathematician, did some of the earliest work with vectors. Hamilton spent many years developing a system of vector-like quantities called quaternions. Although Hamilton was convinced of the benefits of quaternions, the operations he defined did not produce good models for physical phenomena. It was not until the latter half of the nineteenth century that the Scottish physicist James Maxwell (1831–1879) restructured Hamilton's quaternions in a form useful for representing physical quantities such as force, velocity, and acceleration.

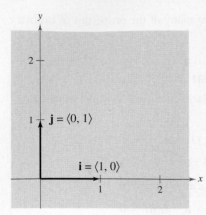

FIGURE 3.25

The unit vectors $\langle 1, 0 \rangle$ and $\langle 0, 1 \rangle$ are called the **standard unit vectors** and are denoted by

$$\mathbf{i} = \langle 1, 0 \rangle \qquad \text{and} \qquad \mathbf{j} = \langle 0, 1 \rangle$$

as shown in Figure 3.25. (Note that the lowercase letter $\mathbf{i}$ is written in boldface to distinguish it from the imaginary number $i = \sqrt{-1}$.) These vectors can be used to represent any vector $\mathbf{v} = \langle v_1, v_2 \rangle$, as follows.

$$\begin{aligned} \mathbf{v} &= \langle v_1, v_2 \rangle \\ &= v_1 \langle 1, 0 \rangle + v_2 \langle 0, 1 \rangle \\ &= v_1 \mathbf{i} + v_2 \mathbf{j} \end{aligned}$$

The scalars v_1 and v_2 are called the **horizontal** and **vertical components of v,** respectively. The vector sum

$$v_1 \mathbf{i} + v_2 \mathbf{j}$$

is called a **linear combination** of the vectors $\mathbf{i}$ and $\mathbf{j}$. Any vector in the plane can be written as a linear combination of the standard unit vectors $\mathbf{i}$ and $\mathbf{j}$.

Example 5 Writing a Linear Combination of Unit Vectors

Let $\mathbf{u}$ be the vector with initial point $(2, -5)$ and terminal point $(-1, 3)$. Write $\mathbf{u}$ as a linear combination of the standard unit vectors $\mathbf{i}$ and $\mathbf{j}$.

Solution

Begin by writing the component form of the vector $\mathbf{u}$.

$$\begin{aligned} \mathbf{u} &= \langle -1 - 2, 3 - (-5) \rangle \\ &= \langle -3, 8 \rangle \\ &= -3\mathbf{i} + 8\mathbf{j} \end{aligned}$$

This result is shown graphically in Figure 3.26.

CHECK*Point* Now try Exercise 53.

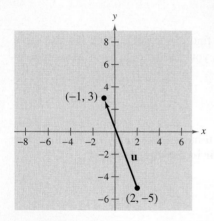

FIGURE 3.26

Example 6 Vector Operations

Let $\mathbf{u} = -3\mathbf{i} + 8\mathbf{j}$ and let $\mathbf{v} = 2\mathbf{i} - \mathbf{j}$. Find $2\mathbf{u} - 3\mathbf{v}$.

Solution

You could solve this problem by converting $\mathbf{u}$ and $\mathbf{v}$ to component form. This, however, is not necessary. It is just as easy to perform the operations in unit vector form.

$$\begin{aligned} 2\mathbf{u} - 3\mathbf{v} &= 2(-3\mathbf{i} + 8\mathbf{j}) - 3(2\mathbf{i} - \mathbf{j}) \\ &= -6\mathbf{i} + 16\mathbf{j} - 6\mathbf{i} + 3\mathbf{j} \\ &= -12\mathbf{i} + 19\mathbf{j} \end{aligned}$$

CHECK*Point* Now try Exercise 59.

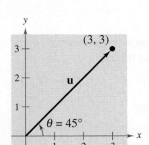

$\|\mathbf{u}\| = 1$

FIGURE 3.27

Direction Angles

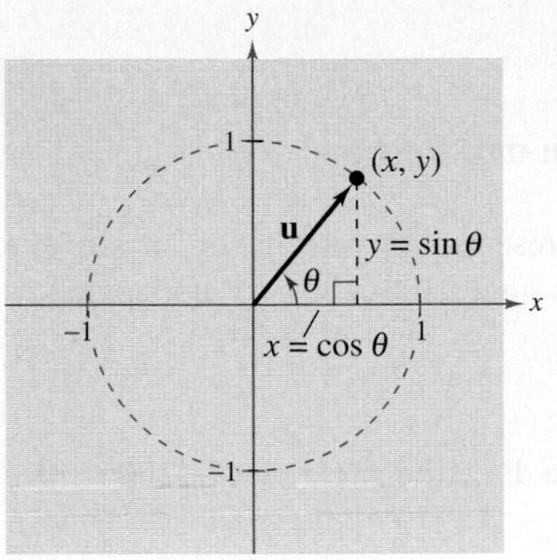

If **u** is a *unit vector* such that θ is the angle (measured counterclockwise) from the positive x-axis to **u**, the terminal point of **u** lies on the unit circle and you have

$$\mathbf{u} = \langle x, y \rangle = \langle \cos\theta, \sin\theta \rangle = (\cos\theta)\mathbf{i} + (\sin\theta)\mathbf{j}$$

as shown in Figure 3.27. The angle θ is the **direction angle** of the vector **u**.

Suppose that **u** is a unit vector with direction angle θ. If $\mathbf{v} = a\mathbf{i} + b\mathbf{j}$ is any vector that makes an angle θ with the positive x-axis, it has the same direction as **u** and you can write

$$\mathbf{v} = \|\mathbf{v}\| \langle \cos\theta, \sin\theta \rangle$$
$$= \|\mathbf{v}\|(\cos\theta)\mathbf{i} + \|\mathbf{v}\|(\sin\theta)\mathbf{j}.$$

Because $\mathbf{v} = a\mathbf{i} + b\mathbf{j} = \|\mathbf{v}\|(\cos\theta)\mathbf{i} + \|\mathbf{v}\|(\sin\theta)\mathbf{j}$, it follows that the direction angle θ for **v** is determined from

$$\tan\theta = \frac{\sin\theta}{\cos\theta} \qquad \text{Quotient identity}$$

$$= \frac{\|\mathbf{v}\|\sin\theta}{\|\mathbf{v}\|\cos\theta} \qquad \text{Multiply numerator and denominator by } \|\mathbf{v}\|.$$

$$= \frac{b}{a}. \qquad \text{Simplify.}$$

Example 7   Finding Direction Angles of Vectors

Find the direction angle of each vector.

a. $\mathbf{u} = 3\mathbf{i} + 3\mathbf{j}$

b. $\mathbf{v} = 3\mathbf{i} - 4\mathbf{j}$

Solution

a. The direction angle is

$$\tan\theta = \frac{b}{a} = \frac{3}{3} = 1.$$

So, $\theta = 45°$, as shown in Figure 3.28.

b. The direction angle is

$$\tan\theta = \frac{b}{a} = \frac{-4}{3}.$$

Moreover, because $\mathbf{v} = 3\mathbf{i} - 4\mathbf{j}$ lies in Quadrant IV, θ lies in Quadrant IV and its reference angle is

$$\theta = \left| \arctan\left(-\frac{4}{3}\right) \right| \approx |-53.13°| = 53.13°.$$

So, it follows that $\theta \approx 360° - 53.13° = 306.87°$, as shown in Figure 3.29.

CHECK Point   Now try Exercise 63.

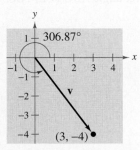

FIGURE 3.28

FIGURE 3.29

Applications of Vectors

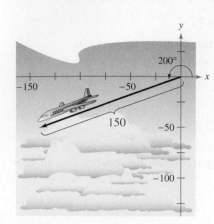

200°

−150   −50

150

−50

−100

FIGURE **3.30**

Example 8   Finding the Component Form of a Vector

Find the component form of the vector that represents the velocity of an airplane descending at a speed of 150 miles per hour at an angle 20° below the horizontal, as shown in Figure 3.30.

Solution

The velocity vector **v** has a magnitude of 150 and a direction angle of $\theta = 200°$.

$$\mathbf{v} = \|\mathbf{v}\|(\cos \theta)\mathbf{i} + \|\mathbf{v}\|(\sin \theta)\mathbf{j}$$
$$= 150(\cos 200°)\mathbf{i} + 150(\sin 200°)\mathbf{j}$$
$$\approx 150(-0.9397)\mathbf{i} + 150(-0.3420)\mathbf{j}$$
$$\approx -140.96\mathbf{i} - 51.30\mathbf{j}$$
$$= \langle -140.96, -51.30 \rangle$$

You can check that **v** has a magnitude of 150, as follows.

$$\|\mathbf{v}\| \approx \sqrt{(-140.96)^2 + (-51.30)^2}$$
$$\approx \sqrt{19{,}869.72 + 2631.69}$$
$$= \sqrt{22{,}501.41} \approx 150$$

CHECK Point   Now try Exercise 83.

Example 9   Using Vectors to Determine Weight

A force of 600 pounds is required to pull a boat and trailer up a ramp inclined at 15° from the horizontal. Find the combined weight of the boat and trailer.

Solution

Based on Figure 3.31, you can make the following observations.

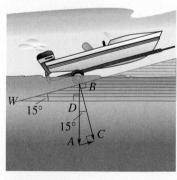

W
15°   D
15°
A   C

FIGURE **3.31**

$\|\overrightarrow{BA}\|$ = force of gravity = combined weight of boat and trailer

$\|\overrightarrow{BC}\|$ = force against ramp

$\|\overrightarrow{AC}\|$ = force required to move boat up ramp = 600 pounds

By construction, triangles *BWD* and *ABC* are similar. Therefore, angle *ABC* is 15°. So, in triangle *ABC* you have

$$\sin 15° = \frac{\|\overrightarrow{AC}\|}{\|\overrightarrow{BA}\|}$$

$$\sin 15° = \frac{600}{\|\overrightarrow{BA}\|}$$

$$\|\overrightarrow{BA}\| = \frac{600}{\sin 15°}$$

$$\|\overrightarrow{BA}\| \approx 2318.$$

Consequently, the combined weight is approximately 2318 pounds. (In Figure 3.31, note that $\overrightarrow{AC}$ is parallel to the ramp.)

CHECK Point   Now try Exercise 95.

Example 10   Using Vectors to Find Speed and Direction

An airplane is traveling at a speed of 500 miles per hour with a bearing of 330° at a fixed altitude with a negligible wind velocity as shown in Figure 3.32(a). When the airplane reaches a certain point, it encounters a wind with a velocity of 70 miles per hour in the direction N 45° E, as shown in Figure 3.32(b). What are the resultant speed and direction of the airplane?

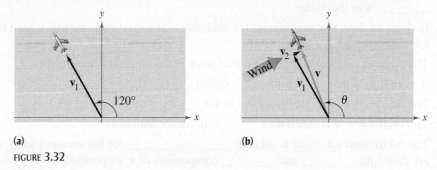

(a)   (b)

FIGURE **3.32**

Solution

Using Figure 3.32, the velocity of the airplane (alone) is

$$\mathbf{v}_1 = 500\langle\cos 120°, \sin 120°\rangle$$

$$= \langle -250, 250\sqrt{3}\rangle$$

and the velocity of the wind is

$$\mathbf{v}_2 = 70\langle\cos 45°, \sin 45°\rangle$$

$$= \langle 35\sqrt{2}, 35\sqrt{2}\rangle.$$

So, the velocity of the airplane (in the wind) is

$$\mathbf{v} = \mathbf{v}_1 + \mathbf{v}_2$$

$$= \langle -250 + 35\sqrt{2}, 250\sqrt{3} + 35\sqrt{2}\rangle$$

$$\approx \langle -200.5, 482.5\rangle$$

and the resultant speed of the airplane is

$$\|\mathbf{v}\| \approx \sqrt{(-200.5)^2 + (482.5)^2}$$

$$\approx 522.5 \text{ miles per hour.}$$

Finally, if θ is the direction angle of the flight path, you have

$$\tan \theta \approx \frac{482.5}{-200.5}$$

$$\approx -2.4065$$

which implies that

$$\theta \approx 180° + \arctan(-2.4065) \approx 180° - 67.4° = 112.6°.$$

So, the true direction of the airplane is approximately

$$270° + (180° - 112.6°) = 337.4°.$$

CHECK*Point*   Now try Exercise 101.

3.3 EXERCISES

See www.CalcChat.com for worked-out solutions to odd-numbered exercises.

VOCABULARY: Fill in the blanks.

1. A _____ _____ _____ can be used to represent a quantity that involves both magnitude and direction.

2. The directed line segment $\overrightarrow{PQ}$ has _____ point P and _____ point Q.

3. The _____ of the directed line segment $\overrightarrow{PQ}$ is denoted by $\|\overrightarrow{PQ}\|$.

4. The set of all directed line segments that are equivalent to a given directed line segment $\overrightarrow{PQ}$ is a _____ **v** in the plane.

5. In order to show that two vectors are equivalent, you must show that they have the same _____ and the same _____ .

6. The directed line segment whose initial point is the origin is said to be in _____ _____ .

7. A vector that has a magnitude of 1 is called a _____ _____ .

8. The two basic vector operations are scalar _____ and vector _____ .

9. The vector **u** + **v** is called the _____ of vector addition.

10. The vector sum $v_1\mathbf{i} + v_2\mathbf{j}$ is called a _____ _____ of the vectors **i** and **j**, and the scalars v_1 and v_2 are called the _____ and _____ components of **v**, respectively.

SKILLS AND APPLICATIONS

In Exercises 11 and 12, show that **u** and **v** are equivalent.

11.
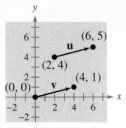

12.
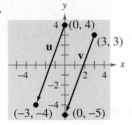

In Exercises 13–24, find the component form and the magnitude of the vector **v**.

13.
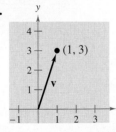

14.

15.
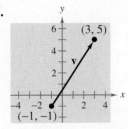

16.
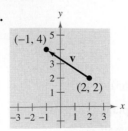

17.
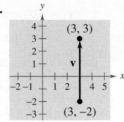

18.
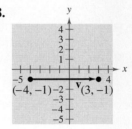

	Initial Point	Terminal Point
19.	$(-3, -5)$	$(5, 1)$
20.	$(-2, 7)$	$(5, -17)$
21.	$(1, 3)$	$(-8, -9)$
22.	$(1, 11)$	$(9, 3)$
23.	$(-1, 5)$	$(15, 12)$
24.	$(-3, 11)$	$(9, 40)$

In Exercises 25–30, use the figure to sketch a graph of the specified vector. To print an enlarged copy of the graph, go to the website *www.mathgraphs.com*.

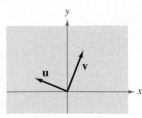

25. $-\mathbf{v}$	**26.** $5\mathbf{v}$
27. $\mathbf{u} + \mathbf{v}$	**28.** $\mathbf{u} + 2\mathbf{v}$
29. $\mathbf{u} - \mathbf{v}$	**30.** $\mathbf{v} - \frac{1}{2}\mathbf{u}$

In Exercises 31–38, find (a) $\mathbf{u} + \mathbf{v}$, (b) $\mathbf{u} - \mathbf{v}$, and (c) $2\mathbf{u} - 3\mathbf{v}$, Then sketch each resultant vector.

31. $\mathbf{u} = \langle 2, 1 \rangle$, $\mathbf{v} = \langle 1, 3 \rangle$ **32.** $\mathbf{u} = \langle 2, 3 \rangle$, $\mathbf{v} = \langle 4, 0 \rangle$

33. $\mathbf{u} = \langle -5, 3 \rangle$, $\mathbf{v} = \langle 0, 0 \rangle$ **34.** $\mathbf{u} = \langle 0, 0 \rangle$, $\mathbf{v} = \langle 2, 1 \rangle$

35. $\mathbf{u} = \mathbf{i} + \mathbf{j}$, $\mathbf{v} = 2\mathbf{i} - 3\mathbf{j}$

36. $\mathbf{u} = -2\mathbf{i} + \mathbf{j}$, $\mathbf{v} = 3\mathbf{j}$

37. $\mathbf{u} = 2\mathbf{i}$, $\mathbf{v} = \mathbf{j}$ **38.** $\mathbf{u} = 2\mathbf{j}$, $\mathbf{v} = 3\mathbf{i}$

In Exercises 39–48, find a unit vector in the direction of the given vector. Verify that the result has a magnitude of 1.

39. $\mathbf{u} = \langle 3, 0 \rangle$ **40.** $\mathbf{u} = \langle 0, -2 \rangle$

41. $\mathbf{v} = \langle -2, 2 \rangle$ **42.** $\mathbf{v} = \langle 5, -12 \rangle$

43. $\mathbf{v} = \mathbf{i} + \mathbf{j}$ **44.** $\mathbf{v} = 6\mathbf{i} - 2\mathbf{j}$

45. $\mathbf{w} = 4\mathbf{j}$ **46.** $\mathbf{w} = -6\mathbf{i}$

47. $\mathbf{w} = \mathbf{i} - 2\mathbf{j}$ **48.** $\mathbf{w} = 7\mathbf{j} - 3\mathbf{i}$

In Exercises 49–52, find the vector $\mathbf{v}$ with the given magnitude and the same direction as $\mathbf{u}$.

	Magnitude	*Direction*
49.	$\|\mathbf{v}\| = 10$	$\mathbf{u} = \langle -3, 4 \rangle$
50.	$\|\mathbf{v}\| = 3$	$\mathbf{u} = \langle -12, -5 \rangle$
51.	$\|\mathbf{v}\| = 9$	$\mathbf{u} = \langle 2, 5 \rangle$
52.	$\|\mathbf{v}\| = 8$	$\mathbf{u} = \langle 3, 3 \rangle$

In Exercises 53–56, the initial and terminal points of a vector are given. Write a linear combination of the standard unit vectors $\mathbf{i}$ and $\mathbf{j}$.

	Initial Point	*Terminal Point*
53.	$(-2, 1)$	$(3, -2)$
54.	$(0, -2)$	$(3, 6)$
55.	$(-6, 4)$	$(0, 1)$
56.	$(-1, -5)$	$(2, 3)$

In Exercises 57–62, find the component form of $\mathbf{v}$ and sketch the specified vector operations geometrically, where $\mathbf{u} = 2\mathbf{i} - \mathbf{j}$, and $\mathbf{w} = \mathbf{i} + 2\mathbf{j}$.

57. $\mathbf{v} = \frac{3}{2}\mathbf{u}$ **58.** $\mathbf{v} = \frac{3}{4}\mathbf{w}$

59. $\mathbf{v} = \mathbf{u} + 2\mathbf{w}$ **60.** $\mathbf{v} = -\mathbf{u} + \mathbf{w}$

61. $\mathbf{v} = \frac{1}{2}(3\mathbf{u} + \mathbf{w})$ **62.** $\mathbf{v} = \mathbf{u} - 2\mathbf{w}$

In Exercises 63–66, find the magnitude and direction angle of the vector $\mathbf{v}$.

63. $\mathbf{v} = 6\mathbf{i} - 6\mathbf{j}$ **64.** $\mathbf{v} = -5\mathbf{i} + 4\mathbf{j}$

65. $\mathbf{v} = 3(\cos 60°\mathbf{i} + \sin 60°\mathbf{j})$

66. $\mathbf{v} = 8(\cos 135°\mathbf{i} + \sin 135°\mathbf{j})$

In Exercises 67–74, find the component form of $\mathbf{v}$ given its magnitude and the angle it makes with the positive x-axis. Sketch $\mathbf{v}$.

	Magnitude	*Angle*
67.	$\|\mathbf{v}\| = 3$	$\theta = 0°$
68.	$\|\mathbf{v}\| = 1$	$\theta = 45°$
69.	$\|\mathbf{v}\| = \frac{7}{2}$	$\theta = 150°$
70.	$\|\mathbf{v}\| = \frac{3}{4}$	$\theta = 150°$
71.	$\|\mathbf{v}\| = 2\sqrt{3}$	$\theta = 45°$
72.	$\|\mathbf{v}\| = 4\sqrt{3}$	$\theta = 90°$
73.	$\|\mathbf{v}\| = 3$	$\mathbf{v}$ in the direction $3\mathbf{i} + 4\mathbf{j}$
74.	$\|\mathbf{v}\| = 2$	$\mathbf{v}$ in the direction $\mathbf{i} + 3\mathbf{j}$

In Exercises 75–78, find the component form of the sum of $\mathbf{u}$ and $\mathbf{v}$ with direction angles $\theta_\mathbf{u}$ and $\theta_\mathbf{v}$.

	Magnitude	*Angle*
75.	$\|\mathbf{u}\| = 5$	$\theta_\mathbf{u} = 0°$
	$\|\mathbf{v}\| = 5$	$\theta_\mathbf{v} = 90°$
76.	$\|\mathbf{u}\| = 4$	$\theta_\mathbf{u} = 60°$
	$\|\mathbf{v}\| = 4$	$\theta_\mathbf{v} = 90°$
77.	$\|\mathbf{u}\| = 20$	$\theta_\mathbf{u} = 45°$
	$\|\mathbf{v}\| = 50$	$\theta_\mathbf{v} = 180°$
78.	$\|\mathbf{u}\| = 50$	$\theta_\mathbf{u} = 30°$
	$\|\mathbf{v}\| = 30$	$\theta_\mathbf{v} = 110°$

In Exercises 79 and 80, use the Law of Cosines to find the angle α between the vectors. (Assume $0° \leq \alpha \leq 180°$.)

79. $\mathbf{v} = \mathbf{i} + \mathbf{j}$, $\mathbf{w} = 2\mathbf{i} - 2\mathbf{j}$

80. $\mathbf{v} = \mathbf{i} + 2\mathbf{j}$, $\mathbf{w} = 2\mathbf{i} - \mathbf{j}$

RESULTANT FORCE In Exercises 81 and 82, find the angle between the forces given the magnitude of their resultant. (*Hint:* Write force 1 as a vector in the direction of the positive x-axis and force 2 as a vector at an angle θ with the positive x-axis.)

	Force 1	*Force 2*	*Resultant Force*
81.	45 pounds	60 pounds	90 pounds
82.	3000 pounds	1000 pounds	3750 pounds

83. VELOCITY A gun with a muzzle velocity of 1200 feet per second is fired at an angle of 6° above the horizontal. Find the vertical and horizontal components of the velocity.

84. Detroit Tigers pitcher Joel Zumaya was recorded throwing a pitch at a velocity of 104 miles per hour. If he threw the pitch at an angle of 35° below the horizontal, find the vertical and horizontal components of the velocity. (Source: Damon Lichtenwalner, Baseball Info Solutions)

85. RESULTANT FORCE Forces with magnitudes of 125 newtons and 300 newtons act on a hook (see figure). The angle between the two forces is 45°. Find the direction and magnitude of the resultant of these forces.

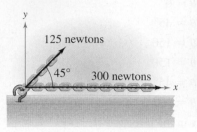

FIGURE FOR 85

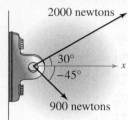

FIGURE FOR 86

86. RESULTANT FORCE Forces with magnitudes of 2000 newtons and 900 newtons act on a machine part at angles of 30° and −45°, respectively, with the x-axis (see figure). Find the direction and magnitude of the resultant of these forces.

87. RESULTANT FORCE Three forces with magnitudes of 75 pounds, 100 pounds, and 125 pounds act on an object at angles of 30°, 45°, and 120°, respectively, with the positive x-axis. Find the direction and magnitude of the resultant of these forces.

88. RESULTANT FORCE Three forces with magnitudes of 70 pounds, 40 pounds, and 60 pounds act on an object at angles of −30°, 45°, and 135°, respectively, with the positive x-axis. Find the direction and magnitude of the resultant of these forces.

89. A traffic light weighing 12 pounds is suspended by two cables (see figure). Find the tension in each cable.

90. Repeat Exercise 89 if $\theta_1 = 40°$ and $\theta_2 = 35°$.

CABLE TENSION In Exercises 91 and 92, use the figure to determine the tension in each cable supporting the load.

91.

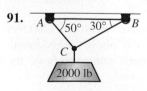

92.

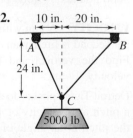

93. TOW LINE TENSION A loaded barge is being towed by two tugboats, and the magnitude of the resultant is 6000 pounds directed along the axis of the barge (see figure). Find the tension in the tow lines if they each make an 18° angle with the axis of the barge.

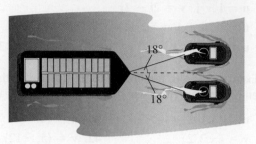

94. ROPE TENSION To carry a 100-pound cylindrical weight, two people lift on the ends of short ropes that are tied to an eyelet on the top center of the cylinder. Each rope makes a 20° angle with the vertical. Draw a figure that gives a visual representation of the situation, and find the tension in the ropes.

In Exercises 95–98, a force of *F* pounds is required to pull an object weighing *W* pounds up a ramp inclined at *θ* degrees from the horizontal.

95. Find *F* if *W* = 100 pounds and $\theta = 12°$.

96. Find *W* if *F* = 600 pounds and $\theta = 14°$.

97. Find *θ* if *F* = 5000 pounds and *W* = 15,000 pounds.

98. Find *F* if *W* = 5000 pounds and $\theta = 26°$.

99. WORK A heavy object is pulled 30 feet across a floor, using a force of 100 pounds. The force is exerted at an angle of 50° above the horizontal (see figure). Find the work done. (Use the formula for work, $W = FD$, where *F* is the component of the force in the direction of motion and *D* is the distance.)

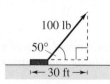

FIGURE FOR 99

FIGURE FOR 100

100. ROPE TENSION A tetherball weighing 1 pound is pulled outward from the pole by a horizontal force **u** until the rope makes a 45° angle with the pole (see figure). Determine the resulting tension in the rope and the magnitude of **u**.

101. NAVIGATION An airplane is flying in the direction of 148°, with an airspeed of 875 kilometers per hour. Because of the wind, its groundspeed and direction are 800 kilometers per hour and 140°, respectively (see figure). Find the direction and speed of the wind.

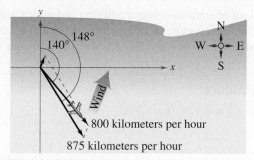

102. NAVIGATION A commercial jet is flying from Miami to Seattle. The jet's velocity with respect to the air is 580 miles per hour, and its bearing is 332°. The wind, at the altitude of the plane, is blowing from the southwest with a velocity of 60 miles per hour.

(a) Draw a figure that gives a visual representation of the situation.

(b) Write the velocity of the wind as a vector in component form.

(c) Write the velocity of the jet relative to the air in component form.

(d) What is the speed of the jet with respect to the ground?

(e) What is the true direction of the jet?

EXPLORATION

TRUE OR FALSE? In Exercises 103–110, use the figure to determine whether the statement is true or false. Justify your answer.

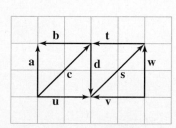

103. $\mathbf{a} = -\mathbf{d}$ 104. $\mathbf{c} = \mathbf{s}$

105. $\mathbf{a} + \mathbf{u} = \mathbf{c}$ 106. $\mathbf{v} + \mathbf{w} = -\mathbf{s}$

107. $\mathbf{a} + \mathbf{w} = -2\mathbf{d}$ 108. $\mathbf{a} + \mathbf{d} = \mathbf{0}$

109. $\mathbf{u} - \mathbf{v} = -2(\mathbf{b} + \mathbf{t})$ 110. $\mathbf{t} - \mathbf{w} = \mathbf{b} - \mathbf{a}$

111. PROOF Prove that $(\cos \theta)\mathbf{i} + (\sin \theta)\mathbf{j}$ is a unit vector for any value of θ.

112. CAPSTONE The initial and terminal points of vector $\mathbf{v}$ are $(3, -4)$ and $(9, 1)$, respectively.

(a) Write $\mathbf{v}$ in component form.

(b) Write $\mathbf{v}$ as the linear combination of the standard unit vectors $\mathbf{i}$ and $\mathbf{j}$.

(c) Sketch $\mathbf{v}$ with its initial point at the origin.

(d) Find the magnitude of $\mathbf{v}$.

113. GRAPHICAL REASONING Consider two forces

$$\mathbf{F}_1 = \langle 10, 0 \rangle \text{ and } \mathbf{F}_2 = 5\langle \cos \theta, \sin \theta \rangle.$$

(a) Find $\|\mathbf{F}_1 + \mathbf{F}_2\|$ as a function of θ.

(b) Use a graphing utility to graph the function in part (a) for $0 \le \theta < 2\pi$.

(c) Use the graph in part (b) to determine the range of the function. What is its maximum, and for what value of θ does it occur? What is its minimum, and for what value of θ does it occur?

(d) Explain why the magnitude of the resultant is never 0.

114. TECHNOLOGY Write a program for your graphing utility that graphs two vectors and their difference given the vectors in component form.

In Exercises 115 and 116, use the program in Exercise 114 to find the difference of the vectors shown in the figure.

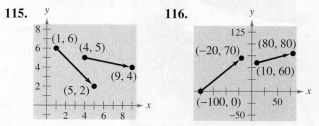

117. WRITING In your own words, state the difference between a scalar and a vector. Give examples of each.

118. WRITING Give geometric descriptions of the operations of addition of vectors and multiplication of a vector by a scalar.

119. WRITING Identify the quantity as a scalar or as a vector. Explain your reasoning.

(a) The muzzle velocity of a bullet

(b) The price of a company's stock

(c) The air temperature in a room

(d) The weight of an automobile

What you should learn

- Find the dot product of two vectors and use the properties of the dot product.
- Find the angle between two vectors and determine whether two vectors are orthogonal.
- Write a vector as the sum of two vector components.
- Use vectors to find the work done by a force.

Why you should learn it

You can use the dot product of two vectors to solve real-life problems involving two vector quantities. For instance, in Exercise 76 on page 320, you can use the dot product to find the force necessary to keep a sport utility vehicle from rolling down a hill.

Joe Raedle/Getty Images

3.4 VECTORS AND DOT PRODUCTS

The Dot Product of Two Vectors

So far you have studied two vector operations—vector addition and multiplication by a scalar—each of which yields another vector. In this section, you will study a third vector operation, the **dot product.** This product yields a scalar, rather than a vector.

Definition of the Dot Product

The **dot product** of $\mathbf{u} = \langle u_1, u_2 \rangle$ and $\mathbf{v} = \langle v_1, v_2 \rangle$ is

$$\mathbf{u} \cdot \mathbf{v} = u_1 v_1 + u_2 v_2.$$

Properties of the Dot Product

Let $\mathbf{u}$, $\mathbf{v}$, and $\mathbf{w}$ be vectors in the plane or in space and let c be a scalar.

1. $\mathbf{u} \cdot \mathbf{v} = \mathbf{v} \cdot \mathbf{u}$

2. $\mathbf{0} \cdot \mathbf{v} = 0$

3. $\mathbf{u} \cdot (\mathbf{v} + \mathbf{w}) = \mathbf{u} \cdot \mathbf{v} + \mathbf{u} \cdot \mathbf{w}$

4. $\mathbf{v} \cdot \mathbf{v} = \|\mathbf{v}\|^2$

5. $c(\mathbf{u} \cdot \mathbf{v}) = c\mathbf{u} \cdot \mathbf{v} = \mathbf{u} \cdot c\mathbf{v}$

For proofs of the properties of the dot product, see Proofs in Mathematics on page 334.

Example 1 Finding Dot Products

Find each dot product.

a. $\langle 4, 5 \rangle \cdot \langle 2, 3 \rangle$ **b.** $\langle 2, -1 \rangle \cdot \langle 1, 2 \rangle$ **c.** $\langle 0, 3 \rangle \cdot \langle 4, -2 \rangle$

Solution

a. $\langle 4, 5 \rangle \cdot \langle 2, 3 \rangle = 4(2) + 5(3)$

$$= 8 + 15$$

$$= 23$$

b. $\langle 2, -1 \rangle \cdot \langle 1, 2 \rangle = 2(1) + (-1)(2)$

$$= 2 - 2 = 0$$

c. $\langle 0, 3 \rangle \cdot \langle 4, -2 \rangle = 0(4) + 3(-2)$

$$= 0 - 6 = -6$$

CHECK*Point* Now try Exercise 7.

In Example 1, be sure you see that the dot product of two vectors is a scalar (a real number), not a vector. Moreover, notice that the dot product can be positive, zero, or negative.

Example 2   Using Properties of Dot Products

Let $\mathbf{u} = \langle -1, 3 \rangle$, $\mathbf{v} = \langle 2, -4 \rangle$, and $\mathbf{w} = \langle 1, -2 \rangle$. Find each dot product.

a. $(\mathbf{u} \cdot \mathbf{v})\mathbf{w}$

b. $\mathbf{u} \cdot 2\mathbf{v}$

Solution

Begin by finding the dot product of $\mathbf{u}$ and $\mathbf{v}$.

$$\mathbf{u} \cdot \mathbf{v} = \langle -1, 3 \rangle \cdot \langle 2, -4 \rangle$$
$$= (-1)(2) + 3(-4)$$
$$= -14$$

a. $(\mathbf{u} \cdot \mathbf{v})\mathbf{w} = -14\langle 1, -2 \rangle$
$$= \langle -14, 28 \rangle$$

b. $\mathbf{u} \cdot 2\mathbf{v} = 2(\mathbf{u} \cdot \mathbf{v})$
$$= 2(-14)$$
$$= -28$$

Notice that the product in part (a) is a vector, whereas the product in part (b) is a scalar. Can you see why?

CHECK*Point*   Now try Exercise 17.

Example 3   Dot Product and Magnitude

The dot product of $\mathbf{u}$ with itself is 5. What is the magnitude of $\mathbf{u}$?

Solution

Because $\|\mathbf{u}\|^2 = \mathbf{u} \cdot \mathbf{u}$ and $\mathbf{u} \cdot \mathbf{u} = 5$, it follows that

$$\|\mathbf{u}\| = \sqrt{\mathbf{u} \cdot \mathbf{u}}$$
$$= \sqrt{5}.$$

CHECK*Point*   Now try Exercise 25.

The Angle Between Two Vectors

The **angle between two nonzero vectors** is the angle θ, $0 \le \theta \le \pi$, between their respective standard position vectors, as shown in Figure 3.33. This angle can be found using the dot product.

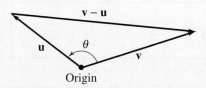

FIGURE **3.33**

Angle Between Two Vectors
If θ is the angle between two nonzero vectors $\mathbf{u}$ and $\mathbf{v}$, then $$\cos \theta = \frac{\mathbf{u} \cdot \mathbf{v}}{\|\mathbf{u}\| \, \|\mathbf{v}\|}.$$

For a proof of the angle between two vectors, see Proofs in Mathematics on page 334.

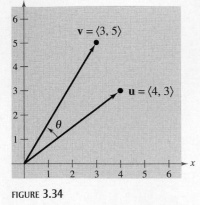

FIGURE **3.34**

Example 4   **Finding the Angle Between Two Vectors**

Find the angle θ between $\mathbf{u} = \langle 4, 3 \rangle$ and $\mathbf{v} = \langle 3, 5 \rangle$.

Solution

The two vectors and θ are shown in Figure 3.34.

$$
\begin{aligned}
\cos \theta &= \frac{\mathbf{u} \cdot \mathbf{v}}{\|\mathbf{u}\| \, \|\mathbf{v}\|} \\
&= \frac{\langle 4, 3 \rangle \cdot \langle 3, 5 \rangle}{\|\langle 4, 3 \rangle\| \, \|\langle 3, 5 \rangle\|} \\
&= \frac{27}{5\sqrt{34}}
\end{aligned}
$$

This implies that the angle between the two vectors is

$$\theta = \arccos \frac{27}{5\sqrt{34}} \approx 22.2°.$$

CHECK*Point*   Now try Exercise 35.

Rewriting the expression for the angle between two vectors in the form

$$\mathbf{u} \cdot \mathbf{v} = \|\mathbf{u}\| \, \|\mathbf{v}\| \cos \theta \qquad \text{Alternative form of dot product}$$

produces an alternative way to calculate the dot product. From this form, you can see that because $\|\mathbf{u}\|$ and $\|\mathbf{v}\|$ are always positive, $\mathbf{u} \cdot \mathbf{v}$ and $\cos \theta$ will always have the same sign. Figure 3.35 shows the five possible orientations of two vectors.

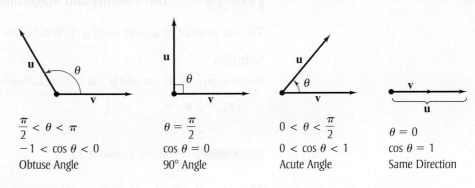

$\theta = \pi$
$\cos \theta = -1$
Opposite Direction

$\dfrac{\pi}{2} < \theta < \pi$
$-1 < \cos \theta < 0$
Obtuse Angle

$\theta = \dfrac{\pi}{2}$
$\cos \theta = 0$
90° Angle

$0 < \theta < \dfrac{\pi}{2}$
$0 < \cos \theta < 1$
Acute Angle

$\theta = 0$
$\cos \theta = 1$
Same Direction

FIGURE **3.35**

Definition of Orthogonal Vectors

The vectors $\mathbf{u}$ and $\mathbf{v}$ are **orthogonal** if $\mathbf{u} \cdot \mathbf{v} = 0$.

The terms *orthogonal* and *perpendicular* mean essentially the same thing—meeting at right angles. Note that the zero vector is orthogonal to every vector $\mathbf{u}$, because $\mathbf{0} \cdot \mathbf{u} = 0$.

Example 5   **Determining Orthogonal Vectors**

Are the vectors $\mathbf{u} = \langle 2, -3 \rangle$ and $\mathbf{v} = \langle 6, 4 \rangle$ orthogonal?

Solution

Find the dot product of the two vectors.

$$\mathbf{u} \cdot \mathbf{v} = \langle 2, -3 \rangle \cdot \langle 6, 4 \rangle = 2(6) + (-3)(4) = 0$$

Because the dot product is 0, the two vectors are orthogonal (see Figure 3.36).

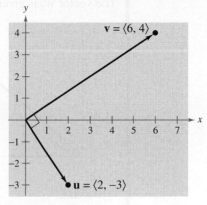

FIGURE **3.36**

CHECK *Point*   Now try Exercise 53.

Finding Vector Components

You have already seen applications in which two vectors are added to produce a resultant vector. Many applications in physics and engineering pose the reverse problem—decomposing a given vector into the sum of two **vector components.**

Consider a boat on an inclined ramp, as shown in Figure 3.37. The force $\mathbf{F}$ due to gravity pulls the boat *down* the ramp and *against* the ramp. These two orthogonal forces, $\mathbf{w}_1$ and $\mathbf{w}_2$, are vector components of $\mathbf{F}$. That is,

$$\mathbf{F} = \mathbf{w}_1 + \mathbf{w}_2. \qquad \text{Vector components of } \mathbf{F}$$

The negative of component $\mathbf{w}_1$ represents the force needed to keep the boat from rolling down the ramp, whereas $\mathbf{w}_2$ represents the force that the tires must withstand against the ramp. A procedure for finding $\mathbf{w}_1$ and $\mathbf{w}_2$ is shown on the following page.

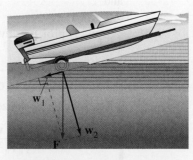

FIGURE **3.37**

Definition of Vector Components

Let $\mathbf{u}$ and $\mathbf{v}$ be nonzero vectors such that

$$\mathbf{u} = \mathbf{w}_1 + \mathbf{w}_2$$

where $\mathbf{w}_1$ and $\mathbf{w}_2$ are orthogonal and $\mathbf{w}_1$ is parallel to (or a scalar multiple of) $\mathbf{v}$, as shown in Figure 3.38. The vectors $\mathbf{w}_1$ and $\mathbf{w}_2$ are called **vector components** of $\mathbf{u}$. The vector $\mathbf{w}_1$ is the **projection** of $\mathbf{u}$ onto $\mathbf{v}$ and is denoted by

$$\mathbf{w}_1 = \text{proj}_{\mathbf{v}}\mathbf{u}.$$

The vector $\mathbf{w}_2$ is given by $\mathbf{w}_2 = \mathbf{u} - \mathbf{w}_1$.

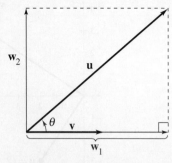

θ is acute.

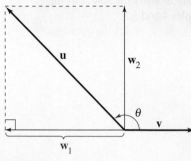

θ is obtuse.

FIGURE **3.38**

From the definition of vector components, you can see that it is easy to find the component $\mathbf{w}_2$ once you have found the projection of $\mathbf{u}$ onto $\mathbf{v}$. To find the projection, you can use the dot product, as follows.

$$\mathbf{u} = \mathbf{w}_1 + \mathbf{w}_2 = c\mathbf{v} + \mathbf{w}_2 \qquad \text{$\mathbf{w}_1$ is a scalar multiple of $\mathbf{v}$.}$$

$$\mathbf{u} \cdot \mathbf{v} = (c\mathbf{v} + \mathbf{w}_2) \cdot \mathbf{v} \qquad \text{Take dot product of each side with $\mathbf{v}$.}$$

$$= c\mathbf{v} \cdot \mathbf{v} + \mathbf{w}_2 \cdot \mathbf{v}$$

$$= c\|\mathbf{v}\|^2 + 0 \qquad \text{$\mathbf{w}_2$ and $\mathbf{v}$ are orthogonal.}$$

So,

$$c = \frac{\mathbf{u} \cdot \mathbf{v}}{\|\mathbf{v}\|^2}$$

and

$$\mathbf{w}_1 = \text{proj}_{\mathbf{v}}\mathbf{u} = c\mathbf{v} = \frac{\mathbf{u} \cdot \mathbf{v}}{\|\mathbf{v}\|^2}\mathbf{v}.$$

Projection of $\mathbf{u}$ onto $\mathbf{v}$

Let $\mathbf{u}$ and $\mathbf{v}$ be nonzero vectors. The projection of $\mathbf{u}$ onto $\mathbf{v}$ is

$$\text{proj}_{\mathbf{v}}\mathbf{u} = \left(\frac{\mathbf{u} \cdot \mathbf{v}}{\|\mathbf{v}\|^2}\right)\mathbf{v}.$$

Example 6   Decomposing a Vector into Components

Find the projection of $\mathbf{u} = \langle 3, -5 \rangle$ onto $\mathbf{v} = \langle 6, 2 \rangle$. Then write $\mathbf{u}$ as the sum of two orthogonal vectors, one of which is $\text{proj}_{\mathbf{v}}\mathbf{u}$.

Solution

The projection of $\mathbf{u}$ onto $\mathbf{v}$ is

$$\mathbf{w}_1 = \text{proj}_{\mathbf{v}}\mathbf{u} = \left(\frac{\mathbf{u} \cdot \mathbf{v}}{\|\mathbf{v}\|^2} \right)\mathbf{v} = \left(\frac{8}{40} \right)\langle 6, 2 \rangle = \left\langle \frac{6}{5}, \frac{2}{5} \right\rangle$$

as shown in Figure 3.39. The other component, $\mathbf{w}_2$, is

$$\mathbf{w}_2 = \mathbf{u} - \mathbf{w}_1 = \langle 3, -5 \rangle - \left\langle \frac{6}{5}, \frac{2}{5} \right\rangle = \left\langle \frac{9}{5}, -\frac{27}{5} \right\rangle.$$

So,

$$\mathbf{u} = \mathbf{w}_1 + \mathbf{w}_2 = \left\langle \frac{6}{5}, \frac{2}{5} \right\rangle + \left\langle \frac{9}{5}, -\frac{27}{5} \right\rangle = \langle 3, -5 \rangle.$$

CHECK *Point*   Now try Exercise 59.

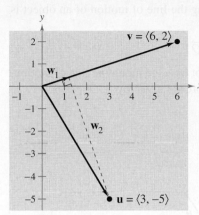

FIGURE **3.39**

Example 7   Finding a Force

A 200-pound cart sits on a ramp inclined at 30°, as shown in Figure 3.40. What force is required to keep the cart from rolling down the ramp?

Solution

Because the force due to gravity is vertical and downward, you can represent the gravitational force by the vector

$$\mathbf{F} = -200\mathbf{j}. \qquad \text{Force due to gravity}$$

To find the force required to keep the cart from rolling down the ramp, project $\mathbf{F}$ onto a unit vector $\mathbf{v}$ in the direction of the ramp, as follows.

$$\mathbf{v} = (\cos 30°)\mathbf{i} + (\sin 30°)\mathbf{j} = \frac{\sqrt{3}}{2}\mathbf{i} + \frac{1}{2}\mathbf{j} \qquad \text{Unit vector along ramp}$$

Therefore, the projection of $\mathbf{F}$ onto $\mathbf{v}$ is

$$
\begin{aligned}
\mathbf{w}_1 &= \text{proj}_{\mathbf{v}}\mathbf{F} \\
&= \left(\frac{\mathbf{F} \cdot \mathbf{v}}{\|\mathbf{v}\|^2} \right)\mathbf{v} \\
&= (\mathbf{F} \cdot \mathbf{v})\mathbf{v} \\
&= (-200)\left(\frac{1}{2} \right)\mathbf{v} \\
&= -100\left(\frac{\sqrt{3}}{2}\mathbf{i} + \frac{1}{2}\mathbf{j} \right).
\end{aligned}
$$

The magnitude of this force is 100, and so a force of 100 pounds is required to keep the cart from rolling down the ramp.

CHECK *Point*   Now try Exercise 75.

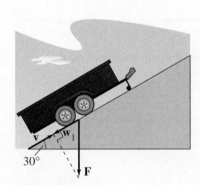

FIGURE **3.40**

Work

The work W done by a *constant* force $\mathbf{F}$ acting along the line of motion of an object is given by

$$W = (\text{magnitude of force})(\text{distance}) = \|\mathbf{F}\|\,\|\overrightarrow{PQ}\|$$

as shown in Figure 3.41. If the constant force $\mathbf{F}$ is not directed along the line of motion, as shown in Figure 3.42, the work W done by the force is given by

$$W = \|\text{proj}_{\overrightarrow{PQ}}\mathbf{F}\|\,\|\overrightarrow{PQ}\| \qquad \text{Projection form for work}$$

$$= (\cos\theta)\|\mathbf{F}\|\,\|\overrightarrow{PQ}\| \qquad \|\text{proj}_{\overrightarrow{PQ}}\mathbf{F}\| = (\cos\theta)\|\mathbf{F}\|$$

$$= \mathbf{F} \cdot \overrightarrow{PQ}. \qquad \text{Alternative form of dot product}$$

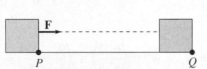

Force acts along the line of motion.
FIGURE 3.41

Force acts at angle θ with the line of motion.
FIGURE 3.42

This notion of work is summarized in the following definition.

Definition of Work

The **work** W done by a constant force $\mathbf{F}$ as its point of application moves along the vector $\overrightarrow{PQ}$ is given by either of the following.

1. $W = \|\text{proj}_{\overrightarrow{PQ}}\mathbf{F}\|\,\|\overrightarrow{PQ}\|$ Projection form

2. $W = \mathbf{F} \cdot \overrightarrow{PQ}$ Dot product form

---

Example 8   Finding Work

To close a sliding barn door, a person pulls on a rope with a constant force of 50 pounds at a constant angle of $60°$, as shown in Figure 3.43. Find the work done in moving the barn door 12 feet to its closed position.

Solution

Using a projection, you can calculate the work as follows.

$$W = \|\text{proj}_{\overrightarrow{PQ}}\mathbf{F}\|\,\|\overrightarrow{PQ}\| \qquad \text{Projection form for work}$$

$$= (\cos 60°)\|\mathbf{F}\|\,\|\overrightarrow{PQ}\|$$

$$= \frac{1}{2}(50)(12) = 300 \text{ foot-pounds}$$

So, the work done is 300 foot-pounds. You can verify this result by finding the vectors $\mathbf{F}$ and $\overrightarrow{PQ}$ and calculating their dot product.

CHECK*Point*   Now try Exercise 79.

FIGURE 3.43

3.4 ‖ EXERCISES

See www.CalcChat.com for worked-out solutions to odd-numbered exercises.

VOCABULARY: Fill in the blanks.

1. The _____ _____ of two vectors yields a scalar, rather than a vector.
2. The dot product of $\mathbf{u} = \langle u_1, u_2 \rangle$ and $\mathbf{v} = \langle v_1, v_2 \rangle$ is $\mathbf{u} \cdot \mathbf{v} =$ _____ .
3. If θ is the angle between two nonzero vectors $\mathbf{u}$ and $\mathbf{v}$, then $\cos \theta =$ _____ .
4. The vectors $\mathbf{u}$ and $\mathbf{v}$ are _____ if $\mathbf{u} \cdot \mathbf{v} = 0$.
5. The projection of $\mathbf{u}$ onto $\mathbf{v}$ is given by $\text{proj}_{\mathbf{v}}\mathbf{u} =$ _____ .
6. The work W done by a constant force $\mathbf{F}$ as its point of application moves along the vector $\overrightarrow{PQ}$ is given by $W =$ _____ or $W =$ _____ .

SKILLS AND APPLICATIONS

In Exercises 7–14, find the dot product of $\mathbf{u}$ and $\mathbf{v}$.

7. $\mathbf{u} = \langle 7, 1 \rangle$
   $\mathbf{v} = \langle -3, 2 \rangle$
8. $\mathbf{u} = \langle 6, 10 \rangle$
   $\mathbf{v} = \langle -2, 3 \rangle$
9. $\mathbf{u} = \langle -4, 1 \rangle$
   $\mathbf{v} = \langle 2, -3 \rangle$
10. $\mathbf{u} = \langle -2, 5 \rangle$
    $\mathbf{v} = \langle -1, -8 \rangle$
11. $\mathbf{u} = 4\mathbf{i} - 2\mathbf{j}$
    $\mathbf{v} = \mathbf{i} - \mathbf{j}$
12. $\mathbf{u} = 3\mathbf{i} + 4\mathbf{j}$
    $\mathbf{v} = 7\mathbf{i} - 2\mathbf{j}$
13. $\mathbf{u} = 3\mathbf{i} + 2\mathbf{j}$
    $\mathbf{v} = -2\mathbf{i} - 3\mathbf{j}$
14. $\mathbf{u} = \mathbf{i} - 2\mathbf{j}$
    $\mathbf{v} = -2\mathbf{i} + \mathbf{j}$

In Exercises 15–24, use the vectors $\mathbf{u} = \langle 3, 3 \rangle$, $\mathbf{v} = \langle -4, 2 \rangle$, and $\mathbf{w} = \langle 3, -1 \rangle$ to find the indicated quantity. State whether the result is a vector or a scalar.

15. $\mathbf{u} \cdot \mathbf{u}$
16. $3\mathbf{u} \cdot \mathbf{v}$
17. $(\mathbf{u} \cdot \mathbf{v})\mathbf{v}$
18. $(\mathbf{v} \cdot \mathbf{u})\mathbf{w}$
19. $(3\mathbf{w} \cdot \mathbf{v})\mathbf{u}$
20. $(\mathbf{u} \cdot 2\mathbf{v})\mathbf{w}$
21. $\|\mathbf{w}\| - 1$
22. $2 - \|\mathbf{u}\|$
23. $(\mathbf{u} \cdot \mathbf{v}) - (\mathbf{u} \cdot \mathbf{w})$
24. $(\mathbf{v} \cdot \mathbf{u}) - (\mathbf{w} \cdot \mathbf{v})$

In Exercises 25–30, use the dot product to find the magnitude of $\mathbf{u}$.

25. $\mathbf{u} = \langle -8, 15 \rangle$
26. $\mathbf{u} = \langle 4, -6 \rangle$
27. $\mathbf{u} = 20\mathbf{i} + 25\mathbf{j}$
28. $\mathbf{u} = 12\mathbf{i} - 16\mathbf{j}$
29. $\mathbf{u} = 6\mathbf{j}$
30. $\mathbf{u} = -21\mathbf{i}$

In Exercises 31–40, find the angle θ between the vectors.

31. $\mathbf{u} = \langle 1, 0 \rangle$
    $\mathbf{v} = \langle 0, -2 \rangle$
32. $\mathbf{u} = \langle 3, 2 \rangle$
    $\mathbf{v} = \langle 4, 0 \rangle$
33. $\mathbf{u} = 3\mathbf{i} + 4\mathbf{j}$
    $\mathbf{v} = -2\mathbf{j}$
34. $\mathbf{u} = 2\mathbf{i} - 3\mathbf{j}$
    $\mathbf{v} = \mathbf{i} - 2\mathbf{j}$
35. $\mathbf{u} = 2\mathbf{i} - \mathbf{j}$
    $\mathbf{v} = 6\mathbf{i} + 4\mathbf{j}$
36. $\mathbf{u} = -6\mathbf{i} - 3\mathbf{j}$
    $\mathbf{v} = -8\mathbf{i} + 4\mathbf{j}$

37. $\mathbf{u} = 5\mathbf{i} + 5\mathbf{j}$
    $\mathbf{v} = -6\mathbf{i} + 6\mathbf{j}$
38. $\mathbf{u} = 2\mathbf{i} - 3\mathbf{j}$
    $\mathbf{v} = 4\mathbf{i} + 3\mathbf{j}$
39. $\mathbf{u} = \cos\left(\dfrac{\pi}{3}\right)\mathbf{i} + \sin\left(\dfrac{\pi}{3}\right)\mathbf{j}$
    $\mathbf{v} = \cos\left(\dfrac{3\pi}{4}\right)\mathbf{i} + \sin\left(\dfrac{3\pi}{4}\right)\mathbf{j}$
40. $\mathbf{u} = \cos\left(\dfrac{\pi}{4}\right)\mathbf{i} + \sin\left(\dfrac{\pi}{4}\right)\mathbf{j}$
    $\mathbf{v} = \cos\left(\dfrac{\pi}{2}\right)\mathbf{i} + \sin\left(\dfrac{\pi}{2}\right)\mathbf{j}$

In Exercises 41–44, graph the vectors and find the degree measure of the angle θ between the vectors.

41. $\mathbf{u} = 3\mathbf{i} + 4\mathbf{j}$
    $\mathbf{v} = -7\mathbf{i} + 5\mathbf{j}$
42. $\mathbf{u} = 6\mathbf{i} + 3\mathbf{j}$
    $\mathbf{v} = -4\mathbf{i} + 4\mathbf{j}$
43. $\mathbf{u} = 5\mathbf{i} + 5\mathbf{j}$
    $\mathbf{v} = -8\mathbf{i} + 8\mathbf{j}$
44. $\mathbf{u} = 2\mathbf{i} - 3\mathbf{j}$
    $\mathbf{v} = 8\mathbf{i} + 3\mathbf{j}$

In Exercises 45–48, use vectors to find the interior angles of the triangle with the given vertices.

45. $(1, 2), (3, 4), (2, 5)$
46. $(-3, -4), (1, 7), (8, 2)$
47. $(-3, 0), (2, 2), (0, 6)$
48. $(-3, 5), (-1, 9), (7, 9)$

In Exercises 49–52, find $\mathbf{u} \cdot \mathbf{v}$, where θ is the angle between $\mathbf{u}$ and $\mathbf{v}$.

49. $\|\mathbf{u}\| = 4, \|\mathbf{v}\| = 10, \theta = \dfrac{2\pi}{3}$
50. $\|\mathbf{u}\| = 100, \|\mathbf{v}\| = 250, \theta = \dfrac{\pi}{6}$
51. $\|\mathbf{u}\| = 9, \|\mathbf{v}\| = 36, \theta = \dfrac{3\pi}{4}$
52. $\|\mathbf{u}\| = 4, \|\mathbf{v}\| = 12, \theta = \dfrac{\pi}{3}$

In Exercises 53–58, determine whether **u** and **v** are orthogonal, parallel, or neither.

53. $\mathbf{u} = \langle -12, 30 \rangle$
$\mathbf{v} = \left\langle \frac{1}{2}, -\frac{5}{4} \right\rangle$

54. $\mathbf{u} = \langle 3, 15 \rangle$
$\mathbf{v} = \langle -1, 5 \rangle$

55. $\mathbf{u} = \frac{1}{4}(3\mathbf{i} - \mathbf{j})$
$\mathbf{v} = 5\mathbf{i} + 6\mathbf{j}$

56. $\mathbf{u} = \mathbf{i}$
$\mathbf{v} = -2\mathbf{i} + 2\mathbf{j}$

57. $\mathbf{u} = 2\mathbf{i} - 2\mathbf{j}$
$\mathbf{v} = -\mathbf{i} - \mathbf{j}$

58. $\mathbf{u} = \langle \cos \theta, \sin \theta \rangle$
$\mathbf{v} = \langle \sin \theta, -\cos \theta \rangle$

In Exercises 59–62, find the projection of **u** onto **v**. Then write **u** as the sum of two orthogonal vectors, one of which is proj$_\mathbf{v}$**u**.

59. $\mathbf{u} = \langle 2, 2 \rangle$
$\mathbf{v} = \langle 6, 1 \rangle$

60. $\mathbf{u} = \langle 4, 2 \rangle$
$\mathbf{v} = \langle 1, -2 \rangle$

61. $\mathbf{u} = \langle 0, 3 \rangle$
$\mathbf{v} = \langle 2, 15 \rangle$

62. $\mathbf{u} = \langle -3, -2 \rangle$
$\mathbf{v} = \langle -4, -1 \rangle$

In Exercises 63–66, use the graph to determine mentally the projection of **u** onto **v**. (The coordinates of the terminal points of the vectors in standard position are given.) Use the formula for the projection of **u** onto **v** to verify your result.

63.

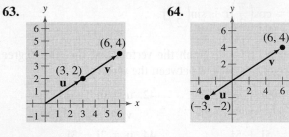

64.

65.

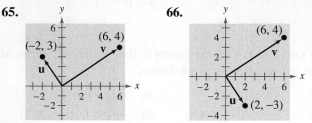

66.

In Exercises 67–70, find two vectors in opposite directions that are orthogonal to the vector **u**. (There are many correct answers.)

67. $\mathbf{u} = \langle 3, 5 \rangle$

68. $\mathbf{u} = \langle -8, 3 \rangle$

69. $\mathbf{u} = \frac{1}{2}\mathbf{i} - \frac{2}{3}\mathbf{j}$

70. $\mathbf{u} = -\frac{5}{2}\mathbf{i} - 3\mathbf{j}$

WORK In Exercises 71 and 72, find the work done in moving a particle from *P* to *Q* if the magnitude and direction of the force are given by **v**.

71. $P(0, 0)$, $Q(4, 7)$, $\mathbf{v} = \langle 1, 4 \rangle$

72. $P(1, 3)$, $Q(-3, 5)$, $\mathbf{v} = -2\mathbf{i} + 3\mathbf{j}$

73. REVENUE The vector $\mathbf{u} = \langle 4600, 5250 \rangle$ gives the numbers of units of two models of cellular phones produced by a telecommunications company. The vector $\mathbf{v} = \langle 79.99, 99.99 \rangle$ gives the prices (in dollars) of the two models of cellular phones, respectively.

(a) Find the dot product $\mathbf{u} \cdot \mathbf{v}$ and interpret the result in the context of the problem.

(b) Identify the vector operation used to increase the prices by 5%.

74. REVENUE The vector $\mathbf{u} = \langle 3140, 2750 \rangle$ gives the numbers of hamburgers and hot dogs, respectively, sold at a fast-food stand in one month. The vector $\mathbf{v} = \langle 2.25, 1.75 \rangle$ gives the prices (in dollars) of the food items.

(a) Find the dot product $\mathbf{u} \cdot \mathbf{v}$ and interpret the result in the context of the problem.

(b) Identify the vector operation used to increase the prices by 2.5%.

75. BRAKING LOAD A truck with a gross weight of 30,000 pounds is parked on a slope of $d°$ (see figure). Assume that the only force to overcome is the force of gravity.

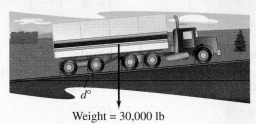

Weight = 30,000 lb

(a) Find the force required to keep the truck from rolling down the hill in terms of the slope *d*.

(b) Use a graphing utility to complete the table.

d	0°	1°	2°	3°	4°	5°
Force						

d	6°	7°	8°	9°	10°
Force					

(c) Find the force perpendicular to the hill when $d = 5°$.

76. BRAKING LOAD A sport utility vehicle with a gross weight of 5400 pounds is parked on a slope of 10°. Assume that the only force to overcome is the force of gravity. Find the force required to keep the vehicle from rolling down the hill. Find the force perpendicular to the hill.

77. WORK   Determine the work done by a person lifting a 245-newton bag of sugar 3 meters.

78. WORK   Determine the work done by a crane lifting a 2400-pound car 5 feet.

79. WORK   A force of 45 pounds exerted at an angle of 30° above the horizontal is required to slide a table across a floor (see figure). The table is dragged 20 feet. Determine the work done in sliding the table.

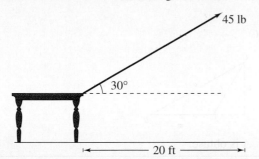

80. WORK   A tractor pulls a log 800 meters, and the tension in the cable connecting the tractor and log is approximately 15,691 newtons. The direction of the force is 35° above the horizontal. Approximate the work done in pulling the log.

81. WORK   One of the events in a local strongman contest is to pull a cement block 100 feet. One competitor pulls the block by exerting a force of 250 pounds on a rope attached to the block at an angle of 30° with the horizontal (see figure). Find the work done in pulling the block.

82. WORK   A toy wagon is pulled by exerting a force of 25 pounds on a handle that makes a 20° angle with the horizontal (see figure). Find the work done in pulling the wagon 50 feet.

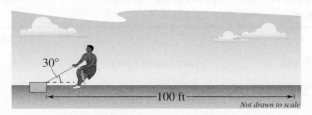

83. PROGRAMMING   Given vectors **u** and **v** in component form, write a program for your graphing utility in which the output is (a) ‖**u**‖, (b) ‖**v**‖, and (c) the angle between **u** and **v**.

84. PROGRAMMING   Use the program you wrote in Exercise 83 to find the angle between the given vectors.

(a) **u** = ⟨8, −4⟩ and **v** = ⟨2, 5⟩

(b) **u** = ⟨2, −6⟩ and **v** = ⟨4, 1⟩

85. PROGRAMMING   Given vectors **u** and **v** in component form, write a program for your graphing utility in which the output is the component form of the projection of **u** onto **v**.

86. PROGRAMMING   Use the program you wrote in Exercise 85 to find the projection of **u** onto **v** for the given vectors.

(a) **u** = ⟨5, 6⟩ and **v** = ⟨−1, 3⟩

(b) **u** = ⟨3, −2⟩ and **v** = ⟨−2, 1⟩

EXPLORATION

TRUE OR FALSE? In Exercises 87 and 88, determine whether the statement is true or false. Justify your answer.

87. The work W done by a constant force **F** acting along the line of motion of an object is represented by a vector.

88. A sliding door moves along the line of vector $\overrightarrow{PQ}$. If a force is applied to the door along a vector that is orthogonal to $\overrightarrow{PQ}$, then no work is done.

89. PROOF   Use vectors to prove that the diagonals of a rhombus are perpendicular.

90. CAPSTONE   What is known about θ, the angle between two nonzero vectors **u** and **v**, under each condition (see figure)?

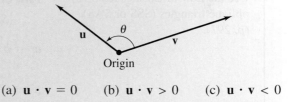

(a) **u** · **v** = 0   (b) **u** · **v** > 0   (c) **u** · **v** < 0

91. THINK ABOUT IT   What can be said about the vectors **u** and **v** under each condition?

(a) The projection of **u** onto **v** equals **u**.

(b) The projection of **u** onto **v** equals **0**.

92. PROOF   Prove the following.

$$\|\mathbf{u} - \mathbf{v}\|^2 = \|\mathbf{u}\|^2 + \|\mathbf{v}\|^2 - 2\mathbf{u} \cdot \mathbf{v}$$

93. PROOF   Prove that if **u** is a unit vector and θ is the angle between **u** and **i**, then $\mathbf{u} = \cos\theta\,\mathbf{i} + \sin\theta\,\mathbf{j}$.

94. PROOF   Prove that if **u** is a unit vector and θ is the angle between **u** and **j**, then

$$\mathbf{u} = \cos\left(\frac{\pi}{2} - \theta\right)\mathbf{i} + \sin\left(\frac{\pi}{2} - \theta\right)\mathbf{j}.$$

3 CHAPTER SUMMARY

What Did You Learn?	Explanation/Examples	Review Exercises
Section 3.1		
Use the Law of Sines to solve oblique triangles (AAS or ASA) (p. 282).	**Law of Sines** If ABC is a triangle with sides a, b, and c, then $$\frac{a}{\sin A} = \frac{b}{\sin B} = \frac{c}{\sin C}.$$ A is acute.　　　　A is obtuse.	1–12
Use the Law of Sines to solve oblique triangles (SSA) (p. 284).	If two sides and one opposite angle are given, three possible situations can occur: (1) no such triangle exists (see Example 4), (2) one such triangle exists (see Example 3), or (3) two distinct triangles may satisfy the conditions (see Example 5).	1–12
Find the areas of oblique triangles (p. 286).	The area of any triangle is one-half the product of the lengths of two sides times the sine of their included angle. That is, $$\text{Area} = \tfrac{1}{2}bc \sin A = \tfrac{1}{2}ab \sin C = \tfrac{1}{2}ac \sin B.$$	13–16
Use the Law of Sines to model and solve real-life problems (p. 287).	The Law of Sines can be used to approximate the total distance of a boat race course. (See Example 7.)	17–20
Section 3.2		
Use the Law of Cosines to solve oblique triangles (SSS or SAS) (p. 291).	**Law of Cosines** *Standard Form*　　　　　*Alternative Form* $a^2 = b^2 + c^2 - 2bc \cos A$　　$\cos A = \dfrac{b^2 + c^2 - a^2}{2bc}$ $b^2 = a^2 + c^2 - 2ac \cos B$　　$\cos B = \dfrac{a^2 + c^2 - b^2}{2ac}$ $c^2 = a^2 + b^2 - 2ab \cos C$　　$\cos C = \dfrac{a^2 + b^2 - c^2}{2ab}$	21–30
Use the Law of Cosines to model and solve real-life problems (p. 293).	The Law of Cosines can be used to find the distance between the pitcher's mound and first base on a women's softball field. (See Example 3.)	35–38
Use Heron's Area Formula to find the area of a triangle (p. 294).	**Heron's Area Formula:** Given any triangle with sides of lengths a, b, and c, the area of the triangle is $$\text{Area} = \sqrt{s(s-a)(s-b)(s-c)}$$ where $s = \dfrac{a+b+c}{2}$.	39–42

What Did You Learn?	Explanation/Examples	Review Exercises
Section 3.3 Represent vectors as directed line segments *(p. 299).*		43, 44
Write the component forms of vectors *(p. 300).*	The component form of the vector with initial point $P(p_1, p_2)$ and terminal point $Q(q_1, q_2)$ is given by $$\overrightarrow{PQ} = \langle q_1 - p_1, q_2 - p_2 \rangle = \langle v_1, v_2 \rangle = \mathbf{v}.$$	45–50
Perform basic vector operations and represent them graphically *(p. 301).*	Let $\mathbf{u} = \langle u_1, u_2 \rangle$ and $\mathbf{v} = \langle v_1, v_2 \rangle$ be vectors and let k be a scalar (a real number). $$\mathbf{u} + \mathbf{v} = \langle u_1 + v_1, u_2 + v_2 \rangle$$ $$k\mathbf{u} = \langle ku_1, ku_2 \rangle$$ $$-\mathbf{v} = \langle -v_1, -v_2 \rangle$$ $$\mathbf{u} - \mathbf{v} = \langle u_1 - v_1, u_2 - v_2 \rangle$$	51–62
Write vectors as linear combinations of unit vectors *(p. 303).*	$$\mathbf{v} = \langle v_1, v_2 \rangle = v_1\langle 1, 0 \rangle + v_2\langle 0, 1 \rangle = v_1\mathbf{i} + v_2\mathbf{j}$$ The scalars v_1 and v_2 are the horizontal and vertical components of $\mathbf{v}$, respectively. The vector sum $v_1\mathbf{i} + v_2\mathbf{j}$ is the linear combination of the vectors $\mathbf{i}$ and $\mathbf{j}$.	63–68
Find the direction angles of vectors *(p. 305).*	If $\mathbf{u} = 2i + 2j$, then the direction angle is $$\tan \theta = \frac{2}{2} = 1.$$ So, $\theta = 45°$.	69–74
Use vectors to model and solve real-life problems *(p. 306).*	Vectors can be used to find the resultant speed and direction of an airplane. (See Example 10.)	75–78
Section 3.4 Find the dot product of two vectors and use the properties of the dot product *(p. 312).*	The dot product of $\mathbf{u} = \langle u_1, u_2 \rangle$ and $\mathbf{v} = \langle v_1, v_2 \rangle$ is $$\mathbf{u} \cdot \mathbf{v} = u_1v_1 + u_2v_2.$$	79–90
Find the angle between two vectors and determine whether two vectors are orthogonal *(p. 313).*	If θ is the angle between two nonzero vectors $\mathbf{u}$ and $\mathbf{v}$, then $$\cos \theta = \frac{\mathbf{u} \cdot \mathbf{v}}{\|\mathbf{u}\| \, \|\mathbf{v}\|}.$$ Vectors $\mathbf{u}$ and $\mathbf{v}$ are orthogonal if $\mathbf{u} \cdot \mathbf{v} = 0$.	91–98
Write a vector as the sum of two vector components *(p. 315).*	Many applications in physics and engineering require the decomposition of a given vector into the sum of two vector components. (See Example 7.)	99–102
Use vectors to find the work done by a force *(p. 318).*	The work W done by a constant force $\mathbf{F}$ as its point of application moves along the vector $\overrightarrow{PQ}$ is given by either of the following. 1. $W = \|\text{proj}_{\overrightarrow{PQ}}\mathbf{F}\| \, \|\overrightarrow{PQ}\|$ 2. $W = \mathbf{F} \cdot \overrightarrow{PQ}$	103–106

3 REVIEW EXERCISES
See www.CalcChat.com for worked-out solutions to odd-numbered exercises.

3.1 In Exercises 1–12, use the Law of Sines to solve (if possible) the triangle. If two solutions exist, find both. Round your answers to two decimal places.

1.
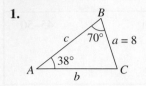

2.
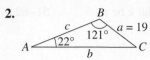

3. $B = 72°$, $C = 82°$, $b = 54$
4. $B = 10°$, $C = 20°$, $c = 33$
5. $A = 16°$, $B = 98°$, $c = 8.4$
6. $A = 95°$, $B = 45°$, $c = 104.8$
7. $A = 24°$, $C = 48°$, $b = 27.5$
8. $B = 64°$, $C = 36°$, $a = 367$
9. $B = 150°$, $b = 30$, $c = 10$
10. $B = 150°$, $a = 10$, $b = 3$
11. $A = 75°$, $a = 51.2$, $b = 33.7$
12. $B = 25°$, $a = 6.2$, $b = 4$

In Exercises 13–16, find the area of the triangle having the indicated angle and sides.

13. $A = 33°$, $b = 7$, $c = 10$
14. $B = 80°$, $a = 4$, $c = 8$
15. $C = 119°$, $a = 18$, $b = 6$
16. $A = 11°$, $b = 22$, $c = 21$

17. **HEIGHT** From a certain distance, the angle of elevation to the top of a building is 17°. At a point 50 meters closer to the building, the angle of elevation is 31°. Approximate the height of the building.

18. **GEOMETRY** Find the length of the side w of the parallelogram.

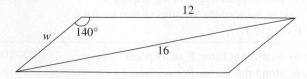

19. **HEIGHT** A tree stands on a hillside of slope 28° from the horizontal. From a point 75 feet down the hill, the angle of elevation to the top of the tree is 45° (see figure). Find the height of the tree.

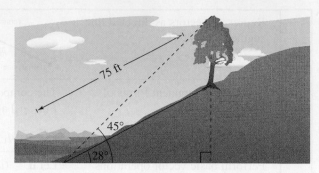

FIGURE FOR **19**

20. **RIVER WIDTH** A surveyor finds that a tree on the opposite bank of a river flowing due east has a bearing of N 22° 30′ E from a certain point and a bearing of N 15° W from a point 400 feet downstream. Find the width of the river.

3.2 In Exercises 21–30, use the Law of Cosines to solve the triangle. Round your answers to two decimal places.

21.
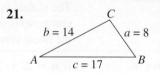

22.
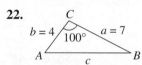

23. $a = 6$, $b = 9$, $c = 14$
24. $a = 75$, $b = 50$, $c = 110$
25. $a = 2.5$, $b = 5.0$, $c = 4.5$
26. $a = 16.4$, $b = 8.8$, $c = 12.2$
27. $B = 108°$, $a = 11$, $c = 11$
28. $B = 150°$, $a = 10$, $c = 20$
29. $C = 43°$, $a = 22.5$, $b = 31.4$
30. $A = 62°$, $b = 11.34$, $c = 19.52$

In Exercises 31–34, determine whether the Law of Sines or the Law of Cosines is needed to solve the triangle. Then solve the triangle.

31. $b = 9$, $c = 13$, $C = 64°$
32. $a = 4$, $c = 5$, $B = 52°$
33. $a = 13$, $b = 15$, $c = 24$
34. $A = 44°$, $B = 31°$, $c = 2.8$

35. GEOMETRY The lengths of the diagonals of a parallelogram are 10 feet and 16 feet. Find the lengths of the sides of the parallelogram if the diagonals intersect at an angle of 28°.

36. GEOMETRY The lengths of the diagonals of a parallelogram are 30 meters and 40 meters. Find the lengths of the sides of the parallelogram if the diagonals intersect at an angle of 34°.

37. SURVEYING To approximate the length of a marsh, a surveyor walks 425 meters from point A to point B. Then the surveyor turns 65° and walks 300 meters to point C (see figure). Approximate the length AC of the marsh.

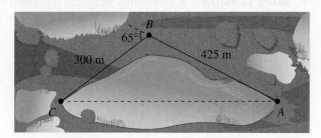

38. NAVIGATION Two planes leave an airport at approximately the same time. One is flying 425 miles per hour at a bearing of 355°, and the other is flying 530 miles per hour at a bearing of 67° (see figure). Determine the distance between the planes after they have flown for 2 hours.

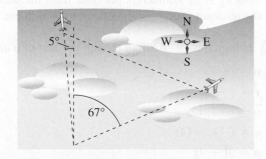

In Exercises 39–42, use Heron's Area Formula to find the area of the triangle.

39. $a = 3$, $b = 6$, $c = 8$

40. $a = 15$, $b = 8$, $c = 10$

41. $a = 12.3$, $b = 15.8$, $c = 3.7$

42. $a = \frac{4}{5}$, $b = \frac{3}{4}$, $c = \frac{5}{8}$

3.3 In Exercises 43 and 44, show that **u** and **v** are equivalent.

43.

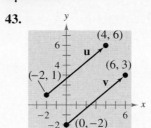

44.

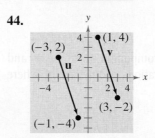

In Exercises 45–50, find the component form of the vector **v** satisfying the conditions.

45.

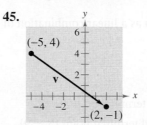

46.

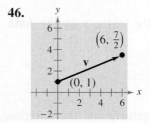

47. Initial point: $(0, 10)$; terminal point: $(7, 3)$

48. Initial point: $(1, 5)$; terminal point: $(15, 9)$

49. $\|\mathbf{v}\| = 8$, $\theta = 120°$

50. $\|\mathbf{v}\| = \frac{1}{2}$, $\theta = 225°$

In Exercises 51–58, find (a) **u** + **v**, (b) **u** − **v**, (c) 4**u**, and (d) 3**v** + 5**u**.

51. **u** = ⟨−1, −3⟩, **v** = ⟨−3, 6⟩

52. **u** = ⟨4, 5⟩, **v** = ⟨0, −1⟩

53. **u** = ⟨−5, 2⟩, **v** = ⟨4, 4⟩

54. **u** = ⟨1, −8⟩, **v** = ⟨3, −2⟩

55. **u** = 2**i** − **j**, **v** = 5**i** + 3**j**

56. **u** = −7**i** − 3**j**, **v** = 4**i** − **j**

57. **u** = 4**i**, **v** = −**i** + 6**j**

58. **u** = −6**j**, **v** = **i** + **j**

In Exercises 59–62, find the component form of **w** and sketch the specified vector operations geometrically, where **u** = 6**i** − 5**j** and **v** = 10**i** + 3**j**.

59. **w** = 2**u** + **v**

60. **w** = 4**u** − 5**v**

61. **w** = 3**v**

62. **w** = ½**v**

In Exercises 63–66, write vector **u** as a linear combination of the standard unit vectors **i** and **j**.

63. **u** = ⟨−1, 5⟩

64. **u** = ⟨−6, −8⟩

65. **u** has initial point (3, 4) and terminal point (9, 8).

66. **u** has initial point (−2, 7) and terminal point (5, −9).

In Exercises 67 and 68, write the vector **v** in the form ‖**v**‖(cos θ)**i** + ‖**v**‖(sin θ)**j**.

67. **v** = −10**i** + 10**j**

68. **v** = 4**i** − **j**

In Exercises 69–74, find the magnitude and the direction angle of the vector **v**.

69. **v** = 7(cos 60°**i** + sin 60°**j**)

70. **v** = 3(cos 150°**i** + sin 150°**j**)

71. **v** = 5**i** + 4**j**

72. **v** = −4**i** + 7**j**

73. **v** = −3**i** − 3**j**

74. **v** = 8**i** − **j**

75. CABLE TENSION In a manufacturing process, an electric hoist lifts 200-pound ingots. Find the tension in the support cables (see figure).

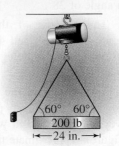

76. ROPE TENSION A 180-pound weight is supported by two ropes, as shown in the figure. Find the tension in each rope.

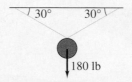

77. NAVIGATION An airplane has an airspeed of 430 miles per hour at a bearing of 135°. The wind velocity is 35 miles per hour in the direction of N 30° E. Find the resultant speed and direction of the airplane.

78. NAVIGATION An airplane has an airspeed of 724 kilometers per hour at a bearing of 30°. The wind velocity is 32 kilometers per hour from the west. Find the resultant speed and direction of the airplane.

3.4 In Exercises 79–82, find the dot product of **u** and **v**.

79. **u** = ⟨6, 7⟩
    v = ⟨−3, 9⟩

80. **u** = ⟨−7, 12⟩
    v = ⟨−4, −14⟩

81. **u** = 3**i** + 7**j**
    v = 11**i** − 5**j**

82. **u** = −7**i** + 2**j**
    v = 16**i** − 12**j**

In Exercises 83–90, use the vectors **u** = ⟨−4, 2⟩ and **v** = ⟨5, 1⟩ to find the indicated quantity. State whether the result is a vector or a scalar.

83. 2**u** · **u**

84. 3**u** · **v**

85. 4 − ‖**u**‖

86. ‖**v**‖²

87. **u**(**u** · **v**)

88. (**u** · **v**)**v**

89. (**u** · **u**) − (**u** · **v**)

90. (**v** · **v**) − (**v** · **u**)

In Exercises 91–94, find the angle θ between the vectors.

91. $\mathbf{u} = \cos\dfrac{7\pi}{4}\mathbf{i} + \sin\dfrac{7\pi}{4}\mathbf{j}$

$\mathbf{v} = \cos\dfrac{5\pi}{6}\mathbf{i} + \sin\dfrac{5\pi}{6}\mathbf{j}$

92. $\mathbf{u} = \cos 45°\mathbf{i} + \sin 45°\mathbf{j}$

$\mathbf{v} = \cos 300°\mathbf{i} + \sin 300°\mathbf{j}$

93. $\mathbf{u} = \langle 2\sqrt{2}, -4\rangle$

$\mathbf{v} = \langle -\sqrt{2}, 1\rangle$

94. $\mathbf{u} = \langle 3, \sqrt{3}\rangle$

$\mathbf{v} = \langle 4, 3\sqrt{3}\rangle$

In Exercises 95–98, determine whether **u** and **v** are orthogonal, parallel, or neither.

95. $\mathbf{u} = \langle -3, 8\rangle$

$\mathbf{v} = \langle 8, 3\rangle$

96. $\mathbf{u} = \langle \frac{1}{4}, -\frac{1}{2}\rangle$

$\mathbf{v} = \langle -2, 4\rangle$

97. $\mathbf{u} = -\mathbf{i}$

$\mathbf{v} = \mathbf{i} + 2\mathbf{j}$

98. $\mathbf{u} = -2\mathbf{i} + \mathbf{j}$

$\mathbf{v} = 3\mathbf{i} + 6\mathbf{j}$

In Exercises 99–102, find the projection of **u** onto **v**. Then write **u** as the sum of two orthogonal vectors, one of which is $\text{proj}_{\mathbf{v}}\mathbf{u}$.

99. $\mathbf{u} = \langle -4, 3\rangle$, $\mathbf{v} = \langle -8, -2\rangle$

100. $\mathbf{u} = \langle 5, 6\rangle$, $\mathbf{v} = \langle 10, 0\rangle$

101. $\mathbf{u} = \langle 2, 7\rangle$, $\mathbf{v} = \langle 1, -1\rangle$

102. $\mathbf{u} = \langle -3, 5\rangle$, $\mathbf{v} = \langle -5, 2\rangle$

WORK In Exercises 103 and 104, find the work done in moving a particle from P to Q if the magnitude and direction of the force are given by **v**.

103. $P(5, 3)$, $Q(8, 9)$, $\mathbf{v} = \langle 2, 7\rangle$

104. $P(-2, -9)$, $Q(-12, 8)$, $\mathbf{v} = 3\mathbf{i} - 6\mathbf{j}$

105. WORK Determine the work done (in foot-pounds) by a crane lifting an 18,000-pound truck 48 inches.

106. WORK A mover exerts a horizontal force of 25 pounds on a crate as it is pushed up a ramp that is 12 feet long and inclined at an angle of 20° above the horizontal. Find the work done in pushing the crate.

EXPLORATION

TRUE OR FALSE? In Exercises 107–110, determine whether the statement is true or false. Justify your answer.

107. The Law of Sines is true if one of the angles in the triangle is a right angle.

108. When the Law of Sines is used, the solution is always unique.

109. If **u** is a unit vector in the direction of **v**, then $\mathbf{v} = \|\mathbf{v}\|\mathbf{u}$.

110. If $\mathbf{v} = a\mathbf{i} + b\mathbf{j} = \mathbf{0}$, then $a = -b$.

111. State the Law of Sines from memory.

112. State the Law of Cosines from memory.

113. What characterizes a vector in the plane?

114. Which vectors in the figure appear to be equivalent?

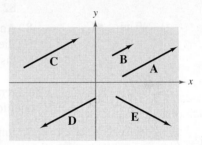

115. The vectors **u** and **v** have the same magnitudes in the two figures. In which figure will the magnitude of the sum be greater? Give a reason for your answer.

(a)

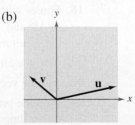

(b)

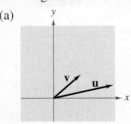

116. Give a geometric description of the scalar multiple $k\mathbf{u}$ of the vector **u**, for $k > 0$ and for $k < 0$.

117. Give a geometric description of the sum of the vectors **u** and **v**.

3 CHAPTER TEST

See www.CalcChat.com for worked-out solutions to odd-numbered exercises.

Take this test as you would take a test in class. When you are finished, check your work against the answers given in the back of the book.

In Exercises 1–6, use the information to solve (if possible) the triangle. If two solutions exist, find both solutions. Round your answers to two decimal places.

1. $A = 24°$, $B = 68°$, $a = 12.2$
2. $B = 110°$, $C = 28°$, $a = 15.6$
3. $A = 24°$, $a = 11.2$, $b = 13.4$
4. $a = 4.0$, $b = 7.3$, $c = 12.4$
5. $B = 100°$, $a = 15$, $b = 23$
6. $C = 121°$, $a = 34$, $b = 55$

7. A triangular parcel of land has borders of lengths 60 meters, 70 meters, and 82 meters. Find the area of the parcel of land.

8. An airplane flies 370 miles from point A to point B with a bearing of 24°. It then flies 240 miles from point B to point C with a bearing of 37° (see figure). Find the distance and bearing from point A to point C.

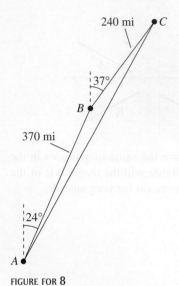

240 mi · C

37°

B ·

370 mi

24°

A ·

FIGURE FOR 8

In Exercises 9 and 10, find the component form of the vector **v** satisfying the given conditions.

9. Initial point of **v**: $(-3, 7)$; terminal point of **v**: $(11, -16)$
10. Magnitude of **v**: $\|\mathbf{v}\| = 12$; direction of **v**: $\mathbf{u} = \langle 3, -5 \rangle$

In Exercises 11–14, $\mathbf{u} = \langle 2, 7 \rangle$ and $\mathbf{v} = \langle -6, 5 \rangle$. Find the resultant vector and sketch its graph.

11. $\mathbf{u} + \mathbf{v}$
12. $\mathbf{u} - \mathbf{v}$
13. $5\mathbf{u} - 3\mathbf{v}$
14. $4\mathbf{u} + 2\mathbf{v}$

15. Find a unit vector in the direction of $\mathbf{u} = \langle 24, -7 \rangle$.

16. Forces with magnitudes of 250 pounds and 130 pounds act on an object at angles of 45° and $-60°$, respectively, with the x-axis. Find the direction and magnitude of the resultant of these forces.

17. Find the angle between the vectors $\mathbf{u} = \langle -1, 5 \rangle$ and $\mathbf{v} = \langle 3, -2 \rangle$.

18. Are the vectors $\mathbf{u} = \langle 6, -10 \rangle$ and $\mathbf{v} = \langle 5, 3 \rangle$ orthogonal?

19. Find the projection of $\mathbf{u} = \langle 6, 7 \rangle$ onto $\mathbf{v} = \langle -5, -1 \rangle$. Then write $\mathbf{u}$ as the sum of two orthogonal vectors.

20. A 500-pound motorcycle is headed up a hill inclined at 12°. What force is required to keep the motorcycle from rolling down the hill when stopped at a red light?

3 CUMULATIVE TEST FOR CHAPTERS 1–3

See www.CalcChat.com for worked-out solutions to odd-numbered exercises.

Take this test as you would take a test in class. When you are finished, check your work against the answers given in the back of the book.

1. Consider the angle $\theta = -120°$.

    (a) Sketch the angle in standard position.

    (b) Determine a coterminal angle in the interval $[0°, 360°)$.

    (c) Convert the angle to radian measure.

    (d) Find the reference angle θ'.

    (e) Find the exact values of the six trigonometric functions of θ.

2. Convert the angle $\theta = -1.45$ radians to degrees. Round the answer to one decimal place.

3. Find $\cos\theta$ if $\tan\theta = -\frac{21}{20}$ and $\sin\theta < 0$.

In Exercises 4–6, sketch the graph of the function. (Include two full periods.)

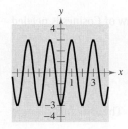

FIGURE FOR 7

4. $f(x) = 3 - 2\sin\pi x$
5. $g(x) = \frac{1}{2}\tan\left(x - \frac{\pi}{2}\right)$
6. $h(x) = -\sec(x + \pi)$

7. Find a, b, and c such that the graph of the function $h(x) = a\cos(bx + c)$ matches the graph in the figure.

8. Sketch the graph of the function $f(x) = \frac{1}{2}x\sin x$ over the interval $-3\pi \le x \le 3\pi$.

In Exercises 9 and 10, find the exact value of the expression without using a calculator.

9. $\tan(\arctan 4.9)$

10. $\tan\left(\arcsin\frac{3}{5}\right)$

11. Write an algebraic expression equivalent to $\sin(\arccos 2x)$.

12. Use the fundamental identities to simplify: $\cos\left(\frac{\pi}{2} - x\right)\csc x$.

13. Subtract and simplify: $\dfrac{\sin\theta - 1}{\cos\theta} - \dfrac{\cos\theta}{\sin\theta - 1}$.

In Exercises 14–16, verify the identity.

14. $\cot^2\alpha(\sec^2\alpha - 1) = 1$

15. $\sin(x + y)\sin(x - y) = \sin^2 x - \sin^2 y$

16. $\sin^2 x\cos^2 x = \frac{1}{8}(1 - \cos 4x)$

In Exercises 17 and 18, find all solutions of the equation in the interval $[0, 2\pi)$.

17. $2\cos^2\beta - \cos\beta = 0$

18. $3\tan\theta - \cot\theta = 0$

19. Use the Quadratic Formula to solve the equation in the interval $[0, 2\pi)$: $\sin^2 x + 2\sin x + 1 = 0$.

20. Given that $\sin u = \frac{12}{13}$, $\cos v = \frac{3}{5}$, and angles u and v are both in Quadrant I, find $\tan(u - v)$.

21. If $\tan\theta = \dfrac{1}{2}$, find the exact value of $\tan(2\theta)$.

22. If $\tan \theta = \dfrac{4}{3}$, find the exact value of $\sin \dfrac{\theta}{2}$.

23. Write the product $5 \sin \dfrac{3\pi}{4} \cdot \cos \dfrac{7\pi}{4}$ as a sum or difference.

24. Write $\cos 9x - \cos 7x$ as a product.

In Exercises 25–28, use the information to solve the triangle shown in the figure. Round your answers to two decimal places.

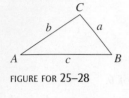

FIGURE FOR 25–28

25. $A = 30°$, $a = 9$, $b = 8$

26. $A = 30°$, $b = 8$, $c = 10$

27. $A = 30°$, $C = 90°$, $b = 10$

28. $a = 4.7$, $b = 8.1$, $c = 10.3$

In Exercises 29 and 30, determine whether the Law of Sines or the Law of Cosines is needed to solve the triangle. Then solve the triangle.

29. $A = 45°$, $B = 26°$, $c = 20$

30. $a = 1.2$, $b = 10$, $C = 80°$

31. Two sides of a triangle have lengths 7 inches and 12 inches. Their included angle measures 99°. Find the area of the triangle.

32. Find the area of a triangle with sides of lengths 30 meters, 41 meters, and 45 meters.

33. Write the vector $\mathbf{u} = \langle 7, 8 \rangle$ as a linear combination of the standard unit vectors $\mathbf{i}$ and $\mathbf{j}$.

34. Find a unit vector in the direction of $\mathbf{v} = \mathbf{i} + \mathbf{j}$.

35. Find $\mathbf{u} \cdot \mathbf{v}$ for $\mathbf{u} = 3\mathbf{i} + 4\mathbf{j}$ and $\mathbf{v} = \mathbf{i} - 2\mathbf{j}$.

36. Find the projection of $\mathbf{u} = \langle 8, -2 \rangle$ onto $\mathbf{v} = \langle 1, 5 \rangle$. Then write $\mathbf{u}$ as the sum of two orthogonal vectors.

37. A ceiling fan with 21-inch blades makes 63 revolutions per minute. Find the angular speed of the fan in radians per minute. Find the linear speed of the tips of the blades in inches per minute.

38. Find the area of the sector of a circle with a radius of 12 yards and a central angle of 105°.

39. From a point 200 feet from a flagpole, the angles of elevation to the bottom and top of the flag are 16° 45′ and 18°, respectively. Approximate the height of the flag to the nearest foot.

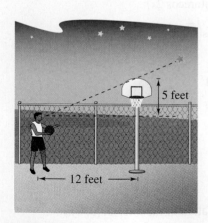

FIGURE FOR 40

40. To determine the angle of elevation of a star in the sky, you get the star in your line of vision with the backboard of a basketball hoop that is 5 feet higher than your eyes (see figure). Your horizontal distance from the backboard is 12 feet. What is the angle of elevation of the star?

41. Write a model for a particle in simple harmonic motion with a displacement of 4 inches and a period of 8 seconds.

42. An airplane's velocity with respect to the air is 500 kilometers per hour, with a bearing of 30°. The wind at the altitude of the plane has a velocity of 50 kilometers per hour with a bearing of N 60° E. What is the true direction of the plane, and what is its speed relative to the ground?

43. A force of 85 pounds exerted at an angle of 60° above the horizontal is required to slide an object across a floor. The object is dragged 10 feet. Determine the work done in sliding the object.

PROOFS IN MATHEMATICS

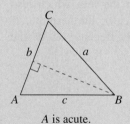

A is acute.

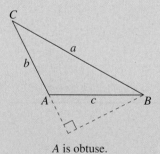

A is obtuse.

Law of Sines   (p. 282)

If ABC is a triangle with sides a, b, and c, then

$$\frac{a}{\sin A} = \frac{b}{\sin B} = \frac{c}{\sin C}.$$

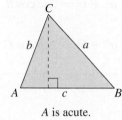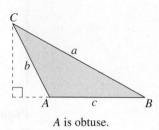

A is acute.   A is obtuse.

Proof

Let h be the altitude of either triangle found in the figure above. Then you have

$$\sin A = \frac{h}{b} \qquad \text{or} \qquad h = b \sin A$$

$$\sin B = \frac{h}{a} \qquad \text{or} \qquad h = a \sin B.$$

Equating these two values of h, you have

$$a \sin B = b \sin A \qquad \text{or} \qquad \frac{a}{\sin A} = \frac{b}{\sin B}.$$

Note that $\sin A \neq 0$ and $\sin B \neq 0$ because no angle of a triangle can have a measure of 0° or 180°. In a similar manner, construct an altitude from vertex B to side AC (extended in the obtuse triangle), as shown at the left. Then you have

$$\sin A = \frac{h}{c} \qquad \text{or} \qquad h = c \sin A$$

$$\sin C = \frac{h}{a} \qquad \text{or} \qquad h = a \sin C.$$

Equating these two values of h, you have

$$a \sin C = c \sin A \qquad \text{or} \qquad \frac{a}{\sin A} = \frac{c}{\sin C}.$$

By the Transitive Property of Equality you know that

$$\frac{a}{\sin A} = \frac{b}{\sin B} = \frac{c}{\sin C}.$$

So, the Law of Sines is established.

PROOFS IN MATH

Law of Cosines *(p. 291)*

Standard Form	*Alternative Form*
$a^2 = b^2 + c^2 - 2bc \cos A$	$\cos A = \dfrac{b^2 + c^2 - a^2}{2bc}$
$b^2 = a^2 + c^2 - 2ac \cos B$	$\cos B = \dfrac{a^2 + c^2 - b^2}{2ac}$
$c^2 = a^2 + b^2 - 2ab \cos C$	$\cos C = \dfrac{a^2 + b^2 - c^2}{2ab}$

Proof

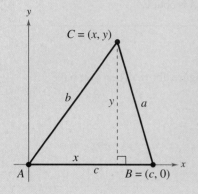

To prove the first formula, consider the top triangle at the left, which has three acute angles. Note that vertex B has coordinates $(c, 0)$. Furthermore, C has coordinates (x, y), where $x = b \cos A$ and $y = b \sin A$. Because a is the distance from vertex C to vertex B, it follows that

$a = \sqrt{(x - c)^2 + (y - 0)^2}$	Distance Formula
$a^2 = (x - c)^2 + (y - 0)^2$	Square each side.
$a^2 = (b \cos A - c)^2 + (b \sin A)^2$	Substitute for x and y.
$a^2 = b^2 \cos^2 A - 2bc \cos A + c^2 + b^2 \sin^2 A$	Expand.
$a^2 = b^2(\sin^2 A + \cos^2 A) + c^2 - 2bc \cos A$	Factor out b^2.
$a^2 = b^2 + c^2 - 2bc \cos A.$	$\sin^2 A + \cos^2 A = 1$

To prove the second formula, consider the bottom triangle at the left, which also has three acute angles. Note that vertex A has coordinates $(c, 0)$. Furthermore, C has coordinates (x, y), where $x = a \cos B$ and $y = a \sin B$. Because b is the distance from vertex C to vertex A, it follows that

$b = \sqrt{(x - c)^2 + (y - 0)^2}$	Distance Formula
$b^2 = (x - c)^2 + (y - 0)^2$	Square each side.
$b^2 = (a \cos B - c)^2 + (a \sin B)^2$	Substitute for x and y.
$b^2 = a^2 \cos^2 B - 2ac \cos B + c^2 + a^2 \sin^2 B$	Expand.
$b^2 = a^2(\sin^2 B + \cos^2 B) + c^2 - 2ac \cos B$	Factor out a^2.
$b^2 = a^2 + c^2 - 2ac \cos B.$	$\sin^2 B + \cos^2 B = 1$

A similar argument is used to establish the third formula.

Heron's Area Formula *(p. 294)*

Given any triangle with sides of lengths a, b, and c, the area of the triangle is

$$\text{Area} = \sqrt{s(s-a)(s-b)(s-c)}$$

where $s = \dfrac{a+b+c}{2}$.

Proof

From Section 3.1, you know that

$$\text{Area} = \frac{1}{2}bc \sin A \qquad \text{Formula for the area of an oblique triangle}$$

$$(\text{Area})^2 = \frac{1}{4}b^2c^2 \sin^2 A \qquad \text{Square each side.}$$

$$\text{Area} = \sqrt{\frac{1}{4}b^2c^2 \sin^2 A} \qquad \text{Take the square root of each side.}$$

$$= \sqrt{\frac{1}{4}b^2c^2(1 - \cos^2 A)} \qquad \text{Pythagorean Identity}$$

$$= \sqrt{\left[\frac{1}{2}bc(1 + \cos A)\right]\left[\frac{1}{2}bc(1 - \cos A)\right]}. \qquad \text{Factor.}$$

Using the Law of Cosines, you can show that

$$\frac{1}{2}bc(1 + \cos A) = \frac{a+b+c}{2} \cdot \frac{-a+b+c}{2}$$

and

$$\frac{1}{2}bc(1 - \cos A) = \frac{a-b+c}{2} \cdot \frac{a+b-c}{2}.$$

Letting $s = (a+b+c)/2$, these two equations can be rewritten as

$$\frac{1}{2}bc(1 + \cos A) = s(s-a)$$

and

$$\frac{1}{2}bc(1 - \cos A) = (s-b)(s-c).$$

By substituting into the last formula for area, you can conclude that

$$\text{Area} = \sqrt{s(s-a)(s-b)(s-c)}.$$

Properties of the Dot Product   *(p. 312)*

Let **u**, **v**, and **w** be vectors in the plane or in space and let c be a scalar.

1. $\mathbf{u} \cdot \mathbf{v} = \mathbf{v} \cdot \mathbf{u}$
2. $\mathbf{0} \cdot \mathbf{v} = 0$
3. $\mathbf{u} \cdot (\mathbf{v} + \mathbf{w}) = \mathbf{u} \cdot \mathbf{v} + \mathbf{u} \cdot \mathbf{w}$
4. $\mathbf{v} \cdot \mathbf{v} = \|\mathbf{v}\|^2$
5. $c(\mathbf{u} \cdot \mathbf{v}) = c\mathbf{u} \cdot \mathbf{v} = \mathbf{u} \cdot c\mathbf{v}$

Proof

Let $\mathbf{u} = \langle u_1, u_2 \rangle$, $\mathbf{v} = \langle v_1, v_2 \rangle$, $\mathbf{w} = \langle w_1, w_2 \rangle$, $\mathbf{0} = \langle 0, 0 \rangle$, and let c be a scalar.

1. $\mathbf{u} \cdot \mathbf{v} = u_1v_1 + u_2v_2 = v_1u_1 + v_2u_2 = \mathbf{v} \cdot \mathbf{u}$

2. $\mathbf{0} \cdot \mathbf{v} = 0 \cdot v_1 + 0 \cdot v_2 = 0$

3. $\mathbf{u} \cdot (\mathbf{v} + \mathbf{w}) = \mathbf{u} \cdot \langle v_1 + w_1, v_2 + w_2 \rangle$

$$= u_1(v_1 + w_1) + u_2(v_2 + w_2)$$

$$= u_1v_1 + u_1w_1 + u_2v_2 + u_2w_2$$

$$= (u_1v_1 + u_2v_2) + (u_1w_1 + u_2w_2) = \mathbf{u} \cdot \mathbf{v} + \mathbf{u} \cdot \mathbf{w}$$

4. $\mathbf{v} \cdot \mathbf{v} = v_1^2 + v_2^2 = \left(\sqrt{v_1^2 + v_2^2}\right)^2 = \|\mathbf{v}\|^2$

5. $c(\mathbf{u} \cdot \mathbf{v}) = c(\langle u_1, u_2 \rangle \cdot \langle v_1, v_2 \rangle)$

$$= c(u_1v_1 + u_2v_2)$$

$$= (cu_1)v_1 + (cu_2)v_2$$

$$= \langle cu_1, cu_2 \rangle \cdot \langle v_1, v_2 \rangle$$

$$= c\mathbf{u} \cdot \mathbf{v}$$

Angle Between Two Vectors   *(p. 313)*

If θ is the angle between two nonzero vectors **u** and **v**, then $\cos \theta = \dfrac{\mathbf{u} \cdot \mathbf{v}}{\|\mathbf{u}\| \, \|\mathbf{v}\|}$.

Proof

Consider the triangle determined by vectors **u**, **v**, and $\mathbf{v} - \mathbf{u}$, as shown in the figure. By the Law of Cosines, you can write

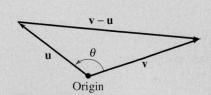

$$\|\mathbf{v} - \mathbf{u}\|^2 = \|\mathbf{u}\|^2 + \|\mathbf{v}\|^2 - 2\|\mathbf{u}\| \, \|\mathbf{v}\| \cos \theta$$

$$(\mathbf{v} - \mathbf{u}) \cdot (\mathbf{v} - \mathbf{u}) = \|\mathbf{u}\|^2 + \|\mathbf{v}\|^2 - 2\|\mathbf{u}\| \, \|\mathbf{v}\| \cos \theta$$

$$(\mathbf{v} - \mathbf{u}) \cdot \mathbf{v} - (\mathbf{v} - \mathbf{u}) \cdot \mathbf{u} = \|\mathbf{u}\|^2 + \|\mathbf{v}\|^2 - 2\|\mathbf{u}\| \, \|\mathbf{v}\| \cos \theta$$

$$\mathbf{v} \cdot \mathbf{v} - \mathbf{u} \cdot \mathbf{v} - \mathbf{v} \cdot \mathbf{u} + \mathbf{u} \cdot \mathbf{u} = \|\mathbf{u}\|^2 + \|\mathbf{v}\|^2 - 2\|\mathbf{u}\| \, \|\mathbf{v}\| \cos \theta$$

$$\|\mathbf{v}\|^2 - 2\mathbf{u} \cdot \mathbf{v} + \|\mathbf{u}\|^2 = \|\mathbf{u}\|^2 + \|\mathbf{v}\|^2 - 2\|\mathbf{u}\| \, \|\mathbf{v}\| \cos \theta$$

$$\cos \theta = \frac{\mathbf{u} \cdot \mathbf{v}}{\|\mathbf{u}\| \, \|\mathbf{v}\|}.$$

PROBLEM SOLVING

This collection of thought-provoking and challenging exercises further explores and expands upon concepts learned in this chapter.

1. In the figure, a beam of light is directed at the blue mirror, reflected to the red mirror, and then reflected back to the blue mirror. Find the distance *PT* that the light travels from the red mirror back to the blue mirror.

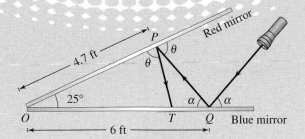

2. A triathlete sets a course to swim S 25° E from a point on shore to a buoy $\frac{3}{4}$ mile away. After swimming 300 yards through a strong current, the triathlete is off course at a bearing of S 35° E. Find the bearing and distance the triathlete needs to swim to correct her course.

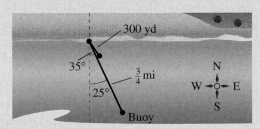

3. A hiking party is lost in a national park. Two ranger stations have received an emergency SOS signal from the party. Station B is 75 miles due east of station A. The bearing from station A to the signal is S 60° E and the bearing from station B to the signal is S 75° W.

   (a) Draw a diagram that gives a visual representation of the problem.

   (b) Find the distance from each station to the SOS signal.

   (c) A rescue party is in the park 20 miles from station A at a bearing of S 80° E. Find the distance and the bearing the rescue party must travel to reach the lost hiking party.

4. You are seeding a triangular courtyard. One side of the courtyard is 52 feet long and another side is 46 feet long. The angle opposite the 52-foot side is 65°.

   (a) Draw a diagram that gives a visual representation of the situation.

   (b) How long is the third side of the courtyard?

   (c) One bag of grass seed covers an area of 50 square feet. How many bags of grass seed will you need to cover the courtyard?

5. For each pair of vectors, find the following.

   (i) $\|\mathbf{u}\|$ (ii) $\|\mathbf{v}\|$ (iii) $\|\mathbf{u} + \mathbf{v}\|$

   (iv) $\left\|\dfrac{\mathbf{u}}{\|\mathbf{u}\|}\right\|$ (v) $\left\|\dfrac{\mathbf{v}}{\|\mathbf{v}\|}\right\|$ (vi) $\left\|\dfrac{\mathbf{u} + \mathbf{v}}{\|\mathbf{u} + \mathbf{v}\|}\right\|$

   (a) $\mathbf{u} = \langle 1, -1 \rangle$ (b) $\mathbf{u} = \langle 0, 1 \rangle$
       $\mathbf{v} = \langle -1, 2 \rangle$ $\mathbf{v} = \langle 3, -3 \rangle$

   (c) $\mathbf{u} = \langle 1, \frac{1}{2} \rangle$ (d) $\mathbf{u} = \langle 2, -4 \rangle$
       $\mathbf{v} = \langle 2, 3 \rangle$ $\mathbf{v} = \langle 5, 5 \rangle$

6. A skydiver is falling at a constant downward velocity of 120 miles per hour. In the figure, vector **u** represents the skydiver's velocity. A steady breeze pushes the skydiver to the east at 40 miles per hour. Vector **v** represents the wind velocity.

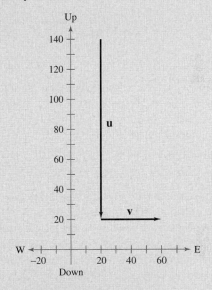

   (a) Write the vectors **u** and **v** in component form.

   (b) Let $\mathbf{s} = \mathbf{u} + \mathbf{v}$. Use the figure to sketch **s**. To print an enlarged copy of the graph, go to the website *www.mathgraphs.com*.

   (c) Find the magnitude of **s**. What information does the magnitude give you about the skydiver's fall?

   (d) If there were no wind, the skydiver would fall in a path perpendicular to the ground. At what angle to the ground is the path of the skydiver when the skydiver is affected by the 40-mile-per-hour wind from due west?

   (e) The skydiver is blown to the west at 30 miles per hour. Draw a new figure that gives a visual representation of the problem and find the skydiver's new velocity.

335

7. Write the vector **w** in terms of **u** and **v**, given that the terminal point of **w** bisects the line segment (see figure).

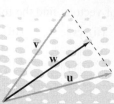

8. Prove that if **u** is orthogonal to **v** and **w**, then **u** is orthogonal to

$$c\mathbf{v} + d\mathbf{w}$$

for any scalars c and d (see figure).

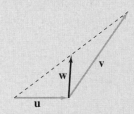

9. Two forces of the same magnitude $\mathbf{F}_1$ and $\mathbf{F}_2$ act at angles θ_1 and θ_2, respectively. Use a diagram to compare the work done by $\mathbf{F}_1$ with the work done by $\mathbf{F}_2$ in moving along the vector PQ if

(a) $\theta_1 = -\theta_2$

(b) $\theta_1 = 60°$ and $\theta_2 = 30°$.

10. Four basic forces are in action during flight: weight, lift, thrust, and drag. To fly through the air, an object must overcome its own *weight*. To do this, it must create an upward force called *lift*. To generate lift, a forward motion called *thrust* is needed. The thrust must be great enough to overcome air resistance, which is called *drag*.

For a commercial jet aircraft, a quick climb is important to maximize efficiency because the performance of an aircraft at high altitudes is enhanced. In addition, it is necessary to clear obstacles such as buildings and mountains and to reduce noise in residential areas. In the diagram, the angle θ is called the climb angle. The velocity of the plane can be represented by a vector **v** with a vertical component $\|\mathbf{v}\| \sin \theta$ (called climb speed) and a horizontal component $\|\mathbf{v}\| \cos \theta$, where $\|\mathbf{v}\|$ is the speed of the plane.

When taking off, a pilot must decide how much of the thrust to apply to each component. The more the thrust is applied to the horizontal component, the faster the airplane will gain speed. The more the thrust is applied to the vertical component, the quicker the airplane will climb.

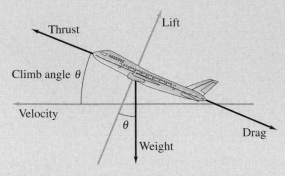

FIGURE FOR 10

(a) Complete the table for an airplane that has a speed of $\|\mathbf{v}\| = 100$ miles per hour.

θ	0.5°	1.0°	1.5°	2.0°	2.5°	3.0°
$\|\mathbf{v}\| \sin \theta$						
$\|\mathbf{v}\| \cos \theta$						

(b) Does an airplane's speed equal the sum of the vertical and horizontal components of its velocity? If not, how could you find the speed of an airplane whose velocity components were known?

(c) Use the result of part (b) to find the speed of an airplane with the given velocity components.

(i) $\|\mathbf{v}\| \sin \theta = 5.235$ miles per hour

$\|\mathbf{v}\| \cos \theta = 149.909$ miles per hour

(ii) $\|\mathbf{v}\| \sin \theta = 10.463$ miles per hour

$\|\mathbf{v}\| \cos \theta = 149.634$ miles per hour

Complex Numbers

4

In Mathematics

The set of complex numbers includes real numbers and imaginary numbers. Complex numbers can be used to solve equations that do not have real solutions.

In Real Life

Complex numbers can be used to create beautiful pictures called fractals. The most famous fractal is called the Mandelbrot Set, named after the mathematician Benoit Mandelbrot. (See Exercise 11, page 374.)

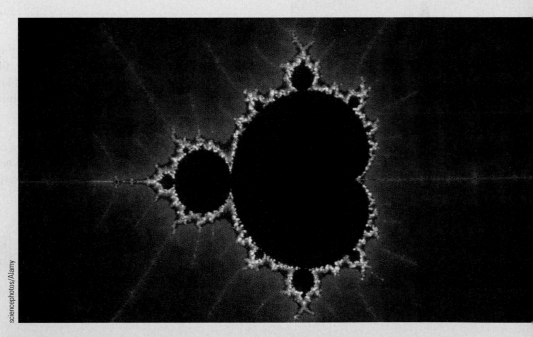

sciencephotos/Alamy

IN CAREERS

There are many careers that use complex numbers. Several are listed below.

- Electrician
  Exercise 89, page 344

- Economist
  Exercise 85, page 352

- Sales Analyst
  Exercise 86, page 352

- Consumer Research Analyst
  Exercise 48, page 368

What you should learn

- Use the imaginary unit i to write complex numbers.
- Add, subtract, and multiply complex numbers.
- Use complex conjugates to write the quotient of two complex numbers in standard form.
- Find complex solutions of quadratic equations.

Why you should learn it

You can use complex numbers to model and solve real-life problems in electronics. For instance, in Exercise 89 on page 344, you will learn how to use complex numbers to find the impedance of an electrical circuit.

© Richard Magna/Fundamental Photographs

4.1 COMPLEX NUMBERS

The Imaginary Unit i

Some quadratic equations have no real solutions. For instance, the quadratic equation $x^2 + 1 = 0$ has no real solution because there is no real number x that can be squared to produce -1. To overcome this deficiency, mathematicians created an expanded system of numbers using the **imaginary unit i**, defined as

$$i = \sqrt{-1} \qquad \text{Imaginary unit}$$

where $i^2 = -1$. By adding real numbers to real multiples of this imaginary unit, the set of **complex numbers** is obtained. Each complex number can be written in the **standard form $a + bi$.** For instance, the standard form of the complex number $-5 + \sqrt{-9}$ is $-5 + 3i$ because

$$-5 + \sqrt{-9} = -5 + \sqrt{3^2(-1)} = -5 + 3\sqrt{-1} = -5 + 3i.$$

In the standard form $a + bi$, the real number a is called the **real part** of the **complex number $a + bi$,** and the number bi (where b is a real number) is called the **imaginary part** of the complex number.

Definition of a Complex Number

If a and b are real numbers, the number $a + bi$ is a **complex number,** and it is said to be written in **standard form.** If $b = 0$, the number $a + bi = a$ is a real number. If $b \neq 0$, the number $a + bi$ is called an **imaginary number.** A number of the form bi, where $b \neq 0$, is called a **pure imaginary number.**

The set of real numbers is a subset of the set of complex numbers, as shown in Figure 4.1. This is true because every real number a can be written as a complex number using $b = 0$. That is, for every real number a, you can write $a = a + 0i$.

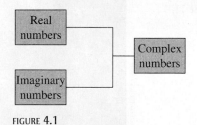

FIGURE 4.1

Equality of Complex Numbers

Two complex numbers $a + bi$ and $c + di$, written in standard form, are equal to each other

$$a + bi = c + di \qquad \text{Equality of two complex numbers}$$

if and only if $a = c$ and $b = d$.

Operations with Complex Numbers

To add (or subtract) two complex numbers, you add (or subtract) the real and imaginary parts of the numbers separately.

Addition and Subtraction of Complex Numbers

If $a + bi$ and $c + di$ are two complex numbers written in standard form, their sum and difference are defined as follows.

Sum: $(a + bi) + (c + di) = (a + c) + (b + d)i$

Difference: $(a + bi) - (c + di) = (a - c) + (b - d)i$

The **additive identity** in the complex number system is zero (the same as in the real number system). Furthermore, the **additive inverse** of the complex number $a + bi$ is

$$-(a + bi) = -a - bi. \qquad \text{Additive inverse}$$

So, you have

$$(a + bi) + (-a - bi) = 0 + 0i = 0.$$

Example 1   Adding and Subtracting Complex Numbers

a. $(4 + 7i) + (1 - 6i) = 4 + 7i + 1 - 6i$ Remove parentheses.

$\qquad\qquad\qquad\quad = (4 + 1) + (7i - 6i)$ Group like terms.

$\qquad\qquad\qquad\quad = 5 + i$ Write in standard form.

b. $(1 + 2i) - (4 + 2i) = 1 + 2i - 4 - 2i$ Remove parentheses.

$\qquad\qquad\qquad\quad = (1 - 4) + (2i - 2i)$ Group like terms.

$\qquad\qquad\qquad\quad = -3 + 0$ Simplify.

$\qquad\qquad\qquad\quad = -3$ Write in standard form.

c. $3i - (-2 + 3i) - (2 + 5i) = 3i + 2 - 3i - 2 - 5i$

$\qquad\qquad\qquad\qquad\qquad = (2 - 2) + (3i - 3i - 5i)$

$\qquad\qquad\qquad\qquad\qquad = 0 - 5i$

$\qquad\qquad\qquad\qquad\qquad = -5i$

d. $(3 + 2i) + (4 - i) - (7 + i) = 3 + 2i + 4 - i - 7 - i$

$\qquad\qquad\qquad\qquad\qquad = (3 + 4 - 7) + (2i - i - i)$

$\qquad\qquad\qquad\qquad\qquad = 0 + 0i$

$\qquad\qquad\qquad\qquad\qquad = 0$

CHECK*Point*   Now try Exercise 21.

Note in Examples 1(b) and 1(d) that the sum of two complex numbers can be a real number.

Many of the properties of real numbers are valid for complex numbers as well. Here are some examples.

Associative Properties of Addition and Multiplication

Commutative Properties of Addition and Multiplication

Distributive Property of Multiplication Over Addition

Notice below how these properties are used when two complex numbers are multiplied.

$$(a + bi)(c + di) = a(c + di) + bi(c + di) \qquad \text{Distributive Property}$$

$$= ac + (ad)i + (bc)i + (bd)i^2 \qquad \text{Distributive Property}$$

$$= ac + (ad)i + (bc)i + (bd)(-1) \qquad i^2 = -1$$

$$= ac - bd + (ad)i + (bc)i \qquad \text{Commutative Property}$$

$$= (ac - bd) + (ad + bc)i \qquad \text{Associative Property}$$

Rather than trying to memorize this multiplication rule, you should simply remember how the Distributive Property is used to multiply two complex numbers.

Example 2   Multiplying Complex Numbers

a. $4(-2 + 3i) = 4(-2) + 4(3i)$ $\qquad$ Distributive Property

$\qquad\qquad\quad = -8 + 12i$ $\qquad$ Simplify.

b. $(2 - i)(4 + 3i) = 2(4 + 3i) - i(4 + 3i)$ $\qquad$ Distributive Property

$\qquad\qquad\qquad = 8 + 6i - 4i - 3i^2$ $\qquad$ Distributive Property

$\qquad\qquad\qquad = 8 + 6i - 4i - 3(-1)$ $\qquad$ $i^2 = -1$

$\qquad\qquad\qquad = (8 + 3) + (6i - 4i)$ $\qquad$ Group like terms.

$\qquad\qquad\qquad = 11 + 2i$ $\qquad$ Write in standard form.

c. $(3 + 2i)(3 - 2i) = 3(3 - 2i) + 2i(3 - 2i)$ $\qquad$ Distributive Property

$\qquad\qquad\qquad = 9 - 6i + 6i - 4i^2$ $\qquad$ Distributive Property

$\qquad\qquad\qquad = 9 - 6i + 6i - 4(-1)$ $\qquad$ $i^2 = -1$

$\qquad\qquad\qquad = 9 + 4$ $\qquad$ Simplify.

$\qquad\qquad\qquad = 13$ $\qquad$ Write in standard form.

d. $(3 + 2i)^2 = (3 + 2i)(3 + 2i)$ $\qquad$ Square of a binomial

$\qquad\qquad\quad = 3(3 + 2i) + 2i(3 + 2i)$ $\qquad$ Distributive Property

$\qquad\qquad\quad = 9 + 6i + 6i + 4i^2$ $\qquad$ Distributive Property

$\qquad\qquad\quad = 9 + 6i + 6i + 4(-1)$ $\qquad$ $i^2 = -1$

$\qquad\qquad\quad = 9 + 12i - 4$ $\qquad$ Simplify.

$\qquad\qquad\quad = 5 + 12i$ $\qquad$ Write in standard form.

CHECK*Point* Now try Exercise 31.

Study Tip

The procedure described above is similar to multiplying two polynomials and combining like terms, as in the FOIL Method. For instance, you can use the FOIL Method to multiply the two complex numbers from Example 2(b).

$\qquad\qquad$ F $\quad$ O $\quad$ I $\quad$ L

$(2 - i)(4 + 3i) = 8 + 6i - 4i - 3i^2$

Complex Conjugates

Notice in Example 2(c) that the product of two complex numbers can be a real number. This occurs with pairs of complex numbers of the form $a + bi$ and $a - bi$, called **complex conjugates**.

$$(a + bi)(a - bi) = a^2 - abi + abi - b^2 i^2$$

$$= a^2 - b^2(-1)$$

$$= a^2 + b^2$$

Example 3   Multiplying Conjugates

Multiply each complex number by its complex conjugate.

a. $1 + i$ **b.** $4 - 3i$

Solution

a. The complex conjugate of $1 + i$ is $1 - i$.

$$(1 + i)(1 - i) = 1^2 - i^2 = 1 - (-1) = 2$$

b. The complex conjugate of $4 - 3i$ is $4 + 3i$.

$$(4 - 3i)(4 + 3i) = 4^2 - (3i)^2 = 16 - 9i^2 = 16 - 9(-1) = 25$$

CHECK*Point*   Now try Exercise 41.

To write the quotient of $a + bi$ and $c + di$ in standard form, where c and d are not both zero, multiply the numerator and denominator by the complex conjugate of the *denominator* to obtain

$$\frac{a + bi}{c + di} = \frac{a + bi}{c + di}\left(\frac{c - di}{c - di}\right)$$

$$= \frac{(ac + bd) + (bc - ad)i}{c^2 + d^2}. \qquad \text{Standard form}$$

Example 4   Writing a Quotient of Complex Numbers in Standard Form

$$\frac{2 + 3i}{4 - 2i} = \frac{2 + 3i}{4 - 2i}\left(\frac{4 + 2i}{4 + 2i}\right) \qquad \text{Multiply numerator and denominator by complex conjugate of denominator.}$$

$$= \frac{8 + 4i + 12i + 6i^2}{16 - 4i^2} \qquad \text{Expand.}$$

$$= \frac{8 - 6 + 16i}{16 + 4} \qquad i^2 = -1$$

$$= \frac{2 + 16i}{20} \qquad \text{Simplify.}$$

$$= \frac{1}{10} + \frac{4}{5}i \qquad \text{Write in standard form.}$$

CHECK*Point*   Now try Exercise 53.

Study Tip

Note that when you multiply the numerator and denominator of a quotient of complex numbers by

$$\frac{c - di}{c - di}$$

you are actually multiplying the quotient by a form of 1. You are not changing the original expression, you are only creating an expression that is equivalent to the original expression.

Complex Solutions of Quadratic Equations

When using the Quadratic Formula to solve a quadratic equation, you often obtain a result such as $\sqrt{-3}$, which you know is not a real number. By factoring out $i = \sqrt{-1}$, you can write this number in standard form.

$$\sqrt{-3} = \sqrt{3(-1)} = \sqrt{3}\sqrt{-1} = \sqrt{3}\,i$$

The number $\sqrt{3}\,i$ is called the *principal square root* of -3.

Principal Square Root of a Negative Number

If a is a positive number, the **principal square root** of the negative number $-a$ is defined as

$$\sqrt{-a} = \sqrt{a}\,i.$$

Example 5 Writing Complex Numbers in Standard Form

a. $\sqrt{-3}\sqrt{-12} = \sqrt{3}\,i\sqrt{12}\,i = \sqrt{36}\,i^2 = 6(-1) = -6$

b. $\sqrt{-48} - \sqrt{-27} = \sqrt{48}\,i - \sqrt{27}\,i = 4\sqrt{3}\,i - 3\sqrt{3}\,i = \sqrt{3}\,i$

c. $\left(-1 + \sqrt{-3}\right)^2 = \left(-1 + \sqrt{3}\,i\right)^2$

$$= (-1)^2 - 2\sqrt{3}\,i + \left(\sqrt{3}\right)^2(i^2)$$

$$= 1 - 2\sqrt{3}\,i + 3(-1)$$

$$= -2 - 2\sqrt{3}\,i$$

CHECK*Point* Now try Exercise 63.

Example 6 Complex Solutions of a Quadratic Equation

Solve (a) $x^2 + 4 = 0$ and (b) $3x^2 - 2x + 5 = 0$.

Solution

a. $x^2 + 4 = 0$ Write original equation.

       $x^2 = -4$ Subtract 4 from each side.

       $x = \pm 2i$ Extract square roots.

b. $3x^2 - 2x + 5 = 0$ Write original equation.

$$x = \frac{-(-2) \pm \sqrt{(-2)^2 - 4(3)(5)}}{2(3)}$$ Quadratic Formula

$$= \frac{2 \pm \sqrt{-56}}{6}$$ Simplify.

$$= \frac{2 \pm 2\sqrt{14}\,i}{6}$$ Write $\sqrt{-56}$ in standard form.

$$= \frac{1}{3} \pm \frac{\sqrt{14}}{3}\,i$$ Write in standard form.

CHECK*Point* Now try Exercise 69.

Algebra Help

You can review the techniques for using the Quadratic Formula in Section P.2.

! WARNING/CAUTION

The definition of principal square root uses the rule

$$\sqrt{ab} = \sqrt{a}\sqrt{b}$$

for $a > 0$ and $b < 0$. This rule is not valid if *both* a and b are negative. For example,

$$\sqrt{-5}\sqrt{-5} = \sqrt{5(-1)}\sqrt{5(-1)}$$

$$= \sqrt{5}\,i\sqrt{5}\,i$$

$$= \sqrt{25}\,i^2$$

$$= 5i^2 = -5$$

whereas

$$\sqrt{(-5)(-5)} = \sqrt{25} = 5.$$

To avoid problems with square roots of negative numbers, be sure to convert complex numbers to standard form *before* multiplying.

4.1   **EXERCISES**

See www.CalcChat.com for worked-out solutions to odd-numbered exercises.

VOCABULARY

1. Match the type of complex number with its definition.

(a) Real number   (i) $a + bi$, $a \neq 0$, $b \neq 0$

(b) Imaginary number   (ii) $a + bi$, $a = 0$, $b \neq 0$

(c) Pure imaginary number   (iii) $a + bi$, $b = 0$

In Exercises 2–4, fill in the blanks.

2. The imaginary unit i is defined as $i =$ _____, where $i^2 =$ _____.

3. If a is a positive number, the _____ _____ root of the negative number $-a$ is defined as $\sqrt{-a} = \sqrt{a}\,i$.

4. The numbers $a + bi$ and $a - bi$ are called _____ _____, and their product is a real number $a^2 + b^2$.

SKILLS AND APPLICATIONS

In Exercises 5–8, find real numbers a and b such that the equation is true.

5. $a + bi = -12 + 7i$

6. $a + bi = 13 + 4i$

7. $(a - 1) + (b + 3)i = 5 + 8i$

8. $(a + 6) + 2bi = 6 - 5i$

In Exercises 9–20, write the complex number in standard form.

9. $8 + \sqrt{-25}$

10. $5 + \sqrt{-36}$

11. $2 - \sqrt{-27}$

12. $1 + \sqrt{-8}$

13. $\sqrt{-80}$

14. $\sqrt{-4}$

15. 14

16. 75

17. $-10i + i^2$

18. $-4i^2 + 2i$

19. $\sqrt{-0.09}$

20. $\sqrt{-0.0049}$

In Exercises 21–30, perform the addition or subtraction and write the result in standard form.

21. $(7 + i) + (3 - 4i)$

22. $(13 - 2i) + (-5 + 6i)$

23. $(9 - i) - (8 - i)$

24. $(3 + 2i) - (6 + 13i)$

25. $\left(-2 + \sqrt{-8}\right) + \left(5 - \sqrt{-50}\right)$

26. $\left(8 + \sqrt{-18}\right) - \left(4 + 3\sqrt{2}i\right)$

27. $13i - (14 - 7i)$

28. $25 + (-10 + 11i) + 15i$

29. $-\left(\frac{3}{2} + \frac{5}{2}i\right) + \left(\frac{5}{3} + \frac{11}{3}i\right)$

30. $(1.6 + 3.2i) + (-5.8 + 4.3i)$

In Exercises 31–40, perform the operation and write the result in standard form.

31. $(1 + i)(3 - 2i)$

32. $(7 - 2i)(3 - 5i)$

33. $12i(1 - 9i)$

34. $-8i(9 + 4i)$

35. $\left(\sqrt{14} + \sqrt{10}i\right)\left(\sqrt{14} - \sqrt{10}i\right)$

36. $\left(\sqrt{3} + \sqrt{15}i\right)\left(\sqrt{3} - \sqrt{15}i\right)$

37. $(6 + 7i)^2$

38. $(5 - 4i)^2$

39. $(2 + 3i)^2 + (2 - 3i)^2$

40. $(1 - 2i)^2 - (1 + 2i)^2$

In Exercises 41–48, write the complex conjugate of the complex number. Then multiply the number by its complex conjugate.

41. $9 + 2i$

42. $8 - 10i$

43. $-1 - \sqrt{5}i$

44. $-3 + \sqrt{2}i$

45. $\sqrt{-20}$

46. $\sqrt{-15}$

47. $\sqrt{6}$

48. $1 + \sqrt{8}$

In Exercises 49–58, write the quotient in standard form.

49. $\dfrac{3}{i}$

50. $-\dfrac{14}{2i}$

51. $\dfrac{2}{4 - 5i}$

52. $\dfrac{13}{1 - i}$

53. $\dfrac{5 + i}{5 - i}$

54. $\dfrac{6 - 7i}{1 - 2i}$

55. $\dfrac{9 - 4i}{i}$

56. $\dfrac{8 + 16i}{2i}$

57. $\dfrac{3i}{(4 - 5i)^2}$

58. $\dfrac{5i}{(2 + 3i)^2}$

In Exercises 59–62, perform the operation and write the result in standard form.

59. $\dfrac{2}{1 + i} - \dfrac{3}{1 - i}$

60. $\dfrac{2i}{2 + i} + \dfrac{5}{2 - i}$

61. $\dfrac{i}{3 - 2i} + \dfrac{2i}{3 + 8i}$

62. $\dfrac{1 + i}{i} - \dfrac{3}{4 - i}$

In Exercises 63–68, write the complex number in standard form.

63. $\sqrt{-6} \cdot \sqrt{-2}$
64. $\sqrt{-5} \cdot \sqrt{-10}$
65. $\left(\sqrt{-15}\right)^2$
66. $\left(\sqrt{-75}\right)^2$
67. $\left(3 + \sqrt{-5}\right)\left(7 - \sqrt{-10}\right)$
68. $\left(2 - \sqrt{-6}\right)^2$

In Exercises 69–78, use the Quadratic Formula to solve the quadratic equation.

69. $x^2 - 2x + 2 = 0$
70. $x^2 + 6x + 10 = 0$
71. $4x^2 + 16x + 17 = 0$
72. $9x^2 - 6x + 37 = 0$
73. $4x^2 + 16x + 15 = 0$
74. $16t^2 - 4t + 3 = 0$
75. $\frac{3}{2}x^2 - 6x + 9 = 0$
76. $\frac{7}{8}x^2 - \frac{3}{4}x + \frac{5}{16} = 0$
77. $1.4x^2 - 2x - 10 = 0$
78. $4.5x^2 - 3x + 12 = 0$

In Exercises 79–88, simplify the complex number and write it in standard form.

79. $-6i^3 + i^2$
80. $4i^2 - 2i^3$
81. $-14i^5$
82. $(-i)^3$
83. $\left(\sqrt{-72}\right)^3$
84. $\left(\sqrt{-2}\right)^6$
85. $\dfrac{1}{i^3}$
86. $\dfrac{1}{(2i)^3}$
87. $(3i)^4$
88. $(-i)^6$

89. IMPEDANCE The opposition to current in an electrical circuit is called its impedance. The impedance z in a parallel circuit with two pathways satisfies the equation

$$\frac{1}{z} = \frac{1}{z_1} + \frac{1}{z_2}$$

where z_1 is the impedance (in ohms) of pathway 1 and z_2 is the impedance of pathway 2.

(a) The impedance of each pathway in a parallel circuit is found by adding the impedances of all components in the pathway. Use the table to find z_1 and z_2.

(b) Find the impedance z.

	Resistor	Inductor	Capacitor
Symbol	─\/\/\─ $a\Omega$	─000─ $b\Omega$	─||─ $c\Omega$
Impedance	a	bi	$-ci$

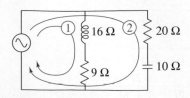

90. Cube each complex number.
(a) 2
(b) $-1 + \sqrt{3}i$
(c) $-1 - \sqrt{3}i$

91. Raise each complex number to the fourth power.
(a) 2
(b) -2
(c) $2i$
(d) $-2i$

92. Write each of the powers of i as i, $-i$, 1, or -1.
(a) i^{40}
(b) i^{25}
(c) i^{50}
(d) i^{67}

EXPLORATION

TRUE OR FALSE? In Exercises 93–96, determine whether the statement is true or false. Justify your answer.

93. There is no complex number that is equal to its complex conjugate.

94. $-i\sqrt{6}$ is a solution of $x^4 - x^2 + 14 = 56$.

95. $i^{44} + i^{150} - i^{74} - i^{109} + i^{61} = -1$

96. The sum of two complex numbers is always a real number.

97. PATTERN RECOGNITION Complete the following.

$i^1 = i$ $i^2 = -1$ $i^3 = -i$ $i^4 = 1$
$i^5 = $ ▨ $i^6 = $ ▨ $i^7 = $ ▨ $i^8 = $ ▨
$i^9 = $ ▨ $i^{10} = $ ▨ $i^{11} = $ ▨ $i^{12} = $ ▨

What pattern do you see? Write a brief description of how you would find i raised to any positive integer power.

98. CAPSTONE Consider the functions

$$f(x) = 2(x - 3)^2 - 4 \text{ and } g(x) = -2(x - 3)^2 - 4.$$

(a) Without graphing either function, determine whether the graph of f and the graph of g have x-intercepts. Explain your reasoning.

(b) Solve $f(x) = 0$ and $g(x) = 0$.

(c) Explain how the zeros of f and g are related to whether their graphs have x-intercepts.

(d) For the function $f(x) = a(x - h)^2 + k$, make a general statement about how a, h, and k affect whether the graph of f has x-intercepts, and whether the zeros of f are real or complex.

99. ERROR ANALYSIS Describe the error.

$$\sqrt{-6}\sqrt{-6} = \sqrt{(-6)(-6)} = \sqrt{36} = 6$$

100. PROOF Prove that the complex conjugate of the product of two complex numbers $a_1 + b_1i$ and $a_2 + b_2i$ is the product of their complex conjugates.

101. PROOF Prove that the complex conjugate of the sum of two complex numbers $a_1 + b_1i$ and $a_2 + b_2i$ is the sum of their complex conjugates.

4.2 COMPLEX SOLUTIONS OF EQUATIONS

What you should learn

- Determine the numbers of solutions of polynomial equations.
- Find solutions of polynomial equations.
- Find zeros of polynomial functions and find polynomial functions given the zeros of the functions.

Why you should learn it

Finding zeros of polynomial functions is an important part of solving real-life problems. For instance, in Exercise 85 on page 352, the zeros of a polynomial function can help you analyze the profit function for a microwave oven.

The Number of Solutions of a Polynomial Equation

The Fundamental Theorem of Algebra implies that a polynomial equation of degree n has precisely n solutions in the complex number system. These solutions can be real or complex and may be repeated. The Fundamental Theorem of Algebra and the Linear Factorization Theorem are listed below for your review. For a proof of the Linear Factorization Theorem, see Proofs in Mathematics on page 372.

The Fundamental Theorem of Algebra

If $f(x)$ is a polynomial of degree n, where $n > 0$, then f has at least one zero in the complex number system.

Note that finding zeros of a polynomial function f is equivalent to finding solutions of the polynomial equation $f(x) = 0$.

Linear Factorization Theorem

If $f(x)$ is a polynomial of degree n, where $n > 0$, then f has precisely n linear factors

$$f(x) = a_n(x - c_1)(x - c_2) \cdots (x - c_n)$$

where $c_1, c_2, \ldots, c_n$ are complex numbers.

Example 1 Solutions of Polynomial Equations

a. The first-degree equation $x - 2 = 0$ has exactly *one* solution: $x = 2$.

b. The second-degree equation

$$x^2 - 6x + 9 = 0 \qquad \text{Second-degree equation}$$
$$(x - 3)(x - 3) = 0 \qquad \text{Factor.}$$

has exactly *two* solutions: $x = 3$ and $x = 3$. (This is called a *repeated solution*.)

c. The third-degree equation

$$x^3 + 4x = 0 \qquad \text{Third-degree equation}$$
$$x(x - 2i)(x + 2i) = 0 \qquad \text{Factor.}$$

has exactly *three* solutions: $x = 0$, $x = 2i$, and $x = -2i$.

d. The fourth-degree equation

$$x^4 - 1 = 0 \qquad \text{Fourth-degree equation}$$
$$(x - 1)(x + 1)(x - i)(x + i) = 0 \qquad \text{Factor.}$$

has exactly *four* solutions: $x = 1$, $x = -1$, $x = i$, and $x = -i$.

CHECK*Point*   Now try Exercise 5.

You can use a graph to check the number of *real* solutions of an equation. As shown in Figure 4.2, the graph of $f(x) = x^4 - 1$ has two x-intercepts, which implies that the equation has two real solutions.

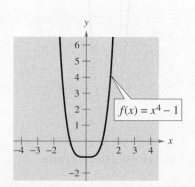

$f(x) = x^4 - 1$

FIGURE 4.2

Every second-degree equation, $ax^2 + bx + c = 0$, has precisely two solutions given by the Quadratic Formula.

$$x = \frac{-b \pm \sqrt{b^2 - 4ac}}{2a}$$

The expression inside the radical, $b^2 - 4ac$, is called the **discriminant,** and can be used to determine whether the solutions are real, repeated, or complex.

1. If $b^2 - 4ac < 0$, the equation has two complex solutions.

2. If $b^2 - 4ac = 0$, the equation has one repeated real solution.

3. If $b^2 - 4ac > 0$, the equation has two distinct real solutions.

Example 2   Using the Discriminant

Use the discriminant to find the number of real solutions of each equation.

a. $4x^2 - 20x + 25 = 0$ **b.** $13x^2 + 7x + 2 = 0$ **c.** $5x^2 - 8x = 0$

Solution

a. For this equation, $a = 4$, $b = -20$, and $c = 25$. So, the discriminant is

$$b^2 - 4ac = (-20)^2 - 4(4)(25) = 400 - 400 = 0.$$

Because the discriminant is zero, there is one repeated real solution.

b. For this equation, $a = 13$, $b = 7$, and $c = 2$. So, the discriminant is

$$b^2 - 4ac = 7^2 - 4(13)(2) = 49 - 104 = -55.$$

Because the discriminant is negative, there are two complex solutions.

c. For this equation, $a = 5$, $b = -8$, and $c = 0$. So, the discriminant is

$$b^2 - 4ac = (-8)^2 - 4(5)(0) = 64 - 0 = 64.$$

Because the discriminant is positive, there are two distinct real solutions.

CHECK Point Now try Exercise 9.

Figure 4.3 shows the graphs of the functions corresponding to the equations in Example 2. Notice that with one repeated solution, the graph *touches* the x-axis at its x-intercept. With two complex solutions, the graph has no x-intercepts. With two real solutions, the graph *crosses* the x-axis at its x-intercepts.

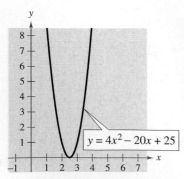

(a) Repeated real solution

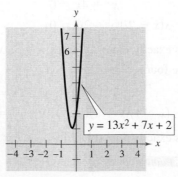

(b) No real solution

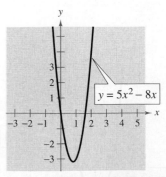

(c) Two distinct real solutions

FIGURE **4.3**

Finding Solutions of Polynomial Equations

| Example 3 | Solving a Quadratic Equation

Solve $x^2 + 2x + 2 = 0$. Write complex solutions in standard form.

Solution

Using $a = 1$, $b = 2$, and $c = 2$, you can apply the Quadratic Formula as follows.

$$x = \frac{-b \pm \sqrt{b^2 - 4ac}}{2a} \qquad \text{Quadratic Formula}$$

$$= \frac{-2 \pm \sqrt{2^2 - 4(1)(2)}}{2(1)} \qquad \text{Substitute 1 for } a, \text{ 2 for } b, \text{ and 2 for } c.$$

$$= \frac{-2 \pm \sqrt{-4}}{2} \qquad \text{Simplify.}$$

$$= \frac{-2 \pm 2i}{2} \qquad \text{Simplify.}$$

$$= -1 \pm i \qquad \text{Write in standard form.}$$

CHECK*Point* Now try Exercise 23.

In Example 3, the two complex solutions are **conjugates.** That is, they are of the form $a \pm bi$. This is not a coincidence, as indicated by the following theorem.

Complex Solutions Occur in Conjugate Pairs

If $a + bi$, $b \neq 0$, is a solution of a polynomial equation with real coefficients, the conjugate $a - bi$ is also a solution of the equation.

Be sure you see that this result is true only if the polynomial has *real* coefficients. For instance, the result applies to the equation $x^2 + 1 = 0$, but not to the equation $x - i = 0$.

| Example 4 | Solving a Polynomial Equation

Solve $x^4 - x^2 - 20 = 0$.

Solution

$$x^4 - x^2 - 20 = 0 \qquad \text{Write original equation.}$$

$$(x^2 - 5)(x^2 + 4) = 0 \qquad \text{Partially factor.}$$

$$\left(x + \sqrt{5}\right)\left(x - \sqrt{5}\right)(x + 2i)(x - 2i) = 0 \qquad \text{Factor completely.}$$

Setting each factor equal to zero yields the solutions $x = -\sqrt{5}$, $x = \sqrt{5}$, $x = -2i$, and $x = 2i$.

CHECK*Point* Now try Exercise 51.

Finding Zeros of Polynomial Functions

The problem of finding the *zeros* of a polynomial function is essentially the same problem as finding the solutions of a polynomial equation. For instance, the zeros of the polynomial function

$$f(x) = 3x^2 - 4x + 5$$

are simply the solutions of the polynomial equation

$$3x^2 - 4x + 5 = 0.$$

Example 5   Finding the Zeros of a Polynomial Function

Find all the zeros of

$$f(x) = x^4 - 3x^3 + 6x^2 + 2x - 60$$

given that $1 + 3i$ is a zero of f.

Algebraic Solution

Because complex zeros occur in conjugate pairs, you know that $1 - 3i$ is also a zero of f. This means that both

$$[x - (1 + 3i)] \quad \text{and} \quad [x - (1 - 3i)]$$

are factors of f. Multiplying these two factors produces

$$[x - (1 + 3i)][x - (1 - 3i)] = [(x - 1) - 3i][(x - 1) + 3i]$$

$$= (x - 1)^2 - 9i^2$$

$$= x^2 - 2x + 10.$$

Using long division, you can divide $x^2 - 2x + 10$ into f to obtain the following.

$$
\begin{array}{r}
x^2 - x - 6 \\
x^2 - 2x + 10 \overline{\smash{)}\, x^4 - 3x^3 + 6x^2 + 2x - 60} \\
\underline{x^4 - 2x^3 + 10x^2} \\
-x^3 - 4x^2 + 2x \\
\underline{-x^3 + 2x^2 - 10x} \\
-6x^2 + 12x - 60 \\
\underline{-6x^2 + 12x - 60} \\
0
\end{array}
$$

So, you have

$$f(x) = (x^2 - 2x + 10)(x^2 - x - 6)$$

$$= (x^2 - 2x + 10)(x - 3)(x + 2)$$

and you can conclude that the zeros of f are $x = 1 + 3i$, $x = 1 - 3i$, $x = 3$, and $x = -2$.

CHECK Point Now try Exercise 53.

Graphical Solution

Complex zeros always occur in conjugate pairs, so you know that $1 - 3i$ is also a zero of f. Because the polynomial is a fourth-degree polynomial, you know that there are two other zeros of the function. Use a graphing utility to graph

$$y = x^4 - 3x^3 + 6x^2 + 2x - 60$$

as shown in Figure 4.4.

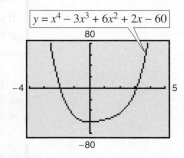

FIGURE **4.4**

You can see that -2 and 3 appear to be x-intercepts of the graph of the function. Use the *zero* or *root* feature or the *zoom* and *trace* features of the graphing utility to confirm that $x = -2$ and $x = 3$ are x-intercepts of the graph. So, you can conclude that the zeros of f are $x = 1 + 3i$, $x = 1 - 3i$, $x = 3$, and $x = -2$.

Example 6   Finding a Polynomial with Given Zeros

Find a fourth-degree polynomial function with real coefficients that has -1, -1, and $3i$ as zeros.

Solution

Because $3i$ is a zero *and* the polynomial is stated to have real coefficients, you know that the conjugate $-3i$ must also be a zero. So, from the Linear Factorization Theorem, $f(x)$ can be written as

$$f(x) = a(x + 1)(x + 1)(x - 3i)(x + 3i).$$

For simplicity, let $a = 1$ to obtain

$$f(x) = (x^2 + 2x + 1)(x^2 + 9)$$
$$= x^4 + 2x^3 + 10x^2 + 18x + 9.$$

CHECKPoint   Now try Exercise 65.

Example 7   Finding a Polynomial with Given Zeros

Find a cubic polynomial function f with real coefficients that has 2 and $1 - i$ as zeros, such that $f(1) = 3$.

Solution

Because $1 - i$ is a zero of f, so is $1 + i$. So,

$$f(x) = a(x - 2)[x - (1 - i)][x - (1 + i)]$$
$$= a(x - 2)[(x - 1) + i][(x - 1) - i]$$
$$= a(x - 2)[(x - 1)^2 - i^2]$$
$$= a(x - 2)(x^2 - 2x + 2)$$
$$= a(x^3 - 4x^2 + 6x - 4).$$

To find the value of a, use the fact that $f(1) = 3$ and obtain

$$f(1) = a[1^3 - 4(1)^2 + 6(1) - 4]$$
$$3 = -a$$
$$-3 = a.$$

So, $a = -3$ and it follows that

$$f(x) = -3(x^3 - 4x^2 + 6x - 4)$$
$$= -3x^3 + 12x^2 - 18x + 12.$$

CHECKPoint   Now try Exercise 71.

4.2   EXERCISES

See www.CalcChat.com for worked-out solutions to odd-numbered exercises.

VOCABULARY: Fill in the blanks.

1. The _____ _____ of _____ states that if $f(x)$ is a polynomial of degree n $(n > 0)$, then f has at least one zero in the complex number system.

2. The _____ _____ _____ states that if $f(x)$ is a polynomial of degree n $(n > 0)$, then f has precisely n linear factors, $f(x) = a_n(x - c_1)(x - c_2) \cdots (x - c_n)$, where $c_1, c_2, \ldots, c_n$ are complex numbers.

3. Two complex solutions of the form $a \pm bi$ of a polynomial equation with real coefficients are called _____.

4. The expression inside the radical of the Quadratic Formula, $b^2 - 4ac$, is called the _____ and is used to determine types of solutions of a quadratic equation.

SKILLS AND APPLICATIONS

In Exercises 5–8, determine the number of solutions of the equation in the complex number system.

5. $2x^3 + 3x + 1 = 0$
6. $x^6 + 4x^2 + 12 = 0$
7. $50 - 2x^4 = 0$
8. $14 - x + 4x^2 - 7x^5 = 0$

In Exercises 9–16, use the discriminant to determine the number of real solutions of the quadratic equation.

9. $2x^2 - 5x + 5 = 0$
10. $2x^2 - x - 1 = 0$
11. $\frac{1}{5}x^2 + \frac{6}{5}x - 8 = 0$
12. $\frac{1}{3}x^2 - 5x + 25 = 0$
13. $2x^2 - x - 15 = 0$
14. $-2x^2 + 11x - 2 = 0$
15. $x^2 + 2x + 10 = 0$
16. $x^2 - 4x + 53 = 0$

In Exercises 17–30, solve the equation. Write complex solutions in standard form.

17. $x^2 - 5 = 0$
18. $3x^2 - 1 = 0$
19. $(x + 5)^2 - 6 = 0$
20. $16 - (x - 1)^2 = 0$
21. $x^2 - 8x + 16 = 0$
22. $4x^2 + 4x + 1 = 0$
23. $x^2 + 2x + 5 = 0$
24. $54 + 16x - x^2 = 0$
25. $4x^2 - 4x + 5 = 0$
26. $4x^2 - 4x + 21 = 0$
27. $230 + 20x - 0.5x^2 = 0$
28. $125 - 30x + 0.4x^2 = 0$
29. $8 + (x + 3)^2 = 0$
30. $(x - 1)^2 + 12 = 0$

GRAPHICAL AND ANALYTICAL ANALYSIS In Exercises 31–34, (a) use a graphing utility to graph the function, (b) find all the zeros of the function, and (c) describe the relationship between the number of real zeros and the number of x-intercepts of the graph.

31. $f(x) = x^3 - 4x^2 + x - 4$
32. $f(x) = x^3 - 4x^2 - 4x + 16$
33. $f(x) = x^4 + 4x^2 + 4$
34. $f(x) = x^4 - 3x^2 - 4$

In Exercises 35–52, find all the zeros of the function and write the polynomial as a product of linear factors.

35. $f(x) = x^2 + 36$
36. $f(x) = x^2 - x + 56$
37. $h(x) = x^2 - 2x + 17$
38. $g(x) = x^2 + 10x + 17$
39. $f(x) = x^4 - 81$
40. $f(y) = y^4 - 256$
41. $f(z) = z^2 - 2z + 2$
42. $h(x) = x^2 - 6x - 10$
43. $g(x) = x^3 + 3x^2 - 3x - 9$
44. $f(x) = x^3 - 8x^2 - 12x + 96$
45. $h(x) = x^3 - 4x^2 + 16x - 64$
46. $h(x) = x^3 + 5x^2 + 2x + 10$
47. $f(x) = 2x^3 - x^2 + 36x - 18$
48. $g(x) = 4x^3 + 3x^2 + 96x + 72$
49. $g(x) = x^4 - 6x^3 + 16x^2 - 96x$
50. $h(x) = x^4 + x^3 + 100x^2 + 100x$
51. $f(x) = x^4 + 10x^2 + 9$
52. $f(x) = x^4 + 29x^2 + 100$

In Exercises 53–62, use the given zero to find all the zeros of the function.

Function	Zero
53. $f(x) = 2x^3 + 3x^2 + 50x + 75$	$5i$
54. $f(x) = x^3 + x^2 + 9x + 9$	$3i$
55. $f(x) = 2x^4 - x^3 + 7x^2 - 4x - 4$	$2i$
56. $f(x) = x^4 - 4x^3 + 6x^2 - 4x + 5$	i
57. $g(x) = 4x^3 + 23x^2 + 34x - 10$	$-3 + i$
58. $g(x) = x^3 - 7x^2 - x + 87$	$5 + 2i$
59. $f(x) = x^3 - 2x^2 - 14x + 40$	$3 - i$
60. $f(x) = x^3 + 4x^2 + 14x + 20$	$-1 - 3i$
61. $f(x) = x^4 + 3x^3 - 5x^2 - 21x + 22$	$-3 + \sqrt{2}i$
62. $h(x) = 3x^3 - 4x^2 + 8x + 8$	$1 - \sqrt{3}i$

In Exercises 63–68, find a polynomial function with real coefficients that has the given zeros. (There are many correct answers.)

63. $1, 5i$

64. $4, -3i$

65. $2, 5 + i$

66. $5, 3 - 2i$

67. $\frac{2}{3}, -1, 3 + \sqrt{2}i$

68. $-5, -5, 1 + \sqrt{3}i$

In Exercises 69–74, find a cubic polynomial function f with real coefficients that has the given zeros and the given function value.

	Zeros	Function Value
69.	$1, 2i$	$f(-1) = 10$
70.	$2, i$	$f(-1) = 6$
71.	$-1, 2 + i$	$f(2) = -9$
72.	$-2, 1 - 2i$	$f(2) = -10$
73.	$\frac{1}{2}, 1 + \sqrt{3}i$	$f(1) = -3$
74.	$\frac{3}{2}, 2 + \sqrt{2}i$	$f(1) = -6$

In Exercises 75–80, find a cubic polynomial function f with real coefficients that has the given complex zeros and x-intercept. (There are many correct answers.)

	Complex Zeros	x-Intercept
75.	$x = 4 \pm 2i$	$(-2, 0)$
76.	$x = 3 \pm i$	$(1, 0)$
77.	$x = 2 \pm \sqrt{6}i$	$(-1, 0)$
78.	$x = 2 \pm \sqrt{5}i$	$(2, 0)$
79.	$x = -1 \pm \sqrt{3}i$	$(4, 0)$
80.	$x = -3 \pm \sqrt{2}i$	$(-2, 0)$

81. Find the fourth-degree polynomial function f with real coefficients that has the zeros $x = \pm \sqrt{2}i$ and the x-intercepts shown in the graph.

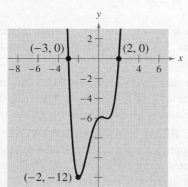

82. Find the fourth-degree polynomial function f with real coefficients that has the zeros $x = \pm \sqrt{5}i$ and the x-intercepts shown in the graph.

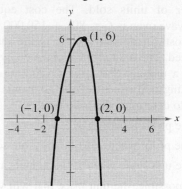

83. HEIGHT OF A BALL A ball is kicked upward from ground level with an initial velocity of 48 feet per second. The height h (in feet) of the ball is given by $h(t) = -16t^2 + 48t$, $0 \le t \le 3$, where t is the time (in seconds).

(a) Complete the table to find the heights h of the ball for the given times t.

t	0	0.5	1	1.5	2	2.5	3
h							

(b) From the table in part (a), does it appear that the ball reaches a height of 64 feet?

(c) Determine algebraically if the ball reaches a height of 64 feet.

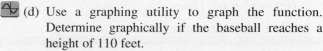(d) Use a graphing utility to graph the function. Determine graphically if the ball reaches a height of 64 feet.

(e) Compare your results from parts (b), (c), and (d).

84. HEIGHT OF A BASEBALL A baseball is thrown upward from a height of 5 feet with an initial velocity of 79 feet per second. The height h (in feet) of the baseball is given by $h = -16t^2 + 79t + 5$, $0 \le t \le 5$, where t is the time (in seconds).

(a) Complete the table to find the heights h of the baseball for the given times t.

t	0	1	2	3	4	5
h						

(b) From the table in part (a), does it appear that the baseball reaches a height of 110 feet?

(c) Determine algebraically if the baseball reaches a height of 110 feet.

(d) Use a graphing utility to graph the function. Determine graphically if the baseball reaches a height of 110 feet.

(e) Compare your results from parts (b), (c), and (d).

85. PROFIT The demand equation for a microwave oven is given by $p = 140 - 0.0001x$, where p is the unit price (in dollars) of the microwave oven and x is the number of units sold. The cost equation for the microwave oven is $C = 80x + 150{,}000$, where C is the total cost (in dollars) and x is the number of units produced. The total profit P obtained by producing and selling x units is $P = xp - C$. You are working in the marketing department of the company and have been asked to determine the following.

(a) The profit function

(b) The profit when 250,000 units are sold

(c) The unit price when 250,000 units are sold

(d) If possible, the unit price that will yield a profit of 10 million dollars.

86. DATA ANALYSIS: SALES The sales S (in billions of dollars) for Texas Instruments, Inc. for the years 2003 through 2008 are shown in the table. (Source: Texas Instruments, Inc.)

Year	Sales, S
2003	9.8
2004	12.6
2005	13.4
2006	14.3
2007	13.8
2008	12.5

(a) Use the *regression* feature of a graphing utility to find a quadratic model for the data. Let t represent the year, with $t = 3$ corresponding to 2003.

(b) Use a graphing utility to graph the model you found in part (a).

(c) Use your graph from part (b) to determine the year in which sales reached $15 billion. Is this possible?

(d) Determine algebraically the year in which sales reached $15 billion. Is this possible? Explain.

EXPLORATION

TRUE OR FALSE? In Exercises 87 and 88, decide whether the statement is true or false. Justify your answer.

87. It is possible for a third-degree polynomial function with integer coefficients to have no real zeros.

88. If $x = -i$ is a zero of the function given by
$$f(x) = x^3 + ix^2 + ix - 1$$
then $x = i$ must also be a zero of f.

89. From each graph, can you tell whether the discriminant is positive, zero, or negative? Explain your reasoning. Find each discriminant to verify your answers.

(a) $x^2 - 2x = 0$

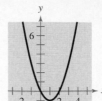

(b) $x^2 - 2x + 1 = 0$

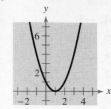

(c) $x^2 - 2x + 2 = 0$

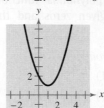

How many solutions would part (c) have if the linear term was $2x$? If the constant was -2?

90. CAPSTONE Write a paragraph explaining the relationships among the solutions of a polynomial equation, the zeros of a polynomial function, and the x-intercepts of the graph of a polynomial function. Include examples in your paragraph.

THINK ABOUT IT In Exercises 91–96, determine (if possible) the zeros of the function g if the function f has zeros at $x = r_1$, $x = r_2$, and $x = r_3$.

91. $g(x) = -f(x)$

92. $g(x) = 3f(x)$

93. $g(x) = f(x - 5)$

94. $g(x) = f(2x)$

95. $g(x) = 3 + f(x)$

96. $g(x) = f(-x)$

97. Find a quadratic function f (with integer coefficients) that has $\pm\sqrt{b}\,i$ as zeros. Assume that b is a positive integer.

98. Find a quadratic function f (with integer coefficients) that has $a \pm bi$ as zeros. Assume that b is a positive integer and a is an integer not equal to zero.

PROJECT: HEAD START ENROLLMENT To work an extended application analyzing Head Start enrollment in the United States from 1988 through 2007, visit this text's website at *academic.cengage.com*. (Data Source: U.S. Department of Health and Human Services)

4.3 TRIGONOMETRIC FORM OF A COMPLEX NUMBER

What you should learn

- Plot complex numbers in the complex plane and find absolute values of complex numbers.
- Write the trigonometric forms of complex numbers.
- Multiply and divide complex numbers written in trigonometric form.

Why you should learn it

You can perform the operations of multiplication and division on complex numbers by learning to write complex numbers in trigonometric form. For instance, in Exercises 63–70 on page 359, you can multiply and divide complex numbers in trigonometric form and standard form.

The Complex Plane

Just as real numbers can be represented by points on the real number line, you can represent a complex number

$$z = a + bi$$

as the point (a, b) in a coordinate plane (the **complex plane**). The horizontal axis is called the **real axis** and the vertical axis is called the **imaginary axis,** as shown in Figure 4.5.

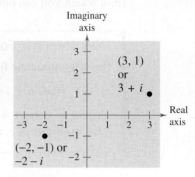

FIGURE **4.5**

The **absolute value** of the complex number $a + bi$ is defined as the distance between the origin $(0, 0)$ and the point (a, b).

Definition of the Absolute Value of a Complex Number

The **absolute value** of the complex number $z = a + bi$ is

$$|a + bi| = \sqrt{a^2 + b^2}.$$

If the complex number $a + bi$ is a real number (that is, if $b = 0$), then this definition agrees with that given for the absolute value of a real number

$$|a + 0i| = \sqrt{a^2 + 0^2} = |a|.$$

Example 1   Finding the Absolute Value of a Complex Number

Plot $z = -2 + 5i$ and find its absolute value.

Solution

The number is plotted in Figure 4.6. It has an absolute value of

$$|z| = \sqrt{(-2)^2 + 5^2}$$

$$= \sqrt{29}.$$

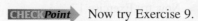 Now try Exercise 9.

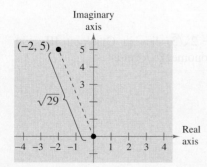

FIGURE **4.6**

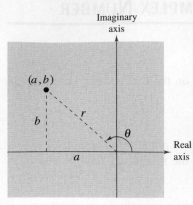

FIGURE **4.7**

Trigonometric Form of a Complex Number

In Section 4.1, you learned how to add, subtract, multiply, and divide complex numbers. To work effectively with *powers* and *roots* of complex numbers, it is helpful to write complex numbers in trigonometric form. In Figure 4.7, consider the nonzero complex number $a + bi$. By letting θ be the angle from the positive real axis (measured counterclockwise) to the line segment connecting the origin and the point (a, b), you can write

$$a = r \cos \theta \quad \text{and} \quad b = r \sin \theta$$

where $r = \sqrt{a^2 + b^2}$. Consequently, you have

$$a + bi = (r \cos \theta) + (r \sin \theta)i$$

from which you can obtain the **trigonometric form of a complex number.**

Trigonometric Form of a Complex Number

The **trigonometric form** of the complex number $z = a + bi$ is

$$z = r(\cos \theta + i \sin \theta)$$

where $a = r \cos \theta$, $b = r \sin \theta$, $r = \sqrt{a^2 + b^2}$, and $\tan \theta = b/a$. The number r is the **modulus** of z, and θ is called an **argument** of z.

The trigonometric form of a complex number is also called the *polar form*. Because there are infinitely many choices for θ, the trigonometric form of a complex number is not unique. Normally, θ is restricted to the interval $0 \leq \theta < 2\pi$, although on occasion it is convenient to use $\theta < 0$.

Example 2   Writing a Complex Number in Trigonometric Form

Write the complex number $z = -2 - 2\sqrt{3}i$ in trigonometric form.

Solution

The absolute value of z is

$$r = \left| -2 - 2\sqrt{3}i \right| = \sqrt{(-2)^2 + \left(-2\sqrt{3}\right)^2} = \sqrt{16} = 4$$

and the reference angle θ' is given by

$$\tan \theta' = \frac{b}{a} = \frac{-2\sqrt{3}}{-2} = \sqrt{3}.$$

Because $\tan(\pi/3) = \sqrt{3}$ and because $z = -2 - 2\sqrt{3}i$ lies in Quadrant III, you choose θ to be $\theta = \pi + \pi/3 = 4\pi/3$. So, the trigonometric form is

$$z = r(\cos \theta + i \sin \theta)$$

$$= 4\left(\cos \frac{4\pi}{3} + i \sin \frac{4\pi}{3} \right).$$

See Figure 4.8.

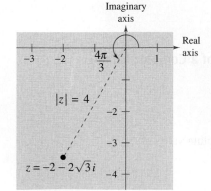

FIGURE **4.8**

CHECK*Point* Now try Exercise 17.

Example 3   Writing a Complex Number in Trigonometric Form

Write the complex number in trigonometric form.

$$z = 6 + 2i$$

Solution

The absolute value of z is

$$r = |6 + 2i|$$

$$= \sqrt{6^2 + 2^2}$$

$$= \sqrt{40}$$

$$= 2\sqrt{10}$$

and the angle θ is

$$\tan \theta = \frac{b}{a} = \frac{2}{6} = \frac{1}{3}.$$

Because $z = 6 + 2i$ is in Quadrant I, you can conclude that

$$\theta = \arctan \frac{1}{3} \approx 0.32175 \text{ radian} \approx 18.4°.$$

So, the trigonometric form of z is

$$z = r(\cos \theta + i \sin \theta)$$

$$= 2\sqrt{10}\left[\cos\left(\arctan \frac{1}{3} \right) + i \sin\left(\arctan \frac{1}{3} \right) \right]$$

$$\approx 2\sqrt{10}(\cos 18.4° + i \sin 18.4°).$$

This result is illustrated graphically in Figure 4.9.

CHECK*Point*   Now try Exercise 23.

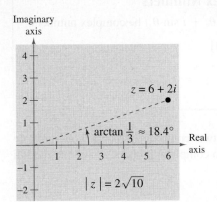

FIGURE 4.9

Example 4   Writing a Complex Number in Standard Form

Write the complex number in standard form $a + bi$.

$$z = \sqrt{8}\left[\cos\left(-\frac{\pi}{3} \right) + i \sin\left(-\frac{\pi}{3} \right) \right]$$

Solution

Because $\cos(-\pi/3) = \frac{1}{2}$ and $\sin(-\pi/3) = -\sqrt{3}/2$, you can write

$$z = \sqrt{8}\left[\cos\left(-\frac{\pi}{3} \right) + i \sin\left(-\frac{\pi}{3} \right) \right]$$

$$= 2\sqrt{2}\left(\frac{1}{2} - \frac{\sqrt{3}}{2}i \right)$$

$$= \sqrt{2} - \sqrt{6}i.$$

CHECK*Point*   Now try Exercise 37.

TECHNOLOGY

A graphing utility can be used to convert a complex number in trigonometric (or polar) form to standard form. For specific keystrokes, see the user's manual for your graphing utility.

Multiplication and Division of Complex Numbers

The trigonometric form adapts nicely to multiplication and division of complex numbers. Suppose you are given two complex numbers

$$z_1 = r_1(\cos \theta_1 + i \sin \theta_1) \qquad \text{and} \qquad z_2 = r_2(\cos \theta_2 + i \sin \theta_2).$$

The product of z_1 and z_2 is given by

$$z_1 z_2 = r_1 r_2 (\cos \theta_1 + i \sin \theta_1)(\cos \theta_2 + i \sin \theta_2)$$
$$= r_1 r_2 [(\cos \theta_1 \cos \theta_2 - \sin \theta_1 \sin \theta_2) + i(\sin \theta_1 \cos \theta_2 + \cos \theta_1 \sin \theta_2)].$$

Using the sum and difference formulas for cosine and sine, you can rewrite this equation as

$$z_1 z_2 = r_1 r_2 [\cos(\theta_1 + \theta_2) + i \sin(\theta_1 + \theta_2)].$$

This establishes the first part of the following rule. The second part is left for you to verify (see Exercise 83).

Product and Quotient of Two Complex Numbers

Let $z_1 = r_1(\cos \theta_1 + i \sin \theta_1)$ and $z_2 = r_2(\cos \theta_2 + i \sin \theta_2)$ be complex numbers.

$$z_1 z_2 = r_1 r_2 [\cos(\theta_1 + \theta_2) + i \sin(\theta_1 + \theta_2)] \qquad \text{Product}$$

$$\frac{z_1}{z_2} = \frac{r_1}{r_2}[\cos(\theta_1 - \theta_2) + i \sin(\theta_1 - \theta_2)], \quad z_2 \neq 0 \qquad \text{Quotient}$$

Note that this rule says that to *multiply* two complex numbers you multiply moduli and add arguments, whereas to *divide* two complex numbers you divide moduli and subtract arguments.

Example 5 Dividing Complex Numbers

Find the quotient z_1/z_2 of the complex numbers.

$$z_1 = 24(\cos 300° + i \sin 300°) \qquad z_2 = 8(\cos 75° + i \sin 75°)$$

Solution

$$\frac{z_1}{z_2} = \frac{24(\cos 300° + i \sin 300°)}{8(\cos 75° + i \sin 75°)}$$

$$= \frac{24}{8}[\cos(300° - 75°) + i \sin(300° - 75°)] \qquad \text{Divide moduli and subtract arguments.}$$

$$= 3(\cos 225° + i \sin 225°)$$

$$= 3\left[\left(-\frac{\sqrt{2}}{2}\right) + i\left(-\frac{\sqrt{2}}{2}\right)\right]$$

$$= -\frac{3\sqrt{2}}{2} - \frac{3\sqrt{2}}{2}i$$

CHECK *Point* Now try Exercise 57.

Example 6   Multiplying Complex Numbers

Find the product z_1z_2 of the complex numbers.

$$z_1 = 2\left(\cos\frac{2\pi}{3} + i\sin\frac{2\pi}{3}\right) \qquad z_2 = 8\left(\cos\frac{11\pi}{6} + i\sin\frac{11\pi}{6}\right)$$

Solution

$$z_1z_2 = 2\left(\cos\frac{2\pi}{3} + i\sin\frac{2\pi}{3}\right) \cdot 8\left(\cos\frac{11\pi}{6} + i\sin\frac{11\pi}{6}\right)$$

$$= 16\left[\cos\left(\frac{2\pi}{3} + \frac{11\pi}{6}\right) + i\sin\left(\frac{2\pi}{3} + \frac{11\pi}{6}\right)\right] \qquad \text{Multiply moduli and add arguments.}$$

$$= 16\left(\cos\frac{5\pi}{2} + i\sin\frac{5\pi}{2}\right)$$

$$= 16\left(\cos\frac{\pi}{2} + i\sin\frac{\pi}{2}\right)$$

$$= 16[0 + i(1)]$$

$$= 16i$$

CHECK *Point*   Now try Exercise 51.

You can check the result in Example 6 by first converting the complex numbers to the standard forms $z_1 = -1 + \sqrt{3}i$ and $z_2 = 4\sqrt{3} - 4i$ and then multiplying algebraically, as in Section 4.1.

$$z_1z_2 = \left(-1 + \sqrt{3}i\right)\left(4\sqrt{3} - 4i\right)$$

$$= -4\sqrt{3} + 4i + 12i + 4\sqrt{3}$$

$$= 16i$$

CLASSROOM DISCUSSION

Multiplying Complex Numbers Graphically   Discuss how you can graphically approximate the product of the complex numbers. Then, approximate the values of the products and check your answers analytically.

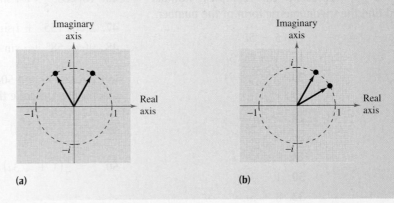

(a)   (b)

4.3   EXERCISES

See www.CalcChat.com for worked-out solutions to odd-numbered exercises.

VOCABULARY: Fill in the blanks.

1. In the complex plane, the horizontal axis is called the _____ axis and the vertical axis is called the _____ axis.

2. The _____ _____ of a complex number $a + bi$ is the distance between the origin $(0, 0)$ and the point (a, b).

3. The _____ _____ of a complex number $z = a + bi$ is given by $z = r(\cos \theta + i \sin \theta)$, where r is the _____ of z and θ is the _____ of z.

4. Let $z_1 = r_1(\cos \theta_1 + i \sin \theta_1)$ and $z_2 = r_2(\cos \theta_2 + i \sin \theta_2)$ be complex numbers, then the product $z_1 z_2 =$ _____ and the quotient $z_1/z_2 =$ _____ $(z_2 \neq 0)$.

SKILLS AND APPLICATIONS

In Exercises 5–10, plot the complex number and find its absolute value.

5. $-6 + 8i$

6. $5 - 12i$

7. $-7i$

8. -7

9. $4 - 6i$

10. $-8 + 3i$

In Exercises 11–14, write the complex number in trigonometric form.

11.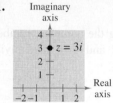

12.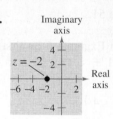

13.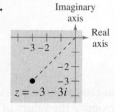

14.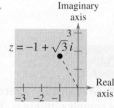

In Exercises 15–34, represent the complex number graphically, and find the trigonometric form of the number.

15. $1 + i$

16. $5 - 5i$

17. $1 - \sqrt{3}i$

18. $4 - 4\sqrt{3}i$

19. $-2(1 + \sqrt{3}i)$

20. $\frac{5}{2}(\sqrt{3} - i)$

21. $-5i$

22. $12i$

23. $-7 + 4i$

24. $3 - i$

25. 2

26. 4

27. $3 + \sqrt{3}i$

28. $2\sqrt{2} - i$

29. $-3 - i$

30. $1 + 3i$

31. $5 + 2i$

32. $8 + 3i$

33. $-8 - 5\sqrt{3}i$

34. $-9 - 2\sqrt{10}i$

In Exercises 35–44, find the standard form of the complex number. Then represent the complex number graphically.

35. $2(\cos 60° + i \sin 60°)$

36. $5(\cos 135° + i \sin 135°)$

37. $\sqrt{48}[\cos(-30°) + i \sin(-30°)]$

38. $\sqrt{8}(\cos 225° + i \sin 225°)$

39. $\frac{9}{4}\left(\cos \frac{3\pi}{4} + i \sin \frac{3\pi}{4}\right)$

40. $6\left(\cos \frac{5\pi}{12} + i \sin \frac{5\pi}{12}\right)$

41. $7(\cos 0 + i \sin 0)$

42. $8\left(\cos \frac{\pi}{2} + i \sin \frac{\pi}{2}\right)$

43. $5[\cos(198° \, 45') + i \sin(198° \, 45')]$

44. $9.75[\cos(280° \, 30') + i \sin(280° \, 30')]$

In Exercises 45–48, use a graphing utility to represent the complex number in standard form.

45. $5\left(\cos \frac{\pi}{9} + i \sin \frac{\pi}{9}\right)$

46. $10\left(\cos \frac{2\pi}{5} + i \sin \frac{2\pi}{5}\right)$

47. $2(\cos 155° + i \sin 155°)$

48. $9(\cos 58° + i \sin 58°)$

In Exercises 49 and 50, represent the powers z, z^2, z^3, and z^4 graphically. Describe the pattern.

49. $z = \frac{\sqrt{2}}{2}(1 + i)$

50. $z = \frac{1}{2}(1 + \sqrt{3}i)$

In Exercises 51–62, perform the operation and leave the result in trigonometric form.

51. $\left[2\left(\cos \dfrac{\pi}{4} + i \sin \dfrac{\pi}{4} \right) \right] \left[6\left(\cos \dfrac{\pi}{12} + i \sin \dfrac{\pi}{12} \right) \right]$

52. $\left[\dfrac{3}{4}\left(\cos \dfrac{\pi}{3} + i \sin \dfrac{\pi}{3} \right) \right] \left[4\left(\cos \dfrac{3\pi}{4} + i \sin \dfrac{3\pi}{4} \right) \right]$

53. $\left[\dfrac{5}{3}(\cos 120° + i \sin 120°) \right] \left[\dfrac{2}{3}(\cos 30° + i \sin 30°) \right]$

54. $\left[\dfrac{1}{2}(\cos 100° + i \sin 100°) \right] \left[\dfrac{4}{5}(\cos 300° + i \sin 300°) \right]$

55. $(\cos 80° + i \sin 80°)(\cos 330° + i \sin 330°)$

56. $(\cos 5° + i \sin 5°)(\cos 20° + i \sin 20°)$

57. $\dfrac{3(\cos 50° + i \sin 50°)}{9(\cos 20° + i \sin 20°)}$

58. $\dfrac{\cos 120° + i \sin 120°}{2(\cos 40° + i \sin 40°)}$

59. $\dfrac{\cos \pi + i \sin \pi}{\cos(\pi/3) + i \sin(\pi/3)}$

60. $\dfrac{5(\cos 4.3 + i \sin 4.3)}{4(\cos 2.1 + i \sin 2.1)}$

61. $\dfrac{12(\cos 92° + i \sin 92°)}{2(\cos 122° + i \sin 122°)}$

62. $\dfrac{6(\cos 40° + i \sin 40°)}{7(\cos 100° + i \sin 100°)}$

In Exercises 63–70, (a) write the trigonometric forms of the complex numbers, (b) perform the indicated operation using the trigonometric forms, and (c) perform the indicated operation using the standard forms, and check your result with that of part (b).

63. $(2 + 2i)(1 - i)$

64. $(\sqrt{3} + i)(1 + i)$

65. $-2i(1 + i)$

66. $3i(1 - \sqrt{2}i)$

67. $\dfrac{3 + 4i}{1 - \sqrt{3}i}$

68. $\dfrac{1 + \sqrt{3}i}{6 - 3i}$

69. $\dfrac{5}{2 + 3i}$

70. $\dfrac{4i}{-4 + 2i}$

In Exercises 71–74, sketch the graph of all complex numbers z satisfying the given condition.

71. $|z| = 2$

72. $|z| = 3$

73. $\theta = \dfrac{\pi}{6}$

74. $\theta = \dfrac{5\pi}{4}$

ELECTRICAL ENGINEERING   In Exercises 75–80, use the formula to find the missing quantity for the given conditions. The formula

$$E = I \cdot Z$$

where E represents voltage, I represents current, and Z represents impedance (a measure of opposition to a sinusoidal electric current), is used in electrical engineering. Each variable is a complex number.

75. $I = 10 + 2i$
$Z = 4 + 3i$

76. $I = 12 + 2i$
$Z = 3 + 5i$

77. $I = 2 + 4i$
$E = 5 + 5i$

78. $I = 10 + 2i$
$E = 4 + 5i$

79. $E = 12 + 24i$
$Z = 12 + 20i$

80. $E = 15 + 12i$
$Z = 25 + 24i$

EXPLORATION

TRUE OR FALSE?   In Exercises 81 and 82, determine whether the statement is true or false. Justify your answer.

81. Although the square of the complex number bi is given by $(bi)^2 = -b^2$, the absolute value of the complex number $z = a + bi$ is defined as
$$|a + bi| = \sqrt{a^2 + b^2}.$$

82. The product of two complex numbers
$$z_1 = r_1(\cos \theta_1 + i \sin \theta_1)$$
and
$$z_2 = r_2(\cos \theta_2 + i \sin \theta_2)$$
is zero only when $r_1 = 0$ and/or $r_2 = 0$.

83. Given two complex numbers $z_1 = r_1(\cos \theta_1 + i \sin \theta_1)$ and $z_2 = r_2(\cos \theta_2 + i \sin \theta_2)$, $z_2 \neq 0$, show that
$$\dfrac{z_1}{z_2} = \dfrac{r_1}{r_2}\left[\cos(\theta_1 - \theta_2) + i \sin(\theta_1 - \theta_2) \right].$$

84. Show that $\bar{z} = r[\cos(-\theta) + i \sin(-\theta)]$ is the complex conjugate of $z = r(\cos \theta + i \sin \theta)$.

85. Use the trigonometric forms of z and $\bar{z}$ in Exercise 84 to find (a) $z\bar{z}$ and (b) $z/\bar{z}$, $\bar{z} \neq 0$.

86. CAPSTONE   Given two complex numbers z_1 and z_2, discuss the advantages and disadvantages of using the trigonometric forms of these numbers (versus the standard forms) when performing the following operations.

(a) $z_1 + z_2$

(b) $z_1 - z_2$

(c) $z_1 \cdot z_2$

(d) z_1/z_2

What you should learn

- Use DeMoivre's Theorem to find powers of complex numbers.
- Find *n*th roots of complex numbers.

Why you should learn it

You can use the trigonometric form of a complex number to perform operations with complex numbers. For instance, in Exercises 55–70 on page 365, you can use the trigonometric forms of complex numbers to help you solve polynomial equations.

Powers of Complex Numbers

The trigonometric form of a complex number is used to raise a complex number to a power. To accomplish this, consider repeated use of the multiplication rule.

$$z = r(\cos \theta + i \sin \theta)$$

$$z^2 = r(\cos \theta + i \sin \theta)r(\cos \theta + i \sin \theta) = r^2(\cos 2\theta + i \sin 2\theta)$$

$$z^3 = r^2(\cos 2\theta + i \sin 2\theta)r(\cos \theta + i \sin \theta) = r^3(\cos 3\theta + i \sin 3\theta)$$

$$z^4 = r^4(\cos 4\theta + i \sin 4\theta)$$

$$z^5 = r^5(\cos 5\theta + i \sin 5\theta)$$

$$\vdots$$

This pattern leads to DeMoivre's Theorem, which is named after the French mathematician Abraham DeMoivre (1667–1754).

DeMoivre's Theorem

If $z = r(\cos \theta + i \sin \theta)$ is a complex number and n is a positive integer, then

$$z^n = [r(\cos \theta + i \sin \theta)]^n$$

$$= r^n(\cos n\theta + i \sin n\theta).$$

Example 1   Finding a Power of a Complex Number

Use DeMoivre's Theorem to find $\left(-1 + \sqrt{3}i\right)^{12}$.

Solution

First convert the complex number to trigonometric form using

$$r = \sqrt{(-1)^2 + \left(\sqrt{3}\right)^2} = 2 \quad \text{and} \quad \theta = \arctan \frac{\sqrt{3}}{-1} = \frac{2\pi}{3}.$$

So, the trigonometric form is

$$z = -1 + \sqrt{3}i = 2\left(\cos \frac{2\pi}{3} + i \sin \frac{2\pi}{3}\right).$$

Then, by DeMoivre's Theorem, you have

$$\left(-1 + \sqrt{3}i\right)^{12} = \left[2\left(\cos \frac{2\pi}{3} + i \sin \frac{2\pi}{3}\right)\right]^{12}$$

$$= 2^{12}\left[\cos \frac{12(2\pi)}{3} + i \sin \frac{12(2\pi)}{3}\right]$$

$$= 4096(\cos 8\pi + i \sin 8\pi)$$

$$= 4096(1 + 0)$$

$$= 4096.$$

CHECK*Point* ▶ Now try Exercise 5.

Roots of Complex Numbers

Recall that a consequence of the Fundamental Theorem of Algebra is that a polynomial equation of degree n has n solutions in the complex number system. So, the equation $x^6 = 1$ has six solutions, and in this particular case you can find the six solutions by factoring and using the Quadratic Formula.

$$x^6 - 1 = (x^3 - 1)(x^3 + 1)$$
$$= (x - 1)(x^2 + x + 1)(x + 1)(x^2 - x + 1) = 0$$

Consequently, the solutions are

$$x = \pm 1, \qquad x = \frac{-1 \pm \sqrt{3}\,i}{2}, \qquad \text{and} \qquad x = \frac{1 \pm \sqrt{3}\,i}{2}.$$

Each of these numbers is a sixth root of 1. In general, an ***n*th root of a complex number** is defined as follows.

Definition of an *n*th Root of a Complex Number

The complex number $u = a + bi$ is an ***n*th root** of the complex number z if

$$z = u^n = (a + bi)^n.$$

To find a formula for an nth root of a complex number, let u be an nth root of z, where

$$u = s(\cos \beta + i \sin \beta)$$

and

$$z = r(\cos \theta + i \sin \theta).$$

By DeMoivre's Theorem and the fact that $u^n = z$, you have

$$s^n(\cos n\beta + i \sin n\beta) = r(\cos \theta + i \sin \theta).$$

Taking the absolute value of each side of this equation, it follows that $s^n = r$. Substituting back into the previous equation and dividing by r, you get

$$\cos n\beta + i \sin n\beta = \cos \theta + i \sin \theta.$$

So, it follows that

$$\cos n\beta = \cos \theta \qquad \text{and} \qquad \sin n\beta = \sin \theta.$$

Because both sine and cosine have a period of 2π, these last two equations have solutions if and only if the angles differ by a multiple of 2π. Consequently, there must exist an integer k such that

$$n\beta = \theta + 2\pi k$$

$$\beta = \frac{\theta + 2\pi k}{n}.$$

By substituting this value of β into the trigonometric form of u, you get the result stated on the following page.

Finding nth Roots of a Complex Number

For a positive integer n, the complex number $z = r(\cos\theta + i\sin\theta)$ has exactly n distinct nth roots given by

$$\sqrt[n]{r}\left(\cos\frac{\theta + 2\pi k}{n} + i\sin\frac{\theta + 2\pi k}{n}\right)$$

where $k = 0, 1, 2, \ldots, n - 1$.

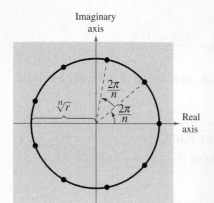

Imaginary
axis

$\dfrac{2\pi}{n}$

$\sqrt[n]{r}$

$\dfrac{2\pi}{n}$

Real
axis

FIGURE 4.10

When k exceeds $n - 1$, the roots begin to repeat. For instance, if $k = n$, the angle

$$\frac{\theta + 2\pi n}{n} = \frac{\theta}{n} + 2\pi$$

is coterminal with θ/n, which is also obtained when $k = 0$.

The formula for the nth roots of a complex number z has a nice geometrical interpretation, as shown in Figure 4.10. Note that because the nth roots of z all have the same magnitude $\sqrt[n]{r}$, they all lie on a circle of radius $\sqrt[n]{r}$ with center at the origin. Furthermore, because successive nth roots have arguments that differ by $2\pi/n$, the n roots are equally spaced around the circle.

You have already found the sixth roots of 1 by factoring and by using the Quadratic Formula. Example 2 shows how you can solve the same problem with the formula for nth roots.

Example 2   Finding the nth Roots of a Real Number

Find all sixth roots of 1.

Solution

First write 1 in the trigonometric form $1 = 1(\cos 0 + i\sin 0)$. Then, by the nth root formula, with $n = 6$ and $r = 1$, the roots have the form

$$\sqrt[6]{1}\left(\cos\frac{0 + 2\pi k}{6} + i\sin\frac{0 + 2\pi k}{6}\right) = \cos\frac{\pi k}{3} + i\sin\frac{\pi k}{3}.$$

So, for $k = 0, 1, 2, 3, 4,$ and 5, the sixth roots are as follows. (See Figure 4.11.)

$$\cos 0 + i\sin 0 = 1$$

$$\cos\frac{\pi}{3} + i\sin\frac{\pi}{3} = \frac{1}{2} + \frac{\sqrt{3}}{2}i \qquad \text{Increment by } \frac{2\pi}{n} = \frac{2\pi}{6} = \frac{\pi}{3}$$

$$\cos\frac{2\pi}{3} + i\sin\frac{2\pi}{3} = -\frac{1}{2} + \frac{\sqrt{3}}{2}i$$

$$\cos\pi + i\sin\pi = -1$$

$$\cos\frac{4\pi}{3} + i\sin\frac{4\pi}{3} = -\frac{1}{2} - \frac{\sqrt{3}}{2}i$$

$$\cos\frac{5\pi}{3} + i\sin\frac{5\pi}{3} = \frac{1}{2} - \frac{\sqrt{3}}{2}i$$

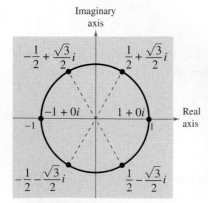

Imaginary
axis

$-\dfrac{1}{2} + \dfrac{\sqrt{3}}{2}i$ $\dfrac{1}{2} + \dfrac{\sqrt{3}}{2}i$

$-1 + 0i$ $1 + 0i$

Real
axis

$-\dfrac{1}{2} - \dfrac{\sqrt{3}}{2}i$ $\dfrac{1}{2} - \dfrac{\sqrt{3}}{2}i$

FIGURE 4.11

CHECK*Point*   Now try Exercise 47.

In Figure 4.11, notice that the roots obtained in Example 2 all have a magnitude of 1 and are equally spaced around the unit circle. Also notice that the complex roots occur in conjugate pairs, as discussed in Section 4.2. The n distinct nth roots of 1 are called the **nth roots of unity.**

Example 3   Finding the nth Roots of a Complex Number

Find the three cube roots of $z = -2 + 2i$.

Solution

Because z lies in Quadrant II, the trigonometric form of z is

$$z = -2 + 2i$$

$$= \sqrt{8}\,(\cos 135° + i \sin 135°). \qquad \theta = \arctan\!\left(\frac{2}{-2}\right) = 135°$$

By the formula for nth roots, the cube roots have the form

$$\sqrt[6]{8}\left(\cos\frac{135° + 360°k}{3} + i \sin\frac{135° + 360°k}{3}\right).$$

Finally, for $k = 0, 1,$ and 2, you obtain the roots

$$\sqrt[6]{8}\left(\cos\frac{135° + 360°(0)}{3} + i \sin\frac{135° + 360°(0)}{3}\right) = \sqrt{2}(\cos 45° + i \sin 45°)$$

$$= 1 + i$$

$$\sqrt[6]{8}\left(\cos\frac{135° + 360°(1)}{3} + i \sin\frac{135° + 360°(1)}{3}\right) = \sqrt{2}(\cos 165° + i \sin 165°)$$

$$\approx -1.3660 + 0.3660i$$

$$\sqrt[6]{8}\left(\cos\frac{135° + 360°(2)}{3} + i \sin\frac{135° + 360°(2)}{3}\right) = \sqrt{2}(\cos 285° + i \sin 285°)$$

$$\approx 0.3660 - 1.3660i.$$

See Figure 4.12.

CHECK*Point*   Now try Exercise 53.

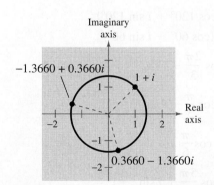

FIGURE **4.12**

Study Tip

Note in Example 3 that the absolute value of z is

$$r = |-2 + 2i|$$

$$= \sqrt{(-2)^2 + 2^2}$$

$$= \sqrt{8}$$

and the angle θ is given by

$$\tan\theta = \frac{b}{a} = \frac{2}{-2} = -1.$$

CLASSROOM DISCUSSION

A Famous Mathematical Formula   The famous formula

$$e^{a+bi} = e^a(\cos b + i \sin b)$$

is called Euler's Formula, after the Swiss mathematician Leonhard Euler (1707–1783). Although the interpretation of this formula is beyond the scope of this text, we decided to include it because it gives rise to one of the most wonderful equations in mathematics.

$$e^{\pi i} + 1 = 0$$

This elegant equation relates the five most famous numbers in mathematics—0, 1, π, e, and i—in a single equation (e is called the natural base and is discussed in Section 5.1). Show how Euler's Formula can be used to derive this equation.

4.4   EXERCISES

See www.CalcChat.com for worked-out solutions to odd-numbered exercises.

VOCABULARY: Fill in the blanks.

1. _____ Theorem states that if $z = r(\cos\theta + i\sin\theta)$ is a complex number and n is a positive integer, then $z^n = r^n(\cos n\theta + i\sin n\theta)$.

2. The complex number $u = a + bi$ is an _____ _____ of the complex number z if $z = u^n = (a + bi)^n$.

3. For a positive integer n, the complex number $z = r(\cos\theta + i\sin\theta)$ has exactly n distinct nth roots given by _____, where $k = 0, 1, 2, \ldots, n - 1$.

4. The n distinct nth roots of 1 are called the nth roots of _____.

SKILLS AND APPLICATIONS

In Exercises 5–28, use DeMoivre's Theorem to find the indicated power of the complex number. Write the result in standard form.

5. $(1 + i)^5$

6. $(2 + 2i)^6$

7. $(-1 + i)^6$

8. $(3 - 2i)^8$

9. $2\left(\sqrt{3} + i\right)^{10}$

10. $4\left(1 - \sqrt{3}i\right)^3$

11. $[5(\cos 20° + i\sin 20°)]^3$

12. $[3(\cos 60° + i\sin 60°)]^4$

13. $\left(\cos\dfrac{\pi}{4} + i\sin\dfrac{\pi}{4}\right)^{12}$

14. $\left[2\left(\cos\dfrac{\pi}{2} + i\sin\dfrac{\pi}{2}\right)\right]^8$

15. $[5(\cos 3.2 + i\sin 3.2)]^4$

16. $(\cos 0 + i\sin 0)^{20}$

17. $(3 - 2i)^5$

18. $(2 + 5i)^6$

19. $\left(\sqrt{5} - 4i\right)^3$

20. $\left(\sqrt{3} + 2i\right)^4$

21. $[3(\cos 15° + i\sin 15°)]^4$

22. $[2(\cos 10° + i\sin 10°)]^8$

23. $[5(\cos 95° + i\sin 95°)]^3$

24. $[4(\cos 110° + i\sin 110°)]^4$

25. $\left[2\left(\cos\dfrac{\pi}{10} + i\sin\dfrac{\pi}{10}\right)\right]^5$

26. $\left[2\left(\cos\dfrac{\pi}{8} + i\sin\dfrac{\pi}{8}\right)\right]^6$

27. $\left[3\left(\cos\dfrac{2\pi}{3} + i\sin\dfrac{2\pi}{3}\right)\right]^3$

28. $\left[3\left(\cos\dfrac{\pi}{12} + i\sin\dfrac{\pi}{12}\right)\right]^5$

In Exercises 29–36, find the square roots of the complex number.

29. $2i$

30. $5i$

31. $-3i$

32. $-6i$

33. $2 - 2i$

34. $2 + 2i$

35. $1 + \sqrt{3}i$

36. $1 - \sqrt{3}i$

In Exercises 37–54, (a) use the theorem on page 362 to find the indicated roots of the complex number, (b) represent each of the roots graphically, and (c) write each of the roots in standard form.

37. Square roots of $5(\cos 120° + i\sin 120°)$

38. Square roots of $16(\cos 60° + i\sin 60°)$

39. Cube roots of $8\left(\cos\dfrac{2\pi}{3} + i\sin\dfrac{2\pi}{3}\right)$

40. Cube roots of $64\left(\cos\dfrac{\pi}{3} + i\sin\dfrac{\pi}{3}\right)$

41. Fifth roots of $243\left(\cos\dfrac{\pi}{6} + i\sin\dfrac{\pi}{6}\right)$

42. Fifth roots of $32\left(\cos\dfrac{5\pi}{6} + i\sin\dfrac{5\pi}{6}\right)$

43. Fourth roots of $81i$

44. Fourth roots of $625i$

45. Cube roots of $-\dfrac{125}{2}\left(1 + \sqrt{3}i\right)$

46. Cube roots of $-4\sqrt{2}(-1 + i)$

47. Fourth roots of 16

48. Fourth roots of i

49. Fifth roots of 1

50. Cube roots of 1000

51. Cube roots of -125

52. Fourth roots of -4

53. Fifth roots of $4(1 - i)$

54. Sixth roots of $64i$

In Exercises 55–70, use the theorem on page 362 to find all the solutions of the equation and represent the solutions graphically.

55. $x^4 + i = 0$

56. $x^3 - i = 0$

57. $x^6 + 1 = 0$

58. $x^3 + 1 = 0$

59. $x^5 + 243 = 0$

60. $x^3 + 125 = 0$

61. $x^3 - 64 = 0$

62. $x^3 - 27 = 0$

63. $x^4 + 16i = 0$

64. $x^3 + 27i = 0$

65. $x^4 - 16i = 0$

66. $x^6 + 64i = 0$

67. $x^3 - (1 - i) = 0$

68. $x^5 - (1 - i) = 0$

69. $x^6 + (1 + i) = 0$

70. $x^4 + (1 + i) = 0$

EXPLORATION

TRUE OR FALSE? In Exercises 71–73, determine whether the statement is true or false. Justify your answer.

71. Geometrically, the nth roots of any complex number z are all equally spaced around the unit circle centered at the origin.

72. By DeMoivre's Theorem,

$$\left(4 + \sqrt{6}i\right)^8 = \cos(32) + i\sin\left(8\sqrt{6}\right).$$

73. $\sqrt{3} + i$ is a solution of the equation $x^2 - 8i = 0$.

74. THINK ABOUT IT Explain how you can use DeMoivre's Theorem to solve the polynomial equation $x^4 + 16 = 0$. [*Hint:* Write -16 as $16(\cos \pi + i \sin \pi)$.]

75. Show that $\frac{1}{2}\left(1 - \sqrt{3}i\right)$ is a ninth root of -1.

76. Show that $2^{-1/4}(1 - i)$ is a fourth root of -2.

77. Use the Quadratic Formula and, if necessary, the theorem on page 362 to solve each equation.

(a) $x^2 + ix + 2 = 0$

(b) $x^2 + 2ix + 1 = 0$

(c) $x^2 + 2ix + \sqrt{3}i = 0$

78. CAPSTONE Use the graph of the roots of a complex number.

(a) Write each of the roots in trigonometric form.

(b) Identify the complex number whose roots are given. Use a graphing utility to verify your results.

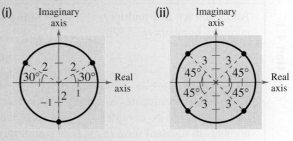

In Exercises 79 and 80, (a) show that the given value of x is a solution of the quadratic equation, (b) find the other solution and write it in trigonometric form, (c) explain how you obtained your answer to part (b), and (d) show that the solution in part (b) satisfies the quadratic equation.

79. $x^2 - 4x + 8 = 0$; $x = 2\sqrt{2}(\cos 45° + i \sin 45°)$

80. $x^2 + 2x + 4 = 0$; $x = 2\left(\cos\dfrac{2\pi}{3} + i \sin\dfrac{2\pi}{3}\right)$

81. Show that $2\left(\cos\dfrac{2\pi}{5} + i \sin\dfrac{2\pi}{5}\right)$ is a fifth root of 32. Then find the other fifth roots of 32, and verify your results.

82. Show that $\sqrt{2}(\cos 7.5° + i \sin 7.5°)$ is a fourth root of $2\sqrt{3} + 2i$. Then find the other fourth roots of $2\sqrt{3} + 2i$, and verify your results.

4 CHAPTER SUMMARY

	What Did You Learn?	Explanation/Examples	Review Exercises
Section 4.1	Use the imaginary unit i to write complex numbers *(p. 338)*.	If a and b are real numbers, the number $a + bi$ is a complex number, and it is said to be written in standard form. **Equality of Complex Numbers** Two complex numbers $a + bi$ and $c + di$, written in standard form, are equal to each other, $a + bi = c + di$, if and only if $a = c$ and $b = d$.	1–6, 27–30
	Add, subtract, and multiply complex numbers *(p. 339)*.	**Sum:** $(a + bi) + (c + di) = (a + c) + (b + d)i$ **Difference:** $(a + bi) - (c + di) = (a - c) + (b - d)i$ The Distributive Property can be used to multiply two complex numbers.	7–16
	Use complex conjugates to write the quotient of two complex numbers in standard form *(p. 341)*.	Complex numbers of the form $a + bi$ and $a - bi$ are complex conjugates. To write $(a + bi)/(c + di)$ in standard form, multiply the numerator and denominator by the complex conjugate of the denominator, $c - di$.	17–22
	Find complex solutions of quadratic equations *(p. 342)*.	**Principal Square Root of a Negative Number** If a is a positive number, the principal square root of the negative number $-a$ is defined as $\sqrt{-a} = \sqrt{a}i$.	23–26
Section 4.2	Determine the numbers of solutions of polynomial equations *(p. 345)*.	**The Fundamental Theorem of Algebra** If $f(x)$ is a polynomial of degree n, where $n > 0$, then f has at least one zero in the complex number system. **Linear Factorization Theorem** If $f(x)$ is a polynomial of degree n, where $n > 0$, then f has precisely n linear factors $f(x) = a_n(x - c_1)(x - c_2) \cdots (x - c_n)$ where $c_1, c_2, \ldots, c_n$ are complex numbers. Every second-degree equation, $ax^2 + bx + c = 0$, has precisely two solutions given by the Quadratic Formula. The expression inside the radical of the Quadratic Formula, $b^2 - 4ac$, is the discriminant, and can be used to determine whether the solutions are real, repeated, or complex. 1. $b^2 - 4ac < 0$: two complex solutions 2. $b^2 - 4ac = 0$: one repeated real solution 3. $b^2 - 4ac > 0$: two distinct real solutions	31–38
	Find solutions of polynomial equations *(p. 347)*.	**Complex Solutions Occur in Conjugate Pairs** If $a + bi$, $b \neq 0$, is a solution of a polynomial equation with real coefficients, the conjugate $a - bi$ is also a solution of the equation.	39–48
	Find zeros of polynomial functions and find polynomial functions given the zeros of the functions *(p. 348)*.	Finding the zeros of a polynomial function is essentially the same as finding the solutions of a polynomial equation.	49–72

What Did You Learn?	**Explanation/Examples**	**Review Exercises**
Section 4.3 — Plot complex numbers in the complex plane and find absolute values of complex numbers (*p. 353*).	A complex number $z = a + bi$ can be represented as the point (a, b) in the complex plane. The horizontal axis is the real axis and the vertical axis is the imaginary axis. The absolute value of $z = a + bi$ is $$\lvert a + bi \rvert = \sqrt{a^2 + b^2}.$$	73–78
Write the trigonometric forms of complex numbers (*p. 354*).	The trigonometric form of the complex number $z = a + bi$ is $$z = r(\cos\theta + i\sin\theta)$$ where $a = r\cos\theta$, $b = r\sin\theta$, $r = \sqrt{a^2 + b^2}$, and $\tan\theta = b/a$. The number r is the modulus of z, and θ is called an argument of z.	79–86
Multiply and divide complex numbers written in trigonometric form (*p. 356*).	**Product and Quotient of Two Complex Numbers** Let $z_1 = r_1(\cos\theta_1 + i\sin\theta_1)$ and $z_2 = r_2(\cos\theta_2 + i\sin\theta_2)$ be complex numbers. $$z_1 z_2 = r_1 r_2[\cos(\theta_1 + \theta_2) + i\sin(\theta_1 + \theta_2)]$$ $$\frac{z_1}{z_2} = \frac{r_1}{r_2}[\cos(\theta_1 - \theta_2) + i\sin(\theta_1 - \theta_2)], \quad z_2 \neq 0$$	87–94
Section 4.4 — Use DeMoivre's Theorem to find powers of complex numbers (*p. 360*).	**DeMoivre's Theorem** If $z = r(\cos\theta + i\sin\theta)$ is a complex number and n is a positive integer, then $$z^n = [r(\cos\theta + i\sin\theta)]^n = r^n(\cos n\theta + i\sin n\theta).$$	95–100
Find nth roots of complex numbers (*p. 361*).	**Definition of an nth Root of a Complex Number** The complex number $u = a + bi$ is an nth root of the complex number z if $$z = u^n = (a + bi)^n.$$ **Finding nth Roots of a Complex Number** For a positive integer n, the complex number $z = r(\cos\theta + i\sin\theta)$ has exactly n distinct nth roots given by $$\sqrt[n]{r}\left(\cos\frac{\theta + 2\pi k}{n} + i\sin\frac{\theta + 2\pi k}{n}\right)$$ where $k = 0, 1, 2, \ldots, n - 1$.	101–108

4.1 In Exercises 1–6, write the complex number in standard form.

1. $6 + \sqrt{-4}$

2. $3 - \sqrt{-25}$

3. $\sqrt{-48}$

4. 27

5. $i^2 + 3i$

6. $-5i + i^2$

In Exercises 7–16, perform the operation and write the result in standard form.

7. $(7 + 5i) + (-4 + 2i)$

8. $\left(\dfrac{\sqrt{2}}{2} - \dfrac{\sqrt{2}}{2}i\right) - \left(\dfrac{\sqrt{2}}{2} + \dfrac{\sqrt{2}}{2}i\right)$

9. $14 + (-3 + 11i) + 33i$

10. $-\left(\dfrac{1}{4} + \dfrac{7}{4}i\right) + \left(\dfrac{5}{2} + \dfrac{9}{2}i\right)$

11. $5i(13 - 8i)$

12. $(1 + 6i)(5 - 2i)$

13. $(10 - 8i)(2 - 3i)$

14. $i(6 + i)(3 - 2i)$

15. $(2 + 7i)^2$

16. $(3 + 6i)^2 + (3 - 6i)^2$

In Exercises 17–20, write the quotient in standard form.

17. $-\dfrac{10}{3i}$

18. $\dfrac{8}{12 - i}$

19. $\dfrac{6 + i}{4 - i}$

20. $\dfrac{3 + 2i}{5 + i}$

In Exercises 21 and 22, perform the operation and write the result in standard form.

21. $\dfrac{4}{2 - 3i} + \dfrac{2}{1 + i}$

22. $\dfrac{1}{2 + i} - \dfrac{5}{1 + 4i}$

In Exercises 23–26, find all solutions of the equation.

23. $3x^2 + 1 = 0$

24. $2 + 8x^2 = 0$

25. $x^2 - 2x + 10 = 0$

26. $6x^2 + 3x + 27 = 0$

In Exercises 27–30, simplify the complex number and write the result in standard form.

27. $10i^2 - i^3$

28. $-8i^6 + i^2$

29. $\dfrac{1}{i^7}$

30. $\dfrac{1}{(4i)^3}$

4.2 In Exercises 31–34, determine the number of solutions of the equation in the complex number system.

31. $x^5 - 2x^4 + 3x^2 - 5 = 0$

32. $-2x^6 + 7x^3 + x^2 + 4x - 19 = 0$

33. $\frac{1}{2}x^4 + \frac{2}{3}x^3 - x^2 + \frac{3}{10} = 0$

34. $\frac{3}{4}x^3 + \frac{1}{2}x^2 + \frac{3}{2}x + 2 = 0$

In Exercises 35–38, use the discriminant to determine the number of real solutions of the quadratic equation.

35. $6x^2 + x - 2 = 0$

36. $9x^2 - 12x + 4 = 0$

37. $0.13x^2 - 0.45x + 0.65 = 0$

38. $4x^2 + \frac{4}{3}x + \frac{1}{9} = 0$

In Exercises 39–46, solve the equation. Write complex solutions in standard form.

39. $x^2 - 2x = 0$

40. $6x - x^2 = 0$

41. $x^2 - 3x + 5 = 0$

42. $x^2 - 4x + 9 = 0$

43. $x^2 + 8x + 10 = 0$

44. $3 + 4x - x^2 = 0$

45. $2x^2 + 3x + 6 = 0$

46. $4x^2 - x + 10 = 0$

47. **PROFIT** The demand equation for a DVD player is $p = 140 - 0.0001x$, where p is the unit price (in dollars) of the DVD player and x is the number of units produced and sold. The cost equation for the DVD player is $C = 75x + 100{,}000$, where C is the total cost (in dollars) and x is the number of units produced. The total profit obtained by producing and selling x units is

$$P = xp - C.$$

You work in the marketing department of the company that produces this DVD player and are asked to determine a price p that would yield a profit of 9 million dollars. Is this possible? Explain.

48. **CONSUMER AWARENESS** The average monthly bill b (in dollars) for a cellular phone in the United States from 1998 through 2007 can be modeled by

$$b = -0.24t^2 + 7.2t - 3, \quad 8 \le t \le 17$$

where t represents the year, with $t = 8$ corresponding to 1998. According to this model, will the average monthly bill for a cellular phone rise to \$52? Explain your reasoning. (Source: CTIA-The Wireless Association)

In Exercises 49–54, find all the zeros of the function and write the polynomial as a product of linear factors.

49. $r(x) = 2x^2 + 2x + 3$

50. $s(x) = 2x^2 + 5x + 4$

51. $f(x) = 2x^3 - 3x^2 + 50x - 75$

52. $f(x) = 4x^3 - x^2 + 128x - 32$

53. $f(x) = 4x^4 + 3x^2 - 10$

54. $f(x) = 5x^4 + 126x^2 + 25$

In Exercises 55–62, use the given zero to find all the zeros of the function. Write the polynomial as a product of linear factors.

Function	Zero
55. $f(x) = x^3 + 3x^2 - 24x + 28$	2
56. $f(x) = 10x^3 + 21x^2 - x - 6$	-2
57. $f(x) = x^3 + 3x^2 - 5x + 25$	-5
58. $g(x) = x^3 - 8x^2 + 29x - 52$	4
59. $h(x) = 2x^3 - 19x^2 + 58x + 34$	$5 + 3i$
60. $f(x) = 5x^3 - 4x^2 + 20x - 16$	$2i$
61. $f(x) = x^4 + 5x^3 + 2x^2 - 50x - 84$	$-3 + \sqrt{5}i$
62. $g(x) = x^4 - 6x^3 + 18x^2 - 26x + 21$	$2 + \sqrt{3}i$

In Exercises 63–70, find a polynomial function with real coefficients that has the given zeros. (There are many correct answers.)

63. $1, 1, \frac{1}{4}, -\frac{2}{3}$

64. $-2, 2, 3, 3$

65. $3, 2 - \sqrt{3}, 2 + \sqrt{3}$

66. $5, 1 - \sqrt{2}, 1 + \sqrt{2}$

67. $\frac{2}{3}, 4, \sqrt{3}i, -\sqrt{3}i$

68. $2, -3, 1 - 2i, 1 + 2i$

69. $-\sqrt{2}i, \sqrt{2}i, -5i, 5i$

70. $-2i, 2i, -4i, 4i$

In Exercises 71 and 72, find a cubic polynomial function f with real coefficients that has the given zeros and the given function value.

Zeros	Function Value
71. $5, 1 - i$	$f(1) = -8$
72. $2, 4 + i$	$f(3) = 4$

4.3 In Exercises 73–78, plot the complex number and find its absolute value.

73. $8i$

74. $-6i$

75. -5

76. $-\sqrt{6}$

77. $5 + 3i$

78. $-10 - 4i$

In Exercises 79–86, write the complex number in trigonometric form.

79.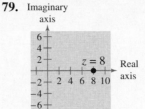

80.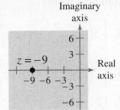

81.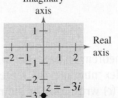

82.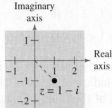

83. $5 - 5i$

84. $5 + 12i$

85. $-3\sqrt{3} + 3i$

86. $-\sqrt{2} + \sqrt{2}i$

In Exercises 87–90, perform the operation and leave the result in trigonometric form.

87. $\left[7\left(\cos\frac{\pi}{3} + i\sin\frac{\pi}{3} \right) \right]\left[4\left(\cos\frac{\pi}{4} + i\sin\frac{\pi}{4} \right) \right]$

88. $[1.5(\cos 25° + i\sin 25°)][5.5(\cos 34° + i\sin 34°)]$

89. $\dfrac{3\left(\cos\dfrac{2\pi}{3} + i\sin\dfrac{2\pi}{3} \right)}{6\left(\cos\dfrac{\pi}{6} + i\sin\dfrac{\pi}{6} \right)}$

90. $\dfrac{8(\cos 50° + i\sin 50°)}{2(\cos 105° + i\sin 105°)}$

In Exercises 91–94, (a) write the two complex numbers in trigonometric form, and (b) use the trigonometric form to find $z_1 z_2$ and $z_1/z_2, z_2 \neq 0$.

91. $z_1 = 1 + i, \quad z_2 = 1 - i$

92. $z_1 = 4 + 4i, \quad z_2 = -5 - 5i$

93. $z_1 = 2\sqrt{3} - 2i, \quad z_2 = -10i$

94. $z_1 = -3(1 + i), \quad z_2 = 2(\sqrt{3} + i)$

4.4 In Exercises 95–100, use DeMoivre's Theorem to find the indicated power of the complex number. Write the result in standard form.

95. $\left[5\left(\cos\dfrac{\pi}{12} + i\sin\dfrac{\pi}{12}\right)\right]^4$

96. $\left[2\left(\cos\dfrac{4\pi}{15} + i\sin\dfrac{4\pi}{15}\right)\right]^5$

97. $(2 + 3i)^6$

98. $(1 - i)^8$

99. $(-1 + i)^7$

100. $\left(\sqrt{3} - i\right)^4$

In Exercises 101–104, (a) use the theorem on page 362 to find the indicated roots of the complex number, (b) represent each of the roots graphically, and (c) write each of the roots in standard form.

101. Sixth roots of $-729i$

102. Fourth roots of 256

103. Fourth roots of -16

104. Fifth roots of -1

In Exercises 105–108, use the theorem on page 362 to find all solutions of the equation and represent the solutions graphically.

105. $x^4 + 81 = 0$

106. $x^5 - 243 = 0$

107. $x^3 + 8i = 0$

108. $(x^3 - 1)(x^2 + 1) = 0$

EXPLORATION

TRUE OR FALSE? In Exercises 109–111, determine whether the statement is true or false. Justify your answer.

109. $\sqrt{-18}\sqrt{-2} = \sqrt{(-18)(-2)}$

110. The equation $325x^2 - 717x + 398 = 0$ has no solution.

111. A fourth-degree polynomial with real coefficients can have -5, $128i$, $4i$, and 5 as its zeros.

112. Write quadratic equations that have (a) two distinct real solutions, (b) two complex solutions, and (c) no real solution.

GRAPHICAL REASONING In Exercises 113 and 114, use the graph of the roots of a complex number.

(a) Write each of the roots in trigonometric form.

(b) Identify the complex number whose roots are given. Use a graphing utility to verify your results.

113.

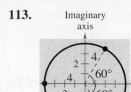

114.

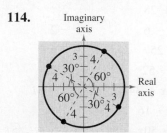

115. The figure shows z_1 and z_2. Describe z_1z_2 and z_1/z_2.

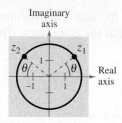

116. One of the fourth roots of a complex number z is shown in the figure.

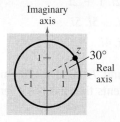

(a) How many roots are not shown?

(b) Describe the other roots.

4 CHAPTER TEST

See www.CalcChat.com for worked-out solutions to odd[...]

Take this test as you would take a test in class. When you are finished, che[...] against the answers given in the back of the book.

1. Write the complex number $-5 + \sqrt{-100}$ in standard form.

In Exercises 2–4, perform the operations and write the result in standard form.

2. $10i - (3 + \sqrt{-25})$ 3. $(4 + 9i)^2$ 4. $(6 + \sqrt{7}i)(6 - \sqrt{7}i)$

5. Write the quotient in standard form: $\dfrac{4}{8 - 3i}$.

6. Use the Quadratic Formula to solve the equation $2x^2 - 2x + 3 = 0$.

In Exercises 7 and 8, determine the number of solutions of the equation in the complex number system.

7. $x^5 + x^3 - x + 1 = 0$ 8. $x^4 - 3x^3 + 2x^2 - 4x - 5 = 0$

In Exercises 9 and 10, find all the zeros of the function.

9. $f(x) = x^3 - 6x^2 + 5x - 30$ 10. $f(x) = x^4 - 2x^2 - 24$

In Exercises 11 and 12, use the given zero(s) to find all the zeros of the function. Write the polynomial as a product of linear factors.

Function	Zero(s)
11. $h(x) = x^4 - 2x^2 - 8$	$-2, 2$
12. $g(v) = 2v^3 - 11v^2 + 22v - 15$	$3/2$

In Exercises 13 and 14, find a polynomial function with real coefficients that has the given zeros. (There are many correct answers.)

13. $0, 7, 4 + i, 4 - i$ 14. $1 + \sqrt{6}i, 1 - \sqrt{6}i, 3, 3$

15. Is it possible for a polynomial function with integer coefficients to have exactly one complex zero? Explain.

16. Write the complex number $z = 5 - 5i$ in trigonometric form.

17. Write the complex number $z = 6(\cos 120° + i \sin 120°)$ in standard form.

In Exercises 18 and 19, use DeMoivre's Theorem to find the indicated power of the complex number. Write the result in standard form.

18. $\left[3\left(\cos \dfrac{7\pi}{6} + i \sin \dfrac{7\pi}{6}\right)\right]^8$ 19. $(3 - 3i)^6$

20. Find the fourth roots of $256(1 + \sqrt{3}i)$.

21. Find all solutions of the equation $x^3 - 27i = 0$ and represent the solutions graphically.

22. A projectile is fired upward from ground level with an initial velocity of 88 feet per second. The height h (in feet) of the projectile is given by

$$h = -16t^2 + 88t, \quad 0 \le t \le 5.5$$

where t is the time (in seconds). You are told that the projectile reaches a height of 125 feet. Is this possible? Explain.

PROOFS IN MATHEMATICS

The Linear Factorization Theorem is closely related to the Fundamental Theorem of Algebra. The Fundamental Theorem of Algebra has a long and interesting history. In the early work with polynomial equations, The Fundamental Theorem of Algebra was thought to have been not true, because imaginary solutions were not considered. In fact, in the very early work by mathematicians such as Abu al-Khwarizmi (c. 800 A.D.), negative solutions were also not considered.

Once imaginary numbers were accepted, several mathematicians attempted to give a general proof of the Fundamental Theorem of Algebra. These mathematicians included Gottfried von Leibniz (1702), Jean D'Alembert (1746), Leonhard Euler (1749), Joseph-Louis Lagrange (1772), and Pierre Simon Laplace (1795). The mathematician usually credited with the first correct proof of the Fundamental Theorem of Algebra is Carl Friedrich Gauss, who published the proof in his doctoral thesis in 1799.

Linear Factorization Theorem *(p. 345)*

If $f(x)$ is a polynomial of degree n, where $n > 0$, then f has precisely n linear factors

$$f(x) = a_n(x - c_1)(x - c_2) \cdots (x - c_n)$$

where $c_1, c_2, \ldots, c_n$ are complex numbers.

Proof

Using the Fundamental Theorem of Algebra, you know that f must have at least one zero, c_1. Consequently, $(x - c_1)$ is a factor of $f(x)$, and you have

$$f(x) = (x - c_1)f_1(x).$$

If the degree of $f_1(x)$ is greater than zero, you again apply the Fundamental Theorem to conclude that f_1 must have a zero c_2, which implies that

$$f(x) = (x - c_1)(x - c_2)f_2(x).$$

It is clear that the degree of $f_1(x)$ is $n - 1$, that the degree of $f_2(x)$ is $n - 2$, and that you can repeatedly apply the Fundamental Theorem n times until you obtain

$$f(x) = a_n(x - c_1)(x - c_2) \cdots (x - c_n)$$

where a_n is the leading coefficient of the polynomial $f(x)$.

PROBLEM SOLVING

This collection of thought-provoking and challenging exercises further explores and expands upon concepts learned in this chapter.

1. (a) The complex numbers

$$z = 2, \; z = \frac{-2 + 2\sqrt{3}i}{2}, \quad \text{and} \quad z = \frac{-2 - 2\sqrt{3}i}{2}$$

are represented graphically (see figure). Evaluate the expression z^3 for each complex number. What do you observe?

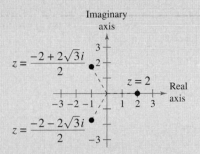

(b) The complex numbers

$$z = 3, \; z = \frac{-3 + 3\sqrt{3}i}{2}, \quad \text{and} \quad z = \frac{-3 - 3\sqrt{3}i}{2}$$

are represented graphically (see figure). Evaluate the expression z^3 for each complex number. What do you observe?

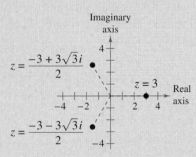

(c) Use your results from parts (a) and (b) to generalize your findings.

2. The multiplicative inverse of z is a complex number z_m such that $z \cdot z_m = 1$. Find the multiplicative inverse of each complex number.

(a) $z = 1 + i$

(b) $z = 3 - i$

(c) $z = -2 + 8i$

3. Show that the product of a complex number $a + bi$ and its conjugate is a real number.

4. Let

$z = a + bi, \bar{z} = a - bi, w = c + di,$ and $\bar{w} = c - di.$

Prove each statement.

(a) $\overline{z + w} = \bar{z} + \bar{w}$

(b) $\overline{z - w} = \bar{z} - \bar{w}$

(c) $\overline{zw} = \bar{z} \cdot \bar{w}$

(d) $\overline{z/w} = \bar{z}/\bar{w}$

(e) $(\bar{z})^2 = \overline{z^2}$

(f) $\bar{\bar{z}} = z$

(g) $\bar{z} = z$ if z is real.

5. Find the values of k such that the equation

$$x^2 - 2kx + k = 0$$

has (a) two real solutions and (b) two complex solutions.

6. Use a graphing utility to graph the function given by

$$f(x) = x^4 - 4x^2 + k$$

for different values of k. Find values of k such that the zeros of f satisfy the specified characteristics. (Some parts do not have unique answers.)

(a) Four real zeros

(b) Two real zeros and two complex zeros

(c) Four complex zeros

7. Will the answers to Exercise 6 change for the function g?

(a) $g(x) = f(x - 2)$

(b) $g(x) = f(2x)$

8. A third-degree polynomial function f has real zeros -2, $\frac{1}{2}$, and 3, and its leading coefficient is negative.

(a) Write an equation for f.

(b) Sketch the graph of f.

(c) How many different polynomial functions are possible for f?

9. The graph of one of the following functions is shown below. Identify the function shown in the graph. Explain why each of the others is not the correct function. Use a graphing utility to verify your result.

   (a) $f(x) = x^2(x + 2)(x - 3.5)$
   (b) $g(x) = (x + 2)(x - 3.5)$
   (c) $h(x) = (x + 2)(x - 3.5)(x^2 + 1)$
   (d) $k(x) = (x + 1)(x + 2)(x - 3.5)$

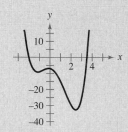

10. Use the information in the table to answer each question.

Interval	Value of $f(x)$
$(-\infty, -2)$	Positive
$(-2, 1)$	Negative
$(1, 4)$	Negative
$(4, \infty)$	Positive

    (a) What are the three real zeros of the polynomial function f?
    (b) What can be said about the behavior of the graph of f at $x = 1$?
    (c) What is the least possible degree of f? Explain. Can the degree of f ever be odd? Explain.
    (d) Is the leading coefficient of f positive or negative? Explain.
    (e) Write an equation for f.
    (f) Sketch a graph of the function you wrote in part (e).

11. A **fractal** is a geometric figure that consists of a pattern that is repeated infinitely on a smaller and smaller scale. The most famous fractal is called the **Mandelbrot Set,** named after the Polish-born mathematician Benoit Mandelbrot. To draw the Mandelbrot Set, consider the following sequence of numbers.

    $$c, c^2 + c, (c^2 + c)^2 + c, [(c^2 + c)^2 + c]^2 + c, \dots$$

    The behavior of this sequence depends on the value of the complex number c. If the sequence is bounded (the absolute value of each number in the sequence, $|a + bi| = \sqrt{a^2 + b^2}$, is less than some fixed number N), the complex number c is in the Mandelbrot Set, and if the sequence is unbounded (the absolute value of the terms of the sequence become infinitely large), the complex number c is not in the Mandelbrot Set. Determine whether the complex number c is in the Mandelbrot Set.

    (a) $c = i$ (b) $c = 1 + i$ (c) $c = -2$

12. (a) Complete the table.

Function	Zeros	Sum of zeros	Product of zeros
$f_1(x) = x^2 - 5x + 6$			
$f_2(x) = x^3 - 7x + 6$			
$f_3(x) = x^4 + 2x^3 + x^2 + 8x - 12$			
$f_4(x) = x^5 - 3x^4 - 9x^3 + 25x^2 - 6x$			

    (b) Use the table to make a conjecture relating the sum of the zeros of a polynomial function to the coefficients of the polynomial function.
    (c) Use the table to make a conjecture relating the product of the zeros of a polynomial function to the coefficients of the polynomial function.

13. Use the Quadratic Formula and, if necessary, DeMoivre's Theorem to solve each equation with complex coefficients.

    (a) $x^2 - (4 + 2i)x + 2 + 4i = 0$
    (b) $x^2 - (3 + 2i)x + 5 + i = 0$
    (c) $2x^2 + (5 - 8i)x - 13 - i = 0$
    (d) $3x^2 - (11 + 14i)x + 1 - 9i = 0$

14. Show that the solutions to

    $$|z - 1| \cdot |\bar{z} - 1| = 1$$

    are the points (x, y) in the complex plane such that $(x - 1)^2 + y^2 = 1$. Identify the graph of the solution set. $\bar{z}$ is the conjugate of z. (*Hint:* Let $z = x + yi$.)

15. Let $z = a + bi$ and $\bar{z} = a - bi$, where $a \neq 0$. Show that the equation

    $$z^2 - \bar{z}^2 = 0$$

    has only real solutions, whereas the equation

    $$z^2 + \bar{z}^2 = 0$$

    has complex solutions.

Exponential and Logarithmic Functions

5

In Mathematics

Exponential functions involve a constant base and a variable exponent. The inverse of an exponential function is a logarithmic function.

In Real Life

Exponential and logarithmic functions are widely used in describing economic and physical phenomena such as compound interest, population growth, memory retention, and decay of radioactive material. For instance, a logarithmic function can be used to relate an animal's weight and its lowest galloping speed. (See Exercise 95, page 402.)

Juniors Bildarchiv / Alamy

IN CAREERS

There are many careers that use exponential and logarithmic functions. Several are listed below.

- Astronomer
  Example 7, page 400

- Psychologist
  Exercise 136, page 413

- Archeologist
  Example 3, page 418

- Forensic Scientist
  Exercise 75, page 426

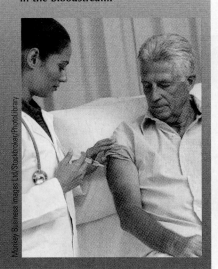

5.1   EXPONENTIAL FUNCTIONS AND THEIR GRAPHS

What you should learn

- Recognize and evaluate exponential functions with base *a*.
- Graph exponential functions and use the One-to-One Property.
- Recognize, evaluate, and graph exponential functions with base *e*.
- Use exponential functions to model and solve real-life problems.

Why you should learn it

Exponential functions can be used to model and solve real-life problems. For instance, in Exercise 76 on page 386, an exponential function is used to model the concentration of a drug in the bloodstream.

Exponential Functions

So far, this text has dealt mainly with **algebraic functions,** which include polynomial functions and rational functions. In this chapter, you will study two types of nonalgebraic functions—*exponential functions* and *logarithmic functions*. These functions are examples of **transcendental functions.**

> **Definition of Exponential Function**
>
> The **exponential function** *f* **with base** *a* is denoted by
>
> $$f(x) = a^x$$
>
> where $a > 0$, $a \neq 1$, and x is any real number.

The base $a = 1$ is excluded because it yields $f(x) = 1^x = 1$. This is a constant function, not an exponential function.

You have evaluated a^x for integer and rational values of x. For example, you know that $4^3 = 64$ and $4^{1/2} = 2$. However, to evaluate 4^x for any real number x, you need to interpret forms with *irrational* exponents. For the purposes of this text, it is sufficient to think of

$$a^{\sqrt{2}} \quad (\text{where } \sqrt{2} \approx 1.41421356)$$

as the number that has the successively closer approximations

$$a^{1.4}, a^{1.41}, a^{1.414}, a^{1.4142}, a^{1.41421}, \ldots .$$

Example 1   Evaluating Exponential Functions

Use a calculator to evaluate each function at the indicated value of x.

Function	Value
a. $f(x) = 2^x$	$x = -3.1$
b. $f(x) = 2^{-x}$	$x = \pi$
c. $f(x) = 0.6^x$	$x = \frac{3}{2}$

Solution

Function Value	Graphing Calculator Keystrokes	Display
a. $f(-3.1) = 2^{-3.1}$	2 ^ (−) 3.1 ENTER	0.1166291
b. $f(\pi) = 2^{-\pi}$	2 ^ (−) π ENTER	0.1133147
c. $f\left(\frac{3}{2}\right) = (0.6)^{3/2}$	.6 ^ (3 ÷ 2) ENTER	0.4647580

CHECK*Point*   Now try Exercise 7.

When evaluating exponential functions with a calculator, remember to enclose fractional exponents in parentheses. Because the calculator follows the order of operations, parentheses are crucial in order to obtain the correct result.

Graphs of Exponential Functions

The graphs of all exponential functions have similar characteristics, as shown in Examples 2, 3, and 5.

| Example 2 | **Graphs of $y = a^x$** |

In the same coordinate plane, sketch the graph of each function.

a. $f(x) = 2^x$ **b.** $g(x) = 4^x$

Solution

The table below lists some values for each function, and Figure 5.1 shows the graphs of the two functions. Note that both graphs are increasing. Moreover, the graph of $g(x) = 4^x$ is increasing more rapidly than the graph of $f(x) = 2^x$.

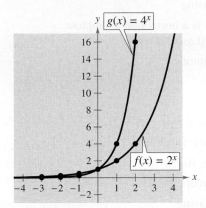

FIGURE 5.1

x	-3	-2	-1	0	1	2
2^x	$\frac{1}{8}$	$\frac{1}{4}$	$\frac{1}{2}$	1	2	4
4^x	$\frac{1}{64}$	$\frac{1}{16}$	$\frac{1}{4}$	1	4	16

CHECK Point   Now try Exercise 17.

The table in Example 2 was evaluated by hand. You could, of course, use a graphing utility to construct tables with even more values.

| Example 3 | **Graphs of $y = a^{-x}$** |

In the same coordinate plane, sketch the graph of each function.

a. $F(x) = 2^{-x}$ **b.** $G(x) = 4^{-x}$

Solution

The table below lists some values for each function, and Figure 5.2 shows the graphs of the two functions. Note that both graphs are decreasing. Moreover, the graph of $G(x) = 4^{-x}$ is decreasing more rapidly than the graph of $F(x) = 2^{-x}$.

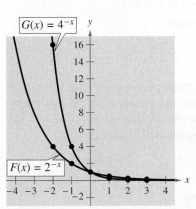

FIGURE 5.2

x	-2	-1	0	1	2	3
2^{-x}	4	2	1	$\frac{1}{2}$	$\frac{1}{4}$	$\frac{1}{8}$
4^{-x}	16	4	1	$\frac{1}{4}$	$\frac{1}{16}$	$\frac{1}{64}$

CHECK Point   Now try Exercise 19.

In Example 3, note that by using one of the properties of exponents, the functions $F(x) = 2^{-x}$ and $G(x) = 4^{-x}$ can be rewritten with positive exponents.

$$F(x) = 2^{-x} = \frac{1}{2^x} = \left(\frac{1}{2}\right)^x \quad \text{and} \quad G(x) = 4^{-x} = \frac{1}{4^x} = \left(\frac{1}{4}\right)^x$$

Comparing the functions in Examples 2 and 3, observe that

$$F(x) = 2^{-x} = f(-x) \qquad \text{and} \qquad G(x) = 4^{-x} = g(-x).$$

Consequently, the graph of F is a reflection (in the y-axis) of the graph of f. The graphs of G and g have the same relationship. The graphs in Figures 5.1 and 5.2 are typical of the exponential functions $y = a^x$ and $y = a^{-x}$. They have one y-intercept and one horizontal asymptote (the x-axis), and they are continuous. The basic characteristics of these exponential functions are summarized in Figures 5.3 and 5.4.

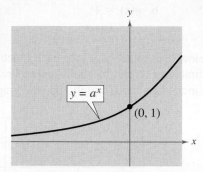

FIGURE 5.3

Graph of $y = a^x$, $a > 1$
- Domain: $(-\infty, \infty)$
- Range: $(0, \infty)$
- y-intercept: $(0, 1)$
- Increasing
- x-axis is a horizontal asymptote $(a^x \to 0$ as $x \to -\infty)$.
- Continuous

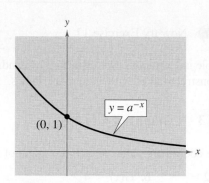

FIGURE 5.4

Graph of $y = a^{-x}$, $a > 1$
- Domain: $(-\infty, \infty)$
- Range: $(0, \infty)$
- y-intercept: $(0, 1)$
- Decreasing
- x-axis is a horizontal asymptote $(a^{-x} \to 0$ as $x \to \infty)$.
- Continuous

From Figures 5.3 and 5.4, you can see that the graph of an exponential function is always increasing or always decreasing. As a result, the graphs pass the Horizontal Line Test, and therefore the functions are one-to-one functions. You can use the following **One-to-One Property** to solve simple exponential equations.

For $a > 0$ and $a \neq 1$, $a^x = a^y$ if and only if $x = y$.   One-to-One Property

Example 4   Using the One-to-One Property

a. $9 = 3^{x+1}$ Original equation

$3^2 = 3^{x+1}$ $9 = 3^2$

$2 = x + 1$ One-to-One Property

$1 = x$ Solve for x.

b. $\left(\frac{1}{2}\right)^x = 8 \Longrightarrow 2^{-x} = 2^3 \Longrightarrow x = -3$

CHECK*Point*   Now try Exercise 51.

In the following example, notice how the graph of $y = a^x$ can be used to sketch the graphs of functions of the form $f(x) = b \pm a^{x+c}$.

Algebra Help

You can review the techniques for transforming the graph of a function in Section P.8.

Example 5   Transformations of Graphs of Exponential Functions

Each of the following graphs is a transformation of the graph of $f(x) = 3^x$.

a. Because $g(x) = 3^{x+1} = f(x + 1)$, the graph of g can be obtained by shifting the graph of f one unit to the *left*, as shown in Figure 5.5.

b. Because $h(x) = 3^x - 2 = f(x) - 2$, the graph of h can be obtained by shifting the graph of f *downward* two units, as shown in Figure 5.6.

c. Because $k(x) = -3^x = -f(x)$, the graph of k can be obtained by *reflecting* the graph of f in the *x*-axis, as shown in Figure 5.7.

d. Because $j(x) = 3^{-x} = f(-x)$, the graph of j can be obtained by *reflecting* the graph of f in the *y*-axis, as shown in Figure 5.8.

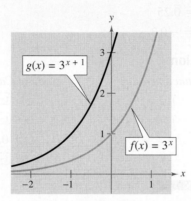

FIGURE **5.5**   Horizontal shift

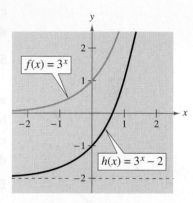

FIGURE **5.6**   Vertical shift

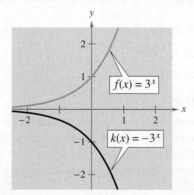

FIGURE **5.7**   Reflection in *x*-axis

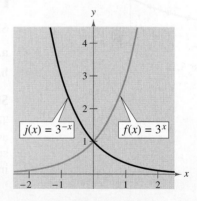

FIGURE **5.8**   Reflection in *y*-axis

CHECK*Point*   Now try Exercise 23.

Notice that the transformations in Figures 5.5, 5.7, and 5.8 keep the *x*-axis as a horizontal asymptote, but the transformation in Figure 5.6 yields a new horizontal asymptote of $y = -2$. Also, be sure to note how the *y*-intercept is affected by each transformation.

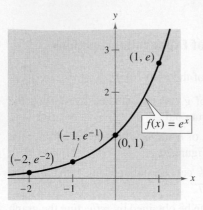

FIGURE 5.9

The Natural Base e

In many applications, the most convenient choice for a base is the irrational number

$$e \approx 2.718281828 \ldots .$$

This number is called the **natural base**. The function given by $f(x) = e^x$ is called the **natural exponential function.** Its graph is shown in Figure 5.9. Be sure you see that for the exponential function $f(x) = e^x$, e is the constant $2.718281828 \ldots$, whereas x is the variable.

Example 6   Evaluating the Natural Exponential Function

Use a calculator to evaluate the function given by $f(x) = e^x$ at each indicated value of x.

a. $x = -2$

b. $x = -1$

c. $x = 0.25$

d. $x = -0.3$

Solution

Function Value	Graphing Calculator Keystrokes	Display
a. $f(-2) = e^{-2}$	e^x (−) 2 ENTER	0.1353353
b. $f(-1) = e^{-1}$	e^x (−) 1 ENTER	0.3678794
c. $f(0.25) = e^{0.25}$	e^x 0.25 ENTER	1.2840254
d. $f(-0.3) = e^{-0.3}$	e^x (−) 0.3 ENTER	0.7408182

CHECK Point Now try Exercise 33.

Example 7   Graphing Natural Exponential Functions

Sketch the graph of each natural exponential function.

a. $f(x) = 2e^{0.24x}$

b. $g(x) = \frac{1}{2}e^{-0.58x}$

Solution

To sketch these two graphs, you can use a graphing utility to construct a table of values, as shown below. After constructing the table, plot the points and connect them with smooth curves, as shown in Figures 5.10 and 5.11. Note that the graph in Figure 5.10 is increasing, whereas the graph in Figure 5.11 is decreasing.

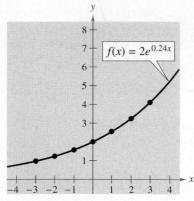

FIGURE 5.10

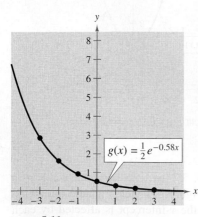

FIGURE 5.11

x	-3	-2	-1	0	1	2	3
$f(x)$	0.974	1.238	1.573	2.000	2.542	3.232	4.109
$g(x)$	2.849	1.595	0.893	0.500	0.280	0.157	0.088

CHECK Point Now try Exercise 41.

Applications

One of the most familiar examples of exponential growth is an investment earning *continuously compounded interest*. Using exponential functions, you can *develop* a formula for interest compounded n times per year and show how it leads to continuous compounding.

Suppose a principal P is invested at an annual interest rate r, compounded once per year. If the interest is added to the principal at the end of the year, the new balance P_1 is

$$P_1 = P + Pr$$

$$= P(1 + r).$$

This pattern of multiplying the previous principal by $1 + r$ is then repeated each successive year, as shown below.

Year	*Balance After Each Compounding*
0	$P = P$
1	$P_1 = P(1 + r)$
2	$P_2 = P_1(1 + r) = P(1 + r)(1 + r) = P(1 + r)^2$
3	$P_3 = P_2(1 + r) = P(1 + r)^2(1 + r) = P(1 + r)^3$
$\vdots$	$\vdots$
t	$P_t = P(1 + r)^t$

To accommodate more frequent (quarterly, monthly, or daily) compounding of interest, let n be the number of compoundings per year and let t be the number of years. Then the rate per compounding is r/n and the account balance after t years is

$$A = P\left(1 + \frac{r}{n}\right)^{nt}. \qquad \text{Amount (balance) with } n \text{ compoundings per year}$$

If you let the number of compoundings n increase without bound, the process approaches what is called **continuous compounding.** In the formula for n compoundings per year, let $m = n/r$. This produces

$$A = P\left(1 + \frac{r}{n}\right)^{nt} \qquad \text{Amount with } n \text{ compoundings per year}$$

$$= P\left(1 + \frac{r}{mr}\right)^{mrt} \qquad \text{Substitute } mr \text{ for } n.$$

$$= P\left(1 + \frac{1}{m}\right)^{mrt} \qquad \text{Simplify.}$$

$$= P\left[\left(1 + \frac{1}{m}\right)^{m}\right]^{rt}. \qquad \text{Property of exponents}$$

As m increases without bound, the table at the left shows that $[1 + (1/m)]^m \to e$ as $m \to \infty$. From this, you can conclude that the formula for continuous compounding is

$$A = Pe^{rt}. \qquad \text{Substitute } e \text{ for } (1 + 1/m)^m.$$

m	$\left(1 + \dfrac{1}{m}\right)^m$
1	2
10	2.59374246
100	2.704813829
1,000	2.716923932
10,000	2.718145927
100,000	2.718268237
1,000,000	2.718280469
10,000,000	2.718281693
$\downarrow$	$\downarrow$
∞	e

Be sure you see that the annual interest rate must be written in decimal form. For instance, 6% should be written as 0.06.

Formulas for Compound Interest

After t years, the balance A in an account with principal P and annual interest rate r (in decimal form) is given by the following formulas.

1. For n compoundings per year: $A = P\left(1 + \dfrac{r}{n}\right)^{nt}$

2. For continuous compounding: $A = Pe^{rt}$

Example 8 Compound Interest

A total of $12,000 is invested at an annual interest rate of 9%. Find the balance after 5 years if it is compounded

a. quarterly.

b. monthly.

c. continuously.

Solution

a. For quarterly compounding, you have $n = 4$. So, in 5 years at 9%, the balance is

$$A = P\left(1 + \frac{r}{n}\right)^{nt} \qquad \text{Formula for compound interest}$$

$$= 12,000\left(1 + \frac{0.09}{4}\right)^{4(5)} \qquad \text{Substitute for } P, r, n, \text{ and } t.$$

$$\approx \$18,726.11. \qquad \text{Use a calculator.}$$

b. For monthly compounding, you have $n = 12$. So, in 5 years at 9%, the balance is

$$A = P\left(1 + \frac{r}{n}\right)^{nt} \qquad \text{Formula for compound interest}$$

$$= 12,000\left(1 + \frac{0.09}{12}\right)^{12(5)} \qquad \text{Substitute for } P, r, n, \text{ and } t.$$

$$\approx \$18,788.17. \qquad \text{Use a calculator.}$$

c. For continuous compounding, the balance is

$$A = Pe^{rt} \qquad \text{Formula for continuous compounding}$$

$$= 12,000e^{0.09(5)} \qquad \text{Substitute for } P, r, \text{ and } t.$$

$$\approx \$18,819.75. \qquad \text{Use a calculator.}$$

CHECK Point Now try Exercise 59.

In Example 8, note that continuous compounding yields more than quarterly or monthly compounding. This is typical of the two types of compounding. That is, for a given principal, interest rate, and time, continuous compounding will always yield a larger balance than compounding n times per year.

Example 9	**Radioactive Decay**

The *half-life* of radioactive radium (^{226}Ra) is about 1599 years. That is, for a given amount of radium, *half* of the original amount will remain after 1599 years. After another 1599 years, one-quarter of the original amount will remain, and so on. Let y represent the mass, in grams, of a quantity of radium. The quantity present after t years, then, is $y = 25\left(\frac{1}{2}\right)^{t/1599}$.

a. What is the initial mass (when $t = 0$)?

b. How much of the initial mass is present after 2500 years?

Algebraic Solution

a. $y = 25\left(\frac{1}{2}\right)^{t/1599}$ Write original equation.

$= 25\left(\frac{1}{2}\right)^{0/1599}$ Substitute 0 for t.

$= 25$ Simplify.

So, the initial mass is 25 grams.

b. $y = 25\left(\frac{1}{2}\right)^{t/1599}$ Write original equation.

$= 25\left(\frac{1}{2}\right)^{2500/1599}$ Substitute 2500 for t.

$\approx 25\left(\frac{1}{2}\right)^{1.563}$ Simplify.

≈ 8.46 Use a calculator.

So, about 8.46 grams is present after 2500 years.

CHECK Point   Now try Exercise 73.

Graphical Solution

Use a graphing utility to graph $y = 25\left(\frac{1}{2}\right)^{t/1599}$.

a. Use the *value* feature or the *zoom* and *trace* features of the graphing utility to determine that when $x = 0$, the value of y is 25, as shown in Figure 5.12. So, the initial mass is 25 grams.

b. Use the *value* feature or the *zoom* and *trace* features of the graphing utility to determine that when $x = 2500$, the value of y is about 8.46, as shown in Figure 5.13. So, about 8.46 grams is present after 2500 years.

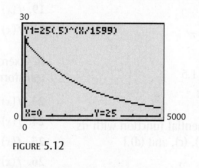

FIGURE 5.12

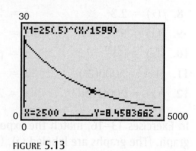

FIGURE 5.13

CLASSROOM DISCUSSION

Identifying Exponential Functions   Which of the following functions generated the two tables below? Discuss how you were able to decide. What do these functions have in common? Are any of them the same? If so, explain why.

a. $f_1(x) = 2^{(x+3)}$

b. $f_2(x) = 8\left(\frac{1}{2}\right)^x$

c. $f_3(x) = \left(\frac{1}{2}\right)^{(x-3)}$

d. $f_4(x) = \left(\frac{1}{2}\right)^x + 7$

e. $f_5(x) = 7 + 2^x$

f. $f_6(x) = 8(2^x)$

x	-1	0	1	2	3
$g(x)$	7.5	8	9	11	15

x	-2	-1	0	1	2
$h(x)$	32	16	8	4	2

Create two different exponential functions of the forms $y = a(b)^x$ and $y = c^x + d$ with y-intercepts of $(0, -3)$.

5.1 EXERCISES

See www.CalcChat.com for worked-out solutions to odd-numbered exercises.

VOCABULARY: Fill in the blanks.

1. Polynomial and rational functions are examples of _____ functions.

2. Exponential and logarithmic functions are examples of nonalgebraic functions, also called _____ functions.

3. You can use the _____ Property to solve simple exponential equations.

4. The exponential function given by $f(x) = e^x$ is called the _____ _____ function, and the base e is called the _____ base.

5. To find the amount A in an account after t years with principal P and an annual interest rate r compounded n times per year, you can use the formula _____.

6. To find the amount A in an account after t years with principal P and an annual interest rate r compounded continuously, you can use the formula _____.

SKILLS AND APPLICATIONS

In Exercises 7–12, evaluate the function at the indicated value of x. Round your result to three decimal places.

Function	Value
7. $f(x) = 0.9^x$	$x = 1.4$
8. $f(x) = 2.3^x$	$x = \frac{3}{2}$
9. $f(x) = 5^x$	$x = -\pi$
10. $f(x) = \left(\frac{2}{3}\right)^{5x}$	$x = \frac{3}{10}$
11. $g(x) = 5000(2^x)$	$x = -1.5$
12. $f(x) = 200(1.2)^{12x}$	$x = 24$

In Exercises 13–16, match the exponential function with its graph. [The graphs are labeled (a), (b), (c), and (d).]

(a)
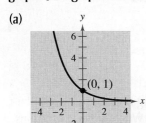

(b)
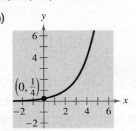

(c)
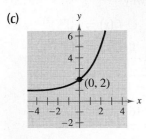

(d)
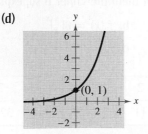

13. $f(x) = 2^x$

14. $f(x) = 2^x + 1$

15. $f(x) = 2^{-x}$

16. $f(x) = 2^{x-2}$

In Exercises 17–22, use a graphing utility to construct a table of values for the function. Then sketch the graph of the function.

17. $f(x) = \left(\frac{1}{2}\right)^x$

18. $f(x) = \left(\frac{1}{2}\right)^{-x}$

19. $f(x) = 6^{-x}$

20. $f(x) = 6^x$

21. $f(x) = 2^{x-1}$

22. $f(x) = 4^{x-3} + 3$

In Exercises 23–28, use the graph of f to describe the transformation that yields the graph of g.

23. $f(x) = 3^x$, $g(x) = 3^x + 1$

24. $f(x) = 4^x$, $g(x) = 4^{x-3}$

25. $f(x) = 2^x$, $g(x) = 3 - 2^x$

26. $f(x) = 10^x$, $g(x) = 10^{-x+3}$

27. $f(x) = \left(\frac{7}{2}\right)^x$, $g(x) = -\left(\frac{7}{2}\right)^{-x}$

28. $f(x) = 0.3^x$, $g(x) = -0.3^x + 5$

In Exercises 29–32, use a graphing utility to graph the exponential function.

29. $y = 2^{-x^2}$

30. $y = 3^{-|x|}$

31. $y = 3^{x-2} + 1$

32. $y = 4^{x+1} - 2$

In Exercises 33–38, evaluate the function at the indicated value of x. Round your result to three decimal places.

Function	Value
33. $h(x) = e^{-x}$	$x = \frac{3}{4}$
34. $f(x) = e^x$	$x = 3.2$
35. $f(x) = 2e^{-5x}$	$x = 10$
36. $f(x) = 1.5e^{x/2}$	$x = 240$
37. $f(x) = 5000e^{0.06x}$	$x = 6$
38. $f(x) = 250e^{0.05x}$	$x = 20$

In Exercises 39–44, use a graphing utility to construct a table of values for the function. Then sketch the graph of the function.

39. $f(x) = e^x$

40. $f(x) = e^{-x}$

41. $f(x) = 3e^{x+4}$

42. $f(x) = 2e^{-0.5x}$

43. $f(x) = 2e^{x-2} + 4$

44. $f(x) = 2 + e^{x-5}$

 In Exercises 45–50, use a graphing utility to graph the exponential function.

45. $y = 1.08^{-5x}$

46. $y = 1.08^{5x}$

47. $s(t) = 2e^{0.12t}$

48. $s(t) = 3e^{-0.2t}$

49. $g(x) = 1 + e^{-x}$

50. $h(x) = e^{x-2}$

In Exercises 51–58, use the One-to-One Property to solve the equation for *x*.

51. $3^{x+1} = 27$

52. $2^{x-3} = 16$

53. $\left(\frac{1}{2}\right)^x = 32$

54. $5^{x-2} = \frac{1}{125}$

55. $e^{3x+2} = e^3$

56. $e^{2x-1} = e^4$

57. $e^{x^2-3} = e^{2x}$

58. $e^{x^2+6} = e^{5x}$

COMPOUND INTEREST In Exercises 59–62, complete the table to determine the balance *A* for *P* dollars invested at rate *r* for *t* years and compounded *n* times per year.

n	1	2	4	12	365	Continuous
A						

59. $P = \$1500, r = 2\%, t = 10$ years

60. $P = \$2500, r = 3.5\%, t = 10$ years

61. $P = \$2500, r = 4\%, t = 20$ years

62. $P = \$1000, r = 6\%, t = 40$ years

COMPOUND INTEREST In Exercises 63–66, complete the table to determine the balance *A* for \$12,000 invested at rate *r* for *t* years, compounded continuously.

t	10	20	30	40	50
A					

63. $r = 4\%$

64. $r = 6\%$

65. $r = 6.5\%$

66. $r = 3.5\%$

67. TRUST FUND On the day of a child's birth, a deposit of \$30,000 is made in a trust fund that pays 5% interest, compounded continuously. Determine the balance in this account on the child's 25th birthday.

68. TRUST FUND A deposit of \$5000 is made in a trust fund that pays 7.5% interest, compounded continuously. It is specified that the balance will be given to the college from which the donor graduated after the money has earned interest for 50 years. How much will the college receive?

69. INFLATION If the annual rate of inflation averages 4% over the next 10 years, the approximate costs *C* of goods or services during any year in that decade will be modeled by $C(t) = P(1.04)^t$, where *t* is the time in years and *P* is the present cost. The price of an oil change for your car is presently \$23.95. Estimate the price 10 years from now.

70. COMPUTER VIRUS The number *V* of computers infected by a computer virus increases according to the model $V(t) = 100e^{4.6052t}$, where *t* is the time in hours. Find the number of computers infected after (a) 1 hour, (b) 1.5 hours, and (c) 2 hours.

 71. POPULATION GROWTH The projected populations of California for the years 2015 through 2030 can be modeled by $P = 34.696e^{0.0098t}$, where *P* is the population (in millions) and *t* is the time (in years), with $t = 15$ corresponding to 2015. (Source: U.S. Census Bureau)

(a) Use a graphing utility to graph the function for the years 2015 through 2030.

(b) Use the *table* feature of a graphing utility to create a table of values for the same time period as in part (a).

(c) According to the model, when will the population of California exceed 50 million?

72. POPULATION The populations *P* (in millions) of Italy from 1990 through 2008 can be approximated by the model $P = 56.8e^{0.0015t}$, where *t* represents the year, with $t = 0$ corresponding to 1990. (Source: U.S. Census Bureau, International Data Base)

(a) According to the model, is the population of Italy increasing or decreasing? Explain.

(b) Find the populations of Italy in 2000 and 2008.

(c) Use the model to predict the populations of Italy in 2015 and 2020.

73. RADIOACTIVE DECAY Let *Q* represent a mass of radioactive plutonium (^{239}Pu) (in grams), whose half-life is 24,100 years. The quantity of plutonium present after *t* years is $Q = 16\left(\frac{1}{2}\right)^{t/24,100}$.

(a) Determine the initial quantity (when $t = 0$).

(b) Determine the quantity present after 75,000 years.

(c) Use a graphing utility to graph the function over the interval $t = 0$ to $t = 150,000$.

74. RADIOACTIVE DECAY Let Q represent a mass of carbon 14 (^{14}C) (in grams), whose half-life is 5715 years. The quantity of carbon 14 present after t years is $Q = 10\left(\frac{1}{2}\right)^{t/5715}$.

(a) Determine the initial quantity (when $t = 0$).

(b) Determine the quantity present after 2000 years.

(c) Sketch the graph of this function over the interval $t = 0$ to $t = 10,000$.

75. DEPRECIATION After t years, the value of a wheelchair conversion van that originally cost $30,500 depreciates so that each year it is worth $\frac{7}{8}$ of its value for the previous year.

(a) Find a model for $V(t)$, the value of the van after t years.

(b) Determine the value of the van 4 years after it was purchased.

76. DRUG CONCENTRATION Immediately following an injection, the concentration of a drug in the bloodstream is 300 milligrams per milliliter. After t hours, the concentration is 75% of the level of the previous hour.

(a) Find a model for $C(t)$, the concentration of the drug after t hours.

(b) Determine the concentration of the drug after 8 hours.

EXPLORATION

TRUE OR FALSE? In Exercises 77 and 78, determine whether the statement is true or false. Justify your answer.

77. The line $y = -2$ is an asymptote for the graph of $f(x) = 10^x - 2$.

78. $e = \dfrac{271,801}{99,990}$

THINK ABOUT IT In Exercises 79–82, use properties of exponents to determine which functions (if any) are the same.

79. $f(x) = 3^{x-2}$
$g(x) = 3^x - 9$
$h(x) = \frac{1}{9}(3^x)$

80. $f(x) = 4^x + 12$
$g(x) = 2^{2x+6}$
$h(x) = 64(4^x)$

81. $f(x) = 16(4^{-x})$
$g(x) = \left(\frac{1}{4}\right)^{x-2}$
$h(x) = 16(2^{-2x})$

82. $f(x) = e^{-x} + 3$
$g(x) = e^{3-x}$
$h(x) = -e^{x-3}$

83. Graph the functions given by $y = 3^x$ and $y = 4^x$ and use the graphs to solve each inequality.

(a) $4^x < 3^x$

(b) $4^x > 3^x$

 84. Use a graphing utility to graph each function. Use the graph to find where the function is increasing and decreasing, and approximate any relative maximum or minimum values.

(a) $f(x) = x^2 e^{-x}$

(b) $g(x) = x2^{3-x}$

 85. GRAPHICAL ANALYSIS Use a graphing utility to graph $y_1 = (1 + 1/x)^x$ and $y_2 = e$ in the same viewing window. Using the *trace* feature, explain what happens to the graph of y_1 as x increases.

86. GRAPHICAL ANALYSIS Use a graphing utility to graph

$$f(x) = \left(1 + \frac{0.5}{x}\right)^x \quad \text{and} \quad g(x) = e^{0.5}$$

in the same viewing window. What is the relationship between f and g as x increases and decreases without bound?

 87. GRAPHICAL ANALYSIS Use a graphing utility to graph each pair of functions in the same viewing window. Describe any similarities and differences in the graphs.

(a) $y_1 = 2^x, y_2 = x^2$

(b) $y_1 = 3^x, y_2 = x^3$

88. THINK ABOUT IT Which functions are exponential?

(a) $3x$ (b) $3x^2$ (c) 3^x (d) 2^{-x}

89. COMPOUND INTEREST Use the formula

$$A = P\left(1 + \frac{r}{n}\right)^{nt}$$

to calculate the balance of an account when $P = \$3000$, $r = 6\%$, and $t = 10$ years, and compounding is done (a) by the day, (b) by the hour, (c) by the minute, and (d) by the second. Does increasing the number of compoundings per year result in unlimited growth of the balance of the account? Explain.

90. CAPSTONE The figure shows the graphs of $y = 2^x$, $y = e^x$, $y = 10^x$, $y = 2^{-x}$, $y = e^{-x}$, and $y = 10^{-x}$. Match each function with its graph. [The graphs are labeled (a) through (f).] Explain your reasoning.

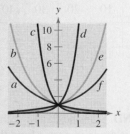

PROJECT: POPULATION PER SQUARE MILE To work an extended application analyzing the population per square mile of the United States, visit this text's website at *academic.cengage.com*. (Data Source: U.S. Census Bureau)

5.2 LOGARITHMIC FUNCTIONS AND THEIR GRAPHS

What you should learn

- Recognize and evaluate logarithmic functions with base a.
- Graph logarithmic functions.
- Recognize, evaluate, and graph natural logarithmic functions.
- Use logarithmic functions to model and solve real-life problems.

Why you should learn it

Logarithmic functions are often used to model scientific observations. For instance, in Exercise 97 on page 396, a logarithmic function is used to model human memory.

Logarithmic Functions

In Section P.10, you studied the concept of an inverse function. There, you learned that if a function is one-to-one—that is, if the function has the property that no horizontal line intersects the graph of the function more than once—the function must have an inverse function. By looking back at the graphs of the exponential functions introduced in Section 5.1, you will see that every function of the form $f(x) = a^x$ passes the Horizontal Line Test and therefore must have an inverse function. This inverse function is called the **logarithmic function with base a.**

> ### Definition of Logarithmic Function with Base a
>
> For $x > 0$, $a > 0$, and $a \neq 1$,
>
> $$y = \log_a x \text{ if and only if } x = a^y.$$
>
> The function given by
>
> $$f(x) = \log_a x \qquad \text{Read as "log base } a \text{ of } x.\text{"}$$
>
> is called the **logarithmic function with base a.**

The equations

$$y = \log_a x \qquad \text{and} \qquad x = a^y$$

are equivalent. The first equation is in logarithmic form and the second is in exponential form. For example, the logarithmic equation $2 = \log_3 9$ can be rewritten in exponential form as $9 = 3^2$. The exponential equation $5^3 = 125$ can be rewritten in logarithmic form as $\log_5 125 = 3$.

When evaluating logarithms, remember that *a logarithm is an exponent*. This means that $\log_a x$ is the exponent to which a must be raised to obtain x. For instance, $\log_2 8 = 3$ because 2 must be raised to the third power to get 8.

Example 1   Evaluating Logarithms

Use the definition of logarithmic function to evaluate each logarithm at the indicated value of x.

a. $f(x) = \log_2 x, \quad x = 32$ **b.** $f(x) = \log_3 x, \quad x = 1$

c. $f(x) = \log_4 x, \quad x = 2$ **d.** $f(x) = \log_{10} x, \quad x = \frac{1}{100}$

Solution

a. $f(32) = \log_2 32 = 5$ because $2^5 = 32$.

b. $f(1) = \log_3 1 = 0$ because $3^0 = 1$.

c. $f(2) = \log_4 2 = \frac{1}{2}$ because $4^{1/2} = \sqrt{4} = 2$.

d. $f\left(\frac{1}{100}\right) = \log_{10} \frac{1}{100} = -2$ because $10^{-2} = \frac{1}{10^2} = \frac{1}{100}$.

CHECK Point ▶ Now try Exercise 23.

The logarithmic function with base 10 is called the **common logarithmic function.** It is denoted by $\log_{10}$ or simply by log. On most calculators, this function is denoted by $\boxed{\text{LOG}}$. Example 2 shows how to use a calculator to evaluate common logarithmic functions. You will learn how to use a calculator to calculate logarithms to any base in the next section.

Example 2   Evaluating Common Logarithms on a Calculator

Use a calculator to evaluate the function given by $f(x) = \log x$ at each value of x.

a. $x = 10$ **b.** $x = \frac{1}{3}$ **c.** $x = 2.5$ **d.** $x = -2$

Solution

Function Value	Graphing Calculator Keystrokes	Display
a. $f(10) = \log 10$	$\boxed{\text{LOG}}$ 10 $\boxed{\text{ENTER}}$	1
b. $f\left(\frac{1}{3}\right) = \log \frac{1}{3}$	$\boxed{\text{LOG}}$ $\boxed{(}$ 1 $\boxed{\div}$ 3 $\boxed{)}$ $\boxed{\text{ENTER}}$	-0.4771213
c. $f(2.5) = \log 2.5$	$\boxed{\text{LOG}}$ 2.5 $\boxed{\text{ENTER}}$	0.3979400
d. $f(-2) = \log(-2)$	$\boxed{\text{LOG}}$ $\boxed{-}$ 2 $\boxed{\text{ENTER}}$	ERROR

Note that the calculator displays an error message (or a complex number) when you try to evaluate $\log(-2)$. The reason for this is that there is no real number power to which 10 can be raised to obtain -2.

CHECK *Point*   Now try Exercise 29. ◼

The following properties follow directly from the definition of the logarithmic function with base a.

Properties of Logarithms

1. $\log_a 1 = 0$ because $a^0 = 1$.

2. $\log_a a = 1$ because $a^1 = a$.

3. $\log_a a^x = x$ and $a^{\log_a x} = x$ Inverse Properties

4. If $\log_a x = \log_a y$, then $x = y$.   One-to-One Property

---

Example 3   Using Properties of Logarithms

a. Simplify: $\log_4 1$ **b.** Simplify: $\log_{\sqrt{7}} \sqrt{7}$ **c.** Simplify: $6^{\log_6 20}$

Solution

a. Using Property 1, it follows that $\log_4 1 = 0$.

b. Using Property 2, you can conclude that $\log_{\sqrt{7}} \sqrt{7} = 1$.

c. Using the Inverse Property (Property 3), it follows that $6^{\log_6 20} = 20$.

CHECK *Point*   Now try Exercise 33. ◼

You can use the One-to-One Property (Property 4) to solve simple logarithmic equations, as shown in Example 4.

Example 4   Using the One-to-One Property

a. $\log_3 x = \log_3 12$ Original equation

   $x = 12$ One-to-One Property

b. $\log(2x + 1) = \log 3x \implies 2x + 1 = 3x \implies 1 = x$

c. $\log_4(x^2 - 6) = \log_4 10 \implies x^2 - 6 = 10 \implies x^2 = 16 \implies x = \pm 4$

CHECK*Point*   Now try Exercise 85.

Graphs of Logarithmic Functions

To sketch the graph of $y = \log_a x$, you can use the fact that the graphs of inverse functions are reflections of each other in the line $y = x$.

Example 5   Graphs of Exponential and Logarithmic Functions

In the same coordinate plane, sketch the graph of each function.

a. $f(x) = 2^x$ **b.** $g(x) = \log_2 x$

Solution

a. For $f(x) = 2^x$, construct a table of values. By plotting these points and connecting them with a smooth curve, you obtain the graph shown in Figure 5.14.

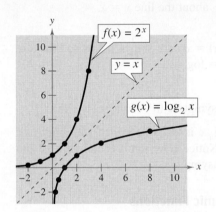

x	-2	-1	0	1	2	3
$f(x) = 2^x$	$\frac{1}{4}$	$\frac{1}{2}$	1	2	4	8

b. Because $g(x) = \log_2 x$ is the inverse function of $f(x) = 2^x$, the graph of g is obtained by plotting the points $(f(x), x)$ and connecting them with a smooth curve. The graph of g is a reflection of the graph of f in the line $y = x$, as shown in Figure 5.14.

FIGURE 5.14

CHECK*Point*   Now try Exercise 37.

Example 6   Sketching the Graph of a Logarithmic Function

Sketch the graph of the common logarithmic function $f(x) = \log x$. Identify the vertical asymptote.

Solution

Begin by constructing a table of values. Note that some of the values can be obtained without a calculator by using the Inverse Property of Logarithms. Others require a calculator. Next, plot the points and connect them with a smooth curve, as shown in Figure 5.15. The vertical asymptote is $x = 0$ (y-axis).

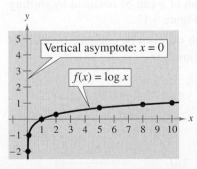

FIGURE 5.15

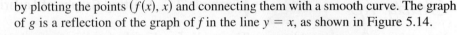

	Without calculator				With calculator		
x	$\frac{1}{100}$	$\frac{1}{10}$	1	10	2	5	8
$f(x) = \log x$	-2	-1	0	1	0.301	0.699	0.903

CHECK*Point*   Now try Exercise 43.

The nature of the graph in Figure 5.15 is typical of functions of the form $f(x) = \log_a x, a > 1$. They have one x-intercept and one vertical asymptote. Notice how slowly the graph rises for $x > 1$. The basic characteristics of logarithmic graphs are summarized in Figure 5.16.

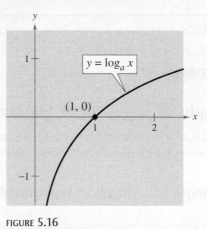

FIGURE 5.16

Graph of $y = \log_a x, a > 1$
- Domain: $(0, \infty)$
- Range: $(-\infty, \infty)$
- x-intercept: $(1, 0)$
- Increasing
- One-to-one, therefore has an inverse function
- y-axis is a vertical asymptote $(\log_a x \to -\infty$ as $x \to 0^+)$.
- Continuous
- Reflection of graph of $y = a^x$ about the line $y = x$

The basic characteristics of the graph of $f(x) = a^x$ are shown below to illustrate the inverse relation between $f(x) = a^x$ and $g(x) = \log_a x$.

- Domain: $(-\infty, \infty)$ • Range: $(0, \infty)$
- y-intercept: $(0, 1)$ • x-axis is a horizontal asymptote $(a^x \to 0$ as $x \to -\infty)$.

In the next example, the graph of $y = \log_a x$ is used to sketch the graphs of functions of the form $f(x) = b \pm \log_a(x + c)$. Notice how a horizontal shift of the graph results in a horizontal shift of the vertical asymptote.

| Example 7 | **Shifting Graphs of Logarithmic Functions** |

The graph of each of the functions is similar to the graph of $f(x) = \log x$.

a. Because $g(x) = \log(x - 1) = f(x - 1)$, the graph of g can be obtained by shifting the graph of f one unit to the right, as shown in Figure 5.17.

b. Because $h(x) = 2 + \log x = 2 + f(x)$, the graph of h can be obtained by shifting the graph of f two units upward, as shown in Figure 5.18.

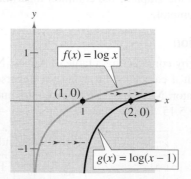

FIGURE 5.17

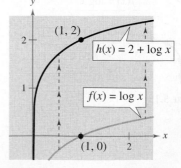

FIGURE 5.18

CHECK Point   Now try Exercise 45.

The Natural Logarithmic Function

By looking back at the graph of the natural exponential function introduced on page 380 in Section 5.1, you will see that $f(x) = e^x$ is one-to-one and so has an inverse function. This inverse function is called the **natural logarithmic function** and is denoted by the special symbol ln x, read as "the natural log of x" or "el en of x." Note that the natural logarithm is written without a base. The base is understood to be e.

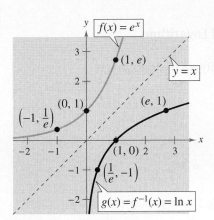

Reflection of graph of $f(x) = e^x$ about the line $y = x$

FIGURE **5.19**

> ### The Natural Logarithmic Function
>
> The function defined by
>
> $$f(x) = \log_e x = \ln x, \quad x > 0$$
>
> is called the **natural logarithmic function.**

The definition above implies that the natural logarithmic function and the natural exponential function are inverse functions of each other. So, every logarithmic equation can be written in an equivalent exponential form, and every exponential equation can be written in logarithmic form. That is, $y = \ln x$ and $x = e^y$ are equivalent equations.

Because the functions given by $f(x) = e^x$ and $g(x) = \ln x$ are inverse functions of each other, their graphs are reflections of each other in the line $y = x$. This reflective property is illustrated in Figure 5.19.

On most calculators, the natural logarithm is denoted by LN, as illustrated in Example 8.

Example 8   Evaluating the Natural Logarithmic Function

Use a calculator to evaluate the function given by $f(x) = \ln x$ for each value of x.

a. $x = 2$

b. $x = 0.3$

c. $x = -1$

d. $x = 1 + \sqrt{2}$

Solution

	Function Value	Graphing Calculator Keystrokes	Display
a.	$f(2) = \ln 2$	LN 2 ENTER	0.6931472
b.	$f(0.3) = \ln 0.3$	LN .3 ENTER	–1.2039728
c.	$f(-1) = \ln(-1)$	LN (–) 1 ENTER	ERROR
d.	$f(1 + \sqrt{2}) = \ln(1 + \sqrt{2})$	LN (1 + √ 2) ENTER	0.8813736

CHECK**Point**   Now try Exercise 67.

In Example 8, be sure you see that $\ln(-1)$ gives an error message on most calculators. (Some calculators may display a complex number.) This occurs because the domain of ln x is the set of positive real numbers (see Figure 5.19). So, $\ln(-1)$ is undefined.

The four properties of logarithms listed on page 388 are also valid for natural logarithms.

WARNING / CAUTION

Notice that as with every other logarithmic function, the domain of the natural logarithmic function is the set of *positive real numbers*—be sure you see that ln x is not defined for zero or for negative numbers.

Properties of Natural Logarithms

1. $\ln 1 = 0$ because $e^0 = 1$.

2. $\ln e = 1$ because $e^1 = e$.

3. $\ln e^x = x$ and $e^{\ln x} = x$ Inverse Properties

4. If $\ln x = \ln y$, then $x = y$.   One-to-One Property

Example 9   Using Properties of Natural Logarithms

Use the properties of natural logarithms to simplify each expression.

a. $\ln \dfrac{1}{e}$ **b.** $e^{\ln 5}$ **c.** $\dfrac{\ln 1}{3}$ **d.** $2 \ln e$

Solution

a. $\ln \dfrac{1}{e} = \ln e^{-1} = -1$ Inverse Property   **b.** $e^{\ln 5} = 5$ Inverse Property

c. $\dfrac{\ln 1}{3} = \dfrac{0}{3} = 0$ Property 1   **d.** $2 \ln e = 2(1) = 2$ Property 2

CHECK Point Now try Exercise 71.

Example 10   Finding the Domains of Logarithmic Functions

Find the domain of each function.

a. $f(x) = \ln(x - 2)$ **b.** $g(x) = \ln(2 - x)$ **c.** $h(x) = \ln x^2$

Solution

a. Because $\ln(x - 2)$ is defined only if $x - 2 > 0$, it follows that the domain of f is $(2, \infty)$. The graph of f is shown in Figure 5.20.

b. Because $\ln(2 - x)$ is defined only if $2 - x > 0$, it follows that the domain of g is $(-\infty, 2)$. The graph of g is shown in Figure 5.21.

c. Because $\ln x^2$ is defined only if $x^2 > 0$, it follows that the domain of h is all real numbers except $x = 0$. The graph of h is shown in Figure 5.22.

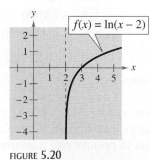

FIGURE **5.20**

FIGURE **5.21**

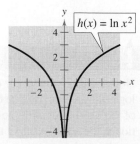

FIGURE **5.22**

  Now try Exercise 75.

Application

Example 11   Human Memory Model

Students participating in a psychology experiment attended several lectures on a subject and were given an exam. Every month for a year after the exam, the students were retested to see how much of the material they remembered. The average scores for the group are given by the *human memory model* $f(t) = 75 - 6 \ln(t + 1)$, $0 \le t \le 12$, where t is the time in months.

a. What was the average score on the original ($t = 0$) exam?

b. What was the average score at the end of $t = 2$ months?

c. What was the average score at the end of $t = 6$ months?

Algebraic Solution

a. The original average score was

$$f(0) = 75 - 6 \ln(0 + 1) \qquad \text{Substitute 0 for } t.$$
$$= 75 - 6 \ln 1 \qquad \text{Simplify.}$$
$$= 75 - 6(0) \qquad \text{Property of natural logarithms}$$
$$= 75. \qquad \text{Solution}$$

b. After 2 months, the average score was

$$f(2) = 75 - 6 \ln(2 + 1) \qquad \text{Substitute 2 for } t.$$
$$= 75 - 6 \ln 3 \qquad \text{Simplify.}$$
$$\approx 75 - 6(1.0986) \qquad \text{Use a calculator.}$$
$$\approx 68.4. \qquad \text{Solution}$$

c. After 6 months, the average score was

$$f(6) = 75 - 6 \ln(6 + 1) \qquad \text{Substitute 6 for } t.$$
$$= 75 - 6 \ln 7 \qquad \text{Simplify.}$$
$$\approx 75 - 6(1.9459) \qquad \text{Use a calculator.}$$
$$\approx 63.3. \qquad \text{Solution}$$

Graphical Solution

Use a graphing utility to graph the model $y = 75 - 6 \ln(x + 1)$. Then use the *value* or *trace* feature to approximate the following.

a. When $x = 0$, $y = 75$ (see Figure 5.23). So, the original average score was 75.

b. When $x = 2$, $y \approx 68.4$ (see Figure 5.24). So, the average score after 2 months was about 68.4.

c. When $x = 6$, $y \approx 63.3$ (see Figure 5.25). So, the average score after 6 months was about 63.3.

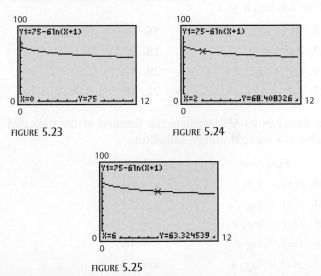

FIGURE 5.23   FIGURE 5.24

FIGURE 5.25

CHECK *Point*   Now try Exercise 97.

CLASSROOM DISCUSSION

Analyzing a Human Memory Model   Use a graphing utility to determine the time in months when the average score in Example 11 was 60. Explain your method of solving the problem. Describe another way that you can use a graphing utility to determine the answer.

5.2 EXERCISES

See www.CalcChat.com for worked-out solutions to odd-numbered exercises.

VOCABULARY: Fill in the blanks.

1. The inverse function of the exponential function given by $f(x) = a^x$ is called the _____ function with base a.

2. The common logarithmic function has base _____ .

3. The logarithmic function given by $f(x) = \ln x$ is called the _____ logarithmic function and has base _____.

4. The Inverse Properties of logarithms and exponentials state that $\log_a a^x = x$ and _____.

5. The One-to-One Property of natural logarithms states that if $\ln x = \ln y$, then _____.

6. The domain of the natural logarithmic function is the set of _____ _____ _____ .

SKILLS AND APPLICATIONS

In Exercises 7–14, write the logarithmic equation in exponential form. For example, the exponential form of $\log_5 25 = 2$ is $5^2 = 25$.

7. $\log_4 16 = 2$ **8.** $\log_7 343 = 3$

9. $\log_9 \frac{1}{81} = -2$ **10.** $\log \frac{1}{1000} = -3$

11. $\log_{32} 4 = \frac{2}{5}$ **12.** $\log_{16} 8 = \frac{3}{4}$

13. $\log_{64} 8 = \frac{1}{2}$ **14.** $\log_8 4 = \frac{2}{3}$

In Exercises 15–22, write the exponential equation in logarithmic form. For example, the logarithmic form of $2^3 = 8$ is $\log_2 8 = 3$.

15. $5^3 = 125$ **16.** $13^2 = 169$

17. $81^{1/4} = 3$ **18.** $9^{3/2} = 27$

19. $6^{-2} = \frac{1}{36}$ **20.** $4^{-3} = \frac{1}{64}$

21. $24^0 = 1$ **22.** $10^{-3} = 0.001$

In Exercises 23–28, evaluate the function at the indicated value of x without using a calculator.

Function	Value
23. $f(x) = \log_2 x$	$x = 64$
24. $f(x) = \log_{25} x$	$x = 5$
25. $f(x) = \log_8 x$	$x = 1$
26. $f(x) = \log x$	$x = 10$
27. $g(x) = \log_a x$	$x = a^2$
28. $g(x) = \log_b x$	$x = b^{-3}$

 In Exercises 29–32, use a calculator to evaluate $f(x) = \log x$ at the indicated value of x. Round your result to three decimal places.

29. $x = \frac{7}{8}$ **30.** $x = \frac{1}{500}$

31. $x = 12.5$ **32.** $x = 96.75$

In Exercises 33–36, use the properties of logarithms to simplify the expression.

33. $\log_{11} 11^7$ **34.** $\log_{3.2} 1$

35. $\log_\pi \pi$ **36.** $9^{\log_9 15}$

In Exercises 37–44, find the domain, x-intercept, and vertical asymptote of the logarithmic function and sketch its graph.

37. $f(x) = \log_4 x$ **38.** $g(x) = \log_6 x$

39. $y = -\log_3 x + 2$ **40.** $h(x) = \log_4(x - 3)$

41. $f(x) = -\log_6(x + 2)$ **42.** $y = \log_5(x - 1) + 4$

43. $y = \log\left(\dfrac{x}{7}\right)$ **44.** $y = \log(-x)$

In Exercises 45–50, use the graph of $g(x) = \log_3 x$ to match the given function with its graph. Then describe the relationship between the graphs of f and g. [The graphs are labeled (a), (b), (c), (d), (e), and (f).]

(a)
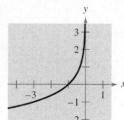

(b)
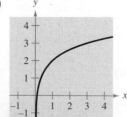

(c)
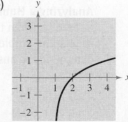

(d)
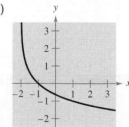

(e)
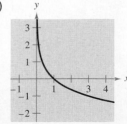

(f)
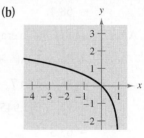

45. $f(x) = \log_3 x + 2$

46. $f(x) = -\log_3 x$

47. $f(x) = -\log_3(x + 2)$

48. $f(x) = \log_3(x - 1)$

49. $f(x) = \log_3(1 - x)$

50. $f(x) = -\log_3(-x)$

In Exercises 51–58, write the logarithmic equation in exponential form.

51. $\ln \frac{1}{2} = -0.693 \ldots$

52. $\ln \frac{2}{5} = -0.916 \ldots$

53. $\ln 7 = 1.945 \ldots$

54. $\ln 10 = 2.302 \ldots$

55. $\ln 250 = 5.521 \ldots$

56. $\ln 1084 = 6.988 \ldots$

57. $\ln 1 = 0$

58. $\ln e = 1$

In Exercises 59–66, write the exponential equation in logarithmic form.

59. $e^4 = 54.598 \ldots$

60. $e^2 = 7.3890 \ldots$

61. $e^{1/2} = 1.6487 \ldots$

62. $e^{1/3} = 1.3956 \ldots$

63. $e^{-0.9} = 0.406 \ldots$

64. $e^{-4.1} = 0.0165 \ldots$

65. $e^x = 4$

66. $e^{2x} = 3$

 In Exercises 67–70, use a calculator to evaluate the function at the indicated value of x. Round your result to three decimal places.

Function	Value
67. $f(x) = \ln x$	$x = 18.42$
68. $f(x) = 3 \ln x$	$x = 0.74$
69. $g(x) = 8 \ln x$	$x = 0.05$
70. $g(x) = -\ln x$	$x = \frac{1}{2}$

In Exercises 71–74, evaluate $g(x) = \ln x$ at the indicated value of x without using a calculator.

71. $x = e^5$

72. $x = e^{-4}$

73. $x = e^{-5/6}$

74. $x = e^{-5/2}$

In Exercises 75–78, find the domain, x-intercept, and vertical asymptote of the logarithmic function and sketch its graph.

75. $f(x) = \ln(x - 4)$

76. $h(x) = \ln(x + 5)$

77. $g(x) = \ln(-x)$

78. $f(x) = \ln(3 - x)$

 In Exercises 79–84, use a graphing utility to graph the function. Be sure to use an appropriate viewing window.

79. $f(x) = \log(x + 9)$

80. $f(x) = \log(x - 6)$

81. $f(x) = \ln(x - 1)$

82. $f(x) = \ln(x + 2)$

83. $f(x) = \ln x + 8$

84. $f(x) = 3 \ln x - 1$

In Exercises 85–92, use the One-to-One Property to solve the equation for x.

85. $\log_5(x + 1) = \log_5 6$

86. $\log_2(x - 3) = \log_2 9$

87. $\log(2x + 1) = \log 15$

88. $\log(5x + 3) = \log 12$

89. $\ln(x + 4) = \ln 12$

90. $\ln(x - 7) = \ln 7$

91. $\ln(x^2 - 2) = \ln 23$

92. $\ln(x^2 - x) = \ln 6$

93. MONTHLY PAYMENT The model

$$t = 16.625 \ln\left(\frac{x}{x - 750}\right), \quad x > 750$$

approximates the length of a home mortgage of $150,000 at 6% in terms of the monthly payment. In the model, t is the length of the mortgage in years and x is the monthly payment in dollars.

(a) Use the model to approximate the lengths of a $150,000 mortgage at 6% when the monthly payment is $897.72 and when the monthly payment is $1659.24.

(b) Approximate the total amounts paid over the term of the mortgage with a monthly payment of $897.72 and with a monthly payment of $1659.24.

(c) Approximate the total interest charges for a monthly payment of $897.72 and for a monthly payment of $1659.24.

(d) What is the vertical asymptote for the model? Interpret its meaning in the context of the problem.

94. COMPOUND INTEREST A principal P, invested at $5\frac{1}{2}\%$ and compounded continuously, increases to an amount K times the original principal after t years, where t is given by $t = (\ln K)/0.055$.

(a) Complete the table and interpret your results.

K	1	2	4	6	8	10	12
t							

(b) Sketch a graph of the function.

95. CABLE TELEVISION The numbers of cable television systems C (in thousands) in the United States from 2001 through 2006 can be approximated by the model

$$C = 10.355 - 0.298t \ln t, \quad 1 \le t \le 6$$

where t represents the year, with $t = 1$ corresponding to 2001. (Source: Warren Communication News)

(a) Complete the table.

t	1	2	3	4	5	6
C						

 (b) Use a graphing utility to graph the function.

(c) Can the model be used to predict the numbers of cable television systems beyond 2006? Explain.

96. POPULATION The time t in years for the world population to double if it is increasing at a continuous rate of r is given by $t = (\ln 2)/r$.

(a) Complete the table and interpret your results.

r	0.005	0.010	0.015	0.020	0.025	0.030
t						

(b) Use a graphing utility to graph the function.

97. HUMAN MEMORY MODEL Students in a mathematics class were given an exam and then retested monthly with an equivalent exam. The average scores for the class are given by the human memory model $f(t) = 80 - 17\log(t + 1)$, $0 \le t \le 12$, where t is the time in months.

(a) Use a graphing utility to graph the model over the specified domain.

(b) What was the average score on the original exam ($t = 0$)?

(c) What was the average score after 4 months?

(d) What was the average score after 10 months?

98. SOUND INTENSITY The relationship between the number of decibels β and the intensity of a sound I in watts per square meter is

$$\beta = 10 \log\left(\frac{I}{10^{-12}}\right).$$

(a) Determine the number of decibels of a sound with an intensity of 1 watt per square meter.

(b) Determine the number of decibels of a sound with an intensity of 10^{-2} watt per square meter.

(c) The intensity of the sound in part (a) is 100 times as great as that in part (b). Is the number of decibels 100 times as great? Explain.

EXPLORATION

TRUE OR FALSE? In Exercises 99 and 100, determine whether the statement is true or false. Justify your answer.

99. You can determine the graph of $f(x) = \log_6 x$ by graphing $g(x) = 6^x$ and reflecting it about the x-axis.

100. The graph of $f(x) = \log_3 x$ contains the point $(27, 3)$.

In Exercises 101–104, sketch the graphs of f and g and describe the relationship between the graphs of f and g. What is the relationship between the functions f and g?

101. $f(x) = 3^x$, $\quad g(x) = \log_3 x$

102. $f(x) = 5^x$, $\quad g(x) = \log_5 x$

103. $f(x) = e^x$, $\quad g(x) = \ln x$

104. $f(x) = 8^x$, $\quad g(x) = \log_8 x$

105. THINK ABOUT IT Complete the table for $f(x) = 10^x$.

x	-2	-1	0	1	2
$f(x)$					

Complete the table for $f(x) = \log x$.

x	$\frac{1}{100}$	$\frac{1}{10}$	1	10	100
$f(x)$					

Compare the two tables. What is the relationship between $f(x) = 10^x$ and $f(x) = \log x$?

106. GRAPHICAL ANALYSIS Use a graphing utility to graph f and g in the same viewing window and determine which is increasing at the greater rate as x approaches $+\infty$. What can you conclude about the rate of growth of the natural logarithmic function?

(a) $f(x) = \ln x$, $\quad g(x) = \sqrt{x}$

(b) $f(x) = \ln x$, $\quad g(x) = \sqrt[4]{x}$

107. (a) Complete the table for the function given by $f(x) = (\ln x)/x$.

x	1	5	10	10^2	10^4	10^6
$f(x)$						

(b) Use the table in part (a) to determine what value $f(x)$ approaches as x increases without bound.

(c) Use a graphing utility to confirm the result of part (b).

108. CAPSTONE The table of values was obtained by evaluating a function. Determine which of the statements may be true and which must be false.

x	y
1	0
2	1
8	3

(a) y is an exponential function of x.

(b) y is a logarithmic function of x.

(c) x is an exponential function of y.

(d) y is a linear function of x.

109. WRITING Explain why $\log_a x$ is defined only for $0 < a < 1$ and $a > 1$.

In Exercises 110 and 111, (a) use a graphing utility to graph the function, (b) use the graph to determine the intervals in which the function is increasing and decreasing, and (c) approximate any relative maximum or minimum values of the function.

110. $f(x) = |\ln x|$ $\qquad$ **111.** $h(x) = \ln(x^2 + 1)$

5.3   PROPERTIES OF LOGARITHMS

What you should learn

- Use the change-of-base formula to rewrite and evaluate logarithmic expressions.
- Use properties of logarithms to evaluate or rewrite logarithmic expressions.
- Use properties of logarithms to expand or condense logarithmic expressions.
- Use logarithmic functions to model and solve real-life problems.

Why you should learn it

Logarithmic functions can be used to model and solve real-life problems. For instance, in Exercises 87–90 on page 402, a logarithmic function is used to model the relationship between the number of decibels and the intensity of a sound.

Dynamic Graphics/ Jupiter Images

Change of Base

Most calculators have only two types of log keys, one for common logarithms (base 10) and one for natural logarithms (base e). Although common logarithms and natural logarithms are the most frequently used, you may occasionally need to evaluate logarithms with other bases. To do this, you can use the following **change-of-base formula.**

> **Change-of-Base Formula**
>
> Let a, b, and x be positive real numbers such that $a \neq 1$ and $b \neq 1$. Then $\log_a x$ can be converted to a different base as follows.
>
> | *Base b* | *Base 10* | *Base e* |
> |---|---|---|
> | $\log_a x = \dfrac{\log_b x}{\log_b a}$ | $\log_a x = \dfrac{\log x}{\log a}$ | $\log_a x = \dfrac{\ln x}{\ln a}$ |

One way to look at the change-of-base formula is that logarithms with base a are simply *constant multiples* of logarithms with base b. The constant multiplier is $1/(\log_b a)$.

Example 1   Changing Bases Using Common Logarithms

a. $\log_4 25 = \dfrac{\log 25}{\log 4}$ $\log_a x = \dfrac{\log x}{\log a}$

$\approx \dfrac{1.39794}{0.60206}$ Use a calculator.

≈ 2.3219 Simplify.

b. $\log_2 12 = \dfrac{\log 12}{\log 2} \approx \dfrac{1.07918}{0.30103} \approx 3.5850$

CHECK *Point*   Now try Exercise 7(a).

Example 2   Changing Bases Using Natural Logarithms

a. $\log_4 25 = \dfrac{\ln 25}{\ln 4}$ $\log_a x = \dfrac{\ln x}{\ln a}$

$\approx \dfrac{3.21888}{1.38629}$ Use a calculator.

≈ 2.3219 Simplify.

b. $\log_2 12 = \dfrac{\ln 12}{\ln 2} \approx \dfrac{2.48491}{0.69315} \approx 3.5850$

CHECK *Point*   Now try Exercise 7(b).

Properties of Logarithms

You know from the preceding section that the logarithmic function with base a is the *inverse function* of the exponential function with base a. So, it makes sense that the properties of exponents should have corresponding properties involving logarithms. For instance, the exponential property $a^0 = 1$ has the corresponding logarithmic property $\log_a 1 = 0$.

> ⚠ **WARNING / CAUTION**
>
> There is no general property that can be used to rewrite $\log_a(u \pm v)$. Specifically, $\log_a(u + v)$ is *not* equal to $\log_a u + \log_a v$.

Properties of Logarithms

Let a be a positive number such that $a \neq 1$, and let n be a real number. If u and v are positive real numbers, the following properties are true.

	Logarithm with Base a	*Natural Logarithm*
1. Product Property:	$\log_a(uv) = \log_a u + \log_a v$	$\ln(uv) = \ln u + \ln v$
2. Quotient Property:	$\log_a \dfrac{u}{v} = \log_a u - \log_a v$	$\ln \dfrac{u}{v} = \ln u - \ln v$
3. Power Property:	$\log_a u^n = n \log_a u$	$\ln u^n = n \ln u$

For proofs of the properties listed above, see Proofs in Mathematics on page 434.

Example 3   Using Properties of Logarithms

Write each logarithm in terms of $\ln 2$ and $\ln 3$.

a. $\ln 6$ **b.** $\ln \dfrac{2}{27}$

Solution

a. $\ln 6 = \ln(2 \cdot 3)$ Rewrite 6 as $2 \cdot 3$.

$\qquad = \ln 2 + \ln 3$ Product Property

b. $\ln \dfrac{2}{27} = \ln 2 - \ln 27$ Quotient Property

$\qquad\quad = \ln 2 - \ln 3^3$ Rewrite 27 as 3^3.

$\qquad\quad = \ln 2 - 3 \ln 3$ Power Property

CHECK *Point* Now try Exercise 27.

Example 4   Using Properties of Logarithms

Find the exact value of each expression without using a calculator.

a. $\log_5 \sqrt[3]{5}$ **b.** $\ln e^6 - \ln e^2$

Solution

a. $\log_5 \sqrt[3]{5} = \log_5 5^{1/3} = \frac{1}{3} \log_5 5 = \frac{1}{3}(1) = \frac{1}{3}$

b. $\ln e^6 - \ln e^2 = \ln \dfrac{e^6}{e^2} = \ln e^4 = 4 \ln e = 4(1) = 4$

CHECK *Point* Now try Exercise 29.

HISTORICAL NOTE

John Napier, a Scottish mathematician, developed logarithms as a way to simplify some of the tedious calculations of his day. Beginning in 1594, Napier worked about 20 years on the invention of logarithms. Napier was only partially successful in his quest to simplify tedious calculations. Nonetheless, the development of logarithms was a step forward and received immediate recognition.

Rewriting Logarithmic Expressions

The properties of logarithms are useful for rewriting logarithmic expressions in forms that simplify the operations of algebra. This is true because these properties convert complicated products, quotients, and exponential forms into simpler sums, differences, and products, respectively.

Example 5   Expanding Logarithmic Expressions

Expand each logarithmic expression.

a. $\log_4 5x^3y$ **b.** $\ln \dfrac{\sqrt{3x-5}}{7}$

Solution

a. $\log_4 5x^3y = \log_4 5 + \log_4 x^3 + \log_4 y$ Product Property

$\qquad\qquad\; = \log_4 5 + 3\log_4 x + \log_4 y$ Power Property

b. $\ln \dfrac{\sqrt{3x-5}}{7} = \ln \dfrac{(3x-5)^{1/2}}{7}$ Rewrite using rational exponent.

$\qquad\qquad\qquad = \ln(3x-5)^{1/2} - \ln 7$ Quotient Property

$\qquad\qquad\qquad = \dfrac{1}{2}\ln(3x-5) - \ln 7$ Power Property

CHECK Point Now try Exercise 53.

In Example 5, the properties of logarithms were used to *expand* logarithmic expressions. In Example 6, this procedure is reversed and the properties of logarithms are used to *condense* logarithmic expressions.

Example 6   Condensing Logarithmic Expressions

Condense each logarithmic expression.

a. $\frac{1}{2}\log x + 3\log(x+1)$ **b.** $2\ln(x+2) - \ln x$

c. $\frac{1}{3}[\log_2 x + \log_2(x+1)]$

Solution

a. $\frac{1}{2}\log x + 3\log(x+1) = \log x^{1/2} + \log(x+1)^3$ Power Property

$\qquad\qquad\qquad\qquad\quad = \log\left[\sqrt{x}(x+1)^3\right]$ Product Property

b. $2\ln(x+2) - \ln x = \ln(x+2)^2 - \ln x$ Power Property

$\qquad\qquad\qquad\quad = \ln \dfrac{(x+2)^2}{x}$ Quotient Property

c. $\frac{1}{3}[\log_2 x + \log_2(x+1)] = \frac{1}{3}\{\log_2[x(x+1)]\}$ Product Property

$\qquad\qquad\qquad\qquad\quad = \log_2[x(x+1)]^{1/3}$ Power Property

$\qquad\qquad\qquad\qquad\quad = \log_2 \sqrt[3]{x(x+1)}$ Rewrite with a radical.

CHECK Point Now try Exercise 75.

Application

One method of determining how the x- and y-values for a set of nonlinear data are related is to take the natural logarithm of each of the x- and y-values. If the points are graphed and fall on a line, then you can determine that the x- and y-values are related by the equation

$$\ln y = m \ln x$$

where m is the slope of the line.

Example 7   Finding a Mathematical Model

The table shows the mean distance from the sun x and the period y (the time it takes a planet to orbit the sun) for each of the six planets that are closest to the sun. In the table, the mean distance is given in terms of astronomical units (where Earth's mean distance is defined as 1.0), and the period is given in years. Find an equation that relates y and x.

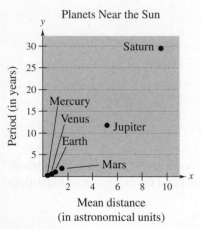

Planets Near the Sun

FIGURE 5.26

Planet	Mean distance, x	Period, y
Mercury	0.387	0.241
Venus	0.723	0.615
Earth	1.000	1.000
Mars	1.524	1.881
Jupiter	5.203	11.860
Saturn	9.537	29.460

Solution

The points in the table above are plotted in Figure 5.26. From this figure it is not clear how to find an equation that relates y and x. To solve this problem, take the natural logarithm of each of the x- and y-values in the table. This produces the following results.

Planet	Mercury	Venus	Earth	Mars	Jupiter	Saturn
$\ln x$	−0.949	−0.324	0.000	0.421	1.649	2.255
$\ln y$	−1.423	−0.486	0.000	0.632	2.473	3.383

Now, by plotting the points in the second table, you can see that all six of the points appear to lie in a line (see Figure 5.27). Choose any two points to determine the slope of the line. Using the two points $(0.421, 0.632)$ and $(0, 0)$, you can determine that the slope of the line is

$$m = \frac{0.632 - 0}{0.421 - 0} \approx 1.5 = \frac{3}{2}.$$

By the point-slope form, the equation of the line is $Y = \frac{3}{2}X$, where $Y = \ln y$ and $X = \ln x$. You can therefore conclude that $\ln y = \frac{3}{2} \ln x$.

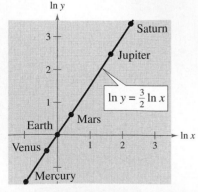

FIGURE 5.27

CHECK Point ▶ Now try Exercise 91.

5.3   EXERCISES

See www.CalcChat.com for worked-out solutions to odd-numbered exercises.

VOCABULARY

In Exercises 1–3, fill in the blanks.

1. To evaluate a logarithm to any base, you can use the _____ formula.

2. The change-of-base formula for base e is given by $\log_a x =$ _____.

3. You can consider $\log_a x$ to be a constant multiple of $\log_b x$; the constant multiplier is _____.

In Exercises 4–6, match the property of logarithms with its name.

4. $\log_a(uv) = \log_a u + \log_a v$ (a) Power Property

5. $\ln u^n = n \ln u$ (b) Quotient Property

6. $\log_a \dfrac{u}{v} = \log_a u - \log_a v$ (c) Product Property

SKILLS AND APPLICATIONS

In Exercises 7–14, rewrite the logarithm as a ratio of (a) common logarithms and (b) natural logarithms.

7. $\log_5 16$ **8.** $\log_3 47$

9. $\log_{1/5} x$ **10.** $\log_{1/3} x$

11. $\log_x \frac{3}{10}$ **12.** $\log_x \frac{3}{4}$

13. $\log_{2.6} x$ **14.** $\log_{7.1} x$

In Exercises 15–22, evaluate the logarithm using the change-of-base formula. Round your result to three decimal places.

15. $\log_3 7$ **16.** $\log_7 4$

17. $\log_{1/2} 4$ **18.** $\log_{1/4} 5$

19. $\log_9 0.1$ **20.** $\log_{20} 0.25$

21. $\log_{15} 1250$ **22.** $\log_3 0.015$

In Exercises 23–28, use the properties of logarithms to rewrite and simplify the logarithmic expression.

23. $\log_4 8$ **24.** $\log_2(4^2 \cdot 3^4)$

25. $\log_5 \frac{1}{250}$ **26.** $\log \frac{9}{300}$

27. $\ln(5e^6)$ **28.** $\ln \dfrac{6}{e^2}$

In Exercises 29–44, find the exact value of the logarithmic expression without using a calculator. (If this is not possible, state the reason.)

29. $\log_3 9$ **30.** $\log_5 \frac{1}{125}$

31. $\log_2 \sqrt[4]{8}$ **32.** $\log_6 \sqrt[3]{6}$

33. $\log_4 16^2$ **34.** $\log_3 81^{-3}$

35. $\log_2(-2)$ **36.** $\log_3(-27)$

37. $\ln e^{4.5}$ **38.** $3 \ln e^4$

39. $\ln \dfrac{1}{\sqrt{e}}$ **40.** $\ln \sqrt[4]{e^3}$

41. $\ln e^2 + \ln e^5$ **42.** $2 \ln e^6 - \ln e^5$

43. $\log_5 75 - \log_5 3$ **44.** $\log_4 2 + \log_4 32$

In Exercises 45–66, use the properties of logarithms to expand the expression as a sum, difference, and/or constant multiple of logarithms. (Assume all variables are positive.)

45. $\ln 4x$ **46.** $\log_3 10z$

47. $\log_8 x^4$ **48.** $\log_{10} \dfrac{y}{2}$

49. $\log_5 \dfrac{5}{x}$ **50.** $\log_6 \dfrac{1}{z^3}$

51. $\ln \sqrt{z}$ **52.** $\ln \sqrt[3]{t}$

53. $\ln xyz^2$ **54.** $\log 4x^2 y$

55. $\ln z(z-1)^2,\ z > 1$ **56.** $\ln\left(\dfrac{x^2-1}{x^3}\right),\ x > 1$

57. $\log_2 \dfrac{\sqrt{a-1}}{9},\ a > 1$ **58.** $\ln \dfrac{6}{\sqrt{x^2+1}}$

59. $\ln \sqrt[3]{\dfrac{x}{y}}$ **60.** $\ln \sqrt{\dfrac{x^2}{y^3}}$

61. $\ln x^2 \sqrt{\dfrac{y}{z}}$ **62.** $\log_2 x^4 \sqrt{\dfrac{y}{z^3}}$

63. $\log_5 \dfrac{x^2}{y^2 z^3}$ **64.** $\log_{10} \dfrac{xy^4}{z^5}$

65. $\ln \sqrt[4]{x^3(x^2+3)}$ **66.** $\ln \sqrt{x^2(x+2)}$

In Exercises 67–84, condense the expression to the logarithm of a single quantity.

67. $\ln 2 + \ln x$

68. $\ln y + \ln t$

69. $\log_4 z - \log_4 y$

70. $\log_5 8 - \log_5 t$

71. $2 \log_2 x + 4 \log_2 y$

72. $\frac{2}{3} \log_7 (z - 2)$

73. $\frac{1}{4} \log_3 5x$

74. $-4 \log_6 2x$

75. $\log x - 2 \log(x + 1)$

76. $2 \ln 8 + 5 \ln(z - 4)$

77. $\log x - 2 \log y + 3 \log z$

78. $3 \log_3 x + 4 \log_3 y - 4 \log_3 z$

79. $\ln x - [\ln(x + 1) + \ln(x - 1)]$

80. $4[\ln z + \ln(z + 5)] - 2 \ln(z - 5)$

81. $\frac{1}{3}[2 \ln(x + 3) + \ln x - \ln(x^2 - 1)]$

82. $2[3 \ln x - \ln(x + 1) - \ln(x - 1)]$

83. $\frac{1}{3}[\log_8 y + 2 \log_8(y + 4)] - \log_8(y - 1)$

84. $\frac{1}{2}[\log_4(x + 1) + 2 \log_4(x - 1)] + 6 \log_4 x$

In Exercises 85 and 86, compare the logarithmic quantities. If two are equal, explain why.

85. $\dfrac{\log_2 32}{\log_2 4}$, $\log_2 \dfrac{32}{4}$, $\log_2 32 - \log_2 4$

86. $\log_7 \sqrt{70}$, $\log_7 35$, $\frac{1}{2} + \log_7 \sqrt{10}$

SOUND INTENSITY In Exercises 87–90, use the following information. The relationship between the number of decibels β and the intensity of a sound I in watts per square meter is given by

$$\beta = 10 \log\left(\frac{I}{10^{-12}}\right).$$

87. Use the properties of logarithms to write the formula in simpler form, and determine the number of decibels of a sound with an intensity of 10^{-6} watt per square meter.

88. Find the difference in loudness between an average office with an intensity of 1.26×10^{-7} watt per square meter and a broadcast studio with an intensity of 3.16×10^{-10} watt per square meter.

89. Find the difference in loudness between a vacuum cleaner with an intensity of 10^{-4} watt per square meter and rustling leaves with an intensity of 10^{-11} watt per square meter.

90. You and your roommate are playing your stereos at the same time and at the same intensity. How much louder is the music when both stereos are playing compared with just one stereo playing?

CURVE FITTING In Exercises 91–94, find a logarithmic equation that relates y and x. Explain the steps used to find the equation.

91.

x	1	2	3	4	5	6
y	1	1.189	1.316	1.414	1.495	1.565

92.

x	1	2	3	4	5	6
y	1	1.587	2.080	2.520	2.924	3.302

93.

x	1	2	3	4	5	6
y	2.5	2.102	1.9	1.768	1.672	1.597

94.

x	1	2	3	4	5	6
y	0.5	2.828	7.794	16	27.951	44.091

95. GALLOPING SPEEDS OF ANIMALS Four-legged animals run with two different types of motion: trotting and galloping. An animal that is trotting has at least one foot on the ground at all times, whereas an animal that is galloping has all four feet off the ground at some point in its stride. The number of strides per minute at which an animal breaks from a trot to a gallop depends on the weight of the animal. Use the table to find a logarithmic equation that relates an animal's weight x (in pounds) and its lowest galloping speed y (in strides per minute).

Weight, x	Galloping speed, y
25	191.5
35	182.7
50	173.8
75	164.2
500	125.9
1000	114.2

96. NAIL LENGTH The approximate lengths and diameters (in inches) of common nails are shown in the table. Find a logarithmic equation that relates the diameter y of a common nail to its length x.

Length, x	Diameter, y	Length, x	Diameter, y
1	0.072	4	0.203
2	0.120	5	0.238
3	0.148	6	0.284

97. **COMPARING MODELS** A cup of water at an initial temperature of 78°C is placed in a room at a constant temperature of 21°C. The temperature of the water is measured every 5 minutes during a half-hour period. The results are recorded as ordered pairs of the form (t, T), where t is the time (in minutes) and T is the temperature (in degrees Celsius).

(0, 78.0°), (5, 66.0°), (10, 57.5°), (15, 51.2°),
(20, 46.3°), (25, 42.4°), (30, 39.6°)

(a) The graph of the model for the data should be asymptotic with the graph of the temperature of the room. Subtract the room temperature from each of the temperatures in the ordered pairs. Use a graphing utility to plot the data points (t, T) and $(t, T - 21)$.

(b) An exponential model for the data $(t, T - 21)$ is given by $T - 21 = 54.4(0.964)^t$. Solve for T and graph the model. Compare the result with the plot of the original data.

(c) Take the natural logarithms of the revised temperatures. Use a graphing utility to plot the points $(t, \ln(T - 21))$ and observe that the points appear to be linear. Use the *regression* feature of the graphing utility to fit a line to these data. This resulting line has the form $\ln(T - 21) = at + b$. Solve for T, and verify that the result is equivalent to the model in part (b).

(d) Fit a rational model to the data. Take the reciprocals of the y-coordinates of the revised data points to generate the points

$$\left(t, \frac{1}{T - 21}\right).$$

Use a graphing utility to graph these points and observe that they appear to be linear. Use the *regression* feature of a graphing utility to fit a line to these data. The resulting line has the form

$$\frac{1}{T - 21} = at + b.$$

Solve for T, and use a graphing utility to graph the rational function and the original data points.

(e) Why did taking the logarithms of the temperatures lead to a linear scatter plot? Why did taking the reciprocals of the temperatures lead to a linear scatter plot?

EXPLORATION

98. **PROOF** Prove that $\log_b \dfrac{u}{v} = \log_b u - \log_b v$.

99. **PROOF** Prove that $\log_b u^n = n \log_b u$.

100. **CAPSTONE** A classmate claims that the following are true.

(a) $\ln(u + v) = \ln u + \ln v = \ln(uv)$

(b) $\ln(u - v) = \ln u - \ln v = \ln \dfrac{u}{v}$

(c) $(\ln u)^n = n(\ln u) = \ln u^n$

Discuss how you would demonstrate that these claims are not true.

TRUE OR FALSE? In Exercises 101–106, determine whether the statement is true or false given that $f(x) = \ln x$. Justify your answer.

101. $f(0) = 0$

102. $f(ax) = f(a) + f(x)$, $a > 0, x > 0$

103. $f(x - 2) = f(x) - f(2)$, $x > 2$

104. $\sqrt{f(x)} = \frac{1}{2}f(x)$

105. If $f(u) = 2f(v)$, then $v = u^2$.

106. If $f(x) < 0$, then $0 < x < 1$.

 In Exercises 107–112, use the change-of-base formula to rewrite the logarithm as a ratio of logarithms. Then use a graphing utility to graph the ratio.

107. $f(x) = \log_2 x$

108. $f(x) = \log_4 x$

109. $f(x) = \log_{1/2} x$

110. $f(x) = \log_{1/4} x$

111. $f(x) = \log_{11.8} x$

112. $f(x) = \log_{12.4} x$

113. **THINK ABOUT IT** Consider the functions below.

$$f(x) = \ln \frac{x}{2}, \quad g(x) = \frac{\ln x}{\ln 2}, \quad h(x) = \ln x - \ln 2$$

Which two functions should have identical graphs? Verify your answer by sketching the graphs of all three functions on the same set of coordinate axes.

 114. **GRAPHICAL ANALYSIS** Use a graphing utility to graph the functions given by $y_1 = \ln x - \ln(x - 3)$ and $y_2 = \ln \dfrac{x}{x - 3}$ in the same viewing window. Does the graphing utility show the functions with the same domain? If so, should it? Explain your reasoning.

115. **THINK ABOUT IT** For how many integers between 1 and 20 can the natural logarithms be approximated given the values $\ln 2 \approx 0.6931$, $\ln 3 \approx 1.0986$, and $\ln 5 \approx 1.6094$? Approximate these logarithms (do not use a calculator).

What you should learn

- Solve simple exponential and logarithmic equations.
- Solve more complicated exponential equations.
- Solve more complicated logarithmic equations.
- Use exponential and logarithmic equations to model and solve real-life problems.

Why you should learn it

Exponential and logarithmic equations are used to model and solve life science applications. For instance, in Exercise 132 on page 413, an exponential function is used to model the number of trees per acre given the average diameter of the trees.

5.4 EXPONENTIAL AND LOGARITHMIC EQUATIONS

Introduction

So far in this chapter, you have studied the definitions, graphs, and properties of exponential and logarithmic functions. In this section, you will study procedures for *solving equations* involving these exponential and logarithmic functions.

There are two basic strategies for solving exponential or logarithmic equations. The first is based on the One-to-One Properties and was used to solve simple exponential and logarithmic equations in Sections 5.1 and 5.2. The second is based on the Inverse Properties. For $a > 0$ and $a \neq 1$, the following properties are true for all x and y for which $\log_a x$ and $\log_a y$ are defined.

One-to-One Properties

$$a^x = a^y \text{ if and only if } x = y.$$

$$\log_a x = \log_a y \text{ if and only if } x = y.$$

Inverse Properties

$$a^{\log_a x} = x$$

$$\log_a a^x = x$$

Example 1 Solving Simple Equations

	Original Equation	*Rewritten Equation*	*Solution*	*Property*
a.	$2^x = 32$	$2^x = 2^5$	$x = 5$	One-to-One
b.	$\ln x - \ln 3 = 0$	$\ln x = \ln 3$	$x = 3$	One-to-One
c.	$\left(\frac{1}{3}\right)^x = 9$	$3^{-x} = 3^2$	$x = -2$	One-to-One
d.	$e^x = 7$	$\ln e^x = \ln 7$	$x = \ln 7$	Inverse
e.	$\ln x = -3$	$e^{\ln x} = e^{-3}$	$x = e^{-3}$	Inverse
f.	$\log x = -1$	$10^{\log x} = 10^{-1}$	$x = 10^{-1} = \frac{1}{10}$	Inverse
g.	$\log_3 x = 4$	$3^{\log_3 x} = 3^4$	$x = 81$	Inverse

CHECK *Point* Now try Exercise 17.

The strategies used in Example 1 are summarized as follows.

Strategies for Solving Exponential and Logarithmic Equations

1. Rewrite the original equation in a form that allows the use of the One-to-One Properties of exponential or logarithmic functions.

2. Rewrite an *exponential* equation in logarithmic form and apply the Inverse Property of logarithmic functions.

3. Rewrite a *logarithmic* equation in exponential form and apply the Inverse Property of exponential functions.

Solving Exponential Equations

| Example 2 | Solving Exponential Equations

Solve each equation and approximate the result to three decimal places, if necessary.

a. $e^{-x^2} = e^{-3x-4}$

b. $3(2^x) = 42$

Solution

a.

$e^{-x^2} = e^{-3x-4}$	Write original equation.
$-x^2 = -3x - 4$	One-to-One Property
$x^2 - 3x - 4 = 0$	Write in general form.
$(x + 1)(x - 4) = 0$	Factor.
$(x + 1) = 0 \Rightarrow x = -1$	Set 1st factor equal to 0.
$(x - 4) = 0 \Rightarrow x = 4$	Set 2nd factor equal to 0.

The solutions are $x = -1$ and $x = 4$. Check these in the original equation.

b.

$3(2^x) = 42$	Write original equation.
$2^x = 14$	Divide each side by 3.
$\log_2 2^x = \log_2 14$	Take log (base 2) of each side.
$x = \log_2 14$	Inverse Property
$x = \dfrac{\ln 14}{\ln 2} \approx 3.807$	Change-of-base formula

The solution is $x = \log_2 14 \approx 3.807$. Check this in the original equation.

CHECK *Point*   Now try Exercise 29.

In Example 2(b), the exact solution is $x = \log_2 14$ and the approximate solution is $x \approx 3.807$. An exact answer is preferred when the solution is an intermediate step in a larger problem. For a final answer, an approximate solution is easier to comprehend.

| Example 3 | Solving an Exponential Equation

Solve $e^x + 5 = 60$ and approximate the result to three decimal places.

Solution

$e^x + 5 = 60$	Write original equation.
$e^x = 55$	Subtract 5 from each side.
$\ln e^x = \ln 55$	Take natural log of each side.
$x = \ln 55 \approx 4.007$	Inverse Property

The solution is $x = \ln 55 \approx 4.007$. Check this in the original equation.

CHECK *Point*   Now try Exercise 55.

Study Tip

Another way to solve Example 2(b) is by taking the natural log of each side and then applying the Power Property, as follows.

$$3(2^x) = 42$$

$$2^x = 14$$

$$\ln 2^x = \ln 14$$

$$x \ln 2 = \ln 14$$

$$x = \frac{\ln 14}{\ln 2} \approx 3.807$$

As you can see, you obtain the same result as in Example 2(b).

Study Tip

Remember that the natural logarithmic function has a base of e.

Example 4 Solving an Exponential Equation

Solve $2(3^{2t-5}) - 4 = 11$ and approximate the result to three decimal places.

Solution

$$2(3^{2t-5}) - 4 = 11 \qquad \text{Write original equation.}$$

$$2(3^{2t-5}) = 15 \qquad \text{Add 4 to each side.}$$

$$3^{2t-5} = \frac{15}{2} \qquad \text{Divide each side by 2.}$$

$$\log_3 3^{2t-5} = \log_3 \frac{15}{2} \qquad \text{Take log (base 3) of each side.}$$

$$2t - 5 = \log_3 \frac{15}{2} \qquad \text{Inverse Property}$$

$$2t = 5 + \log_3 7.5 \qquad \text{Add 5 to each side.}$$

$$t = \frac{5}{2} + \frac{1}{2} \log_3 7.5 \qquad \text{Divide each side by 2.}$$

$$t \approx 3.417 \qquad \text{Use a calculator.}$$

The solution is $t = \frac{5}{2} + \frac{1}{2} \log_3 7.5 \approx 3.417$. Check this in the original equation.

CHECK Point Now try Exercise 57.

When an equation involves two or more exponential expressions, you can still use a procedure similar to that demonstrated in Examples 2, 3, and 4. However, the algebra is a bit more complicated.

Example 5 Solving an Exponential Equation of Quadratic Type

Solve $e^{2x} - 3e^x + 2 = 0$.

Algebraic Solution

$$e^{2x} - 3e^x + 2 = 0 \qquad \text{Write original equation.}$$

$$(e^x)^2 - 3e^x + 2 = 0 \qquad \text{Write in quadratic form.}$$

$$(e^x - 2)(e^x - 1) = 0 \qquad \text{Factor.}$$

$$e^x - 2 = 0 \qquad \text{Set 1st factor equal to 0.}$$

$$x = \ln 2 \qquad \text{Solution}$$

$$e^x - 1 = 0 \qquad \text{Set 2nd factor equal to 0.}$$

$$x = 0 \qquad \text{Solution}$$

The solutions are $x = \ln 2 \approx 0.693$ and $x = 0$. Check these in the original equation.

CHECK Point Now try Exercise 59.

Graphical Solution

Use a graphing utility to graph $y = e^{2x} - 3e^x + 2$. Use the *zero* or *root* feature or the *zoom* and *trace* features of the graphing utility to approximate the values of x for which $y = 0$. In Figure 5.28, you can see that the zeros occur at $x = 0$ and at $x \approx 0.693$. So, the solutions are $x = 0$ and $x \approx 0.693$.

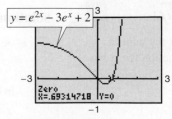

FIGURE 5.28

Solving Logarithmic Equations

To solve a logarithmic equation, you can write it in exponential form.

$$\ln x = 3 \qquad \text{Logarithmic form}$$

$$e^{\ln x} = e^3 \qquad \text{Exponentiate each side.}$$

$$x = e^3 \qquad \text{Exponential form}$$

This procedure is called *exponentiating* each side of an equation.

Example 6   Solving Logarithmic Equations

WARNING/CAUTION

Remember to check your solutions in the original equation when solving equations to verify that the answer is correct and to make sure that the answer lies in the domain of the original equation.

a. $\ln x = 2$ Original equation

$\qquad e^{\ln x} = e^2$ Exponentiate each side.

$\qquad x = e^2$ Inverse Property

b. $\log_3(5x - 1) = \log_3(x + 7)$ Original equation

$\qquad 5x - 1 = x + 7$ One-to-One Property

$\qquad 4x = 8$ Add $-x$ and 1 to each side.

$\qquad x = 2$ Divide each side by 4.

c. $\log_6(3x + 14) - \log_6 5 = \log_6 2x$ Original equation

$$\log_6\left(\frac{3x + 14}{5}\right) = \log_6 2x \qquad \text{Quotient Property of Logarithms}$$

$$\frac{3x + 14}{5} = 2x \qquad \text{One-to-One Property}$$

$$3x + 14 = 10x \qquad \text{Cross multiply.}$$

$$-7x = -14 \qquad \text{Isolate } x.$$

$$x = 2 \qquad \text{Divide each side by } -7.$$

CHECK Point   Now try Exercise 83.

Example 7   Solving a Logarithmic Equation

Solve $5 + 2 \ln x = 4$ and approximate the result to three decimal places.

Algebraic Solution

$5 + 2 \ln x = 4$ Write original equation.

$2 \ln x = -1$ Subtract 5 from each side.

$\ln x = -\dfrac{1}{2}$ Divide each side by 2.

$e^{\ln x} = e^{-1/2}$ Exponentiate each side.

$x = e^{-1/2}$ Inverse Property

$x \approx 0.607$ Use a calculator.

Graphical Solution

Use a graphing utility to graph $y_1 = 5 + 2 \ln x$ and $y_2 = 4$ in the same viewing window. Use the *intersect* feature or the *zoom* and *trace* features to approximate the intersection point, as shown in Figure 5.29. So, the solution is $x \approx 0.607$.

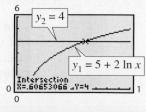

FIGURE 5.29

CHECK Point   Now try Exercise 93.

| **Example 8** | **Solving a Logarithmic Equation** |

Solve $2 \log_5 3x = 4$.

Solution

$2 \log_5 3x = 4$	Write original equation.
$\log_5 3x = 2$	Divide each side by 2.
$5^{\log_5 3x} = 5^2$	Exponentiate each side (base 5).
$3x = 25$	Inverse Property
$x = \dfrac{25}{3}$	Divide each side by 3.

The solution is $x = \frac{25}{3}$. Check this in the original equation.

CHECK Point Now try Exercise 97.

Because the domain of a logarithmic function generally does not include all real numbers, you should be sure to check for extraneous solutions of logarithmic equations.

| **Example 9** | **Checking for Extraneous Solutions** |

Solve $\log 5x + \log(x - 1) = 2$.

Algebraic Solution

$\log 5x + \log(x - 1) = 2$	Write original equation.
$\log[5x(x - 1)] = 2$	Product Property of Logarithms
$10^{\log(5x^2 - 5x)} = 10^2$	Exponentiate each side (base 10).
$5x^2 - 5x = 100$	Inverse Property
$x^2 - x - 20 = 0$	Write in general form.
$(x - 5)(x + 4) = 0$	Factor.
$x - 5 = 0$	Set 1st factor equal to 0.
$x = 5$	Solution
$x + 4 = 0$	Set 2nd factor equal to 0.
$x = -4$	Solution

The solutions appear to be $x = 5$ and $x = -4$. However, when you check these in the original equation, you can see that $x = 5$ is the only solution.

CHECK Point Now try Exercise 109.

Graphical Solution

Use a graphing utility to graph $y_1 = \log 5x + \log(x - 1)$ and $y_2 = 2$ in the same viewing window. From the graph shown in Figure 5.30, it appears that the graphs intersect at one point. Use the *intersect* feature or the *zoom* and *trace* features to determine that the graphs intersect at approximately $(5, 2)$. So, the solution is $x = 5$. Verify that 5 is an exact solution algebraically.

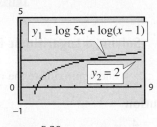

FIGURE **5.30**

In Example 9, the domain of $\log 5x$ is $x > 0$ and the domain of $\log(x - 1)$ is $x > 1$, so the domain of the original equation is $x > 1$. Because the domain is all real numbers greater than 1, the solution $x = -4$ is extraneous. The graph in Figure 5.30 verifies this conclusion.

Applications

| Example 10   Doubling an Investment

You have deposited $500 in an account that pays 6.75% interest, compounded continuously. How long will it take your money to double?

Solution

Using the formula for continuous compounding, you can find that the balance in the account is

$$A = Pe^{rt}$$

$$A = 500e^{0.0675t}.$$

To find the time required for the balance to double, let $A = 1000$ and solve the resulting equation for t.

$500e^{0.0675t} = 1000$	Let $A = 1000$.
$e^{0.0675t} = 2$	Divide each side by 500.
$\ln e^{0.0675t} = \ln 2$	Take natural log of each side.
$0.0675t = \ln 2$	Inverse Property
$t = \dfrac{\ln 2}{0.0675}$	Divide each side by 0.0675.
$t \approx 10.27$	Use a calculator.

The balance in the account will double after approximately 10.27 years. This result is demonstrated graphically in Figure 5.31.

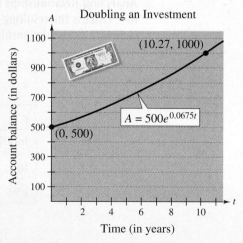

Doubling an Investment

(10.27, 1000)

$A = 500e^{0.0675t}$

(0, 500)

FIGURE **5.31**

CHECK*Point*   Now try Exercise 117.

In Example 10, an approximate answer of 10.27 years is given. Within the context of the problem, the exact solution, $(\ln 2)/0.0675$ years, does not make sense as an answer.

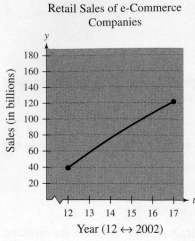

Retail Sales of e-Commerce
Companies

Sales (in billions)

Year (12 ↔ 2002)

FIGURE 5.32

Example 11 Retail Sales

The retail sales y (in billions) of e-commerce companies in the United States from 2002 through 2007 can be modeled by

$$y = -549 + 236.7 \ln t, \quad 12 \le t \le 17$$

where t represents the year, with $t = 12$ corresponding to 2002 (see Figure 5.32). During which year did the sales reach \$108 billion? (Source: U.S. Census Bureau)

Solution

$-549 + 236.7 \ln t = y$	Write original equation.
$-549 + 236.7 \ln t = 108$	Substitute 108 for y.
$236.7 \ln t = 657$	Add 549 to each side.
$\ln t = \dfrac{657}{236.7}$	Divide each side by 236.7.
$e^{\ln t} = e^{657/236.7}$	Exponentiate each side.
$t = e^{657/236.7}$	Inverse Property
$t \approx 16$	Use a calculator.

The solution is $t \approx 16$. Because $t = 12$ represents 2002, it follows that the sales reached \$108 billion in 2006.

CHECK Point Now try Exercise 133.

CLASSROOM DISCUSSION

Analyzing Relationships Numerically Use a calculator to fill in the table row-by-row. Discuss the resulting pattern. What can you conclude? Find two equations that summarize the relationships you discovered.

x	$\dfrac{1}{2}$	1	2	10	25	50
e^x						
$\ln(e^x)$						
$\ln x$						
$e^{\ln x}$						

5.4   EXERCISES

See www.CalcChat.com for worked-out solutions to odd-numbered exercises.

VOCABULARY: Fill in the blanks.

1. To _____ an equation in x means to find all values of x for which the equation is true.

2. To solve exponential and logarithmic equations, you can use the following One-to-One and Inverse Properties.

    (a) $a^x = a^y$ if and only if _____.
    (b) $\log_a x = \log_a y$ if and only if _____.
    (c) $a^{\log_a x} =$ _____
    (d) $\log_a a^x =$ _____

3. To solve exponential and logarithmic equations, you can use the following strategies.

    (a) Rewrite the original equation in a form that allows the use of the _____ Properties of exponential or logarithmic functions.

    (b) Rewrite an exponential equation in _____ form and apply the Inverse Property of _____ functions.

    (c) Rewrite a logarithmic equation in _____ form and apply the Inverse Property of _____ functions.

4. An _____ solution does not satisfy the original equation.

SKILLS AND APPLICATIONS

In Exercises 5–12, determine whether each x-value is a solution (or an approximate solution) of the equation.

5. $4^{2x-7} = 64$
    (a) $x = 5$
    (b) $x = 2$

6. $2^{3x+1} = 32$
    (a) $x = -1$
    (b) $x = 2$

7. $3e^{x+2} = 75$
    (a) $x = -2 + e^{25}$
    (b) $x = -2 + \ln 25$
    (c) $x \approx 1.219$

8. $4e^{x-1} = 60$
    (a) $x = 1 + \ln 15$
    (b) $x \approx 3.7081$
    (c) $x = \ln 16$

9. $\log_4(3x) = 3$
    (a) $x \approx 21.333$
    (b) $x = -4$
    (c) $x = \frac{64}{3}$

10. $\log_2(x + 3) = 10$
    (a) $x = 1021$
    (b) $x = 17$
    (c) $x = 10^2 - 3$

11. $\ln(2x + 3) = 5.8$
    (a) $x = \frac{1}{2}(-3 + \ln 5.8)$
    (b) $x = \frac{1}{2}(-3 + e^{5.8})$
    (c) $x \approx 163.650$

12. $\ln(x - 1) = 3.8$
    (a) $x = 1 + e^{3.8}$
    (b) $x \approx 45.701$
    (c) $x = 1 + \ln 3.8$

In Exercises 13–24, solve for x.

13. $4^x = 16$
14. $3^x = 243$
15. $\left(\frac{1}{2}\right)^x = 32$
16. $\left(\frac{1}{4}\right)^x = 64$
17. $\ln x - \ln 2 = 0$
18. $\ln x - \ln 5 = 0$
19. $e^x = 2$
20. $e^x = 4$
21. $\ln x = -1$
22. $\log x = -2$
23. $\log_4 x = 3$
24. $\log_5 x = \frac{1}{2}$

In Exercises 25–28, approximate the point of intersection of the graphs of f and g. Then solve the equation $f(x) = g(x)$ algebraically to verify your approximation.

25. $f(x) = 2^x$
    $g(x) = 8$

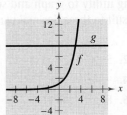

26. $f(x) = 27^x$
    $g(x) = 9$

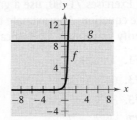

27. $f(x) = \log_3 x$
    $g(x) = 2$

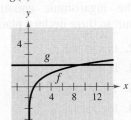

28. $f(x) = \ln(x - 4)$
    $g(x) = 0$

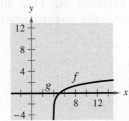

In Exercises 29–70, solve the exponential equation algebraically. Approximate the result to three decimal places.

29. $e^x = e^{x^2 - 2}$
30. $e^{2x} = e^{x^2 - 8}$
31. $e^{x^2 - 3} = e^{x - 2}$
32. $e^{-x^2} = e^{x^2 - 2x}$
33. $4(3^x) = 20$
34. $2(5^x) = 32$
35. $2e^x = 10$
36. $4e^x = 91$
37. $e^x - 9 = 19$
38. $6^x + 10 = 47$
39. $3^{2x} = 80$
40. $6^{5x} = 3000$
41. $5^{-t/2} = 0.20$
42. $4^{-3t} = 0.10$
43. $3^{x-1} = 27$
44. $2^{x-3} = 32$
45. $2^{3-x} = 565$
46. $8^{-2-x} = 431$

47. $8(10^{3x}) = 12$

48. $5(10^{x-6}) = 7$

49. $3(5^{x-1}) = 21$

50. $8(3^{6-x}) = 40$

51. $e^{3x} = 12$

52. $e^{2x} = 50$

53. $500e^{-x} = 300$

54. $1000e^{-4x} = 75$

55. $7 - 2e^x = 5$

56. $-14 + 3e^x = 11$

57. $6(2^{3x-1}) - 7 = 9$

58. $8(4^{6-2x}) + 13 = 41$

59. $e^{2x} - 4e^x - 5 = 0$

60. $e^{2x} - 5e^x + 6 = 0$

61. $e^{2x} - 3e^x - 4 = 0$

62. $e^{2x} + 9e^x + 36 = 0$

63. $\dfrac{500}{100 - e^{x/2}} = 20$

64. $\dfrac{400}{1 + e^{-x}} = 350$

65. $\dfrac{3000}{2 + e^{2x}} = 2$

66. $\dfrac{119}{e^{6x} - 14} = 7$

67. $\left(1 + \dfrac{0.065}{365}\right)^{365t} = 4$

68. $\left(4 - \dfrac{2.471}{40}\right)^{9t} = 21$

69. $\left(1 + \dfrac{0.10}{12}\right)^{12t} = 2$

70. $\left(16 - \dfrac{0.878}{26}\right)^{3t} = 30$

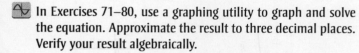 In Exercises 71–80, use a graphing utility to graph and solve the equation. Approximate the result to three decimal places. Verify your result algebraically.

71. $7 = 2^x$

72. $5^x = 212$

73. $6e^{1-x} = 25$

74. $-4e^{-x-1} + 15 = 0$

75. $3e^{3x/2} = 962$

76. $8e^{-2x/3} = 11$

77. $e^{0.09t} = 3$

78. $-e^{1.8x} + 7 = 0$

79. $e^{0.125t} - 8 = 0$

80. $e^{2.724x} = 29$

In Exercises 81–112, solve the logarithmic equation algebraically. Approximate the result to three decimal places.

81. $\ln x = -3$

82. $\ln x = 1.6$

83. $\ln x - 7 = 0$

84. $\ln x + 1 = 0$

85. $\ln 2x = 2.4$

86. $2.1 = \ln 6x$

87. $\log x = 6$

88. $\log 3z = 2$

89. $3 \ln 5x = 10$

90. $2 \ln x = 7$

91. $\ln \sqrt{x + 2} = 1$

92. $\ln \sqrt{x - 8} = 5$

93. $7 + 3 \ln x = 5$

94. $2 - 6 \ln x = 10$

95. $-2 + 2 \ln 3x = 17$

96. $2 + 3 \ln x = 12$

97. $6 \log_3(0.5x) = 11$

98. $4 \log(x - 6) = 11$

99. $\ln x - \ln(x + 1) = 2$

100. $\ln x + \ln(x + 1) = 1$

101. $\ln x + \ln(x - 2) = 1$

102. $\ln x + \ln(x + 3) = 1$

103. $\ln(x + 5) = \ln(x - 1) - \ln(x + 1)$

104. $\ln(x + 1) - \ln(x - 2) = \ln x$

105. $\log_2(2x - 3) = \log_2(x + 4)$

106. $\log(3x + 4) = \log(x - 10)$

107. $\log(x + 4) - \log x = \log(x + 2)$

108. $\log_2 x + \log_2(x + 2) = \log_2(x + 6)$

109. $\log_4 x - \log_4(x - 1) = \frac{1}{2}$

110. $\log_3 x + \log_3(x - 8) = 2$

111. $\log 8x - \log\left(1 + \sqrt{x}\right) = 2$

112. $\log 4x - \log\left(12 + \sqrt{x}\right) = 2$

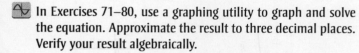 In Exercises 113–116, use a graphing utility to graph and solve the equation. Approximate the result to three decimal places. Verify your result algebraically.

113. $3 - \ln x = 0$

114. $10 - 4 \ln(x - 2) = 0$

115. $2 \ln(x + 3) = 3$

116. $\ln(x + 1) = 2 - \ln x$

COMPOUND INTEREST In Exercises 117–120, $2500 is invested in an account at interest rate r, compounded continuously. Find the time required for the amount to (a) double and (b) triple.

117. $r = 0.05$

118. $r = 0.045$

119. $r = 0.025$

120. $r = 0.0375$

In Exercises 121–128, solve the equation algebraically. Round the result to three decimal places. Verify your answer using a graphing utility.

121. $2x^2 e^{2x} + 2x e^{2x} = 0$

122. $-x^2 e^{-x} + 2x e^{-x} = 0$

123. $-xe^{-x} + e^{-x} = 0$

124. $e^{-2x} - 2x e^{-2x} = 0$

125. $2x \ln x + x = 0$

126. $\dfrac{1 - \ln x}{x^2} = 0$

127. $\dfrac{1 + \ln x}{2} = 0$

128. $2x \ln\left(\dfrac{1}{x}\right) - x = 0$

129. DEMAND The demand equation for a limited edition coin set is

$$p = 1000\left(1 - \frac{5}{5 + e^{-0.001x}}\right).$$

Find the demand x for a price of (a) $p = \$139.50$ and (b) $p = \$99.99$.

130. DEMAND The demand equation for a hand-held electronic organizer is

$$p = 5000\left(1 - \frac{4}{4 + e^{-0.002x}}\right).$$

Find the demand x for a price of (a) $p = \$600$ and (b) $p = \$400$.

131. FOREST YIELD The yield V (in millions of cubic feet per acre) for a forest at age t years is given by $V = 6.7e^{-48.1/t}$.

 (a) Use a graphing utility to graph the function.

(b) Determine the horizontal asymptote of the function. Interpret its meaning in the context of the problem.

(c) Find the time necessary to obtain a yield of 1.3 million cubic feet.

132. TREES PER ACRE The number N of trees of a given species per acre is approximated by the model $N = 68(10^{-0.04x})$, $5 \le x \le 40$, where x is the average diameter of the trees (in inches) 3 feet above the ground. Use the model to approximate the average diameter of the trees in a test plot when $N = 21$.

133. U.S. CURRENCY The values y (in billions of dollars) of U.S. currency in circulation in the years 2000 through 2007 can be modeled by $y = -451 + 444 \ln t$, $10 \le t \le 17$, where t represents the year, with $t = 10$ corresponding to 2000. During which year did the value of U.S. currency in circulation exceed \$690 billion? (Source: Board of Governors of the Federal Reserve System)

 134. MEDICINE The numbers y of freestanding ambulatory care surgery centers in the United States from 2000 through 2007 can be modeled by

$$y = 2875 + \frac{2635.11}{1 + 14.215e^{-0.8038t}}, \quad 0 \le t \le 7$$

where t represents the year, with $t = 0$ corresponding to 2000. (Source: Verispan)

(a) Use a graphing utility to graph the model.

(b) Use the *trace* feature of the graphing utility to estimate the year in which the number of surgery centers exceeded 3600.

135. AVERAGE HEIGHTS The percent m of American males between the ages of 18 and 24 who are no more than x inches tall is modeled by

$$m(x) = \frac{100}{1 + e^{-0.6114(x - 69.71)}}$$

and the percent f of American females between the ages of 18 and 24 who are no more than x inches tall is modeled by

$$f(x) = \frac{100}{1 + e^{-0.66607(x - 64.51)}}.$$

(Source: U.S. National Center for Health Statistics)

(a) Use the graph to determine any horizontal asymptotes of the graphs of the functions. Interpret the meaning in the context of the problem.

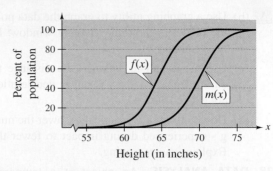

(b) What is the average height of each sex?

 136. LEARNING CURVE In a group project in learning theory, a mathematical model for the proportion P of correct responses after n trials was found to be $P = 0.83/(1 + e^{-0.2n})$.

(a) Use a graphing utility to graph the function.

(b) Use the graph to determine any horizontal asymptotes of the graph of the function. Interpret the meaning of the upper asymptote in the context of this problem.

(c) After how many trials will 60% of the responses be correct?

137. AUTOMOBILES Automobiles are designed with crumple zones that help protect their occupants in crashes. The crumple zones allow the occupants to move short distances when the automobiles come to abrupt stops. The greater the distance moved, the fewer g's the crash victims experience. (One g is equal to the acceleration due to gravity. For very short periods of time, humans have withstood as much as 40 g's.) In crash tests with vehicles moving at 90 kilometers per hour, analysts measured the numbers of g's experienced during deceleration by crash dummies that were permitted to move x meters during impact. The data are shown in the table. A model for the data is given by $y = -3.00 + 11.88 \ln x + (36.94/x)$, where y is the number of g's.

x	g's
0.2	158
0.4	80
0.6	53
0.8	40
1.0	32

(a) Complete the table using the model.

x	0.2	0.4	0.6	0.8	1.0
y					

(b) Use a graphing utility to graph the data points and the model in the same viewing window. How do they compare?

(c) Use the model to estimate the distance traveled during impact if the passenger deceleration must not exceed 30 g's.

(d) Do you think it is practical to lower the number of g's experienced during impact to fewer than 23? Explain your reasoning.

138. DATA ANALYSIS An object at a temperature of 160°C was removed from a furnace and placed in a room at 20°C. The temperature T of the object was measured each hour h and recorded in the table. A model for the data is given by $T = 20[1 + 7(2^{-h})]$. The graph of this model is shown in the figure.

Hour, h	Temperature, T
0	160°
1	90°
2	56°
3	38°
4	29°
5	24°

(a) Use the graph to identify the horizontal asymptote of the model and interpret the asymptote in the context of the problem.

(b) Use the model to approximate the time when the temperature of the object was 100°C.

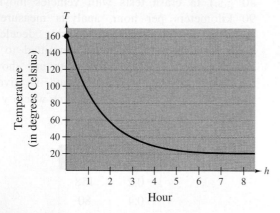

Hour

EXPLORATION

TRUE OR FALSE? In Exercises 139–142, rewrite each verbal statement as an equation. Then decide whether the statement is true or false. Justify your answer.

139. The logarithm of the product of two numbers is equal to the sum of the logarithms of the numbers.

140. The logarithm of the sum of two numbers is equal to the product of the logarithms of the numbers.

141. The logarithm of the difference of two numbers is equal to the difference of the logarithms of the numbers.

142. The logarithm of the quotient of two numbers is equal to the difference of the logarithms of the numbers.

143. THINK ABOUT IT Is it possible for a logarithmic equation to have more than one extraneous solution? Explain.

144. FINANCE You are investing P dollars at an annual interest rate of r, compounded continuously, for t years. Which of the following would result in the highest value of the investment? Explain your reasoning.

(a) Double the amount you invest.

(b) Double your interest rate.

(c) Double the number of years.

145. THINK ABOUT IT Are the times required for the investments in Exercises 117–120 to quadruple twice as long as the times for them to double? Give a reason for your answer and verify your answer algebraically.

146. The *effective yield* of a savings plan is the percent increase in the balance after 1 year. Find the effective yield for each savings plan when $1000 is deposited in a savings account. Which savings plan has the greatest effective yield? Which savings plan will have the highest balance after 5 years?

(a) 7% annual interest rate, compounded annually

(b) 7% annual interest rate, compounded continuously

(c) 7% annual interest rate, compounded quarterly

(d) 7.25% annual interest rate, compounded quarterly

147. GRAPHICAL ANALYSIS Let $f(x) = \log_a x$ and $g(x) = a^x$, where $a > 1$.

(a) Let $a = 1.2$ and use a graphing utility to graph the two functions in the same viewing window. What do you observe? Approximate any points of intersection of the two graphs.

(b) Determine the value(s) of a for which the two graphs have one point of intersection.

(c) Determine the value(s) of a for which the two graphs have two points of intersection.

148. CAPSTONE Write two or three sentences stating the general guidelines that you follow when solving (a) exponential equations and (b) logarithmic equations.

5.5 # EXPONENTIAL AND LOGARITHMIC MODELS

What you should learn

- Recognize the five most common types of models involving exponential and logarithmic functions.
- Use exponential growth and decay functions to model and solve real-life problems.
- Use Gaussian functions to model and solve real-life problems.
- Use logistic growth functions to model and solve real-life problems.
- Use logarithmic functions to model and solve real-life problems.

Why you should learn it

Exponential growth and decay models are often used to model the populations of countries. For instance, in Exercise 44 on page 423, you will use exponential growth and decay models to compare the populations of several countries.

Introduction

The five most common types of mathematical models involving exponential functions and logarithmic functions are as follows.

1. **Exponential growth model:** $y = ae^{bx}, \quad b > 0$

2. **Exponential decay model:** $y = ae^{-bx}, \quad b > 0$

3. **Gaussian model:** $y = ae^{-(x-b)^2/c}$

4. **Logistic growth model:** $y = \dfrac{a}{1 + be^{-rx}}$

5. **Logarithmic models:** $y = a + b \ln x, \quad y = a + b \log x$

The basic shapes of the graphs of these functions are shown in Figure 5.33.

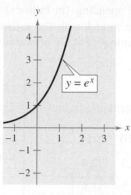

Exponential growth model

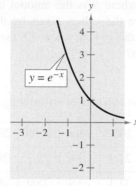

Exponential decay model

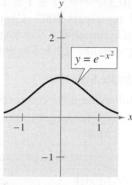

Gaussian model

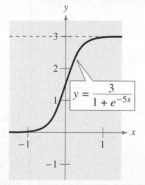

Logistic growth model

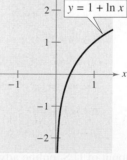

Natural logarithmic model

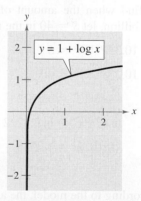

Common logarithmic model

FIGURE 5.33

You can often gain quite a bit of insight into a situation modeled by an exponential or logarithmic function by identifying and interpreting the function's asymptotes. Use the graphs in Figure 5.33 to identify the asymptotes of the graph of each function.

Exponential Growth and Decay

Example 1   Online Advertising

Estimates of the amounts (in billions of dollars) of U.S. online advertising spending from 2007 through 2011 are shown in the table. A scatter plot of the data is shown in Figure 5.34.   (Source: eMarketer)

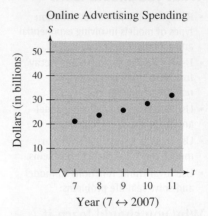

Online Advertising Spending

FIGURE 5.34

Year	Advertising spending
2007	21.1
2008	23.6
2009	25.7
2010	28.5
2011	32.0

An exponential growth model that approximates these data is given by $S = 10.33e^{0.1022t}$, $7 \le t \le 11$, where S is the amount of spending (in billions) and $t = 7$ represents 2007. Compare the values given by the model with the estimates shown in the table. According to this model, when will the amount of U.S. online advertising spending reach $40 billion?

Algebraic Solution

The following table compares the two sets of advertising spending figures.

Year	2007	2008	2009	2010	2011
Advertising spending	21.1	23.6	25.7	28.5	32.0
Model	21.1	23.4	25.9	28.7	31.8

To find when the amount of U.S. online advertising spending will reach $40 billion, let $S = 40$ in the model and solve for t.

$10.33e^{0.1022t} = S$	Write original model.
$10.33e^{0.1022t} = 40$	Substitute 40 for S.
$e^{0.1022t} \approx 3.8722$	Divide each side by 10.33.
$\ln e^{0.1022t} \approx \ln 3.8722$	Take natural log of each side.
$0.1022t \approx 1.3538$	Inverse Property
$t \approx 13.2$	Divide each side by 0.1022.

According to the model, the amount of U.S. online advertising spending will reach $40 billion in 2013.

CHECK*Point*   Now try Exercise 43.

Graphical Solution

Use a graphing utility to graph the model $y = 10.33e^{0.1022x}$ and the data in the same viewing window. You can see in Figure 5.35 that the model appears to fit the data closely.

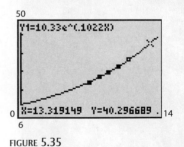

FIGURE 5.35

Use the *zoom* and *trace* features of the graphing utility to find that the approximate value of x for $y = 40$ is $x \approx 13.2$. So, according to the model, the amount of U.S. online advertising spending will reach $40 billion in 2013.

TECHNOLOGY

Some graphing utilities have an *exponential regression* feature that can be used to find exponential models that represent data. If you have such a graphing utility, try using it to find an exponential model for the data given in Example 1. How does your model compare with the model given in Example 1?

In Example 1, you were given the exponential growth model. But suppose this model were not given; how could you find such a model? One technique for doing this is demonstrated in Example 2.

Example 2   Modeling Population Growth

In a research experiment, a population of fruit flies is increasing according to the law of exponential growth. After 2 days there are 100 flies, and after 4 days there are 300 flies. How many flies will there be after 5 days?

Solution

Let y be the number of flies at time t. From the given information, you know that $y = 100$ when $t = 2$ and $y = 300$ when $t = 4$. Substituting this information into the model $y = ae^{bt}$ produces

$$100 = ae^{2b} \quad \text{and} \quad 300 = ae^{4b}.$$

To solve for b, solve for a in the first equation.

$$100 = ae^{2b} \quad \Longrightarrow \quad a = \frac{100}{e^{2b}} \qquad \text{Solve for } a \text{ in the first equation.}$$

Then substitute the result into the second equation.

$300 = ae^{4b}$	Write second equation.
$300 = \left(\dfrac{100}{e^{2b}}\right)e^{4b}$	Substitute $\dfrac{100}{e^{2b}}$ for a.
$\dfrac{300}{100} = e^{2b}$	Divide each side by 100.
$\ln 3 = 2b$	Take natural log of each side.
$\dfrac{1}{2}\ln 3 = b$	Solve for b.

Using $b = \frac{1}{2}\ln 3$ and the equation you found for a, you can determine that

$a = \dfrac{100}{e^{2[(1/2)\ln 3]}}$	Substitute $\frac{1}{2}\ln 3$ for b.
$= \dfrac{100}{e^{\ln 3}}$	Simplify.
$= \dfrac{100}{3}$	Inverse Property
$\approx 33.33.$	Simplify.

So, with $a \approx 33.33$ and $b = \frac{1}{2}\ln 3 \approx 0.5493$, the exponential growth model is

$$y = 33.33e^{0.5493t}$$

as shown in Figure 5.36. This implies that, after 5 days, the population will be

$$y = 33.33e^{0.5493(5)} \approx 520 \text{ flies.}$$

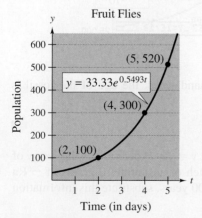

Fruit Flies

$y = 33.33e^{0.5493t}$

(5, 520)

(4, 300)

(2, 100)

Time (in days)

FIGURE 5.36

CHECK **Point**   Now try Exercise 49.

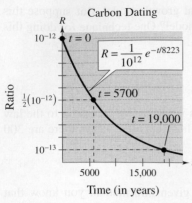

Carbon Dating

$R = \dfrac{1}{10^{12}}e^{-t/8223}$

FIGURE 5.37

In living organic material, the ratio of the number of radioactive carbon isotopes (carbon 14) to the number of nonradioactive carbon isotopes (carbon 12) is about 1 to 10^{12}. When organic material dies, its carbon 12 content remains fixed, whereas its radioactive carbon 14 begins to decay with a half-life of about 5700 years. To estimate the age of dead organic material, scientists use the following formula, which denotes the ratio of carbon 14 to carbon 12 present at any time t (in years).

$$R = \frac{1}{10^{12}}e^{-t/8223} \qquad \text{Carbon dating model}$$

The graph of R is shown in Figure 5.37. Note that R decreases as t increases.

Example 3   Carbon Dating

Estimate the age of a newly discovered fossil in which the ratio of carbon 14 to carbon 12 is

$R = 1/10^{13}$.

Algebraic Solution

In the carbon dating model, substitute the given value of R to obtain the following.

$$\frac{1}{10^{12}}e^{-t/8223} = R \qquad \text{Write original model.}$$

$$\frac{e^{-t/8223}}{10^{12}} = \frac{1}{10^{13}} \qquad \text{Let } R = \frac{1}{10^{13}}.$$

$$e^{-t/8223} = \frac{1}{10} \qquad \text{Multiply each side by } 10^{12}.$$

$$\ln e^{-t/8223} = \ln\frac{1}{10} \qquad \text{Take natural log of each side.}$$

$$-\frac{t}{8223} \approx -2.3026 \qquad \text{Inverse Property}$$

$$t \approx 18{,}934 \qquad \text{Multiply each side by } -8223.$$

So, to the nearest thousand years, the age of the fossil is about 19,000 years.

Graphical Solution

Use a graphing utility to graph the formula for the ratio of carbon 14 to carbon 12 at any time t as

$$y_1 = \frac{1}{10^{12}}e^{-x/8223}.$$

In the same viewing window, graph $y_2 = 1/(10^{13})$. Use the *intersect* feature or the *zoom* and *trace* features of the graphing utility to estimate that $x \approx 18{,}934$ when $y = 1/(10^{13})$, as shown in Figure 5.38.

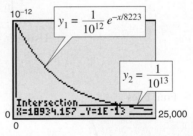

$y_1 = \dfrac{1}{10^{12}}e^{-x/8223}$

$y_2 = \dfrac{1}{10^{13}}$

FIGURE 5.38

So, to the nearest thousand years, the age of the fossil is about 19,000 years.

CHECK Point   Now try Exercise 51.

The value of b in the exponential decay model $y = ae^{-bt}$ determines the *decay* of radioactive isotopes. For instance, to find how much of an initial 10 grams of ^{226}Ra isotope with a half-life of 1599 years is left after 500 years, substitute this information into the model $y = ae^{-bt}$.

$$\frac{1}{2}(10) = 10e^{-b(1599)} \quad \Longrightarrow \quad \ln\frac{1}{2} = -1599b \quad \Longrightarrow \quad b = -\frac{\ln\frac{1}{2}}{1599}$$

Using the value of b found above and $a = 10$, the amount left is

$$y = 10e^{-[-\ln(1/2)/1599](500)} \approx 8.05 \text{ grams.}$$

Gaussian Models

As mentioned at the beginning of this section, Gaussian models are of the form

$$y = ae^{-(x-b)^2/c}.$$

This type of model is commonly used in probability and statistics to represent populations that are **normally distributed.** The graph of a Gaussian model is called a **bell-shaped curve.** Try graphing the normal distribution with a graphing utility. Can you see why it is called a bell-shaped curve?

For *standard* normal distributions, the model takes the form

$$y = \frac{1}{\sqrt{2\pi}}e^{-x^2/2}.$$

The **average value** of a population can be found from the bell-shaped curve by observing where the maximum y-value of the function occurs. The x-value corresponding to the maximum y-value of the function represents the average value of the independent variable—in this case, x.

Example 4   SAT Scores

In 2008, the Scholastic Aptitude Test (SAT) math scores for college-bound seniors roughly followed the normal distribution given by

$$y = 0.0034e^{-(x-515)^2/26,912}, \quad 200 \le x \le 800$$

where x is the SAT score for mathematics. Sketch the graph of this function. From the graph, estimate the average SAT score.   (Source: College Board)

Solution

The graph of the function is shown in Figure 5.39. On this bell-shaped curve, the maximum value of the curve represents the average score. From the graph, you can estimate that the average mathematics score for college-bound seniors in 2008 was 515.

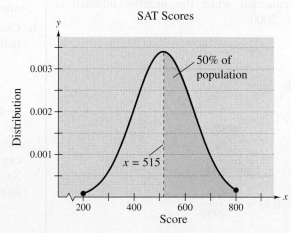

SAT Scores

FIGURE 5.39

CHECK Point ▶   Now try Exercise 57.

Logistic Growth Models

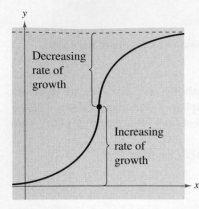

FIGURE 5.40

Some populations initially have rapid growth, followed by a declining rate of growth, as indicated by the graph in Figure 5.40. One model for describing this type of growth pattern is the **logistic curve** given by the function

$$y = \frac{a}{1 + be^{-rx}}$$

where y is the population size and x is the time. An example is a bacteria culture that is initially allowed to grow under ideal conditions, and then under less favorable conditions that inhibit growth. A logistic growth curve is also called a **sigmoidal curve.**

Example 5 Spread of a Virus

On a college campus of 5000 students, one student returns from vacation with a contagious and long-lasting flu virus. The spread of the virus is modeled by

$$y = \frac{5000}{1 + 4999e^{-0.8t}}, \quad t \geq 0$$

where y is the total number of students infected after t days. The college will cancel classes when 40% or more of the students are infected.

a. How many students are infected after 5 days?

b. After how many days will the college cancel classes?

Algebraic Solution

a. After 5 days, the number of students infected is

$$y = \frac{5000}{1 + 4999e^{-0.8(5)}} = \frac{5000}{1 + 4999e^{-4}} \approx 54.$$

b. Classes are canceled when the number infected is $(0.40)(5000) = 2000$.

$$2000 = \frac{5000}{1 + 4999e^{-0.8t}}$$

$$1 + 4999e^{-0.8t} = 2.5$$

$$e^{-0.8t} = \frac{1.5}{4999}$$

$$\ln e^{-0.8t} = \ln \frac{1.5}{4999}$$

$$-0.8t = \ln \frac{1.5}{4999}$$

$$t = -\frac{1}{0.8} \ln \frac{1.5}{4999}$$

$$t \approx 10.1$$

So, after about 10 days, at least 40% of the students will be infected, and the college will cancel classes.

Graphical Solution

a. Use a graphing utility to graph $y = \dfrac{5000}{1 + 4999e^{-0.8x}}$. Use the *value* feature or the *zoom* and *trace* features of the graphing utility to estimate that $y \approx 54$ when $x = 5$. So, after 5 days, about 54 students will be infected.

b. Classes are canceled when the number of infected students is $(0.40)(5000) = 2000$. Use a graphing utility to graph

$$y_1 = \frac{5000}{1 + 4999e^{-0.8x}} \quad \text{and} \quad y_2 = 2000$$

in the same viewing window. Use the *intersect* feature or the *zoom* and *trace* features of the graphing utility to find the point of intersection of the graphs. In Figure 5.41, you can see that the point of intersection occurs near $x \approx 10.1$. So, after about 10 days, at least 40% of the students will be infected, and the college will cancel classes.

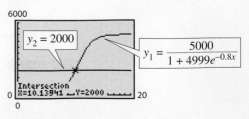

FIGURE 5.41

CHECK*Point* Now try Exercise 59.

On May 12, 2008, an earthquake of magnitude 7.9 struck Eastern Sichuan Province, China. The total economic loss was estimated at 86 billion U.S. dollars.

Logarithmic Models

| Example 6 | Magnitudes of Earthquakes |

On the Richter scale, the magnitude R of an earthquake of intensity I is given by

$$R = \log \frac{I}{I_0}$$

where $I_0 = 1$ is the minimum intensity used for comparison. Find the intensity of each earthquake. (Intensity is a measure of the wave energy of an earthquake.)

a. Nevada in 2008: $R = 6.0$

b. Eastern Sichuan, China in 2008: $R = 7.9$

Solution

a. Because $I_0 = 1$ and $R = 6.0$, you have

$$6.0 = \log \frac{I}{1}$$ Substitute 1 for I_0 and 6.0 for R.

$$10^{6.0} = 10^{\log I}$$ Exponentiate each side.

$$I = 10^{6.0} = 1{,}000{,}000.$$ Inverse Property

b. For $R = 7.9$, you have

$$7.9 = \log \frac{I}{1}$$ Substitute 1 for I_0 and 7.9 for R.

$$10^{7.9} = 10^{\log I}$$ Exponentiate each side.

$$I = 10^{7.9} \approx 79{,}400{,}000.$$ Inverse Property

Note that an increase of 1.9 units on the Richter scale (from 6.0 to 7.9) represents an increase in intensity by a factor of

$$\frac{79{,}400{,}000}{1{,}000{,}000} = 79.4.$$

In other words, the intensity of the earthquake in Eastern Sichuan was about 79 times as great as that of the earthquake in Nevada.

CHECK Point Now try Exercise 63.

CLASSROOM DISCUSSION

Comparing Population Models The populations P (in millions) of the United States for the census years from 1910 to 2000 are shown in the table at the left. Least squares regression analysis gives the best quadratic model for these data as $P = 1.0328t^2 + 9.607t + 81.82$, and the best exponential model for these data as $P = 82.677e^{0.124t}$. Which model better fits the data? Describe how you reached your conclusion.   (Source: U.S. Census Bureau)

t	Year	Population, P
1	1910	92.23
2	1920	106.02
3	1930	123.20
4	1940	132.16
5	1950	151.33
6	1960	179.32
7	1970	203.30
8	1980	226.54
9	1990	248.72
10	2000	281.42

5.5 EXERCISES

See www.CalcChat.com for worked-out solutions to odd-numbered exercises.

VOCABULARY: Fill in the blanks.

1. An exponential growth model has the form _____ and an exponential decay model has the form _____.
2. A logarithmic model has the form _____ or _____.
3. Gaussian models are commonly used in probability and statistics to represent populations that are _____ _____.
4. The graph of a Gaussian model is _____ shaped, where the _____ _____ is the maximum y-value of the graph.
5. A logistic growth model has the form _____.
6. A logistic curve is also called a _____ curve.

SKILLS AND APPLICATIONS

In Exercises 7–12, match the function with its graph. [The graphs are labeled (a), (b), (c), (d), (e), and (f).]

(a)
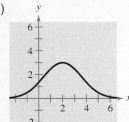

(b)
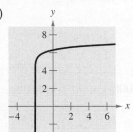

(c)
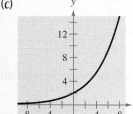

(d)
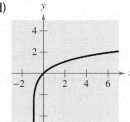

(e)
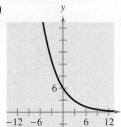

(f)
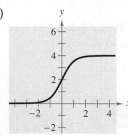

7. $y = 2e^{x/4}$
8. $y = 6e^{-x/4}$
9. $y = 6 + \log(x + 2)$
10. $y = 3e^{-(x-2)^2/5}$
11. $y = \ln(x + 1)$
12. $y = \dfrac{4}{1 + e^{-2x}}$

In Exercises 13 and 14, (a) solve for P and (b) solve for t.

13. $A = Pe^{rt}$
14. $A = P\left(1 + \dfrac{r}{n}\right)^{nt}$

COMPOUND INTEREST In Exercises 15–22, complete the table for a savings account in which interest is compounded continuously.

	Initial Investment	Annual % Rate	Time to Double	Amount After 10 Years
15.	$1000	3.5%		
16.	$750	$10\frac{1}{2}$%		
17.	$750		$7\frac{3}{4}$ yr	
18.	$10,000		12 yr	
19.	$500			$1505.00
20.	$600			$19,205.00
21.		4.5%		$10,000.00
22.		2%		$2000.00

COMPOUND INTEREST In Exercises 23 and 24, determine the principal P that must be invested at rate r, compounded monthly, so that $500,000 will be available for retirement in t years.

23. $r = 5\%, t = 10$
24. $r = 3\frac{1}{2}\%, t = 15$

COMPOUND INTEREST In Exercises 25 and 26, determine the time necessary for $1000 to double if it is invested at interest rate r compounded (a) annually, (b) monthly, (c) daily, and (d) continuously.

25. $r = 10\%$
26. $r = 6.5\%$

27. **COMPOUND INTEREST** Complete the table for the time t (in years) necessary for P dollars to triple if interest is compounded continuously at rate r.

r	2%	4%	6%	8%	10%	12%
t						

 28. **MODELING DATA** Draw a scatter plot of the data in Exercise 27. Use the *regression* feature of a graphing utility to find a model for the data.

29. COMPOUND INTEREST Complete the table for the time t (in years) necessary for P dollars to triple if interest is compounded annually at rate r.

r	2%	4%	6%	8%	10%	12%
t						

30. MODELING DATA Draw a scatter plot of the data in Exercise 29. Use the *regression* feature of a graphing utility to find a model for the data.

31. COMPARING MODELS If \$1 is invested in an account over a 10-year period, the amount in the account, where t represents the time in years, is given by $A = 1 + 0.075[\![t]\!]$ or $A = e^{0.07t}$ depending on whether the account pays simple interest at $7\frac{1}{2}\%$ or continuous compound interest at 7%. Graph each function on the same set of axes. Which grows at a higher rate? (Remember that $[\![t]\!]$ is the greatest integer function discussed in Section 2.4.)

32. COMPARING MODELS If \$1 is invested in an account over a 10-year period, the amount in the account, where t represents the time in years, is given by $A = 1 + 0.06[\![t]\!]$ or $A = [1 + (0.055/365)]^{[\![365t]\!]}$ depending on whether the account pays simple interest at 6% or compound interest at $5\frac{1}{2}\%$ compounded daily. Use a graphing utility to graph each function in the same viewing window. Which grows at a higher rate?

RADIOACTIVE DECAY In Exercises 33–38, complete the table for the radioactive isotope.

Isotope	Half-life (years)	Initial Quantity	Amount After 1000 Years
33. ^{226}Ra	1599	10 g	
34. ^{14}C	5715	6.5 g	
35. ^{239}Pu	24,100	2.1 g	
36. ^{226}Ra	1599		2 g
37. ^{14}C	5715		2 g
38. ^{239}Pu	24,100		0.4 g

In Exercises 39–42, find the exponential model $y = ae^{bx}$ that fits the points shown in the graph or table.

39.
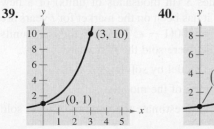

40.
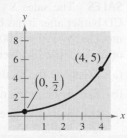

41.

x	0	4
y	5	1

42.

x	0	3
y	1	$\frac{1}{4}$

43. POPULATION The populations P (in thousands) of Horry County, South Carolina from 1970 through 2007 can be modeled by

$$P = -18.5 + 92.2e^{0.0282t}$$

where t represents the year, with $t = 0$ corresponding to 1970. (Source: U.S. Census Bureau)

(a) Use the model to complete the table.

Year	1970	1980	1990	2000	2007
Population					

(b) According to the model, when will the population of Horry County reach 300,000?

(c) Do you think the model is valid for long-term predictions of the population? Explain.

44. POPULATION The table shows the populations (in millions) of five countries in 2000 and the projected populations (in millions) for the year 2015. (Source: U.S. Census Bureau)

Country	2000	2015
Bulgaria	7.8	6.9
Canada	31.1	35.1
China	1268.9	1393.4
United Kingdom	59.5	62.2
United States	282.2	325.5

(a) Find the exponential growth or decay model $y = ae^{bt}$ or $y = ae^{-bt}$ for the population of each country by letting $t = 0$ correspond to 2000. Use the model to predict the population of each country in 2030.

(b) You can see that the populations of the United States and the United Kingdom are growing at different rates. What constant in the equation $y = ae^{bt}$ is determined by these different growth rates? Discuss the relationship between the different growth rates and the magnitude of the constant.

(c) You can see that the population of China is increasing while the population of Bulgaria is decreasing. What constant in the equation $y = ae^{bt}$ reflects this difference? Explain.

45. WEBSITE GROWTH The number y of hits a new search-engine website receives each month can be modeled by $y = 4080e^{kt}$, where t represents the number of months the website has been operating. In the website's third month, there were 10,000 hits. Find the value of k, and use this value to predict the number of hits the website will receive after 24 months.

46. VALUE OF A PAINTING The value V (in millions of dollars) of a famous painting can be modeled by $V = 10e^{kt}$, where t represents the year, with $t = 0$ corresponding to 2000. In 2008, the same painting was sold for $65 million. Find the value of k, and use this value to predict the value of the painting in 2014.

47. POPULATION The populations P (in thousands) of Reno, Nevada from 2000 through 2007 can be modeled by $P = 346.8e^{kt}$, where t represents the year, with $t = 0$ corresponding to 2000. In 2005, the population of Reno was about 395,000. (Source: U.S. Census Bureau)

(a) Find the value of k. Is the population increasing or decreasing? Explain.

(b) Use the model to find the populations of Reno in 2010 and 2015. Are the results reasonable? Explain.

(c) According to the model, during what year will the population reach 500,000?

48. POPULATION The populations P (in thousands) of Orlando, Florida from 2000 through 2007 can be modeled by $P = 1656.2e^{kt}$, where t represents the year, with $t = 0$ corresponding to 2000. In 2005, the population of Orlando was about 1,940,000. (Source: U.S. Census Bureau)

(a) Find the value of k. Is the population increasing or decreasing? Explain.

(b) Use the model to find the populations of Orlando in 2010 and 2015. Are the results reasonable? Explain.

(c) According to the model, during what year will the population reach 2.2 million?

49. BACTERIA GROWTH The number of bacteria in a culture is increasing according to the law of exponential growth. After 3 hours, there are 100 bacteria, and after 5 hours, there are 400 bacteria. How many bacteria will there be after 6 hours?

50. BACTERIA GROWTH The number of bacteria in a culture is increasing according to the law of exponential growth. The initial population is 250 bacteria, and the population after 10 hours is double the population after 1 hour. How many bacteria will there be after 6 hours?

51. CARBON DATING

(a) The ratio of carbon 14 to carbon 12 in a piece of wood discovered in a cave is $R = 1/8^{14}$. Estimate the age of the piece of wood.

(b) The ratio of carbon 14 to carbon 12 in a piece of paper buried in a tomb is $R = 1/13^{11}$. Estimate the age of the piece of paper.

52. RADIOACTIVE DECAY Carbon 14 dating assumes that the carbon dioxide on Earth today has the same radioactive content as it did centuries ago. If this is true, the amount of ^{14}C absorbed by a tree that grew several centuries ago should be the same as the amount of ^{14}C absorbed by a tree growing today. A piece of ancient charcoal contains only 15% as much radioactive carbon as a piece of modern charcoal. How long ago was the tree burned to make the ancient charcoal if the half-life of ^{14}C is 5715 years?

53. DEPRECIATION A sport utility vehicle that costs $23,300 new has a book value of $12,500 after 2 years.

(a) Find the linear model $V = mt + b$.

(b) Find the exponential model $V = ae^{kt}$.

(c) Use a graphing utility to graph the two models in the same viewing window. Which model depreciates faster in the first 2 years?

(d) Find the book values of the vehicle after 1 year and after 3 years using each model.

(e) Explain the advantages and disadvantages of using each model to a buyer and a seller.

54. DEPRECIATION A laptop computer that costs $1150 new has a book value of $550 after 2 years.

(a) Find the linear model $V = mt + b$.

(b) Find the exponential model $V = ae^{kt}$.

(c) Use a graphing utility to graph the two models in the same viewing window. Which model depreciates faster in the first 2 years?

(d) Find the book values of the computer after 1 year and after 3 years using each model.

(e) Explain the advantages and disadvantages of using each model to a buyer and a seller.

55. SALES The sales S (in thousands of units) of a new CD burner after it has been on the market for t years are modeled by $S(t) = 100(1 - e^{kt})$. Fifteen thousand units of the new product were sold the first year.

(a) Complete the model by solving for k.

(b) Sketch the graph of the model.

(c) Use the model to estimate the number of units sold after 5 years.

56. LEARNING CURVE The management at a plastics factory has found that the maximum number of units a worker can produce in a day is 30. The learning curve for the number N of units produced per day after a new employee has worked t days is modeled by $N = 30(1 - e^{kt})$. After 20 days on the job, a new employee produces 19 units.

(a) Find the learning curve for this employee (first, find the value of k).

(b) How many days should pass before this employee is producing 25 units per day?

57. IQ SCORES The IQ scores for a sample of a class of returning adult students at a small northeastern college roughly follow the normal distribution $y = 0.0266e^{-(x-100)^2/450}$, $70 \leq x \leq 115$, where x is the IQ score.

(a) Use a graphing utility to graph the function.

(b) From the graph in part (a), estimate the average IQ score of an adult student.

58. EDUCATION The amount of time (in hours per week) a student utilizes a math-tutoring center roughly follows the normal distribution $y = 0.7979e^{-(x-5.4)^2/0.5}$, $4 \leq x \leq 7$, where x is the number of hours.

(a) Use a graphing utility to graph the function.

(b) From the graph in part (a), estimate the average number of hours per week a student uses the tutoring center.

59. CELL SITES A cell site is a site where electronic communications equipment is placed in a cellular network for the use of mobile phones. The numbers y of cell sites from 1985 through 2008 can be modeled by

$$y = \frac{237{,}101}{1 + 1950e^{-0.355t}}$$

where t represents the year, with $t = 5$ corresponding to 1985. (Source: CTIA-The Wireless Association)

(a) Use the model to find the numbers of cell sites in the years 1985, 2000, and 2006.

(b) Use a graphing utility to graph the function.

(c) Use the graph to determine the year in which the number of cell sites will reach 235,000.

(d) Confirm your answer to part (c) algebraically.

60. POPULATION The populations P (in thousands) of Pittsburgh, Pennsylvania from 2000 through 2007 can be modeled by

$$P = \frac{2632}{1 + 0.083e^{0.0500t}}$$

where t represents the year, with $t = 0$ corresponding to 2000. (Source: U.S. Census Bureau)

(a) Use the model to find the populations of Pittsburgh in the years 2000, 2005, and 2007.

(b) Use a graphing utility to graph the function.

(c) Use the graph to determine the year in which the population will reach 2.2 million.

(d) Confirm your answer to part (c) algebraically.

61. POPULATION GROWTH A conservation organization releases 100 animals of an endangered species into a game preserve. The organization believes that the preserve has a carrying capacity of 1000 animals and that the growth of the pack will be modeled by the logistic curve

$$p(t) = \frac{1000}{1 + 9e^{-0.1656t}}$$

where t is measured in months (see figure).

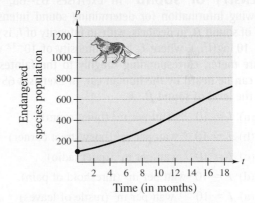

(a) Estimate the population after 5 months.

(b) After how many months will the population be 500?

(c) Use a graphing utility to graph the function. Use the graph to determine the horizontal asymptotes, and interpret the meaning of the asymptotes in the context of the problem.

62. SALES After discontinuing all advertising for a tool kit in 2004, the manufacturer noted that sales began to drop according to the model

$$S = \frac{500{,}000}{1 + 0.4e^{kt}}$$

where S represents the number of units sold and $t = 4$ represents 2004. In 2008, the company sold 300,000 units.

(a) Complete the model by solving for k.

(b) Estimate sales in 2012.

GEOLOGY In Exercises 63 and 64, use the Richter scale

$$R = \log \frac{I}{I_0}$$

for measuring the magnitudes of earthquakes.

63. Find the intensity I of an earthquake measuring R on the Richter scale (let $I_0 = 1$).

(a) Southern Sumatra, Indonesia in 2007, $R = 8.5$

(b) Illinois in 2008, $R = 5.4$

(c) Costa Rica in 2009, $R = 6.1$

64. Find the magnitude R of each earthquake of intensity I (let $I_0 = 1$).

(a) $I = 199,500,000$ (b) $I = 48,275,000$

(c) $I = 17,000$

INTENSITY OF SOUND In Exercises 65–68, use the following information for determining sound intensity. The level of sound β, in decibels, with an intensity of I, is given by $\beta = 10 \log(I/I_0)$, where I_0 is an intensity of 10^{-12} watt per square meter, corresponding roughly to the faintest sound that can be heard by the human ear. In Exercises 65 and 66, find the level of sound β.

65. (a) $I = 10^{-10}$ watt per m^2 (quiet room)

(b) $I = 10^{-5}$ watt per m^2 (busy street corner)

(c) $I = 10^{-8}$ watt per m^2 (quiet radio)

(d) $I = 10^{0}$ watt per m^2 (threshold of pain)

66. (a) $I = 10^{-11}$ watt per m^2 (rustle of leaves)

(b) $I = 10^{2}$ watt per m^2 (jet at 30 meters)

(c) $I = 10^{-4}$ watt per m^2 (door slamming)

(d) $I = 10^{-2}$ watt per m^2 (siren at 30 meters)

67. Due to the installation of noise suppression materials, the noise level in an auditorium was reduced from 93 to 80 decibels. Find the percent decrease in the intensity level of the noise as a result of the installation of these materials.

68. Due to the installation of a muffler, the noise level of an engine was reduced from 88 to 72 decibels. Find the percent decrease in the intensity level of the noise as a result of the installation of the muffler.

pH LEVELS In Exercises 69–74, use the acidity model given by pH $= -\log[\text{H}^+]$, where acidity (pH) is a measure of the hydrogen ion concentration $[\text{H}^+]$ (measured in moles of hydrogen per liter) of a solution.

69. Find the pH if $[\text{H}^+] = 2.3 \times 10^{-5}$.

70. Find the pH if $[\text{H}^+] = 1.13 \times 10^{-5}$.

71. Compute $[\text{H}^+]$ for a solution in which pH $= 5.8$.

72. Compute $[\text{H}^+]$ for a solution in which pH $= 3.2$.

73. Apple juice has a pH of 2.9 and drinking water has a pH of 8.0. The hydrogen ion concentration of the apple juice is how many times the concentration of drinking water?

74. The pH of a solution is decreased by one unit. The hydrogen ion concentration is increased by what factor?

75. **FORENSICS** At 8:30 A.M., a coroner was called to the home of a person who had died during the night. In order to estimate the time of death, the coroner took the person's temperature twice. At 9:00 A.M. the temperature was 85.7°F, and at 11:00 A.M. the temperature was 82.8°F. From these two temperatures, the coroner was able to determine that the time elapsed since death and the body temperature were related by the formula

$$t = -10 \ln \frac{T - 70}{98.6 - 70}$$

where t is the time in hours elapsed since the person died and T is the temperature (in degrees Fahrenheit) of the person's body. (This formula is derived from a general cooling principle called *Newton's Law of Cooling*. It uses the assumptions that the person had a normal body temperature of 98.6°F at death, and that the room temperature was a constant 70°F.) Use the formula to estimate the time of death of the person.

 76. **HOME MORTGAGE** A $120,000 home mortgage for 30 years at $7\frac{1}{2}\%$ has a monthly payment of $839.06. Part of the monthly payment is paid toward the interest charge on the unpaid balance, and the remainder of the payment is used to reduce the principal. The amount that is paid toward the interest is

$$u = M - \left(M - \frac{Pr}{12} \right)\left(1 + \frac{r}{12} \right)^{12t}$$

and the amount that is paid toward the reduction of the principal is

$$v = \left(M - \frac{Pr}{12} \right)\left(1 + \frac{r}{12} \right)^{12t}.$$

In these formulas, P is the size of the mortgage, r is the interest rate, M is the monthly payment, and t is the time (in years).

(a) Use a graphing utility to graph each function in the same viewing window. (The viewing window should show all 30 years of mortgage payments.)

(b) In the early years of the mortgage, is the larger part of the monthly payment paid toward the interest or the principal? Approximate the time when the monthly payment is evenly divided between interest and principal reduction.

(c) Repeat parts (a) and (b) for a repayment period of 20 years ($M = \$966.71$). What can you conclude?

77. HOME MORTGAGE The total interest u paid on a home mortgage of P dollars at interest rate r for t years is

$$u = P\left[\frac{rt}{1 - \left(\dfrac{1}{1 + r/12}\right)^{12t}} - 1\right].$$

Consider a \$120,000 home mortgage at $7\frac{1}{2}\%$.

 (a) Use a graphing utility to graph the total interest function.

(b) Approximate the length of the mortgage for which the total interest paid is the same as the size of the mortgage. Is it possible that some people are paying twice as much in interest charges as the size of the mortgage?

 78. DATA ANALYSIS The table shows the time t (in seconds) required for a car to attain a speed of s miles per hour from a standing start.

Speed, s	Time, t
30	3.4
40	5.0
50	7.0
60	9.3
70	12.0
80	15.8
90	20.0

Two models for these data are as follows.

$$t_1 = 40.757 + 0.556s - 15.817 \ln s$$

$$t_2 = 1.2259 + 0.0023s^2$$

(a) Use the *regression* feature of a graphing utility to find a linear model t_3 and an exponential model t_4 for the data.

(b) Use a graphing utility to graph the data and each model in the same viewing window.

(c) Create a table comparing the data with estimates obtained from each model.

(d) Use the results of part (c) to find the sum of the absolute values of the differences between the data and the estimated values given by each model. Based on the four sums, which model do you think best fits the data? Explain.

EXPLORATION

TRUE OR FALSE? In Exercises 79–82, determine whether the statement is true or false. Justify your answer.

79. The domain of a logistic growth function cannot be the set of real numbers.

80. A logistic growth function will always have an x-intercept.

81. The graph of $f(x) = \dfrac{4}{1 + 6e^{-2x}} + 5$ is the graph of $g(x) = \dfrac{4}{1 + 6e^{-2x}}$ shifted to the right five units.

82. The graph of a Gaussian model will never have an x-intercept.

83. WRITING Use your school's library, the Internet, or some other reference source to write a paper describing John Napier's work with logarithms.

84. CAPSTONE Identify each model as exponential, Gaussian, linear, logarithmic, logistic, quadratic, or none of the above. Explain your reasoning.

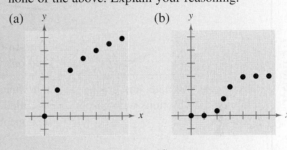

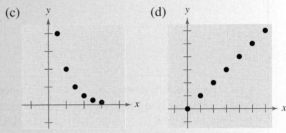

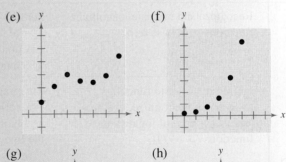

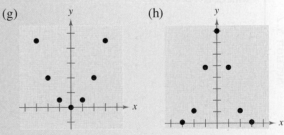

PROJECT: SALES PER SHARE To work an extended application analyzing the sales per share for Kohl's Corporation from 1992 through 2007, visit this text's website at *academic.cengage.com*. (Data Source: Kohl's Corporation)

5 CHAPTER SUMMARY

What Did You Learn?	Explanation/Examples	Review Exercises
Recognize and evaluate exponential functions with base a (p. 376).	The exponential function f with base a is denoted by $f(x) = a^x$ where $a > 0$, $a \neq 1$, and x is any real number.	1–6
Graph exponential functions and use the One-to-One Property (p. 377).	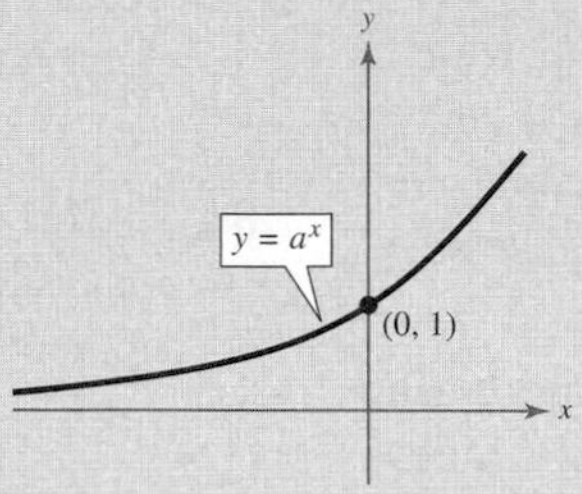 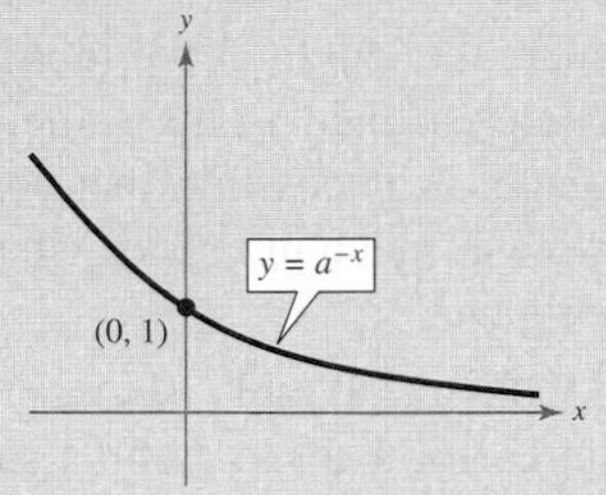 **One-to-One Property:** For $a > 0$ and $a \neq 1$, $a^x = a^y$ if and only if $x = y$.	7–24
Recognize, evaluate, and graph exponential functions with base e (p. 380).	The function $f(x) = e^x$ is called the natural exponential function. 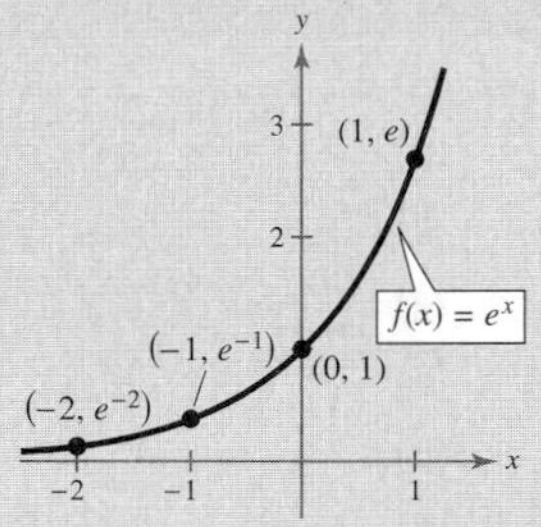	25–32
Use exponential functions to model and solve real-life problems (p. 381).	Exponential functions are used in compound interest formulas (See Example 8.) and in radioactive decay models. (See Example 9.)	33–36
Recognize and evaluate logarithmic functions with base a (p. 387).	For $x > 0$, $a > 0$, and $a \neq 1$, $y = \log_a x$ if and only if $x = a^y$. The function $f(x) = \log_a x$ is called the logarithmic function with base a. The logarithmic function with base 10 is the common logarithmic function. It is denoted by $\log_{10}$ or log.	37–48
Graph logarithmic functions (p. 389) and recognize, evaluate, and graph natural logarithmic functions (p. 391).	The graph of $y = \log_a x$ is a reflection of the graph of $y = a^x$ about the line $y = x$. The function defined by $f(x) = \ln x$, $x > 0$, is called the natural logarithmic function. Its graph is a reflection of the graph of $f(x) = e^x$ about the line $y = x$. 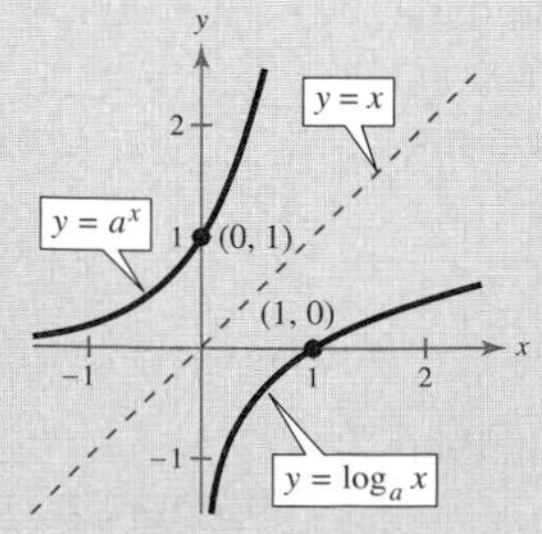 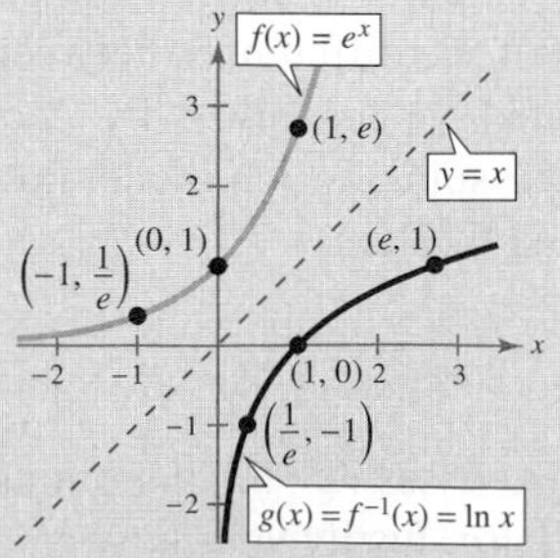	49–52 53–58
Use logarithmic functions to model and solve real-life problems (p. 393).	A logarithmic function is used in the human memory model. (See Example 11.)	59, 60

Section 5.1

Section 5.2

What Did You Learn?	Explanation/Examples	Review Exercises
Section 5.3 — Use the change-of-base formula to rewrite and evaluate logarithmic expressions (*p. 397*).	Let a, b, and x be positive real numbers such that $a \neq 1$ and $b \neq 1$. Then $\log_a x$ can be converted to a different base as follows. *Base b* $\log_a x = \dfrac{\log_b x}{\log_b a}$ *Base 10* $\log_a x = \dfrac{\log x}{\log a}$ *Base e* $\log_a x = \dfrac{\ln x}{\ln a}$	61–64
Use properties of logarithms to evaluate, rewrite, expand, or condense logarithmic expressions (*p. 398*).	Let a be a positive number ($a \neq 1$), n be a real number, and u and v be positive real numbers. **1. Product Property:** $\log_a(uv) = \log_a u + \log_a v$ $\ln(uv) = \ln u + \ln v$ **2. Quotient Property:** $\log_a(u/v) = \log_a u - \log_a v$ $\ln(u/v) = \ln u - \ln v$ **3. Power Property:** $\log_a u^n = n \log_a u$, $\ln u^n = n \ln u$	65–80
Use logarithmic functions to model and solve real-life problems (*p. 400*).	Logarithmic functions can be used to find an equation that relates the periods of several planets and their distances from the sun. (See Example 7.)	81, 82
Section 5.4 — Solve simple exponential and logarithmic equations (*p. 404*).	One-to-One Properties and Inverse Properties of exponential or logarithmic functions can be used to help solve exponential or logarithmic equations.	83–88
Solve more complicated exponential equations (*p. 405*) and logarithmic equations (*p. 407*).	To solve more complicated equations, rewrite the equations so that the One-to-One Properties and Inverse Properties of exponential or logarithmic functions can be used. (See Examples 2–8.)	89–108
Use exponential and logarithmic equations to model and solve real-life problems (*p. 409*).	Exponential and logarithmic equations can be used to find how long it will take to double an investment (see Example 10) and to find the year in which companies reached a given amount of sales. (See Example 11.)	109, 110
Section 5.5 — Recognize the five most common types of models involving exponential and logarithmic functions (*p. 415*).	**1. Exponential growth model:** $y = ae^{bx}$, $b > 0$ **2. Exponential decay model:** $y = ae^{-bx}$, $b > 0$ **3. Gaussian model:** $y = ae^{-(x-b)^2/c}$ **4. Logistic growth model:** $y = \dfrac{a}{1 + be^{-rx}}$ **5. Logarithmic models:** $y = a + b \ln x$, $y = a + b \log x$	111–116
Use exponential growth and decay functions to model and solve real-life problems (*p. 416*).	An exponential growth function can be used to model a population of fruit flies (see Example 2) and an exponential decay function can be used to find the age of a fossil (see Example 3).	117–120
Use Gaussian functions (*p. 419*), logistic growth functions (*p. 420*), and logarithmic functions (*p. 421*) to model and solve real-life problems.	A Gaussian function can be used to model SAT math scores for college-bound seniors. (See Example 4.) A logistic growth function can be used to model the spread of a flu virus. (See Example 5.) A logarithmic function can be used to find the intensity of an earthquake using its magnitude. (See Example 6.)	121–123

5 REVIEW EXERCISES

See www.CalcChat.com for worked-out solutions to odd-numbered exercises.

5.1 In Exercises 1–6, evaluate the function at the indicated value of x. Round your result to three decimal places.

1. $f(x) = 0.3^x$, $x = 1.5$
2. $f(x) = 30^x$, $x = \sqrt{3}$
3. $f(x) = 2^{-0.5x}$, $x = \pi$
4. $f(x) = 1278^{x/5}$, $x = 1$
5. $f(x) = 7(0.2^x)$, $x = -\sqrt{11}$
6. $f(x) = -14(5^x)$, $x = -0.8$

In Exercises 7–14, use the graph of f to describe the transformation that yields the graph of g.

7. $f(x) = 2^x$, $g(x) = 2^x - 2$
8. $f(x) = 5^x$, $g(x) = 5^x + 1$
9. $f(x) = 4^x$, $g(x) = 4^{-x+2}$
10. $f(x) = 6^x$, $g(x) = 6^{x+1}$
11. $f(x) = 3^x$, $g(x) = 1 - 3^x$
12. $f(x) = 0.1^x$, $g(x) = -0.1^x$
13. $f(x) = \left(\frac{1}{2}\right)^x$, $g(x) = -\left(\frac{1}{2}\right)^{x+2}$
14. $f(x) = \left(\frac{2}{3}\right)^x$, $g(x) = 8 - \left(\frac{2}{3}\right)^x$

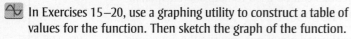 In Exercises 15–20, use a graphing utility to construct a table of values for the function. Then sketch the graph of the function.

15. $f(x) = 4^{-x} + 4$
16. $f(x) = 2.65^{x-1}$
17. $f(x) = 5^{x-2} + 4$
18. $f(x) = 2^{x-6} - 5$
19. $f(x) = \left(\frac{1}{2}\right)^{-x} + 3$
20. $f(x) = \left(\frac{1}{8}\right)^{x+2} - 5$

In Exercises 21–24, use the One-to-One Property to solve the equation for x.

21. $\left(\frac{1}{3}\right)^{x-3} = 9$
22. $3^{x+3} = \frac{1}{81}$
23. $e^{3x-5} = e^7$
24. $e^{8-2x} = e^{-3}$

In Exercises 25–28, evaluate $f(x) = e^x$ at the indicated value of x. Round your result to three decimal places.

25. $x = 8$
26. $x = \frac{5}{8}$
27. $x = -1.7$
28. $x = 0.278$

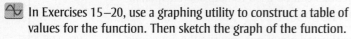 In Exercises 29–32, use a graphing utility to construct a table of values for the function. Then sketch the graph of the function.

29. $h(x) = e^{-x/2}$
30. $h(x) = 2 - e^{-x/2}$
31. $f(x) = e^{x+2}$
32. $s(t) = 4e^{-2/t}$, $t > 0$

COMPOUND INTEREST In Exercises 33 and 34, complete the table to determine the balance A for P dollars invested at rate r for t years and compounded n times per year.

n	1	2	4	12	365	Continuous
A						

TABLE FOR 33 AND 34

33. $P = \$5000$, $r = 3\%$, $t = 10$ years
34. $P = \$4500$, $r = 2.5\%$, $t = 30$ years

35. **WAITING TIMES** The average time between incoming calls at a switchboard is 3 minutes. The probability F of waiting less than t minutes until the next incoming call is approximated by the model $F(t) = 1 - e^{-t/3}$. A call has just come in. Find the probability that the next call will be within

(a) $\frac{1}{2}$ minute. (b) 2 minutes. (c) 5 minutes.

36. **DEPRECIATION** After t years, the value V of a car that originally cost \$23,970 is given by $V(t) = 23,970\left(\frac{3}{4}\right)^t$.

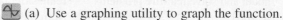 (a) Use a graphing utility to graph the function.

(b) Find the value of the car 2 years after it was purchased.

(c) According to the model, when does the car depreciate most rapidly? Is this realistic? Explain.

(d) According to the model, when will the car have no value?

5.2 In Exercises 37–40, write the exponential equation in logarithmic form. For example, the logarithmic form of $2^3 = 8$ is $\log_2 8 = 3$.

37. $3^3 = 27$
38. $25^{3/2} = 125$
39. $e^{0.8} = 2.2255\ldots$
40. $e^0 = 1$

In Exercises 41–44, evaluate the function at the indicated value of x without using a calculator.

41. $f(x) = \log x$, $x = 1000$
42. $g(x) = \log_9 x$, $x = 3$
43. $g(x) = \log_2 x$, $x = \frac{1}{4}$
44. $f(x) = \log_3 x$, $x = \frac{1}{81}$

In Exercises 45–48, use the One-to-One Property to solve the equation for x.

45. $\log_4(x + 7) = \log_4 14$
46. $\log_8(3x - 10) = \log_8 5$
47. $\ln(x + 9) = \ln 4$
48. $\ln(2x - 1) = \ln 11$

In Exercises 49–52, find the domain, x-intercept, and vertical asymptote of the logarithmic function and sketch its graph.

49. $g(x) = \log_7 x$
50. $f(x) = \log\left(\frac{x}{3}\right)$
51. $f(x) = 4 - \log(x + 5)$
52. $f(x) = \log(x - 3) + 1$

53. Use a calculator to evaluate $f(x) = \ln x$ at (a) $x = 22.6$ and (b) $x = 0.98$. Round your results to three decimal places if necessary.

 54. Use a calculator to evaluate $f(x) = 5 \ln x$ at (a) $x = e^{-12}$ and (b) $x = \sqrt{3}$. Round your results to three decimal places if necessary.

In Exercises 55–58, find the domain, *x*-intercept, and vertical asymptote of the logarithmic function and sketch its graph.

55. $f(x) = \ln x + 3$ **56.** $f(x) = \ln(x - 3)$

57. $h(x) = \ln(x^2)$ **58.** $f(x) = \frac{1}{4} \ln x$

59. ANTLER SPREAD The antler spread a (in inches) and shoulder height h (in inches) of an adult male American elk are related by the model $h = 116 \log(a + 40) - 176$. Approximate the shoulder height of a male American elk with an antler spread of 55 inches.

60. SNOW REMOVAL The number of miles s of roads cleared of snow is approximated by the model

$$s = 25 - \frac{13 \ln(h/12)}{\ln 3}, \quad 2 \le h \le 15$$

where h is the depth of the snow in inches. Use this model to find s when $h = 10$ inches.

5.3 In Exercises 61–64, evaluate the logarithm using the change-of-base formula. Do each exercise twice, once with common logarithms and once with natural logarithms. Round the results to three decimal places.

61. $\log_2 6$ **62.** $\log_{12} 200$

63. $\log_{1/2} 5$ **64.** $\log_3 0.28$

In Exercises 65–68, use the properties of logarithms to rewrite and simplify the logarithmic expression.

65. $\log 18$ **66.** $\log_2\left(\frac{1}{12}\right)$

67. $\ln 20$ **68.** $\ln(3e^{-4})$

In Exercises 69–74, use the properties of logarithms to expand the expression as a sum, difference, and/or constant multiple of logarithms. (Assume all variables are positive.)

69. $\log_5 5x^2$ **70.** $\log 7x^4$

71. $\log_3 \dfrac{9}{\sqrt{x}}$ **72.** $\log_7 \dfrac{\sqrt[3]{x}}{14}$

73. $\ln x^2 y^2 z$ **74.** $\ln\left(\dfrac{y-1}{4}\right)^2, \quad y > 1$

In Exercises 75–80, condense the expression to the logarithm of a single quantity.

75. $\log_2 5 + \log_2 x$ **76.** $\log_6 y - 2 \log_6 z$

77. $\ln x - \frac{1}{4} \ln y$ **78.** $3 \ln x + 2 \ln(x + 1)$

79. $\frac{1}{2} \log_3 x - 2 \log_3(y + 8)$

80. $5 \ln(x - 2) - \ln(x + 2) - 3 \ln x$

81. CLIMB RATE The time t (in minutes) for a small plane to climb to an altitude of h feet is modeled by $t = 50 \log[18,000/(18,000 - h)]$, where 18,000 feet is the plane's absolute ceiling.

(a) Determine the domain of the function in the context of the problem.

 (b) Use a graphing utility to graph the function and identify any asymptotes.

(c) As the plane approaches its absolute ceiling, what can be said about the time required to increase its altitude?

(d) Find the time for the plane to climb to an altitude of 4000 feet.

82. HUMAN MEMORY MODEL Students in a learning theory study were given an exam and then retested monthly for 6 months with an equivalent exam. The data obtained in the study are given as the ordered pairs (t, s), where t is the time in months after the initial exam and s is the average score for the class. Use these data to find a logarithmic equation that relates t and s.

(1, 84.2), (2, 78.4), (3, 72.1),
(4, 68.5), (5, 67.1), (6, 65.3)

5.4 In Exercises 83–88, solve for *x*.

83. $5^x = 125$ **84.** $6^x = \frac{1}{216}$

85. $e^x = 3$ **86.** $\log_6 x = -1$

87. $\ln x = 4$ **88.** $\ln x = -1.6$

In Exercises 89–92, solve the exponential equation algebraically. Approximate your result to three decimal places.

89. $e^{4x} = e^{x^2 + 3}$ **90.** $e^{3x} = 25$

91. $2^x - 3 = 29$ **92.** $e^{2x} - 6e^x + 8 = 0$

 In Exercises 93 and 94, use a graphing utility to graph and solve the equation. Approximate the result to three decimal places.

93. $25e^{-0.3x} = 12$ **94.** $2^x = 3 + x - e^x$

In Exercises 95–104, solve the logarithmic equation algebraically. Approximate the result to three decimal places.

95. $\ln 3x = 8.2$ **96.** $4 \ln 3x = 15$

97. $\ln x - \ln 3 = 2$ **98.** $\ln x - \ln 5 = 4$

99. $\ln \sqrt{x} = 4$ **100.** $\ln \sqrt{x + 8} = 3$

101. $\log_8(x - 1) = \log_8(x - 2) - \log_8(x + 2)$

102. $\log_6(x + 2) - \log_6 x = \log_6(x + 5)$

103. $\log(1 - x) = -1$ **104.** $\log(-x - 4) = 2$

In Exercises 105–108, use a graphing utility to graph and solve the equation. Approximate the result to three decimal places.

105. $2 \ln(x + 3) - 3 = 0$ **106.** $x - 2 \log(x + 4) = 0$

107. $6 \log(x^2 + 1) - x = 0$

108. $3 \ln x + 2 \log x = e^x - 25$

109. COMPOUND INTEREST You deposit $8500 in an account that pays 3.5% interest, compounded continuously. How long will it take for the money to triple?

110. METEOROLOGY The speed of the wind S (in miles per hour) near the center of a tornado and the distance d (in miles) the tornado travels are related by the model $S = 93 \log d + 65$. On March 18, 1925, a large tornado struck portions of Missouri, Illinois, and Indiana with a wind speed at the center of about 283 miles per hour. Approximate the distance traveled by this tornado.

5.5 In Exercises 111–116, match the function with its graph. [The graphs are labeled (a), (b), (c), (d), (e), and (f).]

(a)
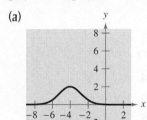

(b)
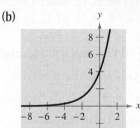

(c)
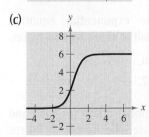

(d)
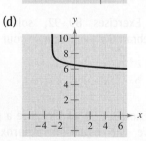

(e)
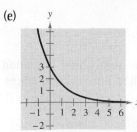

(f)
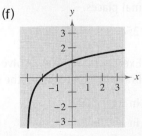

111. $y = 3e^{-2x/3}$ **112.** $y = 4e^{2x/3}$

113. $y = \ln(x + 3)$ **114.** $y = 7 - \log(x + 3)$

115. $y = 2e^{-(x+4)^2/3}$ **116.** $y = \dfrac{6}{1 + 2e^{-2x}}$

In Exercises 117 and 118, find the exponential model $y = ae^{bx}$ that passes through the points.

117. $(0, 2), (4, 3)$ **118.** $\left(0, \frac{1}{2}\right), (5, 5)$

119. POPULATION In 2007, the population of Florida residents aged 65 and over was about 3.10 million. In 2015 and 2020, the populations of Florida residents aged 65 and over are projected to be about 4.13 million and 5.11 million, respectively. An exponential growth model that approximates these data is given by $P = 2.36e^{0.0382t}$, $7 \le t \le 20$, where P is the population (in millions) and $t = 7$ represents 2007. (Source: U.S. Census Bureau)

 (a) Use a graphing utility to graph the model and the data in the same viewing window. Is the model a good fit for the data? Explain.

(b) According to the model, when will the population of Florida residents aged 65 and over reach 5.5 million? Does your answer seem reasonable? Explain.

120. WILDLIFE POPULATION A species of bat is in danger of becoming extinct. Five years ago, the total population of the species was 2000. Two years ago, the total population of the species was 1400. What was the total population of the species one year ago?

 121. TEST SCORES The test scores for a biology test follow a normal distribution modeled by $y = 0.0499e^{-(x-71)^2/128}$, $40 \le x \le 100$, where x is the test score. Use a graphing utility to graph the equation and estimate the average test score.

122. TYPING SPEED In a typing class, the average number N of words per minute typed after t weeks of lessons was found to be $N = 157/(1 + 5.4e^{-0.12t})$. Find the time necessary to type (a) 50 words per minute and (b) 75 words per minute.

123. SOUND INTENSITY The relationship between the number of decibels β and the intensity of a sound I in watts per square meter is $\beta = 10 \log(I/10^{-12})$. Find I for each decibel level β.

(a) $\beta = 60$ (b) $\beta = 135$ (c) $\beta = 1$

EXPLORATION

124. Consider the graph of $y = e^{kt}$. Describe the characteristics of the graph when k is positive and when k is negative.

TRUE OR FALSE? In Exercises 125 and 126, determine whether the equation is true or false. Justify your answer.

125. $\log_b b^{2x} = 2x$ **126.** $\ln(x + y) = \ln x + \ln y$

5 CHAPTER TEST

See www.CalcChat.com for worked-out solutions to odd-numbered exercises.

Take this test as you would take a test in class. When you are finished, check your work against the answers given in the back of the book.

In Exercises 1–4, evaluate the expression. Approximate your result to three decimal places.

1. $4.2^{0.6}$ **2.** $4^{3\pi/2}$ **3.** $e^{-7/10}$ **4.** $e^{3.1}$

In Exercises 5–7, construct a table of values. Then sketch the graph of the function.

5. $f(x) = 10^{-x}$ **6.** $f(x) = -6^{x-2}$ **7.** $f(x) = 1 - e^{2x}$

8. Evaluate (a) $\log_7 7^{-0.89}$ and (b) $4.6 \ln e^2$.

In Exercises 9–11, construct a table of values. Then sketch the graph of the function. Identify any asymptotes.

9. $f(x) = -\log x - 6$ **10.** $f(x) = \ln(x - 4)$ **11.** $f(x) = 1 + \ln(x + 6)$

In Exercises 12–14, evaluate the logarithm using the change-of-base formula. Round your result to three decimal places.

12. $\log_7 44$ **13.** $\log_{16} 0.63$ **14.** $\log_{3/4} 24$

In Exercises 15–17, use the properties of logarithms to expand the expression as a sum, difference, and/or constant multiple of logarithms.

15. $\log_2 3a^4$ **16.** $\ln \dfrac{5\sqrt{x}}{6}$ **17.** $\log \dfrac{(x-1)^3}{y^2 z}$

In Exercises 18–20, condense the expression to the logarithm of a single quantity.

18. $\log_3 13 + \log_3 y$ **19.** $4 \ln x - 4 \ln y$

20. $3 \ln x - \ln(x + 3) + 2 \ln y$

In Exercises 21–26, solve the equation algebraically. Approximate your result to three decimal places.

21. $5^x = \dfrac{1}{25}$ **22.** $3e^{-5x} = 132$

23. $\dfrac{1025}{8 + e^{4x}} = 5$ **24.** $\ln x = \dfrac{1}{2}$

25. $18 + 4 \ln x = 7$ **26.** $\log x + \log(x - 15) = 2$

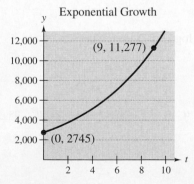

Exponential Growth

(9, 11,277)

(0, 2745)

FIGURE FOR 27

27. Find an exponential growth model for the graph shown in the figure.

28. The half-life of radioactive actinium (^{227}Ac) is 21.77 years. What percent of a present amount of radioactive actinium will remain after 19 years?

29. A model that can be used for predicting the height H (in centimeters) of a child based on his or her age is $H = 70.228 + 5.104x + 9.222 \ln x$, $\frac{1}{4} \le x \le 6$, where x is the age of the child in years. (Source: Snapshots of Applications in Mathematics)

(a) Construct a table of values. Then sketch the graph of the model.

(b) Use the graph from part (a) to estimate the height of a four-year-old child. Then calculate the actual height using the model.

PROOFS IN MATHEMATICS

Each of the following three properties of logarithms can be proved by using properties of exponential functions.

Properties of Logarithms (p. 398)

Let a be a positive number such that $a \neq 1$, and let n be a real number. If u and v are positive real numbers, the following properties are true.

	Logarithm with Base a	Natural Logarithm
1. Product Property:	$\log_a(uv) = \log_a u + \log_a v$	$\ln(uv) = \ln u + \ln v$
2. Quotient Property:	$\log_a \dfrac{u}{v} = \log_a u - \log_a v$	$\ln \dfrac{u}{v} = \ln u - \ln v$
3. Power Property:	$\log_a u^n = n \log_a u$	$\ln u^n = n \ln u$

Proof

Let

$$x = \log_a u \quad \text{and} \quad y = \log_a v.$$

The corresponding exponential forms of these two equations are

$$a^x = u \quad \text{and} \quad a^y = v.$$

To prove the Product Property, multiply u and v to obtain

$$uv = a^x a^y = a^{x+y}.$$

The corresponding logarithmic form of $uv = a^{x+y}$ is $\log_a(uv) = x + y$. So,

$$\log_a(uv) = \log_a u + \log_a v.$$

To prove the Quotient Property, divide u by v to obtain

$$\frac{u}{v} = \frac{a^x}{a^y} = a^{x-y}.$$

The corresponding logarithmic form of $\dfrac{u}{v} = a^{x-y}$ is $\log_a \dfrac{u}{v} = x - y$. So,

$$\log_a \frac{u}{v} = \log_a u - \log_a v.$$

To prove the Power Property, substitute a^x for u in the expression $\log_a u^n$, as follows.

$$\log_a u^n = \log_a (a^x)^n \qquad \text{Substitute } a^x \text{ for } u.$$
$$= \log_a a^{nx} \qquad \text{Property of Exponents}$$
$$= nx \qquad \text{Inverse Property of Logarithms}$$
$$= n \log_a u \qquad \text{Substitute } \log_a u \text{ for } x.$$

So, $\log_a u^n = n \log_a u.$

PROBLEM SOLVING

This collection of thought-provoking and challenging exercises further explores and expands upon concepts learned in this chapter.

1. Graph the exponential function given by $y = a^x$ for $a = 0.5, 1.2,$ and 2.0. Which of these curves intersects the line $y = x$? Determine all positive numbers a for which the curve $y = a^x$ intersects the line $y = x$.

2. Use a graphing utility to graph $y_1 = e^x$ and each of the functions $y_2 = x^2$, $y_3 = x^3$, $y_4 = \sqrt{x}$, and $y_5 = |x|$. Which function increases at the greatest rate as x approaches $+\infty$?

3. Use the result of Exercise 2 to make a conjecture about the rate of growth of $y_1 = e^x$ and $y = x^n$, where n is a natural number and x approaches $+\infty$.

4. Use the results of Exercises 2 and 3 to describe what is implied when it is stated that a quantity is growing exponentially.

5. Given the exponential function

   $$f(x) = a^x$$

   show that

   (a) $f(u + v) = f(u) \cdot f(v)$. (b) $f(2x) = [f(x)]^2$.

6. Given that

   $$f(x) = \frac{e^x + e^{-x}}{2} \text{ and } g(x) = \frac{e^x - e^{-x}}{2}$$

   show that

   $$[f(x)]^2 - [g(x)]^2 = 1.$$

7. Use a graphing utility to compare the graph of the function given by $y = e^x$ with the graph of each given function. [$n!$ (read "n factorial") is defined as $n! = 1 \cdot 2 \cdot 3 \cdots (n - 1) \cdot n$.]

   (a) $y_1 = 1 + \dfrac{x}{1!}$

   (b) $y_2 = 1 + \dfrac{x}{1!} + \dfrac{x^2}{2!}$

   (c) $y_3 = 1 + \dfrac{x}{1!} + \dfrac{x^2}{2!} + \dfrac{x^3}{3!}$

8. Identify the pattern of successive polynomials given in Exercise 7. Extend the pattern one more term and compare the graph of the resulting polynomial function with the graph of $y = e^x$. What do you think this pattern implies?

9. Graph the function given by

   $$f(x) = e^x - e^{-x}.$$

   From the graph, the function appears to be one-to-one. Assuming that the function has an inverse function, find $f^{-1}(x)$.

10. Find a pattern for $f^{-1}(x)$ if

    $$f(x) = \frac{a^x + 1}{a^x - 1}$$

    where $a > 0, a \neq 1$.

11. By observation, identify the equation that corresponds to the graph. Explain your reasoning.

    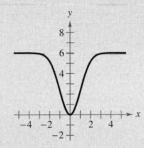

    (a) $y = 6e^{-x^2/2}$

    (b) $y = \dfrac{6}{1 + e^{-x/2}}$

    (c) $y = 6(1 - e^{-x^2/2})$

12. You have two options for investing $500. The first earns 7% compounded annually and the second earns 7% simple interest. The figure shows the growth of each investment over a 30-year period.

    (a) Identify which graph represents each type of investment. Explain your reasoning.

    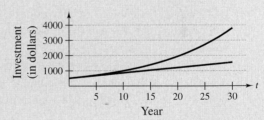

    (b) Verify your answer in part (a) by finding the equations that model the investment growth and graphing the models.

    (c) Which option would you choose? Explain your reasoning.

13. Two different samples of radioactive isotopes are decaying. The isotopes have initial amounts of c_1 and c_2, as well as half-lives of k_1 and k_2, respectively. Find the time t required for the samples to decay to equal amounts.

435

14. A lab culture initially contains 500 bacteria. Two hours later, the number of bacteria has decreased to 200. Find the exponential decay model of the form

$$B = B_0 a^{kt}$$

that can be used to approximate the number of bacteria after t hours.

15. The table shows the colonial population estimates of the American colonies from 1700 to 1780. (Source: U.S. Census Bureau)

Year	Population
1700	250,900
1710	331,700
1720	466,200
1730	629,400
1740	905,600
1750	1,170,800
1760	1,593,600
1770	2,148,100
1780	2,780,400

In each of the following, let y represent the population in the year t, with $t = 0$ corresponding to 1700.

(a) Use the *regression* feature of a graphing utility to find an exponential model for the data.

(b) Use the *regression* feature of the graphing utility to find a quadratic model for the data.

(c) Use the graphing utility to plot the data and the models from parts (a) and (b) in the same viewing window.

(d) Which model is a better fit for the data? Would you use this model to predict the population of the United States in 2015? Explain your reasoning.

16. Show that $\dfrac{\log_a x}{\log_{a/b} x} = 1 + \log_a \dfrac{1}{b}$.

17. Solve $(\ln x)^2 = \ln x^2$.

18. Use a graphing utility to compare the graph of the function $y = \ln x$ with the graph of each given function.

(a) $y_1 = x - 1$

(b) $y_2 = (x - 1) - \frac{1}{2}(x - 1)^2$

(c) $y_3 = (x - 1) - \frac{1}{2}(x - 1)^2 + \frac{1}{3}(x - 1)^3$

19. Identify the pattern of successive polynomials given in Exercise 18. Extend the pattern one more term and compare the graph of the resulting polynomial function with the graph of $y = \ln x$. What do you think the pattern implies?

20. Using

$$y = ab^x \quad \text{and} \quad y = ax^b$$

take the natural logarithm of each side of each equation. What are the slope and y-intercept of the line relating x and $\ln y$ for $y = ab^x$? What are the slope and y-intercept of the line relating $\ln x$ and $\ln y$ for $y = ax^b$?

In Exercises 21 and 22, use the model

$$y = 80.4 - 11 \ln x, \quad 100 \le x \le 1500$$

which approximates the minimum required ventilation rate in terms of the air space per child in a public school classroom. In the model, x is the air space per child in cubic feet and y is the ventilation rate per child in cubic feet per minute.

21. Use a graphing utility to graph the model and approximate the required ventilation rate if there is 300 cubic feet of air space per child.

22. A classroom is designed for 30 students. The air conditioning system in the room has the capacity of moving 450 cubic feet of air per minute.

(a) Determine the ventilation rate per child, assuming that the room is filled to capacity.

(b) Estimate the air space required per child.

(c) Determine the minimum number of square feet of floor space required for the room if the ceiling height is 30 feet.

In Exercises 23–26, (a) use a graphing utility to create a scatter plot of the data, (b) decide whether the data could best be modeled by a linear model, an exponential model, or a logarithmic model, (c) explain why you chose the model you did in part (b), (d) use the *regression* feature of a graphing utility to find the model you chose in part (b) for the data and graph the model with the scatter plot, and (e) determine how well the model you chose fits the data.

23. $(1, 2.0), (1.5, 3.5), (2, 4.0), (4, 5.8), (6, 7.0), (8, 7.8)$

24. $(1, 4.4), (1.5, 4.7), (2, 5.5), (4, 9.9), (6, 18.1), (8, 33.0)$

25. $(1, 7.5), (1.5, 7.0), (2, 6.8), (4, 5.0), (6, 3.5), (8, 2.0)$

26. $(1, 5.0), (1.5, 6.0), (2, 6.4), (4, 7.8), (6, 8.6), (8, 9.0)$

Topics in Analytic Geometry

6

In Mathematics

A conic is a collection of points satisfying a geometric property.

In Real Life

Conics are used as models in construction, planetary orbits, radio navigation, and projectile motion. For instance, you can use conics to model the orbits of the planets as they move about the sun. Using the techniques presented in this chapter, you can determine the distances between the planets and the center of the sun. (See Exercises 55–62, page 508.)

Mike Agliolo/Photo Researchers, Inc.

IN CAREERS

There are many careers that use conics and other topics in analytic geometry. Several are listed below.

- Home Contractor
  Exercise 69, page 444

- Civil Engineer
  Exercises 73 and 74, page 452

- Artist
  Exercise 51, page 471

- Astronomer
  Exercises 63 and 64, page 508

6.1 LINES

What you should learn

- Find the inclination of a line.
- Find the angle between two lines.
- Find the distance between a point and a line.

Why you should learn it

The inclination of a line can be used to measure heights indirectly. For instance, in Exercise 70 on page 444, the inclination of a line can be used to determine the change in elevation from the base to the top of the Falls Incline Railway in Niagara Falls, Ontario, Canada.

Inclination of a Line

In Section P.4, you learned that the graph of the linear equation

$$y = mx + b$$

is a nonvertical line with slope m and y-intercept $(0, b)$. There, the slope of a line was described as the rate of change in y with respect to x. In this section, you will look at the slope of a line in terms of the angle of inclination of the line.

Every nonhorizontal line must intersect the x-axis. The angle formed by such an intersection determines the **inclination** of the line, as specified in the following definition.

Definition of Inclination

The **inclination** of a nonhorizontal line is the positive angle θ (less than π) measured counterclockwise from the x-axis to the line. (See Figure 6.1.)

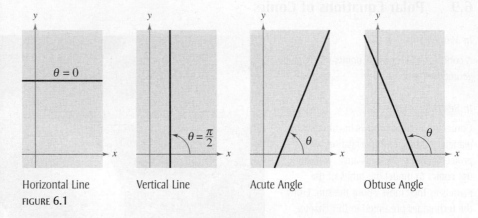

Horizontal Line   Vertical Line   Acute Angle   Obtuse Angle

FIGURE **6.1**

The inclination of a line is related to its slope in the following manner.

Inclination and Slope

If a nonvertical line has inclination θ and slope m, then

$$m = \tan \theta.$$

For a proof of this relation between inclination and slope, see Proofs in Mathematics on page 518.

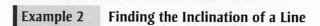

Example 1   Finding the Inclination of a Line

Find the inclination of the line $x - y = 2$.

Solution

The slope of this line is $m = 1$. So, its inclination is determined from the equation

$$\tan \theta = 1.$$

From Figure 6.2, it follows that $0 < \theta < \dfrac{\pi}{2}$. This means that

$$\theta = \arctan 1$$

$$= \frac{\pi}{4}.$$

The angle of inclination is $\dfrac{\pi}{4}$ radian or 45°.

CHECK*Point* ▶ Now try Exercise 27.

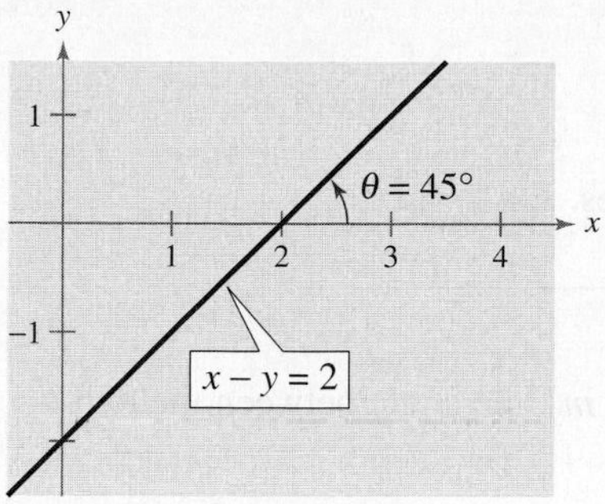

FIGURE 6.2

Example 2   Finding the Inclination of a Line

Find the inclination of the line $2x + 3y = 6$.

Solution

The slope of this line is $m = -\frac{2}{3}$. So, its inclination is determined from the equation

$$\tan \theta = -\frac{2}{3}.$$

From Figure 6.3, it follows that $\dfrac{\pi}{2} < \theta < \pi$. This means that

$$\theta = \pi + \arctan\left(-\frac{2}{3}\right)$$

$$\approx \pi + (-0.588)$$

$$= \pi - 0.588$$

$$\approx 2.554.$$

The angle of inclination is about 2.554 radians or about 146.3°.

CHECK*Point* ▶ Now try Exercise 33.

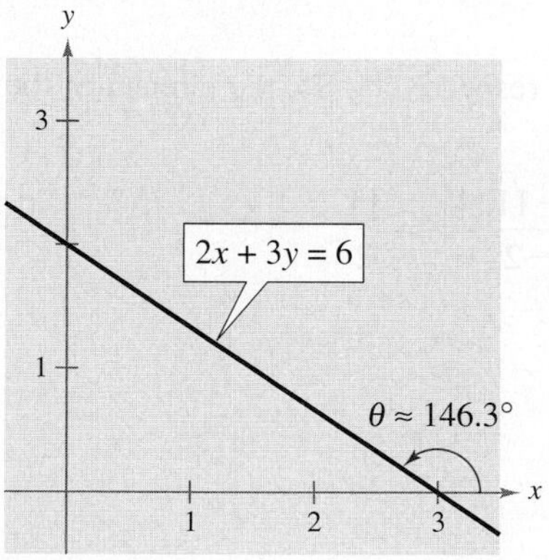

FIGURE 6.3

The Angle Between Two Lines

Two distinct lines in a plane are either parallel or intersecting. If they intersect and are nonperpendicular, their intersection forms two pairs of opposite angles. One pair is acute and the other pair is obtuse. The smaller of these angles is called the **angle between the two lines.** As shown in Figure 6.4, you can use the inclinations of the two lines to find the angle between the two lines. If two lines have inclinations θ_1 and θ_2, where $\theta_1 < \theta_2$ and $\theta_2 - \theta_1 < \pi/2$, the angle between the two lines is

$$\theta = \theta_2 - \theta_1.$$

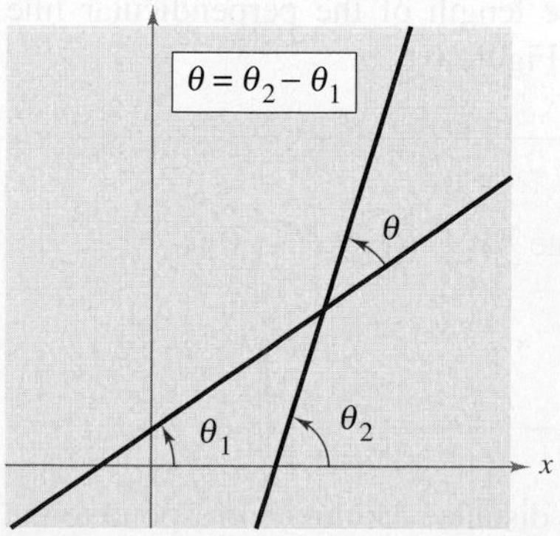

FIGURE 6.4

You can use the formula for the tangent of the difference of two angles

$$\tan \theta = \tan(\theta_2 - \theta_1)$$

$$= \frac{\tan \theta_2 - \tan \theta_1}{1 + \tan \theta_1 \tan \theta_2}$$

to obtain the formula for the angle between two lines.

Angle Between Two Lines

If two nonperpendicular lines have slopes m_1 and m_2, the angle between the two lines is

$$\tan \theta = \left| \frac{m_2 - m_1}{1 + m_1 m_2} \right|.$$

Example 3   Finding the Angle Between Two Lines

Find the angle between the two lines.

Line 1: $2x - y - 4 = 0$ *Line 2:* $3x + 4y - 12 = 0$

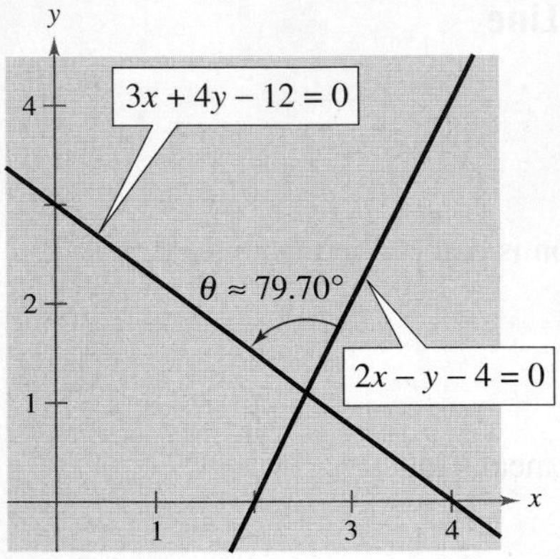

Solution

The two lines have slopes of $m_1 = 2$ and $m_2 = -\frac{3}{4}$, respectively. So, the tangent of the angle between the two lines is

$$\tan \theta = \left| \frac{m_2 - m_1}{1 + m_1 m_2} \right| = \left| \frac{(-3/4) - 2}{1 + (2)(-3/4)} \right| = \left| \frac{-11/4}{-2/4} \right| = \frac{11}{2}.$$

Finally, you can conclude that the angle is

$$\theta = \arctan \frac{11}{2} \approx 1.391 \text{ radians} \approx 79.70°$$

as shown in Figure 6.5.

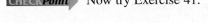   **CHECK***Point*   Now try Exercise 41.

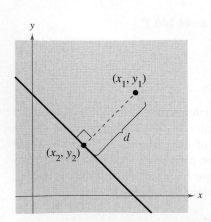

FIGURE 6.5

The Distance Between a Point and a Line

Finding the distance between a line and a point not on the line is an application of perpendicular lines. This distance is defined as the length of the perpendicular line segment joining the point and the line, as shown in Figure 6.6.

Distance Between a Point and a Line

The distance between the point (x_1, y_1) and the line $Ax + By + C = 0$ is

$$d = \frac{|Ax_1 + By_1 + C|}{\sqrt{A^2 + B^2}}.$$

Remember that the values of A, B, and C in this distance formula correspond to the general equation of a line, $Ax + By + C = 0$. For a proof of this formula for the distance between a point and a line, see Proofs in Mathematics on page 518.

FIGURE 6.6

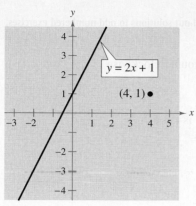

FIGURE **6.7**

Example 4   Finding the Distance Between a Point and a Line

Find the distance between the point $(4, 1)$ and the line $y = 2x + 1$.

Solution

The general form of the equation is $-2x + y - 1 = 0$. So, the distance between the point and the line is

$$d = \frac{|-2(4) + 1(1) + (-1)|}{\sqrt{(-2)^2 + 1^2}} = \frac{8}{\sqrt{5}} \approx 3.58 \text{ units.}$$

The line and the point are shown in Figure 6.7.

CHECK Point Now try Exercise 53.

Example 5   An Application of Two Distance Formulas

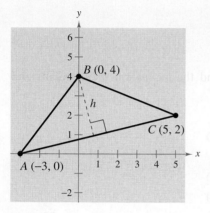

FIGURE **6.8**

Figure 6.8 shows a triangle with vertices $A(-3, 0)$, $B(0, 4)$, and $C(5, 2)$.

a. Find the altitude h from vertex B to side AC.

b. Find the area of the triangle.

Solution

a. To find the altitude, use the formula for the distance between line AC and the point $(0, 4)$. The equation of line AC is obtained as follows.

$$\textit{Slope: } \quad m = \frac{2 - 0}{5 - (-3)} = \frac{2}{8} = \frac{1}{4}$$

$$\textit{Equation: } \qquad y - 0 = \frac{1}{4}(x + 3) \qquad \text{Point-slope form}$$

$$4y = x + 3 \qquad \text{Multiply each side by 4.}$$

$$x - 4y + 3 = 0 \qquad \text{General form}$$

So, the distance between this line and the point $(0, 4)$ is

$$\text{Altitude} = h = \frac{|1(0) + (-4)(4) + 3|}{\sqrt{1^2 + (-4)^2}} = \frac{13}{\sqrt{17}} \text{ units.}$$

b. Using the formula for the distance between two points, you can find the length of the base AC to be

$$b = \sqrt{[5 - (-3)]^2 + (2 - 0)^2} \qquad \text{Distance Formula}$$

$$= \sqrt{8^2 + 2^2} \qquad \text{Simplify.}$$

$$= 2\sqrt{17} \text{ units.} \qquad \text{Simplify.}$$

Finally, the area of the triangle in Figure 6.8 is

$$A = \frac{1}{2}bh \qquad \text{Formula for the area of a triangle}$$

$$= \frac{1}{2}\left(2\sqrt{17}\right)\left(\frac{13}{\sqrt{17}}\right) \qquad \text{Substitute for } b \text{ and } h.$$

$$= 13 \text{ square units.} \qquad \text{Simplify.}$$

CHECK Point Now try Exercise 59.

6.1 EXERCISES

See www.CalcChat.com for worked-out solutions to odd-numbered exercises.

VOCABULARY: Fill in the blanks.

1. The _____ of a nonhorizontal line is the positive angle θ (less than π) measured counterclockwise from the x-axis to the line.

2. If a nonvertical line has inclination θ and slope m, then $m = $ _____ .

3. If two nonperpendicular lines have slopes m_1 and m_2, the angle between the two lines is $\tan \theta = $ _____ .

4. The distance between the point (x_1, y_1) and the line $Ax + By + C = 0$ is given by $d = $ _____ .

SKILLS AND APPLICATIONS

In Exercises 5–12, find the slope of the line with inclination θ.

5.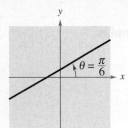

6.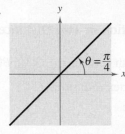

7.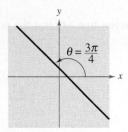

8.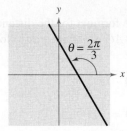

9. $\theta = \dfrac{\pi}{3}$ radians

10. $\theta = \dfrac{5\pi}{6}$ radians

11. $\theta = 1.27$ radians

12. $\theta = 2.88$ radians

In Exercises 13–18, find the inclination θ (in radians and degrees) of the line with a slope of m.

13. $m = -1$
14. $m = -2$
15. $m = 1$
16. $m = 2$
17. $m = \dfrac{3}{4}$
18. $m = -\dfrac{5}{2}$

In Exercises 19–26, find the inclination θ (in radians and degrees) of the line passing through the points.

19. $\left(\sqrt{3}, 2\right), (0, 1)$
20. $\left(1, 2\sqrt{3}\right), \left(0, \sqrt{3}\right)$

21. $\left(-\sqrt{3}, -1\right), (0, -2)$
22. $\left(3, \sqrt{3}\right), \left(6, -2\sqrt{3}\right)$
23. $(6, 1), (10, 8)$
24. $(12, 8), (-4, -3)$
25. $(-2, 20), (10, 0)$
26. $(0, 100), (50, 0)$

In Exercises 27–36, find the inclination θ (in radians and degrees) of the line.

27. $2x + 2y - 5 = 0$
28. $x - \sqrt{3}y + 1 = 0$
29. $3x - 3y + 1 = 0$
30. $\sqrt{3}x - y + 2 = 0$
31. $x + \sqrt{3}y + 2 = 0$
32. $-2\sqrt{3}x - 2y = 0$
33. $6x - 2y + 8 = 0$
34. $4x + 5y - 9 = 0$
35. $5x + 3y = 0$
36. $2x - 6y - 12 = 0$

In Exercises 37–46, find the angle θ (in radians and degrees) between the lines.

37. $3x + y = 3$
    $x - y = 2$

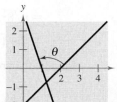

38. $x + 3y = \quad 2$
    $x - 2y = -3$

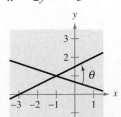

39. $x - y = 0$
$3x - 2y = -1$

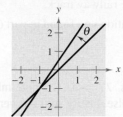

40. $2x - y = 2$
$4x + 3y = 24$

41. $x - 2y = 7$
$6x + 2y = 5$

42. $5x + 2y = 16$
$3x - 5y = -1$

43. $x + 2y = 8$
$x - 2y = 2$

44. $3x - 5y = 3$
$3x + 5y = 12$

45. $0.05x - 0.03y = 0.21$
$0.07x + 0.02y = 0.16$

46. $0.02x - 0.05y = -0.19$
$0.03x + 0.04y = 0.52$

ANGLE MEASUREMENT In Exercises 47–50, find the slope of each side of the triangle and use the slopes to find the measures of the interior angles.

47.

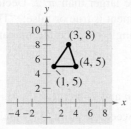

48.

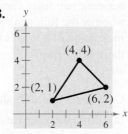

49.

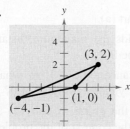

50.

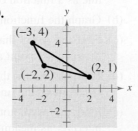

In Exercises 51–58, find the distance between the point and the line.

Point	Line
51. $(0, 0)$	$4x + 3y = 0$
52. $(0, 0)$	$2x - y = 4$

Point	Line
53. $(2, 3)$	$3x + y = 1$
54. $(-2, 1)$	$x - y = 2$
55. $(6, 2)$	$x + 1 = 0$
56. $(2, 1)$	$-2x + y - 2 = 0$
57. $(0, 8)$	$6x - y = 0$
58. $(4, 2)$	$x - y = 20$

In Exercises 59–62, the points represent the vertices of a triangle. (a) Draw triangle ABC in the coordinate plane, (b) find the altitude from vertex B of the triangle to side AC, and (c) find the area of the triangle.

59. $A = (0, 0), B = (1, 4), C = (4, 0)$
60. $A = (0, 0), B = (4, 5), C = (5, -2)$
61. $A = \left(-\frac{1}{2}, \frac{1}{2}\right), B = (2, 3), C = \left(\frac{5}{2}, 0\right)$
62. $A = (-4, -5), B = (3, 10), C = (6, 12)$

In Exercises 63 and 64, find the distance between the parallel lines.

63. $x + y = 1$
$x + y = 5$

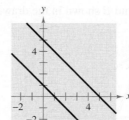

64. $3x - 4y = 1$
$3x - 4y = 10$

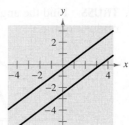

65. ROAD GRADE A straight road rises with an inclination of 0.10 radian from the horizontal (see figure). Find the slope of the road and the change in elevation over a two-mile stretch of the road.

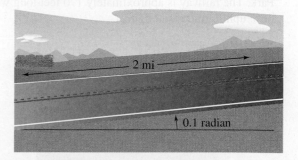

66. ROAD GRADE A straight road rises with an inclination of 0.20 radian from the horizontal. Find the slope of the road and the change in elevation over a one-mile stretch of the road.

67. PITCH OF A ROOF A roof has a rise of 3 feet for every horizontal change of 5 feet (see figure). Find the inclination of the roof.

68. CONVEYOR DESIGN A moving conveyor is built so that it rises 1 meter for each 3 meters of horizontal travel.

(a) Draw a diagram that gives a visual representation of the problem.

(b) Find the inclination of the conveyor.

(c) The conveyor runs between two floors in a factory. The distance between the floors is 5 meters. Find the length of the conveyor.

69. TRUSS Find the angles α and β shown in the drawing of the roof truss.

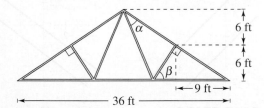

70. The Falls Incline Railway in Niagara Falls, Ontario, Canada is an inclined railway that was designed to carry people from the City of Niagara Falls to Queen Victoria Park. The railway is approximately 170 feet long with a 36% uphill grade (see figure).

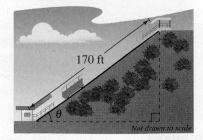

(a) Find the inclination θ of the railway.

(b) Find the change in elevation from the base to the top of the railway.

(c) Using the origin of a rectangular coordinate system as the base of the inclined plane, find the equation of the line that models the railway track.

(d) Sketch a graph of the equation you found in part (c).

EXPLORATION

TRUE OR FALSE? In Exercises 71 and 72, determine whether the statement is true or false. Justify your answer.

71. A line that has an inclination greater than $\pi/2$ radians has a negative slope.

72. To find the angle between two lines whose angles of inclination θ_1 and θ_2 are known, substitute θ_1 and θ_2 for m_1 and m_2, respectively, in the formula for the angle between two lines.

73. Consider a line with slope m and y-intercept $(0, 4)$.

(a) Write the distance d between the origin and the line as a function of m.

(b) Graph the function in part (a).

(c) Find the slope that yields the maximum distance between the origin and the line.

(d) Find the asymptote of the graph in part (b) and interpret its meaning in the context of the problem.

74. CAPSTONE Discuss why the inclination of a line can be an angle that is larger than $\pi/2$, but the angle between two lines cannot be larger than $\pi/2$. Decide whether the following statement is true or false: "The inclination of a line is the angle between the line and the x-axis." Explain.

75. Consider a line with slope m and y-intercept $(0, 4)$.

(a) Write the distance d between the point $(3, 1)$ and the line as a function of m.

(b) Graph the function in part (a).

(c) Find the slope that yields the maximum distance between the point and the line.

(d) Is it possible for the distance to be 0? If so, what is the slope of the line that yields a distance of 0?

(e) Find the asymptote of the graph in part (b) and interpret its meaning in the context of the problem.

6.2 INTRODUCTION TO CONICS: PARABOLAS

What you should learn

- Recognize a conic as the intersection of a plane and a double-napped cone.
- Write equations of parabolas in standard form and graph parabolas.
- Use the reflective property of parabolas to solve real-life problems.

Why you should learn it

Parabolas can be used to model and solve many types of real-life problems. For instance, in Exercise 71 on page 451, a parabola is used to model the cables of the Golden Gate Bridge.

Conics

Conic sections were discovered during the classical Greek period, 600 to 300 B.C. The early Greeks were concerned largely with the geometric properties of conics. It was not until the 17th century that the broad applicability of conics became apparent and played a prominent role in the early development of calculus.

A **conic section** (or simply **conic**) is the intersection of a plane and a double-napped cone. Notice in Figure 6.9 that in the formation of the four basic conics, the intersecting plane does not pass through the vertex of the cone. When the plane does pass through the vertex, the resulting figure is a **degenerate conic,** as shown in Figure 6.10.

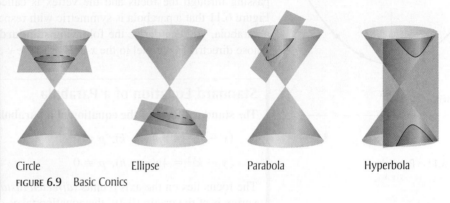

Circle   Ellipse   Parabola   Hyperbola
FIGURE 6.9   Basic Conics

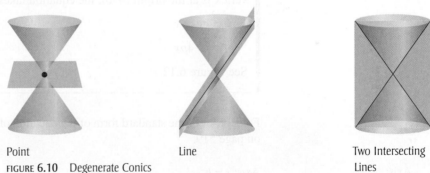

Point   Line   Two Intersecting
FIGURE 6.10   Degenerate Conics   Lines

There are several ways to approach the study of conics. You could begin by defining conics in terms of the intersections of planes and cones, as the Greeks did, or you could define them algebraically, in terms of the general second-degree equation

$$Ax^2 + Bxy + Cy^2 + Dx + Ey + F = 0.$$

However, you will study a third approach, in which each of the conics is defined as a **locus** (collection) of points satisfying a geometric property. For example, in Section P.3, you learned that a circle is defined as the collection of all points (x, y) that are equidistant from a fixed point (h, k). This leads to the standard form of the equation of a circle

$$(x - h)^2 + (y - k)^2 = r^2. \qquad \text{Equation of circle}$$

Parabolas

In Section P.3, you learned that the graph of the quadratic function

$$f(x) = ax^2 + bx + c$$

is a parabola that opens upward or downward. The following definition of a parabola is more general in the sense that it is independent of the orientation of the parabola.

Definition of Parabola

A **parabola** is the set of all points (x, y) in a plane that are equidistant from a fixed line **(directrix)** and a fixed point **(focus)** not on the line.

The midpoint between the focus and the directrix is called the **vertex,** and the line passing through the focus and the vertex is called the **axis** of the parabola. Note in Figure 6.11 that a parabola is symmetric with respect to its axis. Using the definition of a parabola, you can derive the following **standard form** of the equation of a parabola whose directrix is parallel to the x-axis or to the y-axis.

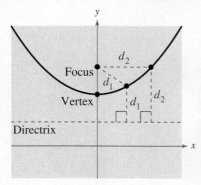

FIGURE 6.11   Parabola

Standard Equation of a Parabola

The **standard form of the equation of a parabola** with vertex at (h, k) is as follows.

$$(x - h)^2 = 4p(y - k), \quad p \neq 0 \qquad \text{Vertical axis, directrix: } y = k - p$$

$$(y - k)^2 = 4p(x - h), \quad p \neq 0 \qquad \text{Horizontal axis, directrix: } x = h - p$$

The focus lies on the axis p units (*directed distance*) from the vertex. If the vertex is at the origin $(0, 0)$, the equation takes one of the following forms.

$$x^2 = 4py \qquad\qquad \text{Vertical axis}$$

$$y^2 = 4px \qquad\qquad \text{Horizontal axis}$$

See Figure 6.12.

For a proof of the standard form of the equation of a parabola, see Proofs in Mathematics on page 519.

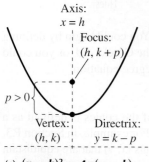

(a) $(x - h)^2 = 4p(y - k)$
Vertical axis: $p > 0$

(b) $(x - h)^2 = 4p(y - k)$
Vertical axis: $p < 0$

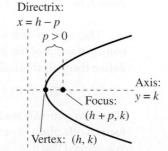

(c) $(y - k)^2 = 4p(x - h)$
Horizontal axis: $p > 0$

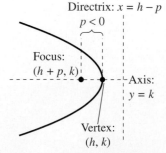

(d) $(y - k)^2 = 4p(x - h)$
Horizontal axis: $p < 0$

FIGURE 6.12

Example 1   Vertex at the Origin

Find the standard equation of the parabola with vertex at the origin and focus $(2, 0)$.

Solution

The axis of the parabola is horizontal, passing through $(0, 0)$ and $(2, 0)$, as shown in Figure 6.13.

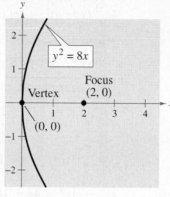

FIGURE **6.13**

The standard form is $y^2 = 4px$, where $h = 0$, $k = 0$, and $p = 2$. So, the equation is $y^2 = 8x$.

CHECK_Point_   Now try Exercise 23.

Example 2   Finding the Focus of a Parabola

Find the focus of the parabola given by $y = -\frac{1}{2}x^2 - x + \frac{1}{2}$.

Solution

To find the focus, convert to standard form by completing the square.

$$y = -\tfrac{1}{2}x^2 - x + \tfrac{1}{2} \qquad \text{Write original equation.}$$

$$-2y = x^2 + 2x - 1 \qquad \text{Multiply each side by } -2.$$

$$1 - 2y = x^2 + 2x \qquad \text{Add 1 to each side.}$$

$$1 + 1 - 2y = x^2 + 2x + 1 \qquad \text{Complete the square.}$$

$$2 - 2y = x^2 + 2x + 1 \qquad \text{Combine like terms.}$$

$$-2(y - 1) = (x + 1)^2 \qquad \text{Standard form}$$

Comparing this equation with

$$(x - h)^2 = 4p(y - k)$$

you can conclude that $h = -1$, $k = 1$, and $p = -\frac{1}{2}$. Because p is negative, the parabola opens downward, as shown in Figure 6.14. So, the focus of the parabola is $\left(h, k + p\right) = \left(-1, \frac{1}{2}\right)$.

CHECK_Point_   Now try Exercise 43.

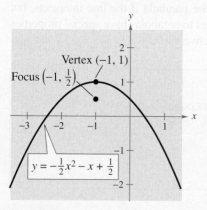

FIGURE 6.14

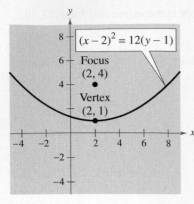

FIGURE **6.15**

Example 3 Finding the Standard Equation of a Parabola

Find the standard form of the equation of the parabola with vertex $(2, 1)$ and focus $(2, 4)$. Then write the quadratic form of the equation.

Solution

Because the axis of the parabola is vertical, passing through $(2, 1)$ and $(2, 4)$, consider the equation

$$(x - h)^2 = 4p(y - k)$$

where $h = 2$, $k = 1$, and $p = 4 - 1 = 3$. So, the standard form is

$$(x - 2)^2 = 12(y - 1).$$

You can obtain the more common quadratic form as follows.

$$(x - 2)^2 = 12(y - 1)$$ Write original equation.

$$x^2 - 4x + 4 = 12y - 12$$ Multiply.

$$x^2 - 4x + 16 = 12y$$ Add 12 to each side.

$$\frac{1}{12}(x^2 - 4x + 16) = y$$ Divide each side by 12.

The graph of this parabola is shown in Figure 6.15.

CHECK Point Now try Exercise 55.

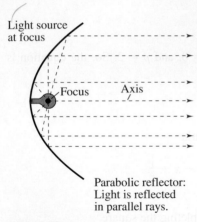

Parabolic reflector:
Light is reflected
in parallel rays.

FIGURE **6.16**

Application

A line segment that passes through the focus of a parabola and has endpoints on the parabola is called a **focal chord.** The specific focal chord perpendicular to the axis of the parabola is called the **latus rectum.**

Parabolas occur in a wide variety of applications. For instance, a parabolic reflector can be formed by revolving a parabola around its axis. The resulting surface has the property that all incoming rays parallel to the axis are reflected through the focus of the parabola. This is the principle behind the construction of the parabolic mirrors used in reflecting telescopes. Conversely, the light rays emanating from the focus of a parabolic reflector used in a flashlight are all parallel to one another, as shown in Figure 6.16.

A line is **tangent** to a parabola at a point on the parabola if the line intersects, but does not cross, the parabola at the point. Tangent lines to parabolas have special properties related to the use of parabolas in constructing reflective surfaces.

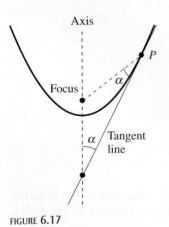

FIGURE **6.17**

Reflective Property of a Parabola

The tangent line to a parabola at a point P makes equal angles with the following two lines (see Figure 6.17).

1. The line passing through P and the focus

2. The axis of the parabola

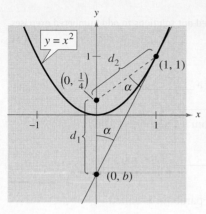

FIGURE **6.18**

TECHNOLOGY

Use a graphing utility to confirm the result of Example 4. By graphing

$$y_1 = x^2 \quad \text{and} \quad y_2 = 2x - 1$$

in the same viewing window, you should be able to see that the line touches the parabola at the point $(1, 1)$.

Algebra Help

You can review techniques for writing linear equations in Section P.4.

| Example 4 | **Finding the Tangent Line at a Point on a Parabola** |

Find the equation of the tangent line to the parabola given by $y = x^2$ at the point $(1, 1)$.

Solution

For this parabola, $p = \frac{1}{4}$ and the focus is $\left(0, \frac{1}{4}\right)$, as shown in Figure 6.18. You can find the y-intercept $(0, b)$ of the tangent line by equating the lengths of the two sides of the isosceles triangle shown in Figure 6.18:

$$d_1 = \frac{1}{4} - b$$

and

$$d_2 = \sqrt{(1 - 0)^2 + \left[1 - \left(\frac{1}{4}\right)\right]^2} = \frac{5}{4}.$$

Note that $d_1 = \frac{1}{4} - b$ rather than $b - \frac{1}{4}$. The order of subtraction for the distance is important because the distance must be positive. Setting $d_1 = d_2$ produces

$$\frac{1}{4} - b = \frac{5}{4}$$

$$b = -1.$$

So, the slope of the tangent line is

$$m = \frac{1 - (-1)}{1 - 0} = 2$$

and the equation of the tangent line in slope-intercept form is

$$y = 2x - 1.$$

CHECKPoint   Now try Exercise 65.

CLASSROOM DISCUSSION

Satellite Dishes   Cross sections of satellite dishes are parabolic in shape. Use the figure shown to write a paragraph explaining why satellite dishes are parabolic.

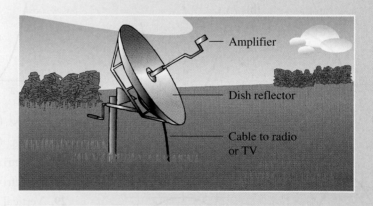

6.2 EXERCISES

See www.CalcChat.com for worked-out solutions to odd-numbered exercises.

VOCABULARY: Fill in the blanks.

1. A _____ is the intersection of a plane and a double-napped cone.

2. When a plane passes through the vertex of a double-napped cone, the intersection is a _____ _____.

3. A collection of points satisfying a geometric property can also be referred to as a _____ of points.

4. A _____ is defined as the set of all points (x, y) in a plane that are equidistant from a fixed line, called the _____, and a fixed point, called the _____, not on the line.

5. The line that passes through the focus and the vertex of a parabola is called the _____ of the parabola.

6. The _____ of a parabola is the midpoint between the focus and the directrix.

7. A line segment that passes through the focus of a parabola and has endpoints on the parabola is called a _____ _____ .

8. A line is _____ to a parabola at a point on the parabola if the line intersects, but does not cross, the parabola at the point.

SKILLS AND APPLICATIONS

In Exercises 9–12, describe in words how a plane could intersect with the double-napped cone shown to form the conic section.

9. Circle
10. Ellipse
11. Parabola
12. Hyperbola

In Exercises 13–18, match the equation with its graph. [The graphs are labeled (a), (b), (c), (d), (e), and (f).]

(a)
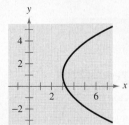

(b)
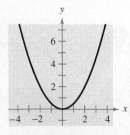

(c)
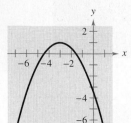

(d)
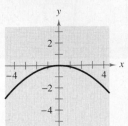

(e)
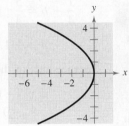

(f)
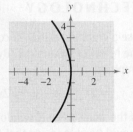

13. $y^2 = -4x$
14. $x^2 = 2y$
15. $x^2 = -8y$
16. $y^2 = -12x$
17. $(y - 1)^2 = 4(x - 3)$
18. $(x + 3)^2 = -2(y - 1)$

In Exercises 19–32, find the standard form of the equation of the parabola with the given characteristic(s) and vertex at the origin.

19.
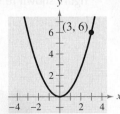

20.
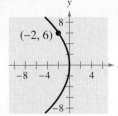

21. Focus: $\left(0, \frac{1}{2}\right)$
22. Focus: $\left(-\frac{3}{2}, 0\right)$
23. Focus: $(-2, 0)$
24. Focus: $(0, -2)$
25. Directrix: $y = 1$
26. Directrix: $y = -2$
27. Directrix: $x = -1$
28. Directrix: $x = 3$
29. Vertical axis and passes through the point $(4, 6)$
30. Vertical axis and passes through the point $(-3, -3)$
31. Horizontal axis and passes through the point $(-2, 5)$
32. Horizontal axis and passes through the point $(3, -2)$

In Exercises 33–46, find the vertex, focus, and directrix of the parabola, and sketch its graph.

33. $y = \frac{1}{2}x^2$ **34.** $y = -2x^2$

35. $y^2 = -6x$ **36.** $y^2 = 3x$

37. $x^2 + 6y = 0$ **38.** $x + y^2 = 0$

39. $(x-1)^2 + 8(y+2) = 0$

40. $(x+5) + (y-1)^2 = 0$

41. $(x+3)^2 = 4\left(y - \frac{3}{2}\right)$ **42.** $\left(x + \frac{1}{2}\right)^2 = 4(y-1)$

43. $y = \frac{1}{4}(x^2 - 2x + 5)$ **44.** $x = \frac{1}{4}(y^2 + 2y + 33)$

45. $y^2 + 6y + 8x + 25 = 0$

46. $y^2 - 4y - 4x = 0$

 In Exercises 47–50, find the vertex, focus, and directrix of the parabola. Use a graphing utility to graph the parabola.

47. $x^2 + 4x + 6y - 2 = 0$ **48.** $x^2 - 2x + 8y + 9 = 0$

49. $y^2 + x + y = 0$ **50.** $y^2 - 4x - 4 = 0$

In Exercises 51–60, find the standard form of the equation of the parabola with the given characteristics.

51.

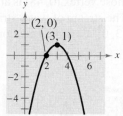

52.

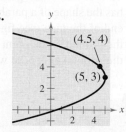

53.

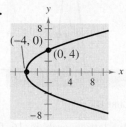

54.

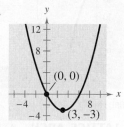

55. Vertex: $(4, 3)$; focus: $(6, 3)$

56. Vertex: $(-1, 2)$; focus: $(-1, 0)$

57. Vertex: $(0, 2)$; directrix: $y = 4$

58. Vertex: $(1, 2)$; directrix: $y = -1$

59. Focus: $(2, 2)$; directrix: $x = -2$

60. Focus: $(0, 0)$; directrix: $y = 8$

In Exercises 61 and 62, change the equation of the parabola so that its graph matches the description.

61. $(y-3)^2 = 6(x+1)$; upper half of parabola

62. $(y+1)^2 = 2(x-4)$; lower half of parabola

 In Exercises 63 and 64, the equations of a parabola and a tangent line to the parabola are given. Use a graphing utility to graph both equations in the same viewing window. Determine the coordinates of the point of tangency.

Parabola	*Tangent Line*
63. $y^2 - 8x = 0$	$x - y + 2 = 0$
64. $x^2 + 12y = 0$	$x + y - 3 = 0$

In Exercises 65–68, find an equation of the tangent line to the parabola at the given point, and find the x-intercept of the line.

65. $x^2 = 2y$, $(4, 8)$ **66.** $x^2 = 2y$, $\left(-3, \frac{9}{2}\right)$

67. $y = -2x^2$, $(-1, -2)$ **68.** $y = -2x^2$, $(2, -8)$

 69. REVENUE   The revenue R (in dollars) generated by the sale of x units of a patio furniture set is given by

$$(x - 106)^2 = -\frac{4}{5}(R - 14{,}045).$$

Use a graphing utility to graph the function and approximate the number of sales that will maximize revenue.

 70. REVENUE   The revenue R (in dollars) generated by the sale of x units of a digital camera is given by

$$(x - 135)^2 = -\frac{5}{7}(R - 25{,}515).$$

Use a graphing utility to graph the function and approximate the number of sales that will maximize revenue.

71. SUSPENSION BRIDGE   Each cable of the Golden Gate Bridge is suspended (in the shape of a parabola) between two towers that are 1280 meters apart. The top of each tower is 152 meters above the roadway. The cables touch the roadway midway between the towers.

(a) Draw a sketch of the bridge. Locate the origin of a rectangular coordinate system at the center of the roadway. Label the coordinates of the known points.

(b) Write an equation that models the cables.

(c) Complete the table by finding the height y of the suspension cables over the roadway at a distance of x meters from the center of the bridge.

Distance, x	Height, y
0	
100	
250	
400	
500	

72. SATELLITE DISH The receiver in a parabolic satellite dish is 4.5 feet from the vertex and is located at the focus (see figure). Write an equation for a cross section of the reflector. (Assume that the dish is directed upward and the vertex is at the origin.)

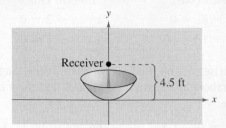

73. ROAD DESIGN Roads are often designed with parabolic surfaces to allow rain to drain off. A particular road that is 32 feet wide is 0.4 foot higher in the center than it is on the sides (see figure).

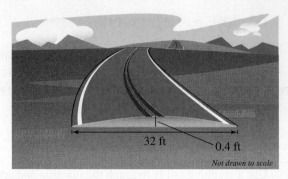

Cross section of road surface

(a) Find an equation of the parabola that models the road surface. (Assume that the origin is at the center of the road.)

(b) How far from the center of the road is the road surface 0.1 foot lower than in the middle?

74. HIGHWAY DESIGN Highway engineers design a parabolic curve for an entrance ramp from a straight street to an interstate highway (see figure). Find an equation of the parabola.

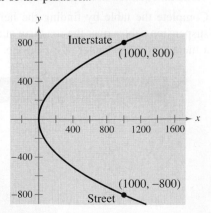

75. BEAM DEFLECTION A simply supported beam is 12 meters long and has a load at the center (see figure). The deflection of the beam at its center is 2 centimeters. Assume that the shape of the deflected beam is parabolic.

(a) Write an equation of the parabola. (Assume that the origin is at the center of the deflected beam.)

(b) How far from the center of the beam is the deflection equal to 1 centimeter?

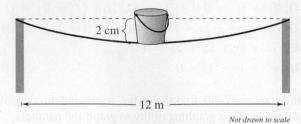

Not drawn to scale

76. BEAM DEFLECTION Repeat Exercise 75 if the length of the beam is 16 meters and the deflection of the beam at the center is 3 centimeters.

77. FLUID FLOW Water is flowing from a horizontal pipe 48 feet above the ground. The falling stream of water has the shape of a parabola whose vertex $(0, 48)$ is at the end of the pipe (see figure). The stream of water strikes the ground at the point $(10\sqrt{3}, 0)$. Find the equation of the path taken by the water.

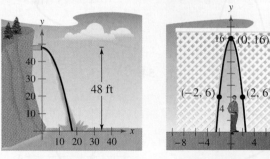

FIGURE FOR **77** FIGURE FOR **78**

78. LATTICE ARCH A parabolic lattice arch is 16 feet high at the vertex. At a height of 6 feet, the width of the lattice arch is 4 feet (see figure). How wide is the lattice arch at ground level?

79. SATELLITE ORBIT A satellite in a 100-mile-high circular orbit around Earth has a velocity of approximately 17,500 miles per hour. If this velocity is multiplied by $\sqrt{2}$, the satellite will have the minimum velocity necessary to escape Earth's gravity and it will follow a parabolic path with the center of Earth as the focus (see figure on the next page).

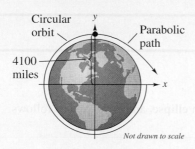

FIGURE FOR 79

(a) Find the escape velocity of the satellite.

(b) Find an equation of the parabolic path of the satel-lite (assume that the radius of Earth is 4000 miles).

80. PATH OF A SOFTBALL The path of a softball is modeled by $-12.5(y - 7.125) = (x - 6.25)^2$, where the coordinates x and y are measured in feet, with $x = 0$ corresponding to the position from which the ball was thrown.

(a) Use a graphing utility to graph the trajectory of the softball.

(b) Use the *trace* feature of the graphing utility to approximate the highest point and the range of the trajectory.

PROJECTILE MOTION In Exercises 81 and 82, consider the path of a projectile projected horizontally with a velocity of v feet per second at a height of s feet, where the model for the path is

$$x^2 = -\frac{v^2}{16}(y - s).$$

In this model (in which air resistance is disregarded), y is the height (in feet) of the projectile and x is the horizontal distance (in feet) the projectile travels.

81. A ball is thrown from the top of a 100-foot tower with a velocity of 28 feet per second.

(a) Find the equation of the parabolic path.

(b) How far does the ball travel horizontally before striking the ground?

82. A cargo plane is flying at an altitude of 30,000 feet and a speed of 540 miles per hour. A supply crate is dropped from the plane. How many *feet* will the crate travel horizontally before it hits the ground?

EXPLORATION

TRUE OR FALSE? In Exercises 83 and 84, determine whether the statement is true or false. Justify your answer.

83. It is possible for a parabola to intersect its directrix.

84. If the vertex and focus of a parabola are on a horizontal line, then the directrix of the parabola is vertical.

85. Let (x_1, y_1) be the coordinates of a point on the parabola $x^2 = 4py$. The equation of the line tangent to the parabola at the point is

$$y - y_1 = \frac{x_1}{2p}(x - x_1).$$

What is the slope of the tangent line?

86. CAPSTONE Explain what each of the following equations represents, and how equations (a) and (b) are equivalent.

(a) $y = a(x - h)^2 + k$, $a \neq 0$

(b) $(x - h)^2 = 4p(y - k)$, $p \neq 0$

(c) $(y - k)^2 = 4p(x - h)$, $p \neq 0$

87. GRAPHICAL REASONING Consider the parabola $x^2 = 4py$.

(a) Use a graphing utility to graph the parabola for $p = 1$, $p = 2$, $p = 3$, and $p = 4$. Describe the effect on the graph when p increases.

(b) Locate the focus for each parabola in part (a).

(c) For each parabola in part (a), find the length of the latus rectum (see figure). How can the length of the latus rectum be determined directly from the standard form of the equation of the parabola?

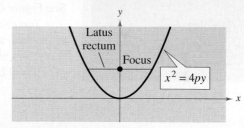

(d) Explain how the result of part (c) can be used as a sketching aid when graphing parabolas.

88. GEOMETRY The area of the shaded region in the figure is $A = \frac{8}{3}p^{1/2}b^{3/2}$.

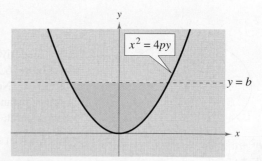

(a) Find the area when $p = 2$ and $b = 4$.

(b) Give a geometric explanation of why the area approaches 0 as p approaches 0.

6.3 ELLIPSES

What you should learn

- Write equations of ellipses in standard form and graph ellipses.
- Use properties of ellipses to model and solve real-life problems.
- Find eccentricities of ellipses.

Why you should learn it

Ellipses can be used to model and solve many types of real-life problems. For instance, in Exercise 65 on page 461, an ellipse is used to model the orbit of Halley's comet.

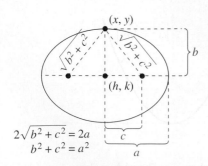

Introduction

The second type of conic is called an **ellipse,** and is defined as follows.

> #### Definition of Ellipse
> An **ellipse** is the set of all points (x, y) in a plane, the sum of whose distances from two distinct fixed points **(foci)** is constant. See Figure 6.19.

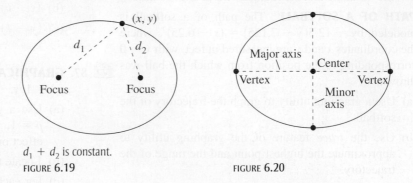

$d_1 + d_2$ is constant.

FIGURE 6.19   FIGURE 6.20

The line through the foci intersects the ellipse at two points called **vertices.** The chord joining the vertices is the **major axis,** and its midpoint is the **center** of the ellipse. The chord perpendicular to the major axis at the center is the **minor axis** of the ellipse. See Figure 6.20.

You can visualize the definition of an ellipse by imagining two thumbtacks placed at the foci, as shown in Figure 6.21. If the ends of a fixed length of string are fastened to the thumbtacks and the string is *drawn taut* with a pencil, the path traced by the pencil will be an ellipse.

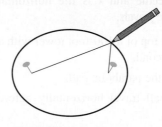

FIGURE 6.21

To derive the standard form of the equation of an ellipse, consider the ellipse in Figure 6.22 with the following points: center, (h, k); vertices, $(h \pm a, k)$; foci, $(h \pm c, k)$. Note that the center is the midpoint of the segment joining the foci. The sum of the distances from any point on the ellipse to the two foci is constant. Using a vertex point, this constant sum is

$$(a + c) + (a - c) = 2a \qquad \text{Length of major axis}$$

or simply the length of the major axis. Now, if you let (x, y) be *any* point on the ellipse, the sum of the distances between (x, y) and the two foci must also be $2a$.

$2\sqrt{b^2 + c^2} = 2a$
$b^2 + c^2 = a^2$

FIGURE 6.22

That is,

$$\sqrt{[x - (h - c)]^2 + (y - k)^2} + \sqrt{[x - (h + c)]^2 + (y - k)^2} = 2a$$

which, after expanding and regrouping, reduces to

$$(a^2 - c^2)(x - h)^2 + a^2(y - k)^2 = a^2(a^2 - c^2).$$

Finally, in Figure 6.22, you can see that

$$b^2 = a^2 - c^2$$

which implies that the equation of the ellipse is

$$b^2(x - h)^2 + a^2(y - k)^2 = a^2b^2$$

$$\frac{(x - h)^2}{a^2} + \frac{(y - k)^2}{b^2} = 1.$$

You would obtain a similar equation in the derivation by starting with a vertical major axis. Both results are summarized as follows.

Standard Equation of an Ellipse

The **standard form of the equation of an ellipse,** with center (h, k) and major and minor axes of lengths $2a$ and $2b$, respectively, where $0 < b < a$, is

$$\frac{(x - h)^2}{a^2} + \frac{(y - k)^2}{b^2} = 1 \qquad \text{Major axis is horizontal.}$$

$$\frac{(x - h)^2}{b^2} + \frac{(y - k)^2}{a^2} = 1. \qquad \text{Major axis is vertical.}$$

The foci lie on the major axis, c units from the center, with $c^2 = a^2 - b^2$. If the center is at the origin $(0, 0)$, the equation takes one of the following forms.

$$\frac{x^2}{a^2} + \frac{y^2}{b^2} = 1 \qquad \text{Major axis is horizontal.} \qquad\qquad \frac{x^2}{b^2} + \frac{y^2}{a^2} = 1 \qquad \text{Major axis is vertical.}$$

Figure 6.23 shows both the horizontal and vertical orientations for an ellipse.

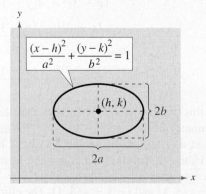

Major axis is horizontal.
FIGURE **6.23**

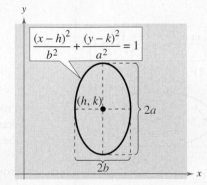

Major axis is vertical.

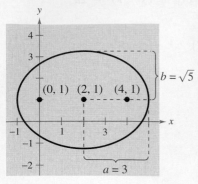

FIGURE **6.24**

Example 1   Finding the Standard Equation of an Ellipse

Find the standard form of the equation of the ellipse having foci at $(0, 1)$ and $(4, 1)$ and a major axis of length 6, as shown in Figure 6.24.

Solution

Because the foci occur at $(0, 1)$ and $(4, 1)$, the center of the ellipse is $(2, 1)$ and the distance from the center to one of the foci is $c = 2$. Because $2a = 6$, you know that $a = 3$. Now, from $c^2 = a^2 - b^2$, you have

$$b = \sqrt{a^2 - c^2} = \sqrt{3^2 - 2^2} = \sqrt{5}.$$

Because the major axis is horizontal, the standard equation is

$$\frac{(x - 2)^2}{3^2} + \frac{(y - 1)^2}{\left(\sqrt{5}\right)^2} = 1.$$

This equation simplifies to

$$\frac{(x - 2)^2}{9} + \frac{(y - 1)^2}{5} = 1.$$

CHECK *Point*   Now try Exercise 23.

Example 2   Sketching an Ellipse

Sketch the ellipse given by $x^2 + 4y^2 + 6x - 8y + 9 = 0$.

Solution

Begin by writing the original equation in standard form. In the fourth step, note that 9 and 4 are added to *both* sides of the equation when completing the squares.

$$x^2 + 4y^2 + 6x - 8y + 9 = 0 \qquad \text{Write original equation.}$$

$$\left(x^2 + 6x + \rule{1cm}{0.3cm}\right) + \left(4y^2 - 8y + \rule{1cm}{0.3cm}\right) = -9 \qquad \text{Group terms.}$$

$$\left(x^2 + 6x + \rule{1cm}{0.3cm}\right) + 4\left(y^2 - 2y + \rule{1cm}{0.3cm}\right) = -9 \qquad \text{Factor 4 out of } y\text{-terms.}$$

$$(x^2 + 6x + 9) + 4(y^2 - 2y + 1) = -9 + 9 + 4(1)$$

$$(x + 3)^2 + 4(y - 1)^2 = 4 \qquad \text{Write in completed square form.}$$

$$\frac{(x + 3)^2}{4} + \frac{(y - 1)^2}{1} = 1 \qquad \text{Divide each side by 4.}$$

$$\frac{(x + 3)^2}{2^2} + \frac{(y - 1)^2}{1^2} = 1 \qquad \text{Write in standard form.}$$

From this standard form, it follows that the center is $(h, k) = (-3, 1)$. Because the denominator of the x-term is $a^2 = 2^2$, the endpoints of the major axis lie two units to the right and left of the center. Similarly, because the denominator of the y-term is $b^2 = 1^2$, the endpoints of the minor axis lie one unit up and down from the center. Now, from $c^2 = a^2 - b^2$, you have $c = \sqrt{2^2 - 1^2} = \sqrt{3}$. So, the foci of the ellipse are $\left(-3 - \sqrt{3}, 1\right)$ and $\left(-3 + \sqrt{3}, 1\right)$. The ellipse is shown in Figure 6.25.

CHECK *Point*   Now try Exercise 47.

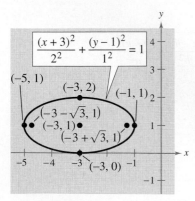

FIGURE **6.25**

Example 3   Analyzing an Ellipse

Find the center, vertices, and foci of the ellipse $4x^2 + y^2 - 8x + 4y - 8 = 0$.

Solution

By completing the square, you can write the original equation in standard form.

$$4x^2 + y^2 - 8x + 4y - 8 = 0 \qquad \text{Write original equation.}$$

$$\left(4x^2 - 8x + \boxed{}\right) + \left(y^2 + 4y + \boxed{}\right) = 8 \qquad \text{Group terms.}$$

$$4\left(x^2 - 2x + \boxed{}\right) + \left(y^2 + 4y + \boxed{}\right) = 8 \qquad \text{Factor 4 out of } x\text{-terms.}$$

$$4(x^2 - 2x + 1) + (y^2 + 4y + 4) = 8 + 4(1) + 4$$

$$4(x - 1)^2 + (y + 2)^2 = 16 \qquad \text{Write in completed square form.}$$

$$\frac{(x - 1)^2}{4} + \frac{(y + 2)^2}{16} = 1 \qquad \text{Divide each side by 16.}$$

$$\frac{(x - 1)^2}{2^2} + \frac{(y + 2)^2}{4^2} = 1 \qquad \text{Write in standard form.}$$

The major axis is vertical, where $h = 1$, $k = -2$, $a = 4$, $b = 2$, and

$$c = \sqrt{a^2 - b^2} = \sqrt{16 - 4} = \sqrt{12} = 2\sqrt{3}.$$

So, you have the following.

Center: $(1, -2)$ Vertices: $(1, -6)$ Foci: $\left(1, -2 - 2\sqrt{3}\right)$

$(1, 2)$ $\left(1, -2 + 2\sqrt{3}\right)$

The graph of the ellipse is shown in Figure 6.26.

CHECK Point   Now try Exercise 51.

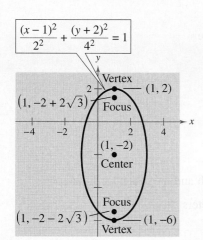

$$\frac{(x - 1)^2}{2^2} + \frac{(y + 2)^2}{4^2} = 1$$

FIGURE **6.26**

TECHNOLOGY

You can use a graphing utility to graph an ellipse by graphing the upper and lower portions in the same viewing window. For instance, to graph the ellipse in Example 3, first solve for y to get

$$y_1 = -2 + 4\sqrt{1 - \frac{(x - 1)^2}{4}} \quad \text{and} \quad y_2 = -2 - 4\sqrt{1 - \frac{(x - 1)^2}{4}}.$$

Use a viewing window in which $-6 \le x \le 9$ and $-7 \le y \le 3$. You should obtain the graph shown below.

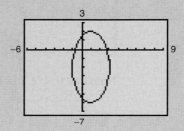

Application

Ellipses have many practical and aesthetic uses. For instance, machine gears, supporting arches, and acoustic designs often involve elliptical shapes. The orbits of satellites and planets are also ellipses. Example 4 investigates the elliptical orbit of the moon about Earth.

Example 4   An Application Involving an Elliptical Orbit

The moon travels about Earth in an elliptical orbit with Earth at one focus, as shown in Figure 6.27. The major and minor axes of the orbit have lengths of 768,800 kilometers and 767,640 kilometers, respectively. Find the greatest and smallest distances (the *apogee* and *perigee*, respectively) from Earth's center to the moon's center.

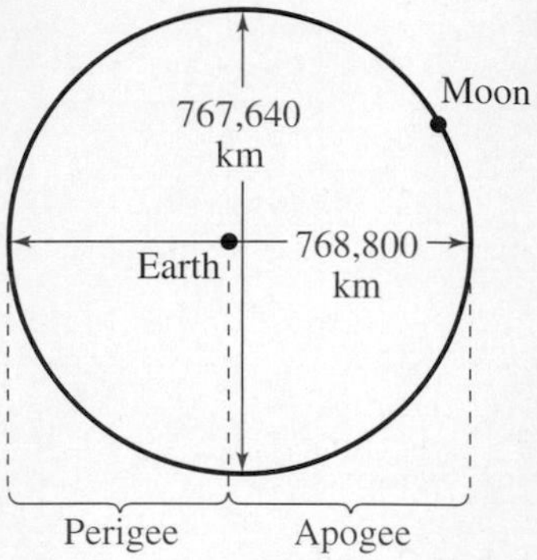

Solution

Because $2a = 768,800$ and $2b = 767,640$, you have

$$a = 384,400 \text{ and } b = 383,820$$

which implies that

$$c = \sqrt{a^2 - b^2}$$
$$= \sqrt{384,400^2 - 383,820^2}$$
$$\approx 21,108.$$

So, the greatest distance between the center of Earth and the center of the moon is

$$a + c \approx 384,400 + 21,108 = 405,508 \text{ kilometers}$$

and the smallest distance is

$$a - c \approx 384,400 - 21,108 = 363,292 \text{ kilometers}.$$

CHECK Point Now try Exercise 65. ▮

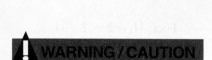

767,640 km

Moon

Earth

768,800 km

Perigee   Apogee

FIGURE **6.27**

⚠ **WARNING / CAUTION**

Note in Example 4 and Figure 6.27 that Earth *is not* the center of the moon's orbit.

Eccentricity

One of the reasons it was difficult for early astronomers to detect that the orbits of the planets are ellipses is that the foci of the planetary orbits are relatively close to their centers, and so the orbits are nearly circular. To measure the ovalness of an ellipse, you can use the concept of **eccentricity.**

Definition of Eccentricity

The **eccentricity** e of an ellipse is given by the ratio

$$e = \frac{c}{a}.$$

Note that $0 < e < 1$ for *every* ellipse.

To see how this ratio is used to describe the shape of an ellipse, note that because the foci of an ellipse are located along the major axis between the vertices and the center, it follows that

$$0 < c < a.$$

For an ellipse that is nearly circular, the foci are close to the center and the ratio c/a is small, as shown in Figure 6.28. On the other hand, for an elongated ellipse, the foci are close to the vertices and the ratio c/a is close to 1, as shown in Figure 6.29.

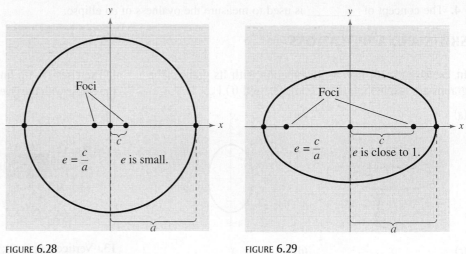

FIGURE **6.28**   FIGURE **6.29**

The orbit of the moon has an eccentricity of $e \approx 0.0549$, and the eccentricities of the eight planetary orbits are as follows.

Mercury:	$e \approx 0.2056$	Jupiter:	$e \approx 0.0484$
Venus:	$e \approx 0.0068$	Saturn:	$e \approx 0.0542$
Earth:	$e \approx 0.0167$	Uranus:	$e \approx 0.0472$
Mars:	$e \approx 0.0934$	Neptune:	$e \approx 0.0086$

The time it takes Saturn to orbit the sun is about 29.4 Earth years.

CLASSROOM DISCUSSION

Ellipses and Circles

a. Show that the equation of an ellipse can be written as

$$\frac{(x - h)^2}{a^2} + \frac{(y - k)^2}{a^2(1 - e^2)} = 1.$$

b. For the equation in part (a), let $a = 4$, $h = 1$, and $k = 2$, and use a graphing utility to graph the ellipse for $e = 0.95$, $e = 0.75$, $e = 0.5$, $e = 0.25$, and $e = 0.1$. Discuss the changes in the shape of the ellipse as e approaches 0.

c. Make a conjecture about the shape of the graph in part (b) when $e = 0$. What is the equation of this ellipse? What is another name for an ellipse with an eccentricity of 0?

6.3 EXERCISES

See www.CalcChat.com for worked-out solutions to odd-numbered exercises.

VOCABULARY: Fill in the blanks.

1. An _____ is the set of all points (x, y) in a plane, the sum of whose distances from two distinct fixed points, called _____, is constant.

2. The chord joining the vertices of an ellipse is called the _____ _____, and its midpoint is the _____ of the ellipse.

3. The chord perpendicular to the major axis at the center of the ellipse is called the _____ _____ of the ellipse.

4. The concept of _____ is used to measure the ovalness of an ellipse.

SKILLS AND APPLICATIONS

In Exercises 5–10, match the equation with its graph. [The graphs are labeled (a), (b), (c), (d), (e), and (f).]

(a)
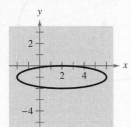

(b)
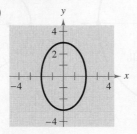

(c)
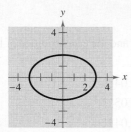

(d)
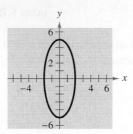

(e)
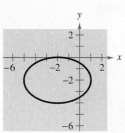

(f)
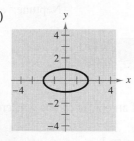

5. $\dfrac{x^2}{4} + \dfrac{y^2}{9} = 1$

6. $\dfrac{x^2}{9} + \dfrac{y^2}{4} = 1$

7. $\dfrac{x^2}{4} + \dfrac{y^2}{25} = 1$

8. $\dfrac{x^2}{4} + y^2 = 1$

9. $\dfrac{(x-2)^2}{16} + (y+1)^2 = 1$

10. $\dfrac{(x+2)^2}{9} + \dfrac{(y+2)^2}{4} = 1$

In Exercises 11–18, find the standard form of the equation of the ellipse with the given characteristics and center at the origin.

11.
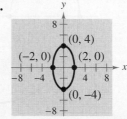

12.
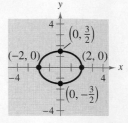

13. Vertices: $(\pm 7, 0)$; foci: $(\pm 2, 0)$

14. Vertices: $(0, \pm 8)$; foci: $(0, \pm 4)$

15. Foci: $(\pm 5, 0)$; major axis of length 14

16. Foci: $(\pm 2, 0)$; major axis of length 10

17. Vertices: $(0, \pm 5)$; passes through the point $(4, 2)$

18. Vertical major axis; passes through the points $(0, 6)$ and $(3, 0)$

In Exercises 19–28, find the standard form of the equation of the ellipse with the given characteristics.

19.
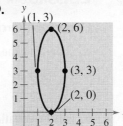

20.
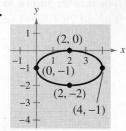

21. Vertices: $(0, 2)$, $(8, 2)$; minor axis of length 2

22. Foci: $(0, 0)$, $(4, 0)$; major axis of length 6

23. Foci: $(0, 0)$, $(0, 8)$; major axis of length 16

24. Center: $(2, -1)$; vertex: $\left(2, \frac{1}{2}\right)$; minor axis of length 2

25. Center: $(0, 4)$; $a = 2c$; vertices: $(-4, 4)$, $(4, 4)$

26. Center: $(3, 2)$; $a = 3c$; foci: $(1, 2)$, $(5, 2)$

27. Vertices: $(0, 2), (4, 2)$; endpoints of the minor axis: $(2, 3), (2, 1)$

28. Vertices: $(5, 0), (5, 12)$; endpoints of the minor axis: $(1, 6), (9, 6)$

In Exercises 29–52, identify the conic as a circle or an ellipse. Then find the center, radius, vertices, foci, and eccentricity of the conic (if applicable), and sketch its graph.

29. $\dfrac{x^2}{25} + \dfrac{y^2}{16} = 1$ **30.** $\dfrac{x^2}{16} + \dfrac{y^2}{81} = 1$

31. $\dfrac{x^2}{25} + \dfrac{y^2}{25} = 1$ **32.** $\dfrac{x^2}{9} + \dfrac{y^2}{9} = 1$

33. $\dfrac{x^2}{5} + \dfrac{y^2}{9} = 1$ **34.** $\dfrac{x^2}{64} + \dfrac{y^2}{28} = 1$

35. $\dfrac{(x - 4)^2}{16} + \dfrac{(y + 1)^2}{25} = 1$

36. $\dfrac{(x + 3)^2}{12} + \dfrac{(y - 2)^2}{16} = 1$

37. $\dfrac{x^2}{4/9} + \dfrac{(y + 1)^2}{4/9} = 1$

38. $\dfrac{(x + 5)^2}{9/4} + (y - 1)^2 = 1$

39. $(x + 2)^2 + \dfrac{(y + 4)^2}{1/4} = 1$

40. $\dfrac{(x - 3)^2}{25/4} + \dfrac{(y - 1)^2}{25/4} = 1$

41. $9x^2 + 4y^2 + 36x - 24y + 36 = 0$

42. $9x^2 + 4y^2 - 54x + 40y + 37 = 0$

43. $x^2 + y^2 - 2x + 4y - 31 = 0$

44. $x^2 + 5y^2 - 8x - 30y - 39 = 0$

45. $3x^2 + y^2 + 18x - 2y - 8 = 0$

46. $6x^2 + 2y^2 + 18x - 10y + 2 = 0$

47. $x^2 + 4y^2 - 6x + 20y - 2 = 0$

48. $x^2 + y^2 - 4x + 6y - 3 = 0$

49. $9x^2 + 9y^2 + 18x - 18y + 14 = 0$

50. $16x^2 + 25y^2 - 32x + 50y + 16 = 0$

51. $9x^2 + 25y^2 - 36x - 50y + 60 = 0$

52. $16x^2 + 16y^2 - 64x + 32y + 55 = 0$

In Exercises 53–56, use a graphing utility to graph the ellipse. Find the center, foci, and vertices. (Recall that it may be necessary to solve the equation for y and obtain two equations.)

53. $5x^2 + 3y^2 = 15$ **54.** $3x^2 + 4y^2 = 12$

55. $12x^2 + 20y^2 - 12x + 40y - 37 = 0$

56. $36x^2 + 9y^2 + 48x - 36y - 72 = 0$

In Exercises 57–60, find the eccentricity of the ellipse.

57. $\dfrac{x^2}{4} + \dfrac{y^2}{9} = 1$ **58.** $\dfrac{x^2}{25} + \dfrac{y^2}{36} = 1$

59. $x^2 + 9y^2 - 10x + 36y + 52 = 0$

60. $4x^2 + 3y^2 - 8x + 18y + 19 = 0$

61. Find an equation of the ellipse with vertices $(\pm 5, 0)$ and eccentricity $e = \frac{3}{5}$.

62. Find an equation of the ellipse with vertices $(0, \pm 8)$ and eccentricity $e = \frac{1}{2}$.

63. ARCHITECTURE   A semielliptical arch over a tunnel for a one-way road through a mountain has a major axis of 50 feet and a height at the center of 10 feet.

(a) Draw a rectangular coordinate system on a sketch of the tunnel with the center of the road entering the tunnel at the origin. Identify the coordinates of the known points.

(b) Find an equation of the semielliptical arch.

(c) You are driving a moving truck that has a width of 8 feet and a height of 9 feet. Will the moving truck clear the opening of the arch?

64. ARCHITECTURE   A fireplace arch is to be constructed in the shape of a semiellipse. The opening is to have a height of 2 feet at the center and a width of 6 feet along the base (see figure). The contractor draws the outline of the ellipse using tacks as described at the beginning of this section. Determine the required positions of the tacks and the length of the string.

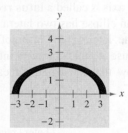

65. COMET ORBIT   Halley's comet has an elliptical orbit, with the sun at one focus. The eccentricity of the orbit is approximately 0.967. The length of the major axis of the orbit is approximately 35.88 astronomical units. (An astronomical unit is about 93 million miles.)

(a) Find an equation of the orbit. Place the center of the orbit at the origin, and place the major axis on the x-axis.

(b) Use a graphing utility to graph the equation of the orbit.

(c) Find the greatest (aphelion) and smallest (perihelion) distances from the sun's center to the comet's center.

66. SATELLITE ORBIT The first artificial satellite to orbit Earth was Sputnik I (launched by the former Soviet Union in 1957). Its highest point above Earth's surface was 947 kilometers, and its lowest point was 228 kilometers (see figure). The center of Earth was at one focus of the elliptical orbit, and the radius of Earth is 6378 kilometers. Find the eccentricity of the orbit.

228 km —   — 947 km

67. MOTION OF A PENDULUM The relation between the velocity y (in radians per second) of a pendulum and its angular displacement θ from the vertical can be modeled by a semiellipse. A 12-centimeter pendulum crests ($y = 0$) when the angular displacement is -0.2 radian and 0.2 radian. When the pendulum is at equilibrium ($\theta = 0$), the velocity is -1.6 radians per second.

(a) Find an equation that models the motion of the pendulum. Place the center at the origin.

(b) Graph the equation from part (a).

(c) Which half of the ellipse models the motion of the pendulum?

68. GEOMETRY A line segment through a focus of an ellipse with endpoints on the ellipse and perpendicular to the major axis is called a **latus rectum** of the ellipse. Therefore, an ellipse has two latera recta. Knowing the length of the latera recta is helpful in sketching an ellipse because it yields other points on the curve (see figure). Show that the length of each latus rectum is $2b^2/a$.

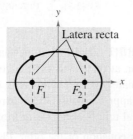

In Exercises 69–72, sketch the graph of the ellipse, using latera recta (see Exercise 68).

69. $\dfrac{x^2}{9} + \dfrac{y^2}{16} = 1$ **70.** $\dfrac{x^2}{4} + \dfrac{y^2}{1} = 1$

71. $5x^2 + 3y^2 = 15$ **72.** $9x^2 + 4y^2 = 36$

EXPLORATION

TRUE OR FALSE? In Exercises 73 and 74, determine whether the statement is true or false. Justify your answer.

73. The graph of $x^2 + 4y^4 - 4 = 0$ is an ellipse.

74. It is easier to distinguish the graph of an ellipse from the graph of a circle if the eccentricity of the ellipse is large (close to 1).

75. Consider the ellipse

$$\frac{x^2}{a^2} + \frac{y^2}{b^2} = 1, \quad a + b = 20.$$

(a) The area of the ellipse is given by $A = \pi ab$. Write the area of the ellipse as a function of a.

(b) Find the equation of an ellipse with an area of 264 square centimeters.

(c) Complete the table using your equation from part (a), and make a conjecture about the shape of the ellipse with maximum area.

a	8	9	10	11	12	13
A						

(d) Use a graphing utility to graph the area function and use the graph to support your conjecture in part (c).

76. THINK ABOUT IT At the beginning of this section it was noted that an ellipse can be drawn using two thumbtacks, a string of fixed length (greater than the distance between the two tacks), and a pencil. If the ends of the string are fastened at the tacks and the string is drawn taut with a pencil, the path traced by the pencil is an ellipse.

(a) What is the length of the string in terms of a?

(b) Explain why the path is an ellipse.

77. THINK ABOUT IT Find the equation of an ellipse such that for any point on the ellipse, the sum of the distances from the point (2, 2) and (10, 2) is 36.

78. CAPSTONE Describe the relationship between circles and ellipses. How are they similar? How do they differ?

79. PROOF Show that $a^2 = b^2 + c^2$ for the ellipse

$$\frac{x^2}{a^2} + \frac{y^2}{b^2} = 1$$

where $a > 0$, $b > 0$, and the distance from the center of the ellipse (0, 0) to a focus is c.

6.4 HYPERBOLAS

What you should learn

- Write equations of hyperbolas in standard form.
- Find asymptotes of and graph hyperbolas.
- Use properties of hyperbolas to solve real-life problems.
- Classify conics from their general equations.

Why you should learn it

Hyperbolas can be used to model and solve many types of real-life problems. For instance, in Exercise 54 on page 471, hyperbolas are used in long distance radio navigation for aircraft and ships.

U.S. Navy, William Lipski/AP Photo

Introduction

The third type of conic is called a **hyperbola.** The definition of a hyperbola is similar to that of an ellipse. The difference is that for an ellipse the *sum* of the distances between the foci and a point on the ellipse is fixed, whereas for a hyperbola the *difference* of the distances between the foci and a point on the hyperbola is fixed.

> ### Definition of Hyperbola
>
> A **hyperbola** is the set of all points (x, y) in a plane, the difference of whose distances from two distinct fixed points (**foci**) is a positive constant. See Figure 6.30.

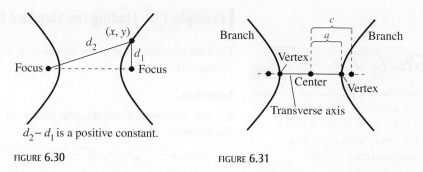

FIGURE **6.30**

FIGURE **6.31**

The graph of a hyperbola has two disconnected **branches.** The line through the two foci intersects the hyperbola at its two **vertices.** The line segment connecting the vertices is the **transverse axis,** and the midpoint of the transverse axis is the **center** of the hyperbola. See Figure 6.31. The development of the standard form of the equation of a hyperbola is similar to that of an ellipse. Note in the definition below that a, b, and c are related differently for hyperbolas than for ellipses.

> ### Standard Equation of a Hyperbola
>
> The **standard form of the equation of a hyperbola** with center (h, k) is
>
> $$\frac{(x - h)^2}{a^2} - \frac{(y - k)^2}{b^2} = 1 \qquad \text{Transverse axis is horizontal.}$$
>
> $$\frac{(y - k)^2}{a^2} - \frac{(x - h)^2}{b^2} = 1. \qquad \text{Transverse axis is vertical.}$$
>
> The vertices are a units from the center, and the foci are c units from the center. Moreover, $c^2 = a^2 + b^2$. If the center of the hyperbola is at the origin $(0, 0)$, the equation takes one of the following forms.
>
> $$\frac{x^2}{a^2} - \frac{y^2}{b^2} = 1 \qquad \text{Transverse axis is horizontal.} \qquad \frac{y^2}{a^2} - \frac{x^2}{b^2} = 1 \qquad \text{Transverse axis is vertical.}$$

Figure 6.32 shows both the horizontal and vertical orientations for a hyperbola.

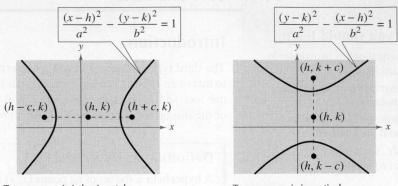

$$\frac{(x-h)^2}{a^2} - \frac{(y-k)^2}{b^2} = 1$$

$$\frac{(y-k)^2}{a^2} - \frac{(x-h)^2}{b^2} = 1$$

Transverse axis is horizontal.   Transverse axis is vertical.

FIGURE **6.32**

Example 1 Finding the Standard Equation of a Hyperbola

Find the standard form of the equation of the hyperbola with foci $(-1, 2)$ and $(5, 2)$ and vertices $(0, 2)$ and $(4, 2)$.

Solution

By the Midpoint Formula, the center of the hyperbola occurs at the point $(2, 2)$. Furthermore, $c = 5 - 2 = 3$ and $a = 4 - 2 = 2$, and it follows that

$$b = \sqrt{c^2 - a^2} = \sqrt{3^2 - 2^2} = \sqrt{9 - 4} = \sqrt{5}.$$

So, the hyperbola has a horizontal transverse axis and the standard form of the equation is

$$\frac{(x-2)^2}{2^2} - \frac{(y-2)^2}{\left(\sqrt{5}\right)^2} = 1. \qquad \text{See Figure 6.33.}$$

This equation simplifies to

$$\frac{(x-2)^2}{4} - \frac{(y-2)^2}{5} = 1.$$

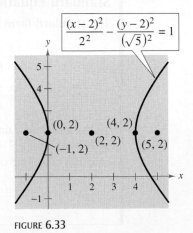

$$\frac{(x-2)^2}{2^2} - \frac{(y-2)^2}{(\sqrt{5})^2} = 1$$

FIGURE **6.33**

CHECK *Point* Now try Exercise 35.

Asymptotes of a Hyperbola

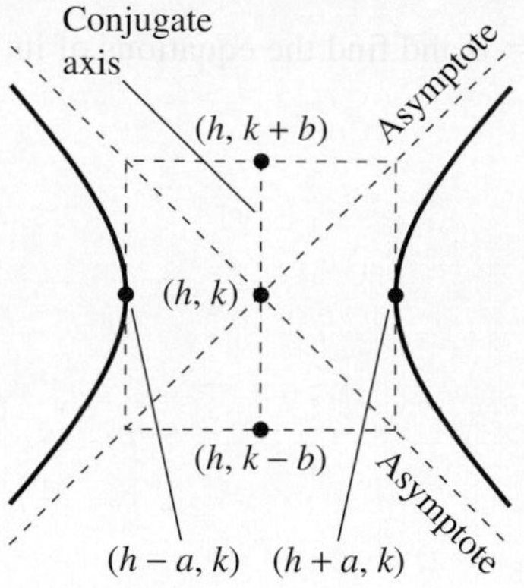

Each hyperbola has two **asymptotes** that intersect at the center of the hyperbola, as shown in Figure 6.34. The asymptotes pass through the vertices of a rectangle of dimensions $2a$ by $2b$, with its center at (h, k). The line segment of length $2b$ joining $(h, k + b)$ and $(h, k - b)$ [or $(h + b, k)$ and $(h - b, k)$] is the **conjugate axis** of the hyperbola.

Conjugate axis

$(h, k + b)$

Asymptote

(h, k)

$(h, k - b)$

$(h - a, k)$ $(h + a, k)$

Asymptote

FIGURE **6.34**

Asymptotes of a Hyperbola

The equations of the asymptotes of a hyperbola are

$$y = k \pm \frac{b}{a}(x - h)$$ Transverse axis is horizontal.

$$y = k \pm \frac{a}{b}(x - h).$$ Transverse axis is vertical.

Example 2   Using Asymptotes to Sketch a Hyperbola

Sketch the hyperbola whose equation is $4x^2 - y^2 = 16$.

Algebraic Solution

Divide each side of the original equation by 16, and rewrite the equation in standard form.

$$\frac{x^2}{2^2} - \frac{y^2}{4^2} = 1$$ Write in standard form.

From this, you can conclude that $a = 2$, $b = 4$, and the transverse axis is horizontal. So, the vertices occur at $(-2, 0)$ and $(2, 0)$, and the endpoints of the conjugate axis occur at $(0, -4)$ and $(0, 4)$. Using these four points, you are able to sketch the rectangle shown in Figure 6.35. Now, from $c^2 = a^2 + b^2$, you have $c = \sqrt{2^2 + 4^2} = \sqrt{20} = 2\sqrt{5}$. So, the foci of the hyperbola are $(-2\sqrt{5}, 0)$ and $(2\sqrt{5}, 0)$. Finally, by drawing the asymptotes through the corners of this rectangle, you can complete the sketch shown in Figure 6.36. Note that the asymptotes are $y = 2x$ and $y = -2x$.

Graphical Solution

Solve the equation of the hyperbola for y as follows.

$$4x^2 - y^2 = 16$$
$$4x^2 - 16 = y^2$$
$$\pm\sqrt{4x^2 - 16} = y$$

Then use a graphing utility to graph $y_1 = \sqrt{4x^2 - 16}$ and $y_2 = -\sqrt{4x^2 - 16}$ in the same viewing window. Be sure to use a square setting. From the graph in Figure 6.37, you can see that the transverse axis is horizontal. You can use the *zoom* and *trace* features to approximate the vertices to be $(-2, 0)$ and $(2, 0)$.

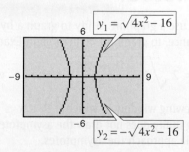

$y_1 = \sqrt{4x^2 - 16}$

$y_2 = -\sqrt{4x^2 - 16}$

FIGURE **6.37**

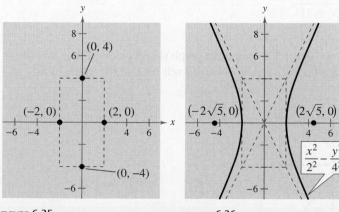

FIGURE **6.35**   FIGURE **6.36**

CHECK *Point*   Now try Exercise 11.

Example 3 **Finding the Asymptotes of a Hyperbola**

Sketch the hyperbola given by $4x^2 - 3y^2 + 8x + 16 = 0$ and find the equations of its asymptotes and the foci.

Solution

$4x^2 - 3y^2 + 8x + 16 = 0$	Write original equation.
$(4x^2 + 8x) - 3y^2 = -16$	Group terms.
$4(x^2 + 2x) - 3y^2 = -16$	Factor 4 from x-terms.
$4(x^2 + 2x + 1) - 3y^2 = -16 + 4$	Add 4 to each side.
$4(x + 1)^2 - 3y^2 = -12$	Write in completed square form.
$-\dfrac{(x + 1)^2}{3} + \dfrac{y^2}{4} = 1$	Divide each side by -12.
$\dfrac{y^2}{2^2} - \dfrac{(x + 1)^2}{(\sqrt{3})^2} = 1$	Write in standard form.

From this equation you can conclude that the hyperbola has a vertical transverse axis, centered at $(-1, 0)$, has vertices $(-1, 2)$ and $(-1, -2)$, and has a conjugate axis with endpoints $(-1 - \sqrt{3}, 0)$ and $(-1 + \sqrt{3}, 0)$. To sketch the hyperbola, draw a rectangle through these four points. The asymptotes are the lines passing through the corners of the rectangle. Using $a = 2$ and $b = \sqrt{3}$, you can conclude that the equations of the asymptotes are

$$y = \frac{2}{\sqrt{3}}(x + 1) \qquad \text{and} \qquad y = -\frac{2}{\sqrt{3}}(x + 1).$$

Finally, you can determine the foci by using the equation $c^2 = a^2 + b^2$. So, you have $c = \sqrt{2^2 + (\sqrt{3})^2} = \sqrt{7}$, and the foci are $(-1, \sqrt{7})$ and $(-1, -\sqrt{7})$. The hyperbola is shown in Figure 6.38.

CHECK *Point* Now try Exercise 19.

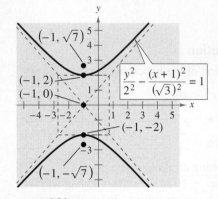

FIGURE 6.38

TECHNOLOGY

You can use a graphing utility to graph a hyperbola by graphing the upper and lower portions in the same viewing window. For instance, to graph the hyperbola in Example 3, first solve for y to get

$$y_1 = 2\sqrt{1 + \frac{(x + 1)^2}{3}} \qquad \text{and} \qquad y_2 = -2\sqrt{1 + \frac{(x + 1)^2}{3}}.$$

Use a viewing window in which $-9 \leq x \leq 9$ and $-6 \leq y \leq 6$. You should obtain the graph shown below. Notice that the graphing utility does not draw the asymptotes. However, if you trace along the branches, you will see that the values of the hyperbola approach the asymptotes.

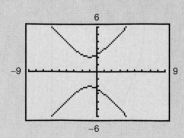

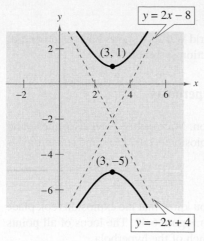

$y = 2x - 8$

$(3, 1)$

$(3, -5)$

$y = -2x + 4$

FIGURE **6.39**

Example 4   Using Asymptotes to Find the Standard Equation

Find the standard form of the equation of the hyperbola having vertices $(3, -5)$ and $(3, 1)$ and having asymptotes

$$y = 2x - 8 \qquad \text{and} \qquad y = -2x + 4$$

as shown in Figure 6.39.

Solution

By the Midpoint Formula, the center of the hyperbola is $(3, -2)$. Furthermore, the hyperbola has a vertical transverse axis with $a = 3$. From the original equations, you can determine the slopes of the asymptotes to be

$$m_1 = 2 = \frac{a}{b} \qquad \text{and} \qquad m_2 = -2 = -\frac{a}{b}$$

and, because $a = 3$, you can conclude

$$2 = \frac{a}{b} \implies 2 = \frac{3}{b} \implies b = \frac{3}{2}.$$

So, the standard form of the equation is

$$\frac{(y + 2)^2}{3^2} - \frac{(x - 3)^2}{\left(\dfrac{3}{2}\right)^2} = 1.$$

CHECK Point   Now try Exercise 43.

As with ellipses, the *eccentricity* of a hyperbola is

$$e = \frac{c}{a} \qquad \text{Eccentricity}$$

and because $c > a$, it follows that $e > 1$. If the eccentricity is large, the branches of the hyperbola are nearly flat, as shown in Figure 6.40. If the eccentricity is close to 1, the branches of the hyperbola are more narrow, as shown in Figure 6.41.

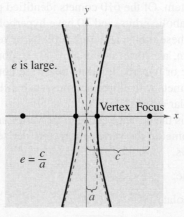

e is large.

Vertex Focus

$e = \dfrac{c}{a}$

FIGURE **6.40**

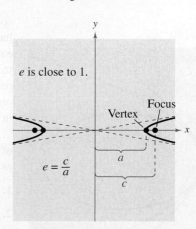

e is close to 1.

Focus

Vertex

$e = \dfrac{c}{a}$

FIGURE **6.41**

Applications

The following application was developed during World War II. It shows how the properties of hyperbolas can be used in radar and other detection systems.

Example 5   An Application Involving Hyperbolas

Two microphones, 1 mile apart, record an explosion. Microphone A receives the sound 2 seconds before microphone B. Where did the explosion occur? (Assume sound travels at 1100 feet per second.)

Solution

Assuming sound travels at 1100 feet per second, you know that the explosion took place 2200 feet farther from B than from A, as shown in Figure 6.42. The locus of all points that are 2200 feet closer to A than to B is one branch of the hyperbola

$$\frac{x^2}{a^2} - \frac{y^2}{b^2} = 1$$

where

$$c = \frac{5280}{2} = 2640$$

and

$$a = \frac{2200}{2} = 1100.$$

So, $b^2 = c^2 - a^2 = 2640^2 - 1100^2 = 5{,}759{,}600$, and you can conclude that the explosion occurred somewhere on the right branch of the hyperbola

$$\frac{x^2}{1{,}210{,}000} - \frac{y^2}{5{,}759{,}600} = 1.$$

CHECKPoint   Now try Exercise 53. ∎

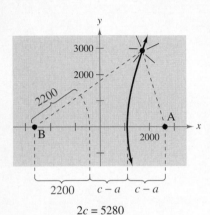

2c = 5280
2200 + 2(c − a) = 5280

FIGURE 6.42

Another interesting application of conic sections involves the orbits of comets in our solar system. Of the 610 comets identified prior to 1970, 245 have elliptical orbits, 295 have parabolic orbits, and 70 have hyperbolic orbits. The center of the sun is a focus of each of these orbits, and each orbit has a vertex at the point where the comet is closest to the sun, as shown in Figure 6.43. Undoubtedly, there have been many comets with parabolic or hyperbolic orbits that were not identified. We only get to see such comets *once*. Comets with elliptical orbits, such as Halley's comet, are the only ones that remain in our solar system.

If p is the distance between the vertex and the focus (in meters), and v is the velocity of the comet at the vertex (in meters per second), then the type of orbit is determined as follows.

1. Ellipse: $v < \sqrt{2GM/p}$

2. Parabola: $v = \sqrt{2GM/p}$

3. Hyperbola: $v > \sqrt{2GM/p}$

In each of these relations, $M = 1.989 \times 10^{30}$ kilograms (the mass of the sun) and $G \approx 6.67 \times 10^{-11}$ cubic meter per kilogram-second squared (the universal gravitational constant).

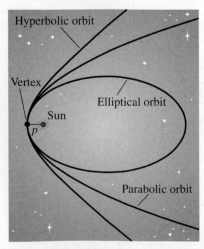

FIGURE 6.43

General Equations of Conics

> ### Classifying a Conic from Its General Equation
> The graph of $Ax^2 + Cy^2 + Dx + Ey + F = 0$ is one of the following.
>
> **1.** *Circle:* $A = C$
>
> **2.** *Parabola:* $AC = 0$ $A = 0$ or $C = 0$, but not both.
>
> **3.** *Ellipse:* $AC > 0$ A and C have like signs.
>
> **4.** *Hyperbola:* $AC < 0$ A and C have unlike signs.

The test above is valid *if* the graph is a conic. The test does not apply to equations such as $x^2 + y^2 = -1$, whose graph is not a conic.

Example 6   Classifying Conics from General Equations

Classify the graph of each equation.

a. $4x^2 - 9x + y - 5 = 0$

b. $4x^2 - y^2 + 8x - 6y + 4 = 0$

c. $2x^2 + 4y^2 - 4x + 12y = 0$

d. $2x^2 + 2y^2 - 8x + 12y + 2 = 0$

Solution

a. For the equation $4x^2 - 9x + y - 5 = 0$, you have

$$AC = 4(0) = 0. \qquad \text{Parabola}$$

So, the graph is a parabola.

b. For the equation $4x^2 - y^2 + 8x - 6y + 4 = 0$, you have

$$AC = 4(-1) < 0. \qquad \text{Hyperbola}$$

So, the graph is a hyperbola.

c. For the equation $2x^2 + 4y^2 - 4x + 12y = 0$, you have

$$AC = 2(4) > 0. \qquad \text{Ellipse}$$

So, the graph is an ellipse.

d. For the equation $2x^2 + 2y^2 - 8x + 12y + 2 = 0$, you have

$$A = C = 2. \qquad \text{Circle}$$

So, the graph is a circle.

CHECK Point Now try Exercise 61.

CLASSROOM DISCUSSION

Sketching Conics Sketch each of the conics described in Example 6. Write a paragraph describing the procedures that allow you to sketch the conics efficiently.

EXERCISES

See www.CalcChat.com for worked-out solutions to odd-numbered exercises.

VOCABULARY: Fill in the blanks.

1. A _____ is the set of all points (x, y) in a plane, the difference of whose distances from two distinct fixed points, called _____, is a positive constant.

2. The graph of a hyperbola has two disconnected parts called _____.

3. The line segment connecting the vertices of a hyperbola is called the _____ _____, and the midpoint of the line segment is the _____ of the hyperbola.

4. Each hyperbola has two _____ that intersect at the center of the hyperbola.

SKILLS AND APPLICATIONS

In Exercises 5–8, match the equation with its graph. [The graphs are labeled (a), (b), (c), and (d).]

(a)
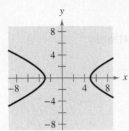

(b)
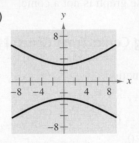

(c)
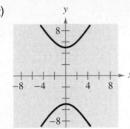

(d)
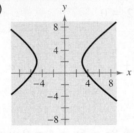

5. $\dfrac{y^2}{9} - \dfrac{x^2}{25} = 1$

6. $\dfrac{y^2}{25} - \dfrac{x^2}{9} = 1$

7. $\dfrac{(x-1)^2}{16} - \dfrac{y^2}{4} = 1$

8. $\dfrac{(x+1)^2}{16} - \dfrac{(y-2)^2}{9} = 1$

In Exercises 9–22, find the center, vertices, foci, and the equations of the asymptotes of the hyperbola, and sketch its graph using the asymptotes as an aid.

9. $x^2 - y^2 = 1$

10. $\dfrac{x^2}{9} - \dfrac{y^2}{25} = 1$

11. $\dfrac{y^2}{25} - \dfrac{x^2}{81} = 1$

12. $\dfrac{x^2}{36} - \dfrac{y^2}{4} = 1$

13. $\dfrac{y^2}{1} - \dfrac{x^2}{4} = 1$

14. $\dfrac{y^2}{9} - \dfrac{x^2}{1} = 1$

15. $\dfrac{(x-1)^2}{4} - \dfrac{(y+2)^2}{1} = 1$

16. $\dfrac{(x+3)^2}{144} - \dfrac{(y-2)^2}{25} = 1$

17. $\dfrac{(y+6)^2}{1/9} - \dfrac{(x-2)^2}{1/4} = 1$

18. $\dfrac{(y-1)^2}{1/4} - \dfrac{(x+3)^2}{1/16} = 1$

19. $9x^2 - y^2 - 36x - 6y + 18 = 0$

20. $x^2 - 9y^2 + 36y - 72 = 0$

21. $x^2 - 9y^2 + 2x - 54y - 80 = 0$

22. $16y^2 - x^2 + 2x + 64y + 63 = 0$

In Exercises 23–28, find the center, vertices, foci, and the equations of the asymptotes of the hyperbola. Use a graphing utility to graph the hyperbola and its asymptotes.

23. $2x^2 - 3y^2 = 6$

24. $6y^2 - 3x^2 = 18$

25. $4x^2 - 9y^2 = 36$

26. $25x^2 - 4y^2 = 100$

27. $9y^2 - x^2 + 2x + 54y + 62 = 0$

28. $9x^2 - y^2 + 54x + 10y + 55 = 0$

In Exercises 29–34, find the standard form of the equation of the hyperbola with the given characteristics and center at the origin.

29. Vertices: $(0, \pm 2)$; foci: $(0, \pm 4)$

30. Vertices: $(\pm 4, 0)$; foci: $(\pm 6, 0)$

31. Vertices: $(\pm 1, 0)$; asymptotes: $y = \pm 5x$

32. Vertices: $(0, \pm 3)$; asymptotes: $y = \pm 3x$

33. Foci: $(0, \pm 8)$; asymptotes: $y = \pm 4x$

34. Foci: $(\pm 10, 0)$; asymptotes: $y = \pm \frac{3}{4}x$

In Exercises 35–46, find the standard form of the equation of the hyperbola with the given characteristics.

35. Vertices: $(2, 0), (6, 0)$; foci: $(0, 0), (8, 0)$

36. Vertices: $(2, 3), (2, -3)$; foci: $(2, 6), (2, -6)$

37. Vertices: $(4, 1), (4, 9)$; foci: $(4, 0), (4, 10)$

38. Vertices: $(-2, 1), (2, 1)$; foci: $(-3, 1), (3, 1)$

39. Vertices: $(2, 3)$, $(2, -3)$;
   passes through the point $(0, 5)$

40. Vertices: $(-2, 1)$, $(2, 1)$;
   passes through the point $(5, 4)$

41. Vertices: $(0, 4)$, $(0, 0)$;
   passes through the point $\left(\sqrt{5}, -1 \right)$

42. Vertices: $(1, 2)$, $(1, -2)$;
   passes through the point $\left(0, \sqrt{5} \right)$

43. Vertices: $(1, 2)$, $(3, 2)$;
   asymptotes: $y = x$, $y = 4 - x$

44. Vertices: $(3, 0)$, $(3, 6)$;
   asymptotes: $y = 6 - x$, $y = x$

45. Vertices: $(0, 2)$, $(6, 2)$;
   asymptotes: $y = \frac{2}{3}x$, $y = 4 - \frac{2}{3}x$

46. Vertices: $(3, 0)$, $(3, 4)$;
   asymptotes: $y = \frac{2}{3}x$, $y = 4 - \frac{2}{3}x$

In Exercises 47–50, write the standard form of the equation of the hyperbola.

47.

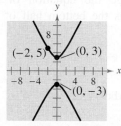

48.

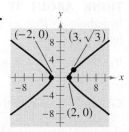

49.

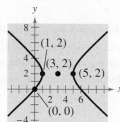

50.

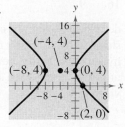

51. ART A sculpture has a hyperbolic cross section (see figure).

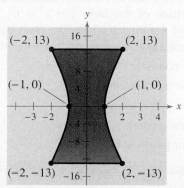

(a) Write an equation that models the curved sides of the sculpture.

(b) Each unit in the coordinate plane represents 1 foot. Find the width of the sculpture at a height of 5 feet.

52. SOUND LOCATION You and a friend live 4 miles apart (on the same "east-west" street) and are talking on the phone. You hear a clap of thunder from lightning in a storm, and 18 seconds later your friend hears the thunder. Find an equation that gives the possible places where the lightning could have occurred. (Assume that the coordinate system is measured in feet and that sound travels at 1100 feet per second.)

53. SOUND LOCATION Three listening stations located at $(3300, 0)$, $(3300, 1100)$, and $(-3300, 0)$ monitor an explosion. The last two stations detect the explosion 1 second and 4 seconds after the first, respectively. Determine the coordinates of the explosion. (Assume that the coordinate system is measured in feet and that sound travels at 1100 feet per second.)

54. LORAN Long distance radio navigation for aircraft and ships uses synchronized pulses transmitted by widely separated transmitting stations. These pulses travel at the speed of light (186,000 miles per second). The difference in the times of arrival of these pulses at an aircraft or ship is constant on a hyperbola having the transmitting stations as foci. Assume that two stations, 300 miles apart, are positioned on the rectangular coordinate system at points with coordinates $(-150, 0)$ and $(150, 0)$, and that a ship is traveling on a hyperbolic path with coordinates $(x, 75)$ (see figure).

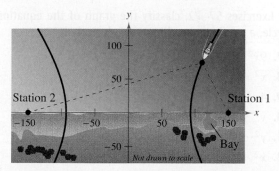

(a) Find the x-coordinate of the position of the ship if the time difference between the pulses from the transmitting stations is 1000 microseconds (0.001 second).

(b) Determine the distance between the ship and station 1 when the ship reaches the shore.

(c) The ship wants to enter a bay located between the two stations. The bay is 30 miles from station 1. What should be the time difference between the pulses?

(d) The ship is 60 miles offshore when the time difference in part (c) is obtained. What is the position of the ship?

55. PENDULUM The base for a pendulum of a clock has the shape of a hyperbola (see figure).

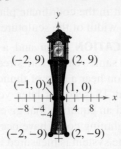

(−2, 9) (2, 9)

(−1, 0) (1, 0)

−8 −4 4 8

(−2, −9) (2, −9)

(a) Write an equation of the cross section of the base.

(b) Each unit in the coordinate plane represents $\frac{1}{2}$ foot. Find the width of the base of the pendulum 4 inches from the bottom.

56. HYPERBOLIC MIRROR A hyperbolic mirror (used in some telescopes) has the property that a light ray directed at a focus will be reflected to the other focus. The focus of a hyperbolic mirror (see figure) has coordinates (24, 0). Find the vertex of the mirror if the mount at the top edge of the mirror has coordinates (24, 24).

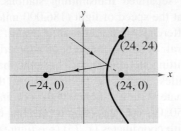

(24, 24)

(−24, 0) (24, 0)

In Exercises 57–72, classify the graph of the equation as a circle, a parabola, an ellipse, or a hyperbola.

57. $9x^2 + 4y^2 - 18x + 16y - 119 = 0$

58. $x^2 + y^2 - 4x - 6y - 23 = 0$

59. $4x^2 - y^2 - 4x - 3 = 0$

60. $y^2 - 6y - 4x + 21 = 0$

61. $y^2 - 4x^2 + 4x - 2y - 4 = 0$

62. $x^2 + y^2 - 4x + 6y - 3 = 0$

63. $y^2 + 12x + 4y + 28 = 0$

64. $4x^2 + 25y^2 + 16x + 250y + 541 = 0$

65. $4x^2 + 3y^2 + 8x - 24y + 51 = 0$

66. $4y^2 - 2x^2 - 4y - 8x - 15 = 0$

67. $25x^2 - 10x - 200y - 119 = 0$

68. $4y^2 + 4x^2 - 24x + 35 = 0$

69. $x^2 - 6x - 2y + 7 = 0$

70. $9x^2 + 4y^2 - 90x + 8y + 228 = 0$

71. $100x^2 + 100y^2 - 100x + 400y + 409 = 0$

72. $4x^2 - y^2 + 4x + 2y - 1 = 0$

EXPLORATION

TRUE OR FALSE? In Exercises 73–76, determine whether the statement is true or false. Justify your answer.

73. In the standard form of the equation of a hyperbola, the larger the ratio of b to a, the larger the eccentricity of the hyperbola.

74. In the standard form of the equation of a hyperbola, the trivial solution of two intersecting lines occurs when $b = 0$.

75. If $D \neq 0$ and $E \neq 0$, then the graph of $x^2 - y^2 + Dx + Ey = 0$ is a hyperbola.

76. If the asymptotes of the hyperbola $\dfrac{x^2}{a^2} - \dfrac{y^2}{b^2} = 1$, where $a, b > 0$, intersect at right angles, then $a = b$.

77. Consider a hyperbola centered at the origin with a horizontal transverse axis. Use the definition of a hyperbola to derive its standard form.

78. WRITING Explain how the central rectangle of a hyperbola can be used to sketch its asymptotes.

79. THINK ABOUT IT Change the equation of the hyperbola so that its graph is the bottom half of the hyperbola.

$$9x^2 - 54x - 4y^2 + 8y + 41 = 0$$

80. CAPSTONE Given the hyperbolas

$$\frac{x^2}{16} - \frac{y^2}{9} = 1 \quad \text{and} \quad \frac{y^2}{9} - \frac{x^2}{16} = 1$$

describe any common characteristics that the hyperbolas share, as well as any differences in the graphs of the hyperbolas. Verify your results by using a graphing utility to graph each of the hyperbolas in the same viewing window.

81. A circle and a parabola can have 0, 1, 2, 3, or 4 points of intersection. Sketch the circle given by $x^2 + y^2 = 4$. Discuss how this circle could intersect a parabola with an equation of the form $y = x^2 + C$. Then find the values of C for each of the five cases described below. Use a graphing utility to verify your results.

(a) No points of intersection

(b) One point of intersection

(c) Two points of intersection

(d) Three points of intersection

(e) Four points of intersection

6.5   ROTATION OF CONICS

What you should learn

- Rotate the coordinate axes to eliminate the *xy*-term in equations of conics.
- Use the discriminant to classify conics.

Why you should learn it

As illustrated in Exercises 13–26 on page 479, rotation of the coordinate axes can help you identify the graph of a general second-degree equation.

Rotation

In the preceding section, you learned that the equation of a conic with axes parallel to one of the coordinate axes has a standard form that can be written in the general form

$$Ax^2 + Cy^2 + Dx + Ey + F = 0. \qquad \text{Horizontal or vertical axis}$$

In this section, you will study the equations of conics whose axes are rotated so that they are not parallel to either the *x*-axis or the *y*-axis. The general equation for such conics contains an *xy*-term.

$$Ax^2 + Bxy + Cy^2 + Dx + Ey + F = 0 \qquad \text{Equation in } xy\text{-plane}$$

To eliminate this *xy*-term, you can use a procedure called **rotation of axes.** The objective is to rotate the *x*- and *y*-axes until they are parallel to the axes of the conic. The rotated axes are denoted as the *x'*-axis and the *y'*-axis, as shown in Figure 6.44.

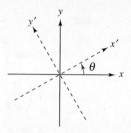

FIGURE 6.44

After the rotation, the equation of the conic in the new *x'y'*-plane will have the form

$$A'(x')^2 + C'(y')^2 + D'x' + E'y' + F' = 0. \qquad \text{Equation in } x'y'\text{-plane}$$

Because this equation has no *xy*-term, you can obtain a standard form by completing the square. The following theorem identifies how much to rotate the axes to eliminate the *xy*-term and also the equations for determining the new coefficients A', C', D', E', and F'.

Rotation of Axes to Eliminate an *xy*-Term

The general second-degree equation

$$Ax^2 + Bxy + Cy^2 + Dx + Ey + F = 0$$

can be rewritten as

$$A'(x')^2 + C'(y')^2 + D'x' + E'y' + F' = 0$$

by rotating the coordinate axes through an angle θ, where

$$\cot 2\theta = \frac{A - C}{B}.$$

The coefficients of the new equation are obtained by making the substitutions $x = x'\cos\theta - y'\sin\theta$ and $y = x'\sin\theta + y'\cos\theta$.

Remember that the substitutions

$$x = x' \cos \theta - y' \sin \theta$$

and

$$y = x' \sin \theta + y' \cos \theta$$

were developed to eliminate the $x'y'$-term in the rotated system. You can use this as a check on your work. In other words, if your final equation contains an $x'y'$-term, you know that you have made a mistake.

Example 1   **Rotation of Axes for a Hyperbola**

Write the equation $xy - 1 = 0$ in standard form.

Solution

Because $A = 0$, $B = 1$, and $C = 0$, you have

$$\cot 2\theta = \frac{A - C}{B} = 0 \quad \Longrightarrow \quad 2\theta = \frac{\pi}{2} \quad \Longrightarrow \quad \theta = \frac{\pi}{4}$$

which implies that

$$x = x' \cos \frac{\pi}{4} - y' \sin \frac{\pi}{4}$$

$$= x'\left(\frac{1}{\sqrt{2}}\right) - y'\left(\frac{1}{\sqrt{2}}\right)$$

$$= \frac{x' - y'}{\sqrt{2}}$$

and

$$y = x' \sin \frac{\pi}{4} + y' \cos \frac{\pi}{4}$$

$$= x'\left(\frac{1}{\sqrt{2}}\right) + y'\left(\frac{1}{\sqrt{2}}\right)$$

$$= \frac{x' + y'}{\sqrt{2}}.$$

The equation in the $x'y'$-system is obtained by substituting these expressions in the equation $xy - 1 = 0$.

$$\left(\frac{x' - y'}{\sqrt{2}}\right)\left(\frac{x' + y'}{\sqrt{2}}\right) - 1 = 0$$

$$\frac{(x')^2 - (y')^2}{2} - 1 = 0$$

$$\frac{(x')^2}{(\sqrt{2})^2} - \frac{(y')^2}{(\sqrt{2})^2} = 1 \qquad \text{Write in standard form.}$$

In the $x'y'$-system, this is a hyperbola centered at the origin with vertices at $(\pm\sqrt{2}, 0)$, as shown in Figure 6.45. To find the coordinates of the vertices in the xy-system, substitute the coordinates $(\pm\sqrt{2}, 0)$ in the equations

$$x = \frac{x' - y'}{\sqrt{2}} \qquad \text{and} \qquad y = \frac{x' + y'}{\sqrt{2}}.$$

This substitution yields the vertices $(1, 1)$ and $(-1, -1)$ in the xy-system. Note also that the asymptotes of the hyperbola have equations $y' = \pm x'$, which correspond to the original x- and y-axes.

CHECK*Point*   Now try Exercise 13.

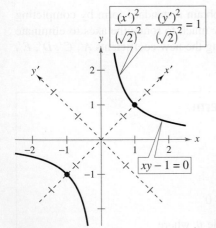

Vertices:
In $x'y'$-system: $(\sqrt{2}, 0), (-\sqrt{2}, 0)$
In xy-system: $(1, 1), (-1, -1)$
FIGURE **6.45**

| **Example 2**   **Rotation of Axes for an Ellipse**

Sketch the graph of $7x^2 - 6\sqrt{3}xy + 13y^2 - 16 = 0$.

Solution

Because $A = 7$, $B = -6\sqrt{3}$, and $C = 13$, you have

$$\cot 2\theta = \frac{A - C}{B} = \frac{7 - 13}{-6\sqrt{3}} = \frac{1}{\sqrt{3}}$$

which implies that $\theta = \pi/6$. The equation in the $x'y'$-system is obtained by making the substitutions

$$x = x' \cos\frac{\pi}{6} - y' \sin\frac{\pi}{6}$$

$$= x'\left(\frac{\sqrt{3}}{2}\right) - y'\left(\frac{1}{2}\right)$$

$$= \frac{\sqrt{3}x' - y'}{2}$$

and

$$y = x' \sin\frac{\pi}{6} + y' \cos\frac{\pi}{6}$$

$$= x'\left(\frac{1}{2}\right) + y'\left(\frac{\sqrt{3}}{2}\right)$$

$$= \frac{x' + \sqrt{3}y'}{2}$$

in the original equation. So, you have

$$7x^2 - 6\sqrt{3}xy + 13y^2 - 16 = 0$$

$$7\left(\frac{\sqrt{3}x' - y'}{2}\right)^2 - 6\sqrt{3}\left(\frac{\sqrt{3}x' - y'}{2}\right)\left(\frac{x' + \sqrt{3}y'}{2}\right)$$

$$+ 13\left(\frac{x' + \sqrt{3}y'}{2}\right)^2 - 16 = 0$$

which simplifies to

$$4(x')^2 + 16(y')^2 - 16 = 0$$

$$4(x')^2 + 16(y')^2 = 16$$

$$\frac{(x')^2}{4} + \frac{(y')^2}{1} = 1$$

$$\frac{(x')^2}{2^2} + \frac{(y')^2}{1^2} = 1. \qquad \text{Write in standard form.}$$

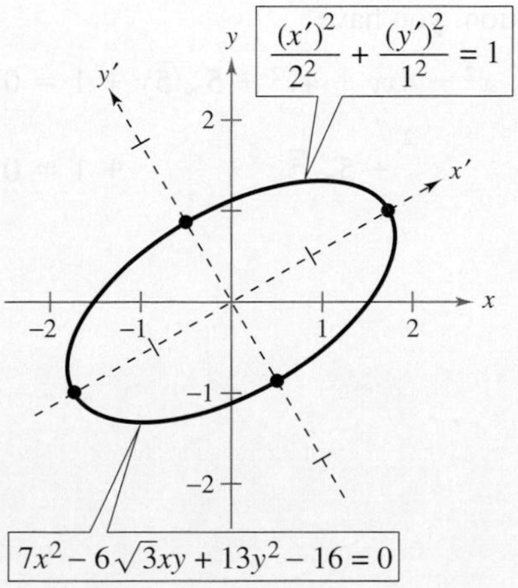

This is the equation of an ellipse centered at the origin with vertices $(\pm 2, 0)$ in the $x'y'$-system, as shown in Figure 6.46.

CHECK*Point*   Now try Exercise 19.

$$\frac{(x')^2}{2^2} + \frac{(y')^2}{1^2} = 1$$

$7x^2 - 6\sqrt{3}xy + 13y^2 - 16 = 0$

Vertices:
In $x'y'$-system: $(\pm 2, 0)$
In xy-system: $(\sqrt{3}, 1), (-\sqrt{3}, -1)$
FIGURE **6.46**

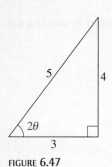

FIGURE **6.47**

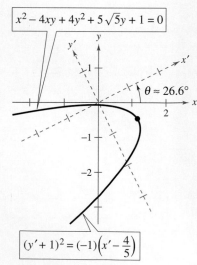

$$x^2 - 4xy + 4y^2 + 5\sqrt{5}y + 1 = 0$$

$\theta \approx 26.6°$

$$(y' + 1)^2 = (-1)\left(x' - \frac{4}{5}\right)$$

Vertex:

In $x'y'$-system: $\left(\frac{4}{5}, -1\right)$

In xy-system: $\left(\frac{13}{5\sqrt{5}}, -\frac{6}{5\sqrt{5}}\right)$

FIGURE **6.48**

Example 3   Rotation of Axes for a Parabola

Sketch the graph of $x^2 - 4xy + 4y^2 + 5\sqrt{5}y + 1 = 0$.

Solution

Because $A = 1$, $B = -4$, and $C = 4$, you have

$$\cot 2\theta = \frac{A - C}{B} = \frac{1 - 4}{-4} = \frac{3}{4}.$$

Using this information, draw a right triangle as shown in Figure 6.47. From the figure, you can see that $\cos 2\theta = \frac{3}{5}$. To find the values of $\sin \theta$ and $\cos \theta$, you can use the half-angle formulas in the forms

$$\sin \theta = \sqrt{\frac{1 - \cos 2\theta}{2}} \quad \text{and} \quad \cos \theta = \sqrt{\frac{1 + \cos 2\theta}{2}}.$$

So,

$$\sin \theta = \sqrt{\frac{1 - \cos 2\theta}{2}} = \sqrt{\frac{1 - \frac{3}{5}}{2}} = \sqrt{\frac{1}{5}} = \frac{1}{\sqrt{5}}$$

$$\cos \theta = \sqrt{\frac{1 + \cos 2\theta}{2}} = \sqrt{\frac{1 + \frac{3}{5}}{2}} = \sqrt{\frac{4}{5}} = \frac{2}{\sqrt{5}}.$$

Consequently, you use the substitutions

$$x = x' \cos \theta - y' \sin \theta$$

$$= x'\left(\frac{2}{\sqrt{5}}\right) - y'\left(\frac{1}{\sqrt{5}}\right) = \frac{2x' - y'}{\sqrt{5}}$$

$$y = x' \sin \theta + y' \cos \theta$$

$$= x'\left(\frac{1}{\sqrt{5}}\right) + y'\left(\frac{2}{\sqrt{5}}\right) = \frac{x' + 2y'}{\sqrt{5}}.$$

Substituting these expressions in the original equation, you have

$$x^2 - 4xy + 4y^2 + 5\sqrt{5}y + 1 = 0$$

$$\left(\frac{2x' - y'}{\sqrt{5}}\right)^2 - 4\left(\frac{2x' - y'}{\sqrt{5}}\right)\left(\frac{x' + 2y'}{\sqrt{5}}\right) + 4\left(\frac{x' + 2y'}{\sqrt{5}}\right)^2 + 5\sqrt{5}\left(\frac{x' + 2y'}{\sqrt{5}}\right) + 1 = 0$$

which simplifies as follows.

$$5(y')^2 + 5x' + 10y' + 1 = 0$$

$$5[(y')^2 + 2y'] = -5x' - 1 \qquad \text{Group terms.}$$

$$5(y' + 1)^2 = -5x' + 4 \qquad \text{Write in completed square form.}$$

$$(y' + 1)^2 = (-1)\left(x' - \frac{4}{5}\right) \qquad \text{Write in standard form.}$$

The graph of this equation is a parabola with vertex $\left(\frac{4}{5}, -1\right)$. Its axis is parallel to the x'-axis in the $x'y'$-system, and because $\sin \theta = 1/\sqrt{5}$, $\theta \approx 26.6°$, as shown in Figure 6.48.

CHECK Point   Now try Exercise 25.

Invariants Under Rotation

In the rotation of axes theorem listed at the beginning of this section, note that the constant term is the same in both equations, $F' = F$. Such quantities are **invariant under rotation.** The next theorem lists some other rotation invariants.

Rotation Invariants

The rotation of the coordinate axes through an angle θ that transforms the equation $Ax^2 + Bxy + Cy^2 + Dx + Ey + F = 0$ into the form

$$A'(x')^2 + C'(y')^2 + D'x' + E'y' + F' = 0$$

has the following rotation invariants.

1. $F = F'$

2. $A + C = A' + C'$

3. $B^2 - 4AC = (B')^2 - 4A'C'$

WARNING/CAUTION

If there is an xy-term in the equation of a conic, you should realize then that the conic is rotated. Before rotating the axes, you should use the discriminant to classify the conic.

You can use the results of this theorem to classify the graph of a second-degree equation *with* an xy-term in much the same way you do for a second-degree equation *without* an xy-term. Note that because $B' = 0$, the invariant $B^2 - 4AC$ reduces to

$$B^2 - 4AC = -4A'C'. \qquad \text{Discriminant}$$

This quantity is called the **discriminant** of the equation

$$Ax^2 + Bxy + Cy^2 + Dx + Ey + F = 0.$$

Now, from the classification procedure given in Section 6.4, you know that the sign of $A'C'$ determines the type of graph for the equation

$$A'(x')^2 + C'(y')^2 + D'x' + E'y' + F' = 0.$$

Consequently, the sign of $B^2 - 4AC$ will determine the type of graph for the original equation, as given in the following classification.

Classification of Conics by the Discriminant

The graph of the equation $Ax^2 + Bxy + Cy^2 + Dx + Ey + F = 0$ is, except in degenerate cases, determined by its discriminant as follows.

1. *Ellipse or circle:* $B^2 - 4AC < 0$

2. *Parabola:* $B^2 - 4AC = 0$

3. *Hyperbola:* $B^2 - 4AC > 0$

For example, in the general equation

$$3x^2 + 7xy + 5y^2 - 6x - 7y + 15 = 0$$

you have $A = 3$, $B = 7$, and $C = 5$. So the discriminant is

$$B^2 - 4AC = 7^2 - 4(3)(5) = 49 - 60 = -11.$$

Because $-11 < 0$, the graph of the equation is an ellipse or a circle.

Example 4 Rotation and Graphing Utilities

For each equation, classify the graph of the equation, use the Quadratic Formula to solve for y, and then use a graphing utility to graph the equation.

a. $2x^2 - 3xy + 2y^2 - 2x = 0$

b. $x^2 - 6xy + 9y^2 - 2y + 1 = 0$

c. $3x^2 + 8xy + 4y^2 - 7 = 0$

Solution

a. Because $B^2 - 4AC = 9 - 16 < 0$, the graph is a circle or an ellipse. Solve for y as follows.

$$2x^2 - 3xy + 2y^2 - 2x = 0 \qquad \text{Write original equation.}$$

$$2y^2 - 3xy + (2x^2 - 2x) = 0 \qquad \text{Quadratic form } ay^2 + by + c = 0$$

$$y = \frac{-(-3x) \pm \sqrt{(-3x)^2 - 4(2)(2x^2 - 2x)}}{2(2)}$$

$$y = \frac{3x \pm \sqrt{x(16 - 7x)}}{4}$$

Graph both of the equations to obtain the ellipse shown in Figure 6.49.

$$y_1 = \frac{3x + \sqrt{x(16 - 7x)}}{4} \qquad \text{Top half of ellipse}$$

$$y_2 = \frac{3x - \sqrt{x(16 - 7x)}}{4} \qquad \text{Bottom half of ellipse}$$

b. Because $B^2 - 4AC = 36 - 36 = 0$, the graph is a parabola.

$$x^2 - 6xy + 9y^2 - 2y + 1 = 0 \qquad \text{Write original equation.}$$

$$9y^2 - (6x + 2)y + (x^2 + 1) = 0 \qquad \text{Quadratic form } ay^2 + by + c = 0$$

$$y = \frac{(6x + 2) \pm \sqrt{(6x + 2)^2 - 4(9)(x^2 + 1)}}{2(9)}$$

Graphing both of the equations to obtain the parabola shown in Figure 6.50.

c. Because $B^2 - 4AC = 64 - 48 > 0$, the graph is a hyperbola.

$$3x^2 + 8xy + 4y^2 - 7 = 0 \qquad \text{Write original equation.}$$

$$4y^2 + 8xy + (3x^2 - 7) = 0 \qquad \text{Quadratic form } ay^2 + by + c = 0$$

$$y = \frac{-8x \pm \sqrt{(8x)^2 - 4(4)(3x^2 - 7)}}{2(4)}$$

The graphs of these two equations yield the hyperbola shown in Figure 6.51.

CHECK Point ▶ Now try Exercise 43.

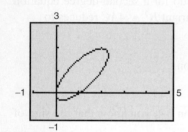

FIGURE 6.49

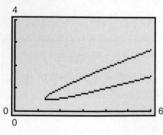

FIGURE 6.50

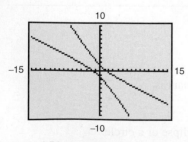

FIGURE 6.51

CLASSROOM DISCUSSION

Classifying a Graph as a Hyperbola The graph of $f(x) = 1/x$ is a hyperbola. Use the techniques in this section to verify this, and justify each step. Compare your results with those of another student.

6.5 EXERCISES

See www.CalcChat.com for worked-out solutions to odd-numbered exercises.

VOCABULARY: Fill in the blanks.

1. The procedure used to eliminate the xy-term in a general second-degree equation is called _____ of _____.

2. After rotating the coordinate axes through an angle θ, the general second-degree equation in the new $x'y'$-plane will have the form _____.

3. Quantities that are equal in both the original equation of a conic and the equation of the rotated conic are _____ _____ _____.

4. The quantity $B^2 - 4AC$ is called the _____ of the equation $Ax^2 + Bxy + Cy^2 + Dx + Ey + F = 0$.

SKILLS AND APPLICATIONS

In Exercises 5–12, the $x'y'$-coordinate system has been rotated θ degrees from the xy-coordinate system. The coordinates of a point in the xy-coordinate system are given. Find the coordinates of the point in the rotated coordinate system.

5. $\theta = 90°$, $(0, 3)$
6. $\theta = 90°$, $(2, 2)$
7. $\theta = 30°$, $(1, 3)$
8. $\theta = 30°$, $(2, 4)$
9. $\theta = 45°$, $(2, 1)$
10. $\theta = 45°$, $(4, 4)$
11. $\theta = 60°$, $(1, 2)$
12. $\theta = 60°$, $(3, 1)$

In Exercises 13–26, rotate the axes to eliminate the xy-term in the equation. Then write the equation in standard form. Sketch the graph of the resulting equation, showing both sets of axes.

13. $xy + 1 = 0$
14. $xy - 4 = 0$
15. $x^2 - 2xy + y^2 - 1 = 0$
16. $xy + 2x - y + 4 = 0$
17. $xy - 8x - 4y = 0$
18. $2x^2 - 3xy - 2y^2 + 10 = 0$
19. $5x^2 - 6xy + 5y^2 - 12 = 0$
20. $2x^2 + xy + 2y^2 - 8 = 0$
21. $x^2 + 2xy + y^2 - 4x + 4y = 0$
22. $13x^2 + 6\sqrt{3}xy + 7y^2 - 16 = 0$
23. $3x^2 - 2\sqrt{3}xy + y^2 + 2x + 2\sqrt{3}y = 0$
24. $16x^2 - 24xy + 9y^2 - 60x - 80y + 100 = 0$
25. $9x^2 + 24xy + 16y^2 + 90x - 130y = 0$
26. $9x^2 + 24xy + 16y^2 + 80x - 60y = 0$

In Exercises 27–36, use a graphing utility to graph the conic. Determine the angle θ through which the axes are rotated. Explain how you used the graphing utility to obtain the graph.

27. $x^2 + 2xy + y^2 = 20$
28. $x^2 - 4xy + 2y^2 = 6$
29. $17x^2 + 32xy - 7y^2 = 75$

30. $40x^2 + 36xy + 25y^2 = 52$
31. $32x^2 + 48xy + 8y^2 = 50$
32. $24x^2 + 18xy + 12y^2 = 34$
33. $2x^2 + 4xy + 2y^2 + \sqrt{26}x + 3y = -15$
34. $7x^2 - 2\sqrt{3}xy + 5y^2 = 16$
35. $4x^2 - 12xy + 9y^2 + (4\sqrt{13} - 12)x - (6\sqrt{13} + 8)y = 91$
36. $6x^2 - 4xy + 8y^2 + (5\sqrt{5} - 10)x - (7\sqrt{5} + 5)y = 80$

In Exercises 37–42, match the graph with its equation. [The graphs are labeled (a), (b), (c), (d), (e), and (f).]

(a)
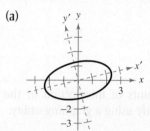

(b)
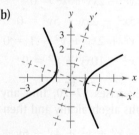

(c)
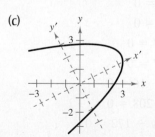

(d)
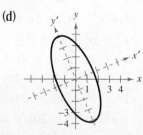

(e)
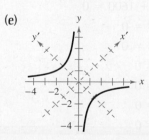

(f)
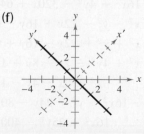

37. $xy + 2 = 0$

38. $x^2 + 2xy + y^2 = 0$

39. $-2x^2 + 3xy + 2y^2 + 3 = 0$

40. $x^2 - xy + 3y^2 - 5 = 0$

41. $3x^2 + 2xy + y^2 - 10 = 0$

42. $x^2 - 4xy + 4y^2 + 10x - 30 = 0$

In Exercises 43–50, (a) use the discriminant to classify the graph, (b) use the Quadratic Formula to solve for y, and (c) use a graphing utility to graph the equation.

43. $16x^2 - 8xy + y^2 - 10x + 5y = 0$

44. $x^2 - 4xy - 2y^2 - 6 = 0$

45. $12x^2 - 6xy + 7y^2 - 45 = 0$

46. $2x^2 + 4xy + 5y^2 + 3x - 4y - 20 = 0$

47. $x^2 - 6xy - 5y^2 + 4x - 22 = 0$

48. $36x^2 - 60xy + 25y^2 + 9y = 0$

49. $x^2 + 4xy + 4y^2 - 5x - y - 3 = 0$

50. $x^2 + xy + 4y^2 + x + y - 4 = 0$

In Exercises 51–56, sketch (if possible) the graph of the degenerate conic.

51. $y^2 - 16x^2 = 0$

52. $x^2 + y^2 - 2x + 6y + 10 = 0$

53. $x^2 - 2xy + y^2 = 0$

54. $5x^2 - 2xy + 5y^2 = 0$

55. $x^2 + 2xy + y^2 - 1 = 0$

56. $x^2 - 10xy + y^2 = 0$

In Exercises 57–70, find any points of intersection of the graphs algebraically and then verify using a graphing utility.

57. $-x^2 + y^2 + 4x - 6y + 4 = 0$

$x^2 + y^2 - 4x - 6y + 12 = 0$

58. $-x^2 - y^2 - 8x + 20y - 7 = 0$

$x^2 + 9y^2 + 8x + 4y + 7 = 0$

59. $-4x^2 - y^2 - 16x + 24y - 16 = 0$

$4x^2 + y^2 + 40x - 24y + 208 = 0$

60. $x^2 - 4y^2 - 20x - 64y - 172 = 0$

$16x^2 + 4y^2 - 320x + 64y + 1600 = 0$

61. $x^2 - y^2 - 12x + 16y - 64 = 0$

$x^2 + y^2 - 12x - 16y + 64 = 0$

62. $x^2 + 4y^2 - 2x - 8y + 1 = 0$

$-x^2 + 2x - 4y - 1 = 0$

63. $-16x^2 - y^2 + 24y - 80 = 0$

$16x^2 + 25y^2 - 400 = 0$

64. $16x^2 - y^2 + 16y - 128 = 0$

$y^2 - 48x - 16y - 32 = 0$

65. $x^2 + y^2 - 4 = 0$

$3x - y^2 = 0$

66. $4x^2 + 9y^2 - 36y = 0$

$x^2 + 9y - 27 = 0$

67. $x^2 + 2y^2 - 4x + 6y - 5 = 0$

$-x + y - 4 = 0$

68. $x^2 + 2y^2 - 4x + 6y - 5 = 0$

$x^2 - 4x - y + 4 = 0$

69. $xy + x - 2y + 3 = 0$

$x^2 + 4y^2 - 9 = 0$

70. $5x^2 - 2xy + 5y^2 - 12 = 0$

$x + y - 1 = 0$

EXPLORATION

TRUE OR FALSE? In Exercises 71 and 72, determine whether the statement is true or false. Justify your answer.

71. The graph of the equation

$$x^2 + xy + ky^2 + 6x + 10 = 0$$

where k is any constant less than $\frac{1}{4}$, is a hyperbola.

72. After a rotation of axes is used to eliminate the xy-term from an equation of the form

$$Ax^2 + Bxy + Cy^2 + Dx + Ey + F = 0$$

the coefficients of the x^2- and y^2-terms remain A and C, respectively.

73. Show that the equation

$$x^2 + y^2 = r^2$$

is invariant under rotation of axes.

74. CAPSTONE Consider the equation

$$6x^2 - 3xy + 6y^2 - 25 = 0.$$

(a) Without calculating, explain how to rewrite the equation so that it does not have an xy-term.

(b) Explain how to identify the graph of the equation.

75. Find the lengths of the major and minor axes of the ellipse graphed in Exercise 22.

6.6 PARAMETRIC EQUATIONS

Plane Curves

Up to this point you have been representing a graph by a single equation involving the *two* variables x and y. In this section, you will study situations in which it is useful to introduce a *third* variable to represent a curve in the plane.

To see the usefulness of this procedure, consider the path followed by an object that is propelled into the air at an angle of 45°. If the initial velocity of the object is 48 feet per second, it can be shown that the object follows the parabolic path

$$y = -\frac{x^2}{72} + x \qquad \text{Rectangular equation}$$

as shown in Figure 6.52. However, this equation does not tell the whole story. Although it does tell you *where* the object has been, it does not tell you *when* the object was at a given point (x, y) on the path. To determine this time, you can introduce a third variable t, called a **parameter.** It is possible to write both x and y as functions of t to obtain the **parametric equations**

$$x = 24\sqrt{2}\,t \qquad \text{Parametric equation for } x$$

$$y = -16t^2 + 24\sqrt{2}\,t. \qquad \text{Parametric equation for } y$$

From this set of equations you can determine that at time $t = 0$, the object is at the point $(0, 0)$. Similarly, at time $t = 1$, the object is at the point $\left(24\sqrt{2}, 24\sqrt{2} - 16\right)$, and so on, as shown in Figure 6.52.

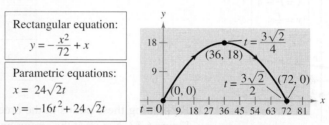

Curvilinear Motion: Two Variables for Position, One Variable for Time
FIGURE **6.52**

For this particular motion problem, x and y are continuous functions of t, and the resulting path is a **plane curve.** (Recall that a *continuous function* is one whose graph can be traced without lifting the pencil from the paper.)

Definition of Plane Curve

If f and g are continuous functions of t on an interval I, the set of ordered pairs $(f(t), g(t))$ is a **plane curve** C. The equations

$$x = f(t) \qquad \text{and} \qquad y = g(t)$$

are **parametric equations** for C, and t is the **parameter.**

Sketching a Plane Curve

When sketching a curve represented by a pair of parametric equations, you still plot points in the xy-plane. Each set of coordinates (x, y) is determined from a value chosen for the parameter t. Plotting the resulting points in the order of *increasing* values of t traces the curve in a specific direction. This is called the **orientation** of the curve.

Example 1   Sketching a Curve

Sketch the curve given by the parametric equations

$$x = t^2 - 4 \quad \text{and} \quad y = \frac{t}{2}, \quad -2 \le t \le 3.$$

Solution

Using values of t in the specified interval, the parametric equations yield the points (x, y) shown in the table.

t	x	y
-2	0	-1
-1	-3	$-\dfrac{1}{2}$
0	-4	0
1	-3	$\dfrac{1}{2}$
2	0	1
3	5	$\dfrac{3}{2}$

By plotting these points in the order of increasing t, you obtain the curve C shown in Figure 6.53. Note that the arrows on the curve indicate its orientation as t increases from -2 to 3. So, if a particle were moving on this curve, it would start at $(0, -1)$ and then move along the curve to the point $\left(5, \frac{3}{2}\right)$.

CHECK*Point*   Now try Exercises 5(a) and (b).

Note that the graph shown in Figure 6.53 does not define y as a function of x. This points out one benefit of parametric equations—they can be used to represent graphs that are more general than graphs of functions.

It often happens that two different sets of parametric equations have the same graph. For example, the set of parametric equations

$$x = 4t^2 - 4 \quad \text{and} \quad y = t, \quad -1 \le t \le \frac{3}{2}$$

has the same graph as the set given in Example 1. However, by comparing the values of t in Figures 6.53 and 6.54, you can see that this second graph is traced out more *rapidly* (considering t as time) than the first graph. So, in applications, different parametric representations can be used to represent various *speeds* at which objects travel along a given path.

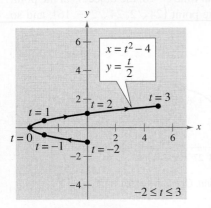

FIGURE **6.53**

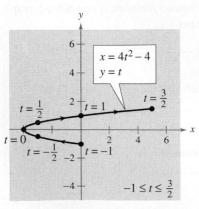

FIGURE **6.54**

Eliminating the Parameter

Example 1 uses simple point plotting to sketch the curve. This tedious process can sometimes be simplified by finding a rectangular equation (in x and y) that has the same graph. This process is called **eliminating the parameter.**

Parametric equations	$\Rightarrow$	Solve for t in one equation.	$\Rightarrow$	Substitute in other equation.	$\Rightarrow$	Rectangular equation

$$x = t^2 - 4 \qquad\qquad t = 2y \qquad\qquad x = (2y)^2 - 4 \qquad x = 4y^2 - 4$$

$$y = \frac{t}{2}$$

Now you can recognize that the equation $x = 4y^2 - 4$ represents a parabola with a horizontal axis and vertex at $(-4, 0)$.

When converting equations from parametric to rectangular form, you may need to alter the domain of the rectangular equation so that its graph matches the graph of the parametric equations. Such a situation is demonstrated in Example 2.

▎Example 2   Eliminating the Parameter

Sketch the curve represented by the equations

$$x = \frac{1}{\sqrt{t + 1}} \quad \text{and} \quad y = \frac{t}{t + 1}$$

by eliminating the parameter and adjusting the domain of the resulting rectangular equation.

Solution

Solving for t in the equation for x produces

$$x = \frac{1}{\sqrt{t + 1}} \quad\Rightarrow\quad x^2 = \frac{1}{t + 1}$$

which implies that

$$t = \frac{1 - x^2}{x^2}.$$

Now, substituting in the equation for y, you obtain the rectangular equation

$$y = \frac{t}{t + 1} = \frac{\dfrac{(1 - x^2)}{x^2}}{\left[\dfrac{(1 - x^2)}{x^2}\right] + 1} = \frac{\dfrac{1 - x^2}{x^2}}{\dfrac{1 - x^2}{x^2} + 1} \cdot \frac{x^2}{x^2} = 1 - x^2.$$

From this rectangular equation, you can recognize that the curve is a parabola that opens downward and has its vertex at $(0, 1)$. Also, this rectangular equation is defined for all values of x, but from the parametric equation for x you can see that the curve is defined only when $t > -1$. This implies that you should restrict the domain of x to positive values, as shown in Figure 6.55.

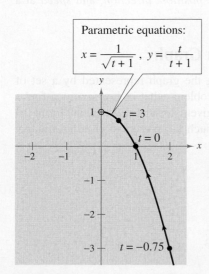

Parametric equations:

$$x = \frac{1}{\sqrt{t + 1}}, \quad y = \frac{t}{t + 1}$$

FIGURE **6.55**

CHECK*Point*   Now try Exercise 5(c).

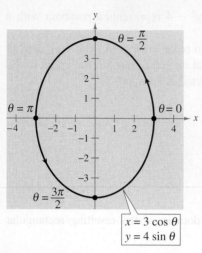

FIGURE 6.56

It is not necessary for the parameter in a set of parametric equations to represent time. The next example uses an *angle* as the parameter.

Example 3 Eliminating an Angle Parameter

Sketch the curve represented by

$$x = 3 \cos \theta \quad \text{and} \quad y = 4 \sin \theta, \quad 0 \le \theta \le 2\pi$$

by eliminating the parameter.

Solution

Begin by solving for $\cos \theta$ and $\sin \theta$ in the equations.

$$\cos \theta = \frac{x}{3} \quad \text{and} \quad \sin \theta = \frac{y}{4} \qquad \text{Solve for } \cos \theta \text{ and } \sin \theta.$$

Use the identity $\sin^2 \theta + \cos^2 \theta = 1$ to form an equation involving only x and y.

$$\cos^2 \theta + \sin^2 \theta = 1 \qquad \text{Pythagorean identity}$$

$$\left(\frac{x}{3}\right)^2 + \left(\frac{y}{4}\right)^2 = 1 \qquad \text{Substitute } \frac{x}{3} \text{ for } \cos \theta \text{ and } \frac{y}{4} \text{ for } \sin \theta.$$

$$\frac{x^2}{9} + \frac{y^2}{16} = 1 \qquad \text{Rectangular equation}$$

From this rectangular equation, you can see that the graph is an ellipse centered at $(0, 0)$, with vertices $(0, 4)$ and $(0, -4)$ and minor axis of length $2b = 6$, as shown in Figure 6.56. Note that the elliptic curve is traced out *counterclockwise* as θ varies from 0 to 2π.

CHECK *Point* Now try Exercise 17.

In Examples 2 and 3, it is important to realize that eliminating the parameter is primarily an *aid to curve sketching*. If the parametric equations represent the path of a moving object, the graph alone is not sufficient to describe the object's motion. You still need the parametric equations to tell you the *position*, *direction*, and *speed* at a given time.

Finding Parametric Equations for a Graph

You have been studying techniques for sketching the graph represented by a set of parametric equations. Now consider the *reverse* problem—that is, how can you find a set of parametric equations for a given graph or a given physical description? From the discussion following Example 1, you know that such a representation is not unique. That is, the equations

$$x = 4t^2 - 4 \quad \text{and} \quad y = t, \ -1 \le t \le \frac{3}{2}$$

produced the same graph as the equations

$$x = t^2 - 4 \quad \text{and} \quad y = \frac{t}{2}, \ -2 \le t \le 3.$$

This is further demonstrated in Example 4.

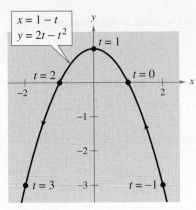

$x = 1 - t$
$y = 2t - t^2$

FIGURE 6.57

Example 4   Finding Parametric Equations for a Graph

Find a set of parametric equations to represent the graph of $y = 1 - x^2$, using the following parameters.

a. $t = x$ **b.** $t = 1 - x$

Solution

a. Letting $t = x$, you obtain the parametric equations

$$x = t \quad \text{and} \quad y = 1 - x^2 = 1 - t^2.$$

b. Letting $t = 1 - x$, you obtain the parametric equations

$$x = 1 - t \quad \text{and} \quad y = 1 - x^2 = 1 - (1 - t)^2 = 2t - t^2.$$

In Figure 6.57, note how the resulting curve is oriented by the increasing values of t. For part (a), the curve would have the opposite orientation.

CHECK*Point* Now try Exercise 45.

Example 5   Parametric Equations for a Cycloid

Describe the **cycloid** traced out by a point P on the circumference of a circle of radius a as the circle rolls along a straight line in a plane.

Solution

As the parameter, let θ be the measure of the circle's rotation, and let the point $P = (x, y)$ begin at the origin. When $\theta = 0$, P is at the origin; when $\theta = \pi$, P is at a maximum point $(\pi a, 2a)$; and when $\theta = 2\pi$, P is back on the x-axis at $(2\pi a, 0)$. From Figure 6.58, you can see that $\angle APC = 180° - \theta$. So, you have

$$\sin \theta = \sin(180° - \theta) = \sin(\angle APC) = \frac{AC}{a} = \frac{BD}{a}$$

$$\cos \theta = -\cos(180° - \theta) = -\cos(\angle APC) = -\frac{AP}{a}$$

which implies that $BD = a \sin \theta$ and $AP = -a \cos \theta$. Because the circle rolls along the x-axis, you know that $OD = \overset{\frown}{PD} = a\theta$. Furthermore, because $BA = DC = a$, you have

$$x = OD - BD = a\theta - a \sin \theta \quad \text{and} \quad y = BA + AP = a - a \cos \theta.$$

So, the parametric equations are $x = a(\theta - \sin \theta)$ and $y = a(1 - \cos \theta)$.

Study Tip

In Example 5, $\overset{\frown}{PD}$ represents the arc of the circle between points P and D.

TECHNOLOGY

You can use a graphing utility in *parametric* mode to obtain a graph similar to Figure 6.58 by graphing the following equations.

$$X_{1T} = T - \sin T$$

$$Y_{1T} = 1 - \cos T$$

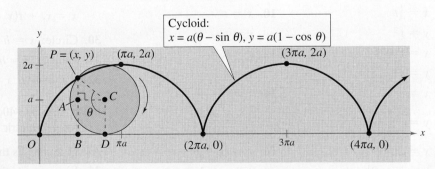

Cycloid:
$x = a(\theta - \sin \theta), y = a(1 - \cos \theta)$

FIGURE 6.58

CHECK*Point* Now try Exercise 67.

6.6 EXERCISES

See www.CalcChat.com for worked-out solutions to odd-numbered exercises.

VOCABULARY: Fill in the blanks.

1. If f and g are continuous functions of t on an interval I, the set of ordered pairs $(f(t), g(t))$ is a _____ _____ C.

2. The _____ of a curve is the direction in which the curve is traced out for increasing values of the parameter.

3. The process of converting a set of parametric equations to a corresponding rectangular equation is called _____ the _____.

4. A curve traced by a point on the circumference of a circle as the circle rolls along a straight line in a plane is called a _____.

SKILLS AND APPLICATIONS

5. Consider the parametric equations $x = \sqrt{t}$ and $y = 3 - t$.

   (a) Create a table of x- and y-values using $t = 0, 1, 2, 3,$ and 4.

   (b) Plot the points (x, y) generated in part (a), and sketch a graph of the parametric equations.

   (c) Find the rectangular equation by eliminating the parameter. Sketch its graph. How do the graphs differ?

6. Consider the parametric equations $x = 4\cos^2 \theta$ and $y = 2\sin \theta$.

   (a) Create a table of x- and y-values using $\theta = -\pi/2, -\pi/4, 0, \pi/4,$ and $\pi/2$.

   (b) Plot the points (x, y) generated in part (a), and sketch a graph of the parametric equations.

   (c) Find the rectangular equation by eliminating the parameter. Sketch its graph. How do the graphs differ?

In Exercises 7–26, (a) sketch the curve represented by the parametric equations (indicate the orientation of the curve) and (b) eliminate the parameter and write the corresponding rectangular equation whose graph represents the curve. Adjust the domain of the resulting rectangular equation if necessary.

7. $x = t - 1$
   $y = 3t + 1$

8. $x = 3 - 2t$
   $y = 2 + 3t$

9. $x = \frac{1}{4}t$
   $y = t^2$

10. $x = t$
    $y = t^3$

11. $x = t + 2$
    $y = t^2$

12. $x = \sqrt{t}$
    $y = 1 - t$

13. $x = t + 1$
    $y = \dfrac{t}{t + 1}$

14. $x = t - 1$
    $y = \dfrac{t}{t - 1}$

15. $x = 2(t + 1)$
    $y = |t - 2|$

16. $x = |t - 1|$
    $y = t + 2$

17. $x = 4\cos \theta$
    $y = 2\sin \theta$

18. $x = 2\cos \theta$
    $y = 3\sin \theta$

19. $x = 6\sin 2\theta$
    $y = 6\cos 2\theta$

20. $x = \cos \theta$
    $y = 2\sin 2\theta$

21. $x = 1 + \cos \theta$
    $y = 1 + 2\sin \theta$

22. $x = 2 + 5\cos \theta$
    $y = -6 + 4\sin \theta$

23. $x = e^{-t}$
    $y = e^{3t}$

24. $x = e^{2t}$
    $y = e^t$

25. $x = t^3$
    $y = 3\ln t$

26. $x = \ln 2t$
    $y = 2t^2$

In Exercises 27 and 28, determine how the plane curves differ from each other.

27. (a) $x = t$
        $y = 2t + 1$

    (b) $x = \cos \theta$
        $y = 2\cos \theta + 1$

    (c) $x = e^{-t}$
        $y = 2e^{-t} + 1$

    (d) $x = e^t$
        $y = 2e^t + 1$

28. (a) $x = t$
        $y = t^2 - 1$

    (b) $x = t^2$
        $y = t^4 - 1$

    (c) $x = \sin t$
        $y = \sin^2 t - 1$

    (d) $x = e^t$
        $y = e^{2t} - 1$

In Exercises 29–32, eliminate the parameter and obtain the standard form of the rectangular equation.

29. Line through (x_1, y_1) and (x_2, y_2):

    $x = x_1 + t(x_2 - x_1)$, $y = y_1 + t(y_2 - y_1)$

30. Circle: $x = h + r\cos \theta$, $y = k + r\sin \theta$

31. Ellipse: $x = h + a\cos \theta$, $y = k + b\sin \theta$

32. Hyperbola: $x = h + a\sec \theta$, $y = k + b\tan \theta$

In Exercises 33–40, use the results of Exercises 29–32 to find a set of parametric equations for the line or conic.

33. Line: passes through $(0, 0)$ and $(3, 6)$

34. Line: passes through $(3, 2)$ and $(-6, 3)$

35. Circle: center: $(3, 2)$; radius: 4

36. Circle: center: $(5, -3)$; radius: 4

37. Ellipse: vertices: $(\pm 5, 0)$; foci: $(\pm 4, 0)$

38. Ellipse: vertices: $(3, 7), (3, -1)$;

   foci: $(3, 5), (3, 1)$

39. Hyperbola: vertices: $(\pm 4, 0)$; foci: $(\pm 5, 0)$

40. Hyperbola: vertices: $(\pm 2, 0)$; foci: $(\pm 4, 0)$

In Exercises 41–48, find a set of parametric equations for the rectangular equation using (a) $t = x$ and (b) $t = 2 - x$.

41. $y = 3x - 2$

42. $x = 3y - 2$

43. $y = 2 - x$

44. $y = x^2 + 1$

45. $y = x^2 - 3$

46. $y = 1 - 2x^2$

47. $y = \dfrac{1}{x}$

48. $y = \dfrac{1}{2x}$

 In Exercises 49–56, use a graphing utility to graph the curve represented by the parametric equations.

49. Cycloid: $x = 4(\theta - \sin \theta), \ y = 4(1 - \cos \theta)$

50. Cycloid: $x = \theta + \sin \theta, \ y = 1 - \cos \theta$

51. Prolate cycloid: $x = \theta - \frac{3}{2} \sin \theta, \ y = 1 - \frac{3}{2} \cos \theta$

52. Prolate cycloid: $x = 2\theta - 4 \sin \theta, \ y = 2 - 4 \cos \theta$

53. Hypocycloid: $x = 3 \cos^3 \theta, \ y = 3 \sin^3 \theta$

54. Curtate cycloid: $x = 8\theta - 4 \sin \theta, \ y = 8 - 4 \cos \theta$

55. Witch of Agnesi: $x = 2 \cot \theta, \ y = 2 \sin^2 \theta$

56. Folium of Descartes: $x = \dfrac{3t}{1 + t^3}, \ y = \dfrac{3t^2}{1 + t^3}$

In Exercises 57–60, match the parametric equations with the correct graph and describe the domain and range. [The graphs are labeled (a), (b), (c), and (d).]

(a)

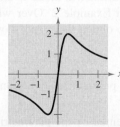

(b)

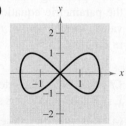

(c)

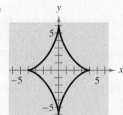

(d)

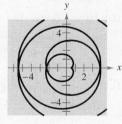

57. Lissajous curve: $x = 2 \cos \theta, \ y = \sin 2\theta$

58. Evolute of ellipse: $x = 4 \cos^3 \theta, \ y = 6 \sin^3 \theta$

59. Involute of circle: $x = \frac{1}{2}(\cos \theta + \theta \sin \theta)$

   $y = \frac{1}{2}(\sin \theta - \theta \cos \theta)$

60. Serpentine curve: $x = \frac{1}{2} \cot \theta, \ y = 4 \sin \theta \cos \theta$

PROJECTILE MOTION   A projectile is launched at a height of h feet above the ground at an angle of θ with the horizontal. The initial velocity is v_0 feet per second, and the path of the projectile is modeled by the parametric equations

$$x = (v_0 \cos \theta)t \quad \text{and} \quad y = h + (v_0 \sin \theta)t - 16t^2.$$

In Exercises 61 and 62, use a graphing utility to graph the paths of a projectile launched from ground level at each value of θ and v_0. For each case, use the graph to approximate the maximum height and the range of the projectile.

61. (a) $\theta = 60°, \quad v_0 = 88$ feet per second

   (b) $\theta = 60°, \quad v_0 = 132$ feet per second

   (c) $\theta = 45°, \quad v_0 = 88$ feet per second

   (d) $\theta = 45°, \quad v_0 = 132$ feet per second

62. (a) $\theta = 15°, \quad v_0 = 50$ feet per second

   (b) $\theta = 15°, \quad v_0 = 120$ feet per second

   (c) $\theta = 10°, \quad v_0 = 50$ feet per second

   (d) $\theta = 10°, \quad v_0 = 120$ feet per second

63. SPORTS   The center field fence in Yankee Stadium is 7 feet high and 408 feet from home plate. A baseball is hit at a point 3 feet above the ground. It leaves the bat at an angle of θ degrees with the horizontal at a speed of 100 miles per hour (see figure).

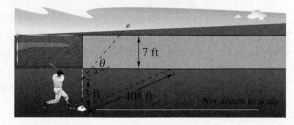

(a) Write a set of parametric equations that model the path of the baseball.

(b) Use a graphing utility to graph the path of the baseball when $\theta = 15°$. Is the hit a home run?

(c) Use the graphing utility to graph the path of the baseball when $\theta = 23°$. Is the hit a home run?

(d) Find the minimum angle required for the hit to be a home run.

64. SPORTS   An archer releases an arrow from a bow at a point 5 feet above the ground. The arrow leaves the bow at an angle of $15°$ with the horizontal and at an initial speed of 225 feet per second.

(a) Write a set of parametric equations that model the path of the arrow.

(b) Assuming the ground is level, find the distance the arrow travels before it hits the ground. (Ignore air resistance.)

(c) Use a graphing utility to graph the path of the arrow and approximate its maximum height.

(d) Find the total time the arrow is in the air.

65. PROJECTILE MOTION   Eliminate the parameter t from the parametric equations

$$x = (v_0 \cos \theta)t \quad \text{and} \quad y = h + (v_0 \sin \theta)t - 16t^2$$

for the motion of a projectile to show that the rectangular equation is

$$y = -\frac{16 \sec^2 \theta}{v_0^2}x^2 + (\tan \theta)x + h.$$

66. PATH OF A PROJECTILE   The path of a projectile is given by the rectangular equation

$$y = 7 + x - 0.02x^2.$$

(a) Use the result of Exercise 65 to find h, v_0, and θ. Find the parametric equations of the path.

(b) Use a graphing utility to graph the rectangular equation for the path of the projectile. Confirm your answer in part (a) by sketching the curve represented by the parametric equations.

(c) Use the graphing utility to approximate the maximum height of the projectile and its range.

67. CURTATE CYCLOID   A wheel of radius a units rolls along a straight line without slipping. The curve traced by a point P that is b units from the center ($b < a$) is called a **curtate cycloid** (see figure). Use the angle θ shown in the figure to find a set of parametric equations for the curve.

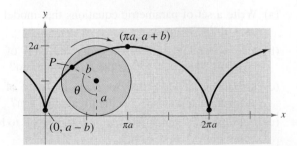

68. EPICYCLOID   A circle of radius one unit rolls around the outside of a circle of radius two units without slipping. The curve traced by a point on the circumference of the smaller circle is called an **epicycloid** (see figure). Use the angle θ shown in the figure to find a set of parametric equations for the curve.

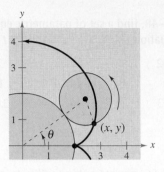

EXPLORATION

TRUE OR FALSE?   In Exercises 69 and 70, determine whether the statement is true or false. Justify your answer.

69. The two sets of parametric equations $x = t$, $y = t^2 + 1$ and $x = 3t$, $y = 9t^2 + 1$ have the same rectangular equation.

70. If y is a function of t, and x is a function of t, then y must be a function of x.

71. WRITING   Write a short paragraph explaining why parametric equations are useful.

72. WRITING   Explain the process of sketching a plane curve given by parametric equations. What is meant by the orientation of the curve?

73. Use a graphing utility set in *parametric* mode to enter the parametric equations from Example 2. Over what values should you let t vary to obtain the graph shown in Figure 6.55?

74. CAPSTONE   Consider the parametric equations $x = 8 \cos t$ and $y = 8 \sin t$.

(a) Describe the curve represented by the parametric equations.

(b) How does the curve represented by the parametric equations $x = 8 \cos t + 3$ and $y = 8 \sin t + 6$ compare with the curve described in part (a)?

(c) How does the original curve change when cosine and sine are interchanged?

6.7 POLAR COORDINATES

Introduction

So far, you have been representing graphs of equations as collections of points (x, y) on the rectangular coordinate system, where x and y represent the directed distances from the coordinate axes to the point (x, y). In this section, you will study a different system called the **polar coordinate system.**

To form the polar coordinate system in the plane, fix a point O, called the **pole** (or **origin**), and construct from O an initial ray called the **polar axis,** as shown in Figure 6.59. Then each point P in the plane can be assigned **polar coordinates** (r, θ) as follows.

1. $r = $ *directed distance* from O to P

2. $\theta = $ *directed angle*, counterclockwise from polar axis to segment $\overline{OP}$

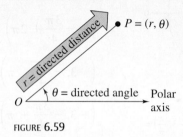

FIGURE 6.59

Example 1   **Plotting Points on the Polar Coordinate System**

a. The point $(r, \theta) = (2, \pi/3)$ lies two units from the pole on the terminal side of the angle $\theta = \pi/3$, as shown in Figure 6.60.

b. The point $(r, \theta) = (3, -\pi/6)$ lies three units from the pole on the terminal side of the angle $\theta = -\pi/6$, as shown in Figure 6.61.

c. The point $(r, \theta) = (3, 11\pi/6)$ coincides with the point $(3, -\pi/6)$, as shown in Figure 6.62.

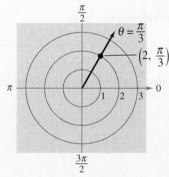

FIGURE 6.60

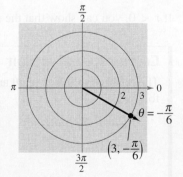

FIGURE 6.61

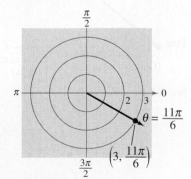

FIGURE 6.62

CHECK*Point* Now try Exercise 7.

In rectangular coordinates, each point (x, y) has a unique representation. This is not true for polar coordinates. For instance, the coordinates (r, θ) and $(r, \theta + 2\pi)$ represent the same point, as illustrated in Example 1. Another way to obtain multiple representations of a point is to use negative values for r. Because r is a *directed distance*, the coordinates (r, θ) and $(-r, \theta + \pi)$ represent the same point. In general, the point (r, θ) can be represented as

$$(r, \theta) = (r, \theta \pm 2n\pi) \qquad \text{or} \qquad (r, \theta) = (-r, \theta \pm (2n + 1)\pi)$$

where n is any integer. Moreover, the pole is represented by $(0, \theta)$, where θ is any angle.

Example 2 Multiple Representations of Points

Plot the point $(3, -3\pi/4)$ and find three additional polar representations of this point, using $-2\pi < \theta < 2\pi$.

Solution

The point is shown in Figure 6.63. Three other representations are as follows.

$$\left(3, -\frac{3\pi}{4} + 2\pi\right) = \left(3, \frac{5\pi}{4}\right) \qquad \text{Add } 2\pi \text{ to } \theta.$$

$$\left(-3, -\frac{3\pi}{4} - \pi\right) = \left(-3, -\frac{7\pi}{4}\right) \qquad \text{Replace } r \text{ by } -r; \text{ subtract } \pi \text{ from } \theta.$$

$$\left(-3, -\frac{3\pi}{4} + \pi\right) = \left(-3, \frac{\pi}{4}\right) \qquad \text{Replace } r \text{ by } -r; \text{ add } \pi \text{ to } \theta.$$

CHECK Point Now try Exercise 13.

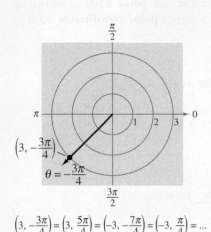

$$\left(3, -\frac{3\pi}{4}\right) = \left(3, \frac{5\pi}{4}\right) = \left(-3, -\frac{7\pi}{4}\right) = \left(-3, \frac{\pi}{4}\right) = \cdots$$

FIGURE 6.63

Coordinate Conversion

To establish the relationship between polar and rectangular coordinates, let the polar axis coincide with the positive x-axis and the pole with the origin, as shown in Figure 6.64. Because (x, y) lies on a circle of radius r, it follows that $r^2 = x^2 + y^2$. Moreover, for $r > 0$, the definitions of the trigonometric functions imply that

$$\tan \theta = \frac{y}{x}, \qquad \cos \theta = \frac{x}{r}, \qquad \text{and} \qquad \sin \theta = \frac{y}{r}.$$

If $r < 0$, you can show that the same relationships hold.

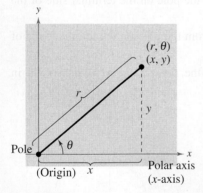

FIGURE 6.64

Coordinate Conversion

The polar coordinates (r, θ) are related to the rectangular coordinates (x, y) as follows.

Polar-to-Rectangular	*Rectangular-to-Polar*
$x = r \cos \theta$	$\tan \theta = \dfrac{y}{x}$
$y = r \sin \theta$	$r^2 = x^2 + y^2$

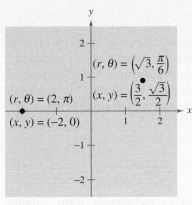

FIGURE **6.65**

Example 3   Polar-to-Rectangular Conversion

Convert each point to rectangular coordinates.

a. $(2, \pi)$ **b.** $\left(\sqrt{3}, \dfrac{\pi}{6} \right)$

Solution

a. For the point $(r, \theta) = (2, \pi)$, you have the following.

$$x = r \cos \theta = 2 \cos \pi = -2$$

$$y = r \sin \theta = 2 \sin \pi = 0$$

The rectangular coordinates are $(x, y) = (-2, 0)$. (See Figure 6.65.)

b. For the point $(r, \theta) = \left(\sqrt{3}, \dfrac{\pi}{6} \right)$, you have the following.

$$x = \sqrt{3} \cos \frac{\pi}{6} = \sqrt{3} \left(\frac{\sqrt{3}}{2} \right) = \frac{3}{2}$$

$$y = \sqrt{3} \sin \frac{\pi}{6} = \sqrt{3} \left(\frac{1}{2} \right) = \frac{\sqrt{3}}{2}$$

The rectangular coordinates are $(x, y) = \left(\dfrac{3}{2}, \dfrac{\sqrt{3}}{2} \right)$.

CHECK *Point*   Now try Exercise 23.

Example 4   Rectangular-to-Polar Conversion

Convert each point to polar coordinates.

a. $(-1, 1)$ **b.** $(0, 2)$

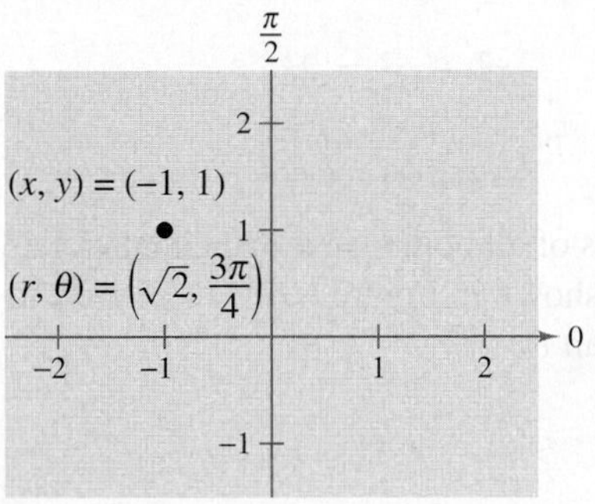

Solution

a. For the second-quadrant point $(x, y) = (-1, 1)$, you have

$$\tan \theta = \frac{y}{x} = -1$$

$$\theta = \frac{3\pi}{4}.$$

Because θ lies in the same quadrant as (x, y), use positive r.

$$r = \sqrt{x^2 + y^2} = \sqrt{(-1)^2 + (1)^2} = \sqrt{2}$$

So, *one* set of polar coordinates is $(r, \theta) = \left(\sqrt{2}, 3\pi/4 \right)$, as shown in Figure 6.66.

b. Because the point $(x, y) = (0, 2)$ lies on the positive y-axis, choose

$$\theta = \frac{\pi}{2} \quad \text{and} \quad r = 2.$$

This implies that *one* set of polar coordinates is $(r, \theta) = (2, \pi/2)$, as shown in Figure 6.67.

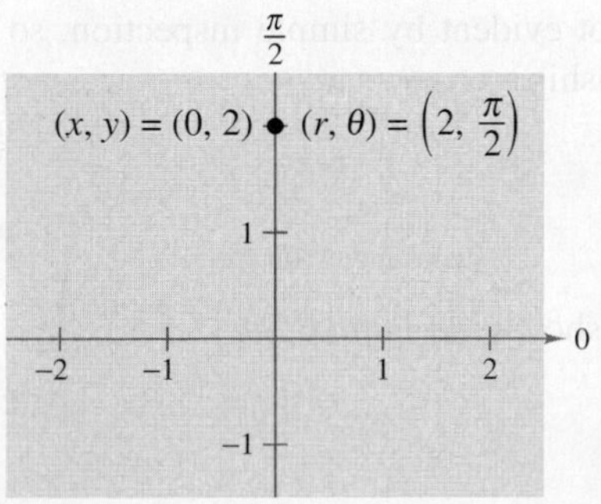

CHECK *Point*   Now try Exercise 41.

FIGURE **6.66**

FIGURE **6.67**

Equation Conversion

By comparing Examples 3 and 4, you can see that point conversion from the polar to the rectangular system is straightforward, whereas point conversion from the rectangular to the polar system is more involved. For equations, the opposite is true. To convert a rectangular equation to polar form, you simply replace x by $r \cos \theta$ and y by $r \sin \theta$. For instance, the rectangular equation $y = x^2$ can be written in polar form as follows.

$$y = x^2 \qquad \text{Rectangular equation}$$

$$r \sin \theta = (r \cos \theta)^2 \qquad \text{Polar equation}$$

$$r = \sec \theta \tan \theta \qquad \text{Simplest form}$$

On the other hand, converting a polar equation to rectangular form requires considerable ingenuity.

Example 5 demonstrates several polar-to-rectangular conversions that enable you to sketch the graphs of some polar equations.

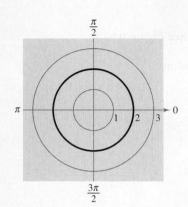

FIGURE **6.68**

| **Example 5** | **Converting Polar Equations to Rectangular Form** |

Describe the graph of each polar equation and find the corresponding rectangular equation.

a. $r = 2$ **b.** $\theta = \dfrac{\pi}{3}$ **c.** $r = \sec \theta$

Solution

a. The graph of the polar equation $r = 2$ consists of all points that are two units from the pole. In other words, this graph is a circle centered at the origin with a radius of 2, as shown in Figure 6.68. You can confirm this by converting to rectangular form, using the relationship $r^2 = x^2 + y^2$.

$$\underbrace{r = 2}_{\text{Polar equation}} \implies r^2 = 2^2 \implies \underbrace{x^2 + y^2 = 2^2}_{\text{Rectangular equation}}$$

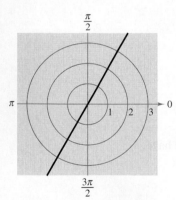

FIGURE **6.69**

b. The graph of the polar equation $\theta = \pi/3$ consists of all points on the line that makes an angle of $\pi/3$ with the positive polar axis, as shown in Figure 6.69. To convert to rectangular form, make use of the relationship $\tan \theta = y/x$.

$$\underbrace{\theta = \dfrac{\pi}{3}}_{\text{Polar equation}} \implies \tan \theta = \sqrt{3} \implies \underbrace{y = \sqrt{3}x}_{\text{Rectangular equation}}$$

c. The graph of the polar equation $r = \sec \theta$ is not evident by simple inspection, so convert to rectangular form by using the relationship $r \cos \theta = x$.

$$\underbrace{r = \sec \theta}_{\text{Polar equation}} \implies r \cos \theta = 1 \implies \underbrace{x = 1}_{\text{Rectangular equation}}$$

Now you see that the graph is a vertical line, as shown in Figure 6.70.

CHECK Point Now try Exercise 109.

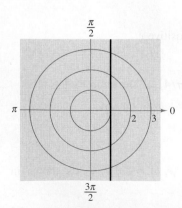

FIGURE **6.70**

6.7   EXERCISES

See www.CalcChat.com for worked-out solutions to odd-numbered exercises.

VOCABULARY: Fill in the blanks.

1. The origin of the polar coordinate system is called the _____.

2. For the point (r, θ), r is the _____ _____ from O to P and θ is the _____ _____, counterclockwise from the polar axis to the line segment $\overline{OP}$.

3. To plot the point (r, θ), use the _____ coordinate system.

4. The polar coordinates (r, θ) are related to the rectangular coordinates (x, y) as follows:

   $x =$ _____ $y =$ _____ $\tan \theta =$ _____ $r^2 =$ _____

SKILLS AND APPLICATIONS

In Exercises 5–18, plot the point given in polar coordinates and find two additional polar representations of the point, using $-2\pi < \theta < 2\pi$.

5. $\left(2, \dfrac{5\pi}{6}\right)$

6. $\left(3, \dfrac{5\pi}{4}\right)$

7. $\left(4, -\dfrac{\pi}{3}\right)$

8. $\left(-1, -\dfrac{3\pi}{4}\right)$

9. $(2, 3\pi)$

10. $\left(4, \dfrac{5\pi}{2}\right)$

11. $\left(-2, \dfrac{2\pi}{3}\right)$

12. $\left(-3, \dfrac{11\pi}{6}\right)$

13. $\left(0, -\dfrac{7\pi}{6}\right)$

14. $\left(0, -\dfrac{7\pi}{2}\right)$

15. $(\sqrt{2}, 2.36)$

16. $(2\sqrt{2}, 4.71)$

17. $(-3, -1.57)$

18. $(-5, -2.36)$

In Exercises 19–28, a point in polar coordinates is given. Convert the point to rectangular coordinates.

19. $(3, \pi/2)$

20. $(3, 3\pi/2)$

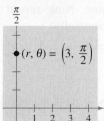

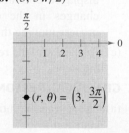

21. $(-1, 5\pi/4)$

22. $(0, -\pi)$

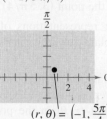

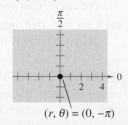

23. $(2, 3\pi/4)$

24. $(1, 5\pi/4)$

25. $(-2, 7\pi/6)$

26. $(-3, 5\pi/6)$

27. $(-2.5, 1.1)$

28. $(-2, 5.76)$

 In Exercises 29–36, use a graphing utility to find the rectangular coordinates of the point given in polar coordinates. Round your results to two decimal places.

29. $(2, 2\pi/9)$

30. $(4, 11\pi/9)$

31. $(-4.5, 1.3)$

32. $(8.25, 3.5)$

33. $(2.5, 1.58)$

34. $(5.4, 2.85)$

35. $(-4.1, -0.5)$

36. $(8.2, -3.2)$

In Exercises 37–54, a point in rectangular coordinates is given. Convert the point to polar coordinates.

37. $(1, 1)$

38. $(2, 2)$

39. $(-3, -3)$

40. $(-4, -4)$

41. $(-6, 0)$

42. $(3, 0)$

43. $(0, -5)$

44. $(0, 5)$

45. $(-3, 4)$

46. $(-4, -3)$

47. $\left(-\sqrt{3}, -\sqrt{3}\right)$

48. $\left(-\sqrt{3}, \sqrt{3}\right)$

49. $\left(\sqrt{3}, -1\right)$

50. $\left(-1, \sqrt{3}\right)$

51. $(6, 9)$

52. $(6, 2)$

53. $(5, 12)$

54. $(7, 15)$

In Exercises 55–64, use a graphing utility to find one set of polar coordinates for the point given in rectangular coordinates.

55. $(3, -2)$

56. $(-4, -2)$

57. $(-5, 2)$

58. $(7, -2)$

59. $\left(\sqrt{3}, 2\right)$

60. $\left(5, -\sqrt{2}\right)$

61. $\left(\dfrac{5}{2}, \dfrac{4}{3}\right)$

62. $\left(\dfrac{9}{5}, \dfrac{11}{2}\right)$

63. $\left(\dfrac{7}{4}, \dfrac{3}{2}\right)$

64. $\left(-\dfrac{7}{9}, -\dfrac{3}{4}\right)$

In Exercises 65–84, convert the rectangular equation to polar form. Assume $a > 0$.

65. $x^2 + y^2 = 9$

66. $x^2 + y^2 = 16$

67. $y = 4$

68. $y = x$

69. $x = 10$

70. $x = 4a$

71. $y = -2$

72. $y = 1$

73. $3x - y + 2 = 0$

74. $3x + 5y - 2 = 0$

75. $xy = 16$

76. $2xy = 1$

77. $y^2 - 8x - 16 = 0$

78. $(x^2 + y^2)^2 = 9(x^2 - y^2)$

79. $x^2 + y^2 = a^2$

80. $x^2 + y^2 = 9a^2$

81. $x^2 + y^2 - 2ax = 0$

82. $x^2 + y^2 - 2ay = 0$

83. $y^3 = x^2$

84. $y^2 = x^3$

In Exercises 85–108, convert the polar equation to rectangular form.

85. $r = 4 \sin \theta$

86. $r = 2 \cos \theta$

87. $r = -2 \cos \theta$

88. $r = -5 \sin \theta$

89. $\theta = 2\pi/3$

90. $\theta = 5\pi/3$

91. $\theta = 11\pi/6$

92. $\theta = 5\pi/6$

93. $r = 4$

94. $r = 10$

95. $r = 4 \csc \theta$

96. $r = 2 \csc \theta$

97. $r = -3 \sec \theta$

98. $r = -\sec \theta$

99. $r^2 = \cos \theta$

100. $r^2 = 2 \sin \theta$

101. $r^2 = \sin 2\theta$

102. $r^2 = \cos 2\theta$

103. $r = 2 \sin 3\theta$

104. $r = 3 \cos 2\theta$

105. $r = \dfrac{2}{1 + \sin \theta}$

106. $r = \dfrac{1}{1 - \cos \theta}$

107. $r = \dfrac{6}{2 - 3 \sin \theta}$

108. $r = \dfrac{6}{2 \cos \theta - 3 \sin \theta}$

In Exercises 109–118, describe the graph of the polar equation and find the corresponding rectangular equation. Sketch its graph.

109. $r = 6$

110. $r = 8$

111. $\theta = \pi/6$

112. $\theta = 3\pi/4$

113. $r = 2 \sin \theta$

114. $r = 4 \cos \theta$

115. $r = -6 \cos \theta$

116. $r = -3 \sin \theta$

117. $r = 3 \sec \theta$

118. $r = 2 \csc \theta$

EXPLORATION

TRUE OR FALSE? In Exercises 119 and 120, determine whether the statement is true or false. Justify your answer.

119. If $\theta_1 = \theta_2 + 2\pi n$ for some integer n, then (r, θ_1) and (r, θ_2) represent the same point on the polar coordinate system.

120. If $|r_1| = |r_2|$, then (r_1, θ) and (r_2, θ) represent the same point on the polar coordinate system.

121. Convert the polar equation $r = 2(h \cos \theta + k \sin \theta)$ to rectangular form and verify that it is the equation of a circle. Find the radius of the circle and the rectangular coordinates of the center of the circle.

122. Convert the polar equation $r = \cos \theta + 3 \sin \theta$ to rectangular form and identify the graph.

123. THINK ABOUT IT

(a) Show that the distance between the points (r_1, θ_1) and (r_2, θ_2) is $\sqrt{r_1^2 + r_2^2 - 2r_1 r_2 \cos(\theta_1 - \theta_2)}$.

(b) Describe the positions of the points relative to each other for $\theta_1 = \theta_2$. Simplify the Distance Formula for this case. Is the simplification what you expected? Explain.

(c) Simplify the Distance Formula for $\theta_1 - \theta_2 = 90°$. Is the simplification what you expected? Explain.

(d) Choose two points on the polar coordinate system and find the distance between them. Then choose different polar representations of the same two points and apply the Distance Formula again. Discuss the result.

124. GRAPHICAL REASONING

(a) Set the window format of your graphing utility on rectangular coordinates and locate the cursor at any position off the coordinate axes. Move the cursor horizontally and observe any changes in the displayed coordinates of the points. Explain the changes in the coordinates. Now repeat the process moving the cursor vertically.

(b) Set the window format of your graphing utility on polar coordinates and locate the cursor at any position off the coordinate axes. Move the cursor horizontally and observe any changes in the displayed coordinates of the points. Explain the changes in the coordinates. Now repeat the process moving the cursor vertically.

(c) Explain why the results of parts (a) and (b) are not the same.

 125. GRAPHICAL REASONING

(a) Use a graphing utility in *polar* mode to graph the equation $r = 3$.

(b) Use the *trace* feature to move the cursor around the circle. Can you locate the point $(3, 5\pi/4)$?

(c) Can you find other polar representations of the point $(3, 5\pi/4)$? If so, explain how you did it.

126. CAPSTONE In the rectangular coordinate system, each point (x, y) has a unique representation. Explain why this is not true for a point (r, θ) in the polar coordinate system.

6.8 GRAPHS OF POLAR EQUATIONS

What you should learn

- Graph polar equations by point plotting.
- Use symmetry to sketch graphs of polar equations.
- Use zeros and maximum *r*-values to sketch graphs of polar equations.
- Recognize special polar graphs.

Why you should learn it

Equations of several common figures are simpler in polar form than in rectangular form. For instance, Exercise 12 on page 501 shows the graph of a circle and its polar equation.

Introduction

In previous chapters, you learned how to sketch graphs on rectangular coordinate systems. You began with the basic point-plotting method. Then you used sketching aids such as symmetry, intercepts, asymptotes, periods, and shifts to further investigate the natures of graphs. This section approaches curve sketching on the polar coordinate system similarly, beginning with a demonstration of point plotting.

Example 1   Graphing a Polar Equation by Point Plotting

Sketch the graph of the polar equation $r = 4 \sin \theta$.

Solution

The sine function is periodic, so you can get a full range of *r*-values by considering values of θ in the interval $0 \le \theta \le 2\pi$, as shown in the following table.

θ	0	$\dfrac{\pi}{6}$	$\dfrac{\pi}{3}$	$\dfrac{\pi}{2}$	$\dfrac{2\pi}{3}$	$\dfrac{5\pi}{6}$	π	$\dfrac{7\pi}{6}$	$\dfrac{3\pi}{2}$	$\dfrac{11\pi}{6}$	2π
r	0	2	$2\sqrt{3}$	4	$2\sqrt{3}$	2	0	-2	-4	-2	0

If you plot these points as shown in Figure 6.71, it appears that the graph is a circle of radius 2 whose center is at the point $(x, y) = (0, 2)$.

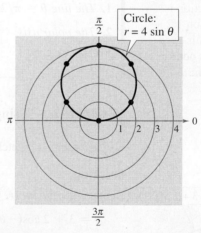

FIGURE **6.71**

CHECK *Point*   Now try Exercise 27.

You can confirm the graph in Figure 6.71 by converting the polar equation to rectangular form and then sketching the graph of the rectangular equation. You can also use a graphing utility set to *polar* mode and graph the polar equation or set the graphing utility to *parametric* mode and graph a parametric representation.

Symmetry

In Figure 6.71 on the preceding page, note that as θ increases from 0 to 2π the graph is traced out twice. Moreover, note that the graph is *symmetric with respect to the line* $\theta = \pi/2$. Had you known about this symmetry and retracing ahead of time, you could have used fewer points.

Symmetry with respect to the line $\theta = \pi/2$ is one of three important types of symmetry to consider in polar curve sketching. (See Figure 6.72.)

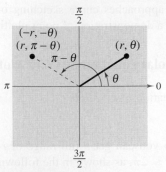

Symmetry with Respect to the Line $\theta = \dfrac{\pi}{2}$

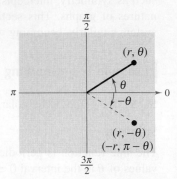

Symmetry with Respect to the Polar Axis

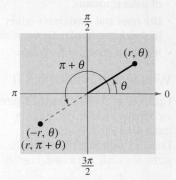

Symmetry with Respect to the Pole

FIGURE 6.72

Note in Example 2 that $\cos(-\theta) = \cos \theta$. This is because the cosine function is *even*. Recall from Section 1.2 that the cosine function is even and the sine function is odd. That is, $\sin(-\theta) = -\sin \theta$.

Tests for Symmetry in Polar Coordinates

The graph of a polar equation is symmetric with respect to the following if the given substitution yields an equivalent equation.

1. *The line $\theta = \pi/2$:*   Replace (r, θ) by $(r, \pi - \theta)$ or $(-r, -\theta)$.

2. *The polar axis:*   Replace (r, θ) by $(r, -\theta)$ or $(-r, \pi - \theta)$.

3. *The pole:*   Replace (r, θ) by $(r, \pi + \theta)$ or $(-r, \theta)$.

Example 2   Using Symmetry to Sketch a Polar Graph

Use symmetry to sketch the graph of $r = 3 + 2 \cos \theta$.

Solution

Replacing (r, θ) by $(r, -\theta)$ produces

$$r = 3 + 2 \cos(-\theta) = 3 + 2 \cos \theta. \qquad \cos(-\theta) = \cos \theta$$

So, you can conclude that the curve is symmetric with respect to the polar axis. Plotting the points in the table and using polar axis symmetry, you obtain the graph shown in Figure 6.73. This graph is called a **limaçon.**

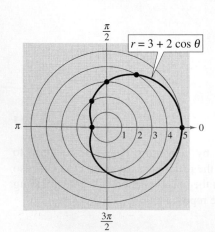

FIGURE 6.73

θ	0	$\dfrac{\pi}{3}$	$\dfrac{\pi}{2}$	$\dfrac{2\pi}{3}$	π
r	5	4	3	2	1

CHECK *Point*   Now try Exercise 33.

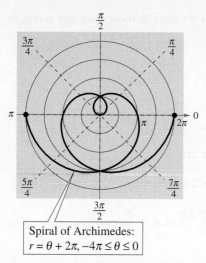

Spiral of Archimedes:
$r = \theta + 2\pi, -4\pi \le \theta \le 0$

FIGURE **6.74**

The three tests for symmetry in polar coordinates listed on page 496 are sufficient to guarantee symmetry, but they are not necessary. For instance, Figure 6.74 shows the graph of $r = \theta + 2\pi$ to be symmetric with respect to the line $\theta = \pi/2$, and yet the tests on page 496 fail to indicate symmetry because neither of the following replacements yields an equivalent equation.

Original Equation	Replacement	New Equation
$r = \theta + 2\pi$	(r, θ) by $(-r, -\theta)$	$-r = -\theta + 2\pi$
$r = \theta + 2\pi$	(r, θ) by $(r, \pi - \theta)$	$r = -\theta + 3\pi$

The equations discussed in Examples 1 and 2 are of the form

$$r = 4 \sin \theta = f(\sin \theta) \quad \text{and} \quad r = 3 + 2 \cos \theta = g(\cos \theta).$$

The graph of the first equation is symmetric with respect to the line $\theta = \pi/2$, and the graph of the second equation is symmetric with respect to the polar axis. This observation can be generalized to yield the following tests.

Quick Tests for Symmetry in Polar Coordinates

1. The graph of $r = f(\sin \theta)$ is symmetric with respect to the line $\theta = \dfrac{\pi}{2}$.

2. The graph of $r = g(\cos \theta)$ is symmetric with respect to the polar axis.

Zeros and Maximum *r*-Values

Two additional aids to graphing of polar equations involve knowing the θ-values for which $|r|$ is maximum and knowing the θ-values for which $r = 0$. For instance, in Example 1, the maximum value of $|r|$ for $r = 4 \sin \theta$ is $|r| = 4$, and this occurs when $\theta = \pi/2$, as shown in Figure 6.71. Moreover, $r = 0$ when $\theta = 0$.

Example 3   Sketching a Polar Graph

Sketch the graph of $r = 1 - 2 \cos \theta$.

Solution

From the equation $r = 1 - 2 \cos \theta$, you can obtain the following.

Symmetry: With respect to the polar axis

Maximum value of $|r|$: $r = 3$ when $\theta = \pi$

Zero of r: $r = 0$ when $\theta = \pi/3$

The table shows several θ-values in the interval $[0, \pi]$. By plotting the corresponding points, you can sketch the graph shown in Figure 6.75.

θ	0	$\dfrac{\pi}{6}$	$\dfrac{\pi}{3}$	$\dfrac{\pi}{2}$	$\dfrac{2\pi}{3}$	$\dfrac{5\pi}{6}$	π
r	-1	-0.73	0	1	2	2.73	3

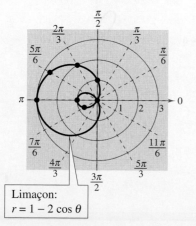

Limaçon:
$r = 1 - 2 \cos \theta$

FIGURE **6.75**

Note how the negative r-values determine the *inner loop* of the graph in Figure 6.75. This graph, like the one in Figure 6.73, is a limaçon.

CHECK*Point* ▶   Now try Exercise 35.

Some curves reach their zeros and maximum r-values at more than one point, as shown in Example 4.

Example 4 Sketching a Polar Graph

Sketch the graph of $r = 2 \cos 3\theta$.

Solution

Symmetry: With respect to the polar axis

Maximum value of $|r|$: $|r| = 2$ when $3\theta = 0, \pi, 2\pi, 3\pi$ or $\theta = 0, \dfrac{\pi}{3}, \dfrac{2\pi}{3}, \pi$

Zeros of r: $r = 0$ when $3\theta = \dfrac{\pi}{2}, \dfrac{3\pi}{2}, \dfrac{5\pi}{2}$ or $\theta = \dfrac{\pi}{6}, \dfrac{\pi}{2}, \dfrac{5\pi}{6}$

θ	0	$\dfrac{\pi}{12}$	$\dfrac{\pi}{6}$	$\dfrac{\pi}{4}$	$\dfrac{\pi}{3}$	$\dfrac{5\pi}{12}$	$\dfrac{\pi}{2}$
r	2	$\sqrt{2}$	0	$-\sqrt{2}$	-2	$-\sqrt{2}$	0

By plotting these points and using the specified symmetry, zeros, and maximum values, you can obtain the graph shown in Figure 6.76. This graph is called a **rose curve,** and each of the loops on the graph is called a *petal* of the rose curve. Note how the entire curve is generated as θ increases from 0 to π.

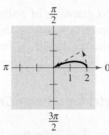

$0 \le \theta \le \dfrac{\pi}{6}$

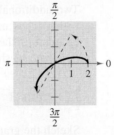

$0 \le \theta \le \dfrac{\pi}{3}$

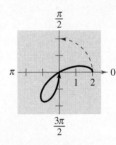

$0 \le \theta \le \dfrac{\pi}{2}$

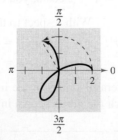

$0 \le \theta \le \dfrac{2\pi}{3}$

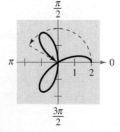

$0 \le \theta \le \dfrac{5\pi}{6}$

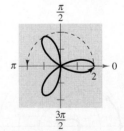

$0 \le \theta \le \pi$

FIGURE 6.76

CHECK *Point* Now try Exercise 39.

TECHNOLOGY

Use a graphing utility in *polar* mode to verify the graph of $r = 2 \cos 3\theta$ shown in Figure 6.76.

Special Polar Graphs

Several important types of graphs have equations that are simpler in polar form than in rectangular form. For example, the circle

$$r = 4 \sin \theta$$

in Example 1 has the more complicated rectangular equation

$$x^2 + (y - 2)^2 = 4.$$

Several other types of graphs that have simple polar equations are shown below.

Limaçons

$r = a \pm b \cos \theta$

$r = a \pm b \sin \theta$

$(a > 0, b > 0)$

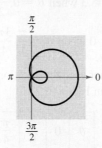

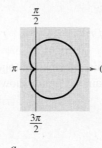

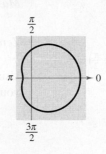

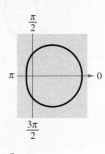

$\dfrac{a}{b} < 1$

Limaçon with inner loop

$\dfrac{a}{b} = 1$

Cardioid (heart-shaped)

$1 < \dfrac{a}{b} < 2$

Dimpled limaçon

$\dfrac{a}{b} \geq 2$

Convex limaçon

Rose Curves

n petals if n is odd,

$2n$ petals if n is even

$(n \geq 2)$.

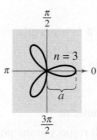

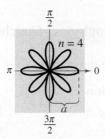

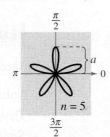

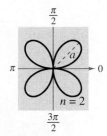

$r = a \cos n\theta$

Rose curve

$r = a \cos n\theta$

Rose curve

$r = a \sin n\theta$

Rose curve

$r = a \sin n\theta$

Rose curve

Circles and Lemniscates

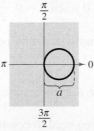

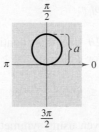

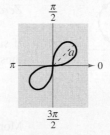

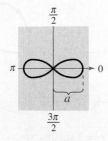

$r = a \cos \theta$

Circle

$r = a \sin \theta$

Circle

$r^2 = a^2 \sin 2\theta$

Lemniscate

$r^2 = a^2 \cos 2\theta$

Lemniscate

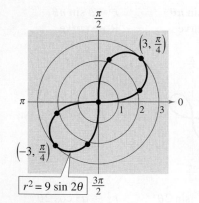

FIGURE **6.77**

Example 5   Sketching a Rose Curve

Sketch the graph of $r = 3 \cos 2\theta$.

Solution

Type of curve:	Rose curve with $2n = 4$ petals
Symmetry:	With respect to polar axis, the line $\theta = \dfrac{\pi}{2}$, and the pole
Maximum value of $\|r\|$:	$\|r\| = 3$ when $\theta = 0, \dfrac{\pi}{2}, \pi, \dfrac{3\pi}{2}$
Zeros of r:	$r = 0$ when $\theta = \dfrac{\pi}{4}, \dfrac{3\pi}{4}$

Using this information together with the additional points shown in the following table, you obtain the graph shown in Figure 6.77.

θ	0	$\dfrac{\pi}{6}$	$\dfrac{\pi}{4}$	$\dfrac{\pi}{3}$
r	3	$\dfrac{3}{2}$	0	$-\dfrac{3}{2}$

CHECK *Point*   Now try Exercise 41.

Example 6   Sketching a Lemniscate

Sketch the graph of $r^2 = 9 \sin 2\theta$.

Solution

Type of curve:	Lemniscate
Symmetry:	With respect to the pole
Maximum value of $\|r\|$:	$\|r\| = 3$ when $\theta = \dfrac{\pi}{4}$
Zeros of r:	$r = 0$ when $\theta = 0, \dfrac{\pi}{2}$

If $\sin 2\theta < 0$, this equation has no solution points. So, you restrict the values of θ to those for which $\sin 2\theta \geq 0$.

$$0 \leq \theta \leq \frac{\pi}{2} \qquad \text{or} \qquad \pi \leq \theta \leq \frac{3\pi}{2}$$

Moreover, using symmetry, you need to consider only the first of these two intervals. By finding a few additional points (see table below), you can obtain the graph shown in Figure 6.78.

θ	0	$\dfrac{\pi}{12}$	$\dfrac{\pi}{4}$	$\dfrac{5\pi}{12}$	$\dfrac{\pi}{2}$
$r = \pm 3\sqrt{\sin 2\theta}$	0	$\dfrac{\pm 3}{\sqrt{2}}$	± 3	$\dfrac{\pm 3}{\sqrt{2}}$	0

CHECK *Point*   Now try Exercise 47.

6.8   EXERCISES

See www.CalcChat.com for worked-out solutions to odd-numbered exercises.

VOCABULARY: Fill in the blanks.

1. The graph of $r = f(\sin \theta)$ is symmetric with respect to the line _____.
2. The graph of $r = g(\cos \theta)$ is symmetric with respect to the _____ _____.
3. The equation $r = 2 + \cos \theta$ represents a _____ _____.
4. The equation $r = 2 \cos \theta$ represents a _____.
5. The equation $r^2 = 4 \sin 2\theta$ represents a _____.
6. The equation $r = 1 + \sin \theta$ represents a _____.

SKILLS AND APPLICATIONS

In Exercises 7–12, identify the type of polar graph.

7.
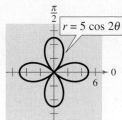
$r = 5 \cos 2\theta$

8.
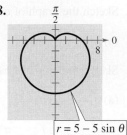
$r = 5 - 5 \sin \theta$

9.
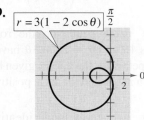
$r = 3(1 - 2 \cos \theta)$

10.
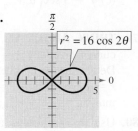
$r^2 = 16 \cos 2\theta$

11.
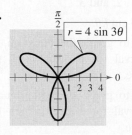
$r = 4 \sin 3\theta$

12.
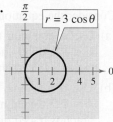
$r = 3 \cos \theta$

In Exercises 13–18, test for symmetry with respect to $\theta = \pi/2$, the polar axis, and the pole.

13. $r = 4 + 3 \cos \theta$
14. $r = 9 \cos 3\theta$
15. $r = \dfrac{2}{1 + \sin \theta}$
16. $r = \dfrac{3}{2 + \cos \theta}$
17. $r^2 = 36 \cos 2\theta$
18. $r^2 = 25 \sin 2\theta$

In Exercises 19–22, find the maximum value of $|r|$ and any zeros of r.

19. $r = 10 - 10 \sin \theta$
20. $r = 6 + 12 \cos \theta$
21. $r = 4 \cos 3\theta$
22. $r = 3 \sin 2\theta$

In Exercises 23–48, sketch the graph of the polar equation using symmetry, zeros, maximum r-values, and any other additional points.

23. $r = 4$
24. $r = -7$
25. $r = \dfrac{\pi}{3}$
26. $r = -\dfrac{3\pi}{4}$
27. $r = \sin \theta$
28. $r = 4 \cos \theta$
29. $r = 3(1 - \cos \theta)$
30. $r = 4(1 - \sin \theta)$
31. $r = 4(1 + \sin \theta)$
32. $r = 2(1 + \cos \theta)$
33. $r = 3 + 6 \sin \theta$
34. $r = 4 - 3 \sin \theta$
35. $r = 1 - 2 \sin \theta$
36. $r = 2 - 4 \cos \theta$
37. $r = 3 - 4 \cos \theta$
38. $r = 4 + 3 \cos \theta$
39. $r = 5 \sin 2\theta$
40. $r = 2 \cos 2\theta$
41. $r = 6 \cos 3\theta$
42. $r = 3 \sin 3\theta$
43. $r = 2 \sec \theta$
44. $r = 5 \csc \theta$
45. $r = \dfrac{3}{\sin \theta - 2 \cos \theta}$
46. $r = \dfrac{6}{2 \sin \theta - 3 \cos \theta}$
47. $r^2 = 9 \cos 2\theta$
48. $r^2 = 4 \sin \theta$

 In Exercises 49–58, use a graphing utility to graph the polar equation. Describe your viewing window.

49. $r = \dfrac{9}{4}$
50. $r = -\dfrac{5}{2}$
51. $r = \dfrac{5\pi}{8}$
52. $r = -\dfrac{\pi}{10}$
53. $r = 8 \cos \theta$
54. $r = \cos 2\theta$
55. $r = 3(2 - \sin \theta)$
56. $r = 2 \cos(3\theta - 2)$
57. $r = 8 \sin \theta \cos^2 \theta$
58. $r = 2 \csc \theta + 5$

In Exercises 59–64, use a graphing utility to graph the polar equation. Find an interval for θ for which the graph is traced *only once*.

59. $r = 3 - 8 \cos \theta$
60. $r = 5 + 4 \cos \theta$
61. $r = 2 \cos\left(\dfrac{3\theta}{2}\right)$
62. $r = 3 \sin\left(\dfrac{5\theta}{2}\right)$

63. $r^2 = 16 \sin 2\theta$ **64.** $r^2 = \dfrac{1}{\theta}$

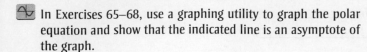

In Exercises 65–68, use a graphing utility to graph the polar equation and show that the indicated line is an asymptote of the graph.

Name of Graph	*Polar Equation*	*Asymptote*
65. Conchoid	$r = 2 - \sec\theta$	$x = -1$
66. Conchoid	$r = 2 + \csc\theta$	$y = 1$
67. Hyperbolic spiral	$r = \dfrac{3}{\theta}$	$y = 3$
68. Strophoid	$r = 2 \cos 2\theta \sec\theta$	$x = -2$

EXPLORATION

TRUE OR FALSE? In Exercises 69 and 70, determine whether the statement is true or false. Justify your answer.

69. In the polar coordinate system, if a graph that has symmetry with respect to the polar axis were folded on the line $\theta = 0$, the portion of the graph above the polar axis would coincide with the portion of the graph below the polar axis.

70. In the polar coordinate system, if a graph that has symmetry with respect to the pole were folded on the line $\theta = 3\pi/4$, the portion of the graph on one side of the fold would coincide with the portion of the graph on the other side of the fold.

71. Sketch the graph of $r = 6 \cos\theta$ over each interval. Describe the part of the graph obtained in each case.

(a) $0 \le \theta \le \dfrac{\pi}{2}$ (b) $\dfrac{\pi}{2} \le \theta \le \pi$

(c) $-\dfrac{\pi}{2} \le \theta \le \dfrac{\pi}{2}$ (d) $\dfrac{\pi}{4} \le \theta \le \dfrac{3\pi}{4}$

72. GRAPHICAL REASONING Use a graphing utility to graph the polar equation $r = 6[1 + \cos(\theta - \phi)]$ for (a) $\phi = 0$, (b) $\phi = \pi/4$, and (c) $\phi = \pi/2$. Use the graphs to describe the effect of the angle ϕ. Write the equation as a function of $\sin\theta$ for part (c).

73. The graph of $r = f(\theta)$ is rotated about the pole through an angle ϕ. Show that the equation of the rotated graph is $r = f(\theta - \phi)$.

74. Consider the graph of $r = f(\sin\theta)$.

(a) Show that if the graph is rotated counterclockwise $\pi/2$ radians about the pole, the equation of the rotated graph is $r = f(-\cos\theta)$.

(b) Show that if the graph is rotated counterclockwise π radians about the pole, the equation of the rotated graph is $r = f(-\sin\theta)$.

(c) Show that if the graph is rotated counterclockwise $3\pi/2$ radians about the pole, the equation of the rotated graph is $r = f(\cos\theta)$.

In Exercises 75–78, use the results of Exercises 73 and 74.

75. Write an equation for the limaçon $r = 2 - \sin\theta$ after it has been rotated through the given angle.

(a) $\dfrac{\pi}{4}$ (b) $\dfrac{\pi}{2}$ (c) π (d) $\dfrac{3\pi}{2}$

76. Write an equation for the rose curve $r = 2 \sin 2\theta$ after it has been rotated through the given angle.

(a) $\dfrac{\pi}{6}$ (b) $\dfrac{\pi}{2}$ (c) $\dfrac{2\pi}{3}$ (d) π

77. Sketch the graph of each equation.

(a) $r = 1 - \sin\theta$ (b) $r = 1 - \sin\left(\theta - \dfrac{\pi}{4}\right)$

78. Sketch the graph of each equation.

(a) $r = 3 \sec\theta$ (b) $r = 3 \sec\left(\theta - \dfrac{\pi}{4}\right)$

(c) $r = 3 \sec\left(\theta + \dfrac{\pi}{3}\right)$ (d) $r = 3 \sec\left(\theta - \dfrac{\pi}{2}\right)$

79. THINK ABOUT IT How many petals do the rose curves given by $r = 2 \cos 4\theta$ and $r = 2 \sin 3\theta$ have? Determine the numbers of petals for the curves given by $r = 2 \cos n\theta$ and $r = 2 \sin n\theta$, where n is a positive integer.

80. Use a graphing utility to graph and identify $r = 2 + k \sin\theta$ for $k = 0$, 1, 2, and 3.

81. Consider the equation $r = 3 \sin k\theta$.

(a) Use a graphing utility to graph the equation for $k = 1.5$. Find the interval for θ over which the graph is traced only once.

(b) Use a graphing utility to graph the equation for $k = 2.5$. Find the interval for θ over which the graph is traced only once.

(c) Is it possible to find an interval for θ over which the graph is traced only once for any rational number k? Explain.

82. CAPSTONE Write a brief paragraph that describes why some polar curves have equations that are simpler in polar form than in rectangular form. Besides a circle, give an example of a curve that is simpler in polar form than in rectangular form. Give an example of a curve that is simpler in rectangular form than in polar form.

6.9 POLAR EQUATIONS OF CONICS

What you should learn

- Define conics in terms of eccentricity.
- Write and graph equations of conics in polar form.
- Use equations of conics in polar form to model real-life problems.

Why you should learn it

The orbits of planets and satellites can be modeled with polar equations. For instance, in Exercise 65 on page 508, a polar equation is used to model the orbit of a satellite.

Corbis

Alternative Definition of Conic

In Sections 6.3 and 6.4, you learned that the rectangular equations of ellipses and hyperbolas take simple forms when the origin lies at their *centers*. As it happens, there are many important applications of conics in which it is more convenient to use one of the *foci* as the origin. In this section, you will learn that polar equations of conics take simple forms if one of the foci lies at the pole.

To begin, consider the following alternative definition of conic that uses the concept of eccentricity.

Alternative Definition of Conic

The locus of a point in the plane that moves so that its distance from a fixed point (focus) is in a constant ratio to its distance from a fixed line (directrix) is a **conic**. The constant ratio is the eccentricity of the conic and is denoted by e. Moreover, the conic is an **ellipse** if $e < 1$, a **parabola** if $e = 1$, and a **hyperbola** if $e > 1$. (See Figure 6.79.)

In Figure 6.79, note that for each type of conic, the focus is at the pole.

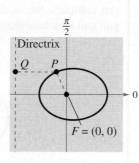

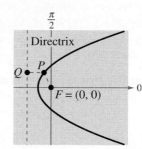

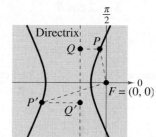

Ellipse: $0 < e < 1$

$\dfrac{PF}{PQ} < 1$

Parabola: $e = 1$

$\dfrac{PF}{PQ} = 1$

Hyperbola $e > 1$

$\dfrac{PF}{PQ} = \dfrac{P'F}{P'Q'} > 1$

FIGURE **6.79**

Polar Equations of Conics

The benefit of locating a focus of a conic at the pole is that the equation of the conic takes on a simpler form. For a proof of the polar equations of conics, see Proofs in Mathematics on page 520.

Polar Equations of Conics

The graph of a polar equation of the form

1. $r = \dfrac{ep}{1 \pm e \cos \theta}$ or   **2.** $r = \dfrac{ep}{1 \pm e \sin \theta}$

is a conic, where $e > 0$ is the eccentricity and $|p|$ is the distance between the focus (pole) and the directrix.

Equations of the form

$$r = \frac{ep}{1 \pm e \cos \theta} = g(\cos \theta) \qquad \text{Vertical directrix}$$

correspond to conics with a vertical directrix and symmetry with respect to the polar axis. Equations of the form

$$r = \frac{ep}{1 \pm e \sin \theta} = g(\sin \theta) \qquad \text{Horizontal directrix}$$

correspond to conics with a horizontal directrix and symmetry with respect to the line $\theta = \pi/2$. Moreover, the converse is also true—that is, any conic with a focus at the pole and having a horizontal or vertical directrix can be represented by one of these equations.

Example 1 Identifying a Conic from Its Equation

Identify the type of conic represented by the equation $r = \dfrac{15}{3 - 2 \cos \theta}$.

Algebraic Solution

To identify the type of conic, rewrite the equation in the form $r = (ep)/(1 \pm e \cos \theta)$.

$$r = \frac{15}{3 - 2 \cos \theta} \qquad \text{Write original equation.}$$

$$= \frac{5}{1 - (2/3) \cos \theta} \qquad \begin{array}{l}\text{Divide numerator and} \\ \text{denominator by 3.}\end{array}$$

Because $e = \frac{2}{3} < 1$, you can conclude that the graph is an ellipse.

Graphical Solution

You can start sketching the graph by plotting points from $\theta = 0$ to $\theta = \pi$. Because the equation is of the form $r = g(\cos \theta)$, the graph of r is symmetric with respect to the polar axis. So, you can complete the sketch, as shown in Figure 6.80. From this, you can conclude that the graph is an ellipse.

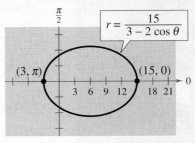

FIGURE **6.80**

CHECK *Point* Now try Exercise 15.

For the ellipse in Figure 6.80, the major axis is horizontal and the vertices lie at $(15, 0)$ and $(3, \pi)$. So, the length of the *major* axis is $2a = 18$. To find the length of the *minor* axis, you can use the equations $e = c/a$ and $b^2 = a^2 - c^2$ to conclude that

$$b^2 = a^2 - c^2$$

$$= a^2 - (ea)^2$$

$$= a^2(1 - e^2). \qquad \text{Ellipse}$$

Because $e = \frac{2}{3}$, you have $b^2 = 9^2\left[1 - \left(\frac{2}{3}\right)^2\right] = 45$, which implies that $b = \sqrt{45} = 3\sqrt{5}$. So, the length of the minor axis is $2b = 6\sqrt{5}$. A similar analysis for hyperbolas yields

$$b^2 = c^2 - a^2$$

$$= (ea)^2 - a^2$$

$$= a^2(e^2 - 1). \qquad \text{Hyperbola}$$

Example 2   Sketching a Conic from Its Polar Equation

Identify the conic $r = \dfrac{32}{3 + 5 \sin \theta}$ and sketch its graph.

Solution

Dividing the numerator and denominator by 3, you have

$$r = \frac{32/3}{1 + (5/3) \sin \theta}.$$

Because $e = \frac{5}{3} > 1$, the graph is a hyperbola. The transverse axis of the hyperbola lies on the line $\theta = \pi/2$, and the vertices occur at $(4, \pi/2)$ and $(-16, 3\pi/2)$. Because the length of the transverse axis is 12, you can see that $a = 6$. To find b, write

$$b^2 = a^2(e^2 - 1) = 6^2\left[\left(\frac{5}{3}\right)^2 - 1\right] = 64.$$

So, $b = 8$. Finally, you can use a and b to determine that the asymptotes of the hyperbola are $y = 10 \pm \frac{3}{4}x$. The graph is shown in Figure 6.81.

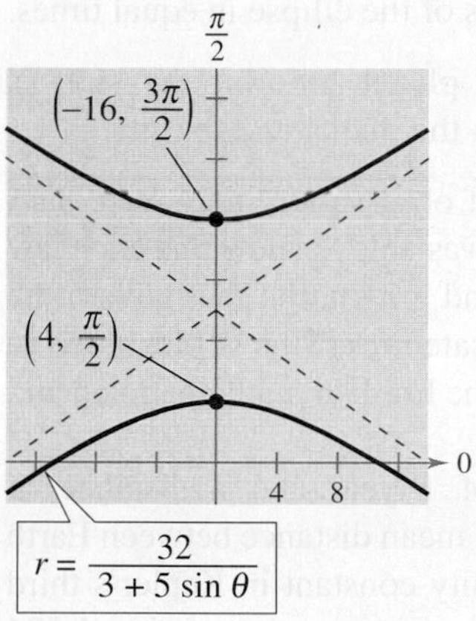

CHECK Point   Now try Exercise 23.

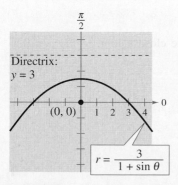

$r = \dfrac{32}{3 + 5 \sin \theta}$

FIGURE **6.81**

In the next example, you are asked to find a polar equation of a specified conic. To do this, let p be the distance between the pole and the directrix.

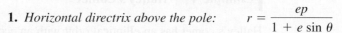

1. *Horizontal directrix above the pole:* $r = \dfrac{ep}{1 + e \sin \theta}$

2. *Horizontal directrix below the pole:* $r = \dfrac{ep}{1 - e \sin \theta}$

3. *Vertical directrix to the right of the pole:* $r = \dfrac{ep}{1 + e \cos \theta}$

4. *Vertical directrix to the left of the pole:* $r = \dfrac{ep}{1 - e \cos \theta}$

TECHNOLOGY

Use a graphing utility set in *polar* mode to verify the four orientations shown at the right. Remember that e must be positive, but p can be positive or negative.

Example 3   Finding the Polar Equation of a Conic

Find the polar equation of the parabola whose focus is the pole and whose directrix is the line $y = 3$.

Solution

From Figure 6.82, you can see that the directrix is horizontal and above the pole, so you can choose an equation of the form

$$r = \frac{ep}{1 + e \sin \theta}.$$

Moreover, because the eccentricity of a parabola is $e = 1$ and the distance between the pole and the directrix is $p = 3$, you have the equation

$$r = \frac{3}{1 + \sin \theta}.$$

CHECK Point   Now try Exercise 39.

Directrix:
$y = 3$

$(0, 0)$

$r = \dfrac{3}{1 + \sin \theta}$

FIGURE **6.82**

Applications

Kepler's Laws (listed below), named after the German astronomer Johannes Kepler (1571–1630), can be used to describe the orbits of the planets about the sun.

1. Each planet moves in an elliptical orbit with the sun at one focus.

2. A ray from the sun to the planet sweeps out equal areas of the ellipse in equal times.

3. The square of the period (the time it takes for a planet to orbit the sun) is proportional to the cube of the mean distance between the planet and the sun.

Although Kepler simply stated these laws on the basis of observation, they were later validated by Isaac Newton (1642–1727). In fact, Newton was able to show that each law can be deduced from a set of universal laws of motion and gravitation that govern the movement of all heavenly bodies, including comets and satellites. This is illustrated in the next example, which involves the comet named after the English mathematician and physicist Edmund Halley (1656–1742).

If you use Earth as a reference with a period of 1 year and a distance of 1 astronomical unit (an *astronomical unit* is defined as the mean distance between Earth and the sun, or about 93 million miles), the proportionality constant in Kepler's third law is 1. For example, because Mars has a mean distance to the sun of $d = 1.524$ astronomical units, its period P is given by $d^3 = P^2$. So, the period of Mars is $P \approx 1.88$ years.

Example 4 Halley's Comet

Halley's comet has an elliptical orbit with an eccentricity of $e \approx 0.967$. The length of the major axis of the orbit is approximately 35.88 astronomical units. Find a polar equation for the orbit. How close does Halley's comet come to the sun?

Solution

Using a vertical axis, as shown in Figure 6.83, choose an equation of the form $r = ep/(1 + e \sin \theta)$. Because the vertices of the ellipse occur when $\theta = \pi/2$ and $\theta = 3\pi/2$, you can determine the length of the major axis to be the sum of the r-values of the vertices. That is,

$$2a = \frac{0.967p}{1 + 0.967} + \frac{0.967p}{1 - 0.967} \approx 29.79p \approx 35.88.$$

So, $p \approx 1.204$ and $ep \approx (0.967)(1.204) \approx 1.164$. Using this value of ep in the equation, you have

$$r = \frac{1.164}{1 + 0.967 \sin \theta}$$

where r is measured in astronomical units. To find the closest point to the sun (the focus), substitute $\theta = \pi/2$ in this equation to obtain

$$r = \frac{1.164}{1 + 0.967 \sin(\pi/2)}$$

$$\approx 0.59 \text{ astronomical unit}$$

$$\approx 55,000,000 \text{ miles.}$$

CHECK*Point* Now try Exercise 63.

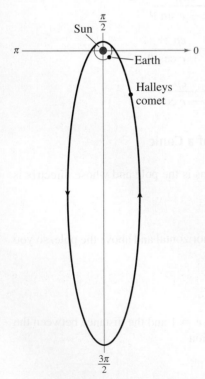

FIGURE 6.83

6.9   EXERCISES

See www.CalcChat.com for worked-out solutions to odd-numbered exercises.

VOCABULARY

In Exercises 1–3, fill in the blanks.

1. The locus of a point in the plane that moves so that its distance from a fixed point (focus) is in a constant ratio to its distance from a fixed line (directrix) is a _____.

2. The constant ratio is the _____ of the conic and is denoted by _____.

3. An equation of the form $r = \dfrac{ep}{1 + e \cos \theta}$ has a _____ directrix to the _____ of the pole.

4. Match the conic with its eccentricity.

   (a) $e < 1$ (b) $e = 1$ (c) $e > 1$

   (i) parabola   (ii) hyperbola   (iii) ellipse

SKILLS AND APPLICATIONS

In Exercises 5–8, write the polar equation of the conic for $e = 1$, $e = 0.5$, and $e = 1.5$. Identify the conic for each equation. Verify your answers with a graphing utility.

5. $r = \dfrac{2e}{1 + e \cos \theta}$ **6.** $r = \dfrac{2e}{1 - e \cos \theta}$

7. $r = \dfrac{2e}{1 - e \sin \theta}$ **8.** $r = \dfrac{2e}{1 + e \sin \theta}$

In Exercises 9–14, match the polar equation with its graph. [The graphs are labeled (a), (b), (c), (d), (e), and (f).]

(a)
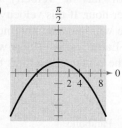

(b)
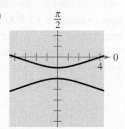

(c)
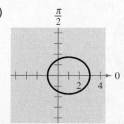

(d)
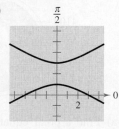

(e)
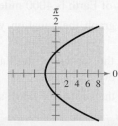

(f)
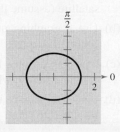

9. $r = \dfrac{4}{1 - \cos \theta}$ **10.** $r = \dfrac{3}{2 - \cos \theta}$

11. $r = \dfrac{3}{1 + 2 \sin \theta}$ **12.** $r = \dfrac{3}{2 + \cos \theta}$

13. $r = \dfrac{4}{1 + \sin \theta}$ **14.** $r = \dfrac{4}{1 - 3 \sin \theta}$

In Exercises 15–28, identify the conic and sketch its graph.

15. $r = \dfrac{3}{1 - \cos \theta}$ **16.** $r = \dfrac{7}{1 + \sin \theta}$

17. $r = \dfrac{5}{1 + \sin \theta}$ **18.** $r = \dfrac{6}{1 + \cos \theta}$

19. $r = \dfrac{2}{2 - \cos \theta}$ **20.** $r = \dfrac{4}{4 + \sin \theta}$

21. $r = \dfrac{6}{2 + \sin \theta}$ **22.** $r = \dfrac{9}{3 - 2 \cos \theta}$

23. $r = \dfrac{3}{2 + 4 \sin \theta}$ **24.** $r = \dfrac{5}{-1 + 2 \cos \theta}$

25. $r = \dfrac{3}{2 - 6 \cos \theta}$ **26.** $r = \dfrac{3}{2 + 6 \sin \theta}$

27. $r = \dfrac{4}{2 - \cos \theta}$ **28.** $r = \dfrac{2}{2 + 3 \sin \theta}$

In Exercises 29–34, use a graphing utility to graph the polar equation. Identify the graph.

29. $r = \dfrac{-1}{1 - \sin \theta}$ **30.** $r = \dfrac{-5}{2 + 4 \sin \theta}$

31. $r = \dfrac{3}{-4 + 2 \cos \theta}$ **32.** $r = \dfrac{4}{1 - 2 \cos \theta}$

33. $r = \dfrac{14}{14 + 17 \sin \theta}$ **34.** $r = \dfrac{12}{2 - \cos \theta}$

In Exercises 35–38, use a graphing utility to graph the rotated conic.

35. $r = \dfrac{3}{1 - \cos(\theta - \pi/4)}$ (See Exercise 15.)

36. $r = \dfrac{4}{4 + \sin(\theta - \pi/3)}$ (See Exercise 20.)

37. $r = \dfrac{6}{2 + \sin(\theta + \pi/6)}$ (See Exercise 21.)

38. $r = \dfrac{5}{-1 + 2\cos(\theta + 2\pi/3)}$ (See Exercise 24.)

In Exercises 39–54, find a polar equation of the conic with its focus at the pole.

Conic	Eccentricity	Directrix
39. Parabola	$e = 1$	$x = -1$
40. Parabola	$e = 1$	$y = -4$
41. Ellipse	$e = \frac{1}{2}$	$y = 1$
42. Ellipse	$e = \frac{3}{4}$	$y = -2$
43. Hyperbola	$e = 2$	$x = 1$
44. Hyperbola	$e = \frac{3}{2}$	$x = -1$

Conic	Vertex or Vertices
45. Parabola	$(1, -\pi/2)$
46. Parabola	$(8, 0)$
47. Parabola	$(5, \pi)$
48. Parabola	$(10, \pi/2)$
49. Ellipse	$(2, 0), (10, \pi)$
50. Ellipse	$(2, \pi/2), (4, 3\pi/2)$
51. Ellipse	$(20, 0), (4, \pi)$
52. Hyperbola	$(2, 0), (8, 0)$
53. Hyperbola	$(1, 3\pi/2), (9, 3\pi/2)$
54. Hyperbola	$(4, \pi/2), (1, \pi/2)$

55. **PLANETARY MOTION** The planets travel in elliptical orbits with the sun at one focus. Assume that the focus is at the pole, the major axis lies on the polar axis, and the length of the major axis is $2a$ (see figure). Show that the polar equation of the orbit is $r = a(1 - e^2)/(1 - e\cos\theta)$, where e is the eccentricity.

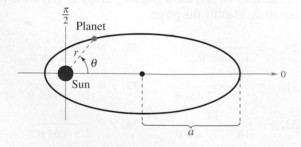

56. **PLANETARY MOTION** Use the result of Exercise 55 to show that the minimum distance (*perihelion distance*) from the sun to the planet is $r = a(1 - e)$ and the maximum distance (*aphelion distance*) is $r = a(1 + e)$.

PLANETARY MOTION In Exercises 57–62, use the results of Exercises 55 and 56 to find the polar equation of the planet's orbit and the perihelion and aphelion distances.

57. Earth $a = 95.956 \times 10^6$ miles, $e = 0.0167$

58. Saturn $a = 1.427 \times 10^9$ kilometers, $e = 0.0542$

59. Venus $a = 108.209 \times 10^6$ kilometers, $e = 0.0068$

60. Mercury $a = 35.98 \times 10^6$ miles, $e = 0.2056$

61. Mars $a = 141.63 \times 10^6$ miles, $e = 0.0934$

62. Jupiter $a = 778.41 \times 10^6$ kilometers, $e = 0.0484$

63. **ASTRONOMY** The comet Encke has an elliptical orbit with an eccentricity of $e \approx 0.847$. The length of the major axis of the orbit is approximately 4.42 astronomical units. Find a polar equation for the orbit. How close does the comet come to the sun?

64. **ASTRONOMY** The comet Hale-Bopp has an elliptical orbit with an eccentricity of $e \approx 0.995$. The length of the major axis of the orbit is approximately 500 astronomical units. Find a polar equation for the orbit. How close does the comet come to the sun?

65. **SATELLITE TRACKING** A satellite in a 100-mile-high circular orbit around Earth has a velocity of approximately 17,500 miles per hour. If this velocity is multiplied by $\sqrt{2}$, the satellite will have the minimum velocity necessary to escape Earth's gravity and will follow a parabolic path with the center of Earth as the focus (see figure).

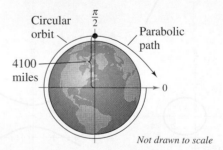

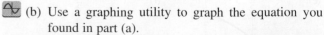

Not drawn to scale

(a) Find a polar equation of the parabolic path of the satellite (assume the radius of Earth is 4000 miles).

(b) Use a graphing utility to graph the equation you found in part (a).

(c) Find the distance between the surface of the Earth and the satellite when $\theta = 30°$.

(d) Find the distance between the surface of Earth and the satellite when $\theta = 60°$.

66. ROMAN COLISEUM The Roman Coliseum is an elliptical amphitheater measuring approximately 188 meters long and 156 meters wide.

(a) Find an equation to model the coliseum that is of the form

$$\frac{x^2}{a^2} + \frac{y^2}{b^2} = 1.$$

(b) Find a polar equation to model the coliseum. (Assume $e \approx 0.5581$ and $p \approx 115.98$.)

(c) Use a graphing utility to graph the equations you found in parts (a) and (b). Are the graphs the same? Why or why not?

(d) In part (c), did you prefer graphing the rectangular equation or the polar equation? Explain.

EXPLORATION

TRUE OR FALSE? In Exercises 67–70, determine whether the statement is true or false. Justify your answer.

67. For a given value of $e > 1$ over the interval $\theta = 0$ to $\theta = 2\pi$, the graph of

$$r = \frac{ex}{1 - e\cos\theta}$$

is the same as the graph of

$$r = \frac{e(-x)}{1 + e\cos\theta}.$$

68. The graph of

$$r = \frac{4}{-3 - 3\sin\theta}$$

has a horizontal directrix above the pole.

69. The conic represented by the following equation is an ellipse.

$$r^2 = \frac{16}{9 - 4\cos\left(\theta + \frac{\pi}{4}\right)}$$

70. The conic represented by the following equation is a parabola.

$$r = \frac{6}{3 - 2\cos\theta}$$

71. WRITING Explain how the graph of each conic differs from the graph of $r = \dfrac{5}{1 + \sin\theta}$. (See Exercise 17.)

(a) $r = \dfrac{5}{1 - \cos\theta}$ (b) $r = \dfrac{5}{1 - \sin\theta}$

(c) $r = \dfrac{5}{1 + \cos\theta}$ (d) $r = \dfrac{5}{1 - \sin[\theta - (\pi/4)]}$

72. CAPSTONE In your own words, define the term *eccentricity* and explain how it can be used to classify conics.

73. Show that the polar equation of the ellipse

$$\frac{x^2}{a^2} + \frac{y^2}{b^2} = 1 \quad \text{is} \quad r^2 = \frac{b^2}{1 - e^2\cos^2\theta}.$$

74. Show that the polar equation of the hyperbola

$$\frac{x^2}{a^2} - \frac{y^2}{b^2} = 1 \quad \text{is} \quad r^2 = \frac{-b^2}{1 - e^2\cos^2\theta}.$$

In Exercises 75–80, use the results of Exercises 73 and 74 to write the polar form of the equation of the conic.

75. $\dfrac{x^2}{169} + \dfrac{y^2}{144} = 1$ **76.** $\dfrac{x^2}{25} + \dfrac{y^2}{16} = 1$

77. $\dfrac{x^2}{9} - \dfrac{y^2}{16} = 1$ **78.** $\dfrac{x^2}{36} - \dfrac{y^2}{4} = 1$

79. Hyperbola   One focus: $(5, 0)$
Vertices: $(4, 0), (4, \pi)$

80. Ellipse   One focus: $(4, 0)$
Vertices: $(5, 0), (5, \pi)$

81. Consider the polar equation

$$r = \frac{4}{1 - 0.4\cos\theta}.$$

(a) Identify the conic without graphing the equation.

(b) Without graphing the following polar equations, describe how each differs from the given polar equation.

$$r_1 = \frac{4}{1 + 0.4\cos\theta} \qquad r_2 = \frac{4}{1 - 0.4\sin\theta}$$

(c) Use a graphing utility to verify your results in part (b).

82. The equation

$$r = \frac{ep}{1 \pm e\sin\theta}$$

is the equation of an ellipse with $e < 1$. What happens to the lengths of both the major axis and the minor axis when the value of e remains fixed and the value of p changes? Use an example to explain your reasoning.

6 CHAPTER SUMMARY

What Did You Learn?	Explanation/Examples	Review Exercises
Section 6.1 Find the inclination of a line (p. 438).	If a nonvertical line has inclination θ and slope m, then $m = \tan \theta$.	1–4
Find the angle between two lines (p. 439).	If two nonperpendicular lines have slopes m_1 and m_2, the angle between the lines is $\tan \theta = \left| (m_2 - m_1)/(1 + m_1 m_2) \right|$.	5–8
Find the distance between a point and a line (p. 440).	The distance between the point (x_1, y_1) and the line $Ax + By + C = 0$ is $d = \left| Ax_1 + By_1 + C \right|/\sqrt{A^2 + B^2}$.	9, 10
Section 6.2 Recognize a conic as the intersection of a plane and a double-napped cone (p. 445).	In the formation of the four basic conics, the intersecting plane does not pass through the vertex of the cone. (See Figure 10.9.)	11, 12
Write equations of parabolas in standard form and graph parabolas (p. 446).	The standard form of the equation of a parabola with vertex at (h, k) is $(x - h)^2 = 4p(y - k)$, $p \neq 0$ (vertical axis), or $(y - k)^2 = 4p(x - h)$, $p \neq 0$ (horizontal axis).	13–16
Use the reflective property of parabolas to solve real-life problems (p. 448).	The tangent line to a parabola at a point P makes equal angles with (1) the line passing through P and the focus and (2) the axis of the parabola.	17–20
Section 6.3 Write equations of ellipses in standard form and graph ellipses (p. 455).	**Horizontal Major Axis** $\dfrac{(x - h)^2}{a^2} + \dfrac{(y - k)^2}{b^2} = 1$ **Vertical Major Axis** $\dfrac{(x - h)^2}{b^2} + \dfrac{(y - k)^2}{a^2} = 1$	21–24
Use properties of ellipses to model and solve real-life problems (p. 458).	The properties of ellipses can be used to find distances from Earth's center to the moon's center in its orbit. (See Example 4.)	25, 26
Find eccentricities (p. 458).	The eccentricity e of an ellipse is given by $e = c/a$.	27–30
Section 6.4 Write equations of hyperbolas in standard form (p. 463) and find asymptotes of and graph hyperbolas (p. 465).	**Horizontal Transverse Axis** $\dfrac{(x - h)^2}{a^2} - \dfrac{(y - k)^2}{b^2} = 1$ **Vertical Transverse Axis** $\dfrac{(y - k)^2}{a^2} - \dfrac{(x - h)^2}{b^2} = 1$ **Asymptotes** $y = k \pm (b/a)(x - h)$ **Asymptotes** $y = k \pm (a/b)(x - h)$	31–38
Use properties of hyperbolas to solve real-life problems (p. 468).	The properties of hyperbolas can be used in radar and other detection systems. (See Example 5.)	39, 40
Classify conics from their general equations (p. 469).	The graph of $Ax^2 + Cy^2 + Dx + Ey + F = 0$ is a circle if $A = C$, a parabola if $AC = 0$, an ellipse if $AC > 0$, and a hyperbola if $AC < 0$.	41–44
Section 6.5 Rotate the coordinate axes to eliminate the xy-term in equations of conics (p. 473).	The equation $Ax^2 + Bxy + Cy^2 + Dx + Ey + F = 0$ can be rewritten as $A'(x')^2 + C'(y')^2 + D'x' + E'y' + F' = 0$ by rotating the coordinate axes through an angle θ, where $\cot 2\theta = (A - C)/B$.	45–48
Use the discriminant to classify conics (p. 477).	The graph of $Ax^2 + Bxy + Cy^2 + Dx + Ey + F = 0$ is, except in degenerate cases, an ellipse or a circle if $B^2 - 4AC < 0$, a parabola if $B^2 - 4AC = 0$, and a hyperbola if $B^2 - 4AC > 0$.	49–52

	What Did You Learn?	**Explanation/Examples**	**Review Exercises**
Section 6.6	Evaluate sets of parametric equations for given values of the parameter *(p. 481)*.	If f and g are continuous functions of t on an interval I, the set of ordered pairs $(f(t), g(t))$ is a plane curve C. The equations $x = f(t)$ and $y = g(t)$ are parametric equations for C, and t is the parameter.	53, 54
	Sketch curves that are represented by sets of parametric equations *(p. 482)*.	Sketching a curve represented by parametric equations requires plotting points in the xy-plane. Each set of coordinates (x, y) is determined from a value chosen for t.	55–60
	Rewrite sets of parametric equations as single rectangular equations by eliminating the parameter *(p. 483)*.	To eliminate the parameter in a pair of parametric equations, solve for t in one equation and substitute that value of t into the other equation. The result is the corresponding rectangular equation.	55–60
	Find sets of parametric equations for graphs *(p. 484)*.	When finding a set of parametric equations for a given graph, remember that the parametric equations are not unique.	61–64
Section 6.7	Plot points on the polar coordinate system *(p. 489)*.		65–68
	Convert points *(p. 490)* and equations *(p. 492)* from rectangular to polar form and vice versa.	**Polar Coordinates (r, θ) and Rectangular Coordinates (x, y)** Polar-to-Rectangular: $x = r \cos \theta$, $y = r \sin \theta$ Rectangular-to-Polar: $\tan \theta = y/x$, $r^2 = x^2 + y^2$ To convert a rectangular equation to polar form, replace x by $r \cos \theta$ and y by $r \sin \theta$. Converting from a polar equation to rectangular form is more complex.	69–88
Section 6.8	Use point plotting *(p. 495)* and symmetry *(p. 496)* to sketch graphs of polar equations.	Graphing a polar equation by point plotting is similar to graphing a rectangular equation. A polar graph is symmetric with respect to the following if the given substitution yields an equivalent equation. **1.** Line $\theta = \pi/2$: Replace (r, θ) by $(r, \pi - \theta)$ or $(-r, -\theta)$. **2.** Polar axis: Replace (r, θ) by $(r, -\theta)$ or $(-r, \pi - \theta)$. **3.** Pole: Replace (r, θ) by $(r, \pi + \theta)$ or $(-r, \theta)$.	89–98
	Use zeros and maximum r-values to sketch graphs of polar equations *(p. 497)*.	Two additional aids to graphing polar equations involve knowing the θ-values for which $|r|$ is maximum and knowing the θ-values for which $r = 0$.	89–98
	Recognize special polar graphs *(p. 499)*.	Several types of graphs, such as limaçons, rose curves, circles, and lemniscates, have equations that are simpler in polar form than in rectangular form. (See page 787.)	99–102
Section 6.9	Define conics in terms of eccentricity *(p. 503)*.	The eccentricity of a conic is denoted by e. **ellipse:** $e < 1$ **parabola:** $e = 1$ **hyperbola:** $e > 1$	103–110
	Write and graph equations of conics in polar form *(p. 503)*.	The graph of a polar equation of the form (1) $r = (ep)/(1 \pm e \cos \theta)$ or (2) $r = (ep)/(1 \pm e \sin \theta)$ is a conic, where $e > 0$ is the eccentricity and $|p|$ is the distance between the focus (pole) and the directrix.	103–110
	Use equations of conics in polar form to model real-life problems *(p. 506)*.	Equations of conics in polar form can be used to model the orbit of Halley's comet. (See Example 4.)	111, 112

6   REVIEW EXERCISES

See www.CalcChat.com for worked-out solutions to odd-numbered exercises.

6.1 In Exercises 1–4, find the inclination θ (in radians and degrees) of the line with the given characteristics.

1. Passes through the points $(-1, 2)$ and $(2, 5)$
2. Passes through the points $(3, 4)$ and $(-2, 7)$
3. Equation: $y = 2x + 4$ 4. Equation: $x - 5y = 7$

In Exercises 5–8, find the angle θ (in radians and degrees) between the lines.

5. $4x + y = 2$
    $-5x + y = -1$

6. $-5x + 3y = 3$
    $-2x + 3y = 1$

7. $2x - 7y = 8$
    $0.4x + y = 0$

8. $0.02x + 0.07y = 0.18$
    $0.09x - 0.04y = 0.17$

In Exercises 9 and 10, find the distance between the point and the line.

	Point	Line
9.	$(5, 3)$	$x - y - 10 = 0$
10.	$(0, 4)$	$x + 2y - 2 = 0$

6.2 In Exercises 11 and 12, state what type of conic is formed by the intersection of the plane and the double-napped cone.

11.   12.

 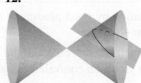

In Exercises 13–16, find the standard form of the equation of the parabola with the given characteristics. Then graph the parabola.

13. Vertex: $(0, 0)$
    Focus: $(4, 0)$

14. Vertex: $(2, 0)$
    Focus: $(0, 0)$

15. Vertex: $(0, 2)$
    Directrix: $x = -3$

16. Vertex: $(-3, -3)$
    Directrix: $y = 0$

In Exercises 17 and 18, find an equation of the tangent line to the parabola at the given point, and find the x-intercept of the line.

17. $y = 2x^2$, $(-1, 2)$ 18. $x^2 = -2y$, $(-4, -8)$

19. **ARCHITECTURE** A parabolic archway is 12 meters high at the vertex. At a height of 10 meters, the width of the archway is 8 meters (see figure). How wide is the archway at ground level?

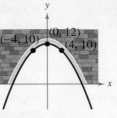

FIGURE FOR 19   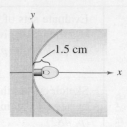

FIGURE FOR 20

20. **FLASHLIGHT** The light bulb in a flashlight is at the focus of its parabolic reflector, 1.5 centimeters from the vertex of the reflector (see figure). Write an equation of a cross section of the flashlight's reflector with its focus on the positive x-axis and its vertex at the origin.

6.3 In Exercises 21–24, find the standard form of the equation of the ellipse with the given characteristics. Then graph the ellipse.

21. Vertices: $(-2, 0), (8, 0)$; foci: $(0, 0), (6, 0)$
22. Vertices: $(4, 3), (4, 7)$; foci: $(4, 4), (4, 6)$
23. Vertices: $(0, 1), (4, 1)$; endpoints of the minor axis: $(2, 0), (2, 2)$
24. Vertices: $(-4, -1), (-4, 11)$; endpoints of the minor axis: $(-6, 5), (-2, 5)$

25. **ARCHITECTURE** A semielliptical archway is to be formed over the entrance to an estate. The arch is to be set on pillars that are 10 feet apart and is to have a height (atop the pillars) of 4 feet. Where should the foci be placed in order to sketch the arch?

26. **WADING POOL** You are building a wading pool that is in the shape of an ellipse. Your plans give an equation for the elliptical shape of the pool measured in feet as

$$\frac{x^2}{324} + \frac{y^2}{196} = 1.$$

Find the longest distance across the pool, the shortest distance, and the distance between the foci.

In Exercises 27–30, find the center, vertices, foci, and eccentricity of the ellipse.

27. $\dfrac{(x + 1)^2}{25} + \dfrac{(y - 2)^2}{49} = 1$

28. $\dfrac{(x - 5)^2}{1} + \dfrac{(y + 3)^2}{36} = 1$

29. $16x^2 + 9y^2 - 32x + 72y + 16 = 0$
30. $4x^2 + 25y^2 + 16x - 150y + 141 = 0$

6.4 In Exercises 31–34, find the standard form of the equation of the hyperbola with the given characteristics.

31. Vertices: $(0, \pm 1)$; foci: $(0, \pm 2)$

32. Vertices: $(3, 3), (-3, 3)$; foci: $(4, 3), (-4, 3)$

33. Foci: $(0, 0), (8, 0)$; asymptotes: $y = \pm 2(x - 4)$

34. Foci: $(3, \pm 2)$; asymptotes: $y = \pm 2(x - 3)$

In Exercises 35–38, find the center, vertices, foci, and the equations of the asymptotes of the hyperbola, and sketch its graph using the asymptotes as an aid.

35. $\dfrac{(x - 5)^2}{36} - \dfrac{(y + 3)^2}{16} = 1$

36. $\dfrac{(y - 1)^2}{4} - x^2 = 1$

37. $9x^2 - 16y^2 - 18x - 32y - 151 = 0$

38. $-4x^2 + 25y^2 - 8x + 150y + 121 = 0$

39. LORAN Radio transmitting station A is located 200 miles east of transmitting station B. A ship is in an area to the north and 40 miles west of station A. Synchronized radio pulses transmitted at 186,000 miles per second by the two stations are received 0.0005 second sooner from station A than from station B. How far north is the ship?

40. LOCATING AN EXPLOSION Two of your friends live 4 miles apart and on the same "east-west" street, and you live halfway between them. You are having a three-way phone conversation when you hear an explosion. Six seconds later, your friend to the east hears the explosion, and your friend to the west hears it 8 seconds after you do. Find equations of two hyperbolas that would locate the explosion. (Assume that the coordinate system is measured in feet and that sound travels at 1100 feet per second.)

In Exercises 41–44, classify the graph of the equation as a circle, a parabola, an ellipse, or a hyperbola.

41. $5x^2 - 2y^2 + 10x - 4y + 17 = 0$

42. $-4y^2 + 5x + 3y + 7 = 0$

43. $3x^2 + 2y^2 - 12x + 12y + 29 = 0$

44. $4x^2 + 4y^2 - 4x + 8y - 11 = 0$

6.5 In Exercises 45–48, rotate the axes to eliminate the xy-term in the equation. Then write the equation in standard form. Sketch the graph of the resulting equation, showing both sets of axes.

45. $xy + 3 = 0$

46. $x^2 - 4xy + y^2 + 9 = 0$

47. $5x^2 - 2xy + 5y^2 - 12 = 0$

48. $4x^2 + 8xy + 4y^2 + 7\sqrt{2}x + 9\sqrt{2}y = 0$

In Exercises 49–52, (a) use the discriminant to classify the graph, (b) use the Quadratic Formula to solve for y, and (c) use a graphing utility to graph the equation.

49. $16x^2 - 24xy + 9y^2 - 30x - 40y = 0$

50. $13x^2 - 8xy + 7y^2 - 45 = 0$

51. $x^2 + y^2 + 2xy + 2\sqrt{2}x - 2\sqrt{2}y + 2 = 0$

52. $x^2 - 10xy + y^2 + 1 = 0$

6.6 In Exercises 53 and 54, (a) create a table of x- and y-values for the parametric equations using $t = -2, -1, 0, 1,$ and 2, and (b) plot the points (x, y) generated in part (a) and sketch a graph of the parametric equations.

53. $x = 3t - 2$ and $y = 7 - 4t$

54. $x = \dfrac{1}{4}t$ and $y = \dfrac{6}{t + 3}$

In Exercises 55–60, (a) sketch the curve represented by the parametric equations (indicate the orientation of the curve) and (b) eliminate the parameter and write the corresponding rectangular equation whose graph represents the curve. Adjust the domain of the resulting rectangular equation, if necessary. (c) Verify your result with a graphing utility.

55. $x = 2t$
$y = 4t$

56. $x = 1 + 4t$
$y = 2 - 3t$

57. $x = t^2$
$y = \sqrt{t}$

58. $x = t + 4$
$y = t^2$

59. $x = 3 \cos \theta$
$y = 3 \sin \theta$

60. $x = 3 + 3 \cos \theta$
$y = 2 + 5 \sin \theta$

61. Find a parametric representation of the line that passes through the points $(-4, 4)$ and $(9, -10)$.

62. Find a parametric representation of the circle with center $(5, 4)$ and radius 6.

63. Find a parametric representation of the ellipse with center $(-3, 4)$, major axis horizontal and eight units in length, and minor axis six units in length.

64. Find a parametric representation of the hyperbola with vertices $(0, \pm 4)$ and foci $(0, \pm 5)$.

6.7 In Exercises 65–68, plot the point given in polar coordinates and find two additional polar representations of the point, using $-2\pi < \theta < 2\pi$.

65. $\left(2, \dfrac{\pi}{4}\right)$

66. $\left(-5, -\dfrac{\pi}{3}\right)$

67. $(-7, 4.19)$

68. $\left(\sqrt{3}, 2.62\right)$

In Exercises 69–72, a point in polar coordinates is given. Convert the point to rectangular coordinates.

69. $\left(-1, \dfrac{\pi}{3}\right)$ **70.** $\left(2, \dfrac{5\pi}{4}\right)$

71. $\left(3, \dfrac{3\pi}{4}\right)$ **72.** $\left(0, \dfrac{\pi}{2}\right)$

In Exercises 73–76, a point in rectangular coordinates is given. Convert the point to polar coordinates.

73. $(0, 1)$ **74.** $\left(-\sqrt{5}, \sqrt{5}\right)$

75. $(4, 6)$ **76.** $(3, -4)$

In Exercises 77–82, convert the rectangular equation to polar form.

77. $x^2 + y^2 = 81$ **78.** $x^2 + y^2 = 48$

79. $x^2 + y^2 - 6y = 0$ **80.** $x^2 + y^2 - 4x = 0$

81. $xy = 5$ **82.** $xy = -2$

In Exercises 83–88, convert the polar equation to rectangular form.

83. $r = 5$ **84.** $r = 12$

85. $r = 3 \cos \theta$ **86.** $r = 8 \sin \theta$

87. $r^2 = \sin \theta$ **88.** $r^2 = 4 \cos 2\theta$

6.8 In Exercises 89–98, determine the symmetry of r, the maximum value of $|r|$, and any zeros of r. Then sketch the graph of the polar equation (plot additional points if necessary).

89. $r = 6$ **90.** $r = 11$

91. $r = 4 \sin 2\theta$ **92.** $r = \cos 5\theta$

93. $r = -2(1 + \cos \theta)$ **94.** $r = 1 - 4 \cos \theta$

95. $r = 2 + 6 \sin \theta$ **96.** $r = 5 - 5 \cos \theta$

97. $r = -3 \cos 2\theta$ **98.** $r^2 = \cos 2\theta$

In Exercises 99–102, identify the type of polar graph and use a graphing utility to graph the equation.

99. $r = 3(2 - \cos \theta)$ **100.** $r = 5(1 - 2 \cos \theta)$

101. $r = 8 \cos 3\theta$ **102.** $r^2 = 2 \sin 2\theta$

6.9 In Exercises 103–106, identify the conic and sketch its graph.

103. $r = \dfrac{1}{1 + 2 \sin \theta}$ **104.** $r = \dfrac{6}{1 + \sin \theta}$

105. $r = \dfrac{4}{5 - 3 \cos \theta}$ **106.** $r = \dfrac{16}{4 + 5 \cos \theta}$

In Exercises 107–110, find a polar equation of the conic with its focus at the pole.

107. Parabola   Vertex: $(2, \pi)$

108. Parabola   Vertex: $(2, \pi/2)$

109. Ellipse   Vertices: $(5, 0), (1, \pi)$

110. Hyperbola   Vertices: $(1, 0), (7, 0)$

111. EXPLORER 18 On November 27, 1963, the United States launched Explorer 18. Its low and high points above the surface of Earth were 119 miles and 122,800 miles, respectively. The center of Earth was at one focus of the orbit (see figure). Find the polar equation of the orbit and find the distance between the surface of Earth (assume Earth has a radius of 4000 miles) and the satellite when $\theta = \pi/3$.

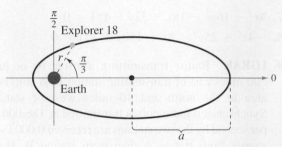

112. ASTEROID An asteroid takes a parabolic path with Earth as its focus. It is about 6,000,000 miles from Earth at its closest approach. Write the polar equation of the path of the asteroid with its vertex at $\theta = \pi/2$. Find the distance between the asteroid and Earth when $\theta = -\pi/3$.

EXPLORATION

TRUE OR FALSE? In Exercises 113–115, determine whether the statement is true or false. Justify your answer.

113. The graph of $\frac{1}{4}x^2 - y^4 = 1$ is a hyperbola.

114. Only one set of parametric equations can represent the line $y = 3 - 2x$.

115. There is a unique polar coordinate representation of each point in the plane.

116. Consider an ellipse with the major axis horizontal and 10 units in length. The number b in the standard form of the equation of the ellipse must be less than what real number? Explain the change in the shape of the ellipse as b approaches this number.

117. What is the relationship between the graphs of the rectangular and polar equations?

(a) $x^2 + y^2 = 25$, $r = 5$

(b) $x - y = 0$, $\theta = \dfrac{\pi}{4}$

See www.CalcChat.com for worked-out solutions to odd-numbered exercises.

Take this test as you would take a test in class. When you are finished, check your work against the answers given in the back of the book.

1. Find the inclination of the line $2x - 5y + 5 = 0$.

2. Find the angle between the lines $3x + 2y - 4 = 0$ and $4x - y + 6 = 0$.

3. Find the distance between the point $(7, 5)$ and the line $y = 5 - x$.

In Exercises 4–7, classify the conic and write the equation in standard form. Identify the center, vertices, foci, and asymptotes (if applicable). Then sketch the graph of the conic.

4. $y^2 - 2x + 2 = 0$

5. $x^2 - 4y^2 - 4x = 0$

6. $9x^2 + 16y^2 + 54x - 32y - 47 = 0$

7. $2x^2 + 2y^2 - 8x - 4y + 9 = 0$

8. Find the standard form of the equation of the parabola with vertex $(2, -3)$, with a vertical axis, and passing through the point $(4, 0)$.

9. Find the standard form of the equation of the hyperbola with foci $(0, 0)$ and $(0, 4)$ and asymptotes $y = \pm\frac{1}{2}x + 2$.

10. (a) Determine the number of degrees the axis must be rotated to eliminate the xy-term of the conic $x^2 + 6xy + y^2 - 6 = 0$.

(b) Graph the conic from part (a) and use a graphing utility to confirm your result.

11. Sketch the curve represented by the parametric equations $x = 2 + 3 \cos \theta$ and $y = 2 \sin \theta$. Eliminate the parameter and write the corresponding rectangular equation.

12. Find a set of parametric equations of the line passing through the points $(2, -3)$ and $(6, 4)$. (There are many correct answers.)

13. Convert the polar coordinate $\left(-2, \dfrac{5\pi}{6}\right)$ to rectangular form.

14. Convert the rectangular coordinate $(2, -2)$ to polar form and find two additional polar representations of this point.

15. Convert the rectangular equation $x^2 + y^2 - 3x = 0$ to polar form.

In Exercises 16–19, sketch the graph of the polar equation. Identify the type of graph.

16. $r = \dfrac{4}{1 + \cos \theta}$

17. $r = \dfrac{4}{2 + \sin \theta}$

18. $r = 2 + 3 \sin \theta$

19. $r = 2 \sin 4\theta$

20. Find a polar equation of the ellipse with focus at the pole, eccentricity $e = \frac{1}{4}$, and directrix $y = 4$.

21. A straight road rises with an inclination of 0.15 radian from the horizontal. Find the slope of the road and the change in elevation over a one-mile stretch of the road.

22. A baseball is hit at a point 3 feet above the ground toward the left field fence. The fence is 10 feet high and 375 feet from home plate. The path of the baseball can be modeled by the parametric equations $x = (115 \cos \theta)t$ and $y = 3 + (115 \sin \theta)t - 16t^2$. Will the baseball go over the fence if it is hit at an angle of $\theta = 30°$? Will the baseball go over the fence if $\theta = 35°$?

6 CUMULATIVE TEST FOR CHAPTERS 4–6

See www.CalcChat.com for worked-out solutions to odd-numbered exercises.

Take this test as you would take a test in class. When you are finished, check your work against the answers given in the back of the book.

1. Write the complex number $6 - \sqrt{-49}$ in standard form.

In Exercises 2–4, perform the operations and write the result in standard form.

2. $6i - \left(2 + \sqrt{-81}\right)$

3. $(5i - 2)^2$

4. $\left(\sqrt{3} + i\right)\left(\sqrt{3} - i\right)$

5. Write the quotient in standard form: $\dfrac{8i}{10 + 2i}$.

In Exercises 6 and 7, find all the zeros of the function.

6. $f(x) = x^3 + 2x^2 + 4x + 8$

7. $f(x) = x^4 + 4x^3 - 21x^2$

8. Find a polynomial with real coefficients that has -6, -3, and $4 + \sqrt{5}i$ as its zeros.

9. Write the complex number $z = -2 + 2i$ in trigonometric form.

10. Find the product $[4(\cos 30° + i \sin 30°)][6(\cos 120° + i \sin 120°)]$. Write the result in standard form.

In Exercises 11 and 12, use DeMoivre's Theorem to find the indicated power of the complex number. Write the result in standard form.

11. $\left[2\left(\cos \dfrac{2\pi}{3} + i \sin \dfrac{2\pi}{3}\right)\right]^4$

12. $\left(-\sqrt{3} - i\right)^6$

13. Find the three cube roots of 1.

14. Write all the solutions of the equation $x^4 - 81i = 0$.

In Exercises 15 and 16, use the graph of f to describe the transformation that yields the graph of g.

15. $f(x) = \left(\frac{2}{5}\right)^x$, $g(x) = -\left(\frac{2}{5}\right)^{-x+3}$

16. $f(x) = 2.2^x$, $g(x) = -2.2^x + 4$

In Exercises 17–20, use a calculator to evaluate each expression. Round your result to three decimal places.

17. $\log_{10} 98$

18. $\log_{10} \frac{6}{7}$

19. $\ln \sqrt{31}$

20. $\ln\left(\sqrt{30} - 4\right)$

In Exercises 21–23, evaluate the logarithm using the change-of-base formula. Round your answer to three decimal places.

21. $\log_5 4.3$

22. $\log_3 0.149$

23. $\log_{1/2} 17$

24. Use the properties of logarithms to expand $\ln\left(\dfrac{x^2 - 16}{x^4}\right)$, where $x > 4$.

25. Write $2 \ln x - \frac{1}{2} \ln(x + 5)$ as a logarithm of a single quantity.

In Exercises 26–29, solve the equation algebraically. Round the result to three decimal places.

26. $6e^{2x} = 72$ **27.** $4^{x-5} + 21 = 30$

28. $\log_2 x + \log_2 5 = 6$ **29.** $\ln 4x - \ln 2 = 8$

30. Use a graphing utility to graph $f(x) = \dfrac{1000}{1 + 4e^{-0.2x}}$ and determine the horizontal asymptotes.

31. The number of bacteria N in a culture is given by the model $N = 175e^{kt}$, where t is the time in hours. If $N = 420$ when $t = 8$, estimate the time required for the population to double in size.

32. The population P of Texas (in thousands) from 2000 through 2007 can be modeled by $P = 20{,}879e^{0.0189t}$, where t represents the year, with $t = 0$ corresponding to 2000. According to this model, when will the population reach 28 million? (Source: U.S. Census Bureau)

33. Find the angle between the lines $2x + y - 3 = 0$ and $x - 3y + 6 = 0$.

34. Find the distance between the point $(6, -3)$ and the line $y = 2x - 4$.

In Exercises 35–38, classify the conic and write the equation in standard form. Identify the center, vertices, foci, and asymptotes (if any). Then sketch the graph.

35. $9x^2 + 4y^2 - 36x + 8y + 4 = 0$ **36.** $4x^2 - y^2 - 4 = 0$

37. $x^2 + y^2 + 2x - 6y - 12 = 0$ **38.** $y^2 + 2x + 2 = 0$

39. Find an equation in rectangular coordinates of the circle with center $(2, -4)$ and passing through the point $(0, 4)$.

40. Find an equation in rectangular coordinates of the hyperbola with foci $(0, 0)$ and $(0, 6)$ and asymptotes $y = \pm \dfrac{2\sqrt{5}}{5}x + 3$.

41. (a) Determine the number of degrees the axes must be rotated to eliminate the xy-term of the conic $x^2 + xy + y^2 + 2x - 3y - 30 = 0$.

(b) Graph the conic and use a graphing utility to confirm your result.

42. Sketch the curve represented by the parametric equations $x = 3 + 4\cos\theta$ and $y = \sin\theta$. Eliminate the parameter and write the corresponding rectangular equation.

43. Find a set of parametric equations of the line passing through the points $(3, -2)$ and $(-3, 4)$. (The answer is not unique.)

44. Plot the point $(-2, -3\pi/4)$ and find three additional polar representations for $-2\pi < \theta < 2\pi$.

45. Convert the rectangular equation $x^2 + y^2 - 16y = 0$ to polar form.

46. Convert the polar equation $r = \dfrac{2}{4 - 5\cos\theta}$ to rectangular form.

In Exercises 47 and 48, sketch the graph of the polar equation.

47. $r = \dfrac{4}{2 + \cos\theta}$ **48.** $r = \dfrac{8}{1 + \sin\theta}$

49. Match each polar equation with its graph at the left.

(a) $r = 2 + 3\sin\theta$ (b) $r = 3\sin\theta$ (c) $r = 3\sin 2\theta$

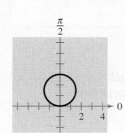

(i)

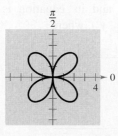

(ii)

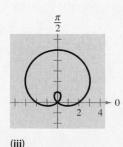

(iii)

FIGURE FOR 49

PROOFS IN MATHEMATICS

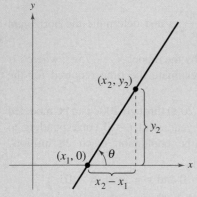

> ## Inclination and Slope   *(p. 438)*
> If a nonvertical line has inclination θ and slope m, then $m = \tan \theta$.

Proof

If $m = 0$, the line is horizontal and $\theta = 0$. So, the result is true for horizontal lines because $m = 0 = \tan 0$.

If the line has a positive slope, it will intersect the x-axis. Label this point $(x_1, 0)$, as shown in the figure. If (x_2, y_2) is a second point on the line, the slope is

$$m = \frac{y_2 - 0}{x_2 - x_1} = \frac{y_2}{x_2 - x_1} = \tan \theta.$$

The case in which the line has a negative slope can be proved in a similar manner.

> ## Distance Between a Point and a Line   *(p. 440)*
> The distance between the point (x_1, y_1) and the line $Ax + By + C = 0$ is
>
> $$d = \frac{|Ax_1 + By_1 + C|}{\sqrt{A^2 + B^2}}.$$

Proof

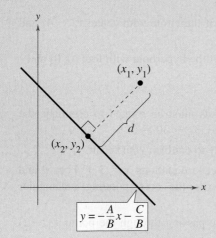

For simplicity, assume that the given line is neither horizontal nor vertical (see figure). By writing the equation $Ax + By + C = 0$ in slope-intercept form

$$y = -\frac{A}{B}x - \frac{C}{B}$$

you can see that the line has a slope of $m = -A/B$. So, the slope of the line passing through (x_1, y_1) and perpendicular to the given line is B/A, and its equation is $y - y_1 = (B/A)(x - x_1)$. These two lines intersect at the point (x_2, y_2), where

$$x_2 = \frac{B(Bx_1 - Ay_1) - AC}{A^2 + B^2} \quad \text{and} \quad y_2 = \frac{A(-Bx_1 + Ay_1) - BC}{A^2 + B^2}.$$

Finally, the distance between (x_1, y_1) and (x_2, y_2) is

$$
\begin{aligned}
d &= \sqrt{(x_2 - x_1)^2 + (y_2 - y_1)^2} \\
&= \sqrt{\left(\frac{B^2 x_1 - ABy_1 - AC}{A^2 + B^2} - x_1\right)^2 + \left(\frac{-ABx_1 + A^2 y_1 - BC}{A^2 + B^2} - y_1\right)^2} \\
&= \sqrt{\frac{A^2(Ax_1 + By_1 + C)^2 + B^2(Ax_1 + By_1 + C)^2}{(A^2 + B^2)^2}} \\
&= \frac{|Ax_1 + By_1 + C|}{\sqrt{A^2 + B^2}}.
\end{aligned}
$$

Parabolic Paths

There are many natural occurrences of parabolas in real life. For instance, the famous astronomer Galileo discovered in the 17th century that an object that is projected upward and obliquely to the pull of gravity travels in a parabolic path. Examples of this are the center of gravity of a jumping dolphin and the path of water molecules in a drinking fountain.

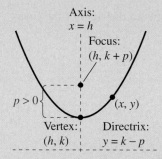

Parabola with vertical axis

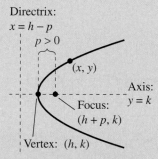

Parabola with horizontal axis

Standard Equation of a Parabola   *(p. 446)*

The standard form of the equation of a parabola with vertex at (h, k) is as follows.

$$(x - h)^2 = 4p(y - k), \quad p \neq 0 \qquad \text{Vertical axis, directrix: } y = k - p$$

$$(y - k)^2 = 4p(x - h), \quad p \neq 0 \qquad \text{Horizontal axis, directrix: } x = h - p$$

The focus lies on the axis p units (*directed distance*) from the vertex. If the vertex is at the origin $(0, 0)$, the equation takes one of the following forms.

$$x^2 = 4py \qquad \text{Vertical axis}$$

$$y^2 = 4px \qquad \text{Horizontal axis}$$

Proof

For the case in which the directrix is parallel to the x-axis and the focus lies above the vertex, as shown in the top figure, if (x, y) is any point on the parabola, then, by definition, it is equidistant from the focus $(h, k + p)$ and the directrix $y = k - p$. So, you have

$$\sqrt{(x - h)^2 + [y - (k + p)]^2} = y - (k - p)$$

$$(x - h)^2 + [y - (k + p)]^2 = [y - (k - p)]^2$$

$$(x - h)^2 + y^2 - 2y(k + p) + (k + p)^2 = y^2 - 2y(k - p) + (k - p)^2$$

$$(x - h)^2 + y^2 - 2ky - 2py + k^2 + 2pk + p^2 = y^2 - 2ky + 2py + k^2 - 2pk + p^2$$

$$(x - h)^2 - 2py + 2pk = 2py - 2pk$$

$$(x - h)^2 = 4p(y - k).$$

For the case in which the directrix is parallel to the y-axis and the focus lies to the right of the vertex, as shown in the bottom figure, if (x, y) is any point on the parabola, then, by definition, it is equidistant from the focus $(h + p, k)$ and the directrix $x = h - p$. So, you have

$$\sqrt{[x - (h + p)]^2 + (y - k)^2} = x - (h - p)$$

$$[x - (h + p)]^2 + (y - k)^2 = [x - (h - p)]^2$$

$$x^2 - 2x(h + p) + (h + p)^2 + (y - k)^2 = x^2 - 2x(h - p) + (h - p)^2$$

$$x^2 - 2hx - 2px + h^2 + 2ph + p^2 + (y - k)^2 = x^2 - 2hx + 2px + h^2 - 2ph + p^2$$

$$-2px + 2ph + (y - k)^2 = 2px - 2ph$$

$$(y - k)^2 = 4p(x - h).$$

Note that if a parabola is centered at the origin, then the two equations above would simplify to $x^2 = 4py$ and $y^2 = 4px$, respectively.

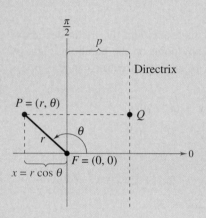

Proof

A proof for $r = \dfrac{ep}{1 + e \cos \theta}$ with $p > 0$ is shown here. The proofs of the other cases are similar. In the figure, consider a vertical directrix, p units to the right of the focus $F = (0, 0)$. If $P = (r, \theta)$ is a point on the graph of

$$r = \frac{ep}{1 + e \cos \theta}$$

the distance between P and the directrix is

$$\begin{aligned}
PQ &= |p - x| \\
&= |p - r \cos \theta| \\
&= \left| p - \left(\frac{ep}{1 + e \cos \theta} \right) \cos \theta \right| \\
&= \left| p \left(1 - \frac{e \cos \theta}{1 + e \cos \theta} \right) \right| \\
&= \left| \frac{p}{1 + e \cos \theta} \right| \\
&= \left| \frac{r}{e} \right|.
\end{aligned}$$

Moreover, because the distance between P and the pole is simply $PF = |r|$, the ratio of PF to PQ is

$$\begin{aligned}
\frac{PF}{PQ} &= \frac{|r|}{\left| \dfrac{r}{e} \right|} \\
&= |e| \\
&= e
\end{aligned}$$

and, by definition, the graph of the equation must be a conic.

PROBLEM SOLVING

This collection of thought-provoking and challenging exercises further explores and expands upon concepts learned in this chapter.

1. Several mountain climbers are located in a mountain pass between two peaks. The angles of elevation to the two peaks are 0.84 radian and 1.10 radians. A range finder shows that the distances to the peaks are 3250 feet and 6700 feet, respectively (see figure).

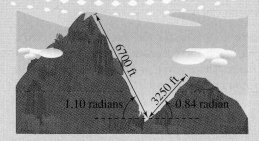

(a) Find the angle between the two lines of sight to the peaks.

(b) Approximate the amount of vertical climb that is necessary to reach the summit of each peak.

2. Statuary Hall is an elliptical room in the United States Capitol in Washington D.C. The room is also called the Whispering Gallery because a person standing at one focus of the room can hear even a whisper spoken by a person standing at the other focus. This occurs because any sound that is emitted from one focus of an ellipse will reflect off the side of the ellipse to the other focus. Statuary Hall is 46 feet wide and 97 feet long.

(a) Find an equation that models the shape of the room.

(b) How far apart are the two foci?

(c) What is the area of the floor of the room? (The area of an ellipse is $A = \pi ab$.)

3. Find the equation(s) of all parabolas that have the x-axis as the axis of symmetry and focus at the origin.

4. Find the area of the square inscribed in the ellipse below.

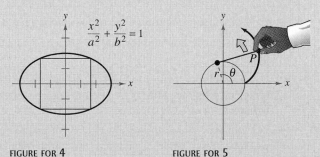

FIGURE FOR 4   FIGURE FOR 5

5. The *involute* of a circle is described by the endpoint P of a string that is held taut as it is unwound from a spool (see figure). The spool does not rotate. Show that

$$x = r(\cos \theta + \theta \sin \theta) \qquad y = r(\sin \theta - \theta \cos \theta)$$

is a parametric representation of the involute of a circle.

6. A tour boat travels between two islands that are 12 miles apart (see figure). For a trip between the islands, there is enough fuel for a 20-mile trip.

Not drawn to scale

(a) Explain why the region in which the boat can travel is bounded by an ellipse.

(b) Let $(0, 0)$ represent the center of the ellipse. Find the coordinates of each island.

(c) The boat travels from one island, straight past the other island to the vertex of the ellipse, and back to the second island. How many miles does the boat travel? Use your answer to find the coordinates of the vertex.

(d) Use the results from parts (b) and (c) to write an equation of the ellipse that bounds the region in which the boat can travel.

7. Find an equation of the hyperbola such that for any point on the hyperbola, the difference between its distances from the points $(2, 2)$ and $(10, 2)$ is 6.

8. Prove that the graph of the equation

$$Ax^2 + Cy^2 + Dx + Ey + F = 0$$

is one of the following (except in degenerate cases).

Conic	Condition
(a) Circle	$A = C$
(b) Parabola	$A = 0$ or $C = 0$ (but not both)
(c) Ellipse	$AC > 0$
(d) Hyperbola	$AC < 0$

9. The following sets of parametric equations model projectile motion.

$$x = (v_0 \cos \theta)t \qquad x = (v_0 \cos \theta)t$$
$$y = (v_0 \sin \theta)t \qquad y = h + (v_0 \sin \theta)t - 16t^2$$

(a) Under what circumstances would you use each model?

(b) Eliminate the parameter for each set of equations.

(c) In which case is the path of the moving object not affected by a change in the velocity v? Explain.

521

10. As t increases, the ellipse given by the parametric equations $x = \cos t$ and $y = 2 \sin t$ is traced out *counterclockwise*. Find a parametric representation for which the same ellipse is traced out *clockwise*.

 11. A **hypocycloid** has the parametric equations

$$x = (a - b) \cos t + b \cos\left(\frac{a - b}{b} t\right)$$

and

$$y = (a - b) \sin t - b \sin\left(\frac{a - b}{b} t\right).$$

Use a graphing utility to graph the hypocycloid for each value of a and b. Describe each graph.

(a) $a = 2, b = 1$ (b) $a = 3, b = 1$

(c) $a = 4, b = 1$ (d) $a = 10, b = 1$

(e) $a = 3, b = 2$ (f) $a = 4, b = 3$

12. The curve given by the parametric equations

$$x = \frac{1 - t^2}{1 + t^2} \quad \text{and} \quad y = \frac{t(1 - t^2)}{1 + t^2}$$

is called a **strophoid.**

(a) Find a rectangular equation of the strophoid.

(b) Find a polar equation of the strophoid.

 (c) Use a graphing utility to graph the strophoid.

 13. The rose curves described in this chapter are of the form

$$r = a \cos n\theta \quad \text{or} \quad r = a \sin n\theta$$

where n is a positive integer that is greater than or equal to 2. Use a graphing utility to graph $r = a \cos n\theta$ and $r = a \sin n\theta$ for some noninteger values of n. Describe the graphs.

14. What conic section is represented by the polar equation

$$r = a \sin \theta + b \cos \theta?$$

15. The graph of the polar equation

$$r = e^{\cos \theta} - 2 \cos 4\theta + \sin^5\left(\frac{\theta}{12}\right)$$

is called the *butterfly curve*, as shown in the figure.

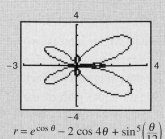

$$r = e^{\cos \theta} - 2 \cos 4\theta + \sin^5\left(\frac{\theta}{12}\right)$$

(a) The graph shown was produced using $0 \le \theta \le 2\pi$. Does this show the entire graph? Explain your reasoning.

(b) Approximate the maximum r-value of the graph. Does this value change if you use $0 \le \theta \le 4\pi$ instead of $0 \le \theta \le 2\pi$? Explain.

 16. Use a graphing utility to graph the polar equation

$$r = \cos 5\theta + n \cos \theta$$

for $0 \le \theta \le \pi$ for the integers $n = -5$ to $n = 5$. As you graph these equations, you should see the graph change shape from a heart to a bell. Write a short paragraph explaining what values of n produce the heart portion of the curve and what values of n produce the bell portion.

17. The planets travel in elliptical orbits with the sun at one focus. The polar equation of the orbit of a planet with one focus at the pole and major axis of length $2a$ (see figure) is

$$r = \frac{(1 - e^2)a}{1 - e \cos \theta}$$

where e is the eccentricity. The minimum distance (perihelion) from the sun to a planet is $r = a(1 - e)$ and the maximum distance (aphelion) is $r = a(1 + e)$. For the planet Neptune, $a = 4.495 \times 10^9$ kilometers and $e = 0.0086$. For the dwarf planet Pluto, $a = 5.906 \times 10^9$ kilometers and $e = 0.2488$.

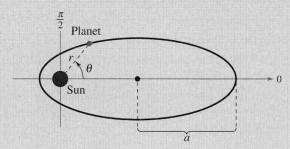

(a) Find the polar equation of the orbit of each planet.

(b) Find the perihelion and aphelion distances for each planet.

 (c) Use a graphing utility to graph the equations of the orbits of Neptune and Pluto in the same viewing window.

(d) Is Pluto ever closer to the sun than Neptune? Until recently, Pluto was considered the ninth planet. Why was Pluto called the ninth planet and Neptune the eighth planet?

(e) Do the orbits of Neptune and Pluto intersect? Will Neptune and Pluto ever collide? Why or why not?

ANSWERS TO ODD-NUMBERED EXERCISES AND TESTS

Chapter P

Section P.1 *(page 12)*

1. rational 3. origin 5. composite
7. variables; constants 9. coefficient
11. (a) $5, 1, 2$ (b) $0, 5, 1, 2$ (c) $-9, 5, 0, 1, -4, 2, -11$
    (d) $-\frac{7}{2}, \frac{2}{3}, -9, 5, 0, 1, -4, 2, -11$ (e) $\sqrt{2}$
13. (a) 1 (b) 1 (c) $-13, 1, -6$
    (d) $2.01, -13, 1, -6, 0.666\ldots$ (e) $0.010110111\ldots$
15. (a) $\frac{6}{3}, 8$ (b) $\frac{6}{3}, 8$ (c) $\frac{6}{3}, -1, 8, -22$
    (d) $-\frac{1}{3}, \frac{6}{3}, -7.5, -1, 8, -22$ (e) $-\pi, \frac{1}{2}\sqrt{2}$
17. (a) (b)

    (c) (d)

19. 0.625 21. $0.\overline{123}$ 23. $-2.5 < 2$
25. 27.

    $-4 > -8$ $\frac{3}{2} < 7$
29.

    $\frac{5}{6} > \frac{2}{3}$
31. (a) $x \le 5$ denotes the set of all real numbers less than or equal to 5.
    (b) (c) Unbounded
33. (a) $x < 0$ denotes the set of all real numbers less than 0.
    (b) (c) Unbounded
35. (a) $[4, \infty)$ denotes the set of all real numbers greater than or equal to 4.
    (b) (c) Unbounded
37. (a) $-2 < x < 2$ denotes the set of all real numbers greater than -2 and less than 2.
    (b) (c) Bounded
39. (a) $-1 \le x < 0$ denotes the set of all real numbers greater than or equal to -1 and less than 0.
    (b) (c) Bounded
41. (a) $[-2, 5)$ denotes the set of all real numbers greater than or equal to -2 and less than 5.
    (b) (c) Bounded

Inequality	*Interval*
43. $y \ge 0$	$[0, \infty)$
45. $-2 < x \le 4$	$(-2, 4]$
47. $10 \le t \le 22$	$[10, 22]$
49. $W > 65$	$(65, \infty)$

51. 10 53. 5 55. -1 57. -1 59. -1
61. $|-3| > -|-3|$ 63. $-5 = -|5|$
65. $-|-2| = -|2|$ 67. 51 69. $\frac{5}{2}$ 71. $\frac{128}{75}$
73. $|x - 5| \le 3$ 75. $|y| \ge 6$
77. $|57 - 236| = 179$ mi
79. $|\$113{,}356 - \$112{,}700| = \$656 > \$500$
    $0.05(\$112{,}700) = \$5635$
    Because the actual expense differs from the budget by more than \$500, there is failure to meet the "budget variance test."
81. $|\$37{,}335 - \$37{,}640| = \$305 < \$500$
    $0.05(\$37{,}640) = \$1882$
    Because the difference between the actual expense and the budget is less than \$500 and less than 5% of the budgeted amount, there is compliance with the "budget variance test."
83. \$1453.2 billion; \$107.4 billion
85. \$2025.5 billion; \$236.3 billion
87. \$1880.3 billion; \$412.7 billion
89. $7x$ and 4 are the terms; 7 is the coefficient.
91. $\sqrt{3}x^2, -8x$, and -11 are the terms; $\sqrt{3}$ and -8 are the coefficients.
93. $4x^3, x/2$, and -5 are the terms; 4 and $\frac{1}{2}$ are the coefficients.
95. (a) -10 (b) -6 97. (a) 14 (b) 2
99. (a) Division by 0 is undefined. (b) 0
101. Commutative Property of Addition
103. Multiplicative Inverse Property
105. Distributive Property
107. Multiplicative Identity Property
109. Associative Property of Addition
111. Distributive Property
113. $\frac{1}{2}$ 115. $\frac{3}{8}$ 117. 48 119. $\frac{5x}{12}$
121. (a) Negative (b) Negative
123. (a)

n	1	0.5	0.01	0.0001	0.000001
$5/n$	5	10	500	50,000	5,000,000

    (b) The value of $5/n$ approaches infinity as n approaches 0.
125. True. Because $b < 0$, $a - b$ subtracts a negative number from (or adds a positive number to) a positive number. The sum of two positive numbers is positive.
127. False. If $a < b$, then $\frac{1}{a} > \frac{1}{b}$, where $a \ne 0$ and $b \ne 0$.
129. (a) No. If one variable is negative and the other is positive, the expressions are unequal.
    (b) No. $|u + v| \le |u| + |v|$
    The expressions are equal when u and v have the same sign. If u and v differ in sign, $|u + v|$ is less than $|u| + |v|$.
131. The only even prime number is 2, because its only factors are itself and 1.
133. Yes. $|a| = -a$ if $a < 0$.

CHAPTER P

Section P.2 *(page 25)*

1. equation   3. extraneous   5. Identity
7. Conditional equation   9. Identity
11. Conditional equation   13. 4   15. -9 17. 5
19. 1   21. No solution   23. $-\frac{96}{23}$ 25. $-\frac{6}{5}$
27. No solution. The x-terms sum to zero, but the constant terms do not.
29. 10   31. 4   33. 0
35. No solution. The solution is extraneous.
37. No solution. The solution is extraneous.
39. No solution. The solution is extraneous.
41. 0   43. $2x^2 + 8x - 3 = 0$
45. $3x^2 - 90x - 10 = 0$ 47. $0, -\frac{1}{2}$ 49. $4, -2$
51. $5, 7$ 53. $3, -\frac{1}{2}$ 55. $2, -6$ 57. $-a$ 59. ± 7
61. $\pm 3\sqrt{3}$ 63. $8, 16$ 65. $-2 \pm \sqrt{14}$
67. $\dfrac{1 \pm 3\sqrt{2}}{2}$ 69. 2   71. $4, -8$ 73. $-6 \pm \sqrt{11}$
75. $2 \pm 2\sqrt{3}$ 77. $\dfrac{-5 \pm \sqrt{89}}{4}$ 79. $\dfrac{15 \pm \sqrt{85}}{10}$
81. $\dfrac{1}{2}, -1$ 83. $1 \pm \sqrt{3}$ 85. $-7 \pm \sqrt{5}$
87. $-4 \pm 2\sqrt{5}$ 89. $\dfrac{2}{3} \pm \dfrac{\sqrt{7}}{3}$ 91. $-\dfrac{4}{3}$ 93. $\dfrac{2}{7}$
95. $2 \pm \dfrac{\sqrt{6}}{2}$ 97. $6 \pm \sqrt{11}$ 99. $-3.449, 1.449$
101. $1.355, -14.071$ 103. $1.687, -0.488$ 105. $1 \pm \sqrt{2}$
107. $6, -12$ 109. $\dfrac{1}{2} \pm \sqrt{3}$ 111. ± 1 113. $0, \pm 5$
115. ± 3 117. -6 119. $3, 1, -1$ 121. ± 1
123. $\pm\sqrt{3}, \pm 1$ 125. $1, -2$ 127. 50   129. 26
131. No solution   133. $-\frac{513}{2}$ 135. $6, 7$ 137. 10
139. $-3 \pm 5\sqrt{5}$ 141. 1   143. $2, -\frac{3}{2}$ 145. $4, -5$
147. $\dfrac{1 \pm \sqrt{31}}{3}$ 149. $3, -2$ 151. $\sqrt{3}, -3$
153. $3, \dfrac{-1 - \sqrt{17}}{2}$ 155. 61.2 in.   157. 23,437.5 mi
159. $\dfrac{5\sqrt{2}}{2} \approx 3.54$ cm   161. 6 in. $\times$ 6 in. $\times$ 2 in.
163. (a) 1998   (b) 2011; Answers will vary.
165. False. See Example 14 on page 24.
167. Equivalent equations have the same solution set, and one is derived from the other by steps for generating equivalent equations.
$2x = 5, \; 2x + 3 = 8$
169. $x^2 - 3x - 18 = 0$ 171. $x^2 - 2x - 1 = 0$
173. Sample answer: $a = 9, b = 9$
175. Sample answer: $a = 20, b = 20$
177. (a) $x = 0, -\dfrac{b}{a}$ (b) $x = 0, 1$

Section P.3 *(page 39)*

1. Cartesian   3. Midpoint Formula   5. graph
7. y-axis
9. A: $(2, 6)$, B: $(-6, -2)$, C: $(4, -4)$, D: $(-3, 2)$

11.

13.

15. $(-3, 4)$ 17. $(-5, -5)$ 19. Quadrant IV
21. Quadrant II   23. Quadrant III or IV   25. Quadrant III
27. Quadrant I or III
29.

31. 8   33. 5   35. 5   37. 5
39. (a) $4, 3, 5$ (b) $4^2 + 3^2 = 5^2$
41. (a)
(b) 10
(c) $(5, 4)$
43. (a)
(b) 17
(c) $\left(0, \frac{5}{2}\right)$
45. (a)
(b) $2\sqrt{10}$
(c) $(2, 3)$

47. (a)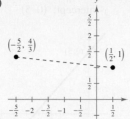

(b) $\dfrac{\sqrt{82}}{3}$

(c) $\left(-1, \dfrac{7}{6}\right)$

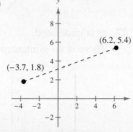

49. (a)

(b) $\sqrt{110.97}$

(c) $(1.25, 3.6)$

51. $\left(\sqrt{5}\right)^2 + \left(\sqrt{45}\right)^2 = \left(\sqrt{50}\right)^2$

53. Distances between the points: $\sqrt{29}, \sqrt{58}, \sqrt{29}$

55. About 9.6%   **57.** $2\sqrt{505} \approx 45$ yd

59. \$4415 million   **61.** (a) Yes   (b) Yes

63. (a) Yes   (b) No

65.

x	-2	0	1	$\frac{4}{3}$	2
y	$-\frac{5}{2}$	-1	$-\frac{1}{4}$	0	$\frac{1}{2}$
(x, y)	$\left(-2, -\frac{5}{2}\right)$	$(0, -1)$	$\left(1, -\frac{1}{4}\right)$	$\left(\frac{4}{3}, 0\right)$	$\left(2, \frac{1}{2}\right)$

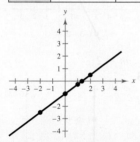

67. x-intercepts: $(\pm 2, 0)$
y-intercept: $(0, 16)$

69. x-intercept: $\left(\frac{6}{5}, 0\right)$
y-intercept: $(0, -6)$

71. x-intercept: $(-4, 0)$
y-intercept: $(0, 2)$

73. x-intercept: $\left(\frac{7}{3}, 0\right)$
y-intercept: $(0, 7)$

75. x-intercepts: $(0, 0), (2, 0)$
y-intercept: $(0, 0)$

77. x-intercept: $(6, 0)$
y-intercepts: $\left(0, \pm\sqrt{6}\right)$

79.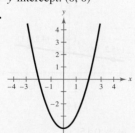

81.

83. y-axis symmetry   **85.** Origin symmetry

87. Origin symmetry   **89.** x-axis symmetry

91.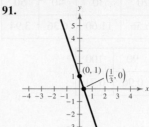

93.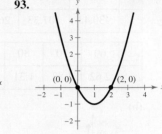

95.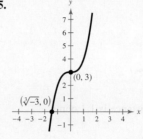

97.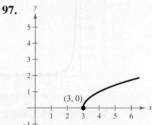

99.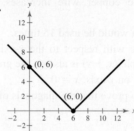

101.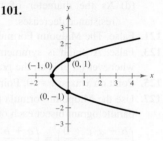

103. $x^2 + y^2 = 36$ **105.** $(x - 2)^2 + (y + 1)^2 = 16$

107. $(x + 1)^2 + (y - 2)^2 = 5$

109. $(x - 3)^2 + (y - 4)^2 = 25$

111. Center: $(0, 0)$; Radius: 5   **113.** Center: $(1, -3)$; Radius: 3

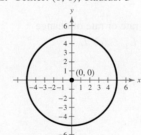

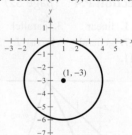

115. Center: $\left(\frac{1}{2}, \frac{1}{2}\right)$; Radius: $\frac{3}{2}$ **117.**

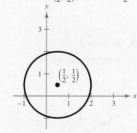

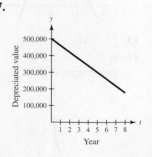

CHAPTER P

119. (a)

x	5	10	20	30	40	50
y	430.43	107.33	26.56	11.60	6.36	3.94

x	60	70	80	90	100
y	2.62	1.83	1.31	0.96	0.71

(b)

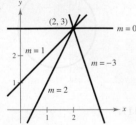

When $x = 85.5$, the resistance is 1.1 ohms.

(c) Answers will vary.

(d) As the diameter of the copper wire increases, the resistance decreases.

121. False. The Midpoint Formula would be used 15 times.

123. False. A graph is symmetric with respect to the x-axis if, whenever (x, y) is on the graph, $(x, -y)$ is also on the graph.

125. Point on x-axis: $y = 0$; Point on y-axis: $x = 0$

127. Use the Midpoint Formula to prove that the diagonals of the parallelogram bisect each other.

$$\left(\frac{b + a}{2}, \frac{c + 0}{2} \right) = \left(\frac{a + b}{2}, \frac{c}{2} \right)$$

$$\left(\frac{a + b + 0}{2}, \frac{c + 0}{2} \right) = \left(\frac{a + b}{2}, \frac{c}{2} \right)$$

Section P.4 (page 52)

1. linear **3.** parallel **5.** rate or rate of change

7. general **9.** (a) L_2 (b) L_3 (c) L_1

11.

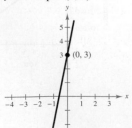

13. $\frac{3}{2}$ **15.** -4

17. $m = 5$
y-intercept: $(0, 3)$

19. $m = -\frac{1}{2}$
y-intercept: $(0, 4)$

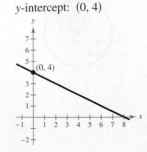

21. m is undefined.
There is no y-intercept.

23. $m = -\frac{7}{6}$
y-intercept: $(0, 5)$

25. $m = 0$
y-intercept: $(0, 3)$

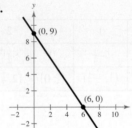

27. m is undefined.
There is no y-intercept.

29.

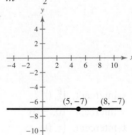

$m = -\frac{3}{2}$

31.

$m = 2$

33.

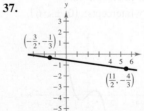

$m = 0$

35.

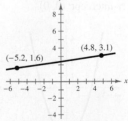

m is undefined.

37.

$m = -\frac{1}{7}$

39.

$m = 0.15$

41. $(0, 1), (3, 1), (-1, 1)$ **43.** $(6, -5), (7, -4), (8, -3)$

45. $(-8, 0), (-8, 2), (-8, 3)$ **47.** $(-4, 6), (-3, 8), (-2, 10)$

49. $(9, -1), (11, 0), (13, 1)$

51. $y = 3x - 2$ **53.** $y = -2x$

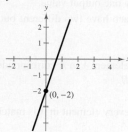 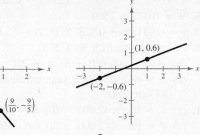

55. $y = -\frac{1}{3}x + \frac{4}{3}$ **57.** $y = -\frac{1}{2}x - 2$

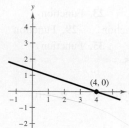

59. $x = 6$ **61.** $y = \frac{5}{2}$

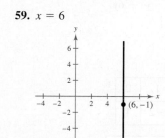

63. $y = 5x + 27.3$ **65.** $y = -\frac{3}{5}x + 2$

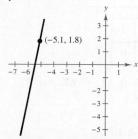 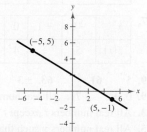

67. $x = -8$ **69.** $y = -\frac{1}{2}x + \frac{3}{2}$

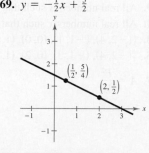

71. $y = -\frac{6}{5}x - \frac{18}{25}$ **73.** $y = 0.4x + 0.2$

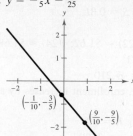

75. $y = -1$ **77.** $x = \frac{7}{3}$

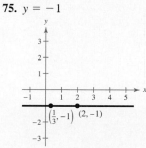

79. Parallel   **81.** Neither   **83.** Perpendicular

85. Parallel   **87.** (a) $y = 2x - 3$ (b) $y = -\frac{1}{2}x + 2$

89. (a) $y = -\frac{3}{4}x + \frac{3}{8}$ (b) $y = \frac{4}{3}x + \frac{127}{72}$

91. (a) $y = 0$ (b) $x = -1$

93. (a) $x = 3$ (b) $y = -2$

95. (a) $y = x + 4.3$ (b) $y = -x + 9.3$

97. $3x + 2y - 6 = 0$ **99.** $12x + 3y + 2 = 0$

101. $x + y - 3 = 0$

103. Line (b) is perpendicular to line (c).

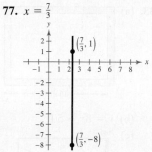

105. Line (a) is parallel to line (b).

Line (c) is perpendicular to line (a) and line (b).

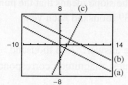

107. $3x - 2y - 1 = 0$ **109.** $80x + 12y + 139 = 0$

111. (a) Sales increasing 135 units/yr

(b) No change in sales

(c) Sales decreasing 40 units/yr

113. (a) The average salary increased the greatest from 2006 to 2008 and increased the least from 2002 to 2004.

(b) $m = 2350.75$

(c) The average salary increased $2350.75 per year over the 12 years between 1996 and 2008.

115. 12 ft   **117.** $V(t) = 3790 - 125t$

119. V-intercept: initial cost; Slope: annual depreciation

121. $V = -175t + 875$ **123.** $S = 0.8L$

125. $W = 0.07S + 2500$

127. $y = 0.03125t + 0.92875$; $y(22) \approx \$1.62$; $y(24) \approx \$1.68$

129. (a) $y(t) = 442.625t + 40{,}571$

(b) $y(10) = 44{,}997$; $y(15) = 47{,}210$

(c) $m = 442.625$; Each year, enrollment increases by about 443 students.

131. (a) $C = 18t + 42{,}000$ (b) $R = 30t$

(c) $P = 12t - 42{,}000$ (d) $t = 3500$ h

133. (a)

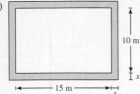

(b) $y = 8x + 50$

(c)

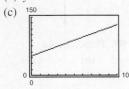

(d) $m = 8$, 8 m

135. (a) and (b)

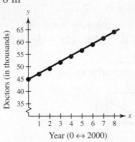

(c) Answers will vary. Sample answer: $y = 2.39x + 44.9$

(d) Answers will vary. Sample answer: The y-intercept indicates that in 2000 there were 44.9 thousand doctors of osteopathic medicine. The slope means that the number of doctors increases by 2.39 thousand each year.

(e) The model is accurate.

(f) Answers will vary. Sample answer: 73.6 thousand

137. False. The slope with the greatest magnitude corresponds to the steepest line.

139. Find the distance between each two points and use the Pythagorean Theorem.

141. No. The slope cannot be determined without knowing the scale on the y-axis. The slopes could be the same.

143. The line $y = 4x$ rises most quickly, and the line $y = -4x$ falls most quickly. The greater the magnitude of the slope (the absolute value of the slope), the faster the line rises or falls.

145. No. The slopes of two perpendicular lines have opposite signs (assume that neither line is vertical or horizontal).

Section P.5   (*page 67*)

1. domain; range; function   **3.** independent; dependent

5. implied domain   **7.** Yes   **9.** No

11. Yes, each input value has exactly one output value.

13. No, the input values 7 and 10 each have two different output values.

15. (a) Function

(b) Not a function, because the element 1 in A corresponds to two elements, -2 and 1, in B.

(c) Function

(d) Not a function, because not every element in A is matched with an element in B.

17. Each is a function. For each year there corresponds one and only one circulation.

19. Not a function   **21.** Function   **23.** Function

25. Not a function   **27.** Not a function   **29.** Function

31. Function   **33.** Not a function   **35.** Function

37. (a) -1 (b) -9 (c) $2x - 5$

39. (a) 36π (b) $\frac{9}{2}\pi$ (c) $\frac{32}{3}\pi r^3$

41. (a) 15   (b) $4t^2 - 19t + 27$ (c) $4t^2 - 3t - 10$

43. (a) 1   (b) 2.5   (c) $3 - 2|x|$

45. (a) $-\dfrac{1}{9}$ (b) Undefined   (c) $\dfrac{1}{y^2 + 6y}$

47. (a) 1   (b) -1 (c) $\dfrac{|x - 1|}{x - 1}$

49. (a) -1 (b) 2   (c) 6

51. (a) -7 (b) 4   (c) 9

53.

x	-2	-1	0	1	2
$f(x)$	1	-2	-3	-2	1

55.

t	-5	-4	-3	-2	-1
$h(t)$	1	$\frac{1}{2}$	0	$\frac{1}{2}$	1

57.

x	-2	-1	0	1	2
$f(x)$	5	$\frac{9}{2}$	4	1	0

59. 5   **61.** $\frac{4}{3}$ **63.** ± 3 **65.** $0, \pm 1$ **67.** $-1, 2$

69. $0, \pm 2$ **71.** All real numbers x

73. All real numbers t except $t = 0$

75. All real numbers y such that $y \geq 10$

77. All real numbers x except $x = 0, -2$

79. All real numbers s such that $s \geq 1$ except $s = 4$

81. All real numbers x such that $x > 0$

83. $\{(-2, 4), (-1, 1), (0, 0), (1, 1), (2, 4)\}$

85. $\{(-2, 4), (-1, 3), (0, 2), (1, 3), (2, 4)\}$

87. $A = \dfrac{P^2}{16}$

89. (a) The maximum volume is 1024 cubic centimeters.

(b) 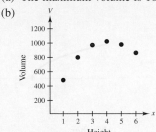 Yes, V is a function of x.

Height

(c) $V = x(24 - 2x)^2, \quad 0 < x < 12$

91. $A = \dfrac{x^2}{2(x - 2)}, \quad x > 2$

93. Yes, the ball will be at a height of 6 feet.

95.
1998: $136,164	2003: $180,419
1999: $140,971	2004: $195,900
2000: $147,800	2005: $216,900
2001: $156,651	2006: $224,000
2002: $167,524	2007: $217,200

97. (a) $C = 12.30x + 98,000$ (b) $R = 17.98x$

(c) $P = 5.68x - 98,000$

99. (a) $R = \dfrac{240n - n^2}{20}, \quad n \geq 80$

(b)

n	90	100	110	120	130	140	150
$R(n)$	$675	$700	$715	$720	$715	$700	$675

The revenue is maximum when 120 people take the trip.

101. (a)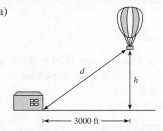

(b) $h = \sqrt{d^2 - 3000^2}, \quad d \geq 3000$

103. $3 + h, \quad h \neq 0$ **105.** $3x^2 + 3xh + h^2 + 3, \quad h \neq 0$

107. $-\dfrac{x + 3}{9x^2}, \quad x \neq 3$ **109.** $\dfrac{\sqrt{5x} - 5}{x - 5}$

111. $g(x) = cx^2; c = -2$ **113.** $r(x) = \dfrac{c}{x}; c = 32$

115. False. A function is a special type of relation.

117. False. The range is $[-1, \infty)$.

119. Domain of $f(x)$: all real numbers $x \geq 1$

Domain of $g(x)$: all real numbers $x > 1$

Notice that the domain of $f(x)$ includes $x = 1$ and the domain of $g(x)$ does not because you cannot divide by 0.

121. No; x is the independent variable, f is the name of the function.

123. (a) Yes. The amount you pay in sales tax will increase as the price of the item purchased increases.

(b) No. The length of time that you study will not necessarily determine how well you do on an exam.

Section P.6 *(page 80)*

1. ordered pairs **3.** zeros **5.** maximum **7.** odd

9. Domain: $(-\infty, -1] \cup [1, \infty)$ **11.** Domain: $[-4, 4]$
Range: $[0, \infty)$ Range: $[0, 4]$

13. Domain: $(-\infty, \infty)$; Range: $[-4, \infty)$
(a) 0 (b) -1 (c) 0 (d) -2

15. Domain: $(-\infty, \infty)$; Range: $(-2, \infty)$
(a) 0 (b) 1 (c) 2 (d) 3

17. Function **19.** Not a function **21.** Function

23. $-\dfrac{5}{2}, 6$ **25.** 0 **27.** $0, \pm\sqrt{2}$ **29.** $\pm\dfrac{1}{2}, 6$ **31.** $\dfrac{1}{2}$

33. 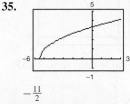 **35.**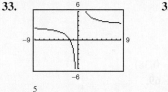

$-\dfrac{5}{3}$ $-\dfrac{11}{2}$

37. 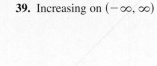 **39.** Increasing on $(-\infty, \infty)$

$\dfrac{1}{3}$

41. Increasing on $(-\infty, 0)$ and $(2, \infty)$
Decreasing on $(0, 2)$

43. Increasing on $(1, \infty)$; Decreasing on $(-\infty, -1)$
Constant on $(-1, 1)$

45. Increasing on $(-\infty, 0)$ and $(2, \infty)$; Constant on $(0, 2)$

47. 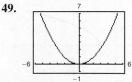 **49.**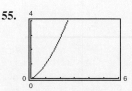

Constant on $(-\infty, \infty)$ Decreasing on $(-\infty, 0)$
Increasing on $(0, \infty)$

51. 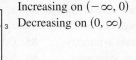 Increasing on $(-\infty, 0)$
Decreasing on $(0, \infty)$

53. 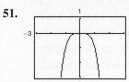 **55.**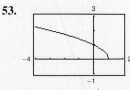

Decreasing on $(-\infty, 1)$ Increasing on $(0, \infty)$

57. 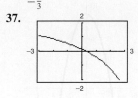 **59.**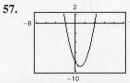

Relative minimum: $(1, -9)$ Relative maximum:
$(1.5, 0.25)$

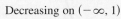

CHAPTER P

61.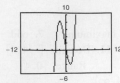

Relative maximum:
$(-1.79, 8.21)$
Relative minimum:
$(1.12, -4.06)$

63.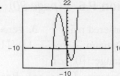

Relative maximum:
$(-2, 20)$
Relative minimum:
$(1, -7)$

65.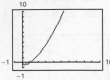

Relative minimum: $(0.33, -0.38)$

67.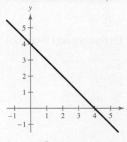

$(-\infty, 4]$

69.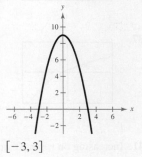

$[-3, 3]$

71.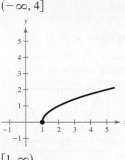

$[1, \infty)$

73.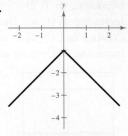

$f(x) < 0$ for all x

75. The average rate of change from $x_1 = 0$ to $x_2 = 3$ is -2.
77. The average rate of change from $x_1 = 1$ to $x_2 = 5$ is 18.
79. The average rate of change from $x_1 = 1$ to $x_2 = 3$ is 0.
81. The average rate of change from $x_1 = 3$ to $x_2 = 11$ is $-\frac{1}{4}$.
83. Even; y-axis symmetry   **85.** Odd; origin symmetry
87. Neither; no symmetry   **89.** Neither; no symmetry
91.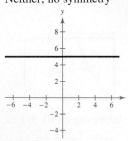

Even

93.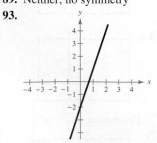

Neither

95.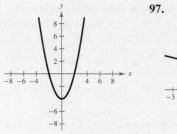

Even

97.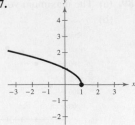

Neither

99.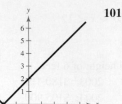

Neither

101. $h = -x^2 + 4x - 3$

103. $h = 2x - x^2$ **105.** $L = \frac{1}{2}y^2$ **107.** $L = 4 - y^2$
109. (a) 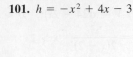   (b) 30 W

111. (a) Ten thousands   (b) Ten millions   (c) Percents
113. (a)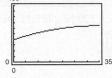

(b) The average rate of change from 1970 to 2005 is 0.705. The enrollment rate of children in preschool has slowly been increasing each year.

115. (a) $s = -16t^2 + 64t + 6$
(b)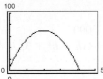

(c) Average rate of change $= 16$
(d) The slope of the secant line is positive.
(e) Secant line: $16t + 6$
(f)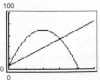

117. (a) $s = -16t^2 + 120t$

(b)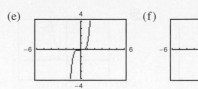

(c) Average rate of change $= -8$

(d) The slope of the secant line is negative.

(e) Secant line: $-8t + 240$

(f)

119. (a) $s = -16t^2 + 120$

(b)

(c) Average rate of change $= -32$

(d) The slope of the secant line is negative.

(e) Secant line: $-32t + 120$

(f)

121. False. The function $f(x) = \sqrt{x^2 + 1}$ has a domain of all real numbers.

123. (a) Even. The graph is a reflection in the x-axis.

(b) Even. The graph is a reflection in the y-axis.

(c) Even. The graph is a vertical translation of f.

(d) Neither. The graph is a horizontal translation of f.

125. (a) $\left(\frac{3}{2}, 4\right)$ (b) $\left(\frac{3}{2}, -4\right)$

127. (a) $(-4, 9)$ (b) $(-4, -9)$

129. (a) $(-x, -y)$ (b) $(-x, y)$

131. (a)   (b)

(c)   (d)

(e)    (f)

All the graphs pass through the origin. The graphs of the odd powers of x are symmetric with respect to the origin, and the graphs of the even powers are symmetric with respect to the y-axis. As the powers increase, the graphs become flatter in the interval $-1 < x < 1$.

133. 60 ft/sec; As the time traveled increases, the distance increases rapidly, causing the average speed to increase with each time increment. From $t = 0$ to $t = 4$, the average speed is less than from $t = 4$ to $t = 9$. Therefore, the overall average from $t = 0$ to $t = 9$ falls below the average found in part (b).

135. Answers will vary.

Section P.7   *(page 90)*

1. g   **2.** i   **3.** h   **4.** a   **5.** b   **6.** e   **7.** f

8. c   **9.** d

11. (a) $f(x) = -2x + 6$ **13.** (a) $f(x) = -3x + 11$

(b) 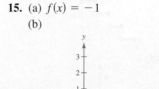   (b)

15. (a) $f(x) = -1$ **17.** (a) $f(x) = \frac{6}{7}x - \frac{45}{7}$

(b) 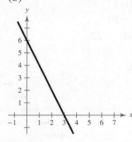   (b)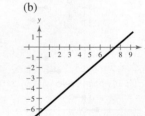

19. 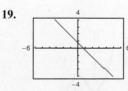   **21.**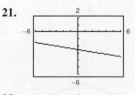

23. 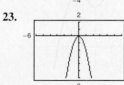   **25.**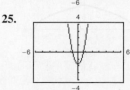

27.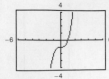

29.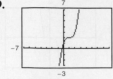

63.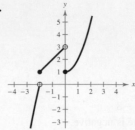

31.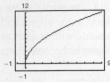

33.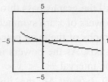

65. (a) 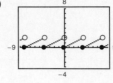   (b) Domain: $(-\infty, \infty)$
Range: $[0, 2)$
(c) Sawtooth pattern

35.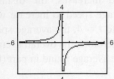

37.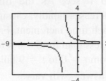

67. (a) 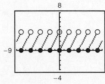   (b) Domain: $(-\infty, \infty)$
Range: $[0, 4)$
(c) Sawtooth pattern

39.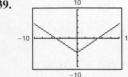

41.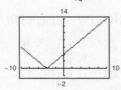

69. (a) 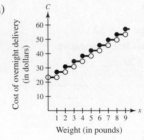   (b) $57.15

43. (a) 2   (b) 2   (c) -4 (d) 3
45. (a) 1   (b) 3   (c) 7   (d) -19
47. (a) 6   (b) -11 (c) 6   (d) -22
49. (a) -10 (b) -4 (c) -1 (d) 41

51.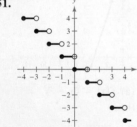

53.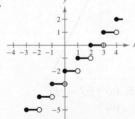

71. (a) $W(30) = 420$; $W(40) = 560$;
$W(45) = 665$; $W(50) = 770$

(b) $W(h) = \begin{cases} 14h, & 0 < h \le 45 \\ 21(h - 45) + 630, & h > 45 \end{cases}$

55.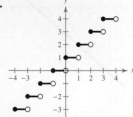

57.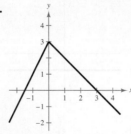

73. (a)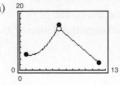

$f(x) = \begin{cases} 0.505x^2 - 1.47x + 6.3, & 1 \le x \le 6 \\ -1.97x + 26.3, & 6 < x \le 12 \end{cases}$

Answers will vary. Sample answer: The domain is determined by inspection of a graph of the data with the two models.

(b) $f(5) = 11.575, f(11) = 4.63$; These values represent the revenue for the months of May and November, respectively.

(c) These values are quite close to the actual data values.

75. False. A linear equation could be a horizontal or vertical line.

59.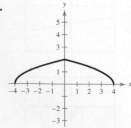

61.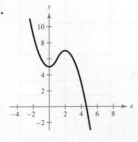

Section P.8   *(page 97)*

1. rigid   **3.** nonrigid
5. vertical stretch; vertical shrink

7. (a)

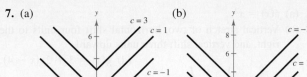

(b)

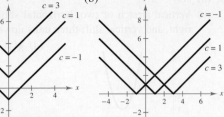

(c)

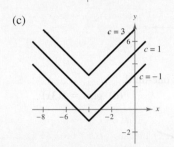

(e)

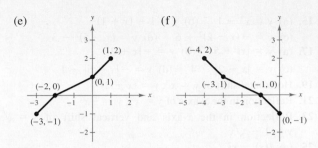

(f)

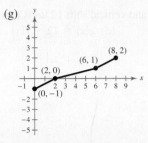

(g)

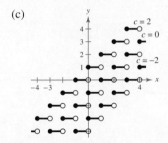

9. (a)

(b)

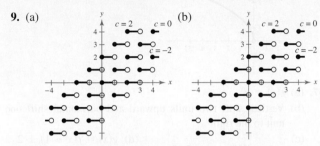

(c)

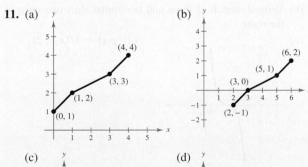

13. (a)

(b)

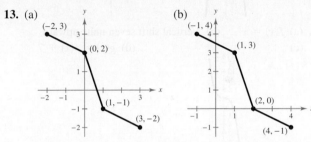

(c)

(d)

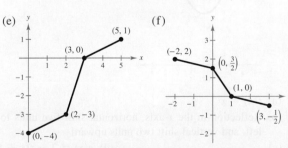

(e)

(f)

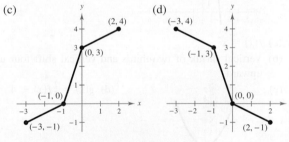

11. (a)

(b)

(c)

(d)

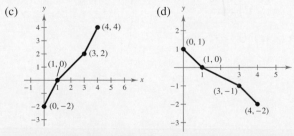

(g)

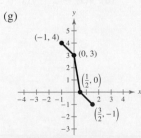

CHAPTER P

15. (a) $y = x^2 - 1$ (b) $y = 1 - (x + 1)^2$
   (c) $y = -(x - 2)^2 + 6$ (d) $y = (x - 5)^2 - 3$

17. (a) $y = |x| + 5$ (b) $y = -|x + 3|$
   (c) $y = |x - 2| - 4$ (d) $y = -|x - 6| - 1$

19. Horizontal shift of $y = x^3$; $y = (x - 2)^3$

21. Reflection in the x-axis of $y = x^2$; $y = -x^2$

23. Reflection in the x-axis and vertical shift of $y = \sqrt{x}$; $y = 1 - \sqrt{x}$

25. (a) $f(x) = x^2$
   (b) Reflection in the x-axis and vertical shift 12 units upward
   (c)   (d) $g(x) = 12 - f(x)$

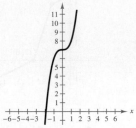

27. (a) $f(x) = x^3$ (b) Vertical shift seven units upward
   (c)   (d) $g(x) = f(x) + 7$

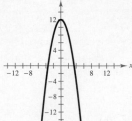

29. (a) $f(x) = x^2$
   (b) Vertical shrink of two-thirds and vertical shift four units upward
   (c)   (d) $g(x) = \frac{2}{3}f(x) + 4$

31. (a) $f(x) = x^2$
   (b) Reflection in the x-axis, horizontal shift five units to the left, and vertical shift two units upward
   (c)   (d) $g(x) = 2 - f(x + 5)$

33. (a) $f(x) = x^2$
   (b) Vertical stretch of two, horizontal shift four units to the right, and vertical shift three units upward
   (c)   (d) $g(x) = 3 + 2f(x - 4)$

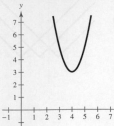

35. (a) $f(x) = \sqrt{x}$
   (b) Horizontal shrink of one-third
   (c)   (d) $g(x) = f(3x)$

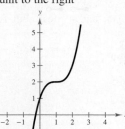

37. (a) $f(x) = x^3$
   (b) Vertical shift two units upward and horizontal shift one unit to the right
   (c)   (d) $g(x) = f(x - 1) + 2$

39. (a) $f(x) = x^3$
   (b) Vertical stretch of three and horizontal shift two units to the right
   (c)   (d) $g(x) = 3f(x - 2)$

41. (a) $f(x) = |x|$
(b) Reflection in the x-axis and vertical shift two units downward
(c)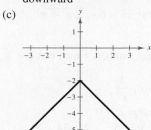
(d) $g(x) = -f(x) - 2$

43. (a) $f(x) = |x|$
(b) Reflection in the x-axis, horizontal shift four units to the left, and vertical shift eight units upward
(c)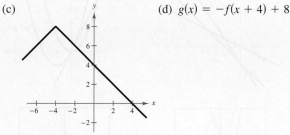
(d) $g(x) = -f(x + 4) + 8$

45. (a) $f(x) = |x|$
(b) Reflection in the x-axis, vertical stretch of two, horizontal shift one unit to the right, and vertical shift four units downward
(c)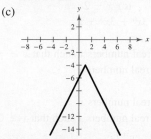
(d) $g(x) = -2f(x - 1) - 4$

47. (a) $f(x) = [\![x]\!]$
(b) Reflection in the x-axis and vertical shift three units upward
(c)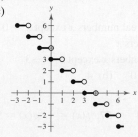
(d) $g(x) = 3 - f(x)$

49. (a) $f(x) = \sqrt{x}$
(b) Horizontal shift nine units to the right
(c)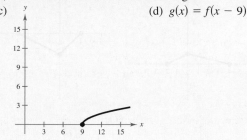
(d) $g(x) = f(x - 9)$

51. (a) $f(x) = \sqrt{x}$
(b) Reflection in the y-axis, horizontal shift seven units to the right, and vertical shift two units downward
(c)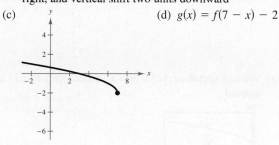
(d) $g(x) = f(7 - x) - 2$

53. (a) $f(x) = \sqrt{x}$
(b) Horizontal stretch and vertical shift four units downward
(c)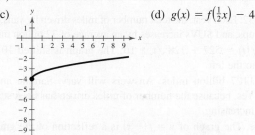
(d) $g(x) = f\left(\tfrac{1}{2}x\right) - 4$

55. $g(x) = (x - 3)^2 - 7$ **57.** $g(x) = (x - 13)^3$
59. $g(x) = -|x| + 12$ **61.** $g(x) = -\sqrt{-x + 6}$
63. (a) $y = -3x^2$ (b) $y = 4x^2 + 3$
65. (a) $y = -\tfrac{1}{2}|x|$ (b) $y = 3|x| - 3$
67. Vertical stretch of $y = x^3$; $y = 2x^3$
69. Reflection in the x-axis and vertical shrink of $y = x^2$; $y = -\tfrac{1}{2}x^2$
71. Reflection in the y-axis and vertical shrink of $y = \sqrt{x}$; $y = \tfrac{1}{2}\sqrt{-x}$
73. $y = -(x - 2)^3 + 2$ **75.** $y = -\sqrt{x} - 3$
77. (a)   (b)
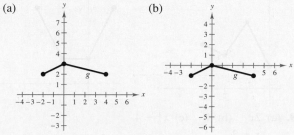

(c) 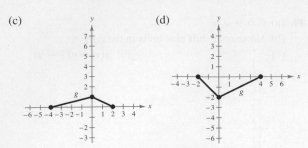   (d)

(e) 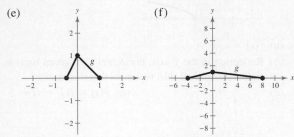   (f)

79. (a) Vertical stretch of 128.0 and a vertical shift of 527 units upward

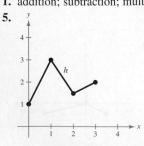

   (b) 32; Each year, the total number of miles driven by vans, pickups, and SUVs increases by an average of 32 billion miles.

   (c) $f(t) = 527 + 128\sqrt{t + 10}$; The graph is shifted 10 units to the left.

   (d) 1127 billion miles; Answers will vary. Sample answer: Yes, because the number of miles driven has been steadily increasing.

81. False. The graph of $y = f(-x)$ is a reflection of the graph of $f(x)$ in the y-axis.

83. True. $|-x| = |x|$

85. (a) $g(t) = \frac{3}{4}f(t)$ (b) $g(t) = f(t) + 10{,}000$

   (c) $g(t) = f(t - 2)$

87. $(-2, 0), (-1, 1), (0, 2)$

89. No. $g(x) = -x^4 - 2$. Yes. $h(x) = -(x - 3)^4$.

Section P.9 (page 107)

1. addition; subtraction; multiplication; division   **3.** $g(x)$

5. 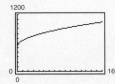   **7.**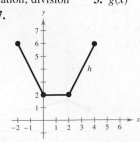

9. (a) $2x$ (b) 4   (c) $x^2 - 4$

   (d) $\dfrac{x + 2}{x - 2}$; all real numbers x except $x = 2$

11. (a) $x^2 + 4x - 5$ (b) $x^2 - 4x + 5$ (c) $4x^3 - 5x^2$

   (d) $\dfrac{x^2}{4x - 5}$; all real numbers x except $x = \dfrac{5}{4}$

13. (a) $x^2 + 6 + \sqrt{1 - x}$ (b) $x^2 + 6 - \sqrt{1 - x}$

   (c) $(x^2 + 6)\sqrt{1 - x}$

   (d) $\dfrac{(x^2 + 6)\sqrt{1 - x}}{1 - x}$; all real numbers x such that $x < 1$

15. (a) $\dfrac{x + 1}{x^2}$ (b) $\dfrac{x - 1}{x^2}$ (c) $\dfrac{1}{x^3}$

   (d) x; all real numbers x except $x = 0$

17. 3   **19.** 5   **21.** $9t^2 - 3t + 5$ **23.** 74

25. 26   **27.** $\frac{3}{5}$

29. 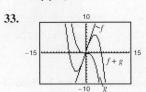   **31.**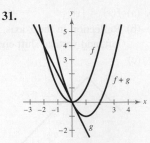

33. 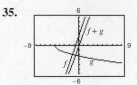   **35.**

   $f(x), g(x)$ $f(x), f(x)$

37. (a) $(x - 1)^2$ (b) $x^2 - 1$ (c) $x - 2$

39. (a) x (b) x (c) $x^9 + 3x^6 + 3x^3 + 2$

41. (a) $\sqrt{x^2 + 4}$ (b) $x + 4$

   Domains of f and $g \circ f$: all real numbers x such that $x \geq -4$

   Domains of g and $f \circ g$: all real numbers x

43. (a) $x + 1$ (b) $\sqrt{x^2 + 1}$

   Domains of f and $g \circ f$: all real numbers x

   Domains of g and $f \circ g$: all real numbers x such that $x \geq 0$

45. (a) $|x + 6|$ (b) $|x| + 6$

   Domains of $f, g, f \circ g$, and $g \circ f$: all real numbers x

47. (a) $\dfrac{1}{x + 3}$ (b) $\dfrac{1}{x} + 3$

   Domains of f and $g \circ f$: all real numbers x except $x = 0$

   Domain of g: all real numbers x

   Domain of $f \circ g$: all real numbers x except $x = -3$

49. (a) 3   (b) 0   **51.** (a) 0   (b) 4

53. $f(x) = x^2$, $g(x) = 2x + 1$

55. $f(x) = \sqrt[3]{x}$, $g(x) = x^2 - 4$ **57.** $f(x) = \dfrac{1}{x}$, $g(x) = x + 2$

59. $f(x) = \dfrac{x + 3}{4 + x}$, $g(x) = -x^2$

61. (a) $T = \frac{3}{4}x + \frac{1}{15}x^2$

(b)

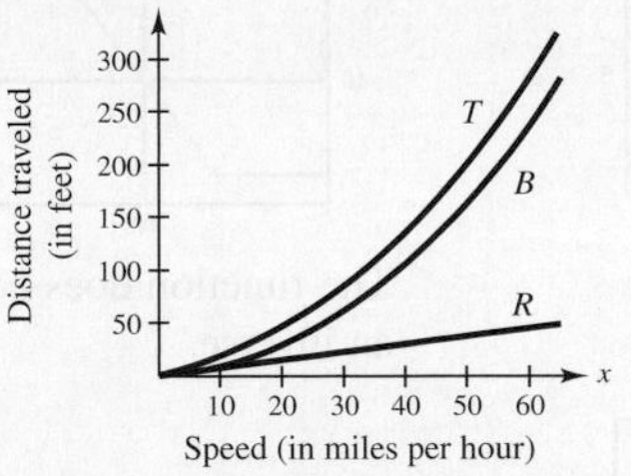

(graph with vertical axis "Distance traveled (in feet)" marked 50, 100, 150, 200, 250, 300; horizontal axis "Speed (in miles per hour)" marked 10 20 30 40 50 60; curves labeled T, B, R)

(c) The braking function $B(x)$. As x increases, $B(x)$ increases at a faster rate than $R(x)$.

63. (a) $c(t) = \dfrac{b(t) - d(t)}{p(t)} \times 100$

(b) $c(5)$ is the percent change in the population due to births and deaths in the year 2005.

65. (a) $(N + M)(t) = 0.227t^3 - 4.11t^2 + 14.6t + 544$, which represents the total number of Navy and Marines personnel combined.

$(N + M)(0) = 544$

$(N + M)(6) \approx 533$

$(N + M)(12) \approx 520$

(b) $(N - M)(t) = 0.157t^3 - 3.65t^2 + 11.2t + 200$, which represents the difference between the number of Navy personnel and the number of Marines personnel.

$(N - M)(0) = 200$

$(N - M)(6) \approx 170$

$(N - M)(12) \approx 80$

67. $(B - D)(t) = -0.197t^3 + 10.17t^2 - 128.0t + 2043$, which represents the change in the United States population.

69. (a) For each time t there corresponds one and only one temperature T.

(b) 60°, 72°

(c) All the temperature changes occur 1 hour later.

(d) The temperature is decreased by 1 degree.

(e) $T(t) = \begin{cases} 60, & 0 \le t \le 6 \\ 12t - 12, & 6 < t < 7 \\ 72, & 7 \le t \le 20 \\ -12t + 312, & 20 < t < 21 \\ 60, & 21 \le t \le 24 \end{cases}$

71. $(A \circ r)(t) = 0.36\pi t^2$; $(A \circ r)(t)$ represents the area of the circle at time t.

73. (a) $N(T(t)) = 30(3t^2 + 2t + 20)$; This represents the number of bacteria in the food as a function of time.

(b) About 653 bacteria   (c) 2.846 h

75. $g(f(x))$ represents 3 percent of an amount over \$500,000.

77. False. $(f \circ g)(x) = 6x + 1$ and $(g \circ f)(x) = 6x + 6$

79. (a) $O(M(Y)) = 2\left(6 + \frac{1}{2}Y\right) = 12 + Y$

(b) Middle child is 8 years old; youngest child is 4 years old.

81. Proof

83. (a) Proof

(b) $\frac{1}{2}[f(x) + f(-x)] + \frac{1}{2}[f(x) - f(-x)]$

$= \frac{1}{2}[f(x) + f(-x) + f(x) - f(-x)]$

$= \frac{1}{2}[2f(x)]$

$= f(x)$

(c) $f(x) = (x^2 + 1) + (-2x)$

$k(x) = \dfrac{-1}{(x + 1)(x - 1)} + \dfrac{x}{(x + 1)(x - 1)}$

Section P.10   *(page 117)*

1. inverse   **3.** range; domain   **5.** one-to-one

7. $f^{-1}(x) = \frac{1}{6}x$ **9.** $f^{-1}(x) = x - 9$

11. $f^{-1}(x) = \dfrac{x - 1}{3}$ **13.** $f^{-1}(x) = x^3$

15. c   **16.** b   **17.** a   **18.** d

19. $f(g(x)) = f\left(-\dfrac{2x + 6}{7}\right) = -\dfrac{7}{2}\left(-\dfrac{2x + 6}{7}\right) - 3 = x$

$g(f(x)) = g\left(-\dfrac{7}{2}x - 3\right) = -\dfrac{2\left(-\frac{7}{2}x - 3\right) + 6}{7} = x$

21. $f(g(x)) = f(\sqrt[3]{x - 5}) = (\sqrt[3]{x - 5})^3 + 5 = x$

$g(f(x)) = g(x^3 + 5) = \sqrt[3]{(x^3 + 5) - 5} = x$

23. (a) $f(g(x)) = f\left(\dfrac{x}{2}\right) = 2\left(\dfrac{x}{2}\right) = x$

$g(f(x)) = g(2x) = \dfrac{(2x)}{2} = x$

(b)

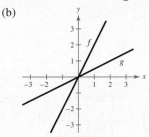

25. (a) $f(g(x)) = f\left(\dfrac{x - 1}{7}\right) = 7\left(\dfrac{x - 1}{7}\right) + 1 = x$

$g(f(x)) = g(7x + 1) = \dfrac{(7x + 1) - 1}{7} = x$

(b)

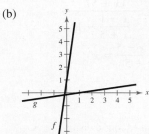

27. (a) $f(g(x)) = f(\sqrt[3]{8x}) = \dfrac{(\sqrt[3]{8x})^3}{8} = x$

$g(f(x)) = g\left(\dfrac{x^3}{8}\right) = \sqrt[3]{8\left(\dfrac{x^3}{8}\right)} = x$

(b)

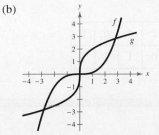

29. (a) $f(g(x)) = f(x^2 + 4), \ x \geq 0$
$$= \sqrt{(x^2 + 4) - 4} = x$$
$g(f(x)) = g(\sqrt{x - 4})$
$$= (\sqrt{x - 4})^2 + 4 = x$$

(b)

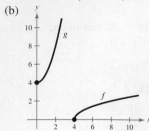

31. (a) $f(g(x)) = f(\sqrt{9 - x}), \ x \leq 9$
$$= 9 - (\sqrt{9 - x})^2 = x$$
$g(f(x)) = g(9 - x^2), \ x \geq 0$
$$= \sqrt{9 - (9 - x^2)} = x$$

(b)

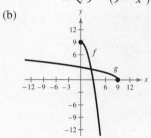

33. (a) $f(g(x)) = f\left(-\dfrac{5x + 1}{x - 1}\right) = \dfrac{-\left(\dfrac{5x + 1}{x - 1}\right) - 1}{-\left(\dfrac{5x + 1}{x - 1}\right) + 5}$
$$= \dfrac{-5x - 1 - x + 1}{-5x - 1 + 5x - 5} = x$$

$g(f(x)) = g\left(\dfrac{x - 1}{x + 5}\right) = \dfrac{-5\left(\dfrac{x - 1}{x + 5}\right) - 1}{\dfrac{x - 1}{x + 5} - 1}$
$$= \dfrac{-5x + 5 - x - 5}{x - 1 - x - 5} = x$$

(b)

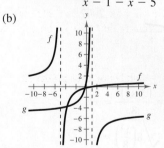

35. No

37.

x	-2	0	2	4	6	8
$f^{-1}(x)$	-2	-1	0	1	2	3

39. Yes **41.** No

43.

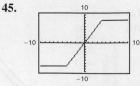

The function has an inverse.

45.

The function does not have an inverse.

47.

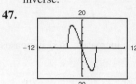

The function does not have an inverse.

49. (a) $f^{-1}(x) = \dfrac{x + 3}{2}$

(b)

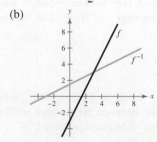

(c) The graph of f^{-1} is the reflection of the graph of f in the line $y = x$.

(d) The domains and ranges of f and f^{-1} are all real numbers.

51. (a) $f^{-1}(x) = \sqrt[5]{x + 2}$

(b)

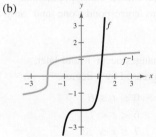

(c) The graph of f^{-1} is the reflection of the graph of f in the line $y = x$.

(d) The domains and ranges of f and f^{-1} are all real numbers.

53. (a) $f^{-1}(x) = \sqrt{4 - x^2}, \ 0 \leq x \leq 2$

(b)

(c) The graph of f^{-1} is the same as the graph of f.

(d) The domains and ranges of f and f^{-1} are all real numbers x such that $0 \leq x \leq 2$.

55. (a) $f^{-1}(x) = \dfrac{4}{x}$

(b)

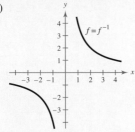

(c) The graph of f^{-1} is the same as the graph of f.

(d) The domains and ranges of f and f^{-1} are all real numbers x except $x = 0$.

57. (a) $f^{-1}(x) = \dfrac{2x + 1}{x - 1}$

(b)

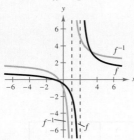

(c) The graph of f^{-1} is the reflection of the graph of f in the line $y = x$.

(d) The domain of f and the range of f^{-1} are all real numbers x except $x = 2$. The domain of f^{-1} and the range of f are all real numbers x except $x = 1$.

59. (a) $f^{-1}(x) = x^3 + 1$

(b)

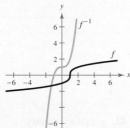

(c) The graph of f^{-1} is the reflection of the graph of f in the line $y = x$.

(d) The domains and ranges of f and f^{-1} are all real numbers.

61. (a) $f^{-1}(x) = \dfrac{5x - 4}{6 - 4x}$

(b)

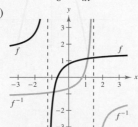

(c) The graph of f^{-1} is the reflection of the graph of f in the line $y = x$.

(d) The domain of f and the range of f^{-1} are all real numbers x except $x = -\frac{5}{4}$. The domain of f^{-1} and the range of f are all real numbers x except $x = \frac{3}{2}$.

63. No inverse　　**65.** $g^{-1}(x) = 8x$　　**67.** No inverse

69. $f^{-1}(x) = \sqrt{x} - 3$　　**71.** No inverse　　**73.** No inverse

75. $f^{-1}(x) = \dfrac{x^2 - 3}{2}, \quad x \geq 0$

77. $f^{-1}(x) = \sqrt{x} + 2$

The domain of f and the range of f^{-1} are all real numbers x such that $x \geq 2$. The domain of f^{-1} and the range of f are all real numbers x such that $x \geq 0$.

79. $f^{-1}(x) = x - 2$

The domain of f and the range of f^{-1} are all real numbers x such that $x \geq -2$. The domain of f^{-1} and the range of f are all real numbers x such that $x \geq 0$.

81. $f^{-1}(x) = \sqrt{x} - 6$

The domain of f and the range of f^{-1} are all real numbers x such that $x \geq -6$. The domain of f^{-1} and the range of f are all real numbers x such that $x \geq 0$.

83. $f^{-1}(x) = \dfrac{\sqrt{-2(x - 5)}}{2}$

The domain of f and the range of f^{-1} are all real numbers x such that $x \geq 0$. The domain of f^{-1} and the range of f are all real numbers x such that $x \leq 5$.

85. $f^{-1}(x) = x + 3$

The domain of f and the range of f^{-1} are all real numbers x such that $x \geq 4$. The domain of f^{-1} and the range of f are all real numbers x such that $x \geq 1$.

87. 32　　**89.** 600　　**91.** $2\sqrt[3]{x + 3}$　　**93.** $\dfrac{x + 1}{2}$

95. $\dfrac{x + 1}{2}$

97. (a) Yes; each European shoe size corresponds to exactly one U.S. shoe size.

(b) 45　　(c) 10　　(d) 41　　(e) 13

99. (a) Yes

(b) S^{-1} represents the time in years for a given sales level.

(c) $S^{-1}(8430) = 6$

(d) No, because then the sales for 2007 and 2009 would be the same, so the function would no longer be one-to-one.

101. (a) $y = \dfrac{x - 10}{0.75}$

x = hourly wage; y = number of units produced

(b) 19 units

103. False. $f(x) = x^2$ has no inverse.

105. Proof

107.

x	1	3	4	6
y	1	2	6	7

x	1	2	6	7
$f^{-1}(x)$	1	3	4	6

109. This situation could be represented by a one-to-one function if the runner does not stop to rest. The inverse function would represent the time in hours for a given number of miles completed.

111. This function could not be represented by a one-to-one function because it oscillates.

113. $k = \frac{1}{4}$

115. 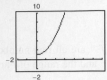 There is an inverse function $f^{-1}(x) = \sqrt{x - 1}$ because the domain of f is equal to the range of f^{-1} and the range of f is equal to the domain of f^{-1}.

Review Exercises *(page 124)*

1. (a) 11   (b) 11   (c) $11, -14$
    (d) $11, -14, -\frac{8}{9}, \frac{5}{2}, 0.4$ (e) $\sqrt{6}$

3. (a) $0.8\overline{3}$ (b) 0.875

$$\frac{5}{6} < \frac{7}{8}$$

5. The set consists of all real numbers less than or equal to 7.

7. 122   **9.** $|x - 7| \geq 4$ **11.** $|y + 30| < 5$
13. (a) -7 (b) -19 **15.** (a) -1 (b) -3
17. Associative Property of Addition
19. Additive Identity Property
21. Commutative Property of Addition
23. -11 **25.** $\frac{1}{12}$ **27.** -144 **29.** Identity
31. 20   **33.** $-\frac{1}{2}$ **35.** -30 **37.** 9   **39.** $-\frac{3}{2}, -1$
41. $\pm\sqrt{2}$ **43.** $-4 \pm 3\sqrt{2}$ **45.** $6 \pm \sqrt{6}$
47. $-\frac{5}{4} \pm \frac{\sqrt{241}}{4}$ **49.** $0, \frac{12}{5}$ **51.** $\pm\sqrt{2}, \pm\sqrt{3}$ **53.** 5
55. No solution   **57.** $-124, 126$ **59.** $-5, 15$ **61.** 1, 3
63.   **65.** Quadrant IV

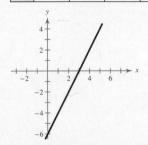

67. (a)   (b) 5
                      (c) $\left(3, \frac{5}{2}\right)$

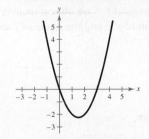

69. (a)   (b) $\sqrt{98.6}$
                          (c) $(2.8, 4.1)$

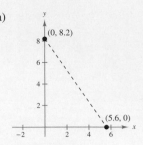

71. \$1460.85 million

73.

x	1	2	3	4	5
y	-4	-2	0	2	4

75.

x	-1	0	1	2	3	4
y	4	0	-2	-2	0	4

77. x-intercepts: $(1, 0), (5, 0)$
    y-intercept: $(0, 5)$

79. No symmetry   **81.** y-axis symmetry

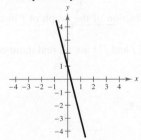

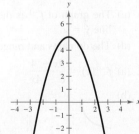

83. No symmetry

85. No symmetry

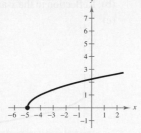

87. Center: $(0, 0)$;
Radius: 3

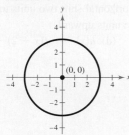

89. Center: $\left(\frac{1}{2}, -1\right)$;
Radius: 6

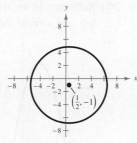

91. $(x - 2)^2 + (y + 3)^2 = 13$

93. $m = -2$
y-intercept: -7

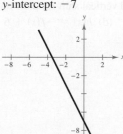

95. $m = 0$
y-intercept: 6

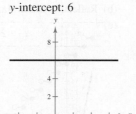

97. $m = 3$
y-intercept: 13

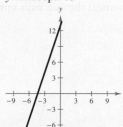

99. $m = -\frac{5}{2}$
y-intercept: -1

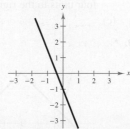

101.

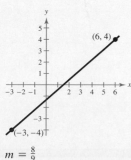

$m = \frac{8}{9}$

103.

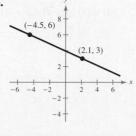

$m = -\frac{5}{11}$

105. $y = \frac{2}{3}x - 2$

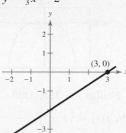

107. $y = -\frac{1}{2}x + 2$

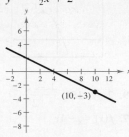

109. $x = 0$ **111.** $y = \frac{2}{7}x + \frac{2}{7}$
113. (a) $y = \frac{5}{4}x - \frac{23}{4}$ (b) $y = -\frac{4}{5}x + \frac{2}{5}$ **115.** $210{,}000$
117. (a) Not a function, because 20 in the domain corresponds to two values in the range and because 10 in A is not matched with any element in B.
  (b) A function, because each input value has exactly one output value.
  (c) A function, because each input value has exactly one output value.
  (d) Not a function, because 30 in A is not matched with any element in B.
119. No   **121.** Yes
123. (a) -3 (b) -1 (c) 2   (d) 6
125. Domain: All real numbers x such that $-5 \le x \le 5$

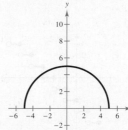

127. Domain: All real numbers s except $s = 3$

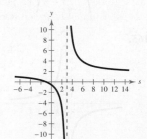

129. Domain: All real numbers x except $x = 3, -2$

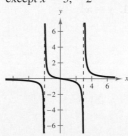

131. (a) 16 ft/sec   (b) 1.5 sec   (c) -16 ft/sec
133. $4x + 2h + 3, \quad h \ne 0$ **135.** Function
137. Not a function   **139.** $\frac{7}{3}, 3$ **141.** $-\frac{3}{8}$
143. Increasing on $(0, \infty)$
  Decreasing on $(-\infty, -1)$
  Constant on $(-1, 0)$

145.
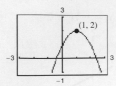

147.
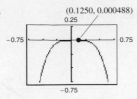

149. 2   **151.** $\dfrac{1 - \sqrt{2}}{2}$ **153.** Neither   **155.** Odd

157. $f(x) = -4x$ **159.** $f(x) = \frac{5}{3}x + \frac{10}{3}$

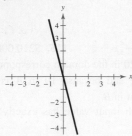
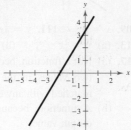

161.
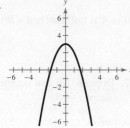

163.
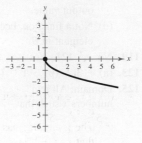

165.
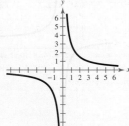

167.
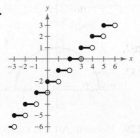

169.
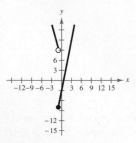

171. $y = x^3$

173. (a) $f(x) = x^2$
(b) Vertical shift nine units downward
(c)   (d) $h(x) = f(x) - 9$

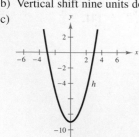

175. (a) $f(x) = \sqrt{x}$
(b) Reflection in the x-axis and vertical shift four units upward
(c)   (d) $h(x) = -f(x) + 4$

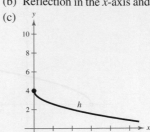

177. (a) $f(x) = x^2$
(b) Reflection in the x-axis, horizontal shift two units to the left, and vertical shift three units upward
(c)   (d) $h(x) = -f(x + 2) + 3$

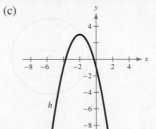

179. (a) $f(x) = [\![x]\!]$
(b) Reflection in the x-axis and vertical shift six units upward
(c)   (d) $h(x) = -f(x) + 6$

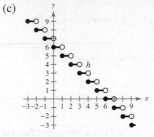

181. (a) $f(x) = |x|$
(b) Reflections in the x-axis and the y-axis, horizontal shift four units to the right, and vertical shift six units upward
(c)

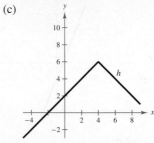

(d) $h(x) = -f(-x + 4) + 6$

183. (a) $f(x) = [\![x]\!]$

(b) Horizontal shift nine units to the right and vertical stretch

(c)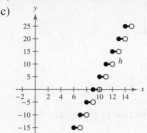

(d) $h(x) = 5f(x - 9)$

185. (a) $f(x) = \sqrt{x}$

(b) Reflection in the x-axis, vertical stretch, and horizontal shift four units to the right

(c)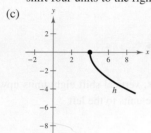

(d) $h(x) = -2f(x - 4)$

187. (a) $x^2 + 2x + 2$ (b) $x^2 - 2x + 4$

(c) $2x^3 - x^2 + 6x - 3$

(d) $\dfrac{x^2 + 3}{2x - 1}$; all real numbers x except $x = \dfrac{1}{2}$

189. (a) $x - \dfrac{8}{3}$ (b) $x - 8$

Domains of $f, g, f \circ g$, and $g \circ f$: all real numbers x

191. $f(x) = x^3, g(x) = 1 - 2x$

193. (a) $(r + c)(t) = 178.8t + 856$; This represents the average annual expenditures for both residential and cellular phone services.

(b)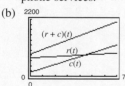

(c) $(r + c)(13) = \$3180.40$

195. $f^{-1}(x) = \dfrac{x - 8}{3}$

$f(f^{-1}(x)) = 3\left(\dfrac{x - 8}{3}\right) + 8 = x$

$f^{-1}(f(x)) = \dfrac{3x + 8 - 8}{3} = x$

197. The function has an inverse.

199. The function has an inverse.

201. The function has an inverse.

203. (a) $f^{-1}(x) = 2x + 6$

(b)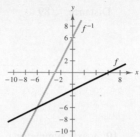

(c) The graph of f^{-1} is the reflection of the graph of f in the line $y = x$.

(d) Both f and f^{-1} have domains and ranges that are all real numbers.

205. (a) $f^{-1}(x) = x^2 - 1, \ x \geq 0$

(b)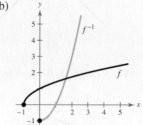

(c) The graph of f^{-1} is the reflection of the graph of f in the line $y = x$.

(d) The graph of f has a domain of all real numbers x such that $x \geq -1$ and a range of $[0, \infty)$. The graph of f^{-1} has a domain of all real numbers x such that $x \geq 0$ and a range of $[-1, \infty)$.

207. $x > 4; \ f^{-1}(x) = \sqrt{\dfrac{x}{2} + 4}, \ x \neq 0$

209. False. The graph is reflected in the x-axis, shifted 9 units to the left, and then shifted 13 units downward.

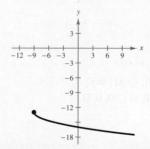

211. Some solutions to certain types of equations may be extraneous solutions, which do not satisfy the original equations. So, checking is crucial.

213. The Vertical Line Test is used to determine if the graph of y is a function of x. The Horizontal Line Test is used to determine if a function has an inverse function.

Chapter Test   *(page 129)*

1. $-\dfrac{10}{3} > -|-4|$ **2.** 9.15

3. Additive Identity Property   **4.** $\dfrac{128}{11}$ **5.** $-4, 5$

6. No solution   **7.** $\pm\sqrt{2}$

CHAPTER P

8.

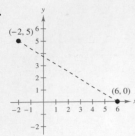

Midpoint: $\left(2, \frac{5}{2}\right)$;
Distance: $\sqrt{89}$

9. No symmetry

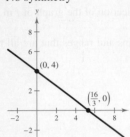

10. y-axis symmetry

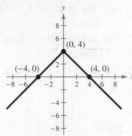

11. Origin symmetry

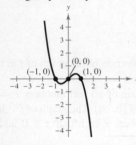

12. Center: $(3, 0)$; Radius: 3

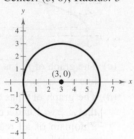

13. $y = -2x + 1$ **14.** $y = -1.7x + 5.9$

15. (a) $y = -\frac{5}{2}x + 4$ (b) $y = \frac{2}{5}x + 4$

16. (a) -9 (b) 1   (c) $|x - 4| - 15$

17. (a)

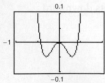

(b) All real numbers x

(c) Increasing on $(-0.31, 0)$, $(0.31, \infty)$

Decreasing on $(-\infty, -0.31)$, $(0, 0.31)$

(d) Even

18. (a)

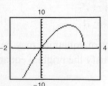

(b) All real numbers x such that $x \le 3$

(c) Increasing on $(-\infty, 2)$

Decreasing on $(2, 3)$

(d) Neither

19. (a)

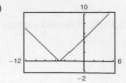

(b) All real numbers x

(c) Increasing on $(-5, \infty)$

Decreasing on $(-\infty, -5)$

(d) Neither

20. (a) $f(x) = [\![x]\!]$ (b) Reflection in the x-axis

(c)

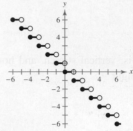

21. (a) $f(x) = \sqrt{x}$

(b) Reflection in the x-axis, vertical shift eight units upward, and horizontal shift five units to the left

(c)

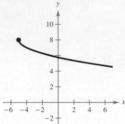

22. (a) $f(x) = x^3$

(b) Reflection in the y-axis, horizontal shift five units to the right, vertical shift three units upward, and vertical stretch

(c)

23. (a) $2x^2 - 4x - 2$ (b) $4x^2 + 4x - 12$

(c) $-3x^4 - 12x^3 + 22x^2 + 28x - 35$

(d) $\dfrac{3x^2 - 7}{-x^2 - 4x + 5}, \quad x \ne 1, -5$

(e) $3x^4 + 24x^3 + 18x^2 - 120x + 68$

(f) $-9x^4 + 30x^2 - 16$

24. (a) $\dfrac{1 + 2x^{3/2}}{x}, \quad x > 0$ (b) $\dfrac{1 - 2x^{3/2}}{x}, \quad x > 0$

(c) $\dfrac{2\sqrt{x}}{x}, \quad x > 0$ (d) $\dfrac{1}{2x^{3/2}}, \quad x > 0$

(e) $\dfrac{\sqrt{x}}{2x}, \quad x > 0$ (f) $\dfrac{2\sqrt{x}}{x}, \quad x > 0$

25. $f^{-1}(x) = \sqrt[3]{x - 8}$ **26.** No inverse

27. $f^{-1}(x) = \left(\dfrac{x}{3}\right)^{2/3}, \; x \geq 0$

Problem Solving (*page 131*)

1. (a) $W_1 = 2000 + 0.07S$ (b) $W_2 = 2300 + 0.05S$

(c)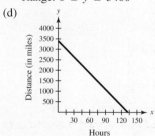

Both jobs pay the same monthly salary if sales equal \$15,000.

(15,000, 3050)

(d) No. Job 1 would pay \$3400 and job 2 would pay \$3300.

3. (a) The function will be even.

(b) The function will be odd.

(c) The function will be neither even nor odd.

5. $f(x) = a_{2n}x^{2n} + a_{2n-2}x^{2n-2} + \cdots + a_2x^2 + a_0$

$f(-x) = a_{2n}(-x)^{2n} + a_{2n-2}(-x)^{2n-2}$

$\qquad\quad + \cdots + a_2(-x)^2 + a_0$

$\qquad = f(x)$

7. (a) $81\frac{2}{3}$ h (b) $25\frac{5}{7}$ mi/h

(c) $y = \dfrac{-180}{7}x + 3400$

Domain: $0 \leq x \leq \dfrac{1190}{9}$

Range: $0 \leq y \leq 3400$

(d)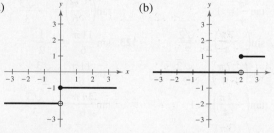

9. (a) $(f \circ g)(x) = 4x + 24$ (b) $(f \circ g)^{-1}(x) = \frac{1}{4}x - 6$

(c) $f^{-1}(x) = \frac{1}{4}x; \; g^{-1}(x) = x - 6$

(d) $(g^{-1} \circ f^{-1})(x) = \frac{1}{4}x - 6$; They are the same.

(e) $(f \circ g)(x) = 8x^3 + 1; \; (f \circ g)^{-1}(x) = \frac{1}{2}\sqrt[3]{x - 1}$;

$f^{-1}(x) = \sqrt[3]{x - 1}; \; g^{-1}(x) = \frac{1}{2}x$;

$(g^{-1} \circ f^{-1})(x) = \frac{1}{2}\sqrt[3]{x - 1}$

(f) Answers will vary.

(g) $(f \circ g)^{-1}(x) = (g^{-1} \circ f^{-1})(x)$

11. (a) (b)

(c) (d)

(e) (f)

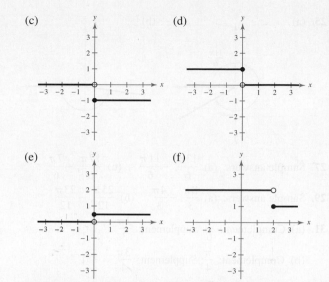

13. Proof

15. (a)

x	-4	-2	0	4
$f(f^{-1}(x))$	-4	-2	0	4

(b)

x	-3	-2	0	1
$(f + f^{-1})(x)$	5	1	-3	-5

(c)

x	-3	-2	0	1
$(f \cdot f^{-1})(x)$	4	0	2	6

(d)

x	-4	-3	0	4
$|f^{-1}(x)|$	2	1	1	3

Chapter 1

Section 1.1 (*page 142*)

1. Trigonometry   **3.** coterminal   **5.** acute; obtuse

7. degree   **9.** linear; angular   **11.** 1 rad

13. 5.5 rad   **15.** -3 rad

17. (a) Quadrant I   (b) Quadrant III

19. (a) Quadrant IV   (b) Quadrant IV

21. (a) Quadrant III   (b) Quadrant II

23. (a) (b)

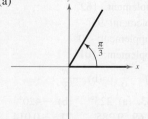

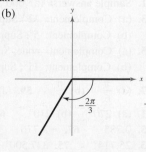

CHAPTER 1

25. (a)   (b)

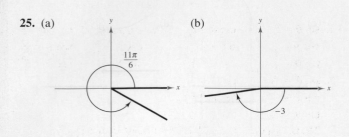

27. Sample answers: (a) $\dfrac{13\pi}{6}, -\dfrac{11\pi}{6}$ (b) $\dfrac{17\pi}{6}, -\dfrac{7\pi}{6}$

29. Sample answers: (a) $\dfrac{8\pi}{3}, -\dfrac{4\pi}{3}$ (b) $\dfrac{25\pi}{12}, -\dfrac{23\pi}{12}$

31. (a) Complement: $\dfrac{\pi}{6}$; Supplement: $\dfrac{2\pi}{3}$

    (b) Complement: $\dfrac{\pi}{4}$; Supplement: $\dfrac{3\pi}{4}$

33. (a) Complement: $\dfrac{\pi}{2} - 1 \approx 0.57$;

    Supplement: $\pi - 1 \approx 2.14$

    (b) Complement: none; Supplement: $\pi - 2 \approx 1.14$

35. $210°$ **37.** $-60°$ **39.** $165°$

41. (a) Quadrant II   (b) Quadrant IV

43. (a) Quadrant III   (b) Quadrant I

45. (a)   (b)

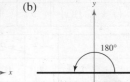

47. (a)   (b)

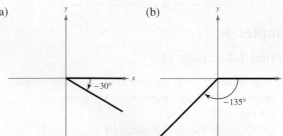

49. Sample answers: (a) $405°, -315°$ (b) $324°, -396°$

51. Sample answers: (a) $600°, -120°$ (b) $180°, -540°$

53. (a) Complement: $72°$; Supplement: $162°$

    (b) Complement: $5°$; Supplement: $95°$

55. (a) Complement: none; Supplement: $30°$

    (b) Complement: $11°$; Supplement: $101°$

57. (a) $\dfrac{\pi}{6}$ (b) $\dfrac{\pi}{4}$ **59.** (a) $-\dfrac{\pi}{9}$ (b) $-\dfrac{\pi}{3}$

61. (a) $270°$ (b) $210°$ **63.** (a) $225°$ (b) $-420°$

65. 0.785 **67.** -3.776 **69.** 9.285 **71.** -0.014

73. $25.714°$ **75.** $337.500°$ **77.** $-756.000°$

79. $-114.592°$ **81.** (a) $54.75°$ (b) $-128.5°$

83. (a) $85.308°$ (b) $330.007°$

85. (a) $240°\,36'$ (b) $-145°\,48'$

87. (a) $2°\,30'$ (b) $-3°\,34'\,48''$

89. 10π in. ≈ 31.42 in.   **91.** 2.5π m ≈ 7.85 m

93. $\dfrac{9}{2}$ rad   **95.** $\dfrac{21}{50}$ rad   **97.** $\dfrac{1}{2}$ rad   **99.** 4 rad

101. 6π in.$^2 \approx 18.85$ in.^{2} **103.** 12.27 ft^2 **105.** 591.3 mi

107. 0.071 rad $\approx 4.04°$ **109.** $\dfrac{5}{12}$ rad

111. (a) $10{,}000\pi$ rad/min $\approx 31{,}415.93$ rad/min

    (b) 9490.23 ft/min

113. (a) $[400\pi, 1000\pi]$ rad/min   (b) $[2400\pi, 6000\pi]$ cm/min

115. (a) 910.37 revolutions/min   (b) 5720 rad/min

117.

$A = 87.5\pi$ m$^2 \approx 274.89$ m^2

119. (a) $\dfrac{14\pi}{3}$ ft/sec ≈ 10 mi/h   (b) $d = \dfrac{7\pi}{7920}n$

    (c) $d = \dfrac{7\pi}{7920}t$ (d) The functions are both linear.

121. False. A measurement of 4π radians corresponds to two complete revolutions from the initial side to the terminal side of an angle.

123. False. The terminal side of the angle lies on the x-axis.

125. Radian. 1 rad $\approx 57.3°$

127. Proof

Section 1.2   *(page 151)*

1. unit circle   **3.** period

5. $\sin t = \dfrac{5}{13}$ $\csc t = \dfrac{13}{5}$

    $\cos t = \dfrac{12}{13}$ $\sec t = \dfrac{13}{12}$

    $\tan t = \dfrac{5}{12}$ $\cot t = \dfrac{12}{5}$

7. $\sin t = -\dfrac{3}{5}$ $\csc t = -\dfrac{5}{3}$

    $\cos t = -\dfrac{4}{5}$ $\sec t = -\dfrac{5}{4}$

    $\tan t = \dfrac{3}{4}$ $\cot t = \dfrac{4}{3}$

9. $(0, 1)$ **11.** $\left(\dfrac{\sqrt{2}}{2}, \dfrac{\sqrt{2}}{2}\right)$ **13.** $\left(-\dfrac{\sqrt{3}}{2}, \dfrac{1}{2}\right)$

15. $\left(-\dfrac{1}{2}, -\dfrac{\sqrt{3}}{2}\right)$

17. $\sin \dfrac{\pi}{4} = \dfrac{\sqrt{2}}{2}$ **19.** $\sin\left(-\dfrac{\pi}{6}\right) = -\dfrac{1}{2}$

    $\cos \dfrac{\pi}{4} = \dfrac{\sqrt{2}}{2}$ $\cos\left(-\dfrac{\pi}{6}\right) = \dfrac{\sqrt{3}}{2}$

    $\tan \dfrac{\pi}{4} = 1$ $\tan\left(-\dfrac{\pi}{6}\right) = -\dfrac{\sqrt{3}}{3}$

21. $\sin\left(-\dfrac{7\pi}{4}\right) = \dfrac{\sqrt{2}}{2}$ **23.** $\sin \dfrac{11\pi}{6} = -\dfrac{1}{2}$

    $\cos\left(-\dfrac{7\pi}{4}\right) = \dfrac{\sqrt{2}}{2}$ $\cos \dfrac{11\pi}{6} = \dfrac{\sqrt{3}}{2}$

    $\tan\left(-\dfrac{7\pi}{4}\right) = 1$ $\tan \dfrac{11\pi}{6} = -\dfrac{\sqrt{3}}{3}$

25. $\sin\left(-\dfrac{3\pi}{2}\right) = 1$

    $\cos\left(-\dfrac{3\pi}{2}\right) = 0$

    $\tan\left(-\dfrac{3\pi}{2}\right)$ is undefined.

27. $\sin \dfrac{2\pi}{3} = \dfrac{\sqrt{3}}{2}$ $\csc \dfrac{2\pi}{3} = \dfrac{2\sqrt{3}}{3}$

$\cos \dfrac{2\pi}{3} = -\dfrac{1}{2}$ $\sec \dfrac{2\pi}{3} = -2$

$\tan \dfrac{2\pi}{3} = -\sqrt{3}$ $\cot \dfrac{2\pi}{3} = -\dfrac{\sqrt{3}}{3}$

29. $\sin \dfrac{4\pi}{3} = -\dfrac{\sqrt{3}}{2}$ $\csc \dfrac{4\pi}{3} = -\dfrac{2\sqrt{3}}{3}$

$\cos \dfrac{4\pi}{3} = -\dfrac{1}{2}$ $\sec \dfrac{4\pi}{3} = -2$

$\tan \dfrac{4\pi}{3} = \sqrt{3}$ $\cot \dfrac{4\pi}{3} = \dfrac{\sqrt{3}}{3}$

31. $\sin \dfrac{3\pi}{4} = \dfrac{\sqrt{2}}{2}$ $\csc \dfrac{3\pi}{4} = \sqrt{2}$

$\cos \dfrac{3\pi}{4} = -\dfrac{\sqrt{2}}{2}$ $\sec \dfrac{3\pi}{4} = -\sqrt{2}$

$\tan \dfrac{3\pi}{4} = -1$ $\cot \dfrac{3\pi}{4} = -1$

33. $\sin\left(-\dfrac{\pi}{2}\right) = -1$ $\csc\left(-\dfrac{\pi}{2}\right) = -1$

$\cos\left(-\dfrac{\pi}{2}\right) = 0$ $\sec\left(-\dfrac{\pi}{2}\right)$ is undefined.

$\tan\left(-\dfrac{\pi}{2}\right)$ is undefined. $\cot\left(-\dfrac{\pi}{2}\right) = 0$

35. $\sin 4\pi = \sin 0 = 0$ **37.** $\cos \dfrac{7\pi}{3} = \cos \dfrac{\pi}{3} = \dfrac{1}{2}$

39. $\cos \dfrac{17\pi}{4} = \cos \dfrac{\pi}{4} = \dfrac{\sqrt{2}}{2}$

41. $\sin\left(-\dfrac{8\pi}{3}\right) = \sin \dfrac{4\pi}{3} = -\dfrac{\sqrt{3}}{2}$

43. (a) $-\dfrac{1}{2}$ (b) -2 **45.** (a) $-\dfrac{1}{5}$ (b) -5

47. (a) $\dfrac{4}{5}$ (b) $-\dfrac{4}{5}$ **49.** 0.7071 **51.** 1.0000

53. -0.1288 **55.** 1.3940 **57.** -1.4486

59. (a) 0.25 ft   (b) 0.02 ft   (c) -0.25 ft

61. False. $\sin(-t) = -\sin(t)$ means that the function is odd, not that the sine of a negative angle is a negative number.

63. False. The real number 0 corresponds to the point $(1, 0)$.

65. (a) y-axis symmetry   (b) $\sin t_1 = \sin(\pi - t_1)$
(c) $\cos(\pi - t_1) = -\cos t_1$

67. Answers will vary.

69. It is an odd function.

71. (a)

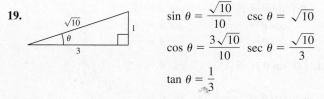

Circle of radius 1 centered at $(0, 0)$

(b) The t-values represent the central angle in radians. The x- and y-values represent the location in the coordinate plane.

(c) $-1 \le x \le 1, -1 \le y \le 1$

Section 1.3   (page 160)

1. (a) v   (b) iv   (c) vi   (d) iii   (e) i   (f) ii

3. complementary

5. $\sin \theta = \dfrac{3}{5}$ $\csc \theta = \dfrac{5}{3}$ **7.** $\sin \theta = \dfrac{9}{41}$ $\csc \theta = \dfrac{41}{9}$

$\cos \theta = \dfrac{4}{5}$ $\sec \theta = \dfrac{5}{4}$ $\cos \theta = \dfrac{40}{41}$ $\sec \theta = \dfrac{41}{40}$

$\tan \theta = \dfrac{3}{4}$ $\cot \theta = \dfrac{4}{3}$ $\tan \theta = \dfrac{9}{40}$ $\cot \theta = \dfrac{40}{9}$

9. $\sin \theta = \dfrac{8}{17}$ $\csc \theta = \dfrac{17}{8}$

$\cos \theta = \dfrac{15}{17}$ $\sec \theta = \dfrac{17}{15}$

$\tan \theta = \dfrac{8}{15}$ $\cot \theta = \dfrac{15}{8}$

The triangles are similar, and corresponding sides are proportional.

11. $\sin \theta = \dfrac{1}{3}$ $\csc \theta = 3$

$\cos \theta = \dfrac{2\sqrt{2}}{3}$ $\sec \theta = \dfrac{3\sqrt{2}}{4}$

$\tan \theta = \dfrac{\sqrt{2}}{4}$ $\cot \theta = 2\sqrt{2}$

The triangles are similar, and corresponding sides are proportional.

13.

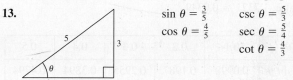

$\sin \theta = \dfrac{3}{5}$ $\csc \theta = \dfrac{5}{3}$

$\cos \theta = \dfrac{4}{5}$ $\sec \theta = \dfrac{5}{4}$

$\cot \theta = \dfrac{4}{3}$

15.

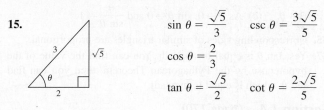

$\sin \theta = \dfrac{\sqrt{5}}{3}$ $\csc \theta = \dfrac{3\sqrt{5}}{5}$

$\cos \theta = \dfrac{2}{3}$

$\tan \theta = \dfrac{\sqrt{5}}{2}$ $\cot \theta = \dfrac{2\sqrt{5}}{5}$

17.

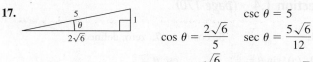

$\csc \theta = 5$

$\cos \theta = \dfrac{2\sqrt{6}}{5}$ $\sec \theta = \dfrac{5\sqrt{6}}{12}$

$\tan \theta = \dfrac{\sqrt{6}}{12}$ $\cot \theta = 2\sqrt{6}$

19.

$\sin \theta = \dfrac{\sqrt{10}}{10}$ $\csc \theta = \sqrt{10}$

$\cos \theta = \dfrac{3\sqrt{10}}{10}$ $\sec \theta = \dfrac{\sqrt{10}}{3}$

$\tan \theta = \dfrac{1}{3}$

21. $\dfrac{\pi}{6}; \dfrac{1}{2}$ **23.** $45°; \sqrt{2}$ **25.** $60°; \dfrac{\pi}{3}$ **27.** $30°; 2$

29. $45°; \dfrac{\pi}{4}$

31. (a) $\dfrac{1}{2}$ (b) $\dfrac{\sqrt{3}}{2}$ (c) $\sqrt{3}$ (d) $\dfrac{\sqrt{3}}{3}$

33. (a) $\dfrac{2\sqrt{2}}{3}$ (b) $2\sqrt{2}$ (c) 3 (d) 3

35. (a) $\dfrac{1}{5}$ (b) $\sqrt{26}$ (c) $\dfrac{1}{5}$ (d) $\dfrac{5\sqrt{26}}{26}$

37–45. Answers will vary.   **47.** (a) 0.1736 (b) 0.1736

49. (a) 0.2815 (b) 3.5523 **51.** (a) 0.9964 (b) 1.0036

53. (a) 5.0273 (b) 0.1989 **55.** (a) 1.8527 (b) 0.9817

CHAPTER 1

57. (a) $30° = \dfrac{\pi}{6}$ (b) $30° = \dfrac{\pi}{6}$

59. (a) $60° = \dfrac{\pi}{3}$ (b) $45° = \dfrac{\pi}{4}$

61. (a) $60° = \dfrac{\pi}{3}$ (b) $45° = \dfrac{\pi}{4}$

63. $9\sqrt{3}$ **65.** $\dfrac{32\sqrt{3}}{3}$ **67.** 443.2 m; 323.3 m

69. $30° = \pi/6$ **71.** (a) 219.9 ft (b) 160.9 ft

73. $(x_1, y_1) = \left(28\sqrt{3},\ 28\right)$
$(x_2, y_2) = \left(28,\ 28\sqrt{3}\right)$

75. $\sin 20° \approx 0.34$, $\cos 20° \approx 0.94$, $\tan 20° \approx 0.36$,
$\csc 20° \approx 2.92$, $\sec 20° \approx 1.06$, $\cot 20° \approx 2.75$

77. True, $\csc x = \dfrac{1}{\sin x}$. **79.** False, $\dfrac{\sqrt{2}}{2} + \dfrac{\sqrt{2}}{2} \neq 1$.

81. False, $1.7321 \neq 0.0349$.

83. (a)

θ	0.1	0.2	0.3	0.4	0.5
$\sin \theta$	0.0998	0.1987	0.2955	0.3894	0.4794

(b) θ (c) As $\theta \to 0$, $\sin \theta \to 0$ and $\dfrac{\theta}{\sin \theta} \to 1$.

85. Corresponding sides of similar triangles are proportional.

87. Yes, $\tan \theta$ is equal to opp/adj. You can find the value of the hypotenuse by the Pythagorean Theorem, then you can find $\sec \theta$, which is equal to hyp/adj.

Section 1.4 (page 170)

1. $\dfrac{y}{r}$ **3.** $\dfrac{y}{x}$ **5.** $\cos \theta$ **7.** zero; defined

9. (a) $\sin \theta = \dfrac{3}{5}$ $\csc \theta = \dfrac{5}{3}$
$\cos \theta = \dfrac{4}{5}$ $\sec \theta = \dfrac{5}{4}$
$\tan \theta = \dfrac{3}{4}$ $\cot \theta = \dfrac{4}{3}$

(b) $\sin \theta = \dfrac{15}{17}$ $\csc \theta = \dfrac{17}{15}$
$\cos \theta = -\dfrac{8}{17}$ $\sec \theta = -\dfrac{17}{8}$
$\tan \theta = -\dfrac{15}{8}$ $\cot \theta = -\dfrac{8}{15}$

11. (a) $\sin \theta = -\dfrac{1}{2}$ $\csc \theta = -2$

$\cos \theta = -\dfrac{\sqrt{3}}{2}$ $\sec \theta = -\dfrac{2\sqrt{3}}{3}$

$\tan \theta = \dfrac{\sqrt{3}}{3}$ $\cot \theta = \sqrt{3}$

(b) $\sin \theta = -\dfrac{\sqrt{17}}{17}$ $\csc \theta = -\sqrt{17}$

$\cos \theta = \dfrac{4\sqrt{17}}{17}$ $\sec \theta = \dfrac{\sqrt{17}}{4}$

$\tan \theta = -\dfrac{1}{4}$ $\cot \theta = -4$

13. $\sin \theta = \dfrac{12}{13}$ $\csc \theta = \dfrac{13}{12}$
$\cos \theta = \dfrac{5}{13}$ $\sec \theta = \dfrac{13}{5}$
$\tan \theta = \dfrac{12}{5}$ $\cot \theta = \dfrac{5}{12}$

15. $\sin \theta = -\dfrac{2\sqrt{29}}{29}$ $\csc \theta = -\dfrac{\sqrt{29}}{2}$

$\cos \theta = -\dfrac{5\sqrt{29}}{29}$ $\sec \theta = -\dfrac{\sqrt{29}}{5}$

$\tan \theta = \dfrac{2}{5}$ $\cot \theta = \dfrac{5}{2}$

17. $\sin \theta = \dfrac{4}{5}$ $\csc \theta = \dfrac{5}{4}$
$\cos \theta = -\dfrac{3}{5}$ $\sec \theta = -\dfrac{5}{3}$
$\tan \theta = -\dfrac{4}{3}$ $\cot \theta = -\dfrac{3}{4}$

19. Quadrant I **21.** Quadrant II

23. $\sin \theta = \dfrac{15}{17}$ $\csc \theta = \dfrac{17}{15}$
$\cos \theta = -\dfrac{8}{17}$ $\sec \theta = -\dfrac{17}{8}$
$\tan \theta = -\dfrac{15}{8}$ $\cot \theta = -\dfrac{8}{15}$

25. $\sin \theta = \dfrac{3}{5}$ $\csc \theta = \dfrac{5}{3}$
$\cos \theta = -\dfrac{4}{5}$ $\sec \theta = -\dfrac{5}{4}$
$\tan \theta = -\dfrac{3}{4}$ $\cot \theta = -\dfrac{4}{3}$

27. $\sin \theta = -\dfrac{\sqrt{10}}{10}$ $\csc \theta = -\sqrt{10}$

$\cos \theta = \dfrac{3\sqrt{10}}{10}$ $\sec \theta = \dfrac{\sqrt{10}}{3}$

$\tan \theta = -\dfrac{1}{3}$ $\cot \theta = -3$

29. $\sin \theta = -\dfrac{\sqrt{3}}{2}$ $\csc \theta = -\dfrac{2\sqrt{3}}{3}$

$\cos \theta = -\dfrac{1}{2}$ $\sec \theta = -2$

$\tan \theta = \sqrt{3}$ $\cot \theta = \dfrac{\sqrt{3}}{3}$

31. $\sin \theta = 0$ $\csc \theta$ is undefined.
$\cos \theta = -1$ $\sec \theta = -1$
$\tan \theta = 0$ $\cot \theta$ is undefined.

33. $\sin \theta = \dfrac{\sqrt{2}}{2}$ $\csc \theta = \sqrt{2}$

$\cos \theta = -\dfrac{\sqrt{2}}{2}$ $\sec \theta = -\sqrt{2}$

$\tan \theta = -1$ $\cot \theta = -1$

35. $\sin \theta = -\dfrac{2\sqrt{5}}{5}$ $\csc \theta = -\dfrac{\sqrt{5}}{2}$

$\cos \theta = -\dfrac{\sqrt{5}}{5}$ $\sec \theta = -\sqrt{5}$

$\tan \theta = 2$ $\cot \theta = \dfrac{1}{2}$

37. 0 **39.** Undefined **41.** 1 **43.** Undefined

45. $\theta' = 20°$ **47.** $\theta' = 55°$

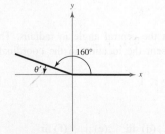

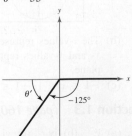

49. $\theta' = \dfrac{\pi}{3}$

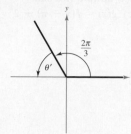

51. $\theta' = 2\pi - 4.8$

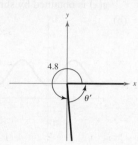

53. $\sin 225° = -\dfrac{\sqrt{2}}{2}$

$\cos 225° = -\dfrac{\sqrt{2}}{2}$

$\tan 225° = 1$

55. $\sin 750° = \dfrac{1}{2}$

$\cos 750° = \dfrac{\sqrt{3}}{2}$

$\tan 750° = \dfrac{\sqrt{3}}{3}$

57. $\sin(-150°) = -\dfrac{1}{2}$

$\cos(-150°) = -\dfrac{\sqrt{3}}{2}$

$\tan(-150°) = \dfrac{\sqrt{3}}{3}$

59. $\sin \dfrac{2\pi}{3} = \dfrac{\sqrt{3}}{2}$

$\cos \dfrac{2\pi}{3} = -\dfrac{1}{2}$

$\tan \dfrac{2\pi}{3} = -\sqrt{3}$

61. $\sin \dfrac{5\pi}{4} = -\dfrac{\sqrt{2}}{2}$

$\cos \dfrac{5\pi}{4} = -\dfrac{\sqrt{2}}{2}$

$\tan \dfrac{5\pi}{4} = 1$

63. $\sin\left(-\dfrac{\pi}{6}\right) = -\dfrac{1}{2}$

$\cos\left(-\dfrac{\pi}{6}\right) = \dfrac{\sqrt{3}}{2}$

$\tan\left(-\dfrac{\pi}{6}\right) = -\dfrac{\sqrt{3}}{3}$

65. $\sin \dfrac{9\pi}{4} = \dfrac{\sqrt{2}}{2}$

$\cos \dfrac{9\pi}{4} = \dfrac{\sqrt{2}}{2}$

$\tan \dfrac{9\pi}{4} = 1$

67. $\sin\left(-\dfrac{3\pi}{2}\right) = 1$

$\cos\left(-\dfrac{3\pi}{2}\right) = 0$

$\tan\left(-\dfrac{3\pi}{2}\right)$ is undefined.

69. $\dfrac{4}{5}$ **71.** $-\dfrac{\sqrt{13}}{2}$ **73.** $\dfrac{8}{5}$ **75.** 0.1736

77. -0.3420 **79.** -1.4826 **81.** 3.2361 **83.** 4.6373

85. 0.3640 **87.** -0.6052 **89.** -0.4142

91. (a) $30° = \dfrac{\pi}{6},\ 150° = \dfrac{5\pi}{6}$ (b) $210° = \dfrac{7\pi}{6},\ 330° = \dfrac{11\pi}{6}$

93. (a) $60° = \dfrac{\pi}{3},\ 120° = \dfrac{2\pi}{3}$ (b) $135° = \dfrac{3\pi}{4},\ 315° = \dfrac{7\pi}{4}$

95. (a) $45° = \dfrac{\pi}{4},\ 225° = \dfrac{5\pi}{4}$ (b) $150° = \dfrac{5\pi}{6},\ 330° = \dfrac{11\pi}{6}$

97. (a) 12 mi (b) 6 mi (c) 6.9 mi

99. (a) $N = 22.099 \sin(0.522t - 2.219) + 55.008$

$F = 36.641 \sin(0.502t - 1.831) + 25.610$

(b) February: $N = 34.6°,\ F = -1.4°$

March: $N = 41.6°,\ F = 13.9°$

May: $N = 63.4°,\ F = 48.6°$

June: $N = 72.5°,\ F = 59.5°$

August: $N = 75.5°,\ F = 55.6°$

September: $N = 68.6°,\ F = 41.7°$

November: $N = 46.8°,\ F = 6.5°$

(c) Answers will vary.

101. (a) 270.63 ft (b) 307.75 ft (c) 270.63 ft

103. False. In each of the four quadrants, the signs of the secant function and the cosine function will be the same, because these functions are reciprocals of each other.

105. As θ increases from $0°$ to $90°$, x decreases from 12 cm to 0 cm and y increases from 0 cm to 12 cm. Therefore, $\sin \theta = y/12$ increases from 0 to 1 and $\cos \theta = x/12$ decreases from 1 to 0. Thus, $\tan \theta = y/x$ increases without bound. When $\theta = 90°$, the tangent is undefined.

107. (a) $\sin t = y$ (b) $r = 1$ because it is a unit circle.

$\cos t = x$

(c) $\sin \theta = y$ (d) $\sin t = \sin \theta$, and $\cos t = \cos \theta$.

$\cos \theta = x$

Section 1.5 (*page 180*)

1. cycle **3.** phase shift **5.** Period: $\dfrac{2\pi}{5}$; Amplitude: 2

7. Period: 4π; Amplitude: $\dfrac{3}{4}$ **9.** Period: 6; Amplitude: $\dfrac{1}{2}$

11. Period: 2π; Amplitude: 4 **13.** Period: $\dfrac{\pi}{5}$; Amplitude: 3

15. Period: $\dfrac{5\pi}{2}$; Amplitude: $\dfrac{5}{3}$ **17.** Period: 1; Amplitude: $\dfrac{1}{4}$

19. g is a shift of f π units to the right.

21. g is a reflection of f in the x-axis.

23. The period of f is twice the period of g.

25. g is a shift of f three units upward.

27. The graph of g has twice the amplitude of the graph of f.

29. The graph of g is a horizontal shift of the graph of f π units to the right.

31.

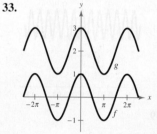

33.

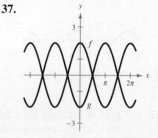

35. **37.**

39. **41.**

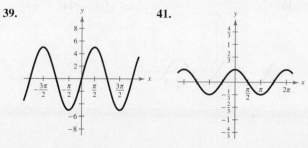

43.

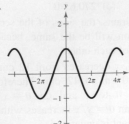

45.

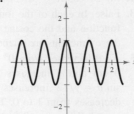

47.

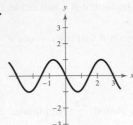

49.

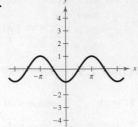

51.

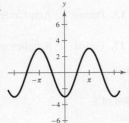

53.

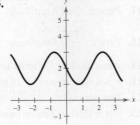

55.

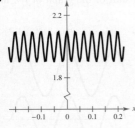

57.

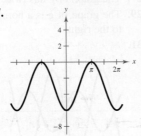

59.

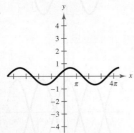

61. (a) $g(x)$ is obtained by a horizontal shrink of four, and one cycle of $g(x)$ corresponds to the interval $[\pi/4, 3\pi/4]$.
(b)

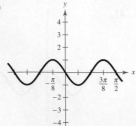

(c) $g(x) = f(4x - \pi)$

63. (a) One cycle of $g(x)$ corresponds to the interval $[\pi, 3\pi]$, and $g(x)$ is obtained by shifting $f(x)$ upward two units.
(b)

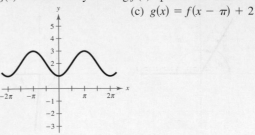

(c) $g(x) = f(x - \pi) + 2$

65. (a) One cycle of $g(x)$ is $[\pi/4, 3\pi/4]$. $g(x)$ is also shifted down three units and has an amplitude of two.
(b)

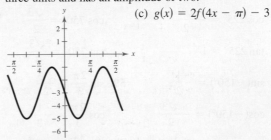

(c) $g(x) = 2f(4x - \pi) - 3$

67.

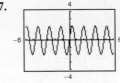

69.

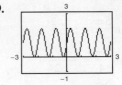

71.

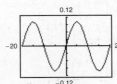

73. $a = 2, d = 1$

75. $a = -4, d = 4$ **77.** $a = -3, b = 2, c = 0$

79. $a = 2, b = 1, c = -\dfrac{\pi}{4}$

81.

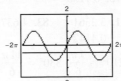

83. $y = 1 + 2\sin(2x - \pi)$

$x = -\dfrac{\pi}{6}, -\dfrac{5\pi}{6}, \dfrac{7\pi}{6}, \dfrac{11\pi}{6}$

85. $y = \cos(2x + 2\pi) - \dfrac{3}{2}$

87. (a) 6 sec   (b) 10 cycles/min
(c)

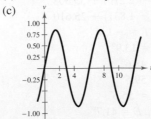

89. (a) $I(t) = 46.2 + 32.4 \cos\left(\dfrac{\pi t}{6} - 3.67\right)$

(b)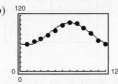

The model fits the data well.

(c)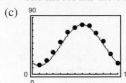

The model fits the data well.

(d) Las Vegas: $80.6°$; International Falls: $46.2°$

The constant term gives the annual average temperature.

(e) 12; yes; One full period is one year.

(f) International Falls; amplitude; The greater the amplitude, the greater the variability in temperature.

91. (a) $\frac{1}{440}$ sec   (b) 440 cycles/sec

93. (a) 365; Yes, because there are 365 days in a year.

(b) 30.3 gal; the constant term

(c) 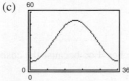 $124 < t < 252$

95. False. The graph of $f(x) = \sin(x + 2\pi)$ translates the graph of $f(x) = \sin x$ exactly one period to the left so that the two graphs look identical.

97. True. Because $\cos x = \sin\left(x + \dfrac{\pi}{2}\right)$, $y = -\cos x$ is a

reflection in the x-axis of $y = \sin\left(x + \dfrac{\pi}{2}\right)$.

99.

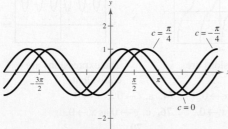

The value of c is a horizontal translation of the graph.

101.

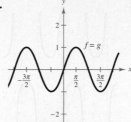

Conjecture: $\sin x = \cos\left(x - \dfrac{\pi}{2}\right)$

103. (a)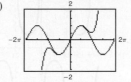

The graphs appear to coincide from $-\dfrac{\pi}{2}$ to $\dfrac{\pi}{2}$.

(b)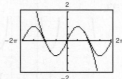

The graphs appear to coincide from $-\dfrac{\pi}{2}$ to $\dfrac{\pi}{2}$.

(c) $-\dfrac{x^7}{7!}, -\dfrac{x^6}{6!}$

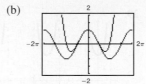   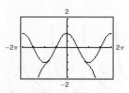

The interval of accuracy increased.

Section 1.6   (page 191)

1. odd; origin   **3.** reciprocal

5. π **7.** $(-\infty, -1] \cup [1, \infty)$

9. e, π **10.** c, 2π **11.** a, 1   **12.** d, 2π

13. f, 4   **14.** b, 4

15.   **17.**

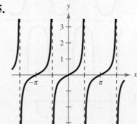

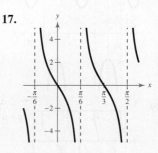

CHAPTER 1

19.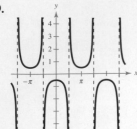

21.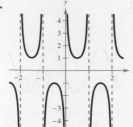

23.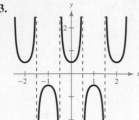

25.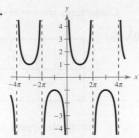

27.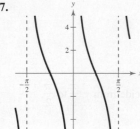

29.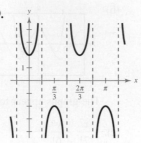

31.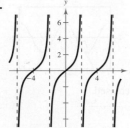

33.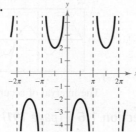

35.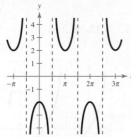

37.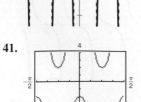

39.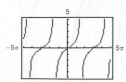

41.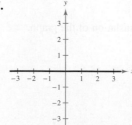

43.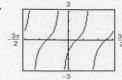

45.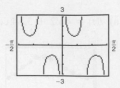

47.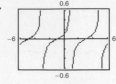

49. $-\dfrac{7\pi}{4}, -\dfrac{3\pi}{4}, \dfrac{\pi}{4}, \dfrac{5\pi}{4}$

51. $-\dfrac{4\pi}{3}, -\dfrac{\pi}{3}, \dfrac{2\pi}{3}, \dfrac{5\pi}{3}$ **53.** $-\dfrac{4\pi}{3}, -\dfrac{2\pi}{3}, \dfrac{2\pi}{3}, \dfrac{4\pi}{3}$

55. $-\dfrac{7\pi}{4}, -\dfrac{5\pi}{4}, \dfrac{\pi}{4}, \dfrac{3\pi}{4}$

57. Even **59.** Odd **61.** Odd **63.** Even

65. (a) 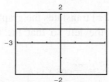 (b) $\dfrac{\pi}{6} < x < \dfrac{5\pi}{6}$

(c) f approaches 0 and g approaches $+\infty$ because the cosecant is the reciprocal of the sine.

67. 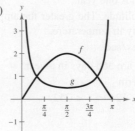 The expressions are equivalent except when $\sin x = 0$, y_1 is undefined.

69. 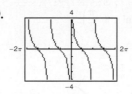 **71.**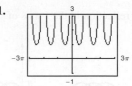

The expressions are equivalent. The expressions are equivalent.

73. d, $f \to 0$ as $x \to 0$. **74.** a, $f \to 0$ as $x \to 0$.

75. b, $g \to 0$ as $x \to 0$. **76.** c, $g \to 0$ as $x \to 0$.

77. 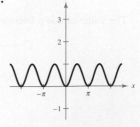 **79.**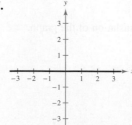

The functions are equal. The functions are equal.

81.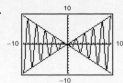

83.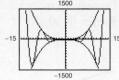

As $x \to \infty$, $g(x)$ oscillates. As $x \to \infty$, $f(x)$ oscillates

85.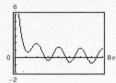

87.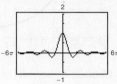

As $x \to 0$, $y \to \infty$. As $x \to 0$, $g(x) \to 1$.

89.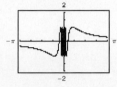

As $x \to 0$, $f(x)$ oscillates between 1 and -1.

91. $d = 7 \cot x$

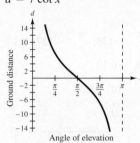

93. (a) Period of $H(t)$: 12 mo
　　　Period of $L(t)$: 12 mo
　　(b) Summer; winter　(c) About 0.5 mo

95. (a) 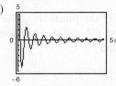　(b) y approaches 0 as t increases.

97. True. $y = \sec x$ is equal to $y = 1/\cos x$, and if the reciprocal of $y = \sin x$ is translated $\pi/2$ units to the left, then
$$\frac{1}{\sin\left(x + \dfrac{\pi}{2}\right)} = \frac{1}{\cos x} = \sec x.$$

99. (a) As $x \to \dfrac{\pi^+}{2}$,　$f(x) \to -\infty$.

　　(b) As $x \to \dfrac{\pi^-}{2}$,　$f(x) \to \infty$.

　　(c) As $x \to -\dfrac{\pi^+}{2}$,　$f(x) \to -\infty$.

　　(d) As $x \to -\dfrac{\pi^-}{2}$,　$f(x) \to \infty$.

101. (a) As $x \to 0^+$,　$f(x) \to \infty$.
　　(b) As $x \to 0^-$,　$f(x) \to -\infty$.
　　(c) As $x \to \pi^+$,　$f(x) \to \infty$.
　　(d) As $x \to \pi^-$,　$f(x) \to -\infty$.

103. (a)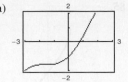

0.7391

　(b) 1, 0.5403, 0.8576, 0.6543, 0.7935, 0.7014, 0.7640, 0.7221, 0.7504, 0.7314, . . . ; 0.7391

105. 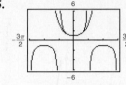　The graphs appear to coincide on the interval $-1.1 \le x \le 1.1$.

Section 1.7 　(page 201)

1. $y = \sin^{-1} x$; $-1 \le x \le 1$

3. $y = \tan^{-1} x$; $-\infty < x < \infty$; $-\dfrac{\pi}{2} < y < \dfrac{\pi}{2}$

5. $\dfrac{\pi}{6}$　**7.** $\dfrac{\pi}{3}$　**9.** $\dfrac{\pi}{6}$　**11.** $\dfrac{5\pi}{6}$　**13.** $-\dfrac{\pi}{3}$

15. $\dfrac{2\pi}{3}$　**17.** $-\dfrac{\pi}{3}$　**19.** 0

21.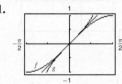

23. 1.19　**25.** -0.85　**27.** -1.25　**29.** 0.32

31. 1.99　**33.** 0.74　**35.** 1.07　**37.** 1.36

39. -1.52　**41.** $-\dfrac{\pi}{3}$, $-\dfrac{\sqrt{3}}{3}$, 1　**43.** $\theta = \arctan \dfrac{x}{4}$

45. $\theta = \arcsin \dfrac{x + 2}{5}$　**47.** $\theta = \arccos \dfrac{x + 3}{2x}$

49. 0.3　**51.** -0.1　**53.** 0　**55.** $\dfrac{3}{5}$　**57.** $\dfrac{\sqrt{5}}{5}$

59. $\dfrac{12}{13}$　**61.** $\dfrac{\sqrt{34}}{5}$　**63.** $\dfrac{\sqrt{5}}{3}$　**65.** 2　**67.** $\dfrac{1}{x}$

69. $\sqrt{1 - 4x^2}$　**71.** $\sqrt{1 - x^2}$　**73.** $\dfrac{\sqrt{9 - x^2}}{x}$

75. $\dfrac{\sqrt{x^2 + 2}}{x}$

77. 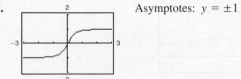　Asymptotes: $y = \pm 1$

79. $\dfrac{9}{\sqrt{x^2 + 81}}$, $x > 0$; $\dfrac{-9}{\sqrt{x^2 + 81}}$, $x < 0$

81. $\dfrac{|x - 1|}{\sqrt{x^2 - 2x + 10}}$

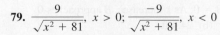

CHAPTER 1

83.

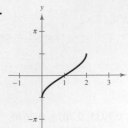

85.

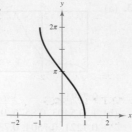

The graph of g is a
horizontal shift one unit
to the right of f.

87.

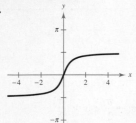

89.

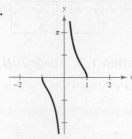

91.

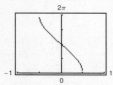

93.

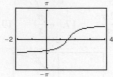

95.

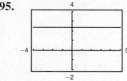

97. $3\sqrt{2}\sin\left(2t + \dfrac{\pi}{4}\right)$

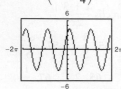

The graph implies that the
identity is true.

99. $\dfrac{\pi}{2}$ **101.** $\dfrac{\pi}{2}$ **103.** π

105. (a) $\theta = \arcsin\dfrac{5}{s}$ (b) $0.13, 0.25$

107. (a)

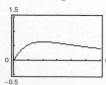

(b) 2 ft   (c) $\beta = 0$; As x increases, β approaches 0.

109. (a) $\theta \approx 26.0°$ (b) 24.4 ft

111. (a) $\theta = \arctan\dfrac{x}{20}$ (b) $14.0°, 31.0°$

113. False. $\dfrac{5\pi}{4}$ is not in the range of the arctangent.

115. Domain: $(-\infty, \infty)$
Range: $(0, \pi)$

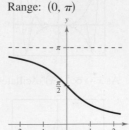

117. Domain: $(-\infty, -1] \cup [1, \infty)$
Range: $[-\pi/2, 0) \cup (0, \pi/2]$

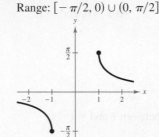

119. $\dfrac{\pi}{4}$ **121.** $\dfrac{3\pi}{4}$ **123.** $\dfrac{\pi}{6}$ **125.** $\dfrac{\pi}{3}$ **127.** 1.17
129. 0.19   **131.** 0.54   **133.** -0.12

135. (a) $\dfrac{\pi}{4}$ (b) $\dfrac{\pi}{2}$ (c) 1.25   (d) 2.03

137. (a) $f \circ f^{-1}$ $f^{-1} \circ f$

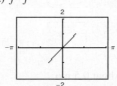   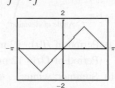

(b) The domains and ranges of the functions are restricted.
The graphs of $f \circ f^{-1}$ and $f^{-1} \circ f$ differ because of the
domains and ranges of f and f^{-1}.

Section 1.8   *(page 211)*

1. bearing   **3.** period
5. $a \approx 1.73$ **7.** $a \approx 8.26$ **9.** $c = 5$
$\quad\;\; c \approx 3.46$ $\quad\;\; c \approx 25.38$ $\quad\;\; A \approx 36.87°$
$\quad\;\; B = 60°$ $\quad\;\; A = 19°$ $\quad\;\; B \approx 53.13°$
11. $a \approx 49.48$ **13.** $a \approx 91.34$
$\quad\;\; A \approx 72.08°$ $\quad\;\;\; b \approx 420.70$
$\quad\;\; B \approx 17.92°$ $\quad\;\;\; B = 77°45'$
15. 3.00   **17.** 2.50   **19.** 214.45 ft   **21.** 19.7 ft
23. 19.9 ft   **25.** 11.8 km   **27.** 56.3°   **29.** 2.06°
31. (a) $\sqrt{h^2 + 34h + 10{,}289}$ (b) $\theta = \arccos\left(\dfrac{100}{l}\right)$
$\quad\;\;$ (c) 53.02 ft
33. (a) $l = 250$ ft, $A \approx 36.87°$, $B \approx 53.13°$ (b) 4.87 sec
35. 554 mi north; 709 mi east
37. (a) 104.95 nautical mi south; 58.18 nautical mi west
$\quad\;\;$ (b) S 36.7° W; distance = 130.9 nautical mi

39. N 56.31° W **41.** (a) N 58° E (b) 68.82 m

43. 78.7° **45.** 35.3° **47.** 29.4 in.

49. $y = \sqrt{3}\,r$ **51.** $a \approx 12.2, b \approx 7$ **53.** $d = 4\sin(\pi t)$

55. $d = 3\cos\left(\dfrac{4\pi t}{3}\right)$ **57.** (a) 9 (b) $\dfrac{3}{5}$ (c) 9 (d) $\dfrac{5}{12}$

59. (a) $\dfrac{1}{4}$ (b) 3 (c) 0 (d) $\dfrac{1}{6}$ **61.** $\omega = 528\pi$

63. (a) (b) $\dfrac{\pi}{8}$ (c) $\dfrac{\pi}{32}$

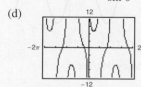

65. (a)

θ	L_1	L_2	$L_1 + L_2$
0.1	$\dfrac{2}{\sin 0.1}$	$\dfrac{3}{\cos 0.1}$	23.0
0.2	$\dfrac{2}{\sin 0.2}$	$\dfrac{3}{\cos 0.2}$	13.1
0.3	$\dfrac{2}{\sin 0.3}$	$\dfrac{3}{\cos 0.3}$	9.9
0.4	$\dfrac{2}{\sin 0.4}$	$\dfrac{3}{\cos 0.4}$	8.4

(b)

θ	L_1	L_2	$L_1 + L_2$
0.5	$\dfrac{2}{\sin 0.5}$	$\dfrac{3}{\cos 0.5}$	7.6
0.6	$\dfrac{2}{\sin 0.6}$	$\dfrac{3}{\cos 0.6}$	7.2
0.7	$\dfrac{2}{\sin 0.7}$	$\dfrac{3}{\cos 0.7}$	7.0
0.8	$\dfrac{2}{\sin 0.8}$	$\dfrac{3}{\cos 0.8}$	7.1

7.0 m

(c) $L = L_1 + L_2 = \dfrac{2}{\sin\theta} + \dfrac{3}{\cos\theta}$

(d) 7.0 m; The answers are the same.

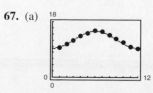

67. (a)

(b) 12; Yes, there are 12 months in a year.

(c) 2.77; The maximum change in the number of hours of daylight

69. False. The scenario does not create a right triangle because the tower is not vertical.

Review Exercises *(page 218)*

1. (a)

(b) Quadrant IV

(c) $\dfrac{23\pi}{4}, -\dfrac{\pi}{4}$

3. (a)

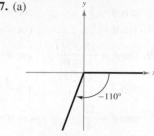

(b) Quadrant II

(c) $\dfrac{2\pi}{3}, -\dfrac{10\pi}{3}$

5. (a)

(b) Quadrant I

(c) 430°, −290°

7. (a)

(b) Quadrant III

(c) 250°, −470°

9. 7.854 **11.** −0.589 **13.** 54.000° **15.** −200.535°

17. 198° 24′ **19.** 0° 39′ **21.** 48.17 in.

23. Area ≈ 339.29 in.^{2} **25.** $\left(-\dfrac{1}{2}, \dfrac{\sqrt{3}}{2}\right)$

27. $\left(-\dfrac{\sqrt{3}}{2}, -\dfrac{1}{2}\right)$

29. $\sin\dfrac{7\pi}{6} = -\dfrac{1}{2}$ $\csc\dfrac{7\pi}{6} = -2$

$\cos\dfrac{7\pi}{6} = -\dfrac{\sqrt{3}}{2}$ $\sec\dfrac{7\pi}{6} = -\dfrac{2\sqrt{3}}{3}$

$\tan\dfrac{7\pi}{6} = \dfrac{\sqrt{3}}{3}$ $\cot\dfrac{7\pi}{6} = \sqrt{3}$

31. $\sin\left(-\dfrac{2\pi}{3}\right) = -\dfrac{\sqrt{3}}{2}$ $\csc\left(-\dfrac{2\pi}{3}\right) = -\dfrac{2\sqrt{3}}{3}$

$\cos\left(-\dfrac{2\pi}{3}\right) = -\dfrac{1}{2}$ $\sec\left(-\dfrac{2\pi}{3}\right) = -2$

$\tan\left(-\dfrac{2\pi}{3}\right) = \sqrt{3}$ $\cot\left(-\dfrac{2\pi}{3}\right) = \dfrac{\sqrt{3}}{3}$

33. $\sin\dfrac{11\pi}{4} = \sin\dfrac{3\pi}{4} = \dfrac{\sqrt{2}}{2}$ **35.** $\sin\left(-\dfrac{17\pi}{6}\right) = \sin\dfrac{7\pi}{6} = -\dfrac{1}{2}$

37. −75.3130 **39.** 3.2361

CHAPTER 1

41. $\sin \theta = \dfrac{4\sqrt{41}}{41}$

$\cos \theta = \dfrac{5\sqrt{41}}{41}$

$\tan \theta = \dfrac{4}{5}$

$\csc \theta = \dfrac{\sqrt{41}}{4}$

$\sec \theta = \dfrac{\sqrt{41}}{5}$

$\cot \theta = \dfrac{5}{4}$

43. (a) 3 (b) $\dfrac{2\sqrt{2}}{3}$ (c) $\dfrac{3\sqrt{2}}{4}$ (d) $\dfrac{\sqrt{2}}{4}$

45. (a) $\dfrac{1}{4}$ (b) $\dfrac{\sqrt{15}}{4}$ (c) $\dfrac{4\sqrt{15}}{15}$ (d) $\dfrac{\sqrt{15}}{15}$

47. 0.6494 **49.** 0.5621 **51.** 3.6722 **53.** 0.6104

55. 71.3 m

57. $\sin \theta = \dfrac{4}{5}$ $\csc \theta = \dfrac{5}{4}$

$\cos \theta = \dfrac{3}{5}$ $\sec \theta = \dfrac{5}{3}$

$\tan \theta = \dfrac{4}{3}$ $\cot \theta = \dfrac{3}{4}$

59. $\sin \theta = \dfrac{15\sqrt{241}}{241}$ $\csc \theta = \dfrac{\sqrt{241}}{15}$

$\cos \theta = \dfrac{4\sqrt{241}}{241}$ $\sec \theta = \dfrac{\sqrt{241}}{4}$

$\tan \theta = \dfrac{15}{4}$ $\cot \theta = \dfrac{4}{15}$

61. $\sin \theta = \dfrac{9\sqrt{82}}{82}$ $\csc \theta = \dfrac{\sqrt{82}}{9}$

$\cos \theta = \dfrac{-\sqrt{82}}{82}$ $\sec \theta = -\sqrt{82}$

$\tan \theta = -9$ $\cot \theta = -\dfrac{1}{9}$

63. $\sin \theta = \dfrac{4\sqrt{17}}{17}$ $\csc \theta = \dfrac{\sqrt{17}}{4}$

$\cos \theta = \dfrac{\sqrt{17}}{17}$ $\sec \theta = \sqrt{17}$

$\tan \theta = 4$ $\cot \theta = \dfrac{1}{4}$

65. $\sin \theta = -\dfrac{\sqrt{11}}{6}$

$\cos \theta = \dfrac{5}{6}$

$\tan \theta = -\dfrac{\sqrt{11}}{5}$

$\csc \theta = -\dfrac{6\sqrt{11}}{11}$

$\cot \theta = -\dfrac{5\sqrt{11}}{11}$

67. $\cos \theta = -\dfrac{\sqrt{55}}{8}$

$\tan \theta = -\dfrac{3\sqrt{55}}{55}$

$\csc \theta = \dfrac{8}{3}$

$\sec \theta = -\dfrac{8\sqrt{55}}{55}$

$\cot \theta = -\dfrac{\sqrt{55}}{3}$

69. $\sin \theta = \dfrac{\sqrt{21}}{5}$

$\tan \theta = -\dfrac{\sqrt{21}}{2}$

$\csc \theta = \dfrac{5\sqrt{21}}{21}$

$\sec \theta = -\dfrac{5}{2}$

$\cot \theta = -\dfrac{2\sqrt{21}}{21}$

71. $\theta' = 84°$ **73.** $\theta' = \dfrac{\pi}{5}$

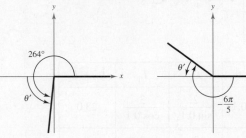

75. $\sin \dfrac{\pi}{3} = \dfrac{\sqrt{3}}{2}$; $\cos \dfrac{\pi}{3} = \dfrac{1}{2}$; $\tan \dfrac{\pi}{3} = \sqrt{3}$

77. $\sin\left(-\dfrac{7\pi}{3}\right) = -\dfrac{\sqrt{3}}{2}$; $\cos\left(-\dfrac{7\pi}{3}\right) = \dfrac{1}{2}$;

$\tan\left(-\dfrac{7\pi}{3}\right) = -\sqrt{3}$

79. $\sin 495° = \dfrac{\sqrt{2}}{2}$; $\cos 495° = -\dfrac{\sqrt{2}}{2}$; $\tan 495° = -1$

81. -0.7568 **83.** 0.9511

85.   **87.**

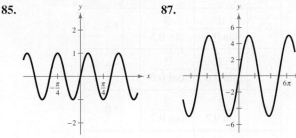

89.   **91.**

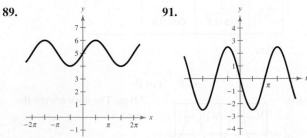

93. (a) $y = 2 \sin 528\pi x$ (b) 264 cycles/sec

95. 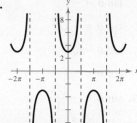 **97.**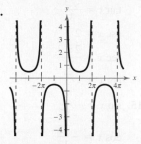

99. 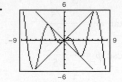 **101.**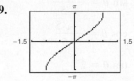

103.
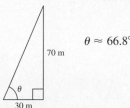

As $x \to +\infty$, $f(x)$ oscillates.

105. $-\dfrac{\pi}{6}$ **107.** 0.41 **109.** -0.46 **111.** $\dfrac{3\pi}{4}$

113. π **115.** 1.24 **117.** -0.98

119. 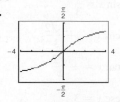 **121.**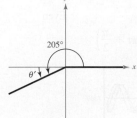

123. $\dfrac{4}{5}$ **125.** $\dfrac{13}{5}$ **127.** $\dfrac{10}{7}$ **129.** $\dfrac{\sqrt{4 - x^2}}{x}$

131. $\dfrac{\pi}{6}$ **133.** $\dfrac{3\pi}{4}$ **135.** 0.09 **137.** 1.98

139.

$\theta \approx 66.8°$

70 m
θ
30 m

141. 1221 mi, 85.6°
143. False. For each θ there corresponds exactly one value of y.
145. The function is undefined because $\sec \theta = 1/\cos \theta$.
147. The ranges of the other four trigonometric functions are $(-\infty, \infty)$ or $(-\infty, -1] \cup [1, \infty)$.
149. Answers will vary.

Chapter Test *(page 221)*

1. (a)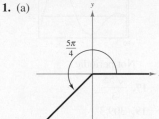

$\dfrac{5\pi}{4}$

(b) $\dfrac{13\pi}{4}$, $-\dfrac{3\pi}{4}$

(c) 225°

2. 3500 rad/min **3.** About 709.04 ft²

4. $\sin \theta = \dfrac{3\sqrt{10}}{10}$ $\csc \theta = \dfrac{\sqrt{10}}{3}$

$\cos \theta = -\dfrac{\sqrt{10}}{10}$ $\sec \theta = -\sqrt{10}$

$\tan \theta = -3$ $\cot \theta = -\dfrac{1}{3}$

5. For $0 \le \theta < \dfrac{\pi}{2}$: For $\pi \le \theta < \dfrac{3\pi}{2}$:

$\sin \theta = \dfrac{3\sqrt{13}}{13}$ $\sin \theta = -\dfrac{3\sqrt{13}}{13}$

$\cos \theta = \dfrac{2\sqrt{13}}{13}$ $\cos \theta = -\dfrac{2\sqrt{13}}{13}$

$\csc \theta = \dfrac{\sqrt{13}}{3}$ $\csc \theta = -\dfrac{\sqrt{13}}{3}$

$\sec \theta = \dfrac{\sqrt{13}}{2}$ $\sec \theta = -\dfrac{\sqrt{13}}{2}$

$\cot \theta = \dfrac{2}{3}$ $\cot \theta = \dfrac{2}{3}$

6. $\theta' = 25°$

205°
θ'

7. Quadrant III **8.** 150°, 210° **9.** 1.33, 1.81

10. $\sin \theta = -\dfrac{4}{5}$ **11.** $\sin \theta = \dfrac{21}{29}$

$\tan \theta = -\dfrac{4}{3}$ $\cos \theta = -\dfrac{20}{29}$

$\csc \theta = -\dfrac{5}{4}$ $\tan \theta = -\dfrac{21}{20}$

$\sec \theta = \dfrac{5}{3}$ $\csc \theta = \dfrac{29}{21}$

$\cot \theta = -\dfrac{3}{4}$ $\cot \theta = -\dfrac{20}{21}$

12. 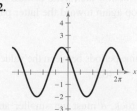 **13.**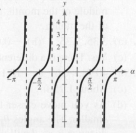

CHAPTER 1

14.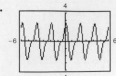

Period: 2

15.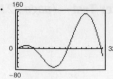

Not periodic

16. $a = -2, b = \dfrac{1}{2}, c = -\dfrac{\pi}{4}$ **17.** $\dfrac{\sqrt{55}}{3}$

18. 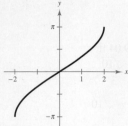   **19.** $309.3°$

20. $d = -6 \cos \pi t$

Problem Solving   (page 223)

1. (a) $\dfrac{11\pi}{2}$ rad or $990°$ (b) About 816.42 ft

3. (a) 4767 ft   (b) 3705 ft

(c) $w = 2183$ ft, $\tan 63° = \dfrac{w + 3705}{3000}$

5. (a) 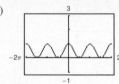   (b)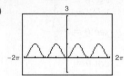

Even   Even

7. $h = 51 - 50 \sin\left(8\pi t + \dfrac{\pi}{2}\right)$

9. (a)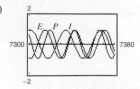

(b)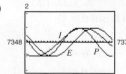

(c) $P(7369) = 0.631$
$E(7369) = 0.901$
$I(7369) = 0.945$

All three drop earlier in the month, then peak toward the middle of the month, and drop again toward the latter part of the month.

11. (a) $3.35, 7.35$ (b) -0.65

(c) Yes. There is a difference of nine periods between the values.

13. (a) $40.5°$ (b) $x \approx 1.71$ ft; $y \approx 3.46$ ft

(c) About 1.75 ft

(d) As you move closer to the rock, d must get smaller and smaller. The angles θ_1 and θ_2 will decrease along with the distance y, so d will decrease. The accuracy improves.

Chapter 2

Section 2.1   (page 231)

1. $\tan u$ **3.** $\cot u$ **5.** $\cot^2 u$ **7.** $\cos u$ **9.** $\cos u$

11. $\sin x = \dfrac{1}{2}$ **13.** $\sin \theta = -\dfrac{\sqrt{2}}{2}$

$\cos x = \dfrac{\sqrt{3}}{2}$ $\cos \theta = \dfrac{\sqrt{2}}{2}$

$\tan x = \dfrac{\sqrt{3}}{3}$ $\tan \theta = -1$

$\csc x = 2$ $\csc \theta = -\sqrt{2}$

$\sec x = \dfrac{2\sqrt{3}}{3}$ $\sec \theta = \sqrt{2}$

$\cot x = \sqrt{3}$ $\cot \theta = -1$

15. $\sin x = -\dfrac{8}{17}$ **17.** $\sin \phi = -\dfrac{\sqrt{5}}{3}$

$\cos x = -\dfrac{15}{17}$ $\cos \phi = \dfrac{2}{3}$

$\tan x = \dfrac{8}{15}$ $\tan \phi = -\dfrac{\sqrt{5}}{2}$

$\csc x = -\dfrac{17}{8}$ $\csc \phi = -\dfrac{3\sqrt{5}}{5}$

$\sec x = -\dfrac{17}{15}$ $\sec \phi = \dfrac{3}{2}$

$\cot x = \dfrac{15}{8}$ $\cot \phi = -\dfrac{2\sqrt{5}}{5}$

19. $\sin x = \dfrac{1}{3}$ **21.** $\sin \theta = -\dfrac{2\sqrt{5}}{5}$

$\cos x = -\dfrac{2\sqrt{2}}{3}$ $\cos \theta = -\dfrac{\sqrt{5}}{5}$

$\tan x = -\dfrac{\sqrt{2}}{4}$ $\tan \theta = 2$

$\csc x = 3$ $\csc \theta = -\dfrac{\sqrt{5}}{2}$

$\sec x = -\dfrac{3\sqrt{2}}{4}$ $\sec \theta = -\sqrt{5}$

$\cot x = -2\sqrt{2}$ $\cot \theta = \dfrac{1}{2}$

23. $\sin \theta = -1$
$\cos \theta = 0$
$\tan \theta$ is undefined.
$\csc \theta = -1$
$\sec \theta$ is undefined.
$\cot \theta = 0$

25. d   **26.** a   **27.** b   **28.** f   **29.** e   **30.** c

31. b   **32.** c   **33.** f   **34.** a   **35.** e   **36.** d

37. $\csc \theta$ **39.** $-\sin x$ **41.** $\cos^2 \phi$ **43.** $\cos x$

45. $\sin^2 x$ **47.** $\cos \theta$ **49.** 1 **51.** $\tan x$

53. $1 + \sin y$ **55.** $\sec \beta$ **57.** $\cos u + \sin u$

59. $\sin^2 x$ **61.** $\sin^2 x \tan^2 x$ **63.** $\sec x + 1$ **65.** $\sec^4 x$

67. $\sin^2 x - \cos^2 x$ **69.** $\cot^2 x(\csc x - 1)$

71. $1 + 2 \sin x \cos x$ **73.** $4 \cot^2 x$ **75.** $2 \csc^2 x$

77. $2 \sec x$ **79.** $\sec x$ **81.** $1 + \cos y$

83. $3(\sec x + \tan x)$

85.

x	0.2	0.4	0.6	0.8	1.0
y_1	0.1987	0.3894	0.5646	0.7174	0.8415
y_2	0.1987	0.3894	0.5646	0.7174	0.8415

x	1.2	1.4
y_1	0.9320	0.9854
y_2	0.9320	0.9854

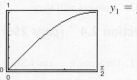
$y_1 = y_2$

87.

x	0.2	0.4	0.6	0.8	1.0
y_1	1.2230	1.5085	1.8958	2.4650	3.4082
y_2	1.2230	1.5085	1.8958	2.4650	3.4082

x	1.2	1.4
y_1	5.3319	11.6814
y_2	5.3319	11.6814

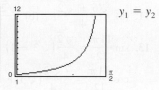
$y_1 = y_2$

89. $\csc x$ **91.** $\tan x$ **93.** $3 \sin \theta$ **95.** $4 \cos \theta$

97. $3 \tan \theta$ **99.** $5 \sec \theta$ **101.** $3 \sec \theta$

103. $\sqrt{2} \cos \theta$ **105.** $3 \cos \theta = 3; \sin \theta = 0; \cos \theta = 1$

107. $4 \sin \theta = 2\sqrt{2}; \sin \theta = \dfrac{\sqrt{2}}{2}; \cos \theta = \dfrac{\sqrt{2}}{2}$

109. $0 \le \theta \le \pi$ **111.** $0 \le \theta < \dfrac{\pi}{2}, \dfrac{3\pi}{2} < \theta < 2\pi$

113. (a) $\csc^2 132° - \cot^2 132° \approx 1.8107 - 0.8107 = 1$

(b) $\csc^2 \dfrac{2\pi}{7} - \cot^2 \dfrac{2\pi}{7} \approx 1.6360 - 0.6360 = 1$

115. (a) $\cos(90° - 80°) = \sin 80° \approx 0.9848$

(b) $\cos\left(\dfrac{\pi}{2} - 0.8\right) = \sin 0.8 \approx 0.7174$

117. $\mu = \tan \theta$ **119.** Answers will vary.

121. True. For example, $\sin(-x) = -\sin x$.

123. $1, 1$ **125.** $\infty, 0$

127. Not an identity because $\cos \theta = \pm \sqrt{1 - \sin^2 \theta}$

129. Not an identity because $\dfrac{\sin k\theta}{\cos k\theta} = \tan k\theta$

131. An identity because $\sin \theta \cdot \dfrac{1}{\sin \theta} = 1$

133. $a \cos \theta$ **135.** $a \tan \theta$

Section 2.2 *(page 239)*

1. identity **3.** $\tan u$ **5.** $\cos^2 u$ **7.** $-\csc u$

9–49. Answers will vary.

51. In the first line, $\cot(x)$ is substituted for $\cot(-x)$, which is incorrect; $\cot(-x) = -\cot(x)$.

53. (a) 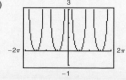 (b)

Identity

(c) Answers will vary.

55. (a) 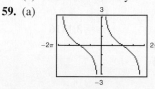 (b)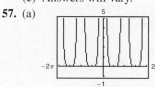

Not an identity

(c) Answers will vary.

57. (a) 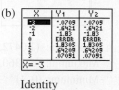 (b)

Identity

(c) Answers will vary.

59. (a) (b)

Identity

(c) Answers will vary.

61–63. Answers will vary. **65.** 1 **67.** 2

69. Answers will vary.

71. True. Many different techniques can be used to verify identities.

73. The equation is not an identity because $\sin \theta = \pm\sqrt{1 - \cos^2 \theta}$.

Possible answer: $\dfrac{7\pi}{4}$

75. The equation is not an identity because $1 - \cos^2 \theta = \sin^2 \theta$.

Possible answer: $-\dfrac{\pi}{2}$

77. The equation is not an identity because $1 + \tan^2 \theta = \sec^2 \theta$.

Possible answer: $\dfrac{\pi}{6}$

Section 2.3 *(page 248)*

1. isolate **3.** quadratic **5–9.** Answers will vary.

11. $\dfrac{2\pi}{3} + 2n\pi, \dfrac{4\pi}{3} + 2n\pi$ **13.** $\dfrac{\pi}{3} + 2n\pi, \dfrac{2\pi}{3} + 2n\pi$

15. $\dfrac{\pi}{6} + n\pi, \dfrac{5\pi}{6} + n\pi$ **17.** $n\pi, \dfrac{3\pi}{2} + 2n\pi$

19. $\dfrac{\pi}{3} + n\pi, \dfrac{2\pi}{3} + n\pi$ **21.** $\dfrac{\pi}{8} + \dfrac{n\pi}{2}, \dfrac{3\pi}{8} + \dfrac{n\pi}{2}$

23. $\dfrac{n\pi}{3}, \dfrac{\pi}{4} + n\pi$ **25.** $0, \dfrac{\pi}{2}, \pi, \dfrac{3\pi}{2}$

27. $0, \pi, \dfrac{\pi}{6}, \dfrac{5\pi}{6}, \dfrac{7\pi}{6}, \dfrac{11\pi}{6}$ **29.** $\dfrac{\pi}{3}, \dfrac{5\pi}{3}, \pi$ **31.** No solution

33. $\pi, \dfrac{\pi}{3}, \dfrac{5\pi}{3}$ **35.** $\dfrac{\pi}{6}, \dfrac{5\pi}{6}, \dfrac{7\pi}{6}, \dfrac{11\pi}{6}$ **37.** $\dfrac{\pi}{2}$

39. $\dfrac{\pi}{6} + n\pi, \dfrac{5\pi}{6} + n\pi$ **41.** $\dfrac{\pi}{12} + \dfrac{n\pi}{3}$

43. $\dfrac{\pi}{2} + 4n\pi, \dfrac{7\pi}{2} + 4n\pi$ **45.** $3 + 4n$

47. $-2 + 6n, 2 + 6n$ **49.** 2.678, 5.820 **51.** 1.047, 5.236

53. 0.860, 3.426 **55.** 0, 2.678, 3.142, 5.820

57. 0.983, 1.768, 4.124, 4.910

59. 0.3398, 0.8481, 2.2935, 2.8018

61. 1.9357, 2.7767, 5.0773, 5.9183

63. $\arctan(-4) + \pi$, $\arctan(-4) + 2\pi$, $\arctan 3$, $\arctan 3 + \pi$

65. $\dfrac{\pi}{4}, \dfrac{5\pi}{4}$, $\arctan 5$, $\arctan 5 + \pi$ **67.** $\dfrac{\pi}{3}, \dfrac{5\pi}{3}$

69. $\arctan\left(\frac{1}{3}\right)$, $\arctan\left(\frac{1}{3}\right) + \pi$, $\arctan\left(-\frac{1}{3}\right) + \pi$, $\arctan\left(-\frac{1}{3}\right) + 2\pi$

71. $\arccos\left(\frac{1}{4}\right)$, $2\pi - \arccos\left(\frac{1}{4}\right)$

73. $\dfrac{\pi}{2}$, $\arcsin\left(-\dfrac{1}{4}\right) + 2\pi$, $\arcsin\left(\dfrac{1}{4}\right) + \pi$

75. $-1.154, 0.534$ **77.** 1.110

79. (a)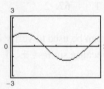

(b) $\dfrac{\pi}{3} \approx 1.0472$

$\dfrac{5\pi}{3} \approx 5.2360$

0

$\pi \approx 3.1416$

Maximum: $(1.0472, 1.25)$
Maximum: $(5.2360, 1.25)$
Minimum: $(0, 1)$
Minimum: $(3.1416, -1)$

81. (a)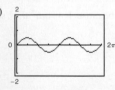

(b) $\dfrac{\pi}{4} \approx 0.7854$

$\dfrac{5\pi}{4} \approx 3.9270$

Maximum: $(0.7854, 1.4142)$
Minimum: $(3.9270, -1.4142)$

83. (a)

(b) $\dfrac{\pi}{4} \approx 0.7854$

$\dfrac{5\pi}{4} \approx 3.9270$

$\dfrac{3\pi}{4} \approx 2.3562$

Maximum: $(0.7854, 0.5)$ $\dfrac{7\pi}{4} \approx 5.4978$
Maximum: $(3.9270, 0.5)$
Minimum: $(2.3562, -0.5)$
Minimum: $(5.4978, -0.5)$

85. 1

87. (a) All real numbers x except $x = 0$
(b) y-axis symmetry; Horizontal asymptote: $y = 1$
(c) Oscillates   (d) Infinitely many solutions; $\dfrac{2}{2n\pi + \pi}$
(e) Yes, 0.6366

89. 0.04 sec, 0.43 sec, 0.83 sec

91. February, March, and April   **93.** $36.9°, 53.1°$

95. (a) $t = 8$ sec and $t = 24$ sec
(b) 5 times: $t = 16, 48, 80, 112, 144$ sec

97. (a)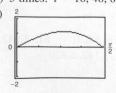

$A \approx 1.12$

(b) $0.6 < x < 1.1$

99. True. The first equation has a smaller period than the second equation, so it will have more solutions in the interval $[0, 2\pi)$.

101. The equation would become $\cos^2 x = 2$; this is not the correct method to use when solving equations.

103. Answers will vary.

Section 2.4   (page 256)

1. $\sin u \cos v - \cos u \sin v$ **3.** $\dfrac{\tan u + \tan v}{1 - \tan u \tan v}$

5. $\cos u \cos v + \sin u \sin v$

7. (a) $\dfrac{\sqrt{2} - \sqrt{6}}{4}$ (b) $\dfrac{\sqrt{2} + 1}{2}$

9. (a) $\dfrac{1}{2}$ (b) $\dfrac{-\sqrt{3} - 1}{2}$

11. (a) $\dfrac{\sqrt{6} + \sqrt{2}}{4}$ (b) $\dfrac{\sqrt{2} - \sqrt{3}}{2}$

13. $\sin \dfrac{11\pi}{12} = \dfrac{\sqrt{2}}{4}\left(\sqrt{3} - 1\right)$

$\cos \dfrac{11\pi}{12} = -\dfrac{\sqrt{2}}{4}\left(\sqrt{3} + 1\right)$

$\tan \dfrac{11\pi}{12} = -2 + \sqrt{3}$

15. $\sin \dfrac{17\pi}{12} = -\dfrac{\sqrt{2}}{4}\left(\sqrt{3} + 1\right)$

$\cos \dfrac{17\pi}{12} = \dfrac{\sqrt{2}}{4}\left(1 - \sqrt{3}\right)$

$\tan \dfrac{17\pi}{12} = 2 + \sqrt{3}$

17. $\sin 105° = \dfrac{\sqrt{2}}{4}\left(\sqrt{3} + 1\right)$

$\cos 105° = \dfrac{\sqrt{2}}{4}\left(1 - \sqrt{3}\right)$

$\tan 105° = -2 - \sqrt{3}$

19. $\sin 195° = \dfrac{\sqrt{2}}{4}\left(1 - \sqrt{3}\right)$

$\cos 195° = -\dfrac{\sqrt{2}}{4}\left(\sqrt{3} + 1\right)$

$\tan 195° = 2 - \sqrt{3}$

21. $\sin \dfrac{13\pi}{12} = \dfrac{\sqrt{2}}{4}\left(1 - \sqrt{3}\right)$

$\cos \dfrac{13\pi}{12} = -\dfrac{\sqrt{2}}{4}\left(1 + \sqrt{3}\right)$

$\tan \dfrac{13\pi}{12} = 2 - \sqrt{3}$

23. $\sin\left(-\dfrac{13\pi}{12}\right) = \dfrac{\sqrt{2}}{4}\left(\sqrt{3} - 1\right)$

$\cos\left(-\dfrac{13\pi}{12}\right) = -\dfrac{\sqrt{2}}{4}\left(\sqrt{3} + 1\right)$

$\tan\left(-\dfrac{13\pi}{12}\right) = -2 + \sqrt{3}$

25. $\sin 285° = -\dfrac{\sqrt{2}}{4}(\sqrt{3}+1)$

$\cos 285° = \dfrac{\sqrt{2}}{4}(\sqrt{3}-1)$

$\tan 285° = -(2+\sqrt{3})$

27. $\sin(-165°) = -\dfrac{\sqrt{2}}{4}(\sqrt{3}-1)$

$\cos(-165°) = -\dfrac{\sqrt{2}}{4}(1+\sqrt{3})$

$\tan(-165°) = 2-\sqrt{3}$

29. $\sin 1.8$ **31.** $\sin 75°$ **33.** $\tan 15°$ **35.** $\tan 3x$

37. $\dfrac{\sqrt{3}}{2}$ **39.** $\dfrac{\sqrt{3}}{2}$ **41.** $-\sqrt{3}$ **43.** $-\dfrac{63}{65}$ **45.** $\dfrac{16}{65}$

47. $-\dfrac{63}{16}$ **49.** $\dfrac{65}{56}$ **51.** $\dfrac{3}{5}$ **53.** $-\dfrac{44}{117}$ **55.** $-\dfrac{125}{44}$

57. 1 **59.** 0 **61–69.** Proofs   **71.** $-\sin x$

73. $-\cos\theta$ **75.** $\dfrac{\pi}{6}, \dfrac{5\pi}{6}$ **77.** $\dfrac{2\pi}{3}, \dfrac{4\pi}{3}$ **79.** $\dfrac{\pi}{3}, \dfrac{5\pi}{3}$

81. $\dfrac{5\pi}{4}, \dfrac{7\pi}{4}$ **83.** $0, \dfrac{\pi}{2}, \dfrac{3\pi}{2}$ **85.** $\dfrac{\pi}{4}, \dfrac{7\pi}{4}$ **87.** $\dfrac{\pi}{2}, \pi, \dfrac{3\pi}{2}$

89. (a) $y = \dfrac{5}{12}\sin(2t+0.6435)$ (b) $\dfrac{5}{12}$ ft   (c) $\dfrac{1}{\pi}$ cycle/sec

91. True. $\sin(u\pm v) = \sin u\cos v \pm \cos u\sin v$

93. False. $\tan\left(x-\dfrac{\pi}{4}\right) = \dfrac{\tan x - 1}{1+\tan x}$

95–97. Answers will vary.

99. (a) $\sqrt{2}\sin\left(\theta+\dfrac{\pi}{4}\right)$ (b) $\sqrt{2}\cos\left(\theta-\dfrac{\pi}{4}\right)$

101. (a) $13\sin(3\theta+0.3948)$ (b) $13\cos(3\theta-1.1760)$

103. $\sqrt{2}\sin\theta + \sqrt{2}\cos\theta$ **105.** Answers will vary.

107. $15°$

109.

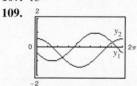

No, $y_1 \neq y_2$ because their graphs are different.

111. (a) and (b) Proofs

Section 2.5   (page 267)

1. $2\sin u\cos u$

3. $\cos^2 u - \sin^2 u = 2\cos^2 u - 1 = 1 - 2\sin^2 u$

5. $\pm\sqrt{\dfrac{1-\cos u}{2}}$ **7.** $\dfrac{1}{2}[\cos(u-v)+\cos(u+v)]$

9. $2\sin\left(\dfrac{u+v}{2}\right)\cos\left(\dfrac{u-v}{2}\right)$ **11.** $\dfrac{15}{17}$ **13.** $\dfrac{8}{15}$

15. $\dfrac{17}{8}$ **17.** $\dfrac{240}{289}$ **19.** $0, \dfrac{\pi}{3}, \pi, \dfrac{5\pi}{3}$

21. $\dfrac{\pi}{12}, \dfrac{5\pi}{12}, \dfrac{13\pi}{12}, \dfrac{17\pi}{12}$ **23.** $0, \dfrac{2\pi}{3}, \dfrac{4\pi}{3}$ **25.** $0, \dfrac{\pi}{2}, \pi, \dfrac{3\pi}{2}$

27. $\dfrac{\pi}{2}, \dfrac{\pi}{6}, \dfrac{5\pi}{6}, \dfrac{7\pi}{6}, \dfrac{3\pi}{2}, \dfrac{11\pi}{6}$ **29.** $3\sin 2x$

31. $3\cos 2x$ **33.** $4\cos 2x$ **35.** $\cos 2x$

37. $\sin 2u = -\dfrac{24}{25}, \cos 2u = \dfrac{7}{25}, \tan 2u = -\dfrac{24}{7}$

39. $\sin 2u = \dfrac{15}{17}, \cos 2u = \dfrac{8}{17}, \tan 2u = \dfrac{15}{8}$

41. $\sin 2u = -\dfrac{\sqrt{3}}{2}, \cos 2u = -\dfrac{1}{2}, \tan 2u = \sqrt{3}$

43. $\dfrac{1}{8}(3 + 4\cos 2x + \cos 4x)$ **45.** $\dfrac{1}{8}(3 + 4\cos 4x + \cos 8x)$

47. $\dfrac{(3 - 4\cos 4x + \cos 8x)}{(3 + 4\cos 4x + \cos 8x)}$ **49.** $\dfrac{1}{8}(1 - \cos 8x)$

51. $\dfrac{1}{16}(1 - \cos 2x - \cos 4x + \cos 2x\cos 4x)$

53. $\dfrac{4\sqrt{17}}{17}$ **55.** $\dfrac{1}{4}$ **57.** $\sqrt{17}$

59. $\sin 75° = \dfrac{1}{2}\sqrt{2+\sqrt{3}}$

$\cos 75° = \dfrac{1}{2}\sqrt{2-\sqrt{3}}$

$\tan 75° = 2 + \sqrt{3}$

61. $\sin 112°30' = \dfrac{1}{2}\sqrt{2+\sqrt{2}}$

$\cos 112°30' = -\dfrac{1}{2}\sqrt{2-\sqrt{2}}$

$\tan 112°30' = -1 - \sqrt{2}$

63. $\sin\dfrac{\pi}{8} = \dfrac{1}{2}\sqrt{2-\sqrt{2}}$ **65.** $\sin\dfrac{3\pi}{8} = \dfrac{1}{2}\sqrt{2+\sqrt{2}}$

$\cos\dfrac{\pi}{8} = \dfrac{1}{2}\sqrt{2+\sqrt{2}}$ $\cos\dfrac{3\pi}{8} = \dfrac{1}{2}\sqrt{2-\sqrt{2}}$

$\tan\dfrac{\pi}{8} = \sqrt{2} - 1$ $\tan\dfrac{3\pi}{8} = \sqrt{2} + 1$

67. (a) Quadrant I

(b) $\sin\dfrac{u}{2} = \dfrac{3}{5}, \cos\dfrac{u}{2} = \dfrac{4}{5}, \tan\dfrac{u}{2} = \dfrac{3}{4}$

69. (a) Quadrant II

(b) $\sin\dfrac{u}{2} = \dfrac{\sqrt{26}}{26}, \cos\dfrac{u}{2} = -\dfrac{5\sqrt{26}}{26}, \tan\dfrac{u}{2} = -\dfrac{1}{5}$

71. (a) Quadrant II

(b) $\sin\dfrac{u}{2} = \dfrac{3\sqrt{10}}{10}, \cos\dfrac{u}{2} = -\dfrac{\sqrt{10}}{10}, \tan\dfrac{u}{2} = -3$

73. $|\sin 3x|$ **75.** $-|\tan 4x|$

77. π **79.** $\dfrac{\pi}{3}, \pi, \dfrac{5\pi}{3}$

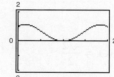

81. $\dfrac{1}{2}\left(\sin\dfrac{\pi}{2} + \sin\dfrac{\pi}{6}\right)$ **83.** $5(\cos 60° + \cos 90°)$

85. $\dfrac{1}{2}(\cos 2\theta - \cos 8\theta)$ **87.** $\dfrac{7}{2}(\sin(-2\beta) - \sin(-8\beta))$

89. $\dfrac{1}{2}(\cos 2y - \cos 2x)$ **91.** $2\sin 2\theta\cos\theta$

93. $2\cos 4x\cos 2x$ **95.** $2\cos\alpha\sin\beta$

97. $-2\sin\theta\sin\dfrac{\pi}{2} = -2\sin\theta$ **99.** $\dfrac{\sqrt{6}}{2}$ **101.** $-\sqrt{2}$

103. $0, \dfrac{\pi}{4}, \dfrac{\pi}{2}, \dfrac{3\pi}{4}, \pi, \dfrac{5\pi}{4}, \dfrac{3\pi}{2}, \dfrac{7\pi}{4}$

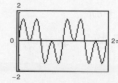

105. $\dfrac{\pi}{6}, \dfrac{5\pi}{6}$ **107.** $\dfrac{120}{169}$

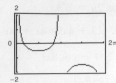

109. $\dfrac{3\sqrt{10}}{10}$ **111–123.** Answers will vary.

125. 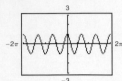 **127.**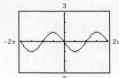

129. 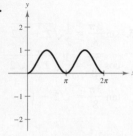 **131.** $2x\sqrt{1-x^2}$

133. $1 - 2x^2$ **135.** $23.85°$

137. (a) π (b) 0.4482 (c) 760 mi/h; 3420 mi/h

   (d) $\theta = 2\sin^{-1}\left(\dfrac{1}{M}\right)$

139. False. For $u < 0$,

   $\sin 2u = -\sin(-2u)$
   $= -2\sin(-u)\cos(-u)$
   $= -2(-\sin u)\cos u$
   $= 2\sin u \cos u.$

Review Exercises *(page 272)*

1. $\tan x$ **3.** $\cos x$ **5.** $|\csc x|$

7. $\tan x = \dfrac{5}{12}$ **9.** $\cos x = \dfrac{\sqrt{2}}{2}$

   $\csc x = \dfrac{13}{5}$ $\tan x = -1$

   $\sec x = \dfrac{13}{12}$ $\csc x = -\sqrt{2}$

   $\cot x = \dfrac{12}{5}$ $\sec x = \sqrt{2}$

   $\cot x = -1$

11. $\sin^2 x$ **13.** 1 **15.** $\cot\theta$ **17.** $\csc\theta$

19. $\cot^2 x$ **21.** $\sec x + 2\sin x$ **23.** $-2\tan^2\theta$

25. $5\cos\theta$ **27–35.** Answers will vary.

37. $\dfrac{\pi}{3} + 2n\pi, \dfrac{2\pi}{3} + 2n\pi$ **39.** $\dfrac{\pi}{6} + n\pi$

41. $\dfrac{\pi}{3} + n\pi, \dfrac{2\pi}{3} + n\pi$ **43.** $0, \dfrac{2\pi}{3}, \dfrac{4\pi}{3}$ **45.** $0, \dfrac{\pi}{2}, \pi$

47. $\dfrac{\pi}{8}, \dfrac{3\pi}{8}, \dfrac{9\pi}{8}, \dfrac{11\pi}{8}$ **49.** $\dfrac{\pi}{2}$

51. $0, \dfrac{\pi}{8}, \dfrac{3\pi}{8}, \dfrac{5\pi}{8}, \dfrac{7\pi}{8}, \dfrac{9\pi}{8}, \dfrac{11\pi}{8}, \dfrac{13\pi}{8}, \dfrac{15\pi}{8}$ **53.** $0, \pi$

55. $\arctan(-3) + \pi, \arctan(-3) + 2\pi, \arctan 2, \arctan 2 + \pi$

57. $\sin 285° = -\dfrac{\sqrt{2}}{4}\left(\sqrt{3} + 1\right)$

   $\cos 285° = \dfrac{\sqrt{2}}{4}\left(\sqrt{3} - 1\right)$

   $\tan 285° = -2 - \sqrt{3}$

59. $\sin\dfrac{25\pi}{12} = \dfrac{\sqrt{2}}{4}\left(\sqrt{3} - 1\right)$

   $\cos\dfrac{25\pi}{12} = \dfrac{\sqrt{2}}{4}\left(\sqrt{3} + 1\right)$

   $\tan\dfrac{25\pi}{12} = 2 - \sqrt{3}$

61. $\sin 15°$ **63.** $\tan 35°$ **65.** $-\dfrac{24}{25}$ **67.** -1

69. $-\dfrac{7}{25}$ **71–75.** Answers will vary.

77. $\dfrac{\pi}{4}, \dfrac{7\pi}{4}$ **79.** $\dfrac{\pi}{6}, \dfrac{11\pi}{6}$

81. $\sin 2u = \dfrac{24}{25}$

   $\cos 2u = -\dfrac{7}{25}$

   $\tan 2u = -\dfrac{24}{7}$

83. $\sin 2u = -\dfrac{4\sqrt{2}}{9}, \cos 2u = -\dfrac{7}{9}, \tan 2u = \dfrac{4\sqrt{2}}{7}$

85. 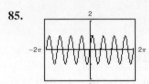 **87.** $\dfrac{1 - \cos 4x}{1 + \cos 4x}$

89. $\dfrac{3 - 4\cos 2x + \cos 4x}{4(1 + \cos 2x)}$

91. $\sin(-75°) = -\dfrac{1}{2}\sqrt{2 + \sqrt{3}}$

   $\cos(-75°) = \dfrac{1}{2}\sqrt{2 - \sqrt{3}}$

   $\tan(-75°) = -2 - \sqrt{3}$

93. $\sin\dfrac{19\pi}{12} = -\dfrac{1}{2}\sqrt{2 + \sqrt{3}}$

   $\cos\dfrac{19\pi}{12} = \dfrac{1}{2}\sqrt{2 - \sqrt{3}}$

   $\tan\dfrac{19\pi}{12} = -2 - \sqrt{3}$

95. (a) Quadrant I

   (b) $\sin\dfrac{u}{2} = \dfrac{\sqrt{2}}{10}, \cos\dfrac{u}{2} = \dfrac{7\sqrt{2}}{10}, \tan\dfrac{u}{2} = \dfrac{1}{7}$

97. (a) Quadrant I

   (b) $\sin\dfrac{u}{2} = \dfrac{3\sqrt{14}}{14}, \cos\dfrac{u}{2} = \dfrac{\sqrt{70}}{14}, \tan\dfrac{u}{2} = \dfrac{3\sqrt{5}}{5}$

99. $-|\cos 5x|$ **101.** $\dfrac{1}{2}\left(\sin\dfrac{\pi}{3} - \sin 0\right) = \dfrac{1}{2}\sin\dfrac{\pi}{3}$

103. $\dfrac{1}{2}[\sin 10\theta - \sin(-2\theta)]$ **105.** $2\cos 6\theta\sin(-2\theta)$

107. $-2\sin x\sin\dfrac{\pi}{6}$ **109.** $\theta = 15°$ or $\dfrac{\pi}{12}$

111.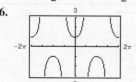

113. $\frac{1}{2}\sqrt{10}$ ft

115. False. If $(\pi/2) < \theta < \pi$, then $\cos(\theta/2) > 0$. The sign of $\cos(\theta/2)$ depends on the quadrant in which $\theta/2$ lies.

117. True. $4\sin(-x)\cos(-x) = 4(-\sin x)\cos x$
$$= -4\sin x \cos x$$
$$= -2(2\sin x \cos x)$$
$$= -2\sin 2x$$

119. Reciprocal identities:

$$\sin\theta = \frac{1}{\csc\theta}, \ \cos\theta = \frac{1}{\sec\theta}, \ \tan\theta = \frac{1}{\cot\theta},$$

$$\csc\theta = \frac{1}{\sin\theta}, \ \sec\theta = \frac{1}{\cos\theta}, \ \cot\theta = \frac{1}{\tan\theta}$$

Quotient identities: $\tan\theta = \dfrac{\sin\theta}{\cos\theta}, \ \cot\theta = \dfrac{\cos\theta}{\sin\theta}$

Pythagorean identities: $\sin^2\theta + \cos^2\theta = 1$,
$1 + \tan^2\theta = \sec^2\theta, \ 1 + \cot^2\theta = \csc^2\theta$

121. $-1 \le \sin x \le 1$ for all x **123.** $y_1 = y_2 + 1$

125. $-1.8431, 2.1758, 3.9903, 8.8935, 9.8820$

Chapter Test   *(page 275)*

1. $\sin\theta = -\dfrac{6\sqrt{61}}{61}$ $\csc\theta = -\dfrac{\sqrt{61}}{6}$

$\cos\theta = -\dfrac{5\sqrt{61}}{61}$ $\sec\theta = -\dfrac{\sqrt{61}}{5}$

$\tan\theta = \dfrac{6}{5}$ $\cot\theta = \dfrac{5}{6}$

2. 1 **3.** 1 **4.** $\csc\theta\sec\theta$

5. $\theta = 0, \dfrac{\pi}{2} < \theta \le \pi, \dfrac{3\pi}{2} < \theta < 2\pi$

6.

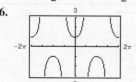 $y_1 = y_2$

7–12. Answers will vary.   **13.** $\frac{1}{8}(3 - 4\cos x + \cos 2x)$

14. $\tan 2\theta$ **15.** $2(\sin 5\theta + \sin \theta)$ **16.** $-2\sin 2\theta \sin\theta$

17. $0, \dfrac{3\pi}{4}, \pi, \dfrac{7\pi}{4}$ **18.** $\dfrac{\pi}{6}, \dfrac{\pi}{2}, \dfrac{5\pi}{6}, \dfrac{3\pi}{2}$ **19.** $\dfrac{\pi}{6}, \dfrac{5\pi}{6}, \dfrac{7\pi}{6}, \dfrac{11\pi}{6}$

20. $\dfrac{\pi}{6}, \dfrac{5\pi}{6}, \dfrac{3\pi}{2}$ **21.** $-2.596, 0, 2.596$ **22.** $\dfrac{\sqrt{2} - \sqrt{6}}{4}$

23. $\sin 2u = -\dfrac{20}{29}, \cos 2u = -\dfrac{21}{29}, \tan 2u = \dfrac{20}{21}$

24. Day 123 to day 223

25. $t = 0.26$ min
0.58 min
0.89 min
1.20 min
1.52 min
1.83 min

Problem Solving   *(page 279)*

1. (a) $\cos\theta = \pm\sqrt{1 - \sin^2\theta}$

$\tan\theta = \pm\dfrac{\sin\theta}{\sqrt{1 - \sin^2\theta}}$

$\cot\theta = \pm\dfrac{\sqrt{1 - \sin^2\theta}}{\sin\theta}$

$\sec\theta = \pm\dfrac{1}{\sqrt{1 - \sin^2\theta}}$

$\csc\theta = \dfrac{1}{\sin\theta}$

(b) $\sin\theta = \pm\sqrt{1 - \cos^2\theta}$

$\tan\theta = \pm\dfrac{\sqrt{1 - \cos^2\theta}}{\cos\theta}$

$\csc\theta = \pm\dfrac{1}{\sqrt{1 - \cos^2\theta}}$

$\sec\theta = \dfrac{1}{\cos\theta}$

$\cot\theta = \pm\dfrac{\cos\theta}{\sqrt{1 - \cos^2\theta}}$

3. Answers will vary.

5. $u + v = w$

7. $\sin\dfrac{\theta}{2} = \sqrt{\dfrac{1 - \cos\theta}{2}}$

$\cos\dfrac{\theta}{2} = \sqrt{\dfrac{1 + \cos\theta}{2}}$

$\tan\dfrac{\theta}{2} = \dfrac{\sin\theta}{1 + \cos\theta}$

9. (a)

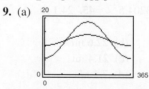

(b) $t \approx 91$ (April 1), $t \approx 274$ (October 1)

(c) Seward; The amplitudes: 6.4 and 1.9

(d) 365.2 days

11. (a) $\dfrac{\pi}{6} \le x \le \dfrac{5\pi}{6}$ (b) $\dfrac{2\pi}{3} \le x \le \dfrac{4\pi}{3}$

(c) $\dfrac{\pi}{2} < x < \pi, \dfrac{3\pi}{2} < x < 2\pi$

(d) $0 \le x \le \dfrac{\pi}{4}, \dfrac{5\pi}{4} \le x \le 2\pi$

13. (a) $\sin(u + v + w)$
$= \sin u \cos v \cos w - \sin u \sin v \sin w$
$+ \cos u \sin v \cos w + \cos u \cos v \sin w$

(b) $\tan(u + v + w)$

$= \dfrac{\tan u + \tan v + \tan w - \tan u \tan v \tan w}{1 - \tan u \tan v - \tan u \tan w - \tan v \tan w}$

15. (a)

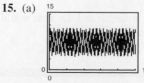   (b) 233.3 times/sec

CHAPTER 2

Chapter 3

Section 3.1 *(page 288)*

1. oblique   3. angles; side
5. $A = 30°, a \approx 14.14, c \approx 27.32$
7. $C = 120°, b \approx 4.75, c \approx 7.17$
9. $B = 60.9°, b \approx 19.32, c \approx 6.36$
11. $B = 42°4', a \approx 22.05, b \approx 14.88$
13. $C = 80°, a \approx 5.82, b \approx 9.20$
15. $C = 83°, a \approx 0.62, b \approx 0.51$
17. $B \approx 21.55°, C \approx 122.45°, c \approx 11.49$
19. $A \approx 10°11', C \approx 154°19', c \approx 11.03$
21. $B \approx 9.43°, C = 25.57°, c \approx 10.53$
23. $B \approx 18°13', C \approx 51°32', c \approx 40.06$
25. $B \approx 48.74°, C \approx 21.26°, c \approx 48.23$
27. No solution
29. Two solutions:
   $B \approx 72.21°, C \approx 49.79°, c \approx 10.27$
   $B \approx 107.79°, C \approx 14.21°, c \approx 3.30$
31. No solution   33. $B = 45°, C = 90°, c \approx 1.41$

35. (a) $b \le 5, b = \dfrac{5}{\sin 36°}$ (b) $5 < b < \dfrac{5}{\sin 36°}$

   (c) $b > \dfrac{5}{\sin 36°}$

37. (a) $b \le 10.8, b = \dfrac{10.8}{\sin 10°}$ (b) $10.8 < b < \dfrac{10.8}{\sin 10°}$

   (c) $b > \dfrac{10.8}{\sin 10°}$

39. 10.4   41. 1675.2   43. 3204.5   45. 24.1 m
47. 16.1°   49. 77 m
51. (a)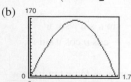
   (b) 22.6 mi
   (c) 21.4 mi
   (d) 7.3 mi

53. 3.2 mi   55. 5.86 mi
57. True. If an angle of a triangle is obtuse (greater than 90°), then the other two angles must be acute and therefore less than 90°. The triangle is oblique.
59. False. If just three angles are known, the triangle cannot be solved.
61. (a) $A = 20\left(15 \sin \dfrac{3\theta}{2} - 4 \sin \dfrac{\theta}{2} - 6 \sin \theta\right)$

   (b)

   [graph with vertical axis labeled 170, horizontal axis from 0 to 1.7]

   (c) Domain: $0 \le \theta \le 1.6690$
   The domain would increase in length and the area would have a greater maximum value.

Section 3.2 *(page 295)*

1. Cosines   3. $b^2 = a^2 + c^2 - 2ac \cos B$
5. $A \approx 38.62°, B \approx 48.51°, C \approx 92.87°$
7. $B \approx 23.79°, C \approx 126.21°, a \approx 18.59$
9. $A \approx 30.11°, B \approx 43.16°, C \approx 106.73°$
11. $A \approx 92.94°, B \approx 43.53°, C \approx 43.53°$
13. $B \approx 27.46°, C \approx 32.54°, a \approx 11.27$
15. $A \approx 141°45', C \approx 27°40', b \approx 11.87$
17. $A \approx 27°10', C \approx 27°10', b \approx 65.84$
19. $A \approx 33.80°, B \approx 103.20°, c \approx 0.54$

	a	b	c	d	θ	ϕ
21.	5	8	12.07	5.69	45°	135°
23.	10	14	20	13.86	68.2°	111.8°
25.	15	16.96	25	20	77.2°	102.8°

27. Law of Cosines; $A \approx 102.44°, C \approx 37.56°, b \approx 5.26$
29. Law of Sines; No solution
31. Law of Sines; $C = 103°, a \approx 0.82, b \approx 0.71$
33. 43.52   35. 10.4   37. 52.11   39. 0.18
41.

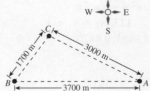

N 37.1° E, S 63.1° E

43. 373.3 m   45. 72.3°   47. 43.3 mi
49. (a) N 58.4° W   (b) S 81.5° W   51. 63.7 ft
53. 24.2 mi   55. $\overline{PQ} \approx 9.4, \overline{QS} = 5, \overline{RS} \approx 12.8$
57.

d (inches)	9	10	12	13	14
θ (degrees)	60.9°	69.5°	88.0°	98.2°	109.6°
s (inches)	20.88	20.28	18.99	18.28	17.48

d (inches)	15	16
θ (degrees)	122.9°	139.8°
s (inches)	16.55	15.37

59. 46,837.5 ft^2 61. \$83,336.37
63. False. For s to be the average of the lengths of the three sides of the triangle, s would be equal to $(a + b + c)/3$.
65. No. The three side lengths do not form a triangle.
67. (a) and (b) Proofs   69. 405.2 ft
71. Either; Because A is obtuse, there is only one solution for B or C.
73. The Law of Cosines can be used to solve the single-solution case of SSA. There is no method that can solve the no-solution case of SSA.
75. Proof

Section 3.3 *(page 308)*

1. directed line segment   3. magnitude
5. magnitude; direction   7. unit vector   9. resultant

11. $\|\mathbf{u}\| = \|\mathbf{v}\| = \sqrt{17}$, slope$_\mathbf{u}$ = slope$_\mathbf{v}$ = $\frac{1}{4}$

u and **v** have the same magnitude and direction, so they are equal.

13. $\mathbf{v} = \langle 1, 3 \rangle$, $\|\mathbf{v}\| = \sqrt{10}$ **15.** $\mathbf{v} = \langle 4, 6 \rangle$; $\|\mathbf{v}\| = 2\sqrt{13}$

17. $\mathbf{v} = \langle 0, 5 \rangle$; $\|\mathbf{v}\| = 5$ **19.** $\mathbf{v} = \langle 8, 6 \rangle$; $\|\mathbf{v}\| = 10$

21. $\mathbf{v} = \langle -9, -12 \rangle$; $\|\mathbf{v}\| = 15$ **23.** $\mathbf{v} = \langle 16, 7 \rangle$; $\|\mathbf{v}\| = \sqrt{305}$

25. 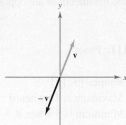 **27.**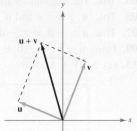

29.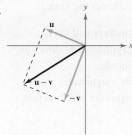

31. (a) $\langle 3, 4 \rangle$ 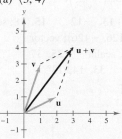 (b) $\langle 1, -2 \rangle$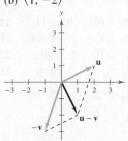

(c) $\langle 1, -7 \rangle$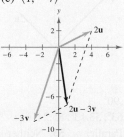

33. (a) $\langle -5, 3 \rangle$ 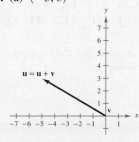 (b) $\langle -5, 3 \rangle$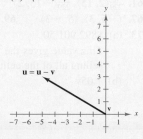

(c) $\langle -10, 6 \rangle$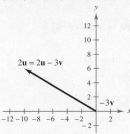

35. (a) $3\mathbf{i} - 2\mathbf{j}$ 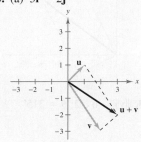 (b) $-\mathbf{i} + 4\mathbf{j}$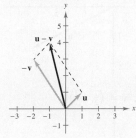

(c) $-4\mathbf{i} + 11\mathbf{j}$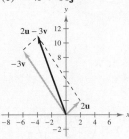

37. (a) $2\mathbf{i} + \mathbf{j}$ 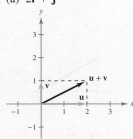 (b) $2\mathbf{i} - \mathbf{j}$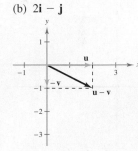

(c) $4\mathbf{i} - 3\mathbf{j}$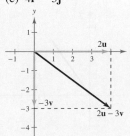

CHAPTER 3

39. $\langle 1, 0 \rangle$ **41.** $\left\langle -\dfrac{\sqrt{2}}{2}, \dfrac{\sqrt{2}}{2} \right\rangle$ **43.** $\dfrac{\sqrt{2}}{2}\mathbf{i} + \dfrac{\sqrt{2}}{2}\mathbf{j}$

45. $\mathbf{j}$ **47.** $\dfrac{\sqrt{5}}{5}\mathbf{i} - \dfrac{2\sqrt{5}}{5}\mathbf{j}$ **49.** $\mathbf{v} = \langle -6, 8 \rangle$

51. $\mathbf{v} = \left\langle \dfrac{18\sqrt{29}}{29}, \dfrac{45\sqrt{29}}{29} \right\rangle$ **53.** $5\mathbf{i} - 3\mathbf{j}$

55. $6\mathbf{i} - 3\mathbf{j}$

57. $\mathbf{v} = \left\langle 3, -\dfrac{3}{2} \right\rangle$ **59.** $\mathbf{v} = \langle 4, 3 \rangle$

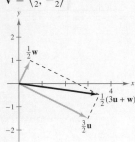   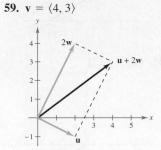

61. $\mathbf{v} = \left\langle \dfrac{7}{2}, -\dfrac{1}{2} \right\rangle$ **63.** $\|\mathbf{v}\| = 6\sqrt{2};\ \theta = 315°$

65. $\|\mathbf{v}\| = 3;\ \theta = 60°$

67. $\mathbf{v} = \langle 3, 0 \rangle$ **69.** $\mathbf{v} = \left\langle -\dfrac{7\sqrt{3}}{4}, \dfrac{7}{4} \right\rangle$

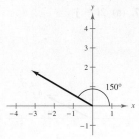

71. $\mathbf{v} = \langle \sqrt{6}, \sqrt{6} \rangle$ **73.** $\mathbf{v} = \left\langle \dfrac{9}{5}, \dfrac{12}{5} \right\rangle$

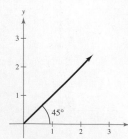   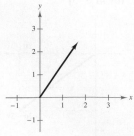

75. $\langle 5, 5 \rangle$ **77.** $\langle 10\sqrt{2} - 50, 10\sqrt{2} \rangle$ **79.** $90°$

81. $62.7°$

83. Vertical ≈ 125.4 ft/sec, Horizontal ≈ 1193.4 ft/sec

85. $12.8°;\ 398.32$ N   **87.** $71.3°;\ 228.5$ lb   **89.** 17.5 lb

91. $T_{AC} \approx 1758.8$ lb
$T_{BC} \approx 1305.4$ lb

93. 3154.4 lb   **95.** 20.8 lb   **97.** $19.5°$

99. 1928.4 ft-lb   **101.** N $21.4°$ E; 138.7 km/h

103. True. The magnitudes are equal and the directions are opposite.

105. True. $\mathbf{a} - \mathbf{b} = \mathbf{c}$ and $\mathbf{u} = -\mathbf{b}$

107. True. $\mathbf{a} = -\mathbf{d},\ \mathbf{w} = -\mathbf{d}$

109. False. $\mathbf{u} - \mathbf{v} = -(\mathbf{b} + \mathbf{t})$ **111.** Proof

113. (a) $5\sqrt{5 + 4\cos\theta}$

(b)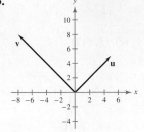

(c) Range: $[5, 15]$
Maximum is 15 when $\theta = 0$.
Minimum is 5 when $\theta = \pi$.

(d) The magnitudes of $\mathbf{F}_1$ and $\mathbf{F}_2$ are not the same.

115. $\langle 1, 3 \rangle$ or $\langle -1, -3 \rangle$ **117.** Answers will vary.

119. (a) Vector. Velocity has both magnitude and direction.

(b) Scalar. Price has only magnitude.

(c) Scalar. Temperature has only magnitude.

(d) Vector. Weight has both magnitude and direction.

Section 3.4   *(page 319)*

1. dot product   **3.** $\dfrac{\mathbf{u} \cdot \mathbf{v}}{\|\mathbf{u}\|\,\|\mathbf{v}\|}$ **5.** $\left(\dfrac{\mathbf{u} \cdot \mathbf{v}}{\|\mathbf{v}\|^2} \right)\mathbf{v}$

7. -19 **9.** -11 **11.** 6 **13.** -12 **15.** 18; scalar

17. $\langle 24, -12 \rangle$; vector   **19.** $\langle -126, -126 \rangle$; vector

21. $\sqrt{10} - 1$; scalar   **23.** -12; scalar   **25.** 17

27. $5\sqrt{41}$ **29.** 6 **31.** $90°$ **33.** $143.13°$

35. $60.26°$ **37.** $90°$ **39.** $\dfrac{5\pi}{12}$

41. 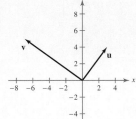   **43.**

About $91.33°$ 90°

45. $26.57°, 63.43°, 90°$ **47.** $41.63°, 53.13°, 85.24°$

49. -20 **51.** -229.1 **53.** Parallel   **55.** Neither

57. Orthogonal   **59.** $\dfrac{1}{37}\langle 84, 14 \rangle, \dfrac{1}{37}\langle -10, 60 \rangle$

61. $\dfrac{45}{229}\langle 2, 15 \rangle, \dfrac{6}{229}\langle -15, 2 \rangle$ **63.** $\langle 3, 2 \rangle$ **65.** $\langle 0, 0 \rangle$

67. $\langle -5, 3 \rangle, \langle 5, -3 \rangle$ **69.** $\dfrac{2}{3}\mathbf{i} + \dfrac{1}{2}\mathbf{j}, -\dfrac{2}{3}\mathbf{i} - \dfrac{1}{2}\mathbf{j}$ **71.** 32

73. (a) $\$892,901.50$

This value gives the total revenue that can be earned by selling all of the cellular phones.

(b) $1.05\mathbf{v}$

75. (a) Force = 30,000 sin d
(b)

d	0°	1°	2°	3°	4°	5°
Force	0	523.6	1047.0	1570.1	2092.7	2614.7

d	6°	7°	8°	9°	10°
Force	3135.9	3656.1	4175.2	4693.0	5209.4

(c) 29,885.8 lb
77. 735 N-m   **79.** 779.4 ft-lb   **81.** 21,650.64 ft-lb
83–85. Answers will vary.
87. False. Work is represented by a scalar.
89. Proof
91. (a) **u** and **v** are parallel.   (b) **u** and **v** are orthogonal.
93. Proof

Review Exercises *(page 324)*

1. $C = 72°$, $b \approx 12.21$, $c \approx 12.36$
3. $A = 26°$, $a \approx 24.89$, $c \approx 56.23$
5. $C = 66°$, $a \approx 2.53$, $b \approx 9.11$
7. $B = 108°$, $a \approx 11.76$, $c \approx 21.49$
9. $A \approx 20.41°$, $C \approx 9.59°$, $a \approx 20.92$
11. $B \approx 39.48°$, $C \approx 65.52°$, $c \approx 48.24$
13. 19.06   **15.** 47.23   **17.** 31.1 m   **19.** 31.01 ft
21. $A \approx 27.81°$, $B \approx 54.75°$, $C \approx 97.44°$
23. $A \approx 16.99°$, $B \approx 26.00°$, $C \approx 137.01°$
25. $A \approx 29.92°$, $B \approx 86.18°$, $C \approx 63.90°$
27. $A = 36°$, $C = 36°$, $b \approx 17.80$
29. $A \approx 45.76°$, $B \approx 91.24°$, $c \approx 21.42$
31. Law of Sines; $A \approx 77.52°$, $B \approx 38.48°$, $a \approx 14.12$
33. Law of Cosines; $A \approx 28.62°$, $B \approx 33.56°$, $C \approx 117.82°$
35. About 4.3 ft, about 12.6 ft   **37.** 615.1 m
39. 7.64   **41.** 8.36
43. $\|\mathbf{u}\| = \|\mathbf{v}\| = \sqrt{61}$, slope$_\mathbf{u}$ = slope$_\mathbf{v}$ = $\frac{5}{6}$
45. $\langle 7, -5\rangle$ **47.** $\langle 7, -7\rangle$ **49.** $\langle -4, 4\sqrt{3}\rangle$
51. (a) $\langle -4, 3\rangle$ (b) $\langle 2, -9\rangle$
(c) $\langle -4, -12\rangle$ (d) $\langle -14, 3\rangle$
53. (a) $\langle -1, 6\rangle$ (b) $\langle -9, -2\rangle$
(c) $\langle -20, 8\rangle$ (d) $\langle -13, 22\rangle$
55. (a) $7\mathbf{i} + 2\mathbf{j}$ (b) $-3\mathbf{i} - 4\mathbf{j}$ (c) $8\mathbf{i} - 4\mathbf{j}$ (d) $25\mathbf{i} + 4\mathbf{j}$
57. (a) $3\mathbf{i} + 6\mathbf{j}$ (b) $5\mathbf{i} - 6\mathbf{j}$ (c) $16\mathbf{i}$ (d) $17\mathbf{i} + 18\mathbf{j}$
59. $\langle 22, -7\rangle$ **61.** $\langle 30, 9\rangle$

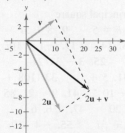

63. $-\mathbf{i} + 5\mathbf{j}$ **65.** $6\mathbf{i} + 4\mathbf{j}$
67. $10\sqrt{2}(\cos 135°\mathbf{i} + \sin 135°\mathbf{j})$ **69.** $\|\mathbf{v}\| = 7$; $\theta = 60°$
71. $\|\mathbf{v}\| = \sqrt{41}$; $\theta = 38.7°$ **73.** $\|\mathbf{v}\| = 3\sqrt{2}$; $\theta = 225°$
75. 115.5 lb each   **77.** 422.30 mi/h; 130.4°
79. 45   **81.** -2 **83.** 40; scalar
85. $4 - 2\sqrt{5}$; scalar   **87.** $\langle 72, -36\rangle$; vector
89. 38; scalar   **91.** $\frac{11\pi}{12}$ **93.** 160.5°   **95.** Orthogonal
97. Neither   **99.** $-\frac{13}{17}\langle 4, 1\rangle$, $\frac{16}{17}\langle -1, 4\rangle$
101. $\frac{5}{2}\langle -1, 1\rangle$, $\frac{9}{2}\langle 1, 1\rangle$ **103.** 48   **105.** 72,000 ft-lb
107. True. Sin 90° is defined in the Law of Sines.
109. True. By definition, $\mathbf{u} = \dfrac{\mathbf{v}}{\|\mathbf{v}\|}$, so $\mathbf{v} = \|\mathbf{v}\|\mathbf{u}$.
111. $\dfrac{a}{\sin A} = \dfrac{b}{\sin B} = \dfrac{c}{\sin C}$ **113.** Direction and magnitude
115. a; The angle between the vectors is acute.
117. The diagonal of the parallelogram with **u** and **v** as its adjacent sides

Chapter Test *(page 328)*

1. $C = 88°$, $b \approx 27.81$, $c \approx 29.98$
2. $A = 42°$, $b \approx 21.91$, $c \approx 10.95$
3. Two solutions:
$B \approx 29.12°$, $C \approx 126.88°$, $c \approx 22.03$
$B \approx 150.88°$, $C \approx 5.12°$, $c \approx 2.46$
4. No solution   **5.** $A \approx 39.96°$, $C \approx 40.04°$, $c \approx 15.02$
6. $A \approx 21.90°$, $B \approx 37.10°$, $c \approx 78.15$ **7.** 2052.5 m^2
8. 606.3 mi; 29.1°   **9.** $\langle 14, -23\rangle$
10. $\left\langle \dfrac{18\sqrt{34}}{17}, -\dfrac{30\sqrt{34}}{17}\right\rangle$
11. $\langle -4, 12\rangle$ **12.** $\langle 8, 2\rangle$

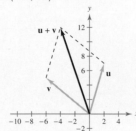   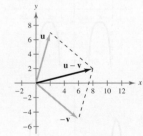

13. $\langle 28, 20\rangle$ **14.** $\langle -4, 38\rangle$

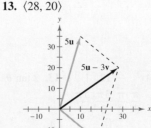   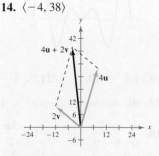

15. $\langle \frac{24}{25}, -\frac{7}{25}\rangle$ **16.** 14.9°; 250.15 lb   **17.** 135°
18. Yes   **19.** $\frac{37}{26}\langle 5, 1\rangle$; $\frac{29}{26}\langle -1, 5\rangle$ **20.** About 104 lb

CHAPTER 3

Cumulative Test for Chapters 1–3 *(page 329)*

1. (a)

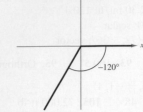

(b) $240°$

(c) $-\dfrac{2\pi}{3}$

(d) $60°$

(e) $\sin(-120°) = -\dfrac{\sqrt{3}}{2}$ $\quad \csc(-120°) = -\dfrac{2\sqrt{3}}{3}$

$\cos(-120°) = -\dfrac{1}{2}$ $\quad \sec(-120°) = -2$

$\tan(-120°) = \sqrt{3}$ $\quad \cot(-120°) = \dfrac{\sqrt{3}}{3}$

2. $-83.1°$ **3.** $\dfrac{20}{29}$

4.

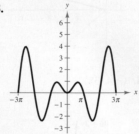

5.

6.

7. $a = -3, b = \pi, c = 0$

8.

9. 4.9 **10.** $\dfrac{3}{4}$ **11.** $\sqrt{1 - 4x^2}$ **12.** 1 **13.** $2 \tan \theta$

14–16. Answers will vary. **17.** $\dfrac{\pi}{3}, \dfrac{\pi}{2}, \dfrac{3\pi}{2}, \dfrac{5\pi}{3}$

18. $\dfrac{\pi}{6}, \dfrac{5\pi}{6}, \dfrac{7\pi}{6}, \dfrac{11\pi}{6}$ **19.** $\dfrac{3\pi}{2}$ **20.** $\dfrac{16}{63}$ **21.** $\dfrac{4}{3}$

22. $\dfrac{\sqrt{5}}{5}, \dfrac{2\sqrt{5}}{5}$ **23.** $\dfrac{5}{2}\left(\sin\dfrac{5\pi}{2} - \sin \pi\right)$

24. $-2 \sin 8x \sin x$ **25.** $B \approx 26.39°, C \approx 123.61°, c \approx 14.99$

26. $B \approx 52.48°, C \approx 97.52°, a \approx 5.04$

27. $B = 60°, a \approx 5.77, c \approx 11.55$

28. $A \approx 26.28°, B \approx 49.74°, C \approx 103.98°$

29. Law of Sines; $C = 109°, a \approx 14.96, b \approx 9.27$

30. Law of Cosines; $A \approx 6.75°, B \approx 93.25°, c \approx 9.86$

31. 41.48 in.^{2} **32.** 599.09 m^2 **33.** $7\mathbf{i} + 8\mathbf{j}$

34. $\left\langle \dfrac{\sqrt{2}}{2}, \dfrac{\sqrt{2}}{2}\right\rangle$ **35.** -5 **36.** $-\dfrac{1}{13}\langle 1, 5\rangle; \dfrac{21}{13}\langle 5, -1\rangle$

37. About 395.8 rad/min; about 8312.7 in./min

38. 42π yd$^2 \approx 131.95$ yd^2 **39.** 5 ft **40.** $22.6°$

41. $d = 4 \cos \dfrac{\pi}{4}t$ **42.** $32.6°; 543.9$ km/h **43.** 425 ft-lb

Problem Solving *(page 335)*

1. 2.01 ft

3. (a)

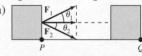

(b) Station A: 27.45 mi; Station B: 53.03 mi

(c) 11.03 mi; S $21.7°$ E

5. (a) (i) $\sqrt{2}$ (ii) $\sqrt{5}$ (iii) 1

(iv) 1 (v) 1 (vi) 1

(b) (i) 1 (ii) $3\sqrt{2}$ (iii) $\sqrt{13}$

(iv) 1 (v) 1 (vi) 1

(c) (i) $\dfrac{\sqrt{5}}{2}$ (ii) $\sqrt{13}$ (iii) $\dfrac{\sqrt{85}}{2}$

(iv) 1 (v) 1 (vi) 1

(d) (i) $2\sqrt{5}$ (ii) $5\sqrt{2}$ (iii) $5\sqrt{2}$

(iv) 1 (v) 1 (vi) 1

7. $\mathbf{w} = \dfrac{1}{2}(\mathbf{u} + \mathbf{v}); \mathbf{w} = \dfrac{1}{2}(\mathbf{v} - \mathbf{u})$

9. (a)

The amount of work done by $\mathbf{F}_1$ is equal to the amount of work done by $\mathbf{F}_2$.

(b)

The amount of work done by $\mathbf{F}_2$ is $\sqrt{3}$ times as great as the amount of work done by $\mathbf{F}_1$.

Chapter 4

Section 4.1 *(page 343)*

1. (a) iii (b) i (c) ii **3.** principal square

5. $a = -12, b = 7$ **7.** $a = 6, b = 5$ **9.** $8 + 5i$

11. $2 - 3\sqrt{3}i$ **13.** $4\sqrt{5}i$ **15.** 14 **17.** $-1 - 10i$

19. $0.3i$ **21.** $10 - 3i$ **23.** 1 **25.** $3 - 3\sqrt{2}i$

27. $-14 + 20i$ **29.** $\dfrac{1}{6} + \dfrac{7}{6}i$ **31.** $5 + i$ **33.** $108 + 12i$

35. 24 **37.** $-13 + 84i$ **39.** -10 **41.** $9 - 2i, 85$

43. $-1 + \sqrt{5}i, 6$ **45.** $-2\sqrt{5}i, 20$ **47.** $\sqrt{6}, 6$

49. $-3i$ **51.** $\dfrac{8}{41} + \dfrac{10}{41}i$ **53.** $\dfrac{12}{13} + \dfrac{5}{13}i$ **55.** $-4 - 9i$

57. $-\dfrac{120}{1681} - \dfrac{27}{1681}i$ **59.** $-\dfrac{1}{2} - \dfrac{5}{2}i$ **61.** $\dfrac{62}{949} + \dfrac{297}{949}i$

63. $-2\sqrt{3}$ **65.** -15

67. $\left(21 + 5\sqrt{2}\right) + \left(7\sqrt{5} - 3\sqrt{10}\right)i$

69. $1 \pm i$ **71.** $-2 \pm \frac{1}{2}i$ **73.** $-\frac{5}{2}, -\frac{3}{2}$ **75.** $2 \pm \sqrt{2}i$

77. $\frac{5}{7} \pm \frac{5\sqrt{15}}{7}$ **79.** $-1 + 6i$ **81.** $-14i$

83. $-432\sqrt{2}i$ **85.** i **87.** 81

89. (a) $z_1 = 9 + 16i$, $z_2 = 20 - 10i$ (b) $z = \frac{11,240}{877} + \frac{4630}{877}i$

91. (a) 16 (b) 16 (c) 16 (d) 16

93. False. If the complex number is real, the number equals its conjugate.

95. False.
$i^{44} + i^{150} - i^{74} - i^{109} + i^{61} = 1 - 1 + 1 - i + i = 1$

97. $i, -1, -i, 1, i, -1, -i, 1$; The pattern repeats the first four results. Divide the exponent by 4.
If the remainder is 1, the result is i.
If the remainder is 2, the result is -1.
If the remainder is 3, the result is $-i$.
If the remainder is 0, the result is 1.

99. $\sqrt{-6}\sqrt{-6} = \sqrt{6}i\sqrt{6}i = 6i^2 = -6$ **101.** Proof

Section 4.2   *(page 350)*

1. Fundamental Theorem; Algebra   **3.** conjugates
5. Three solutions   **7.** Four solutions
9. No real solutions   **11.** Two real solutions
13. Two real solutions   **15.** No real solutions   **17.** $\pm\sqrt{5}$
19. $-5 \pm \sqrt{6}$ **21.** 4   **23.** $-1 \pm 2i$ **25.** $\frac{1}{2} \pm i$
27. $20 \pm 2\sqrt{215}$ **29.** $-3 \pm 2\sqrt{2}i$
31. (a)   (b) $4, \pm i$

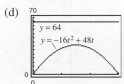

(c) The number of real zeros and the number of x-intercepts are the same.

33. (a)   (b) $\pm\sqrt{2}i$

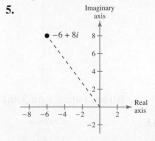

(c) The number of real zeros and the number of x-intercepts are the same.

35. $\pm 6i$; $(x + 6i)(x - 6i)$
37. $1 \pm 4i$; $(x - 1 - 4i)(x - 1 + 4i)$
39. $\pm 3, \pm 3i$; $(x + 3)(x - 3)(x + 3i)(x - 3i)$
41. $1 \pm i$; $(z - 1 + i)(z - 1 - i)$
43. $-3, \pm\sqrt{3}$; $(x + 3)(x + \sqrt{3})(x - \sqrt{3})$
45. $4, \pm 4i$; $(x - 4)(x + 4i)(x - 4i)$
47. $\frac{1}{2}, \pm 3\sqrt{2}i$; $(2x - 1)(x + 3\sqrt{2}i)(x - 3\sqrt{2}i)$
49. $0, 6, \pm 4i$; $x(x - 6)(x + 4i)(x - 4i)$
51. $\pm i, \pm 3i$; $(x + i)(x - i)(x + 3i)(x - 3i)$
53. $-\frac{3}{2}, \pm 5i$ **55.** $\pm 2i, 1, -\frac{1}{2}$ **57.** $-3 \pm i, \frac{1}{4}$
59. $-4, 3 \pm i$ **61.** $2, -3 \pm \sqrt{2}i, 1$
63. $f(x) = x^3 - x^2 + 25x - 25$
65. $f(x) = x^3 - 12x^2 + 46x - 52$
67. $f(x) = 3x^4 - 17x^3 + 25x^2 + 23x - 22$

69. $f(x) = -x^3 + x^2 - 4x + 4$
71. $f(x) = -3x^3 + 9x^2 - 3x - 15$
73. $f(x) = -2x^3 + 5x^2 - 10x + 4$
75. $f(x) = x^3 - 6x^2 + 4x + 40$
77. $f(x) = x^3 - 3x^2 + 6x + 10$
79. $f(x) = x^3 - 2x^2 - 4x - 16$
81. $f(x) = \frac{1}{2}x^4 + \frac{1}{2}x^3 - 2x^2 + x - 6$
83. (a)

t	0	0.5	1	1.5	2	2.5	3
h	0	20	32	36	32	20	0

(b) No

(c) When you set $h = 64$, the resulting equation yields imaginary roots. So, the projectile will not reach a height of 64 feet.

(d)

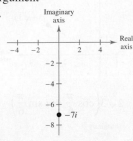

The graphs do not intersect, so the projectile does not reach 64 feet.

(e) The results all show that it is not possible for the projectile to reach a height of 64 feet.

85. (a) $P = -0.0001x^2 + 60x - 150,000$
(b) $\$8,600,000$ (c) $\$115$
(d) It is not possible to have a profit of 10 million dollars.

87. False. The most complex zeros it can have is two, and the Linear Factorization Theorem guarantees that there are three linear factors, so one zero must be real.

89. (a) positive; 4   (b) zero; 0   (c) negative; -4
No real solutions; two real solutions

91. r_1, r_2, r_3 **93.** $5 + r_1, 5 + r_2, 5 + r_3$
95. The zeros cannot be determined.   **97.** $x^2 + b$

Section 4.3   *(page 358)*

1. real; imaginary
3. trigonometric form; modulus; argument
5.

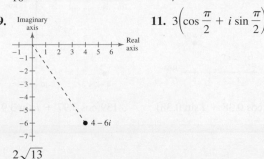

7.

9.

$2\sqrt{13}$

10

7

11. $3\left(\cos\dfrac{\pi}{2} + i\sin\dfrac{\pi}{2}\right)$

13. $3\sqrt{2}\left(\cos\dfrac{5\pi}{4} + i\sin\dfrac{5\pi}{4}\right)$

15.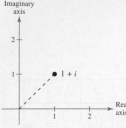

$\sqrt{2}\left(\cos\dfrac{\pi}{4} + i\sin\dfrac{\pi}{4}\right)$

17.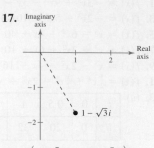

$2\left(\cos\dfrac{5\pi}{3} + i\sin\dfrac{5\pi}{3}\right)$

19.

$4\left(\cos\dfrac{4\pi}{3} + i\sin\dfrac{4\pi}{3}\right)$

21.

$5\left(\cos\dfrac{3\pi}{2} + i\sin\dfrac{3\pi}{2}\right)$

23.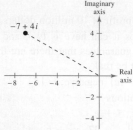

$\sqrt{65}\,(\cos 2.62 + i\sin 2.62)$

25.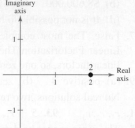

$2(\cos 0 + i\sin 0)$

27.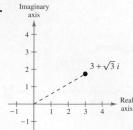

$2\sqrt{3}\left(\cos\dfrac{\pi}{6} + i\sin\dfrac{\pi}{6}\right)$

29.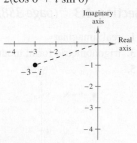

$\sqrt{10}\,(\cos 3.46 + i\sin 3.46)$

31.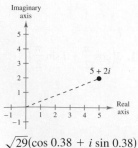

$\sqrt{29}(\cos 0.38 + i\sin 0.38)$

33.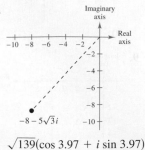

$\sqrt{139}(\cos 3.97 + i\sin 3.97)$

35. $1 + \sqrt{3}i$

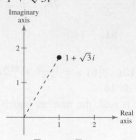

37. $6 - 2\sqrt{3}i$

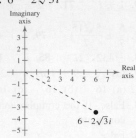

39. $-\dfrac{9\sqrt{2}}{8} + \dfrac{9\sqrt{2}}{8}i$

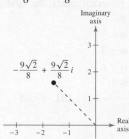

41. 7

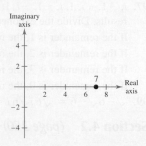

43. $-4.7347 - 1.6072i$

45. $4.6985 + 1.7101i$

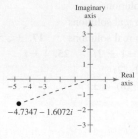

47. $-1.8126 + 0.8452i$

49.

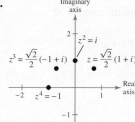

The absolute value of each is 1, and the consecutive powers of z are each 45° apart.

51. $12\left(\cos\dfrac{\pi}{3} + i\sin\dfrac{\pi}{3}\right)$

53. $\dfrac{10}{9}(\cos 150° + i\sin 150°)$

55. $\cos 50° + i\sin 50°$

57. $\frac{1}{3}(\cos 30° + i\sin 30°)$

59. $\cos\dfrac{2\pi}{3} + i\sin\dfrac{2\pi}{3}$

61. $6(\cos 330° + i\sin 330°)$

63. (a) $\left[2\sqrt{2}\left(\cos\dfrac{\pi}{4} + i\sin\dfrac{\pi}{4}\right)\right]\left[\sqrt{2}\left(\cos\dfrac{7\pi}{4} + i\sin\dfrac{7\pi}{4}\right)\right]$

(b) $4(\cos 0 + i\sin 0) = 4$ (c) 4

65. (a) $\left[2\left(\cos\dfrac{3\pi}{2} + i\sin\dfrac{3\pi}{2}\right)\right]\left[\sqrt{2}\left(\cos\dfrac{\pi}{4} + i\sin\dfrac{\pi}{4}\right)\right]$

(b) $2\sqrt{2}\left(\cos\dfrac{7\pi}{4} + i\sin\dfrac{7\pi}{4}\right) = 2 - 2i$

(c) $-2i - 2i^2 = -2i + 2 = 2 - 2i$

67. (a) $[5(\cos 0.93 + i \sin 0.93)] \div \left[2\left(\cos \dfrac{5\pi}{3} + i \sin \dfrac{5\pi}{3}\right)\right]$

(b) $\dfrac{5}{2}(\cos 1.97 + i \sin 1.97) \approx -0.982 + 2.299i$

(c) About $-0.982 + 2.299i$

69. (a) $[5(\cos 0 + i \sin 0)] \div \left[\sqrt{13}(\cos 0.98 + i \sin 0.98)\right]$

(b) $\dfrac{5}{\sqrt{13}}(\cos 5.30 + i \sin 5.30) \approx 0.769 - 1.154i$

(c) $\dfrac{10}{13} - \dfrac{15}{13}i \approx 0.769 - 1.154i$

71.

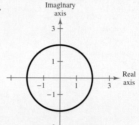

73.

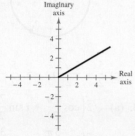

75. $34 + 38i$ **77.** $\dfrac{3}{2} - \dfrac{1}{2}i$ **79.** $\dfrac{39}{34} + \dfrac{3}{34}i$

81. True, by the definition of the absolute value of a complex number.

83. Answers will vary.   **85.** (a) r^2 (b) $\cos 2\theta + i \sin 2\theta$

Section 4.4   (page 364)

1. DeMoivre's   **3.** $\sqrt[n]{r}\left(\cos \dfrac{\theta + 2\pi k}{n} + i \sin \dfrac{\theta + 2\pi k}{n}\right)$

5. $-4 - 4i$ **7.** $8i$ **9.** $1024 - 1024\sqrt{3}i$

11. $\dfrac{125}{2} + \dfrac{125\sqrt{3}}{2}i$ **13.** -1 **15.** $608.0204 + 144.6936i$

17. $-597 - 122i$ **19.** $-43\sqrt{5} + 4i$ **21.** $\dfrac{81}{2} + \dfrac{81\sqrt{3}}{2}i$

23. $32.3524 - 120.7407i$ **25.** $32i$ **27.** 27

29. $1 + i, -1 - i$ **31.** $-\dfrac{\sqrt{6}}{2} + \dfrac{\sqrt{6}}{2}i, \dfrac{\sqrt{6}}{2} - \dfrac{\sqrt{6}}{2}i$

33. $-1.5538 + 0.6436i, 1.5538 - 0.6436i$

35. $\dfrac{\sqrt{6}}{2} + \dfrac{\sqrt{2}}{2}i, -\dfrac{\sqrt{6}}{2} - \dfrac{\sqrt{2}}{2}i$

37. (a) $\sqrt{5}(\cos 60° + i \sin 60°)$

$\sqrt{5}(\cos 240° + i \sin 240°)$

(b)

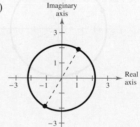

(c) $\dfrac{\sqrt{5}}{2} + \dfrac{\sqrt{15}}{2}i, -\dfrac{\sqrt{5}}{2} - \dfrac{\sqrt{15}}{2}i$

39. (a) $2\left(\cos \dfrac{2\pi}{9} + i \sin \dfrac{2\pi}{9}\right)$

$2\left(\cos \dfrac{8\pi}{9} + i \sin \dfrac{8\pi}{9}\right)$

$2\left(\cos \dfrac{14\pi}{9} + i \sin \dfrac{14\pi}{9}\right)$

(b)

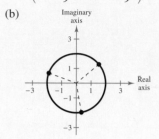

(c) $1.5321 + 1.2856i, -1.8794 + 0.6840i,$
$0.3473 - 1.9696i$

41. (a) $3\left(\cos \dfrac{\pi}{30} + i \sin \dfrac{\pi}{30}\right)$

$3\left(\cos \dfrac{13\pi}{30} + i \sin \dfrac{13\pi}{30}\right)$

$3\left(\cos \dfrac{5\pi}{6} + i \sin \dfrac{5\pi}{6}\right)$

$3\left(\cos \dfrac{37\pi}{30} + i \sin \dfrac{37\pi}{30}\right)$

$3\left(\cos \dfrac{49\pi}{30} + i \sin \dfrac{49\pi}{30}\right)$

(b)

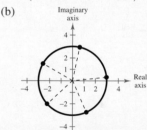

(c) $2.9836 + 0.3136i, 0.6237 + 2.9344i,$
$-2.5981 + 1.5i, -2.2294 - 2.0074i, 1.2202 - 2.7406i$

43. (a) $3\left(\cos \dfrac{\pi}{8} + i \sin \dfrac{\pi}{8}\right)$

$3\left(\cos \dfrac{5\pi}{8} + i \sin \dfrac{5\pi}{8}\right)$

$3\left(\cos \dfrac{9\pi}{8} + i \sin \dfrac{9\pi}{8}\right)$

$3\left(\cos \dfrac{13\pi}{8} + i \sin \dfrac{13\pi}{8}\right)$

(b)

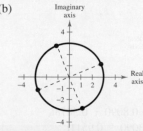

(c) $2.7716 + 1.1481i, -1.1481 + 2.7716i,$
$-2.7716 - 1.1481i, 1.1481 - 2.7716i$

CHAPTER 4

45. (a) $5\left(\cos\dfrac{4\pi}{9} + i\sin\dfrac{4\pi}{9}\right)$

$5\left(\cos\dfrac{10\pi}{9} + i\sin\dfrac{10\pi}{9}\right)$

$5\left(\cos\dfrac{16\pi}{9} + i\sin\dfrac{16\pi}{9}\right)$

(b)

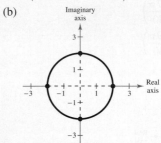

(c) $0.8682 + 4.9240i,\ -4.6985 - 1.7101i,$
$3.8302 - 3.2139i$

47. (a) $2(\cos 0 + i\sin 0)$

$2\left(\cos\dfrac{\pi}{2} + i\sin\dfrac{\pi}{2}\right)$

$2(\cos\pi + i\sin\pi)$

$2\left(\cos\dfrac{3\pi}{2} + i\sin\dfrac{3\pi}{2}\right)$

(b)

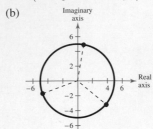

(c) $2, 2i, -2, -2i$

49. (a) $\cos 0 + i\sin 0$

$\cos\dfrac{2\pi}{5} + i\sin\dfrac{2\pi}{5}$

$\cos\dfrac{4\pi}{5} + i\sin\dfrac{4\pi}{5}$

$\cos\dfrac{6\pi}{5} + i\sin\dfrac{6\pi}{5}$

$\cos\dfrac{8\pi}{5} + i\sin\dfrac{8\pi}{5}$

(b)

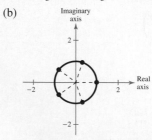

(c) $1, 0.3090 + 0.9511i,\ -0.8090 + 0.5878i,$
$-0.8090 - 0.5878i, 0.3090 - 0.9511i$

51. (a) $5\left(\cos\dfrac{\pi}{3} + i\sin\dfrac{\pi}{3}\right)$

$5(\cos\pi + i\sin\pi)$

$5\left(\cos\dfrac{5\pi}{3} + i\sin\dfrac{5\pi}{3}\right)$

(b)

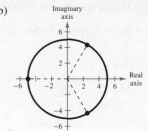

(c) $\dfrac{5}{2} + \dfrac{5\sqrt{3}}{2}i,\ -5,\ \dfrac{5}{2} - \dfrac{5\sqrt{3}}{2}i$

53. (a) $\sqrt{2}\left(\cos\dfrac{7\pi}{20} + i\sin\dfrac{7\pi}{20}\right)$

$\sqrt{2}\left(\cos\dfrac{3\pi}{4} + i\sin\dfrac{3\pi}{4}\right)$

$\sqrt{2}\left(\cos\dfrac{23\pi}{20} + i\sin\dfrac{23\pi}{20}\right)$

$\sqrt{2}\left(\cos\dfrac{31\pi}{20} + i\sin\dfrac{31\pi}{20}\right)$

$\sqrt{2}\left(\cos\dfrac{39\pi}{20} + i\sin\dfrac{39\pi}{20}\right)$

(b)

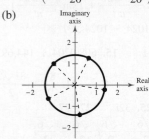

(c) $0.6420 + 1.2601i,\ -1 + i,\ -1.2601 - 0.6420i,$
$0.2212 - 1.3968i,\ 1.3968 - 0.2212i$

55. $\cos\dfrac{3\pi}{8} + i\sin\dfrac{3\pi}{8}$

$\cos\dfrac{7\pi}{8} + i\sin\dfrac{7\pi}{8}$

$\cos\dfrac{11\pi}{8} + i\sin\dfrac{11\pi}{8}$

$\cos\dfrac{15\pi}{8} + i\sin\dfrac{15\pi}{8}$

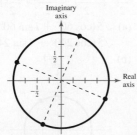

57. $\cos \dfrac{\pi}{6} + i \sin \dfrac{\pi}{6}$

$\cos \dfrac{\pi}{2} + i \sin \dfrac{\pi}{2}$

$\cos \dfrac{5\pi}{6} + i \sin \dfrac{5\pi}{6}$

$\cos \dfrac{7\pi}{6} + i \sin \dfrac{7\pi}{6}$

$\cos \dfrac{3\pi}{2} + i \sin \dfrac{3\pi}{2}$

$\cos \dfrac{11\pi}{6} + i \sin \dfrac{11\pi}{6}$

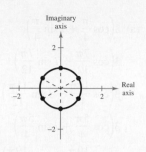

59. $3\left(\cos \dfrac{\pi}{5} + i \sin \dfrac{\pi}{5}\right)$

$3\left(\cos \dfrac{3\pi}{5} + i \sin \dfrac{3\pi}{5}\right)$

$3(\cos \pi + i \sin \pi)$

$3\left(\cos \dfrac{7\pi}{5} + i \sin \dfrac{7\pi}{5}\right)$

$3\left(\cos \dfrac{9\pi}{5} + i \sin \dfrac{9\pi}{5}\right)$

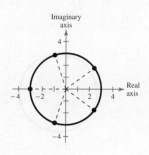

61. $4(\cos 0 + i \sin 0)$

$4\left(\cos \dfrac{2\pi}{3} + i \sin \dfrac{2\pi}{3}\right)$

$4\left(\cos \dfrac{4\pi}{3} + i \sin \dfrac{4\pi}{3}\right)$

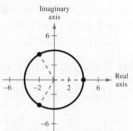

63. $2\left(\cos \dfrac{3\pi}{8} + i \sin \dfrac{3\pi}{8}\right)$

$2\left(\cos \dfrac{7\pi}{8} + i \sin \dfrac{7\pi}{8}\right)$

$2\left(\cos \dfrac{11\pi}{8} + i \sin \dfrac{11\pi}{8}\right)$

$2\left(\cos \dfrac{15\pi}{8} + i \sin \dfrac{15\pi}{8}\right)$

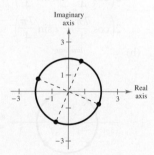

65. $2\left(\cos \dfrac{\pi}{8} + i \sin \dfrac{\pi}{8}\right)$

$2\left(\cos \dfrac{5\pi}{8} + i \sin \dfrac{5\pi}{8}\right)$

$2\left(\cos \dfrac{9\pi}{8} + i \sin \dfrac{9\pi}{8}\right)$

$2\left(\cos \dfrac{13\pi}{8} + i \sin \dfrac{13\pi}{8}\right)$

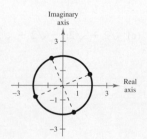

67. $\sqrt[6]{2}\left(\cos \dfrac{7\pi}{12} + i \sin \dfrac{7\pi}{12}\right)$

$\sqrt[6]{2}\left(\cos \dfrac{5\pi}{4} + i \sin \dfrac{5\pi}{4}\right)$

$\sqrt[6]{2}\left(\cos \dfrac{23\pi}{12} + i \sin \dfrac{23\pi}{12}\right)$

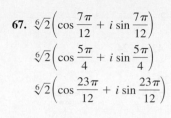

69. $\sqrt[12]{2}\left(\cos \dfrac{5\pi}{24} + i \sin \dfrac{5\pi}{24}\right)$

$\sqrt[12]{2}\left(\cos \dfrac{13\pi}{24} + i \sin \dfrac{13\pi}{24}\right)$

$\sqrt[12]{2}\left(\cos \dfrac{7\pi}{8} + i \sin \dfrac{7\pi}{8}\right)$

$\sqrt[12]{2}\left(\cos \dfrac{29\pi}{24} + i \sin \dfrac{29\pi}{24}\right)$

$\sqrt[12]{2}\left(\cos \dfrac{37\pi}{24} + i \sin \dfrac{37\pi}{24}\right)$

$\sqrt[12]{2}\left(\cos \dfrac{15\pi}{8} + i \sin \dfrac{15\pi}{8}\right)$

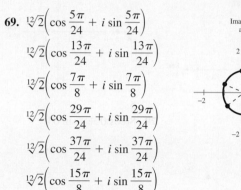

71. False. They are equally spaced around the circle centered at the origin with radius $\sqrt[n]{r}$.

73. False. The solutions are $\pm 2(1 + i)$.

75. Answers will vary.

77. (a) $i, -2i$

(b) $\left(-1 \pm \sqrt{2}\right)i$

(c) $\dfrac{\sqrt{2}}{2} - \left(1 + \dfrac{\sqrt{6}}{2}\right)i, \ -\dfrac{\sqrt{2}}{2} + \left(-1 + \dfrac{\sqrt{6}}{2}\right)i$

79–81. Answers will vary.

Review Exercises   (page 368)

1. $6 + 2i$ **3.** $4\sqrt{3}i$ **5.** $-1 + 3i$ **7.** $3 + 7i$

9. $11 + 44i$ **11.** $40 + 65i$ **13.** $-4 - 46i$

15. $-45 + 28i$ **17.** $\dfrac{10}{3}i$ **19.** $\dfrac{23}{17} + \dfrac{10}{17}i$ **21.** $\dfrac{21}{13} - \dfrac{1}{13}i$

23. $\pm\dfrac{\sqrt{3}}{3}i$ **25.** $1 \pm 3i$ **27.** $-10 + i$ **29.** i

31. Five solutions   **33.** Four solutions

35. Two real solutions   **37.** No real solutions

39. $0, 2$ **41.** $\dfrac{3}{2} \pm \dfrac{\sqrt{11}i}{2}$ **43.** $-4 \pm \sqrt{6}$

45. $-\dfrac{3}{4} \pm \dfrac{\sqrt{39}\,i}{4}$

47. Yes. A price of \$95.41 or \$119.59 per unit would yield a profit of 9 million dollars.

49. $-\dfrac{1}{2} \pm \dfrac{\sqrt{5}}{2}i; \ \dfrac{1}{2}\left(2x + 1 - \sqrt{5}i\right)\left(2x + 1 + \sqrt{5}i\right)$

51. $\dfrac{3}{2}, \pm 5i; \ (2x - 3)(x + 5i)(x - 5i)$

53. $\pm\dfrac{\sqrt{5}}{2}, \pm \sqrt{2}i; \ \left(2x + \sqrt{5}\right)\left(2x - \sqrt{5}\right)\left(x + \sqrt{2}i\right)\left(x - \sqrt{2}i\right)$

55. $-7, 2; \ (x + 7)(x - 2)^2$

57. $-5, 1 \pm 2i; \ (x + 5)(x - 1 - 2i)(x - 1 + 2i)$

59. $-\dfrac{1}{2}, 5 \pm 3i; \ (2x + 1)(x - 5 - 3i)(x - 5 + 3i)$

61. $-2, 3, -3 \pm \sqrt{5}i;$
$(x - 3)(x + 2)\left(x + 3 - \sqrt{5}i\right)\left(x + 3 + \sqrt{5}i\right)$

CHAPTER 4

63. $f(x) = 12x^4 - 19x^3 + 9x - 2$

65. $f(x) = x^3 - 7x^2 + 13x - 3$

67. $f(x) = 3x^4 - 14x^3 + 17x^2 - 42x + 24$

69. $f(x) = x^4 + 27x^2 + 50$

71. $f(x) = 2x^3 - 14x^2 + 24x - 20$

73.

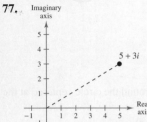

8

75.

Imaginary axis

5

77.

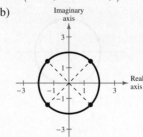

$\sqrt{34}$

79. $8(\cos 0 + i \sin 0)$

81. $3\left(\cos \dfrac{3\pi}{2} + i \sin \dfrac{3\pi}{2}\right)$ **83.** $5\sqrt{2}\left(\cos \dfrac{7\pi}{4} + i \sin \dfrac{7\pi}{4}\right)$

85. $6\left(\cos \dfrac{5\pi}{6} + i \sin \dfrac{5\pi}{6}\right)$ **87.** $28\left(\cos \dfrac{7\pi}{12} + i \sin \dfrac{7\pi}{12}\right)$

89. $\dfrac{1}{2}\left(\cos \dfrac{\pi}{2} + i \sin \dfrac{\pi}{2}\right)$

91. (a) $z_1 = \sqrt{2}\left(\cos \dfrac{\pi}{4} + i \sin \dfrac{\pi}{4}\right)$

$z_2 = \sqrt{2}\left(\cos \dfrac{7\pi}{4} + i \sin \dfrac{7\pi}{4}\right)$

(b) $z_1 z_2 = 2(\cos 0 + i \sin 0)$

$\dfrac{z_1}{z_2} = \left(\cos \dfrac{\pi}{2} + i \sin \dfrac{\pi}{2}\right)$

93. (a) $z_1 = 4\left(\cos \dfrac{11\pi}{6} + i \sin \dfrac{11\pi}{6}\right)$

$z_2 = 10\left(\cos \dfrac{3\pi}{2} + i \sin \dfrac{3\pi}{2}\right)$

(b) $z_1 z_2 = 40\left(\cos \dfrac{4\pi}{3} + i \sin \dfrac{4\pi}{3}\right)$

$\dfrac{z_1}{z_2} = \dfrac{2}{5}\left(\cos \dfrac{\pi}{3} + i \sin \dfrac{\pi}{3}\right)$

95. $\dfrac{625}{2} + \dfrac{625\sqrt{3}}{2}i$ **97.** $2035 - 828i$ **99.** $-8 - 8i$

101. (a) $3\left(\cos \dfrac{\pi}{4} + i \sin \dfrac{\pi}{4}\right)$

$3\left(\cos \dfrac{7\pi}{12} + i \sin \dfrac{7\pi}{12}\right)$

$3\left(\cos \dfrac{11\pi}{12} + i \sin \dfrac{11\pi}{12}\right)$

$3\left(\cos \dfrac{5\pi}{4} + i \sin \dfrac{5\pi}{4}\right)$

$3\left(\cos \dfrac{19\pi}{12} + i \sin \dfrac{19\pi}{12}\right)$

$3\left(\cos \dfrac{23\pi}{12} + i \sin \dfrac{23\pi}{12}\right)$

(b)

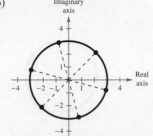

(c) $2.1213 + 2.1213i, -0.7765 + 2.8978i,$

$-2.8978 + 0.7765i, -2.1213 - 2.1213i,$

$0.7765 - 2.8978i, 2.8978 - 0.7765i$

103. (a) $2\left(\cos \dfrac{\pi}{4} + i \sin \dfrac{\pi}{4}\right)$

$2\left(\cos \dfrac{3\pi}{4} + i \sin \dfrac{3\pi}{4}\right)$

$2\left(\cos \dfrac{5\pi}{4} + i \sin \dfrac{5\pi}{4}\right)$

$2\left(\cos \dfrac{7\pi}{4} + i \sin \dfrac{7\pi}{4}\right)$

(b)

Imaginary axis

(c) $\sqrt{2} + \sqrt{2}i, -\sqrt{2} + \sqrt{2}i, -\sqrt{2} - \sqrt{2}i, \sqrt{2} - \sqrt{2}i$

105. $3\left(\cos\dfrac{\pi}{4} + i\sin\dfrac{\pi}{4}\right) = \dfrac{3\sqrt{2}}{2} + \dfrac{3\sqrt{2}}{2}i$

$3\left(\cos\dfrac{3\pi}{4} + i\sin\dfrac{3\pi}{4}\right) = -\dfrac{3\sqrt{2}}{2} + \dfrac{3\sqrt{2}}{2}i$

$3\left(\cos\dfrac{5\pi}{4} + i\sin\dfrac{5\pi}{4}\right) = -\dfrac{3\sqrt{2}}{2} - \dfrac{3\sqrt{2}}{2}i$

$3\left(\cos\dfrac{7\pi}{4} + i\sin\dfrac{7\pi}{4}\right) = \dfrac{3\sqrt{2}}{2} - \dfrac{3\sqrt{2}}{2}i$

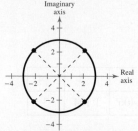

107. $2\left(\cos\dfrac{\pi}{2} + i\sin\dfrac{\pi}{2}\right) = 2i$

$2\left(\cos\dfrac{7\pi}{6} + i\sin\dfrac{7\pi}{6}\right) = -\sqrt{3} - i$

$2\left(\cos\dfrac{11\pi}{6} + i\sin\dfrac{11\pi}{6}\right) = \sqrt{3} - i$

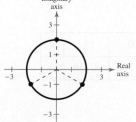

109. False.

$\sqrt{-18}\sqrt{-2} = 3\sqrt{2}i\sqrt{2}i$ and $\sqrt{(-18)(-2)} = \sqrt{36}$

$\qquad\qquad = 3\sqrt{4}i^2 \qquad\qquad\qquad = 6$

$\qquad\qquad = 6i^2$

$\qquad\qquad = -6$

111. False. A fourth-degree polynomial with real coefficients has four zeros, and complex zeros occur in conjugate pairs.

113. (a) $4(\cos 60° + i\sin 60°)$ $\qquad$ (b) -64

$\qquad\quad 4(\cos 180° + i\sin 180°)$

$\qquad\quad 4(\cos 300° + i\sin 300°)$

115. $z_1 z_2 = -4, \dfrac{z_1}{z_2} = -\cos 2\theta - i\sin 2\theta$

Chapter Test *(page 371)*

1. $-5 + 10i$ $\quad$ **2.** $-3 + 5i$ $\quad$ **3.** $-65 + 72i$ $\quad$ **4.** 43

5. $\dfrac{32}{73} + \dfrac{12}{73}i$ $\quad$ **6.** $\dfrac{1}{2} \pm \dfrac{\sqrt{5}}{2}i$ $\quad$ **7.** Five solutions

8. Four solutions $\quad$ **9.** $6, \pm\sqrt{5}i$ $\quad$ **10.** $\pm\sqrt{6}, \pm 2i$

11. $\pm 2, \pm\sqrt{2}i; (x + 2)(x - 2)(x + \sqrt{2}i)(x - \sqrt{2}i)$

12. $\dfrac{3}{2}, 2 \pm i; (2v - 3)(v - 2 - i)(v - 2 + i)$

13. $x^4 - 15x^3 + 73x^2 - 119x$

14. $x^4 - 8x^3 + 28x^2 - 60x + 63$

15. No. If $a + bi, b \neq 0$, is a zero, its conjugate $a - bi$ is also a zero.

16. $5\sqrt{2}\left(\cos\dfrac{7\pi}{4} + i\sin\dfrac{7\pi}{4}\right)$ $\quad$ **17.** $-3 + 3\sqrt{3}i$

18. $-\dfrac{6561}{2} - \dfrac{6561\sqrt{3}}{2}i$ $\quad$ **19.** $5832i$

20. $4\sqrt[4]{2}\left(\cos\dfrac{\pi}{12} + i\sin\dfrac{\pi}{12}\right)$

$\qquad 4\sqrt[4]{2}\left(\cos\dfrac{7\pi}{12} + i\sin\dfrac{7\pi}{12}\right)$

$\qquad 4\sqrt[4]{2}\left(\cos\dfrac{13\pi}{12} + i\sin\dfrac{13\pi}{12}\right)$

$\qquad 4\sqrt[4]{2}\left(\cos\dfrac{19\pi}{12} + i\sin\dfrac{19\pi}{12}\right)$

21. $3\left(\cos\dfrac{\pi}{6} + i\sin\dfrac{\pi}{6}\right)$

$\qquad 3\left(\cos\dfrac{5\pi}{6} + i\sin\dfrac{5\pi}{6}\right)$

$\qquad 3\left(\cos\dfrac{3\pi}{2} + i\sin\dfrac{3\pi}{2}\right)$

22. No. When you set $h = 125$, the resulting equation yields imaginary roots. So, the projectile will not reach a height of 125 feet.

Problem Solving *(page 373)*

1. (a) $z^3 = 8$ for all three complex numbers.

$\quad$ (b) $z^3 = 27$ for all three complex numbers.

$\quad$ (c) The cube roots of a positive real number "a" are:

$\qquad \sqrt[3]{a}, \dfrac{-\sqrt[3]{a} + \sqrt[3]{a}\sqrt{3}i}{2}$, and $\dfrac{-\sqrt[3]{a} - \sqrt[3]{a}\sqrt{3}i}{2}$.

3. $(a + bi)(a - bi) = a^2 + abi - abi - b^2 i^2$

$\qquad\qquad\qquad\quad = a^2 + b^2$

5. (a) $k > 1$ or $k < 0$ $\qquad$ (b) $0 < k < 1$

7. (a) No $\quad$ (b) Yes

9. (a) Not correct because f has $(0, 0)$ as an intercept.

$\quad$ (b) Not correct because the function must be at least a fourth-degree polynomial.

$\quad$ (c) Correct function

$\quad$ (d) Not correct because k has $(-1, 0)$ as an intercept.

11. (a) Yes $\quad$ (b) No $\quad$ (c) Yes

13. (a) $1 + i, 3 + i$ $\quad$ (b) $1 - i, 2 + 3i$

$\quad$ (c) $1 + i, -\dfrac{7}{2} + 3i$ $\quad$ (d) $4 + 5i, -\dfrac{1}{3} - \dfrac{1}{3}i$

15. Answers will vary.

Chapter 5

Section 5.1 *(page 384)*

1. algebraic $\qquad$ **3.** One-to-One $\qquad$ **5.** $A = P\left(1 + \dfrac{r}{n}\right)^{nt}$

7. 0.863 $\qquad$ **9.** 0.006 $\qquad$ **11.** 1767.767

13. d $\qquad$ **14.** c $\qquad$ **15.** a $\qquad$ **16.** b

17.

x	-2	-1	0	1	2
$f(x)$	4	2	1	0.5	0.25

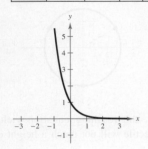

19.

x	-2	-1	0	1	2
$f(x)$	36	6	1	0.167	0.028

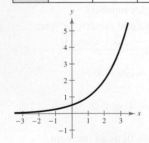

21.

x	-2	-1	0	1	2
$f(x)$	0.125	0.25	0.5	1	2

23. Shift the graph of f one unit upward.

25. Reflect the graph of f in the x-axis and shift three units upward.

27. Reflect the graph of f in the origin.

29. 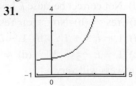 **31.**

33. 0.472   **35.** 3.857×10^{-22} **37.** 7166.647

39.

x	-2	-1	0	1	2
$f(x)$	0.135	0.368	1	2.718	7.389

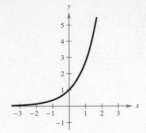

41.

x	-8	-7	-6	-5	-4
$f(x)$	0.055	0.149	0.406	1.104	3

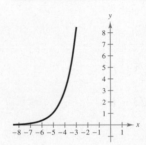

43.

x	-2	-1	0	1	2
$f(x)$	4.037	4.100	4.271	4.736	6

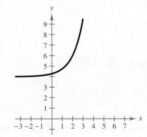

45. 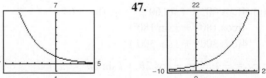 **47.**

49.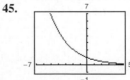

51. $x = 2$ **53.** $x = -5$ **55.** $x = \frac{1}{3}$ **57.** $x = 3, -1$

59.

n	1	2	4	12
A	$1828.49	$1830.29	$1831.19	$1831.80

n	365	Continuous
A	$1832.09	$1832.10

61.

n	1	2	4	12
A	$5477.81	$5520.10	$5541.79	$5556.46

n	365	Continuous
A	$5563.61	$5563.85

63.

t	10	20	30
A	$17,901.90	$26,706.49	$39,841.40

t	40	50
A	$59,436.39	$88,668.67

65.

t	10	20	30
A	$22,986.49	$44,031.56	$84,344.25

t	40	50
A	$161,564.86	$309,484.08

67. $104,710.29   **69.** $35.45

71. (a)

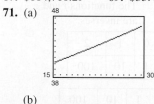

(b)

t	15	16	17	18	19	20
P (in millions)	40.19	40.59	40.99	41.39	41.80	42.21

t	21	22	23	24	25	26
P (in millions)	42.62	43.04	43.47	43.90	44.33	44.77

t	27	28	29	30
P (in millions)	45.21	45.65	46.10	46.56

(c) 2038

73. (a) 16 g   (b) 1.85 g

(c)

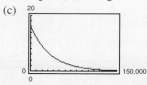

75. (a) $V(t) = 30,500\left(\frac{7}{8}\right)^t$ (b) $17,878.54

77. True. As $x \to -\infty$, $f(x) \to -2$ but never reaches -2.

79. $f(x) = h(x)$ **81.** $f(x) = g(x) = h(x)$

83.

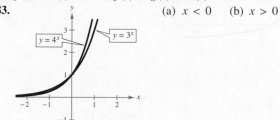

(a) $x < 0$ (b) $x > 0$

85.

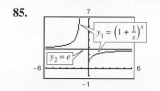

As the x-value increases, y_1 approaches the value of e.

87. (a)   (b)

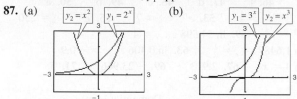

In both viewing windows, the constant raised to a variable power increases more rapidly than the variable raised to a constant power.

89. (a) $A = 5466.09 (b) $A = 5466.35
(c) $A = 5466.36 (d) $A = 5466.38
No. Answers will vary.

Section 5.2   (page 394)

1. logarithmic   **3.** natural; e **5.** $x = y$ **7.** $4^2 = 16$
9. $9^{-2} = \frac{1}{81}$ **11.** $32^{2/5} = 4$ **13.** $64^{1/2} = 8$
15. $\log_5 125 = 3$ **17.** $\log_{81} 3 = \frac{1}{4}$ **19.** $\log_6 \frac{1}{36} = -2$
21. $\log_{24} 1 = 0$ **23.** 6   **25.** 0   **27.** 2   **29.** -0.058
31. 1.097   **33.** 7   **35.** 1
37.

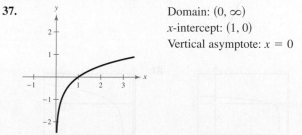

Domain: $(0, \infty)$
x-intercept: $(1, 0)$
Vertical asymptote: $x = 0$

39.

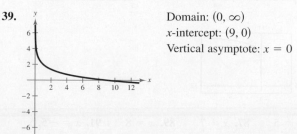

Domain: $(0, \infty)$
x-intercept: $(9, 0)$
Vertical asymptote: $x = 0$

CHAPTER 5

41. 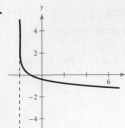 Domain: $(-2, \infty)$
x-intercept: $(-1, 0)$
Vertical asymptote: $x = -2$

43. 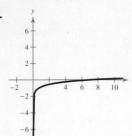 Domain: $(0, \infty)$
x-intercept: $(7, 0)$
Vertical asymptote: $x = 0$

45. c **46.** f **47.** d **48.** e **49.** b **50.** a
51. $e^{-0.693\ldots} = \frac{1}{2}$ **53.** $e^{1.945\ldots} = 7$ **55.** $e^{5.521\ldots} = 250$
57. $e^0 = 1$ **59.** $\ln 54.598\ldots = 4$
61. $\ln 1.6487\ldots = \frac{1}{2}$ **63.** $\ln 0.406\ldots = -0.9$
65. $\ln 4 = x$ **67.** 2.913 **69.** -23.966 **71.** 5
73. $-\frac{5}{6}$
75. 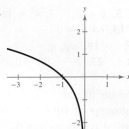 Domain: $(4, \infty)$
x-intercept: $(5, 0)$
Vertical asymptote: $x = 4$

77.  Domain: $(-\infty, 0)$
x-intercept: $(-1, 0)$
Vertical asymptote: $x = 0$

79. 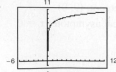 **81.**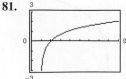

83.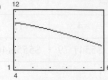

85. $x = 5$ **87.** $x = 7$ **89.** $x = 8$ **91.** $x = -5, 5$

93. (a) 30 yr; 10 yr (b) \$323,179; \$199,109
(c) \$173,179; \$49,109
(d) $x = 750$; The monthly payment must be greater than \$750.
95. (a)

t	1	2	3	4	5	6
C	10.36	9.94	9.37	8.70	7.96	7.15

(b)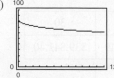

(c) No, the model begins to decrease rapidly, eventually producing negative values.
97. (a) 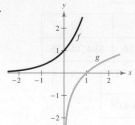 (b) 80 (c) 68.1 (d) 62.3

99. False. Reflecting $g(x)$ about the line $y = x$ will determine the graph of $f(x)$.
101. **103.**

The functions f and g are inverses. The functions f and g are inverses.

105.

x	-2	-1	0	1	2
$f(x) = 10^x$	$\frac{1}{100}$	$\frac{1}{10}$	1	10	100

x	$\frac{1}{100}$	$\frac{1}{10}$	1	10	100
$f(x) = \log x$	-2	-1	0	1	2

The domain of $f(x) = 10^x$ is equal to the range of $f(x) = \log x$ and vice versa. $f(x) = 10^x$ and $f(x) = \log x$ are inverses of each other.
107. (a)

x	1	5	10	10^2
$f(x)$	0	0.322	0.230	0.046

(b) 0

x	10^4	10^6
$f(x)$	0.00092	0.0000138

(c)

109. Answers will vary.

111. (a)

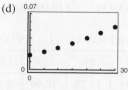

(b) Increasing: $(0, \infty)$
Decreasing: $(-\infty, 0)$
(c) Relative minimum: $(0, 0)$

Section 5.3   (page 401)

1. change-of-base   **3.** $\dfrac{1}{\log_b a}$ **4.** c   **5.** a   **6.** b

7. (a) $\dfrac{\log 16}{\log 5}$ (b) $\dfrac{\ln 16}{\ln 5}$ **9.** (a) $\dfrac{\log x}{\log \frac{1}{5}}$ (b) $\dfrac{\ln x}{\ln \frac{1}{5}}$

11. (a) $\dfrac{\log \frac{3}{10}}{\log x}$ (b) $\dfrac{\ln \frac{3}{10}}{\ln x}$ **13.** (a) $\dfrac{\log x}{\log 2.6}$ (b) $\dfrac{\ln x}{\ln 2.6}$

15. 1.771   **17.** -2.000 **19.** -1.048 **21.** 2.633

23. $\frac{3}{2}$ **25.** $-3 - \log_5 2$ **27.** $6 + \ln 5$ **29.** 2

31. $\frac{3}{4}$ **33.** 4   **35.** -2 is not in the domain of $\log_2 x$.

37. 4.5   **39.** $-\frac{1}{2}$ **41.** 7   **43.** 2   **45.** $\ln 4 + \ln x$

47. $4 \log_8 x$ **49.** $1 - \log_5 x$ **51.** $\frac{1}{2} \ln z$

53. $\ln x + \ln y + 2 \ln z$ **55.** $\ln z + 2 \ln(z - 1)$

57. $\frac{1}{2} \log_2(a - 1) - 2 \log_2 3$ **59.** $\frac{1}{3} \ln x - \frac{1}{3} \ln y$

61. $2 \ln x + \frac{1}{2} \ln y - \frac{1}{2} \ln z$

63. $2 \log_5 x - 2 \log_5 y - 3 \log_5 z$

65. $\frac{3}{4} \ln x + \frac{1}{4} \ln(x^2 + 3)$ **67.** $\ln 2x$ **69.** $\log_4 \dfrac{z}{y}$

71. $\log_2 x^2 y^4$ **73.** $\log_3 \sqrt[4]{5x}$ **75.** $\log \dfrac{x}{(x + 1)^2}$

77. $\log \dfrac{xz^3}{y^2}$ **79.** $\ln \dfrac{x}{(x + 1)(x - 1)}$ **81.** $\ln \sqrt[3]{\dfrac{x(x + 3)^2}{x^2 - 1}}$

83. $\log_8 \dfrac{\sqrt[3]{y(y + 4)^2}}{y - 1}$

85. $\log_2 \frac{32}{4} = \log_2 32 - \log_2 4$; Property 2

87. $\beta = 10(\log I + 12)$; 60 dB   **89.** 70 dB

91. $\ln y = \frac{1}{4} \ln x$ **93.** $\ln y = -\frac{1}{4} \ln x + \ln \frac{5}{2}$

95. $y = 256.24 - 20.8 \ln x$

97. (a) and (b)   (c)

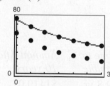

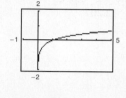

$T = 21 + e^{-0.037t + 3.997}$
The results are similar.

(d)

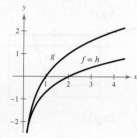

$T = 21 + \dfrac{1}{0.001t + 0.016}$

(e) Answers will vary.

99. Proof

101. False; $\ln 1 = 0$ **103.** False; $\ln(x - 2) \neq \ln x - \ln 2$

105. False; $u = v^2$

107. $f(x) = \dfrac{\log x}{\log 2} = \dfrac{\ln x}{\ln 2}$ **109.** $f(x) = \dfrac{\log x}{\log \frac{1}{2}} = \dfrac{\ln x}{\ln \frac{1}{2}}$

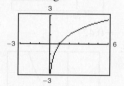   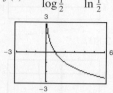

111. $f(x) = \dfrac{\log x}{\log 11.8} = \dfrac{\ln x}{\ln 11.8}$ **113.** $f(x) = h(x)$; Property 2

115.

$\ln 1 = 0$	$\ln 9 \approx 2.1972$
$\ln 2 \approx 0.6931$	$\ln 10 \approx 2.3025$
$\ln 3 \approx 1.0986$	$\ln 12 \approx 2.4848$
$\ln 4 \approx 1.3862$	$\ln 15 \approx 2.7080$
$\ln 5 \approx 1.6094$	$\ln 16 \approx 2.7724$
$\ln 6 \approx 1.7917$	$\ln 18 \approx 2.8903$
$\ln 8 \approx 2.0793$	$\ln 20 \approx 2.9956$

Section 5.4   (page 411)

1. solve

3. (a) One-to-One   (b) logarithmic; logarithmic
(c) exponential; exponential

5. (a) Yes   (b) No

7. (a) No   (b) Yes   (c) Yes, approximate

9. (a) Yes, approximate   (b) No   (c) Yes

11. (a) No   (b) Yes   (c) Yes, approximate

13. 2   **15.** -5 **17.** 2   **19.** $\ln 2 \approx 0.693$

21. $e^{-1} \approx 0.368$ **23.** 64   **25.** $(3, 8)$ **27.** $(9, 2)$

29. $2, -1$ **31.** About 1.618, about -0.618

33. $\dfrac{\ln 5}{\ln 3} \approx 1.465$ **35.** $\ln 5 \approx 1.609$ **37.** $\ln 28 \approx 3.332$

39. $\dfrac{\ln 80}{2 \ln 3} \approx 1.994$ **41.** 2   **43.** 4

45. $3 - \dfrac{\ln 565}{\ln 2} \approx -6.142$ **47.** $\dfrac{1}{3} \log\left(\dfrac{3}{2}\right) \approx 0.059$

49. $1 + \dfrac{\ln 7}{\ln 5} \approx 2.209$ **51.** $\dfrac{\ln 12}{3} \approx 0.828$

53. $-\ln \dfrac{3}{5} \approx 0.511$ **55.** 0   **57.** $\dfrac{\ln \frac{8}{3}}{3 \ln 2} + \dfrac{1}{3} \approx 0.805$

59. $\ln 5 \approx 1.609$ **61.** $\ln 4 \approx 1.386$ **63.** $2 \ln 75 \approx 8.635$

65. $\frac{1}{2} \ln 1498 \approx 3.656$ **67.** $\dfrac{\ln 4}{365 \ln\left(1 + \frac{0.065}{365}\right)} \approx 21.330$

69. $\dfrac{\ln 2}{12 \ln\left(1 + \frac{0.10}{12}\right)} \approx 6.960$

71.
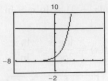
2.807

73.
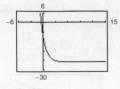
−0.427

75.
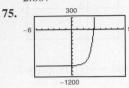
3.847

77.
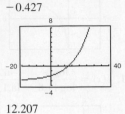
12.207

79.
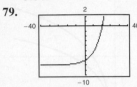
16.636

81. $e^{-3} \approx 0.050$ **83.** $e^7 \approx 1096.633$ **85.** $\dfrac{e^{2.4}}{2} \approx 5.512$

87. 1,000,000   **89.** $\dfrac{e^{10/3}}{5} \approx 5.606$ **91.** $e^2 - 2 \approx 5.389$

93. $e^{-2/3} \approx 0.513$ **95.** $\dfrac{e^{19/2}}{3} \approx 4453.242$

97. $2(3^{11/6}) \approx 14.988$ **99.** No solution

101. $1 + \sqrt{1+e} \approx 2.928$ **103.** No solution   **105.** 7

107. $\dfrac{-1 + \sqrt{17}}{2} \approx 1.562$ **109.** 2

111. $\dfrac{725 + 125\sqrt{33}}{8} \approx 180.384$

113.
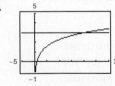
20.086

115.
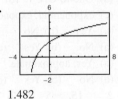
1.482

117. (a) 13.86 yr   (b) 21.97 yr
119. (a) 27.73 yr   (b) 43.94 yr
121. $-1, 0$ **123.** 1   **125.** $e^{-1/2} \approx 0.607$
127. $e^{-1} \approx 0.368$ **129.** (a) 210 coins   (b) 588 coins
131. (a)
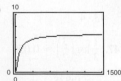

(b) $V = 6.7$; The yield will approach 6.7 million cubic feet per acre.
(c) 29.3 yr
133. 2003
135. (a) $y = 100$ and $y = 0$; The range falls between 0% and 100%.
(b) Males: 69.71 in.   Females: 64.51 in.

137. (a)

x	0.2	0.4	0.6	0.8	1.0
y	162.6	78.5	52.5	40.5	33.9

(b)
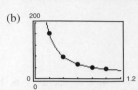

The model appears to fit the data well.
(c) 1.2 m
(d) No. According to the model, when the number of g's is less than 23, x is between 2.276 meters and 4.404 meters, which isn't realistic in most vehicles.

139. $\log_b uv = \log_b u + \log_b v$
True by Property 1 in Section 5.3.
141. $\log_b(u - v) = \log_b u - \log_b v$
False
$1.95 \approx \log(100 - 10) \neq \log 100 - \log 10 = 1$
143. Yes. See Exercise 103.
145. Yes. Time to double: $t = \dfrac{\ln 2}{r}$;

Time to quadruple: $t = \dfrac{\ln 4}{r} = 2\left(\dfrac{\ln 2}{r}\right)$

147. (a)
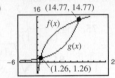
(b) $a = e^{1/e}$
(c) $1 < a < e^{1/e}$

Section 5.5   *(page 422)*

1. $y = ae^{bx}$; $y = ae^{-bx}$ **3.** normally distributed

5. $y = \dfrac{a}{1 + be^{-rx}}$ **7.** c   **8.** e   **9.** b   **10.** a

11. d   **12.** f

13. (a) $P = \dfrac{A}{e^{rt}}$ (b) $t = \dfrac{\ln\left(\dfrac{A}{P}\right)}{r}$

Initial Investment	Annual % Rate	Time to Double	Amount After 10 years
15. $1000	3.5%	19.8 yr	$1419.07
17. $750	8.9438%	7.75 yr	$1834.37
19. $500	11.0%	6.3 yr	$1505.00
21. $6376.28	4.5%	15.4 yr	$10,000.00

23. $303,580.52
25. (a) 7.27 yr   (b) 6.96 yr   (c) 6.93 yr   (d) 6.93 yr
27.

r	2%	4%	6%	8%	10%	12%
t	54.93	27.47	18.31	13.73	10.99	9.16

29.

r	2%	4%	6%	8%	10%	12%
t	55.48	28.01	18.85	14.27	11.53	9.69

31.

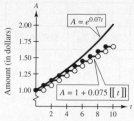

Continuous compounding

Half-life (years)	Initial Quantity	Amount After 1000 Years
33. 1599	10 g	6.48 g
35. 24,100	2.1 g	2.04 g
37. 5715	2.26 g	2 g

39. $y = e^{0.7675x}$ **41.** $y = 5e^{-0.4024x}$

43. (a)

Year	1970	1980	1990	2000	2007
Population	73.7	103.74	143.56	196.35	243.24

(b) 2014

(c) No; The population will not continue to grow at such a quick rate.

45. $k = 0.2988$; About 5,309,734 hits

47. (a) $k = 0.02603$; The population is increasing because $k > 0$.

(b) 449,910; 512,447   (c) 2014

49. About 800 bacteria

51. (a) About 12,180 yr old   (b) About 4797 yr old

53. (a) $V = -5400t + 23,300$ (b) $V = 23,300e^{-0.311t}$

(c)

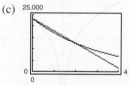

The exponential model depreciates faster.

(d)

t	1 yr	3 yr
$V = -5400t + 23,300$	17,900	7100
$V = 23,300e^{-0.311t}$	17,072	9166

(e) Answers will vary.

55. (a) $S(t) = 100(1 - e^{-0.1625t})$

(b)

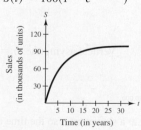

(c) 55,625

57. (a)

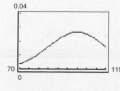

(b) 100

59. (a) 715; 90,880; 199,043

(b)

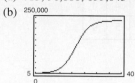

(c) 2014

(d) $235,000 = \dfrac{237,101}{1 + 1950e^{-0.355t}}$

$t \approx 34.63$

61. (a) 203 animals   (b) 13 mo

(c)

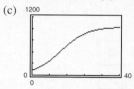

Horizontal asymptotes: $p = 0, p = 1000$. The population size will approach 1000 as time increases.

63. (a) $10^{8.5} \approx 316,227,766$ (b) $10^{5.4} \approx 251,189$

(c) $10^{6.1} \approx 1,258,925$

65. (a) 20 dB   (b) 70 dB   (c) 40 dB   (d) 120 dB

67. 95%   **69.** 4.64   **71.** 1.58×10^{-6} moles/L

73. $10^{5.1}$ **75.** 3:00 A.M.

77. (a)

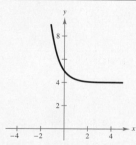

(b) $t \approx 21$ yr; Yes

79. False. The domain can be the set of real numbers for a logistic growth function.

81. False. The graph of $f(x)$ is the graph of $g(x)$ shifted upward five units.

83. Answers will vary.

Review Exercises   (page 430)

1. 0.164   **3.** 0.337   **5.** 1456.529

7. Shift the graph of f two units downward.

9. Reflect f in the y-axis and shift two units to the right.

11. Reflect f in the x-axis and shift one unit upward.

13. Reflect f in the x-axis and shift two units to the left.

15.

x	-1	0	1	2	3
$f(x)$	8	5	4.25	4.063	4.016

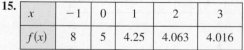

CHAPTER 5

17.

x	-1	0	1	2	3
$f(x)$	4.008	4.04	4.2	5	9

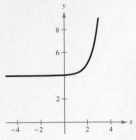

19.

x	-2	-1	0	1	2
$f(x)$	3.25	3.5	4	5	7

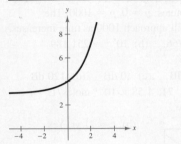

21. $x = 1$ **23.** $x = 4$ **25.** 2980.958   **27.** 0.183

29.

x	-2	-1	0	1	2
$h(x)$	2.72	1.65	1	0.61	0.37

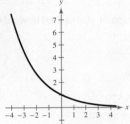

31.

x	-3	-2	-1	0	1
$f(x)$	0.37	1	2.72	7.39	20.09

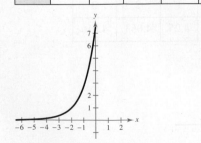

33.

n	1	2	4	12
A	\$6719.58	\$6734.28	\$6741.74	\$6746.77

n	365	Continuous
A	\$6749.21	\$6749.29

35. (a) 0.154   (b) 0.487   (c) 0.811
37. $\log_3 27 = 3$ **39.** $\ln 2.2255 \ldots = 0.8$
41. 3   **43.** -2 **45.** $x = 7$ **47.** $x = -5$
49. Domain: $(0, \infty)$ **51.** Domain: $(-5, \infty)$
x-intercept: $(1, 0)$ x-intercept: $(9995, 0)$
Vertical asymptote: $x = 0$ Vertical asymptote: $x = -5$

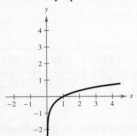   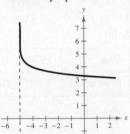

53. (a) 3.118   (b) -0.020
55. Domain: $(0, \infty)$ **57.** Domain: $(-\infty, 0), (0, \infty)$
x-intercept: $(e^{-3}, 0)$ x-intercept: $(\pm 1, 0)$
Vertical asymptote: $x = 0$ Vertical asymptote: $x = 0$

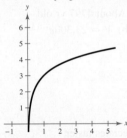   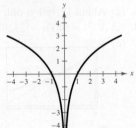

59. 53.4 in.   **61.** 2.585   **63.** -2.322
65. $\log 2 + 2 \log 3 \approx 1.255$ **67.** $2 \ln 2 + \ln 5 \approx 2.996$
69. $1 + 2 \log_5 x$ **71.** $2 - \frac{1}{2} \log_3 x$
73. $2 \ln x + 2 \ln y + \ln z$ **75.** $\log_2 5x$ **77.** $\ln \dfrac{x}{\sqrt[4]{y}}$
79. $\log_3 \dfrac{\sqrt{x}}{(y + 8)^2}$
81. (a) $0 \le h < 18,000$
(b)

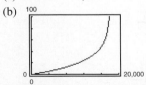

Vertical asymptote: $h = 18,000$

(c) The plane is climbing at a slower rate, so the time required increases.
(d) 5.46 min
83. 3   **85.** $\ln 3 \approx 1.099$ **87.** $e^4 \approx 54.598$
89. $x = 1, 3$ **91.** $\dfrac{\ln 32}{\ln 2} = 5$

93.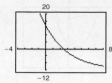

2.447

95. $\frac{1}{3}e^{8.2} \approx 1213.650$

97. $3e^2 \approx 22.167$ **99.** $e^8 \approx 2980.958$

101. No solution **103.** 0.900

105.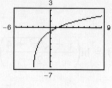

1.482

107.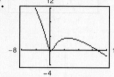

0, 0.416, 13.627

109. 31.4 yr **111.** e **112.** b **113.** f **114.** d

115. a **116.** c **117.** $y = 2e^{0.1014x}$

119. (a)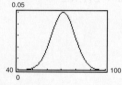

The model fits the data well.

(b) 2022; Answers will vary.

121. (a)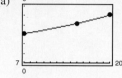

(b) 71

123. (a) 10^{-6} W/m^2 (b) $10\sqrt{10} \text{ W/m}^2$

(c) $1.259 \times 10^{-12} \text{ W/m}^2$

125. True by the inverse properties

Chapter Test *(page 433)*

1. 2.366 **2.** 687.291 **3.** 0.497 **4.** 22.198

5.

x	-1	$-\frac{1}{2}$	0	$\frac{1}{2}$	1
$f(x)$	10	3.162	1	0.316	0.1

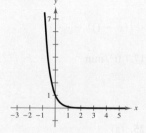

6.

x	-1	0	1	2	3
$f(x)$	-0.005	-0.028	-0.167	-1	-6

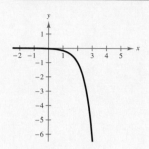

7.

x	-1	$-\frac{1}{2}$	0	$\frac{1}{2}$	1
$f(x)$	0.865	0.632	0	-1.718	-6.389

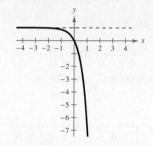

8. (a) -0.89 (b) 9.2

9.

x	$\frac{1}{2}$	1	$\frac{3}{2}$	2	4
$f(x)$	-5.699	-6	-6.176	-6.301	-6.602

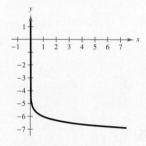

Vertical asymptote: $x = 0$

10.

x	5	7	9	11	13
$f(x)$	0	1.099	1.609	1.946	2.197

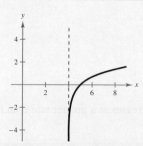

Vertical asymptote: $x = 4$

11.

x	-5	-3	-1	0	1
$f(x)$	1	2.099	2.609	2.792	2.946

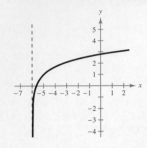

Vertical asymptote: $x = -6$

12. 1.945   **13.** -0.167 **14.** -11.047

15. $\log_2 3 + 4 \log_2 |a|$ **16.** $\ln 5 + \frac{1}{2}\ln x - \ln 6$

17. $3 \log(x - 1) - 2 \log y - \log z$ **18.** $\log_3 13y$

19. $\ln \dfrac{x^4}{y^4}$ **20.** $\ln\left(\dfrac{x^3 y^2}{x + 3}\right)$ **21.** $x = -2$

22. $x = \dfrac{\ln 44}{-5} \approx -0.757$ **23.** $\dfrac{\ln 197}{4} \approx 1.321$

24. $e^{1/2} \approx 1.649$ **25.** $e^{-11/4} \approx 0.0639$ **26.** 20

27. $y = 2745 e^{0.1570t}$ **28.** 55%

29. (a)

x	$\frac{1}{4}$	1	2	4	5	6
H	58.720	75.332	86.828	103.43	110.59	117.38

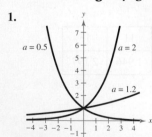

(b) 103 cm; 103.43 cm

Problem Solving   *(page 435)*

1.

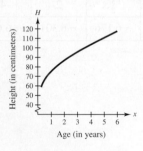

$y = 0.5^x$ and $y = 1.2^x$

$0 < a \le e^{1/e}$

3. As $x \to \infty$, the graph of e^x increases at a greater rate than the graph of x^n.

5. Answers will vary.

7. (a)

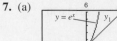
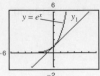

(b)

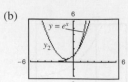

(c)

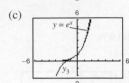

9.

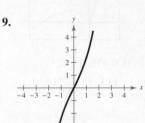

$f^{-1}(x) = \ln\left(\dfrac{x + \sqrt{x^2 + 4}}{2}\right)$

11. c   **13.** $t = \dfrac{\ln c_1 - \ln c_2}{\left(\dfrac{1}{k_2} - \dfrac{1}{k_1}\right)\ln \dfrac{1}{2}}$

15. (a) $y_1 = 252{,}606(1.0310)^t$

(b) $y_2 = 400.88t^2 - 1464.6t + 291{,}782$

(c)

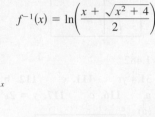

(d) The exponential model is a better fit. No, because the model is rapidly approaching infinity.

17. $1, e^2$

19. $y_4 = (x - 1) - \frac{1}{2}(x - 1)^2 + \frac{1}{3}(x - 1)^3 - \frac{1}{4}(x - 1)^4$

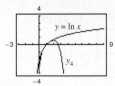

The pattern implies that

$\ln x = (x - 1) - \frac{1}{2}(x - 1)^2 + \frac{1}{3}(x - 1)^3 - \cdots .$

21.

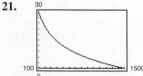

17.7 ft³/min

23. (a)

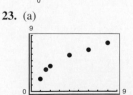

(b)–(e) Answers will vary.

25. (a)

(b)–(e) Answers will vary.

Chapter 6

Section 6.1   *(page 442)*

1. inclination   **3.** $\left|\dfrac{m_2 - m_1}{1 + m_1 m_2}\right|$ **5.** $\dfrac{\sqrt{3}}{3}$ **7.** -1

9. $\sqrt{3}$ **11.** 3.2236   **13.** $\dfrac{3\pi}{4}$ rad, 135°   **15.** $\dfrac{\pi}{4}$ rad, 45°

17. 0.6435 rad, 36.9°   **19.** $\dfrac{\pi}{6}$ rad, 30°   **21.** $\dfrac{5\pi}{6}$ rad, 150°

23. 1.0517 rad, 60.3°   **25.** 2.1112 rad, 121.0°

27. $\dfrac{3\pi}{4}$ rad, 135°   **29.** $\dfrac{\pi}{4}$ rad, 45°   **31.** $\dfrac{5\pi}{6}$ rad, 150°

33. 1.2490 rad, 71.6°   **35.** 2.1112 rad, 121.0°

37. 1.1071 rad, 63.4°   **39.** 0.1974 rad, 11.3°

41. 1.4289 rad, 81.9°   **43.** 0.9273 rad, 53.1°

45. 0.8187 rad, 46.9°

47. $(1, 5) \leftrightarrow (4, 5)$: slope = 0
$(4, 5) \leftrightarrow (3, 8)$: slope = -3
$(3, 8) \leftrightarrow (1, 5)$: slope = $\frac{3}{2}$
$(1, 5)$: 56.3°; $(4, 5)$: 71.6°; $(3, 8)$: 52.1°

49. $(-4, -1) \leftrightarrow (3, 2)$: slope = $\frac{3}{7}$
$(3, 2) \leftrightarrow (1, 0)$: slope = 1
$(1, 0) \leftrightarrow (-4, -1)$: slope = $\frac{1}{5}$
$(-4, -1)$: 11.9°; $(3, 2)$: 21.8°; $(1, 0)$: 146.3°

51. 0   **53.** $\dfrac{4\sqrt{10}}{5} \approx 2.5298$ **55.** 7

57. $\dfrac{8\sqrt{37}}{37} \approx 1.3152$

59. (a)   **61.** (a)

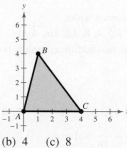

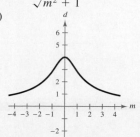

(b) 4   (c) 8

(b) $\dfrac{35\sqrt{37}}{74}$ (c) $\dfrac{35}{8}$

63. $2\sqrt{2}$ **65.** 0.1003, 1054 ft   **67.** 31.0°

69. $\alpha \approx 33.69°$; $\beta \approx 56.31°$

71. True. The inclination of a line is related to its slope by $m = \tan\theta$. If the angle is greater than $\pi/2$ but less than π, then the angle is in the second quadrant, where the tangent function is negative.

73. (a) $d = \dfrac{4}{\sqrt{m^2 + 1}}$

(b)

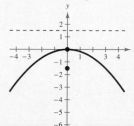

(c) $m = 0$

(d) The graph has a horizontal asymptote of $d = 0$. As the slope becomes larger, the distance between the origin and the line, $y = mx + 4$, becomes smaller and approaches 0.

75. (a) $d = \dfrac{3|m + 1|}{\sqrt{m^2 + 1}}$

(b)

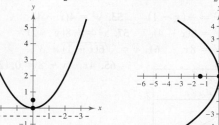

(c) $m = 1$ (d) Yes. $m = -1$

(e) $d = 3$. As the line approaches the vertical, the distance approaches 3.

Section 6.2   *(page 450)*

1. conic   **3.** locus   **5.** axis   **7.** focal chord

9. A circle is formed when a plane intersects the top or bottom half of a double-napped cone and is perpendicular to the axis of the cone.

11. A parabola is formed when a plane intersects the top or bottom half of a double-napped cone, is parallel to the side of the cone, and does not intersect the vertex.

13. e   **14.** b   **15.** d   **16.** f   **17.** a   **18.** c

19. $x^2 = \frac{3}{2}y$ **21.** $x^2 = 2y$ **23.** $y^2 = -8x$

25. $x^2 = -4y$ **27.** $y^2 = 4x$ **29.** $x^2 = \frac{8}{3}y$

31. $y^2 = -\frac{25}{2}x$

33. Vertex: $(0, 0)$
Focus: $\left(0, \frac{1}{2}\right)$
Directrix: $y = -\frac{1}{2}$

35. Vertex: $(0, 0)$
Focus: $\left(-\frac{3}{2}, 0\right)$
Directrix: $x = \frac{3}{2}$

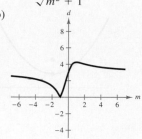

37. Vertex: $(0, 0)$
Focus: $\left(0, -\frac{3}{2}\right)$
Directrix: $y = \frac{3}{2}$

39. Vertex: $(1, -2)$
Focus: $(1, -4)$
Directrix: $y = 0$

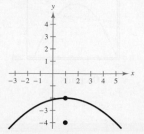

CHAPTER 6

41. Vertex: $\left(-3, \frac{3}{2}\right)$
Focus: $\left(-3, \frac{5}{2}\right)$
Directrix: $y = \frac{1}{2}$

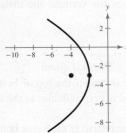

43. Vertex: $(1, 1)$
Focus: $(1, 2)$
Directrix: $y = 0$

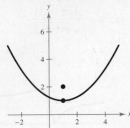

45. Vertex: $(-2, -3)$
Focus: $(-4, -3)$
Directrix: $x = 0$

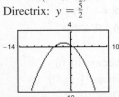

47. Vertex: $(-2, 1)$
Focus: $\left(-2, -\frac{1}{2}\right)$
Directrix: $y = \frac{5}{2}$

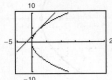

49. Vertex: $\left(\frac{1}{4}, -\frac{1}{2}\right)$
Focus: $\left(0, -\frac{1}{2}\right)$
Directrix: $x = \frac{1}{2}$

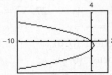

51. $(x - 3)^2 = -(y - 1)$ **53.** $y^2 = 4(x + 4)$
55. $(y - 3)^2 = 8(x - 4)$ **57.** $x^2 = -8(y - 2)$
59. $(y - 2)^2 = 8x$ **61.** $y = \sqrt{6(x + 1)} + 3$
63.

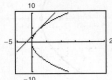

$(2, 4)$

65. $4x - y - 8 = 0$; $(2, 0)$

67. $4x - y + 2 = 0$; $\left(-\frac{1}{2}, 0\right)$
69.

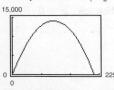

$x = 106$ units

71. (a)

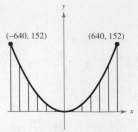

(−640, 152)   (640, 152)

(b) $y = \dfrac{19x^2}{51,200}$

(c)

Distance, x	0	100	250	400	500
Height, y	0	3.71	23.19	59.38	92.77

73. (a) $y = -\frac{1}{640}x^2$ (b) 8 ft
75. (a) $x^2 = 180,000y$ (b) $300\sqrt{2}$ cm ≈ 424.26 cm
77. $x^2 = -\frac{25}{4}(y - 48)$
79. (a) $17,500\sqrt{2}$ mi/h $\approx 24,750$ mi/h
(b) $x^2 = -16,400(y - 4100)$
81. (a) $x^2 = -49(y - 100)$ (b) 70 ft
83. False. If the graph crossed the directrix, there would exist points closer to the directrix than the focus.

85. $m = \dfrac{x_1}{2p}$

87. (a)

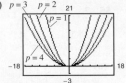

$p = 3$ $p = 2$ 21 $p = 1$ $p = 4$ −18 18 −3

As p increases, the graph becomes wider.
(b) $(0, 1), (0, 2), (0, 3), (0, 4)$ (c) $4, 8, 12, 16$; $4|p|$
(d) It is an easy way to determine two additional points on the graph.

Section 6.3   (page 460)

1. ellipse; foci   **3.** minor axis   **5.** b   **6.** c   **7.** d

8. f   **9.** a   **10.** e   **11.** $\dfrac{x^2}{4} + \dfrac{y^2}{16} = 1$

13. $\dfrac{x^2}{49} + \dfrac{y^2}{45} = 1$ **15.** $\dfrac{x^2}{49} + \dfrac{y^2}{24} = 1$ **17.** $\dfrac{21x^2}{400} + \dfrac{y^2}{25} = 1$

19. $\dfrac{(x - 2)^2}{1} + \dfrac{(y - 3)^2}{9} = 1$ **21.** $\dfrac{(x - 4)^2}{16} + \dfrac{(y - 2)^2}{1} = 1$

23. $\dfrac{x^2}{48} + \dfrac{(y - 4)^2}{64} = 1$ **25.** $\dfrac{x^2}{16} + \dfrac{(y - 4)^2}{12} = 1$

27. $\dfrac{(x - 2)^2}{4} + \dfrac{(y - 2)^2}{1} = 1$

29. Ellipse

Center: $(0, 0)$

Vertices: $(\pm 5, 0)$

Foci: $(\pm 3, 0)$

Eccentricity: $\dfrac{3}{5}$

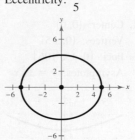

31. Circle

Center: $(0, 0)$

Radius: 5

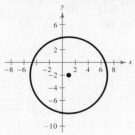

43. Circle

Center: $(1, -2)$

Radius: 6

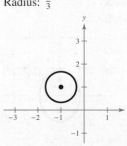

45. Ellipse

Center: $(-3, 1)$

Vertices: $(-3, 7), (-3, -5)$

Foci: $(-3, 1 \pm 2\sqrt{6})$

Eccentricity: $\dfrac{\sqrt{6}}{3}$

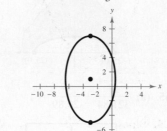

33. Ellipse

Center: $(0, 0)$

Vertices: $(0, \pm 3)$

Foci: $(0, \pm 2)$

Eccentricity: $\dfrac{2}{3}$

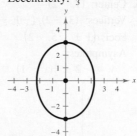

35. Ellipse

Center: $(4, -1)$

Vertices: $(4, -6), (4, 4)$

Foci: $(4, 2), (4, -4)$

Eccentricity: $\dfrac{3}{5}$

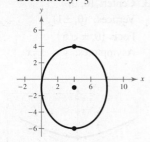

47. Ellipse

Center: $\left(3, -\dfrac{5}{2}\right)$

Vertices: $\left(9, -\dfrac{5}{2}\right), \left(-3, -\dfrac{5}{2}\right)$

Foci: $\left(3 \pm 3\sqrt{3}, -\dfrac{5}{2}\right)$

Eccentricity: $\dfrac{\sqrt{3}}{2}$

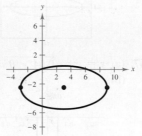

49. Circle

Center: $(-1, 1)$

Radius: $\dfrac{2}{3}$

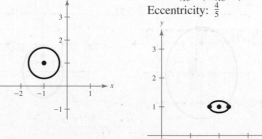

51. Ellipse

Center: $(2, 1)$

Vertices: $\left(\dfrac{7}{3}, 1\right), \left(\dfrac{5}{3}, 1\right)$

Foci: $\left(\dfrac{34}{15}, 1\right), \left(\dfrac{26}{15}, 1\right)$

Eccentricity: $\dfrac{4}{5}$

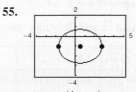

37. Circle

Center: $(0, -1)$

Radius: $\dfrac{2}{3}$

39. Ellipse

Center: $(-2, -4)$

Vertices: $(-3, -4), (-1, -4)$

Foci: $\left(\dfrac{-4 \pm \sqrt{3}}{2}, -4\right)$

Eccentricity: $\dfrac{\sqrt{3}}{2}$

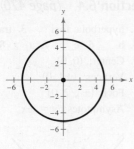

41. Ellipse

Center: $(-2, 3)$

Vertices: $(-2, 6), (-2, 0)$

Foci: $(-2, 3 \pm \sqrt{5})$

Eccentricity: $\dfrac{\sqrt{5}}{3}$

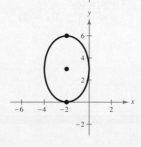

53.

Center: $(0, 0)$

Vertices: $(0, \pm \sqrt{5})$

Foci: $(0, \pm \sqrt{2})$

55.

Center: $\left(\dfrac{1}{2}, -1\right)$

Vertices: $\left(\dfrac{1}{2} \pm \sqrt{5}, -1\right)$

Foci: $\left(\dfrac{1}{2} \pm \sqrt{2}, -1\right)$

57. $\dfrac{\sqrt{5}}{3}$ **59.** $\dfrac{2\sqrt{2}}{3}$ **61.** $\dfrac{x^2}{25} + \dfrac{y^2}{16} = 1$

63. (a)

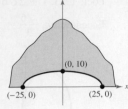

(b) $\dfrac{x^2}{625} + \dfrac{y^2}{100} = 1$

(c) Yes

65. (a) $\dfrac{x^2}{321.84} + \dfrac{y^2}{20.89} = 1$

(b)

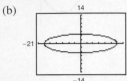

(c) Aphelion:
35.29 astronomical units
Perihelion:
0.59 astronomical unit

67. (a) $y = -8\sqrt{0.04 - \theta^2}$

(b)

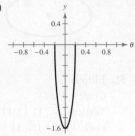

(c) The bottom half

69.

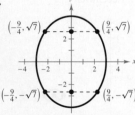

$\left(-\tfrac{9}{4}, \sqrt{7}\right)$ $\left(\tfrac{9}{4}, \sqrt{7}\right)$
$\left(-\tfrac{9}{4}, -\sqrt{7}\right)$ $\left(\tfrac{9}{4}, -\sqrt{7}\right)$

71.

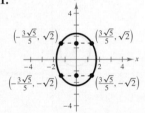

$\left(-\tfrac{3\sqrt{5}}{5}, \sqrt{2}\right)$ $\left(\tfrac{3\sqrt{5}}{5}, \sqrt{2}\right)$
$\left(-\tfrac{3\sqrt{5}}{5}, -\sqrt{2}\right)$ $\left(\tfrac{3\sqrt{5}}{5}, -\sqrt{2}\right)$

73. False. The graph of $(x^2/4) + y^4 = 1$ is not an ellipse. The degree of y is 4, not 2.

75. (a) $A = \pi a(20 - a)$ (b) $\dfrac{x^2}{196} + \dfrac{y^2}{36} = 1$

(c)

a	8	9	10	11	12	13
A	301.6	311.0	314.2	311.0	301.6	285.9

$a = 10$, circle

(d)

The shape of an ellipse with a maximum area is a circle. The maximum area is found when $a = 10$ (verified in part c) and therefore $b = 10$, so the equation produces a circle.

77. $\dfrac{(x - 6)^2}{324} + \dfrac{(y - 2)^2}{308} = 1$ **79.** Proof

Section 6.4 *(page 470)*

1. hyperbola; foci **3.** transverse axis; center

5. b **6.** c **7.** a **8.** d

9. Center: $(0, 0)$
Vertices: $(\pm 1, 0)$
Foci: $\left(\pm \sqrt{2}, 0\right)$
Asymptotes: $y = \pm x$

11. Center: $(0, 0)$
Vertices: $(0, \pm 5)$
Foci: $\left(0, \pm \sqrt{106}\right)$
Asymptotes: $y = \pm \tfrac{5}{9}x$

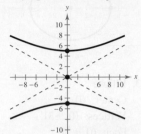

13. Center: $(0, 0)$
Vertices: $(0, \pm 1)$
Foci: $\left(0, \pm \sqrt{5}\right)$
Asymptotes: $y = \pm \tfrac{1}{2}x$

15. Center: $(1, -2)$
Vertices: $(3, -2), (-1, -2)$
Foci: $\left(1 \pm \sqrt{5}, -2\right)$
Asymptotes:
$y = -2 \pm \tfrac{1}{2}(x - 1)$

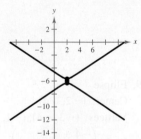

17. Center: $(2, -6)$
Vertices:
$\left(2, -\dfrac{17}{3}\right), \left(2, -\dfrac{19}{3}\right)$
Foci: $\left(2, -6 \pm \dfrac{\sqrt{13}}{6}\right)$
Asymptotes:
$y = -6 \pm \dfrac{2}{3}(x - 2)$

19. Center: $(2, -3)$
Vertices: $(3, -3), (1, -3)$
Foci: $\left(2 \pm \sqrt{10}, -3\right)$
Asymptotes:
$y = -3 \pm 3(x - 2)$

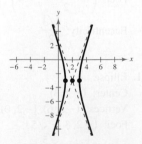

21. The graph of this equation is two lines intersecting at $(-1, -3)$.

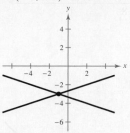

23. Center: $(0, 0)$
Vertices: $\left(\pm\sqrt{3}, 0\right)$
Foci: $\left(\pm\sqrt{5}, 0\right)$
Asymptotes: $y = \pm\dfrac{\sqrt{6}}{3}x$

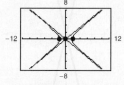

25. Center: $(0, 0)$
Vertices: $(\pm 3, 0)$
Foci: $\left(\pm\sqrt{13}, 0\right)$
Asymptotes: $y = \pm\frac{2}{3}x$

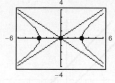

27. Center: $(1, -3)$
Vertices: $\left(1, -3 \pm \sqrt{2}\right)$
Foci: $\left(1, -3 \pm 2\sqrt{5}\right)$
Asymptotes:
$y = -3 \pm \frac{1}{3}(x - 1)$

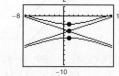

29. $\dfrac{y^2}{4} - \dfrac{x^2}{12} = 1$ **31.** $\dfrac{x^2}{1} - \dfrac{y^2}{25} = 1$

33. $\dfrac{17y^2}{1024} - \dfrac{17x^2}{64} = 1$ **35.** $\dfrac{(x - 4)^2}{4} - \dfrac{y^2}{12} = 1$

37. $\dfrac{(y - 5)^2}{16} - \dfrac{(x - 4)^2}{9} = 1$ **39.** $\dfrac{y^2}{9} - \dfrac{4(x - 2)^2}{9} = 1$

41. $\dfrac{(y - 2)^2}{4} - \dfrac{x^2}{4} = 1$ **43.** $\dfrac{(x - 2)^2}{1} - \dfrac{(y - 2)^2}{1} = 1$

45. $\dfrac{(x - 3)^2}{9} - \dfrac{(y - 2)^2}{4} = 1$ **47.** $\dfrac{y^2}{9} - \dfrac{x^2}{9/4} = 1$

49. $\dfrac{(x - 3)^2}{4} - \dfrac{(y - 2)^2}{16/5} = 1$

51. (a) $\dfrac{x^2}{1} - \dfrac{y^2}{169/3} = 1$ (b) About 2.403 ft

53. $(3300, -2750)$

55. (a) $\dfrac{x^2}{1} - \dfrac{y^2}{27} = 1$; $-9 \leq y \leq 9$ (b) 1.89 ft

57. Ellipse **59.** Hyperbola **61.** Hyperbola
63. Parabola **65.** Ellipse **67.** Parabola
69. Parabola **71.** Circle

73. True. For a hyperbola, $c^2 = a^2 + b^2$. The larger the ratio of b to a, the larger the eccentricity of the hyperbola, $e = c/a$.

75. False. When $D = -E$, the graph is two intersecting lines.

77. Answers will vary.

79. $y = 1 - 3\sqrt{\dfrac{(x - 3)^2}{4} - 1}$

81.

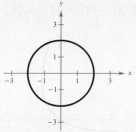

The equation $y = x^2 + C$ is a parabola that could intersect the circle in zero, one, two, three, or four places depending on its location on the y-axis.
(a) $C > 2$ and $C < -\frac{17}{4}$ (b) $C = 2$
(c) $-2 < C < 2$, $C = -\frac{17}{4}$ (d) $C = -2$
(e) $-\frac{17}{4} < C < -2$

Section 6.5 *(page 479)*

1. rotation; axes **3.** invariant under rotation **5.** $(3, 0)$

7. $\left(\dfrac{3 + \sqrt{3}}{2}, \dfrac{3\sqrt{3} - 1}{2}\right)$ **9.** $\left(\dfrac{3\sqrt{2}}{2}, -\dfrac{\sqrt{2}}{2}\right)$

11. $\left(\dfrac{2\sqrt{3} + 1}{2}, \dfrac{2 - \sqrt{3}}{2}\right)$

13. $\dfrac{(y')^2}{2} - \dfrac{(x')^2}{2} = 1$ **15.** $y' = \pm\dfrac{\sqrt{2}}{2}$

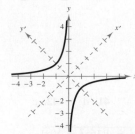

17. $\dfrac{\left(x' - 6\sqrt{2}\right)^2}{64} - \dfrac{\left(y' - 2\sqrt{2}\right)^2}{64} = 1$

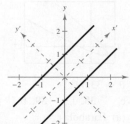

19. $\dfrac{(x')^2}{6} + \dfrac{(y')^2}{3/2} = 1$ **21.** $y' = \dfrac{-\sqrt{2}}{4}(x')^2$

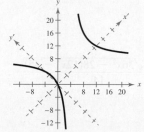

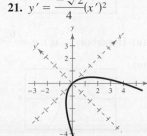

CHAPTER 6

23. $(y')^2 = -x'$ **25.** $(x' - 1)^2 = 6\left(y' + \frac{1}{6}\right)$

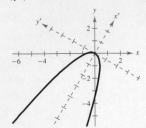   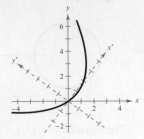

27. 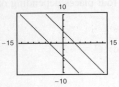   **29.**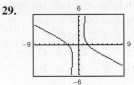

$\theta = 45°$ $\theta \approx 26.57°$

31. 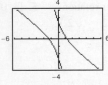   **33.**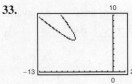

$\theta \approx 31.72°$ $\theta = 45°$

35.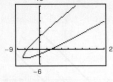

$\theta \approx 33.69°$

37. e   **38.** f   **39.** b   **40.** a   **41.** d   **42.** c

43. (a) Parabola

(b) $y = \dfrac{(8x - 5) \pm \sqrt{(8x - 5)^2 - 4(16x^2 - 10x)}}{2}$

(c)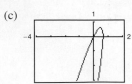

45. (a) Ellipse   (b) $y = \dfrac{6x \pm \sqrt{36x^2 - 28(12x^2 - 45)}}{14}$

(c)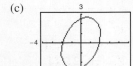

47. (a) Hyperbola   (b) $y = \dfrac{6x \pm \sqrt{36x^2 + 20(x^2 + 4x - 22)}}{-10}$

(c)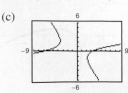

49. (a) Parabola

(b) $y = \dfrac{-(4x - 1) \pm \sqrt{(4x - 1)^2 - 16(x^2 - 5x - 3)}}{8}$

(c)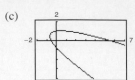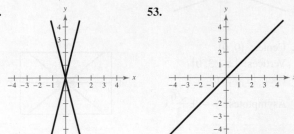

51. 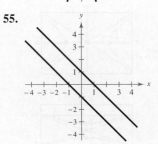   **53.**

55.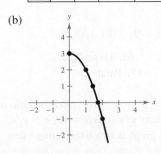

57. $(2, 2), (2, 4)$ **59.** $(-8, 12)$ **61.** $(0, 8), (12, 8)$
63. $(0, 4)$ **65.** $\left(1, \sqrt{3}\right), \left(1, -\sqrt{3}\right)$ **67.** No solution
69. $\left(0, \frac{3}{2}\right), (-3, 0)$

71. True. The graph of the equation can be classified by finding the discriminant. For a graph to be a hyperbola, the discriminant must be greater than zero. If $k \geq \frac{1}{4}$, then the discriminant would be less than or equal to zero.

73. Answers will vary.   **75.** Major axis: 4; Minor axis: 2

Section 6.6   *(page 486)*

1. plane curve   **3.** eliminating; parameter

5. (a)

t	0	1	2	3	4
x	0	1	$\sqrt{2}$	$\sqrt{3}$	2
y	3	2	1	0	-1

(b)

(graph)

(c) $y = 3 - x^2$

The graph of the rectangular equation shows the entire parabola rather than just the right half.

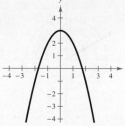

7. (a)

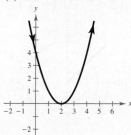

(b) $y = 3x + 4$

9. (a)

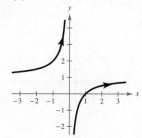

(b) $y = 16x^2$

11. (a)

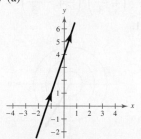

(b) $y = x^2 - 4x + 4$

13. (a)

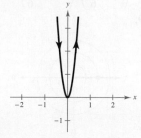

(b) $y = \dfrac{(x - 1)}{x}$

15. (a)

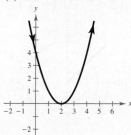

(b) $y = \left| \dfrac{x}{2} - 3 \right|$

17. (a)

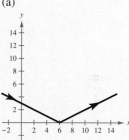

(b) $\dfrac{x^2}{16} + \dfrac{y^2}{4} = 1$

19. (a)

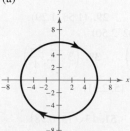

(b) $\dfrac{x^2}{36} + \dfrac{y^2}{36} = 1$

21. (a)

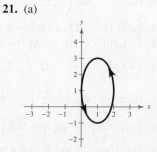

(b) $\dfrac{(x - 1)^2}{1} + \dfrac{(y - 1)^2}{4} = 1$

23. (a)

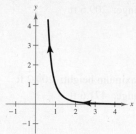

(b) $y = \dfrac{1}{x^3}, \quad x > 0$

25. (a)

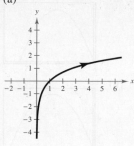

(b) $y = \ln x$

27. Each curve represents a portion of the line $y = 2x + 1$.

Domain	Orientation
(a) $(-\infty, \infty)$	Left to right
(b) $[-1, 1]$	Depends on θ
(c) $(0, \infty)$	Right to left
(d) $(0, \infty)$	Left to right

29. $y - y_1 = m(x - x_1)$

31. $\dfrac{(x - h)^2}{a^2} + \dfrac{(y - k)^2}{b^2} = 1$

33. $x = 3t$
$y = 6t$

35. $x = 3 + 4 \cos \theta$
$y = 2 + 4 \sin \theta$

37. $x = 5 \cos \theta$
$y = 3 \sin \theta$

39. $x = 4 \sec \theta$
$y = 3 \tan \theta$

41. (a) $x = t,\ y = 3t - 2$ (b) $x = -t + 2,\ y = -3t + 4$

43. (a) $x = t,\ y = 2 - t$ (b) $x = -t + 2,\ y = t$

45. (a) $x = t$
$y = t^2 - 3$
(b) $x = 2 - t$
$y = t^2 - 4t + 1$

47. (a) $x = t,\ y = \dfrac{1}{t}$ (b) $x = -t + 2,\ y = -\dfrac{1}{t - 2}$

49.

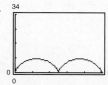

51.

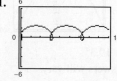

53.

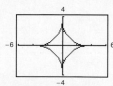

55.

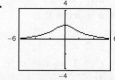

57. b
Domain: $[-2, 2]$
Range: $[-1, 1]$

58. c
Domain: $[-4, 4]$
Range: $[-6, 6]$

59. d
Domain: $(-\infty, \infty)$
Range: $(-\infty, \infty)$

60. a
Domain: $(-\infty, \infty)$
Range: $[-2, 2]$

CHAPTER 6

61. (a)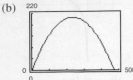

Maximum height: 90.7 ft
Range: 209.6 ft

(b)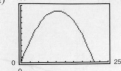

Maximum height: 204.2 ft
Range: 471.6 ft

(c)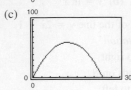

Maximum height: 60.5 ft
Range: 242.0 ft

(d)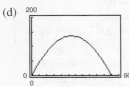

Maximum height: 136.1 ft
Range: 544.5 ft

63. (a) $x = (146.67 \cos \theta)t$
$y = 3 + (146.67 \sin \theta)t - 16t^2$

(b)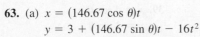

No

(c)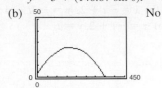

Yes

(d) $19.3°$

65. Answers will vary.

67. $x = a\theta - b \sin \theta$
$y = a - b \cos \theta$

69. True
$x = t$
$y = t^2 + 1 \Longrightarrow y = x^2 + 1$
$x = 3t$
$y = 9t^2 + 1 \Longrightarrow y = x^2 + 1$

71. Parametric equations are useful when graphing two functions simultaneously on the same coordinate system. For example, they are useful when tracking the path of an object so that the position and the time associated with that position can be determined.

73. $-1 < t < \infty$

Section 6.7   (*page 493*)

1. pole   **3.** polar

5.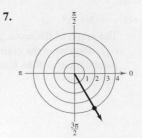

7.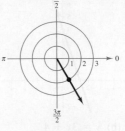

$\left(2, -\dfrac{7\pi}{6}\right), \left(-2, -\dfrac{\pi}{6}\right)$ $\left(4, \dfrac{5\pi}{3}\right), \left(-4, -\dfrac{4\pi}{3}\right)$

9.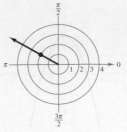

11.

$(2, \pi), (-2, 0)$ $\left(-2, -\dfrac{4\pi}{3}\right), \left(2, \dfrac{5\pi}{3}\right)$

13.

15.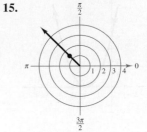

$\left(0, \dfrac{5\pi}{6}\right), \left(0, -\dfrac{\pi}{6}\right)$ $\left(\sqrt{2}, -3.92\right), \left(-\sqrt{2}, -0.78\right)$

17.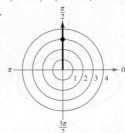

$(-3, 4.71), (3, 1.57)$

19. $(0, 3)$ **21.** $\left(\dfrac{\sqrt{2}}{2}, \dfrac{\sqrt{2}}{2}\right)$ **23.** $\left(-\sqrt{2}, \sqrt{2}\right)$

25. $\left(\sqrt{3}, 1\right)$ **27.** $(-1.1, -2.2)$ **29.** $(1.53, 1.29)$
31. $(-1.20, -4.34)$ **33.** $(-0.02, 2.50)$
35. $(-3.60, 1.97)$ **37.** $\left(\sqrt{2}, \dfrac{\pi}{4}\right)$ **39.** $\left(3\sqrt{2}, \dfrac{5\pi}{4}\right)$
41. $(6, \pi)$ **43.** $\left(5, \dfrac{3\pi}{2}\right)$ **45.** $(5, 2.21)$

47. $\left(\sqrt{6}, \dfrac{5\pi}{4}\right)$ **49.** $\left(2, \dfrac{11\pi}{6}\right)$ **51.** $\left(3\sqrt{13}, 0.98\right)$
53. $(13, 1.18)$ **55.** $\left(\sqrt{13}, 5.70\right)$ **57.** $\left(\sqrt{29}, 2.76\right)$

59. $\left(\sqrt{7}, 0.86\right)$ **61.** $\left(\dfrac{17}{6}, 0.49\right)$ **63.** $\left(\dfrac{\sqrt{85}}{4}, 0.71\right)$

65. $r = 3$ **67.** $r = 4 \csc \theta$ **69.** $r = 10 \sec \theta$

71. $r = -2 \csc \theta$ **73.** $r = \dfrac{-2}{3 \cos \theta - \sin \theta}$

75. $r^2 = 16 \sec \theta \csc \theta = 32 \csc 2\theta$

77. $r = \dfrac{4}{1 - \cos \theta}$ or $-\dfrac{4}{1 + \cos \theta}$ **79.** $r = a$

81. $r = 2a \cos \theta$ **83.** $r = \cot^2 \theta \csc \theta$

85. $x^2 + y^2 - 4y = 0$ **87.** $x^2 + y^2 + 2x = 0$

89. $\sqrt{3}x + y = 0$ **91.** $\dfrac{\sqrt{3}}{3}x + y = 0$

93. $x^2 + y^2 = 16$ **95.** $y = 4$ **97.** $x = -3$

99. $x^2 + y^2 - x^{2/3} = 0$ **101.** $(x^2 + y^2)^2 = 2xy$

103. $(x^2 + y^2)^2 = 6x^2y - 2y^3$ **105.** $x^2 + 4y - 4 = 0$

107. $4x^2 - 5y^2 - 36y - 36 = 0$

109. The graph of the polar
equation consists of all
points that are six units
from the pole.
$x^2 + y^2 = 36$

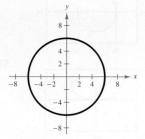

111. The graph of the polar
equation consists of all
points on the line that
makes an angle of $\pi/6$
with the positive polar
axis.
$-\sqrt{3}x + 3y = 0$

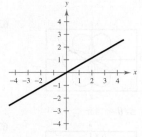

113. The graph of the polar
equation is not evident
by simple inspection, so
convert to rectangular
form.
$x^2 + (y - 1)^2 = 1$

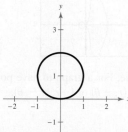

115. The graph of the polar
equation is not evident
by simple inspection, so
convert to rectangular
form.
$(x + 3)^2 + y^2 = 9$

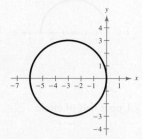

117. The graph of the polar
equation is not evident by
simple inspection, so
convert to rectangular
form.
$x - 3 = 0$

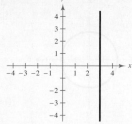

119. True. Because r is a directed distance, the point (r, θ) can be
represented as $(r, \theta \pm 2\pi n)$.

121. $(x - h)^2 + (y - k)^2 = h^2 + k^2$
Radius: $\sqrt{h^2 + k^2}$
Center: (h, k)

123. (a) Answers will vary.
(b) (r_1, θ_1), (r_2, θ_2) and the pole are collinear.
$d = \sqrt{r_1^2 + r_2^2 - 2r_1 r_2} = |r_1 - r_2|$
This represents the distance between two points on the
line $\theta = \theta_1 = \theta_2$.
(c) $d = \sqrt{r_1^2 + r_2^2}$
This is the result of the Pythagorean Theorem.
(d) Answers will vary. For example:
Points: $(3, \pi/6)$, $(4, \pi/3)$
Distance: 2.053
Points: $(-3, 7\pi/6)$, $(-4, 4\pi/3)$
Distance: 2.053

125. (a)

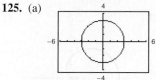

(b) Yes. $\theta \approx 3.927$, $x \approx -2.121$,
$y \approx -2.121$
(c) Yes. Answers will vary.

Section 6.8   *(page 501)*

1. $\theta = \dfrac{\pi}{2}$ **3.** convex limaçon   **5.** lemniscate

7. Rose curve with 4 petals   **9.** Limaçon with inner loop

11. Rose curve with 3 petals   **13.** Polar axis

15. $\theta = \dfrac{\pi}{2}$ **17.** $\theta = \dfrac{\pi}{2}$, polar axis, pole

19. Maximum: $|r| = 20$ when $\theta = \dfrac{3\pi}{2}$
Zero: $r = 0$ when $\theta = \dfrac{\pi}{2}$

21. Maximum: $|r| = 4$ when $\theta = 0, \dfrac{\pi}{3}, \dfrac{2\pi}{3}$
Zeros: $r = 0$ when $\theta = \dfrac{\pi}{6}, \dfrac{\pi}{2}, \dfrac{5\pi}{6}$

CHAPTER 6

23.

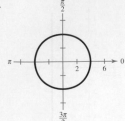

25.

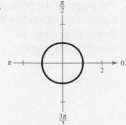

27.

29.

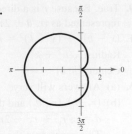

31.

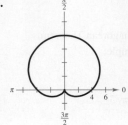

33.

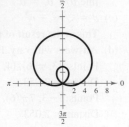

35.

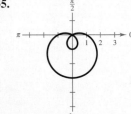

37.

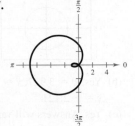

39.

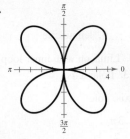

41.

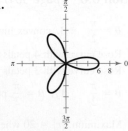

43.

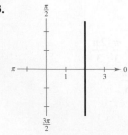

45.

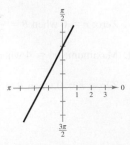

47.

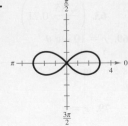

49.

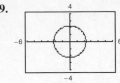

51.

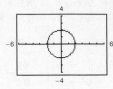

53.

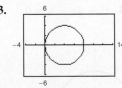

55.

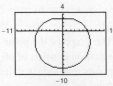

57.

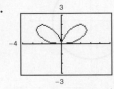

59.

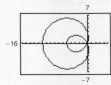

$0 \le \theta < 2\pi$

61.

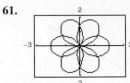

$0 \le \theta < 4\pi$

63.

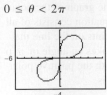

$0 \le \theta < \pi/2$

65.

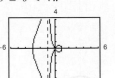

67.

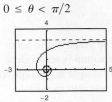

69. True. For a graph to have polar axis symmetry, replace (r, θ) by $(r, -\theta)$ or $(-r, \pi - \theta)$.

71. (a)

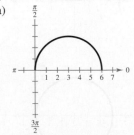

Upper half of circle

(b)

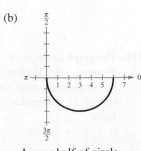

Lower half of circle

(c)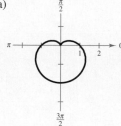

(d)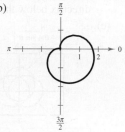

Full circle   Left half of circle

73. Answers will vary.

75. (a) $r = 2 - \dfrac{\sqrt{2}}{2}(\sin \theta - \cos \theta)$ (b) $r = 2 + \cos \theta$
(c) $r = 2 + \sin \theta$ (d) $r = 2 - \cos \theta$

77. (a)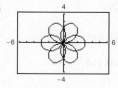

(b)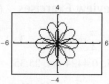

79. 8 petals; 3 petals; For $r = 2 \cos n\theta$ and $r = 2 \sin n\theta$, there are n petals if n is odd, $2n$ petals if n is even.

81. (a)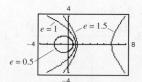

(b)

$0 \le \theta < 4\pi$ $0 \le \theta < 4\pi$

(c) Yes. Explanations will vary.

Section 6.9 *(page 507)*

1. conic   **3.** vertical; right

5. $e = 1$: $r = \dfrac{2}{1 + \cos \theta}$, parabola

$e = 0.5$: $r = \dfrac{1}{1 + 0.5 \cos \theta}$, ellipse

$e = 1.5$: $r = \dfrac{3}{1 + 1.5 \cos \theta}$, hyperbola

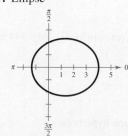

7. $e = 1$: $r = \dfrac{2}{1 - \sin \theta}$, parabola

$e = 0.5$: $r = \dfrac{1}{1 - 0.5 \sin \theta}$, ellipse

$e = 1.5$: $r = \dfrac{3}{1 - 1.5 \sin \theta}$, hyperbola

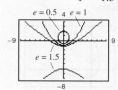

9. e   **10.** c   **11.** d   **12.** f   **13.** a   **14.** b

15. Parabola   **17.** Parabola

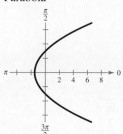
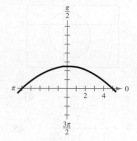

19. Ellipse   **21.** Ellipse

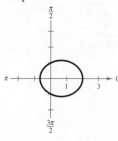
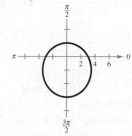

23. Hyperbola   **25.** Hyperbola

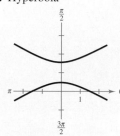
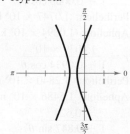

27. Ellipse

CHAPTER 6

29.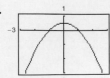

Parabola

31.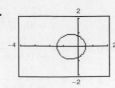

Ellipse

33.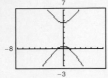

Hyperbola

35.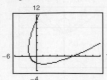

37.

39. $r = \dfrac{1}{1 - \cos\theta}$ **41.** $r = \dfrac{1}{2 + \sin\theta}$

43. $r = \dfrac{2}{1 + 2\cos\theta}$ **45.** $r = \dfrac{2}{1 - \sin\theta}$

47. $r = \dfrac{10}{1 - \cos\theta}$ **49.** $r = \dfrac{10}{3 + 2\cos\theta}$

51. $r = \dfrac{20}{3 - 2\cos\theta}$ **53.** $r = \dfrac{9}{4 - 5\sin\theta}$

55. Answers will vary.

57. $r = \dfrac{9.5929 \times 10^7}{1 - 0.0167\cos\theta}$

Perihelion: 9.4354×10^7 mi

Aphelion: 9.7558×10^7 mi

59. $r = \dfrac{1.0820 \times 10^8}{1 - 0.0068\cos\theta}$

Perihelion: 1.0747×10^8 km

Aphelion: 1.0894×10^8 km

61. $r = \dfrac{1.4039 \times 10^8}{1 - 0.0934\cos\theta}$

Perihelion: 1.2840×10^8 mi

Aphelion: 1.5486×10^8 mi

63. $r = \dfrac{0.624}{1 + 0.847\sin\theta}$;

$r \approx 0.338$ astronomical unit

65. (a) $r = \dfrac{8200}{1 + \sin\theta}$

(b)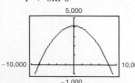

(c) 1467 mi   (d) 394 mi

67. True. The graphs represent the same hyperbola.

69. True. The conic is an ellipse because the eccentricity is less than 1.

71. The original equation graphs as a parabola that opens downward.

(a) The parabola opens to the right.

(b) The parabola opens up.

(c) The parabola opens to the left.

(d) The parabola has been rotated.

73. Answers will vary.

75. $r^2 = \dfrac{24{,}336}{169 - 25\cos^2\theta}$ **77.** $r^2 = \dfrac{144}{25\cos^2\theta - 9}$

79. $r^2 = \dfrac{144}{25\cos^2\theta - 16}$

81. (a) Ellipse

(b) The given polar equation, r, has a vertical directrix to the left of the pole. The equation r_1 has a vertical directrix to the right of the pole, and the equation r_2 has a horizontal directrix below the pole.

(c)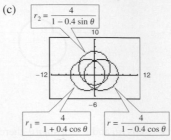

$r_2 = \dfrac{4}{1 - 0.4\sin\theta}$

$r_1 = \dfrac{4}{1 + 0.4\cos\theta}$ $r = \dfrac{4}{1 - 0.4\cos\theta}$

Review Exercises *(page 512)*

1. $\dfrac{\pi}{4}$ rad, $45°$ **3.** 1.1071 rad, $63.43°$

5. 0.4424 rad, $25.35°$ **7.** 0.6588 rad, $37.75°$

9. $4\sqrt{2}$ **11.** Hyperbola

13. $y^2 = 16x$ **15.** $(y - 2)^2 = 12x$

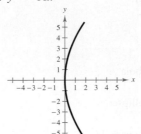   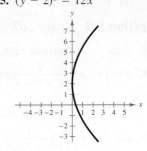

17. $y = -4x - 2; \left(-\frac{1}{2}, 0\right)$ **19.** $8\sqrt{6}$ m

21. $\dfrac{(x - 3)^2}{25} + \dfrac{y^2}{16} = 1$ **23.** $\dfrac{(x - 2)^2}{4} + \dfrac{(y - 1)^2}{1} = 1$

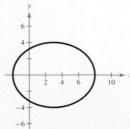   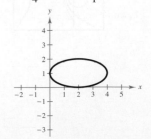

25. The foci occur 3 feet from the center of the arch on a line connecting the tops of the pillars.

27. Center: $(-1, 2)$

Vertices: $(-1, 9), (-1, -5)$

Foci: $\left(-1, 2 \pm 2\sqrt{6}\right)$

Eccentricity: $\dfrac{2\sqrt{6}}{7}$

29. Center: $(1, -4)$

Vertices: $(1, 0), (1, -8)$

Foci: $\left(1, -4 \pm \sqrt{7}\right)$

Eccentricity: $\dfrac{\sqrt{7}}{4}$

31. $\dfrac{y^2}{1} - \dfrac{x^2}{3} = 1$ **33.** $\dfrac{5(x-4)^2}{16} - \dfrac{5y^2}{64} = 1$

35. Center: $(5, -3)$

Vertices: $(11, -3), (-1, -3)$

Foci: $\left(5 \pm 2\sqrt{13}, -3\right)$

Asymptotes:

$y = -3 \pm \frac{2}{3}(x - 5)$

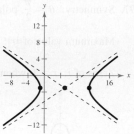

37. Center: $(1, -1)$

Vertices: $(5, -1), (-3, -1)$

Foci: $(6, -1), (-4, -1)$

Asymptotes:

$y = -1 \pm \frac{3}{4}(x - 1)$

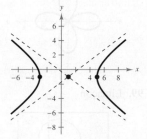

39. 72 mi   **41.** Hyperbola   **43.** Ellipse

45. $\dfrac{(y')^2}{6} - \dfrac{(x')^2}{6} = 1$ **47.** $\dfrac{(x')^2}{3} + \dfrac{(y')^2}{2} = 1$

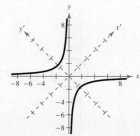

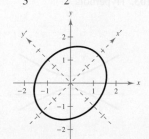

49. (a) Parabola

(b) $y = \dfrac{24x + 40 \pm \sqrt{(24x + 40)^2 - 36(16x^2 - 30x)}}{18}$

(c)

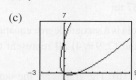

51. (a) Parabola

(b) $y = \dfrac{-\left(2x - 2\sqrt{2}\right) \pm \sqrt{\left(2x - 2\sqrt{2}\right)^2 - 4\left(x^2 + 2\sqrt{2}x + 2\right)}}{2}$

(c)

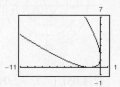

53. (a)

t	-2	-1	0	1	2
x	-8	-5	-2	1	4
y	15	11	7	3	-1

(b)

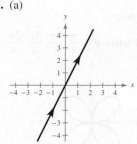

55. (a)

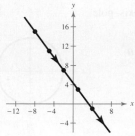

(b) $y = 2x$

57. (a)

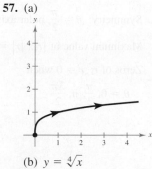

(b) $y = \sqrt[4]{x}$

59. (a)

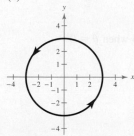

(b) $x^2 + y^2 = 9$

61. $x = -4 + 13t$
   $y = 4 - 14t$

63. $x = -3 + 4\cos\theta$
   $y = 4 + 3\sin\theta$

65.

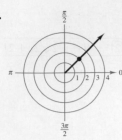

67.

$(2, -7\pi/4), (-2, 5\pi/4)$ $(7, 1.05), (-7, -2.09)$

69. $\left(-\dfrac{1}{2}, -\dfrac{\sqrt{3}}{2}\right)$ **71.** $\left(-\dfrac{3\sqrt{2}}{2}, \dfrac{3\sqrt{2}}{2}\right)$ **73.** $\left(1, \dfrac{\pi}{2}\right)$

75. $\left(2\sqrt{13}, 0.9828\right)$ **77.** $r = 9$ **79.** $r = 6\sin\theta$

81. $r^2 = 10\csc 2\theta$ **83.** $x^2 + y^2 = 25$ **85.** $x^2 + y^2 = 3x$

87. $x^2 + y^2 = y^{2/3}$

89. Symmetry: $\theta = \dfrac{\pi}{2}$, polar axis, pole

Maximum value of $|r|$:

$|r| = 6$ for all values of θ

No zeros of r

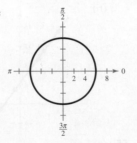

91. Symmetry: $\theta = \dfrac{\pi}{2}$, polar axis, pole

Maximum value of $|r|$: $|r| = 4$ when $\theta = \dfrac{\pi}{4}, \dfrac{3\pi}{4}, \dfrac{5\pi}{4}, \dfrac{7\pi}{4}$

Zeros of r: $r = 0$ when

$\theta = 0, \dfrac{\pi}{2}, \pi, \dfrac{3\pi}{2}$

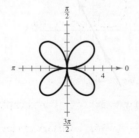

93. Symmetry: polar axis

Maximum value of $|r|$: $|r| = 4$ when $\theta = 0$

Zeros of r: $r = 0$ when $\theta = \pi$

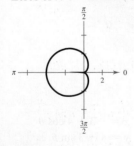

95. Symmetry: $\theta = \dfrac{\pi}{2}$

Maximum value of $|r|$: $|r| = 8$ when $\theta = \dfrac{\pi}{2}$

Zeros of r: $r = 0$ when $\theta = 3.4814, 5.9433$

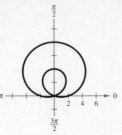

97. Symmetry: $\theta = \dfrac{\pi}{2}$, polar axis, pole

Maximum value of $|r|$: $|r| = 3$ when $\theta = 0, \dfrac{\pi}{2}, \pi, \dfrac{3\pi}{2}$

Zeros of r: $r = 0$ when $\theta = \dfrac{\pi}{4}, \dfrac{3\pi}{4}, \dfrac{5\pi}{4}, \dfrac{7\pi}{4}$

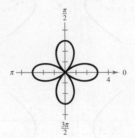

99. Limaçon

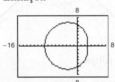

101. Rose curve

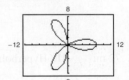

103. Hyperbola

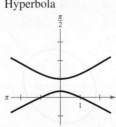

105. Ellipse

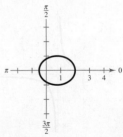

107. $r = \dfrac{4}{1 - \cos\theta}$ **109.** $r = \dfrac{5}{3 - 2\cos\theta}$

111. $r = \dfrac{7978.81}{1 - 0.937\cos\theta}$; 11,011.87 mi

113. False. The equation of a hyperbola is a second-degree equation.

115. False. $(2, \pi/4), (-2, 5\pi/4),$ and $(2, 9\pi/4)$ all represent the same point.

117. (a) The graphs are the same.   (b) The graphs are the same.

Chapter Test   (*page 515*)

1. 0.3805 rad, 21.8°   **2.** 0.8330 rad, 47.7°

3. $\dfrac{7\sqrt{2}}{2}$

4. Parabola: $y^2 = 2(x - 1)$
Vertex: $(1, 0)$
Focus: $\left(\frac{3}{2}, 0\right)$

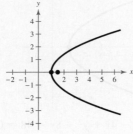

5. Hyperbola: $\dfrac{(x - 2)^2}{4} - y^2 = 1$
Center: $(2, 0)$
Vertices: $(0, 0), (4, 0)$
Foci: $\left(2 \pm \sqrt{5}, 0\right)$
Asymptotes: $y = \pm\dfrac{1}{2}(x - 2)$

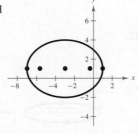

6. Ellipse: $\dfrac{(x + 3)^2}{16} + \dfrac{(y - 1)^2}{9} = 1$
Center: $(-3, 1)$
Vertices: $(1, 1), (-7, 1)$
Foci: $\left(-3 \pm \sqrt{7}, 1\right)$

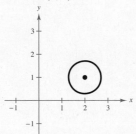

7. Circle: $(x - 2)^2 + (y - 1)^2 = \frac{1}{2}$
Center: $(2, 1)$

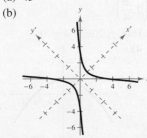

8. $(x - 2)^2 = \dfrac{4}{3}(y + 3)$ **9.** $\dfrac{5(y - 2)^2}{4} - \dfrac{5x^2}{16} = 1$

10. (a) 45°
(b)

11.

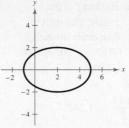

$\dfrac{(x - 2)^2}{9} + \dfrac{y^2}{4} = 1$

12. $x = 6 + 4t$
$y = 4 + 7t$

13. $\left(\sqrt{3}, -1\right)$ **14.** $\left(2\sqrt{2}, \frac{7\pi}{4}\right), \left(-2\sqrt{2}, \frac{3\pi}{4}\right), \left(2\sqrt{2}, -\frac{\pi}{4}\right)$

15. $r = 3\cos\theta$

16.

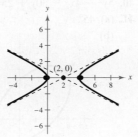

Parabola

17.

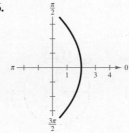

Ellipse

18.

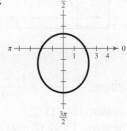

Limaçon with inner loop

19.

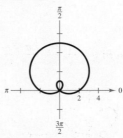

Rose curve

20. Answers will vary. For example: $r = \dfrac{1}{1 + 0.25\sin\theta}$

21. Slope: 0.1511; Change in elevation: 789 ft
22. No; Yes

Cumulative Test for Chapters 4–6   (*page 516*)

1. $6 - 7i$ **2.** $-2 - 3i$ **3.** $-21 - 20i$ **4.** 4

5. $\dfrac{2}{13} + \dfrac{10}{13}i$ **6.** $-2, \pm2i$ **7.** $-7, 0, 3$

8. $x^4 + x^3 - 33x^2 + 45x + 378$

9. $2\sqrt{2}\left(\cos\dfrac{3\pi}{4} + i\sin\dfrac{3\pi}{4}\right)$ **10.** $-12\sqrt{3} + 12i$

11. $-8 + 8\sqrt{3}\,i$ **12.** -64

13. $\cos 0 + i\sin 0$

$\cos\dfrac{2\pi}{3} + i\sin\dfrac{2\pi}{3}$

$\cos\dfrac{4\pi}{3} + i\sin\dfrac{4\pi}{3}$

14. $3\left(\cos\dfrac{\pi}{8} + i\sin\dfrac{\pi}{8}\right)$

$3\left(\cos\dfrac{5\pi}{8} + i\sin\dfrac{5\pi}{8}\right)$

$3\left(\cos\dfrac{9\pi}{8} + i\sin\dfrac{9\pi}{8}\right)$

$3\left(\cos\dfrac{13\pi}{8} + i\sin\dfrac{13\pi}{8}\right)$

CHAPTER 6

15. Reflect f in the x-axis and y-axis, and shift three units to the right.

16. Reflect f in the x-axis, and shift four units upward.

17. 1.991   **18.** -0.067 **19.** 1.717   **20.** 0.390

21. 0.906   **22.** -1.733 **23.** -4.087

24. $\ln(x + 4) + \ln(x - 4) - 4 \ln x, \quad x > 4$

25. $\ln \dfrac{x^2}{\sqrt{x + 5}}, \quad x > 0$ **26.** $\dfrac{\ln 12}{2} \approx 1.242$

27. $\dfrac{\ln 9}{\ln 4} + 5 \approx 6.585$ **28.** $\dfrac{64}{5} = 12.8$

29. $\frac{1}{2}e^8 \approx 1490.479$

30. 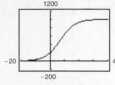 Horizontal asymptotes: $y = 0, y = 1000$

31. 6.3 h   **32.** 2015   **33.** 81.87°   **34.** $\dfrac{11\sqrt{5}}{5}$

35. Ellipse; $\dfrac{(x - 2)^2}{4} + \dfrac{(y + 1)^2}{9} = 1$

Center: $(2, -1)$
Vertices: $(2, 2), (2, -4)$
Foci: $\left(2, -1 \pm \sqrt{5}\right)$

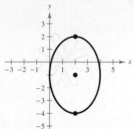

36. Hyperbola; $x^2 - \dfrac{y^2}{4} = 1$

Center: $(0, 0)$
Vertices: $(1, 0), (-1, 0)$
Foci: $\left(\sqrt{5}, 0\right), \left(-\sqrt{5}, 0\right)$
Asymptotes: $y = \pm 2x$

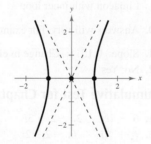

37. Circle; $(x + 1)^2 + (y - 3)^2 = 22$

Center: $(-1, 3)$

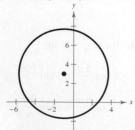

38. Parabola; $y^2 = -2(x + 1)$

Vertex: $(-1, 0)$
Focus: $\left(-\frac{3}{2}, 0\right)$

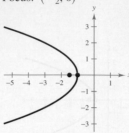

39. $x^2 - 4x + y^2 + 8y - 48 = 0$

40. $5y^2 - 4x^2 - 30y + 25 = 0$

41. (a) 45°

(b)

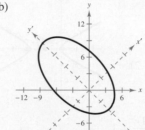

42.

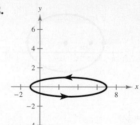

$\dfrac{(x - 3)^2}{16} + y^2 = 1$

43. $x = 3 - 6t$
$ y = -2 + 6t$

44.

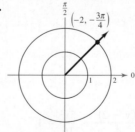

$\left(-2, \dfrac{5\pi}{4}\right), \left(2, -\dfrac{7\pi}{4}\right),$ and $\left(2, \dfrac{\pi}{4}\right)$

45. $r = 16 \sin \theta$ **46.** $9x^2 - 16y^2 + 20x + 4 = 0$

47.   **48.**

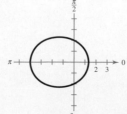

49. (a) iii   (b) i   (c) ii

Problem Solving   *(page 521)*

1. (a) 1.2016 rad   (b) 2420 ft, 5971 ft

3. $y^2 = 4p(x + p)$ **5.** Answers will vary.

7. $\dfrac{(x - 6)^2}{9} - \dfrac{(y - 2)^2}{7} = 1$

9. (a) The first set of parametric equations models projectile motion along a straight line. The second set of parametric equations models projectile motion of an object launched at a height of h units above the ground that will eventually fall back to the ground.

(b) $y = (\tan \theta)x$; $\ y = h + x \tan \theta - \dfrac{16x^2 \sec^2 \theta}{v_0^2}$

(c) In the first case, the path of the moving object is not affected by a change in the velocity because eliminating the parameter removes v_0.

11. (a)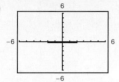

(b)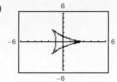

The graph is a line between -2 and 2 on the x-axis.

The graph is a three-sided figure with counterclockwise orientation.

(c)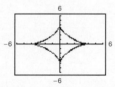

(d)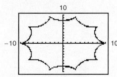

The graph is a four-sided figure with counterclockwise orientation.

The graph is a 10-sided figure with counterclockwise orientation.

(e)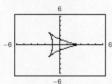

(f)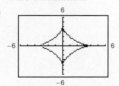

The graph is a three-sided figure with clockwise orientation.

The graph is a four-sided figure with clockwise orientation.

13.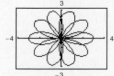

$r = 3 \sin\left(\dfrac{5\theta}{2}\right)$ $r = -\cos\left(\sqrt{2}\theta\right),$
$-2\pi \le \theta \le 2\pi$

Sample answer: If n is a rational number, then the curve has a finite number of petals. If n is an irrational number, then the curve has an infinite number of petals.

15. (a) No. Because of the exponential, the graph will continue to trace the butterfly curve at larger values of r.

(b) $r \approx 4.1$. This value will increase if θ is increased.

17. (a) $r_{\text{Neptune}} = \dfrac{4.4947 \times 10^9}{1 - 0.0086 \cos \theta}$

$r_{\text{Pluto}} = \dfrac{5.54 \times 10^9}{1 - 0.2488 \cos \theta}$

(b) Neptune: Aphelion $= 4.534 \times 10^9$ km
　　　　　　Perihelion $= 4.456 \times 10^9$ km
Pluto: Aphelion $= 7.375 \times 10^9$ km
　　　　Perihelion $= 4.437 \times 10^9$ km

(c)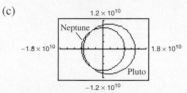

(d) Yes, at times Pluto can be closer to the sun than Neptune. Pluto was called the ninth planet because it has the longest orbit around the sun and therefore also reaches the furthest distance away from the sun.

(e) If the orbits were in the same plane, then they would intersect. Furthermore, since the orbital periods differ (Neptune $=$ 164.79 years, Pluto $=$ 247.68 years), then the two planets would ultimately collide if the orbits intersect. The orbital inclination of Pluto is significantly larger than that of Neptune (17.16° vs. 1.769°), so further analysis is required to determine if the orbits intersect.

CHAPTER 6

INDEX

FORMULAS FROM GEOMETRY

Triangle:

$h = a \sin \theta$

$\text{Area} = \dfrac{1}{2}bh$

$c^2 = a^2 + b^2 - 2ab \cos \theta$ (Law of Cosines)

Right Triangle:

Pythagorean Theorem
$c^2 = a^2 + b^2$

Equilateral Triangle:

$h = \dfrac{\sqrt{3}s}{2}$

$\text{Area} = \dfrac{\sqrt{3}s^2}{4}$

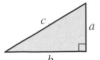

Parallelogram:

$\text{Area} = bh$

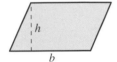

Trapezoid:

$\text{Area} = \dfrac{h}{2}(a + b)$

Circle:

$\text{Area} = \pi r^2$

$\text{Circumference} = 2\pi r$

Sector of Circle:

$\text{Area} = \dfrac{\theta r^2}{2}$

$s = r\theta$

θ in radians

Circular Ring:

$\text{Area} = \pi(R^2 - r^2)$

$\quad\quad = 2\pi pw$

$p = \text{average radius},$

$w = \text{width of ring}$

Sector of Circular Ring:

$\text{Area} = \theta pw$

$p = \text{average radius},$

$w = \text{width of ring},$

θ in radians

Ellipse:

$\text{Area} = \pi ab$

$\text{Circumference} \approx 2\pi \sqrt{\dfrac{a^2 + b^2}{2}}$

Cone:

$\text{Volume} = \dfrac{Ah}{3}$

$A = \text{area of base}$

Right Circular Cone:

$\text{Volume} = \dfrac{\pi r^2 h}{3}$

$\text{Lateral Surface Area} = \pi r \sqrt{r^2 + h^2}$

Frustum of Right Circular Cone:

$\text{Volume} = \dfrac{\pi(r^2 + rR + R^2)h}{3}$

$\text{Lateral Surface Area} = \pi s(R + r)$

Right Circular Cylinder:

$\text{Volume} = \pi r^2 h$

$\text{Lateral Surface Area} = 2\pi rh$

Sphere:

$\text{Volume} = \dfrac{4}{3}\pi r^3$

$\text{Surface Area} = 4\pi r^2$

Wedge:

$A = B \sec \theta$

$A = \text{area of upper face},$

$B = \text{area of base}$

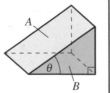

Definition of the Six Trigonometric Functions

Right triangle definitions, where $0 < \theta < \pi/2$

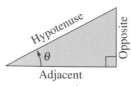

$$\sin \theta = \frac{\text{opp.}}{\text{hyp.}} \qquad \csc \theta = \frac{\text{hyp.}}{\text{opp.}}$$

$$\cos \theta = \frac{\text{adj.}}{\text{hyp.}} \qquad \sec \theta = \frac{\text{hyp.}}{\text{adj.}}$$

$$\tan \theta = \frac{\text{opp.}}{\text{adj.}} \qquad \cot \theta = \frac{\text{adj.}}{\text{opp.}}$$

Circular function definitions, where θ is any angle

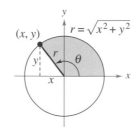

$$\sin \theta = \frac{y}{r} \qquad \csc \theta = \frac{r}{y}$$

$$\cos \theta = \frac{x}{r} \qquad \sec \theta = \frac{r}{x}$$

$$\tan \theta = \frac{y}{x} \qquad \cot \theta = \frac{x}{y}$$

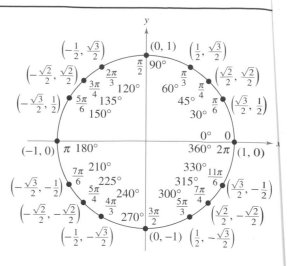

Reciprocal Identities

$$\sin u = \frac{1}{\csc u} \qquad \cos u = \frac{1}{\sec u} \qquad \tan u = \frac{1}{\cot u}$$

$$\csc u = \frac{1}{\sin u} \qquad \sec u = \frac{1}{\cos u} \qquad \cot u = \frac{1}{\tan u}$$

Quotient Identities

$$\tan u = \frac{\sin u}{\cos u} \qquad \cot u = \frac{\cos u}{\sin u}$$

Pythagorean Identities

$$\sin^2 u + \cos^2 u = 1$$

$$1 + \tan^2 u = \sec^2 u \qquad 1 + \cot^2 u = \csc^2 u$$

Cofunction Identities

$$\sin\left(\frac{\pi}{2} - u\right) = \cos u \qquad \cot\left(\frac{\pi}{2} - u\right) = \tan u$$

$$\cos\left(\frac{\pi}{2} - u\right) = \sin u \qquad \sec\left(\frac{\pi}{2} - u\right) = \csc u$$

$$\tan\left(\frac{\pi}{2} - u\right) = \cot u \qquad \csc\left(\frac{\pi}{2} - u\right) = \sec u$$

Even/Odd Identities

$$\sin(-u) = -\sin u \qquad \cot(-u) = -\cot u$$

$$\cos(-u) = \cos u \qquad \sec(-u) = \sec u$$

$$\tan(-u) = -\tan u \qquad \csc(-u) = -\csc u$$

Sum and Difference Formulas

$$\sin(u \pm v) = \sin u \cos v \pm \cos u \sin v$$

$$\cos(u \pm v) = \cos u \cos v \mp \sin u \sin v$$

$$\tan(u \pm v) = \frac{\tan u \pm \tan v}{1 \mp \tan u \tan v}$$

Double-Angle Formulas

$$\sin 2u = 2 \sin u \cos u$$

$$\cos 2u = \cos^2 u - \sin^2 u = 2\cos^2 u - 1 = 1 - 2\sin^2 u$$

$$\tan 2u = \frac{2 \tan u}{1 - \tan^2 u}$$

Power-Reducing Formulas

$$\sin^2 u = \frac{1 - \cos 2u}{2}$$

$$\cos^2 u = \frac{1 + \cos 2u}{2}$$

$$\tan^2 u = \frac{1 - \cos 2u}{1 + \cos 2u}$$

Sum-to-Product Formulas

$$\sin u + \sin v = 2 \sin\left(\frac{u+v}{2}\right) \cos\left(\frac{u-v}{2}\right)$$

$$\sin u - \sin v = 2 \cos\left(\frac{u+v}{2}\right) \sin\left(\frac{u-v}{2}\right)$$

$$\cos u + \cos v = 2 \cos\left(\frac{u+v}{2}\right) \cos\left(\frac{u-v}{2}\right)$$

$$\cos u - \cos v = -2 \sin\left(\frac{u+v}{2}\right) \sin\left(\frac{u-v}{2}\right)$$

Product-to-Sum Formulas

$$\sin u \sin v = \frac{1}{2}[\cos(u - v) - \cos(u + v)]$$

$$\cos u \cos v = \frac{1}{2}[\cos(u - v) + \cos(u + v)]$$

$$\sin u \cos v = \frac{1}{2}[\sin(u + v) + \sin(u - v)]$$

$$\cos u \sin v = \frac{1}{2}[\sin(u + v) - \sin(u - v)]$$

Definition of the Six Trigonometric Functions

Right triangle definitions, where $0 < \theta < \pi/2$

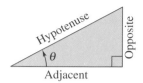

$$\sin\theta = \frac{\text{opp.}}{\text{hyp.}} \qquad \csc\theta = \frac{\text{hyp.}}{\text{opp.}}$$

$$\cos\theta = \frac{\text{adj.}}{\text{hyp.}} \qquad \sec\theta = \frac{\text{hyp.}}{\text{adj.}}$$

$$\tan\theta = \frac{\text{opp.}}{\text{adj.}} \qquad \cot\theta = \frac{\text{adj.}}{\text{opp.}}$$

Circular function definitions, where θ is any angle

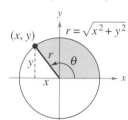

$r = \sqrt{x^2 + y^2}$

$$\sin\theta = \frac{y}{r} \qquad \csc\theta = \frac{r}{y}$$

$$\cos\theta = \frac{x}{r} \qquad \sec\theta = \frac{r}{x}$$

$$\tan\theta = \frac{y}{x} \qquad \cot\theta = \frac{x}{y}$$

Reciprocal Identities

$$\sin u = \frac{1}{\csc u} \qquad \cos u = \frac{1}{\sec u} \qquad \tan u = \frac{1}{\cot u}$$

$$\csc u = \frac{1}{\sin u} \qquad \sec u = \frac{1}{\cos u} \qquad \cot u = \frac{1}{\tan u}$$

Quotient Identities

$$\tan u = \frac{\sin u}{\cos u} \qquad \cot u = \frac{\cos u}{\sin u}$$

Pythagorean Identities

$$\sin^2 u + \cos^2 u = 1$$

$$1 + \tan^2 u = \sec^2 u \qquad 1 + \cot^2 u = \csc^2 u$$

Cofunction Identities

$$\sin\left(\frac{\pi}{2} - u\right) = \cos u \qquad \cot\left(\frac{\pi}{2} - u\right) = \tan u$$

$$\cos\left(\frac{\pi}{2} - u\right) = \sin u \qquad \sec\left(\frac{\pi}{2} - u\right) = \csc u$$

$$\tan\left(\frac{\pi}{2} - u\right) = \cot u \qquad \csc\left(\frac{\pi}{2} - u\right) = \sec u$$

Even/Odd Identities

$$\sin(-u) = -\sin u \qquad \cot(-u) = -\cot u$$

$$\cos(-u) = \cos u \qquad \sec(-u) = \sec u$$

$$\tan(-u) = -\tan u \qquad \csc(-u) = -\csc u$$

Sum and Difference Formulas

$$\sin(u \pm v) = \sin u \cos v \pm \cos u \sin v$$

$$\cos(u \pm v) = \cos u \cos v \mp \sin u \sin v$$

$$\tan(u \pm v) = \frac{\tan u \pm \tan v}{1 \mp \tan u \tan v}$$

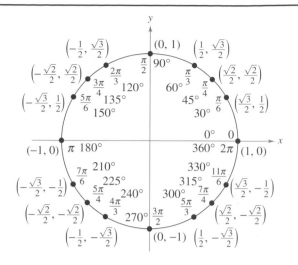

Double-Angle Formulas

$$\sin 2u = 2\sin u \cos u$$

$$\cos 2u = \cos^2 u - \sin^2 u = 2\cos^2 u - 1 = 1 - 2\sin^2 u$$

$$\tan 2u = \frac{2\tan u}{1 - \tan^2 u}$$

Power-Reducing Formulas

$$\sin^2 u = \frac{1 - \cos 2u}{2}$$

$$\cos^2 u = \frac{1 + \cos 2u}{2}$$

$$\tan^2 u = \frac{1 - \cos 2u}{1 + \cos 2u}$$

Sum-to-Product Formulas

$$\sin u + \sin v = 2\sin\left(\frac{u + v}{2}\right)\cos\left(\frac{u - v}{2}\right)$$

$$\sin u - \sin v = 2\cos\left(\frac{u + v}{2}\right)\sin\left(\frac{u - v}{2}\right)$$

$$\cos u + \cos v = 2\cos\left(\frac{u + v}{2}\right)\cos\left(\frac{u - v}{2}\right)$$

$$\cos u - \cos v = -2\sin\left(\frac{u + v}{2}\right)\sin\left(\frac{u - v}{2}\right)$$

Product-to-Sum Formulas

$$\sin u \sin v = \frac{1}{2}[\cos(u - v) - \cos(u + v)]$$

$$\cos u \cos v = \frac{1}{2}[\cos(u - v) + \cos(u + v)]$$

$$\sin u \cos v = \frac{1}{2}[\sin(u + v) + \sin(u - v)]$$

$$\cos u \sin v = \frac{1}{2}[\sin(u + v) - \sin(u - v)]$$

FORMULAS FROM GEOMETRY

Triangle:

$h = a \sin \theta$

$\text{Area} = \dfrac{1}{2} bh$

$c^2 = a^2 + b^2 - 2ab \cos \theta$ (Law of Cosines)

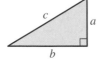

Sector of Circular Ring:

$\text{Area} = \theta p w$

p = average radius,

w = width of ring,

θ in radians

Right Triangle:

Pythagorean Theorem

$c^2 = a^2 + b^2$

Ellipse:

$\text{Area} = \pi a b$

$\text{Circumference} \approx 2\pi \sqrt{\dfrac{a^2 + b^2}{2}}$

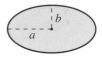

Equilateral Triangle:

$h = \dfrac{\sqrt{3} s}{2}$

$\text{Area} = \dfrac{\sqrt{3} s^2}{4}$

Cone:

$\text{Volume} = \dfrac{Ah}{3}$

A = area of base

Parallelogram:

$\text{Area} = bh$

Right Circular Cone:

$\text{Volume} = \dfrac{\pi r^2 h}{3}$

$\text{Lateral Surface Area} = \pi r \sqrt{r^2 + h^2}$

Trapezoid:

$\text{Area} = \dfrac{h}{2}(a + b)$

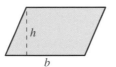

Frustum of Right Circular Cone:

$\text{Volume} = \dfrac{\pi(r^2 + rR + R^2)h}{3}$

$\text{Lateral Surface Area} = \pi s(R + r)$

Circle:

$\text{Area} = \pi r^2$

$\text{Circumference} = 2\pi r$

Right Circular Cylinder:

$\text{Volume} = \pi r^2 h$

$\text{Lateral Surface Area} = 2\pi r h$

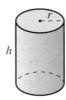

Sector of Circle:

$\text{Area} = \dfrac{\theta r^2}{2}$

$s = r\theta$

θ in radians

Sphere:

$\text{Volume} = \dfrac{4}{3}\pi r^3$

$\text{Surface Area} = 4\pi r^2$

Circular Ring:

$\text{Area} = \pi(R^2 - r^2)$

$\qquad = 2\pi p w$

p = average radius,

w = width of ring

Wedge:

$A = B \sec \theta$

A = area of upper face,

B = area of base

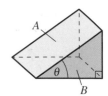